52nd American Association of Swine Veterinarians Annual Meeting 2021

Navigating the Future...Together

Online
27 February – 2 March 2021

ISBN: 978-1-7138-2952-2

Printed from e-media with permission by:

Curran Associates, Inc.
57 Morehouse Lane
Red Hook, NY 12571

Some format issues inherent in the e-media version may also appear in this print version.

American Association of Swine Veterinarians
830 26th Street
Perry, IA 50220-2328
USA

Phone: 515-465-5255
Fax: 515-465-3832

aasv@aasv.org

Additional copies of this publication are available from:

Curran Associates, Inc.
57 Morehouse Lane
Red Hook, NY 12571 USA
Phone: 845-758-0400
Fax: 845-758-2633
Email: curran@proceedings.com
Web: www.proceedings.com

Table of Contents

Howard Dunne Memorial Lecture

Alex Hogg Memorial Lecture

Sunday: Research Topics

Sunday Concurrent Session #1: Student Seminar

Sunday Concurrent Session #2: Industrial Partners

Sunday Concurrent Session #3: Industrial Partners

Sunday Concurrent Session #4: Industrial Partners

Poster Session: Veterinary Students

Poster Session: Research Topics

Poster Session: Industrial Partners

Monday General Session: Navigating the Future … *Together*

Monday Concurrent Session #1: The Same Old Bugs; Not the Same Old Toolkit

Monday Concurrent Session #2: Welfare and the Swine Veterinarian

Monday Concurrent Session #3: African Swine Fever

Tuesday General Session: COVID-19 Lessons Learned

Howard Dunne Memorial Lecture

Navigating the future, together

Jerome O. Geiger, DVM, MS

In remembrance and honor of
Howard W. Dunne

Dr. Howard W. Dunne received his veterinary degree from Iowa State University in 1941. He was in engaged in private practice for one year and industry for four years before entering graduate school at Michigan State University. After receiving his PhD in 1951, he spent two years as deputy chief of veterinary microbiology at the Chemical Corps Biological Laboratories in Camp Detrick, MD. In 1953, he joined the faculty of Pennsylvania State University, where he headed the veterinary science research program and the large animal diagnostic laboratory. Dr. Dunne was internationally known for his work on swine diseases, particularly hog cholera and the swine reproductive viruses then referred to as the SMEDI syndrome. Dr. Dunne authored more than 90 scientific publications, and is perhaps best known for initiating and editing the textbook *Diseases of Swine*, which is still widely considered the "bible" of swine disease.

In 1971, the American Veterinary Medical Association presented Dr. Dunne with the XII International Veterinary Congress Prize to recognize his contributions to the international understanding of veterinary medicine. At the 1974 Congress of the International Pig Veterinary Society (IPVS), Dr. Dunne was elected president of the 1976 IPVS Congress, to be held in Ames, Iowa. His sudden death at the age of 61 in September of 1974 saddened the swine veterinary community, and the 1976 IPVS Congress was dedicated to his memory.

At the suggestion of Dr. Allen Leman, a financial gift made by the 1976 IPVS Congress organizers to the fledgling American Association of Swine Practitioners was combined with additional association funds to endow this prestigious annual lecture in remembrance and honor of Dr. Howard W. Dunne.

HOWARD DUNNE MEMORIAL LECTURERS

1978	Michael Muirhead	1989	Steven Henry	2000	David Reeves	2011	Locke A. Karriker
1979	W. J. (Bill) Smith	1990	Stanley Curtis	2001	Max Rodibaugh	2012	R. Dean Boyd
1980	N.C. Nielsen	1991	Roy Schultz	2002	Tim Blackwell	2013	Craig Rowles
1981	Allen Leman	1992	Joseph Connor	2003	Lonnie King	2014	Daryl Olsen
1982	Michael Wilson	1993	Camille Moore	2004	Robert Desrosiers	2015	Gregory Stevenson
1983	Edward Bohl	1994	Robert Morrison	2005	Scott Dee	2016	John Harding
1984	Richard Ross	1995	Rodney Johnson	2006	Gregg BeVier	2017	Jeff Zimmerman
1985	Kent Kislingbury	1996	Paul Yeske	2007	Tom Burkgren	2018	William R. DuBois
1986	Ralph Vinson	1997	Barbara Straw	2008	Timothy J. Loula	2019	John T. Waddell
1987	Gary Dial	1998	Mike Daniel	2009	John U. Thomson	2020	Bret D. Marsh
1988	Thomas Alexander	1999	Larry Rueff	2010	James F. Lowe		

Navigating the future, together

Jerome O. Geiger, DVM, MS
PIC, Hendersonville, Tennessee

Introduction

It's not easy. Some would say what we do is hard or difficult. Some would ask why we do this. Is it worth the time, commitment, talent and energy we invest? With the multiple and novel challenges we face daily, why do we dedicate so much of ourselves? Why commit with deep conviction to such a profession and industry and lifestyle?

- Why struggle with the consternation of a disease outbreak in a client's herd or the angst of disease eliminations?
- Why put so many mental, physical, financial and emotional resources into holding foreign animal diseases outside our borders?
- Why battle with over-crowded but under-staffed barns due to COVID?
- Why bear the physical, emotional and moral turmoil of large-scale depopulations and then face public scrutiny for doing so?
- Why train and re-train the seemingly endless stream of new caretakers? Why can't they just do their jobs and do them correctly?
- Why endure pressure from animal activists, legal challenges, and an abundance of unfriendly press?
- Indeed, why are you reading this?

The situations described within these questions manifest a mental picture for me, thanks to Dr. John Deen. Often, in my mind I hear his comment, "When you're waist-deep in alligators, it's hard to remember your primary job is to drain the swamp."[1] It is easy to be caught-up in things happening near at hand, the everyday battles and looming challenges both real and imagined.

The primary goal for this lecture is to fulfill Dr. Battrell's vision for this portion of the program. Part of that vision is that we remember the bigger picture, we (for a moment at least) lift our heads and look to the horizon. Navigation is, after all, ascertaining one's current location and then planning and following a route to a destination. If we are consumed by the immediate problems around us, we are not looking at the horizon and we certainly cannot be navigating. If we are merely steering around obstacles, we are not navigating to a destination unseen.

Disclaimers

From the beginning, four disclaimers should be made.

1. Previous Howard Dunne lectures have already presented much of this material. Herein are many quotes from those lectures and references to the same books and articles.
2. There are no slides, no graphs or tables or charts. It is hoped the reader will focus on the thoughts, ideas and philosophies offered rather than decimal points, dendograms and s/p ratios.
3. Overall, this is neither prophetic nor profound, neither elegant nor eloquent. If it's been said before, how can it be prophetic? If someone has already shared it, how is it profound?
4. Many personal examples are used, referring to other AASP/AASV members past and present. They are not included to embarrass or offend; rather these are included to provide relevance and clarity while making the message human and personal.

Start with why. WHY do we do this?

Assuming there is need or desire to navigate the future, it is important to ask and answer WHY one would do such a thing. Have you given serious thought to WHY you do what you do? AASV as an organization cannot know HOW to navigate the future or WHAT we should do, until we know, believe and embrace WHY.

After reading *Start with Why*[2], I've given considerable thought to WHY. One might respond with flippant or half-sincere answers, but if you press further, where does it lead you? In your heart-of-hearts, WHY are you a swine veterinarian, so committed, so devoted to our industry and profession? Personally, the exercise presented me three different paths to three different answers. Without going through the entire length of those convoluted trails, I will summarize.

There is an academic and sometimes physical challenge

Daily, there is both joy and excitement blending science with common sense to solve unique problems. Personally, it's fun enlisting science, technology and my education, combining them with slat-level reality and home-grown common sense and then applying the solution to the real world.

There is the motivation of "it's the right thing to do"

Speaking as a past chair and long-time member of the AASV Pig Welfare committee, we start and finish nearly every conversation with some variation of "Respecting the pig..." This matches a directive passed down to me by generations of animal caretakers. The pig is our charge, our responsibility; it is dependent on us in every way. If we do not represent the pig, who will? Therefore, ultimately, I am compelled "in the interest of the pig."

Daryl Olsen was clear in his Howard Dunne lecture "The Pig Always Wins."[3] He first quoted Dr. Michelle Sprague, "While I understand and respect the fact we all serve our clients, our responsibility as veterinarians is to do what is right for the pig." Daryl went on to write, "If all our decisions are made with the well-being of the pig as the guiding principle, we can defend every and all decisions that we make. When we compromise science or what is best for the pig, we are defenseless."

There is the altruistic aspect

Author Jim Collins has been quoted, "True greatness comes in direct proportion to passionate pursuit of a purpose beyond money."[4] It is my obligation to somehow make the world a better place. In some small part, I feel (perhaps egotistically) we are helping feed the world. There is a need for a higher purpose, even for an organization (e.g. AASV).

Greg Stevenson was bold enough to write "we help feed the world."[5] Craig Rowles wrote, "successful [organizations] of the future won't be those who just sell stuff, but will be those who strive to make a difference in the world."[6]

Agriculture has an important, critical, global role. John Thomson touched on the altruism aspect with two quotes from US presidents:[7]

- John Kennedy. "No group could do more for America, or the world, than farmers if we recognize that food is strength, and food is peace, and food is freedom, and food is a helping hand to people around the world, whose good will and friendship we want."
- Dwight Eisenhower. "This peace we seek, founded upon decent trust and cooperative effort among nations, can be fortified, not by weapons of war, but by wheat and cotton; by milk and wool; by meat and timber; and by rice. These are words that translate into every language on earth."

Embracing this third component of WHY, I must acknowledge there will be no endpoint; the job will never be completed, I will never be done.

Combining those three WHY components, I arrive at an elevator-response that truly answers the WHY question for me. **I combine science and common sense in the interest of the pig, to help feed the world.** That explains my WHY. Does that work for you?

A team is needed. Building the team.

Teamwork, cooperation and coordination have been critical elements in AASV's success to this point, and will be equally important going forward. We are one team, united by a common WHY. We must be one team and we must function as one team. Booker T. Washington observed, "In all things that are purely social, we can be as separate as the fingers, yet one as the hand in all things essential to mutual progress."[8]

"In an organization [that understands its WHY], no one sees himself as any more or any less than anyone else. They all need each other."[2] As with any team, there are different roles to play, different positions to fill. Looking at AASV today, the team has many components:

- Visionaries and doers-implementers. These groups are separate but co-dependent. They go hand-in-hand – one needs the other.
- Truth-tellers unapologetically force the team to face realities, particularly the shortcomings and unfulfilled expectations.
- Stabilizers/grounders that keep the organization and industry focused on science but grounded in common sense.
- Binders/organizers that tie us together to give a common foundation and strength in our numbers. They bind us like corn shocks or sheaves of wheat.
- Statisticians/accountants that scrutinize and analyze the data, crunch the numbers to answer the basic question, "What do the numbers tell us?"
- Guardians that watch over the (financial) health of the organization and offer security for our future.
- Communicators that teach us to effectively communicate with one another and the outside world.
- Promoters that publicly tell the story for AASV and for the industry. Think of Operation Main Street.
- Innovators that bring new technology based in science and pioneer its application.
- Diagnosticians: people with a particular talent for finding agents (sometimes pathogens), identifying and assisting us to put those agents into context.
- Mentors. By virtue of attending this meeting and supporting younger and potential members, one could argue every AASV member qualifies for this group.
- Educators/Teachers. Like mentors, are we not all teachers? This group builds the knowledge base and shares information far and wide.
- Pioneers (geographically speaking)/Scouts/Evangelists. These go to distant lands to share our common knowledge, to explore; but also to bring back ideas and insights.
- Servers/Caretakers that watch over the organization as a whole, giving of themselves.

Consider the AASV organization today, consider the membership. Surely each member can create his or her own list with multiple shining examples of comrades who fit each of these jobs, positions, or descriptions. AASV participates, guides and leads the North American pork production industry in its evolution. We don't do it alone. We need input and support of many other individuals and organizations. Again, this emphasizes "we are one team."

Leadership, be bold

Little is gained by the faint of heart. Navigating the future will require bold leadership, but that should surprise no one. Within this group, Al Leman's adage "we're not making mistakes fast enough" is almost a cliché. Many influential people have similarly advised:

- "Take change by the hand before it takes you by the throat." Stan Curtis quoted Winston Churchill.[9]
- "Our doubts are traitors and make us lose the good we oft might win, by fearing to attempt." Bill Christianson quoting William Shakespeare.[10]
- "The meek ain't gonna inherit nothing west of Chicago." Sam Elliot in The Quick and the Dead, a movie based on a Louis Lamour novel.[11]

As an organization and an industry, we have been bold. AASP changed its name in 2001 to reflect and project the fact it represented more than practitioners alone. Some would say that was bold, AASV's leadership thought it appropriate or even necessary. In 2007, Apple Computers changed its name to Apple to change its image both internally and externally. AASV acted similarly for similar reasons. The change demonstrated commitment and focus.

The history of this organization shows we have been bold in a more tangible way, and that we have built and continue to build a successful team, a team that can effect sweeping change across an entire industry. The conversion from single-site, farrow-finish production typical in the 1980s to the age-segregated production flow, the standard across the industry today, serves as one example of both characteristics.

Steve Henry's 1989 Dunne lecture focused on inefficiencies and inabilities of the grow-finish component of pork production of the time. Dr. Henry wrote that to improve grow-finish performance, we needed to change the system of growing and finishing pigs. In that time, Steve was a truth-teller shedding light on a major area needing significant improvement. He was clear in describing the shortcoming and defining the need. It was our job to improve the situation, in the interest of the pig.

"The superstructure of the swine industry in the United States is housed in a production facility that is 10-20 years old . . . . A finishing pig's needs are simple. They include space, comfort, food, water and good health . . . . Little else matters. Unfortunately, the combination of all five consistently coming up aces is so extremely rare that pigs perform far below the potential we know they have. The pig has done its share in adapting to technology. The challenge now is to more sensitively adapt technology to the pig."[12]

Over time (mid-1980s to present) swine veterinarians developed a technology and physical infrastructure to benefit the growing pig, a change that revolutionized an industry. Tom Alexander identified a seed (a production strategy) that was initially called Medicated Early Weaning (MEW). Hank Harris saw a way to apply that strategy broadly across the industry and planted that seed in North America. Hank was visionary. Others (members of AASV past and present) nurtured, cultivated, watered (and yes, sometimes pruned) the developing growth. They were the doers and implementers. It came to be known and practiced as age segregation, moving newly weaned pigs to biologically separate facilities to break vertical disease transmission. It was a challenge that started a wave that revolutionized an industry, evolving to the production structure we know today.

Reflect a moment on WHY that change occurred. Why did the industry convert its production model so completely? Simply stated, it was done to improve the pig's health, to lower morbidity and mortality. It was done in the interest of the pig.

Leadership, lead with humility

Mike Daniels first introduced AASV to servant leadership in his Dunne lecture.[13] From Wikipedia, "A servant leader shares power, puts the needs of [coworkers] first and helps people develop and perform as highly as possible . . . . Instead of the people working to serve the leader, the leader exists to serve the people."[14] In business, the servant leader focuses on the growth and development of those around him or her. Gregg BeVier taught me you should not expect to be promoted until your successor is fully trained to do your job.

While servant leadership is a timeless concept, the phrase "servant leadership" was coined by Robert K. Greenleaf in "The Servant Leader", an essay first published in 1970.[15] The Bible actually quotes Christ on the topic of a leader acting as a servant five times in three gospels (Matt 23:11, Matt 20:26, Luke 22:26, Mark 10:43, Mark 9:35).

The servant leader exhibits nine characteristics:[16]

1. Values diverse opinions
2. Cultivates a culture of trust
3. Develops other leaders
4. Helps people with life issues
5. Encourages
6. Sells instead of tells
7. Thinks you, not me
8. Thinks long-term
9. Acts with humility

Booker T. Washington said of the servant leader, "Most leaders spend time trying to get others to think highly of them, when instead they should try to get their people to think highly of themselves."[17]

There is a place for servant leadership in AASV. I believe examples abound indicating servant leadership is more than a fiber; it is a rope running through the history (and I predict the future) of AASV.

Can you be a frog kisser?

The true servant leader recognizes "Often the person with the most authority, the leader, is the most removed from the real action and thus has the least information on which to base decisions and strategies."[13] The wise swine consultant/practitioner recognizes many health issues (even the most complex issues) are rooted in the basic and fundamental truth: pigs need feed, water, air and comfort. This is not a new idea, "in 1958 the famous pathologist Ken Jubb made the then novel statement...that scour in piglets was best regarded primarily as a husbandry problem."[18] Dr. Stan Curtis taught this concept in his graduate-level courses at the University of Illinois in 1980s. Dr. David Bechtol applied this to his consultation practice in a method he termed "the frog kisser."

> Dr. David Bechtol was the fourth president of AASV. Scanlon Daniels mentioned Dr. Bechtol in the Alex Hogg Lecture in 2015. John Waddell mentioned David again in the Dunne Lecture in 2019. David has the unique distinction of serving as president of Academy of Veterinary Consultants (1972), AASP (1973), AABP (1979), and American Society of Agricultural Consultants (1991-1992). He was the AABP Practitioner of the Year in 1981 and AVC Consultant of the Year in 1986. Those credentials alone establish his skills and expertise. For those AASV members unfamiliar with David, I challenge you to learn about and from him.

David recognized the success of any health program hinges on timing and execution of basic tasks. As a feedlot consultant, he used the example of a science-based, state-of-the-art receiving/vaccination protocol applied to calves arriving at a feedyard. No matter how brilliant or insightful the program, its success relies heavily on the timely execution/application by the processing crew. The success or failure of the entire feedlot rests every day on the processing crew, typically the least educated, least experienced, lowest ranking, lowest paid segment of the feedlot hierarchy. Realizing that truth, David practiced the art of befriending, knowing, teaching, mentoring and encouraging the individual members of each yard's processing crew. He never missed an opportunity to meet them in the treatment shed to share lunch when he visited a yard. David intentionally and actively sought to raise up each member of the crew (no matter how new he or she might be), "kiss the frog" with education and encouragement, show him or her their value and turn him or her into the prince or princess befitting the job they daily performed.

A parallel can be drawn in pig production. The most successful swine health program starts with the assumption of a healthy pig at birth. The piglet's first health challenge is arguably neonatal diarrheas. The surest way to prevent neonatal diarrheas is proper sanitation of the environment. And who is responsible for the sanitation of the farrowing room pre-loading? The person running the pressure washer, likely the person with the least experience, the most recently hired, the lowest paid (and perhaps least educated), is the individual that can undermine the entire health program before the piglet is even born. The person operating the pressure washer is the sow farm's frog, and it behooves the others involved to lift up (kiss?) the frog, help him or her understand the value, importance and significance of the task, turn that person into a prince or princess befitting their importance and encourage them often.

Consider the observation by Larry Firkins. "It's not an issue needing to motivate employees, it is a case of needing to stop demotivating them."[19] Dr. Firkins also emphasized the effort must be genuine and sincere, "The video must be in synch with the audio to establish trust."

Aside from the farm setting, the servant leader may apply this approach daily. It can be used in the office, in one's practice, at the PTA or church committee, in any organization. The "frog kisser" is a true application of servant leadership.

Can you be a teacher as well as a student?

Without question, AASV is an organization of life-long learners. We embrace the realization, "We're all ignorant, just on different subjects."[20] "And if I claim to be a wise man, that surely means that I don't know."[21] Learning must be a life-long endeavor because we understand age does not bestow wisdom; though perhaps with experience, one might recognize patterns quicker. In our world, "Learning is not compulsory...neither is survival."[22] But our education goes beyond simply learning. We adopt, adapt, adjust and re-direct.

When one accepts the title of DVM or VMD, one is charged with the role and responsibilities of a teacher. The origin of "doctor" traces to the Latin word "docere" meaning to teach. Beginning in the 14th century, "doctor" referred to eminent theologians granted special approval from the Roman Catholic Church as individuals able to explain the doctrines of the church. Over time, its use evolved to reference qualified academic and medical professionals.[23] The title and role of teacher is thus conferred and restricted to those duly ordained.

It is through teaching that we participate, guide, advise and direct our clients specifically and the pork production industry generally on a wide range of topics. "As veterinarians, we have influenced the structure of the industry today, the design and construction of buildings, the dissemination of genetic material, standard operating procedures (SOPs) in place at most farms today, the training and expertise of employees, the nutritional needs and the list goes on and on."[3] Teaching is fundamental to our role in the industry and in agriculture.

We must be teachers. Gordon Spronk closes his emails with "Every day learn, every day teach." Teaching is an active, involved exercise. Teaching is not simply providing a reading list. For some, that may be an unfamiliar or uncomfortable role, but "Every day we should do something that scares us."[24] We must share knowledge, experiences, and insight. Surgeons specifically and veterinarians in general practice the doctrine of "see one, do one, teach one."

And we share

We must not forget, and I emphasize again, we are one team. Harry Truman said, "It is amazing what you can accomplish if you do not care who gets the credit."[25] Barb Straw wrote "We should share information and observations even if they are less than science-based and the practitioner will sort through those values."[26] Why do we share? Why could we, should we, must we share? We share for the benefit of the pig, in the interest of the pig.

As an organization leading an industry, our profession has tackled the unknown with the tools available at the time, and then we shared! The examples are plentiful.

> Howard Dunne himself with SMEDI and Parvo.
> Erwin Kohler and Harley Moon with *E coli*.
> Hank Harris and Bob Glock with dysentery.
> Kelly Lager, Amy Vincent and Marie Culhane with influenza.

We could also add disease elimination, air filtration, and feed and transport biosecurity. The list may not be infinite, but nearly so.

Our history however also gives evidence sharing has not always been practiced within our organization or our industry. Al Leman was direct in his criticism. "For too long, swine practitioners have had 'their clients' and 'their territory' and have failed to utilize the full range of expertise available to them."[27] Al advocated a team approach and with it, greater accountability.

In some respects, no doubt we have fallen short. In 2007, Steve Henry was clear and direct in his admonition. At that time, after nearly 20 years dealing with PRRS, "By all the metrics of disease management, we have not succeeded. The many failings, mistakes, betrayals and inability to confront the brutal facts are a real part of AASV history."[28] Scott Dee observed the lack of a united, focused effort in the early days of Mystery Swine Disease (PRRS). "There was no collaboration. There was tremendous conflict of interest. People were trying to develop vaccines and nobody was sharing. It brought forth the quotation from the late Dr. Mike Murtaugh...in those days 'veterinary research is an oxymoron.' That was clearly true."[29]

More recent events suggest we have collectively changed course. Circovirus and swine enteric coronaviruses (SECs) provide examples of our capabilities when we share. PRRS virus dominated the programs and the hallway conversations of every swine veterinary meeting I can recall from the time it was first reported until the Leman conference in 2006. Then, for a brief moment in history, circovirus seized the swine health spotlight as the industry hastily struggled to understand and mitigate the unpredictable rises in morbidity and mortality. As a professional group, we debated the relevance and significance of circovirus because we understood it was ubiquitous. But when we identified the variability in pathogenicity based on virus type, we moved quickly to develop and disseminate highly effective vaccines. Certainly, porcine circovirus today remains an adversary among us, but collectively we have knowledge and tools to control it.

The coronaviruses provide similar evidence of success through teamwork. The images of dead pigs in those earliest (and undiagnosed) cases are impossible to forget. When Drs. Ackerman and Dufresne described what they saw, it was almost beyond belief. The industry was amazed at the devastation and the speed with which the pathogen traveled. It seemed to ignore biosecurity barriers that worked effectively against other pathogens. We turned collective attention and resources to the problem, identified the pathogen(s), learned and shared effective strategies. SECs had a huge advantage over producers and veterinarians alike, catching the industry by surprise; but the impact was short-lived (albeit devastating in affected sow herds) once we faced the challenge as an industry.

Conclusion

Marie Antoinette has been quoted, "There is nothing new except that which is forgotten."[30] That may not always be true, but the thought has merit. We should navigate the future from a solid foundation built in the past. "Preservation of what has worked in the past will provide the foundation for making the right decisions in the future – even if changes need to be made."[31] We can face the future with a level of confidence, having survived the past.

This lecture has offered nothing new. It shared what was already delivered in previous Dunne lectures, though perhaps it was packaged a bit differently. "I am proud to be a veterinarian and member of this group that, with enthusiasm greets goals as a challenge and each day as an opportunity."[12] These were Steven Henry's words in 1989; but today, humbly, I embrace them as my own. Let us be proud of our past as we navigate our future; but not so focused on our history that we lose sight of our future. Lonnie King warned in 2003. "It is easy to get so distracted looking into our rear-view mirror that we may end up crashing into the future."[32]

Tom Burkgren challenged us, again quoting Jim Collins, "Regardless of past achievements, you will always be merely good relative to what you can become."[31] But, Tim Blackwell countered, "Look around the room, the path to exceeding expectations is all around us."[33]

How will we navigate the future? Gregg BeVier asked, "Are you prepared to invent your future?"[34] Based on history and science, I respond confidently and perhaps a bit defiantly. I do not know every detail, nor can I provide every answer, but we are ready. There will be circumstances and situations that come unexpected and uninvited. We'll get it done. When we talk about navigating the future, we only need remember two things:

One. Start with WHY.

We combine science with common sense, in the interest of the pig, to help feed the world. That's our WHY, our purpose and our core belief.

Two. We are one team.

References

1. Deen J. Prevention and control of emerging and re-emerging zoonotic swine diseases. Proc APVS. Ho Chi Minh City, Vietnam, 2013:K22-24.

2. Sinek S. Start with why. Penguin Books Limited, London. 2009.

3. Olsen D. The pig always wins. Proc AASV. Dallas, TX. 2014:15-19.

4. Collins J. Quoted in Spence R. It's not what you sell, it's what you stand for. Penguin Books Limited, London. 2009.

5. Stevenson G. Because it's the right thing to do. Proc AASV. Orlando, FL. 2015;13-25.

6. Rowles C. Perspective, passion and the purpose inspired practice. Proc AASV. San Diego, CA. 2013:15-20.

7. Thomson J. Securing the future of the swine veterinarian while our nation's food insecurity increases. Proc AASV. Dallas, TX. 2009:13-19.

8. Booker T. Washington quotes. **http://americanradioworks.publicradio.org/features/blackspeech/btwashington.html.** Accessed November 11, 2020.

9. Curtis S. Swine welfare: biology as the basis. Proc AASV. Denver, CO. 1990:165-179.

10. Shakespeare W. "Measure for Measure," Act 1, Scene 4.

11. Day R. The Quick and the Dead. HBO. 1987.

12. Henry S. Assessing grow/finish: Why is the system no better? Proc AASV. Des Moines, IA. 1989:141-170.

13. Daniels M. Leading in the midst of change. Proc AASV. Des Moines, IA. 1998:1-21.

14. Wikipedia servant leadership. **https://en.wikipedia.org/wiki/Servant_leadership.** Accessed November 7, 2020.

15. Greenleaf R. The servant leader pamphlet. **https://www.greenleaf.org/what-is-servant-leadership/. 1970.** Accessed November 7, 2020.

16. Inspired leader concepts. **https://inspiredbusinessconcepts.com/leadership/nine-behaviors-of-servant-leaders/.** Accessed November 7, 2020.

17. Booker T. Washington quote. **https://www.azquotes.com/quote/844250.** Accessed November 11, 2020.

18. Alexander T. Disease control present and future. Proc AASV. St. Louis, MO. 1988:115-126.

19. Firkins L. Getting there from here. Proc AASV. Dallas, TX. 2009:23-35.

20. Clemens S. quote. **https://www.relicsworld.com/mark-twain/we-are-all-ignorant-just-about-different-things-author.** Accessed November 11, 2020.

21. Kansas. Leftoverture, Discogs. 1976, vinyl.

22. Deming W. Brainy quote. **https://www.brainyquote.com/quotes/w_edwards_deming_163061.** Accessed November 7, 2020.

23. The history of 'doctor'. **https://www.merriam-webster.com/words-at-play/the-history-of-doctor.** Accessed November 7, 2020.

24. Lautner B. Classical swine fever: could it be us, could it be you? Proc AASV. St. Louis, MO. 1999:297-300.

25. McCollough D. Truman. Simon and Shuster, New York City. 1993.

26. Straw B. Veterinary practice: art, science and politics. Proc AASV. Quebec City, Quebec. 1997:1-31.

27. Leman A. Preparing for swine practice in the 1980s. Proc AASV. Kansas City, MO. 1981.

28. Henry S. A seat on the bus – lessons from Dr. Alex Hogg. Proc AASV. Orlando, FL. 2007:27-32.

29. Swine research past and future. **https://youtube/qL4ayWjIGYQ.** Accessed November 10, 2020.

30. Antoinette quote. **https://www.goodreads.com/quotes/149557-there-is-nothing-new-except-what-has-been-forgotten.** Accessed November 11, 2020.

31. Burkgren T. AASV: Good to great. Proc AASV. Orlando, FL. 2007:9-25.

32. King L. As the future catches you. Proc AASV. Orlando, FL. 2003:9-21.

33. Blackwell T. Exceeding expectations. Proc AASV. Kansas City, MO. 2002:9-18.

34. BeVier G. Beyond the basics: what will the future bring and how will we get ready? Proc AASV. Kansas City, MO. 2006:9-26.

Alex Hogg Memorial Lecture

Enhancing your brand: the value of lifelong learning, continuing education, and teaching to the swine industry

J.S. Pittman, DVM, MS, DABVP

In remembrance and honor of Alexander Hogg

Alexander Hogg
1920-2006

A native of Scotland, Alex (pronounced "Alec") Hogg immigrated to the United States with his family when he was six years old. After reaching adulthood, he served in the US Navy during World War II before entering veterinary school at Kansas State University. He received his DVM degree in 1950, and was a mixed animal practitioner in Iowa for twenty years. Dr. Hogg returned to college - this time, Iowa State University - to complete a Masters Degree in 1972. He spent the following years serving and educating swine veterinarians and producers in his roles as Nebraska Swine Extension veterinarian and professor at the University of Lincoln-Nebraska. Upon "retirement" from the university in 1990, Alex worked as a technical services consultant for MVP Laboratories. Ever an enthusiastic learner, he graduated from the University of Illinois Executive Veterinary Program in 1995.

An active and longstanding member of the AASV, Dr. Hogg joined what was then the American Association of Swine Practitioners shortly after it was formed. He served as secretary/treasurer of the association from 1974 to 1976, and was elected president in 1979.

The many awards and honors bestowed upon him over the years included several presented by the AASV: The Howard Dunne Memorial Award in 1985, the Meritorious Service Award in 1996, and in 2005, the first-ever AASV Foundation Heritage Award for a lifetime of achievement in veterinary medicine.

ALEX HOGG MEMORIAL LECTURERS

2011 Joseph F. Connor
2012 Lisa Tokach
2013 John T. Waddell
2014 Mark J. Engle
2015 C. Scanlon Daniels
2016 Peggy Anne Hawkins
2017 Matthew Turner
2018 Rodger Main
2019 Deborah Murray
2020 William L. Hollis

Enhancing your brand: the value of lifelong learning, continuing education, and teaching to the swine industry

J.S. Pittman, DVM, MS, DABVP
Smithfield Hog Production – North Region, Waverly, Virginia

Remembering Dr. Alex Hogg

There's two dates in time that they'll carve on your stone
And everyone knows what they mean
What's more important is the time that is known
In that little dash there in between

~Garth Brooks, Pushing Up Daisies

Let us first look back and learn about the remarkable individual for whom this lecture is named; Dr. Alex Hogg. Dr. Hogg was born on April 15th, 1920 in Borgue Scotland. In 1926 he moved with his family to the United States. After graduating high school he spent 2 years attending a business college in Chillicothe, Missouri where he became a student instructor, which would shape his love for teaching. It is also interesting to note that to supplement his education as a young man, he read the entire encyclopedia, demonstrating a drive to continuously learn. Alex joined the US Navy in 1941, serving in many locations during World War II, and most nobly requesting sea duty after the attack on Pearl Harbor. After his service, Alex utilized the G.I. Bill to pursue a degree in veterinary medicine, which he received in 1950 from Kansas State University. For 20 years, Dr. Hogg worked in a mixed animal practice in Coin, Iowa. In 1970, at the age of 50, he attended Iowa State University to obtain a master's degree in veterinary pathology. After graduating Iowa State, Dr. Hogg became the Swine Extension Veterinarian and professor at the University of Nebraska-Lincoln. During this time Dr. Hogg provided bountiful information to the industry through seminars, extension letters, programs and classes; from local instruction to worldwide lectures. The programs were science based and practical, aimed at helping the producer resolve health and production issues. Through these experiences, Dr. Hogg became an international expert in swine production and health. Dr. Hogg was also instrumental in the efforts to eradicate Hog Cholera from the United States.

Upon retiring (used very loosely here!) from the University of Nebraska in 1990, Dr. Hogg became a Technical Services consultant for MVP Laboratories. During this time, Alex authored a monthly newsletter entitled "Hogg Sense" for over 11 years, that was available to the entire industry. This newsletter was a one-pager that highlighted information from Dr. Hogg and other authors on a few key swine topics; short and to the point. With the final issue of "Hogg Sense", and in direct opposition to his character ("shy and humble"), his friends and family encouraged a change in the format (it was NOT about pigs and was TWO pages!) to provide the industry with information about Alex, his life and accomplishments. In 2012, Alex's wife Mary Lou, donated a complete set of the "Hogg Sense" newsletter to the AASV Foundation Auction, which through a consortium of members, was sold for $2,750.00 and then gifted back to Mary Lou, who in turn, graciously donated it back to the AASV.

In the first Alex Hogg Memorial Lecture (2007), appropriately given by Dr. Steve Henry, Dr. Hogg was described as "a builder". "He always continued to learn and to teach. He combined leadership, passion and vison with humility." Alex had the desire to learn, to share and to encourage those around him. The fitting statement that he "elevated those around him" was used by Dr. Henry. Alex always strived to learn more. At 75 years of age, Alex completed the Executive Veterinary Program at the University of Illinois, and as a member of "the Pride of '95" class, was still eager to learn and interact with colleagues. Dr. Larry Firkins in the 2009 Alex Hogg Memorial lecture, remembered Mary Lou telling him "He didn't stop there, Larry, he never stopped." Dr. Firkins also commented that "his enthusiasm as a lifelong learner was uncanny."

Dr. Hogg was a leader in the American Association of Swine Practitioners/Veterinarians, having served as secretary/treasurer of the organization from 1974 to 1976 and President in 1979. Dr. Hogg received numerous accolades including the Howard Dunne Memorial Award in 1985 and the Meritorious Service award in 1996. In 2005, the organization honored Dr. Hogg with the first Heritage Award, presented in recognition of a lifetime of achievement in swine veterinary medicine.

Continuing the legacy

Dr. Alex Hogg passed away on July 18th, 2006, but his influence and commitment to the swine industry continues to this day. In early 2007, the AASV announced the Alex Hogg Memorial Fund that was supported by the Hogg estate and established to recognize and remember Dr. Hogg's contributions to the swine industry and to support lecture and scholarship. The fund is used to support this lecture, the Alex Hogg Memorial Lecture, presented each year at the AASV Annual Meeting and to support a scholarship program for swine practitioners who wish to expand their horizons by pursing an advanced degree.

This is the 15th Alex Hogg Memorial Lecture, and I am both honored and humbled to have been selected to provide this lecture to the association membership and can only hope that I provide value to the industry while representing Dr. Hogg's legacy. I am truly honored to be included on a list of exceptional past speakers, several of whom I know personally, have interacted with closely, and who that have directly influenced my career, but of all the speakers on the list I highly respect and exemplify the character of Dr. Alex Hogg. In addition to this lecture, I have been fortunate to have also benefited from the Alex Hogg Memorial Scholarship, which I co-received in 2011 and used to complete a master's degree in Veterinary Preventative Medicine in 2014. Because of the scholarship, I was fortunate to have met Mary Lou in 2011 and ever since then when we meet at the annual meeting, she is exceptionally kind and interested in my continued involvement with the AASV. I am sure the other scholarship recipients have the same interaction with her and feel the same connectedness to Alex's legacy as I do. Again, I am humbled to be included on a list of so many ambitious and influential colleagues. It is the goal that these individuals will and have, through scholarship, learn new information and skill sets, develop as individuals and professionals, and most importantly share their new-found knowledge with others to benefit the industry worldwide.

I sincerely hope that the following lecture will achieve the goals of the Alex Hogg Memorial Fund, that it will honor the legacy of Alex Hogg, and that it will inspire the audience to embark upon a similar pursuit of lifelong learning and sharing of information. In Alex's memorial tribute it is quoted; "It is for us who remain to be grateful for his influence in our lives and to celebrate a truly remarkable man."

Introduction

The Alex Hogg Memorial Lecture honors his memory by continuing his efforts of lifelong learning, and an ability to educate, encourage and elevate those around him. Therefore, when approached to give this lecture I considered the topics that would most represent the legacy of Dr. Hogg and provide value to the membership. I also wanted to be sure to build on the theme of the 2021 AASV Annual Meeting and the preceding Howard Dunne lecture, given by Dr. Jer Geiger, with the theme of "Navigating the Future Together". For further inspiration I watched Dr. Steve Henry's first Alex Hogg Memorial Lecture (along with the other Hogg Lectures) and Dr. Tom Burkgren's Howard Dunne Memorial Lecture "AASV: Good to Great" from the 2007 AASV Annual Meeting in the AASV's video library [which as a case-in-point for this paper is a great resource to be able to continue to lean on from past lectures] I decided to bring attention to each one of us as the individuals that make up the AASV. How we can each work to improve our abilities to collectively improve ourselves, the organization and the profession. As a graduate of North Carolina State University I must reference a quote from "The Law for the Wolves" from The Second Jungle Book by Rudyard Kipling; *"For the strength of the Pack is the Wolf, and the strength of the Wolf is the Pack."*

Dr. John Waddell said of Alex Hogg's receipt of the AASV's first Heritage Award in 2005; "Throughout his career, Dr. Hogg continually demonstrated his love for learning, teaching, mentoring and promoting the profession of veterinary medicine. He epitomizes the ideal of increasing the knowledge of swine veterinarians." It is imperative that any organization based in science or medicine continue to improve and educate its membership as a basic core value. The AASV's mission is outlined below:

It is the mission of the American Association of Swine Veterinarians to:

- increase the knowledge of swine veterinarians
- protect and promote the health and well-being of pigs
- advocate science-based approaches to veterinary, industry, and public health issues
- promote the development and availability of resources that enhance the effectiveness of professional activities
- create opportunities that inspire personal and professional growth and interaction
- mentor students, encouraging life-long careers as swine veterinarians

It is clear from the mission statement and activities of the AASV that a fundamental pillar of our organization is the education of its' members (and the swine industry), dissemination of knowledge, facilitation of learning and encouragement of the membership to continuously learn. And while the last statement's apparent intent is to students in a conventional sense, I contend that we must all be perpetual "students" regardless of academic matriculation, professional achievements, age, experience or professional seniority.

Please note that while the title of this presentation starts with the catchy "Enhance your brand" statement and the presentation mentions different pathways that an individual may formally advance their education, the intent is by no means to encourage improvement of one's own financial or post-nominal value, but rather to improve their intellectual and personal value as it relates to their professional abilities (their "brand"), thereby improving the value of the AASV collectively.

How much better to acquire wisdom than gold! To gain understanding is more desirable than silver.

[Proverbs 16:16]

Lifelong learning

Never stop learning....Everything you learn you will use somewhere.

~ Marty Sklar, Former President of Disney Imagineering

Increasing your knowledge

We are inherently learners. It is in our nature as humans. Most of us are already programmed to be lifelong learners. The pursuit of a veterinary degree either does that to us, brings that out of our DNA or the profession self-selects for it. Since a young age, we have strived to be part of this profession, and with that desire comes the need to learn, to study and to achieve the next levels of understanding. It shouldn't stop once you receive your veterinary degree. It shouldn't stop once you've become an established veterinarian. It shouldn't stop once you've retired. It should never stop! We are all students and we should always be students. This profession, this industry, this world moves and changes rapidly and information grows exponentially. We accumulate information faster than we can process or implement. We face new challenges all the time and will face more in the future. We have to continuously learn as well as relearn. As you think you have reached the point that you can stop learning, I'd direct you to a quote from Dr. Tom Burkgren's 2007 Howard Dunne Memorial Lecture; "If you think you are great today, tomorrow you're just going to be mediocre."

The goal of lifelong learning and continuing education should be self-motivation and the desire to make one's self better personally or professionally. We should all strive to become better veterinarians than what we are today. We should each take time to critically reflect on our abilities and honestly self-evaluate what could make us better. For example, identify a topic you feel deficient in and seek out instruction on that subject. Reach out to peers or groups that are also interested or actively involved in that topic. Read up on the subject. Search for videos or presentations. Not just the first result in a Google query, but dig deeper. Read a variety of information, historical and current. The last time you had a case or problem, did you explore further? When was the last time you cracked open Diseases of Swine and read the chapter on a pathogen or syndrome? Do you read other sources of information beyond that? What are the 5 written resources you use most? When was the last time you did a literature review on a subject that was plaquing you or your practice? Have you ever done a literature review? What do we know about it and what have others done? Where are the gaps in our knowledge? I challenge you to explore further, dig deeper. Learn how to ask the next question (do you know what question you are actually asking?) and find the next answer. Learn to say; "I don't know!..." but then develop the skill set and say "...but I'll find out!" Continue to develop the skill set that is the real education in veterinary school of how to "find out."

You don't have to pursue formal education to continue to learn. There is plenty of information available to members through the AASV website, the bimonthly JSHAP journal, and other industry publications for those that learn best by reading. For those who prefer to listen to new information, or can take advantage of windshield time, numerous podcasts and lectures are available through many different websites and applications. For the visual learner, there is a growing number of video lectures, recorded webinars and instructional videos being uploaded to places such as AASV, Swine Medicine Education Center, Swine Health Information Center, YouTube or direct from conference organizers. I applaud the Leman Conference organizers who have done this since around 2008. I can tell you from experience, these are great resources to go back and watch the presentations and the slide sets on subjects that have now become relevant to me. We all know that written proceedings may not always provide the same level of detail, contain the most up-to-date information, or convey the same emphasis as the speakers.

I would like to challenge conference and meeting organizers to record more presentations that can be archived, searched and accessed at a later date. These would be valuable resources to the future of our profession as well as provide recorded history of our profession.

Formal education

Opportunities in formal post graduate education can provide individuals with certain skill sets and training not readily available to practicing veterinarians. Depending on the goals, needs and interest of the individual, different programs provide various experiences, different than conventional veterinary school curriculum. There may also be opportunities available to an individual at different times of their career or that work well with current professional requirements. At times an individual may be able to combine a few of the programs together or pursue them concurrently. Many of these programs involve advanced techniques or alternative learning processes that can supplement normal problem-solving strategies of a standard veterinary curriculum. The general veterinary curriculum has gotten cumbersome given all the information accumulation across species. Students interested in swine medicine commonly find less and less instruction on swine specific medicine in their core curricula, and must access this knowledge through other means such as elective courses (if available), special lectures/courses, summer internship programs, externships, conferences (if allowed to attend) or other experience. Some schools provide concurrent degree programs that allow the students to obtain a MS or PhD concurrently with their DVM, or many may continue with these immediately after graduation from vet school. However, many veterinarians pursue practice upon graduation, and then have the ability to return to school for the advanced degrees. This can be a full time return to the university or potentially part-time while continuing to work, possibly even supported by the individual's employer or invested entity. Traditional residency programs for swine medicine are limited but provide another alternative to advance learning in swine medicine post-graduation. Concepts in subjects like statistics, epidemiology, economics, risk assessment, computer programs or machine learning are now commonly used in presentations and health programs across our industry. It would obviously benefit individuals to have formal training in these concepts in order to more effectively use and advance the knowledge of the practice, company or industry. "Classically" trained veterinarians have now moved into management and administration roles in many clinics, production systems, pharmaceutical and affiliated companies, making formal training in business management almost a necessity for these individuals to be successful.

Examples of common formal post-graduate education found among swine veterinarians are listed below:

- Residency or fellowship programs
- Graduate Certificates
- Master's degree: Science (MS), Public Health (MPH), Business Administration (MBA)
- PhD degree
- Post-doc positions
- Executive Veterinary Program – University of Illinois
- Board Certification: Practice based: swine, food animal, other species (ABVP), Pathology (ACVP), Preventative Medicine (ACVPM), Welfare (ACAW), Microbiology (ACVM), Clinical Pharmacology (ACVCP)

While it is the individual that pursued the formal education and received the certificate or diploma, it should not benefit the individual solely, but also benefit the organization and the industry. The information generated from research conducted during the process is usually published, presented and shared, often in conjunction with industry meetings. The knowledge and skill set obtained by the individual is used to make decisions or implement actions that might improve a business or make a client's farm more profitable. Something learned from the experience can be the building block for the next discovery, and the next. The relationships that develop during the process may spawn even further collaborations or ideas that have an impact on the industry. There are many examples of this in the AASV, but one in particular is perfect for this paper. Dr. Roy Schultz, who after 20 years of practice returned to Iowa State to help decipher the cause and develop treatment and control strategies of a new significant respiratory disease that was plaguing one of his client's farm. *Actinobacillus pleuropneumonaie* was identified through Dr. Schultz' work and this information was shared with the industry and had a profound influence on many farmers, veterinarians and the industry worldwide. It should be noted that Dr. Schultz credits Dr. Alex Hogg with influencing his pursuit of lifelong learning. Therefore, all of us as individuals, as an AASV organization and as employers or supervisors should encourage and support the pursuit of formal education by our colleagues.

Continuing education

As with any profession, but especially those in the medical field, there must be a mechanism to continually educate its practitioners. This is often considered an official requirement like Continuing Education (CE) credits, governed by a licensing board, to allow the individual to maintain the ability to practice said profession. The goal is that those engaged in continuing education are up to date on the newest information and what is happening in their fields, thus being more apt to respond appropriately. In professional fields, and the swine industry clearly is not an exception, advances in knowledge and practice evolve rapidly. In swine medicine this has been constantly highlighted with the consistent change in the industry, the role(s) of the veterinarian, and the emergence of new pathogens and challenges. Therefore, our membership cannot remain static in our knowledge and the processes and the processes that we implement for our clients. It is the necessity of the swine veterinarian to learn new information, to adapt and strive to gain more knowledge in order to better serve their client, the profession and the pigs. Whether the subject is a new emergent disease, an old disease with new control strategies, a new diagnostic sampling technique or test, new regulations or simply a review of a "common" disease you haven't had to deal with until recently, the idea is that you have the ability to learn this information from reputable and qualified sources.

However, one should also strive to gain non-requirement based learning. More importantly it should be the effort of the individuals to learn and share new information, concepts and ideas and for the organizations to help formally provide this experience. Information and data move so fast in the world today and there is just too much of it (just ask any veterinary student in the audience). We can't be expected to learn it all, but we should be willing and able to readily gain that knowledge when needed. In addition, many of us have very little time to devote to learning everything that is out there, so the learning requirements very often are reactionary and directed to the closest and hottest "fire(s)" in that moment. As medically minded and trained individuals we have an uncanny ability to absorb volumes of information, organize, compartmentalize and redistribute that information into active tasks. We unfortunately are also human and simply cannot process the vast amount of information that comes constantly, so we elect to focus on information that we feel we need immediately and reserve the right to go look for other information later when needed.

Simply consider the AASV Annual Meeting. There are well over 200 talks and posters, four to six concurrent sessions, over 4-5 days of conference (including those pre-pre-conference meetings many attend), let alone the usual tech table, hallway, breakfast, lunch, dinner and barstool conversations amongst the membership. Thankfully, it is not the intent of the AASV to have their members learn all the information or attend all the sessions, but rather to provide a broad array of information that is valuable to different individuals and situations. If facilitated effectively, the information is still available to the members for review at a later date by way of proceedings, on-line lectures, video capture, etc. I personally feel that if I leave a conference having truly learned something new, been introduced to a different approach or thought process, have a list of to-do's when I get back to work, and had an interaction with someone I previously did not know then it has been a successful conference. I believe that if I leave a conference having a feeling that I wasn't able to "get it all", it was actually a successful conference on behalf of the organizers. There's always more to learn next time!

Teaching

Teaching is the reciprocal of learning. Effective teaching requires the teacher to really learn the material and the audience ("learn one" and "do one" come before "teach one"). In swine medicine we classically think of teaching as a role of professors and clinicians in veterinary school or graduate programs, however, many of us "teach" on a daily basis. We teach clients, farm employees, production servicepersons, managers, students, colleagues, or even laypersons in the community. Teaching, imparting knowledge on others, transferring information or processes is core to what we do as professionals. It's explaining the "Why" and the "How" of a health or management program at different levels. We teach each other as colleagues.

It is the responsibility of the AASV to facilitate ways for their membership to learn from its teachers. We feel it is the responsibility of our industry organizations to provide us with the best teaching available for new and important information. In fact, educating swine veterinarians is what the AASV does. We rely on the organizing committee to decide the key topics to discuss. The topics are timely and important to a large portion of the membership. Regardless of the structure or organization of the topics, it is always important to couple the information with appropriate and effective delivery. For this to happen it is a shared responsibility of the organization to get the right people to speak on the subject, but more importantly for the individual(s) to clearly and effectively "teach" the material to the audience. As the individuals that make up the organization, we should recognize the importance of giving a lecture or talk to members of our organization.

Often it is said of the AASV meeting that the best presentations are in the student session. While that session being a competition helps "up the ante", it is mostly because the students have been well prepared. They have a support system of mentors, professors and peers. The work completed is well thought out and planned. The data is presented with structure and conveys a clear message. The work is timely and usually provided immediate applicability to the industry. And of course, they practice their talks and potential questions, so they are well prepared. What if we all prepared our talks in the same manner or with the same effort? Or attempted to achieve even half the quality of those given in the student session? Interestingly, over the last few years at the AASV Annual Meeting I have noticed several speakers (previous student session participants) giving presentations with the same level of care, attention and preparation. We have all attended lectures where the value of the information was diminished by presentation, structure or poor execution. Anyone asked to speak, should strive to provide the same level of professionalism. This would improve our individual and collective value greatly.

It is the responsibility of university professors and visiting lecturers to provide veterinary students with the most current information available on swine diseases, production and the industry. This is not just in the swine medicine courses (if unique), but in other core curricula where swine pathogens (eg. microbiology) or concepts (eg. pharmacology) are involved. Universities with true swine faculty should work closely with those non-swine professors to update the swine sections of their lectures. If a swine faculty is not available, representatives of the AASV should reach out to help provide this resource. The AASV should continue to explore ways to supplement this education. In addition, those in a position of teaching, should make conscious efforts to update their teaching materials frequently. In my opinion, this is part of the value of continuing education for the teachers in our membership. Attending conferences, meeting with practicing veterinarians and engagement in the field, allows professors to stay abreast of hot topics, new developments and how things are actually being done, that can then be taken back to the university setting. In my experience, students really appreciate being taught with this perspective.

Practitioners! I charge those of you with an interest in teaching and interaction with students to actively reach out to your local university or swine/food animal clubs and request an audience with the students. Go meet and talk to the students; about what you do, the swine industry, an interesting or even a basic case, review differential diseases for syndromes, antibiotics, welfare, etc.

Consider becoming adjunct faculty and help formally teach (there are benefits to this assignment that are not monetary). Students should be considered future colleagues. You should consider that what you are doing is helping to train a future colleague. It is cliché, but you can learn just as much from students as you are teaching to them. You do have to do your due diligence on keeping up with information so that you can successfully teach, which by default, makes you better in the process.

University professors also have an important responsibility in providing many students with their first exposure to swine and the swine industry. While the vast majority of the student body will not become involved with the swine industry, they will all be influenced from what they learn in school. Be part of that positive influence, as others are attempting to influence in a negative direction. Some may become ancillary supporters of our industry, which is a resource we can never have enough of.

University administration must facilitate student's opportunities, either by providing direct opportunities or helping students connect with the industry. Students should be encouraged to attend national association meetings, and when possible, allowed to make up missed work (the vast majority of lectures are recorded and notes are provided digitally). Conference attendance is an important staple of all veterinary disciplines. It is the main way in which veterinarians network with colleagues, obtain required CE, and experience new aspects of the profession. For students interested in swine medicine, what better place to network, gain opportunities, find positions and experience the comradery of our industry than the AASV Annual Meeting and other industry meetings and events? To paraphrase Mark Twain's famous saying, "never let schooling get in the way of your education."

Mentoring

An offering from a Sage and King is worth more than Silver and Gold. It is a seed of hope, a bud of faith.

~Chadwick Boseman

I include under the teaching umbrella, mentorship. I, like all of us, have benefited and still benefit greatly from mentorship in my career. Like many new swine veterinarians and veterinary students, I did not have a background in swine production or medicine (or any agriculture for that matter) and whether by accident or serendipity, I was introduced and immediately drawn into the swine industry. In my case this was primarily because of two things; amazement and mentorship. Learning about modern swine production was "eye-opening." Having not come from an agricultural background I had general misconceptions and plain ignorance on the subject. I've always had a fascination with learning how things worked or were made (I grew up with Mr. Rogers, Reading Rainbow and Richard Scary books). However, more than learning something new, it was the outstanding mentorship that got me hooked. Dr. Billy Flowers taught the swine management course at NC State, and anyone that knows Dr. Flowers immediately knows what a great teacher, mentor and supporter of students he is. That semester was followed with a summer "hands-on" opportunity at the university Swine Unit where the manager and crew continued to teach and support my newly found interest in swine. This mentorship continued through veterinary school with Drs. Glen Almond, the late John Roberts, the late Monte McCaw, Gene Nemechek, Al Scheidt, and a host of others during my 4th year clinical rotations (namely the 3 months spent in Iowa and at Iowa State University). I have no doubt that without the positive mentorship of these individuals, I would not have pursued a career in swine medicine, nor have been as successful.

I strongly believe that a core strength of our profession is the genuine positive experiences we can and do give to students and colleagues. When it comes to "recruiting" new students to the swine industry, I strongly feel that if you can provide a positive hands-on learning experience you will succeed. I encourage all of you to engage with students when we are back meeting face-to-face or reach out through virtual platforms. When you are standing around chatting and you notice a "stalking" student, invite them to join the conversation, introduce them to other veterinarians, learn about them. Remember, many students (and veterinarians) can be intimidated by the aura of older or established veterinarians.

In the organization, there is an overwhelming support of others to mentor even the older student in us. For example, I have always had positive and supportive experiences when I have reached out to another colleague for help with a questions or problem, even if I did not know that person at the time. This is the "Together" that Dr. Jer Geiger references in his 2020 Howard Dunne Memorial Lecture. It is the sense of "family" felt by many in the AASV. We should all have a group of colleagues we consider our core advisors, and our "phone-a-friends." These are the people that you listen to and write down what they say, even if you don't agree or necessarily understand the topic. I have mine, I encourage you to develop yours.

Conclusion

I hope that this lecture has succeeded in honoring Dr. Alex Hogg. It has allowed me to reflect on Dr. Hogg's influence on me and to be introspective about what more I can do, for myself, for my company and for the AASV. However, continuing the goal of this lecture can only be accomplished if this material has also had some influence on the audience; a form of perpetual pay it forward. Dr. Alex Hogg had an influence on Dr. Steve Henry. I have been fortunate to have been influenced by Dr. Henry. I hope that I have influenced others and hope that they, in turn, influence others. I hope that it has inspired you to actively continue lifelong learning. I hope that this has sparked in you the desire to encourage and lift others around you. My hope is that this talk has inspired someone to pursue additional formal education and share their knowledge back with us. My hope is that this talk has encouraged you to start or continue to mentor individuals. If just one person has been influenced by this talk to make themselves a better swine veterinarian, then this has been worth the effort.

Acknowledgements

Thank you to everyone who has interacted with me throughout my career to get me here. I am the product of all those experiences. Thank you to AASV President-elect and 2021 Program Chair Dr. Mary Battrell and the 2021 Program Planning Committee for asking me to give this lecture. Thank you to the AASV for their hard work and direction. Thank you to Mary Lou Hogg for her continued support of the AASV and the scholarship recipients. Most of all thank you Dr. Alex Hogg for the example you set and the legacy you have provided!

Detection and diagnostic trends of five swine endemic bacterial pathogens (2010-2019)

A. P. Poeta Silva[1], DVM, MS; K. Schwartz[1], DVM, PhD; B. Arruda[1], DVM, PhD; E. Burrough[1], DVM, PhD; J. Santos[1], DVM; N. Macedo[1], DVM; O. Sahin[1], DVM, PhD; K. Harmon[1],PhD; P. Gauger[1], DVM PhD; C. Siepker[1], DVM, PhD; P. Sitthicharoenchai[1], DVM, PhD; M. Rahe[1], DVM, PhD; D. Magstadt[1], DVM, PhD; A. Michael[1], DVM, PhD; P. Pineyro[1], DVM, PhD; R. Main[1], DVM, PhD; R. Derscheid[1], DVM, PhD; E. Fano[2], DVM, PhD; M. Clavijo[1,3], DVM, PhD
[1]Veterinary Diagnostic Laboratory, Iowa State University, Ames, Iowa;
[2]Boehringer Ingelheim, Atlanta, Georgia; [3]PIC®, Hendersonville, Tennessee

Introduction

Streptococcus suis (SS), *Glaesserella parasuis* (GPS), *Mycoplasma hyorhinis* (MHR), *Actinobacillus suis* (AS), and *Mycoplasma hyosynoviae* (MHS) are bacterial agents that significantly compromise animal health and welfare in modern swine production, leading to decreased productivity and an over-dependence on antimicrobials. Recently, the Swine Health Information Center[1] released a bacterial swine disease matrix and placed these agents within the top ten bacteria of concern for the swine industry. These agents, usually considered endemic, early colonizers of the upper respiratory tract of pigs, can cause severe systemic disease that often culminates in increased herd morbidity and mortality. While some strains of these organisms can be considered primary disease-causing agents, environmental impacts, co-infections, and/or disorders of the immune system[2] are usually the main drivers of systemic disease outbreaks in swine farms. Accurate diagnosis of disease caused by these agents in the field is challenging due to their commensal ecology and polymicrobial interactions. The goal of this study was to describe the frequency of detection and disease diagnosis using data from the Iowa State University Veterinary Diagnostic Laboratory (ISU VDL).

Materials and methods

Porcine diagnostic submissions with detection of these pathogens via isolation (SS, GPS and AS) and PCR (GPS, MHR and MHS) from preferred sites (central nervous system samples, fibrin, heart, joint, kidney, liver, lung, pleura, and spleen) were selected from the ISU VDL data base over a 10-year period (2010-2019). Thus, temportal patterns in detection of these pathogens was based on one entry for isolation or one PCR result for each agent and for each sample type (preferred sites).

In addition, disease diagnosis data included in this study was based on the pathologic assessment in the context of the clinical history, i.e., a diagnostician reviewed the case and assigned the diagnosis, results of laboratory testing using preferred sites, and associated lesion with disease diagnosis for specific porcine body systems (cardiovascular, musculoskeletal, nervous, respiratory or systemic). Thus, the temporal pattern in disease diagnosis of these pathogens was based on one entry for each body system, for each associated lesion, for each agent and for each sample type (preferred sites).

Results

Temporal patterns in detection and disease diagnosis trends of these 5 agents are shown in Figure 1. Over a 10-year period, a total of 40,676 porcine cases had isolation of any bacterial agent from the preferred sites, from those, 30,808 cases had isolation of one or more of SS (25,258, 81%), GPS (7,236, 23%), and AS (3,119, 10%). PCR testing was performed in 63,269 porcine cases using the preferred sites, from those, PCR detected GPS in 8,661 of 8,832 cases; MHR in 4,014 of 8,162 cases; and MHS in 408 of 2,221 cases.

Based on pathologicassessment in the context of clinical history, a total of 60,000 porcine cases had any disease diagnosis, in which 42,884 cases were identified as having disease in one of the body systems of interest. From those, 10,330 cases received a final disease diagnosis related to at least one of these 5 bacteria. A total of 13% of all cardiovascular cases (2,187) were given either a SS or GPS etiologic diagnosis; 22% of all musculoskeletal cases (1,500) were given either a SS or MHS etiologic diagnosis. From all nervous (1,653) and respiratory cases (28,249), 40% and 9% were given SS etiologic diagnosis, respectively. In systemic cases (15,833), 15%, 15% and 6% were diagnosed with SS, GPS, and MHR-associated disease.

Over a 10-year period, lung (46% of cases) and CNS samples (12%) were frequently used to diagnose SS disease (Figure 2), while lung, fibrin and heart were used to diagnose GPS disease (39%, 19%, 14%), and MHR disease (31%, 24%, 19%).

Disease due to AS was often diagnosed using lung and spleen (60%, 9%). MHS disease was prevalently diagnosed using joints. As showin in Figure 2, the primary lesions associated with SS disease were bronchopneumonia (45%), polyserositis (41%), meningitis (12%), epi/endocarditis, and systemic infection (6% each). Lesions associated with GPS diagnosis (Figure 2) were polyserositis (67%), bronchopneumonia (30%), epi/endocarditis (8%), systemic infection (6%), arthritis (3%) and meningitis (2%). Lesions associated with MHR diagnosis were polyserositis (87%), sepsis (9%), arthritis (8%), and epicarditis (6%). For AS diagnosis, lesions included pleuropneumonia (59%), systemic infection (43%), and polyserositis (5%). MHS diagnosis was only associated with arthritis in 155 cases.

Conclusions and discussion

Results of this study demonstrate a substantial increase in the annual detection and disease diagnoses (Table 1) for all agents, with the exception of MHS. The factors contributing to increased disease diagnostic frequency may be varied and include improved diagnostic assays and protocols, changes in diagnosticians, improved data management system, increased awareness and number of submissions, and changes in antimicrobial use. Lungs were frequently used for SS, GPS, and MHR disease diagnoses (Figure 1 and 2); however, their use for systemic disease diagnosis requires caution due to the commensal nature of these agents in the respiratory system, compared to systemic sites typically targeted by pathologists (joint, brain, spleen, etc). The anatomic location selected for sampling from acutely affected, nonmedicated animals,

Figure 1: Number of porcine cases submitted to the ISU VDL over a 10-year period with detection and disease diagnosis of five endemic bacterial agents.

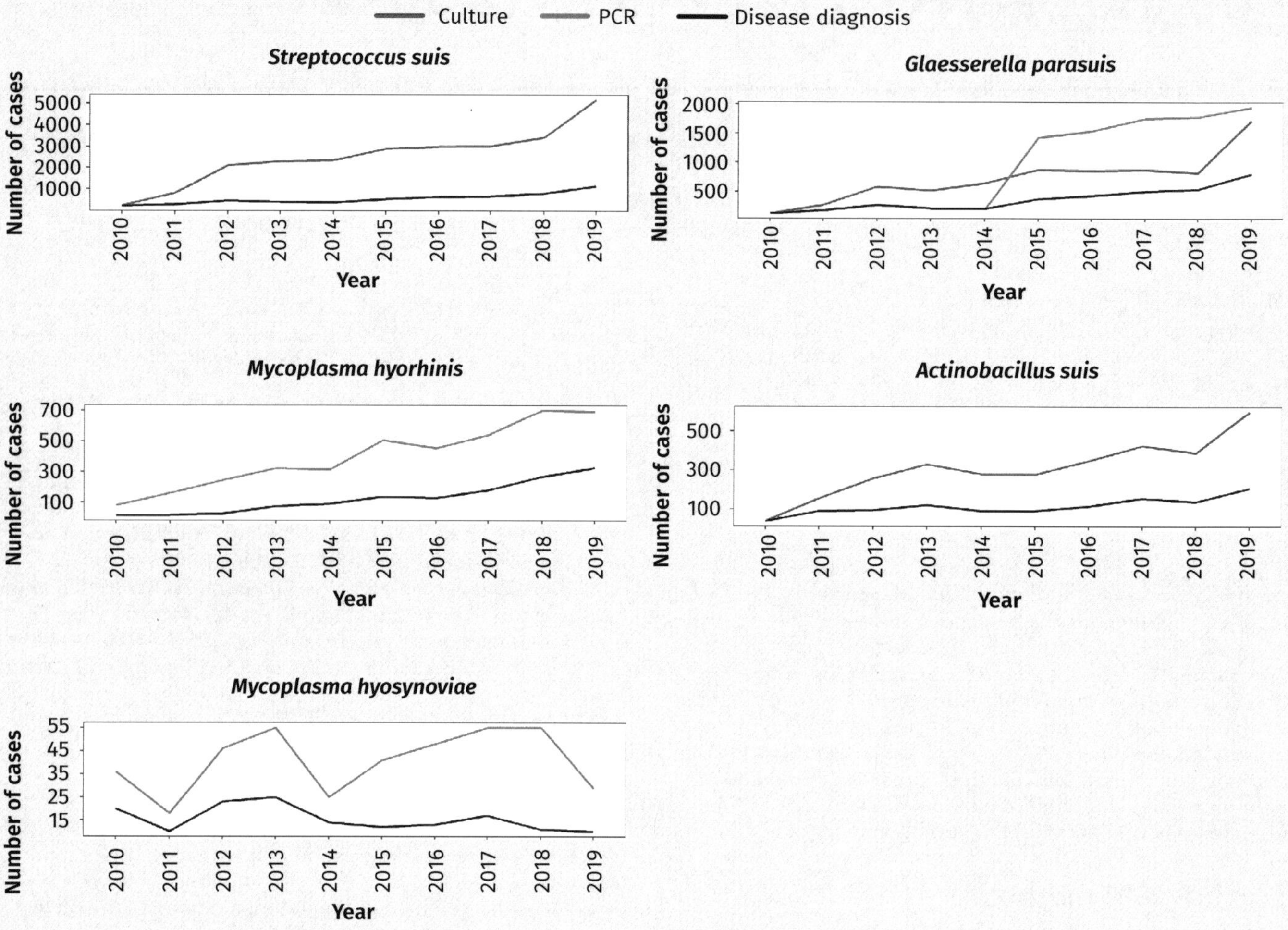

based on a case definition and gross lesions, and verified compatible histopathologic lesions offers a more comprehensive assessment for the contribution of these bacteria to systemic disease.

Although PCR is convenient, highly sensitive, and allows for rapid detection of agents that are difficult to isolate from field samples, e.g. GPS, MHR, and MHS, isolation of pure colonies from preferred sites is warranted for further typing with traditional tests or whole genome sequencing and antimicrobial susceptibility testing. It is critical to refocus attention on these pathogens and begin to close important knowledge gaps to develop improved control and prevention strategies.

References

1. Swine Health Information Center. Access on Ocotober, 15th 2020. **https://www.swinehealth.org/swine-bacterial-disease-matrix/**

2. Burrough E, Baum D, Schwartz K. Collecting Evidence and Establishing Causality. In: Zimmerman JJ, Karriker LA, Ramirez A, Schwartz KJ, Stevenson GW, eds. Diseases of Swine. 11th ed. Ames, IA: Wiley-Blackwell; 2019.

Figure 2: Distribution of preferred sites (tissues) of *Streptococcus suis* and *Glaesserella parasuis* and associated lesions in cases with a disease diagnosis.

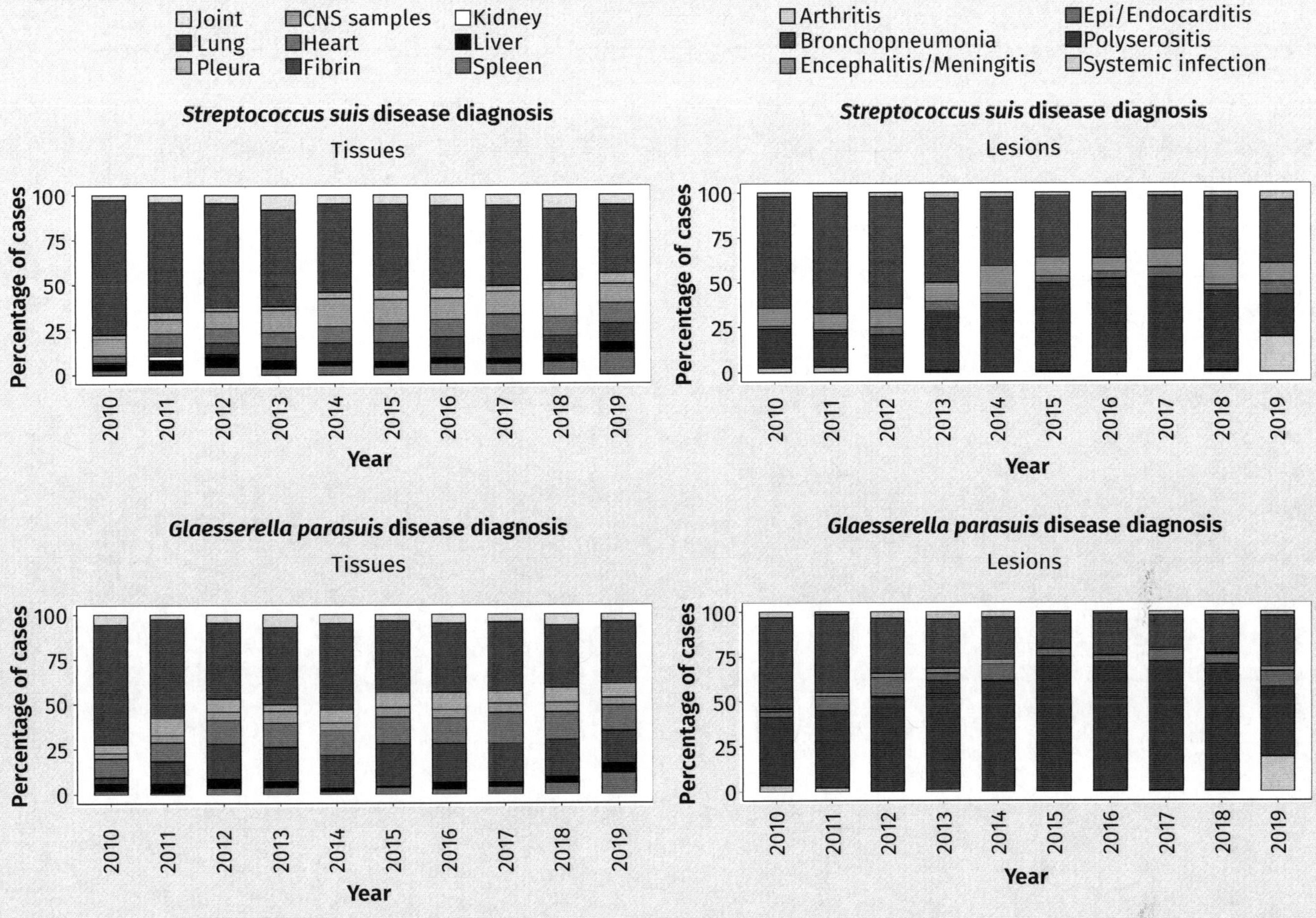

Table 1: Increase (+) or decrease (-) of the number of cases (total cases of the period) with bacterial detection and disease diagnoses over the total cases by a 5- or 10-year period

Agent	2010- 2014	2015 - 2019	2010 - 2019
Streptococcus suis			
Detection	+ 2,112 (7,754)	+ 2,301 (17,504)	+ 4,966 (25,258)
Disease Diagnosis	+ 149 (1,701)	+ 612 (3,748)	+ 923 (54,49)
Glaesserella parasuis			
Detection	+ 718 (2,376)	+ 1,351 (13,638)	+ 3,536 (16,014)
Disease Diagnosis	+ 85 (1,038)	+ 428 (2,684)	+ 677 (3,722)
Mycoplasma hyorhinis			
Detection	+ 234 (1,116)	+ 188 (2,898)	+ 616 (4,014)
Disease Diagnosis	+ 75 (216)	+ 188 (1,036)	+ 311 (1,252)
Actinobacillus suis			
Detection	+ 240 (1,069)	+ 321 (2,050)	+ 560 (3,119)
Disease Diagnosis	+ 52 (444)	+ 116 (717)	+ 169 (1,161)
Mycoplasma hyosynoviae			
Detection	- 11 (180)	- 12 (228)	- 7 (408)
Disease Diagnosis	- 6 (92)	- 2 (63)	- 10 (155)

Figure 3: Distribution of preferred sites (tissues) of *Mycoplasma hyorhinis* and *Actinobacillus suis* and associated lesions in cases with a disease diagnosis.

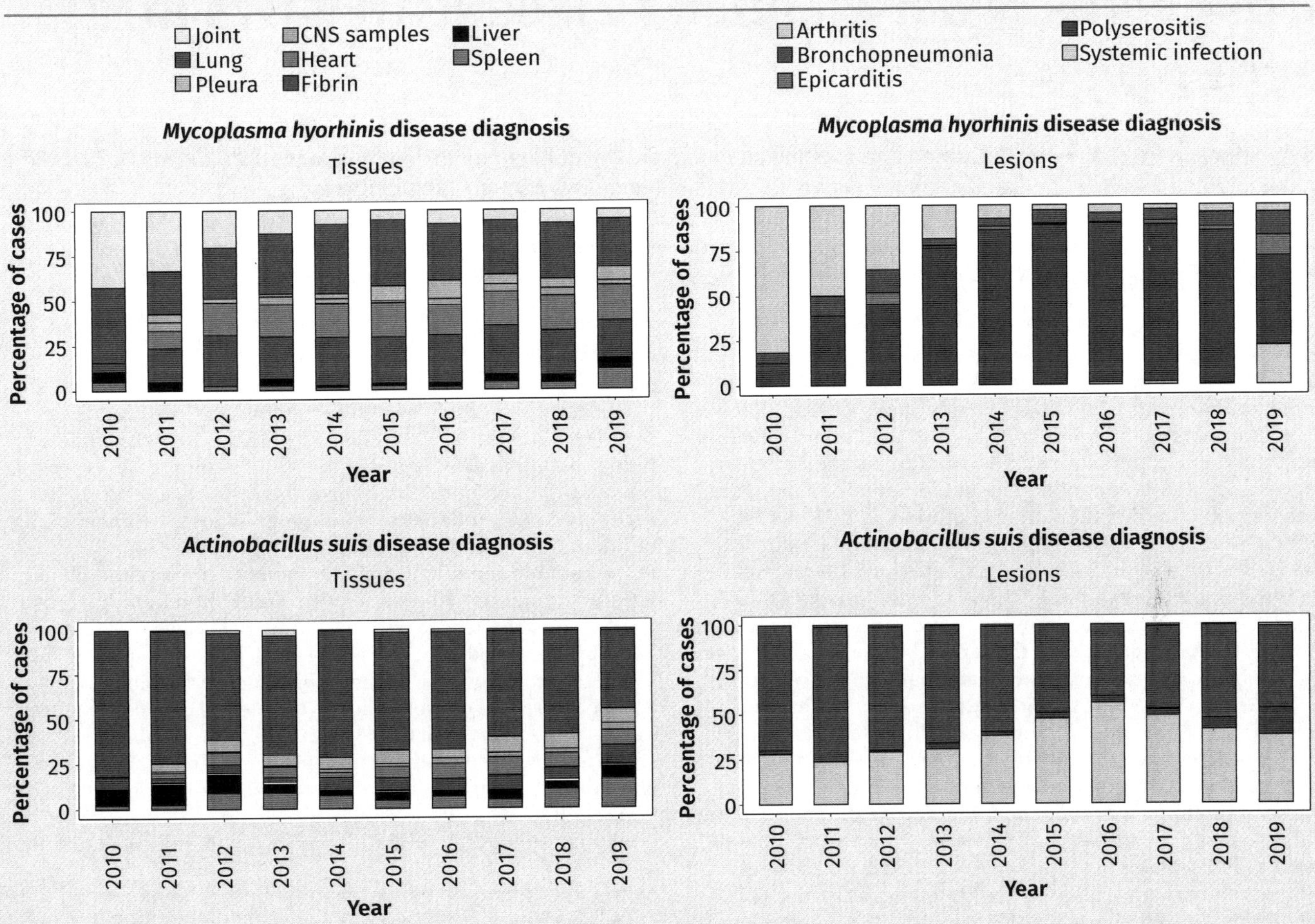

Replication of clinical *Streptococcus equi* subspecies *zooepidemicus* disease in sows and feeder pigs

Samantha J. Hau[1,2]; Kristina Lantz[3]; Keira L. Stuart[3]; Panchan Sitthicharoenchai[4]; Nubia Macedo[4]; Rachel Derscheid[4]; Eric Burrough[4]; Suelee Robbe-Austerman[3]; Susan L. Brockmeier[1]

[1]National Animal Disease Center, USDA, Ames, Iowa; [2]Oak Ridge Institute for Science and Education, Oak Ridge, Tennessee; [3]National Veterinary Services Laboratories, USDA, Ames, Iowa; [4]Department of Veterinary Diagnostic and Production Animal Medicine, College of Veterinary Medicine, Iowa State University, Ames, Iowa

Introduction

Streptococcus equi subspecies *zooepidemicus* (*S zooepidemicus*) is a commensal bacteria of horses that causes opportunistic infections in horses and other mammals, including humans. *S zooepidemicus* has been previously reported to cause severe disease in pigs, with large outbreaks of disease occurring in Indonesia and China. Mortality during the Chinese outbreak in 1975 was estimated at 300,000 animals.[1,2] Reports of clinical disease due to *S zooepidemicus* in swine in the United States have been infrequent with six cases reported by the Iowa State University Veterinary Diagnostic Laboratory over the last 10 years; however, in 2019, high mortality events in cull sows and feeder pigs in assembly yards in the US and Canada were attributed to *S zooepidemicus*.[3] Disease presented with rapid onset of severe depression, lethargy, weakness, and fever and resulted in mortality rates of 30-50%.[3] Isolates from the 2019 mortality events show a high degree of genetic similarity to each other and an isolate from the Chinese outbreak in 1975,[4] which may indicate a genetic adaption of this group of *S zooepidemicus* isolates to swine and the potential threat of these isolates to swine health. To better understand the progression of disease caused by the high mortality isolates and develop a model to replicate disease for future work, this study assessed the susceptibility of sows and 5-month-old feeder pigs to a *S zooepidemicus* isolate from the 2019 high mortality event in Tennessee. Additionally, a genetically distinct horse *S zooepidemicus* isolate and a closely related guinea pig isolate were screened for virulence in pigs to assess adaption of *S zooepidemicus* to pigs.

Materials and methods

Isolates assessed for virulence were from the swine mortality event in Tennessee, a horse with pneumonia and a pleural abscess, and a guinea pig lymph node. For each isolate of *S zooepidemicus* five sows and six 5-month-old feeder pigs were challenged with 5mL or 3mL, respectively, of ~10^8 colony forming units/mL oronasally and monitored for clinical signs of systemic disease including depression/lethargy, weakness, and fever three times daily for 14 days. Pigs were euthanized if disease became severe. Gross and histologic lesions were evaluated and systemic distribution of *S zooepidemicus* was assessed by routine aerobic culture on blood agar plates of the following samples: nasal swab, tonsil swab, joint fluid, serosal swab, cerebrospinal fluid (CSF), bronchoalveolar lavage fluid (BALF), splenic swab, and serum.

Results

S zooepidemicus was isolated from one sow via tonsil swab prior to challenge, all other sows were negative on three samplings prior to challenge. Sows challenged with the swine and guinea pig *S zooepidemicus* isolates showed clinical signs of fever, vomiting, inappetence, purulent nasal discharge, severe lethargy, severe depression, and reluctance or inability to rise. Clinical signs began by 24 hours post-infection (hpi) with severe depression and lethargy being noted by 36 hpi. Sows challenged with the swine isolate succumbed to disease or were euthanized by 56 hpi (Figure 1). Four sows challenged with the guinea pig isolate were euthanized by 96 hpi and one survived through day 14 (Figure 1). Gross lesions at necropsy were minimal with some organ congestion and small areas of consolidation in the lungs. Histologic changes in sows challenged with the swine isolate were consistent with sepsis and included pulmonary, hepatic, and splenic congestion with fibrin thrombi and intralesional bacterial colonies. The sows challenged with the guinea pig isolate had less severe lesions than swine isolate challenged sows with hepatic and splenic

Figure 1: Sow survival curve. Sows challenged with the swine isolate of *S zooepidemicus* succumbed to disease or were euthanized within 56 hpi. Sows challenged with the guinea pig isolate had a longer median survival time and 4/5 sows succumbed to disease or were euthanized by 96 hpi. The horse isolate did not cause clinical disease in sows.

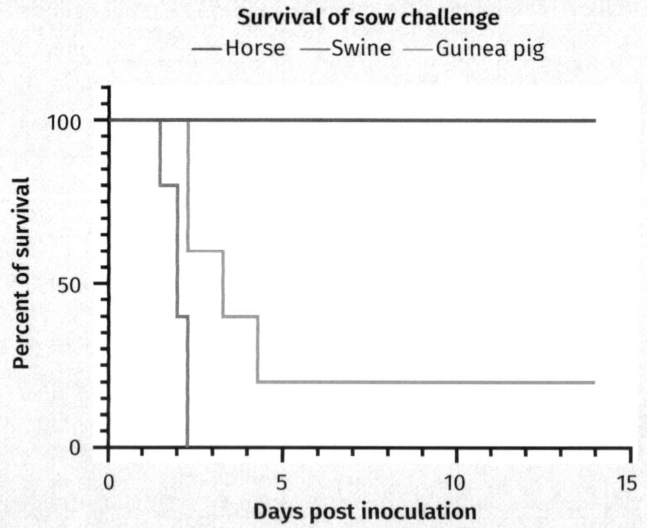

congestion seen most commonly. *S zooepidemicus* was widely distributed in sows challenged with the swine isolate. The spleen and CSF were the most frequently culture positive sample; however, the guinea pig isolate was not readily recovered from systemic sites postmortem in challenged sows. No illness was noted in the sows challenged with the horse *S zooepidemicus* isolate and it was only intermittently recovered from the upper respiratory tract post-challenge. Challenge of feeder pigs resulted in severe systemic disease for swine and guinea pig isolate challenged animals. Clinical signs were similar to those observed in sows (fever, inappetence, lethargy, depression, reluctance to rise); however, 3/6 animals challenged with the swine isolate and 2/6 challenged with the guinea pig isolate developed neurologic signs (twitching, seizures, paddling). Differences in mortality between challenge strains were also noted (Figure 2), with 6/6 swine isolate challenged animals succumbing to disease or being euthanized by 72 hpi and only 2/6 animals challenged with the guinea pig isolate being euthanized by 6 days post challenge. Similar to challenge of sows, feeder pigs were not susceptible to challenge with the horse isolate. Feeder pigs also developed severe lesions following challenge with the swine isolate, including pulmonary congestion and/or hemorrhage, hepatic, renal and splenic congestion, and peritonitis. Fewer and less severe lesions were seen in the feeder pigs challenged with the guinea pig isolate as compared to the swine isolate and no lesions were noted in the feeder pigs challenged with the horse isolate. The distribution of *S zooepidemicus* in feeder pigs was similar to that of sows, with the swine isolate being widely distributed, the guinea pig isolate being isolated only from the CSF, and the horse isolate having no systemic distribution.

Conclusions

Here, we were able to induce clinical disease with a swine isolate of *S zooepidemicus* in pigs with presentation and lesions similar to naturally occurring field cases. Animals were more susceptible to challenge with the *S zooepidemicus* isolate from the mortality event in Tennessee than a genetically distinct horse isolate. The closely related guinea pig isolate also caused severe disease; however, disease severity and systemic distribution of *S zooepidemicus* was more significant with the swine isolate. The difference in susceptibility between the *S zooepidemicus* isolates may be associated with species adaption of the swine isolate or the presence of different virulence factors, which will be important to investigate in future work. The challenge model developed here will enable future studies investigating the importance of different virulence factors to the pathogenicity of this isolate and to assess potential vaccine candidates for prevention of disease.

Figure 2: Feeder pig survival curve. All feeder pigs challenged with the swine isolate of *S zooepidemicus* succumbed to disease or were euthanized within 72 hpi. The guinea pig isolate was less fatal in feeder pigs. Two pigs were euthanized by 6 dpi. The horse isolate did not cause clinical disease in feeder pigs.

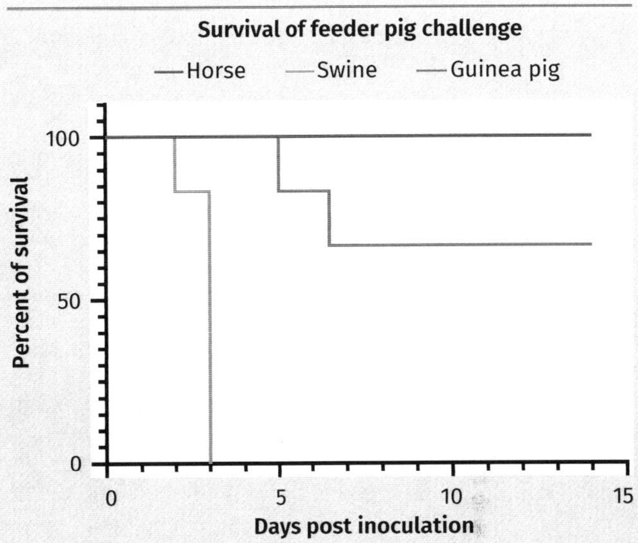

References

1. Soedarmanto I et al. Identification and molecular characterization of serological group C streptococci isolated from diseased pigs and monkeys in Indonesia. *J Clin Microbiol*. 1996; 34:2201-2204.

2. Feng ZJ et al. Outbreak of swine streptococcosis in Sichuan province and identification of pathogen. *Animal Husbandry Vet Med*. 1977; 2:7-12.

3. Sitthicharoenchai P et al. Cases of high mortality in cull sows and feeder pigs associated with *Streptococcus equi* subsp. *zooepidemicus* septicemia. *J Vet Diag Invest*. 2020; 32:565-571.

4. Chen X et al. Genetic characterization of *Streptococcus equi* subspecies *zooepidemicus* associated with high swine mortality in the United States. *Transbound Emerg Dis*. 2020; **doi: 10.1111/tbed.13645.**

Antibiotic susceptibility testing of *Mycoplasma hyopneumoniae* and *Mycoplasma hyorhinis* field isolates from the United States

F. M. A. Leal Zimmer, BVSc; B. Garcia-Morante, DVM, MS, PhD; A. Betlach, DVM; M. Pieters, DVM, PhD
Veterinary Population Medicine Department, College of Veterinary Medicine,
University of Minnesota, St. Paul, Minnesota

Introduction

Mycoplasma hyopneumoniae and *M hyorhinis* are recognized bacteria affecting pigs.[1] *M hyopneumoniae* causes a chronic respiratory disease in swine, known as enzootic pneumonia, which negatively affects swine performance and opens the door to other respiratory pathogens of viral and bacterial origin. Indeed, *M hyopneumoniae* is considered to be one of the most important pathogens involved in the porcine respiratory disease complex (PRDC).[1] On the other hand, *M hyorhinis* is ubiquitous in the swine population worldwide and, under still largely unknown circumstances, can cause polyserositis, arthritis, and less frequently abortion, otitis, and conjunctivitis.[1] In addition, *M hyorhinis* may also be involved in the PRDC.[1]

Both, in the case of *M hyopneumoniae* and *M hyorhinis* infections, the use of antimicrobial agents may be needed when clinical disease occurs. Information about antimicrobial susceptibility is largely limited for both bacterial species, as Mycoplasma spp. are hard to culture (especially *M hyopneumoniae*) and antibiograms are not routinely performed. The most current publicly available data on antimicrobial susceptibility for *M hyopneumoniae* and *M hyorhinis* isolates from the US is dated several decades.[2,3] In this context, the aim of this study was to determine the in vitro susceptibility to different antibiotics of *M hyopneumoniae* and *M hyorhinis* contemporary isolates originated from field clinical cases in the US.

Materials and methods

Eleven *M hyopneumoniae* and nine *M hyorhinis* field isolates were obtained from clinical specimens collected in various US swine farms within the most recent six years. Minimum inhibitory concentration (MIC) values of antimicrobial agents frequently used in the US were determined by the microbroth dilution method.[4] Antimicrobials selected for susceptibility tests were fluoroquinolones (enrofloxacin, marbofloxacin), macrolides (tylosin, tilmicosin, tylvalosin, tulathromycin), lincosamides (lincomycin), tetracyclines (oxytetracycline), pleuromutilins (tiamulin) and phenicols (florfenicol). Briefly, 100 µL of the appropriate antimicrobial solution was distributed into the corresponding well of microtiter plates, with a final range of antimicrobials from 0.001 to 64 µg/mL. The test was accomplished on 10^4 CCU/mL of each isolate. All isolates were tested in three independent replicates. The final MIC recorded was the value obtained in at least two replicates, provided that the third value did not differ more than 2-fold dilutions. For each antimicrobial evaluated, two positive (growth) control wells were included by adding 100 µL of sterile medium instead of antibiotic. Moreover, four negative (uninoculated) control wells were incorporated per compound by filling them with 200 µL of sterile medium. *M hyopneumoniae* ATCC 25095 and *M hyorhinis* ATCC 17981 were used as reference strains for the MIC tests. Plates were incubated at 37°C for up to 14 days for *M hyopneumoniae*, and for up to 7 days in the case of *M hyorhinis*. The MIC_{50} value was defined as the lowest concentration of the antibiotic at which ≥ 50% of the isolates were inhibited.

Results

The MIC_{50} values of antimicrobial agents used against field and type strains of *M hyopneumoniae* and *M hyorhinis* are summarized in Table 1. Compared with ATCC 25095 reference strain, the MIC_{50} values for *M hyopneumoniae* US isolates were lower for all compounds but for enrofloxacin, marbofloxacin, tylvalosin and oxytetracycline. The highest MIC_{50} values were recorded for tilmicosin and oxytetracycline, whereas the lowest value was obtained at 0.016 µg/mL for tylvalosin.

In the case of *M hyorhinis* and by comparing with the ATCC 17981 strain, the US isolate MIC_{50} values were higher for all compounds except for enrofloxacin, marbofloxacin and florfenicol. The highest MIC_{50} values were recorded at ≥ 64 µg/mL for tylosin, tilmicosin, tulathromycin and lincomycin, whereas the lowest MIC_{50} values were obtained at 0.5 µg/mL for oxytetracycline and tiamulin.

Discussion and conclusions

This study presents the antibiotic susceptibility profiles of *M hyopneumoniae* and *M hyorhinis* isolates circulating in the Midwest region of the US. Yet, interpretive categorization (i.e. susceptible or resistant) for the MIC values obtained in *Mycoplasma* spp. affecting pigs could not be assessed due to the lack of data on *in vivo* efficacy and pharmacokinetics. Nevertheless, an overall high *in vitro* efficacy of the tested agents against *M hyopneumoniae* was observed. In contrast, high MIC values (≥ 64 µg/mL) of macrolides and lincomycin were recorded for *M hyorhinis*. Our results are not consistent with former MIC values for *M hyorhinis* US field isolates.[2,3] The discrepancy observed in the present and previous studies in the susceptibility of the isolates to macrolides and lincomycin may suggests the aquisition of specific resistance mechanisms responsible for the observed variation in susceptibility. Thus, underlying mechanisms associated to macrolide and lincosamide resistance in *M hyorhinis* under field conditions deserve to be further characterized. Results from this study represent a renewed step towards appropriate and accurate antibiotic treatment of mycoplasma diseases in swine.

References

1. Pieters M & Maes D. Mycoplasmosis. In: Zimmerman JJ, Karriker LA, Ramirez A, Schwartz KJ, Stevenson GW, Zhang J, editors. *Diseases of Swine*. 11th ed. Ames, IA: John Wiley & Sons.

2. Williams PP. *In vitro* susceptibility of *Mycoplasma hyopneumoniae* and *Mycoplasma hyorhinis* to fifty-one antimicrobial agents. *Antimicrob Agents Chemother.* 1978;14:210-213.

3. Wu CC et al. Antimicrobial susceptibility of *Mycoplasma hyorhinis. Vet Microbiol.* 2000;76:25-30.

4. Klein U et al. Antimicrobial susceptibility monitoring of *Mycoplasma hyopneumoniae* and *Mycoplasma bovis* isolated in Europe. Vet Microbiol. 2007;204:188-193.

Table 1: Minimum inhibitory concentration values (µg/mL) for reference strains (ATCC), and *M hyopneumoniae* and *M hyrohinis* field isolates from the US

	M hyopneumoniae		***M hyorhinis***	
Antimicrobial	**ATCC 25095**	**MIC_{50}**	**ATCC 17981**	**MIC_{50}**
Enrofloxacin	0.062	0.125	1	1
Marbofloxacin	0.062	0.125	1	1
Tylosin	0.25	0.125	0.125	≥ 64
Tilmicosin	8	2	1	≥ 64
Tulathromycin	0.25	0.062	0.125	≥ 64
Tylvalosin	0.008	0.016	0.031	4
Lincomycin	0.5	0.25	0.25	≥ 64
Oxytetracycline	0.031	4	0.062	0.5
Tiamulin	0.125	0.031	0.062	0.5
Florfenicol	2	1	2	2

Effect of tulathromycin treatment on *Mycoplasma hyopneumoniae* detection and infectious potential

A. M. Betlach[1], DVM; D. Baumert[2], DVM; V. Utrera[2], DVM; L. Galina Pantoja[2], DVM, PhD; M. Pieters[1], DVM, PhD
[1]College of Veterinary Medicine, University of Minnesota, St. Paul, Minnesota; [2]Zoetis, Parsippany, New Jersey

Introduction

Mycoplasma hyopneumoniae (M hyopneumoniae), the etiologic agent of enzootic pneumonia,[1,2] continues to cause significant economic losses in the industry worldwide.[3] Control and elimination of *M hyopneumoniae* with the use of management, vaccination, and antibiotics have been well-described.[4] However, challenges in confirming *M hyopneumoniae* elimination exist, due to detection and diagnostic limitations, as well as chronicity and persistence of infection. Due to limitations in *M hyopneumoniae* bacterial culture,[5] veterinarians have been challenged with the detection of *M hyopneumoniae* by PCR long after antibiotic treatment, thus raising the question if the bacterium is still infectious. The study objective was to assess the effect of tulathromycin on *M hyopneumoniae* detection and infectious potential during acute and chronic phases of infection.

Materials and methods

Transmission of *M hyopneumoniae* was evaluated during the acute and chronic phase of infection. For each infection phase (acute and chronic), one age-matched naïve gilt was placed in contact with one *M hyopneumoniae* seeder (1:1 ratio) that was either treated with tulathromycin, treated and *M hyopneumoniae*-vaccinated, or non-treated for two weeks. Four replicates per treatment group were performed for each infection phase (n = 24).

Seeders were intratracheally inoculated with *M hyopneumoniae* strain 232 at a concentration of 1×10^5CCU/mL. Gilts selected to be treated received two doses of tulathromycin (2.5mg/kg, IM) at either 8 and 18 days post-inoculation (dpi) or 64 and 74 dpi. Vaccinated gilts received 2mL of a commercial *M hyopneumoniae* bacterin at 0 and 23 dpi. Blood samples, deep tracheal catheters, bronchial swabs, and lung lesions were collected and evaluated for *M hyopneumoniae* infection. Differences in transmission among groups were estimated by calculating force of infection (λ) and incidence rate.

Results and discussion

Acute phase of infection

A numerical reduction in *M hyopneumoniae* bacterial load was measured in treated seeders (0.5-3 Ct value difference) compared to those non-treated throughout experiment. Treated seeders had 16-18% less lung lesions compared to those non-treated. All naïve gilts remained *M hyopneumoniae* negative when housed with treated seeders. For treated and vaccinated and non-treated groups, 1 and 2 transmission events occurred, respectively. Numerical differences in λ and incidence rate were observed between groups.

Chronic phase of infection

At the end of the experiment, 50% and 33% of treated and treated and vaccinated seeders became *M hyopneumoniae* negative. Moreover, a numerical reduction in *M hyopneumoniae* bacterial load was measured in treated seeders (3 Ct value difference). On postmortem, all non-treated seeders were *M hyopneumoniae* positive via bronchial swabs compared to other groups. Differences in lung lesions were not observed between groups. All naïve gilts remained *M hyopneumoniae* negative in all experimental groups.

Conclusions

Administration of tulathromycin modified *M hyopneumoniae* infectious potential by reducing lung bacterial loads and the rate of new infections. Increased variability in *M hyopneumoniae* detection was evident during chronic phases of infection. Further research focused on the characteristics of *M hyopneumoniae* detection post-treatment is still necessary.

References

1. Mare CJ & Switzer WP. New species: *Mycoplasma hyopneumoniae;* A causative agent of virus pig pneumonia. *Veterinary Medicine, Small Animal Clinician.* 1965;60:841-846.

2. Goodwin RF et al. Production of enzootic pneumonia in pigs with a mycoplasma. *Veterinary Record.* 1965;77:1247-1249.

3. Pieters MG & Maes D. Mycoplasmosis. Diseases of Swine. John Wiley & Sons. 2019;863-883.

4. Maes D et al. Control of *Mycoplasma hyopneumoniae* infections in pigs. *Veterinary Microbiology.* 2008;126:297-309.

5. Sibila M et al. Current perspectives on the diagnosis and epidemiology of *Mycoplasma hyopneumoniae infection. Veterinary Journal.* 2009;181:221-231.

Estimation of pool sensitivity for detection of *Mycoplasma hyopneumoniae* by PCR using deep tracheal catheter field samples

Amanda Sponheim[1,2], DVM; Claudia Munoz-Zanzi[3], DVM, PhD; Eduardo Fano[2], DVM, MS, PhD; Maria Pieters[1], DVM, PhD
[1]Department of Veterinary Population Medicine, College of Veterinary Medicine, University of Minnesota, St. Paul, Minnesota; [2]Boehringer Ingelheim Animal Health USA Inc., Duluth, Georgia;
[3]School of Public Health, Division Environmental Health Sciences, University of Minnesota, Minneapolis, Minnesota

Introduction

Mycoplasma hyopneumoniae (*M hyopneumoniae*) is a cause of major economic losses to the swine industry, due to decreased growth, high feed conversion ratios, and high treatment costs in growing pigs.[2] Breeding herd eradication programs aim to eliminate exposure at the source, but diagnostic programs are needed to measure their success. Deep tracheal catheters have shown the highest reported sensitivity for detection of *M hyopneumoniae* by PCR,[7] but are an individual sample type. Oral fluids and processing fluids are common aggregate samples used in the swine industry.[5,8] However, oral fluid samples have shown low detection of *M hyopneumoniae* by PCR,[4] and more work is needed to validate processing fluids for detection of *M hyopneumoniae*, as the source of the bacterium in the sample type is unknown and detection may be intermittent.[6,9] Pooled-sample testing has been proposed as a more cost-efficient alternative than testing individuals when screening for low prevalence diseases.[1] However, because of the dilution effect, the sensitivity of the test when run on pooled samples is lower than its sensitivity when run on individual samples.[3] Therefore, the objective of this study was to determine the effect of pooling deep tracheal catheter samples on the diagnostic sensitivity of real-time PCR for *M hyopneumoniae* detection.

Materials and methods

Deep tracheal catheter samples from a previous study,[7] collected from pigs prior to inoculation and 113 days post-inoculation with *M hyopneumoniae*, were used in the study. Samples collected from known negative pigs, prior to inoculation, were combined to form one homogenate negative sample. Individual expected positive samples (n = 50) were pooled at ratios of one positive combined with two negatives and one positive combined with four negatives to form pools. Individual positive samples (n = 50), pools of 1:3 (n = 50) and 1:5 (n = 50) were tested using DNA extraction (MagMAX™ Pathogen DNA/RNA Kit and KingFisher™ extraction robot, Life Technologies, Grand Island, NY, USA) a species-specific *M hyopneumoniae* real-time PCR (VetMAXTM qPCR Master Mix and VetMAX™ *M hyopneumoniae* Reagents Kit, Life Technologies, Grand Island, NY, USA). Samples with a Ct value <40 were considered positive for *M hyopneumoniae*.

The sensitivities were calculated as the percentage of positive samples from the total samples, along with Clopper-Pearson (exact) confidence limits.

Results and discussion

DNA detection and pool sensitivity results are presented in Table 1. Pool sensitivity decreased as pool size increased. There is a risk of not detecting one positive animal in a pool when the rest of the pool is negative. These samples represent the latest sensitivity reported after infection in the literature,[7] and are the closest representation of pigs to sample at the end of an eradication program. The data have been applied to sample size calculations to develop point in time diagnostic programs to detect *M hyopneumoniae* in low prevalence scenarios, including at the end of an eradication program.

References

1. Greiner M and Gardner I. 2000. Application of diagnostic tests in veterinary epidemiologic studies. Prev. Vet. Med. 45: 43-45.

2. Maes D et al. 2017. Update on *Mycoplasma hyopnuemoniae* infections in pigs: Knowledge gaps for improved disease control. Transbound. Emerg. Dis. 1-15.

3. Munoz-Zanzi C et al. 2006. Factors affecting sensitivity and specificity of pooled-sample testing for diagnosis of low prevalence infections. Prev. Vet. Med. 74: 309-322.

4. Roos L et al. 2016. A model to investigate the optimal seeder-to-naïve ratio for successful natural Mycoplasma hyopneumoniae gilt exposure prior to entering the breeding herd. Vet. Microbiol. 184: 51-58.

5. Rotolo M et al. 2017. Sampling guidelines for oral fluid-based surveys of group-housed animals. Vet. Microbiol. 209: 20-29.

6. Sponheim A et al. 2018. Detection of *Mycoplasma hyopneumoniae* and *Mycoplasma hyorhinis* by PCR in processing fluids. Allen D. Leman Swine Conference. St. Paul, MN.

7. Sponheim A et al. 2020. Comparison of the sensitivity of laryngeal swabs and deep tracheal catheters for detection of *Mycoplasma hyopneumoniae* in experimentally and naturally infected pigs early and late after infection. Vet Microbiol. 241: 108500

8. Vilalta C et al. 2018. Use of processing fluids and serum samples to characterize porcine reproductive and respiratory syndrome virus dynamics in 3 day-old pigs. Vet. Microbiol. 225: 149-156.

9. Vilalta C et al. 2019. Detection of *Mycoplasma hyopneumoniae* in piglet processing fluids. Vet Record 185: 510.

Table 1: Diagnostic data and mean and 95% confidence interval of the individual and pool sensitivities for seeder deep tracheal samples at 113 dpi for *Mycoplasma hyopneumoniae* PCR

	DNA detection[1]	
Pool size	**# positive/# tested**	**Pool sensitivity (95% CI)**
Individual	49/50	0.98 (089-0.9995)
3	42/50	0.84 (0.71-0.93)
5	41/50	0.82 (0.69-0.91)

1 Real-time PCR. Samples with Ct values < 40 were considered positive.

Distribution of viremic piglets in farrowing rooms violates the homogeneous population assumption – implications for PRRSV detection?

M. N. Almeida[1]; M. Zhang[1]; W. Lopez[2]; C. Vilalta[3]; J. Sanhueza[4]; C. Corzo[5]; J. Zimmerman[1]; D. Linhares[1]
[1]Iowa State University, Ames, Iowa; [2]PIC, Hendersonville, Tennessee; [3]UpNorth Analytics, Arbeca, Spain;
[4]Catholic University of Temuco, Temuco, Chile; [5]University of Minnesota, Minneapolis, Minnesota

Background

Consistently weaning PRRSV-negative piglets is considered an important indicator of significant progress toward eliminating the virus from the breeding herd. Measuring progress toward this goal is based on testing piglets using the sample sizes described by Cannon and Roe[1] in combination with simple random sampling. This approach assumes that PRRSV-viremic piglets are homogenously distributed in farrowing rooms (i.e., following a hypergeometric distribution). Inherently, any potentially useful population risk factor information is ignored, such as the fact that the population (sows and piglets) is distributed among buildings, rooms, and farrowing pens/crates. Given our reliance on the current protocol (30 samples to detect ≥ 1 PRRSV-viremic piglet when prevalence is ≥ 10% with 95% confidence), the aim of this study was to address whether the assumption of homogeneity is valid.

Methods

A cross-sectional study was performed in 12 breed-to-wean PRRSV positive sow farms at different times after a PRRSV outbreak to cover different spectrums of prevalence. Serum was collected from every piglet in 422 due-to-wean litters in 23 farrowing rooms. In total, 4,510 serum samples were tested for PRRSV RNA by RT-PCR. The location of each litter (and piglet) within the farrowing room was recorded using simple (x, y) spatial coordinates. Thereafter, the spatial distribution of PRRSV-viremic piglets among litters within farrowing rooms was evaluated for population homogeneity and spatial clustering. Homogeneous distribution implies that each piglet's probability of viremia is identical and independent, which likewise implies that the proportion of viremic piglets within litters should be similar. The assumption of homogeneity was assessed by comparing the proportions of viremic piglets within litters (Fisher's exact test). The presence of clustering was evaluated by permutation test by first simulating the random allocation of viremic piglets among litters within the room (10,000 iterations). Thereafter, a *P*-value was calculated as the proportion of simulated values as extreme or more extreme than the observed value.

Results and discussion

In 12 PRRSV-endemic farms, 525 of 4,510 piglets (11.6%) were viremic (112 of 422 litters, 26.5%). Among positive litters, 26.8% (n = 30) had one PRRSV-positive piglet; 28.6% (n = 32) had 2 to 4 positive piglets; and 44.6% (n = 50) had 5 to 14 positive piglets. Statistical analysis of the distribution of PRRSV-viremic piglets within rooms/farms revealed a non-homogeneous distribution of PRRSV-viremic piglets in 15 of 16 rooms containing PRRSV-viremic piglets. Cluster analysis revealed spatial clustering of viremic piglets in 11 of 16 rooms (Table 1).

The objective of this project was to understand the pattern of PRRSV distribution within piglets in modern US farrowing using homogeneity and clustering analyses. Homogeneity in PRRSV distribution would imply that each piglet in the farrowing room had an identical and independent chance of being viremic. However, this study demonstrated that the distribution of PRRSV-viremic piglets in modern US farrowing room facilities is rarely homogeneous and often clustered, with clustering of viremic piglets observed in a majority of cases (68.75%).

The idea of heterogeneity in disease distribution in groups of weaned piglets is not new;[2,3] however, the non-homogeneous distribution of PRRSV among litters has not been reported previously. Likewise, PRRSV clustering is not a new concept, but it is most commonly described for larger populations occupying larger spaces, e.g., among farms within regions.[4-7] More recently, clustering within swine wean-to-finish barns has been described as a spatiotemporal phenomenon.[8]

The current American Association of Swine Veterinarians requirement for achieving "PRRSV stable" status relies on the absence of viremic piglets in four consecutive monthly samplings of 30 piglets each.[9] Underlying this approach is the assumption that low prevalence is an indication that PRRSV is being eliminated from the population. However, in one study, the original outbreak virus (based on ORF5 sequencing) was recovered from PRRSV stable herds; thus, low prevalence may not always lead to elimination.[10] Further, a growing body of information suggests that PRRSV can be present in cohorts of piglets at a prevalence < 10%. For example, Cano et al.[11] sampled all piglets in randomly selected litters following whole-herd inoculation with live PRRSV. At 4 and 12 weeks, respectively, prevalence decreased from 8% to 2% in liveborn piglets and 23% to 7% in due-to-wean piglets. Graham et al.[12] tested serum samples from all piglets in 44 litters (n = 454) in a 1,100-sow breed-to-wean farm 60 days after all sows were inoculated with live PRRSV and detected 3 (0.7%) viremic piglets by PRRSV RT-rtPCR. Vilalta et al.[13] tested piglets over time post-outbreak in a 6,000-sow breed-to-wean farm, reporting that prevalence between weeks 11 and 23 ranged from 0.9% to 6.5%. Collectively, these studies demonstrated that it is not uncommon for PRRSV infection to exist in suckling piglet populations at < 10% prevalence, i.e., levels below the level of detection at the current recommended sampling levels.

Cumulatively, heterogeneity, and clustering of viremic piglets in farrowing rooms explain the challenges inherent in PRRSV detection, particularly as the disease prevalence declines during elimination programs.

Table 1: Analysis of population homogeneity and clustering of PRRSV-viremic piglets by farm and room

			Population homogeneity analysis					Clustering analysis	
			Expected positive litters[1]		Observed positive litters				
Farm	Room	Positive piglets	n	Avg. No. of positive piglets per litter	n	Avg. No. of positive piglets in positive litters	*P*-value	Clustered (Y/N)	p (probability)
A	1	90	17	0.57	14	0.66	< 0.01	N	0.684
	2	13	20	0.06	4	0.36	< 0.01	Y	0
	3	29	17	0.19	5	0.58	< 0.01	Y	0
	4	2	5	0.04	1	0.20	> 0.05	N	0.185
C	1	8	7	0.13	1	1.00	< 0.01	Y	0
	3	4	10	0.03	1	0.33	< 0.01	Y	0.001
E		38	13	0.28	8	0.44	< 0.01	Y	0
G	2	3	22	0.01	2	0.14	< 0.05	N	0.219
H		30	19	0.17	8	0.38	< 0.01	Y	0.001
I		66	24	0.24	13	0.49	< 0.01	Y	0
J	1	117	20	0.55	17	0.65	< 0.01	Y	0
	2	58	21	0.36	16	0.46	< 0.01	Y	0.03
K	1	14	7	0.21	4	0.42	< 0.01	Y	0
	2	10	19	0.05	3	0.37	< 0.01	N	0.315
	3	7	4	0.18	2	0.64	< 0.01	Y	0.024
	4	36	17	0.21	13	0.28	< 0.01	N	0.329

[1] Expected based on the assumption of population homogeneity

Conclusion and implications

This study demonstrated that the core assumptions for sampling nursing piglets for PRRSV detection do not hold in modern farrowing room facilities. These results raise important questions regarding the number of piglets and the selection of rooms, litters and piglets within litters for sampling particularly when PRRSV is present at low prevalence. Further studies are needed to define the effect of sample size and sampling approaches on the probability of detecting PRRSV in farrowing rooms.

References

1. Cannon, R.M., Roe, R.T., 1982. A Livestock Disease Surveys: A Field Manual for Veterinarians. Australian Government Publishing Service.

2. Fano, E., Pijoan, C., Dee, S., Deen, J., 2007. Effect of *Mycoplasma hyopneumoniae* colonization at weaning on disease severity in growing pigs. Can J Vet Res 71, 195-200.

3. Pieters, M., Cline, G.S., Payne, B.J., Prado, C., Ertl, J.R., Rendahl, A.K., 2014. Intra-farm risk factors for *Mycoplasma hyopneumoniae* colonization at weaning age. Vet Microbiol 172, 575-580.

4. Arruda, A.G., Friendshpi, R., Carpenter, J., Hand, K., Poljak, Z., 2016. Network, cluster and risk factor analyses for porcine reproductive and respiratory syndrome using data from swine sites participating in a disease control program. Prev Vet Med 128, 41-50.

5. Arruda, A.G., Friendshp, R., Carpenter, J., Hand, K., Ojkic, D., Poljak, Z., 2017. Investigation of the occurrence of porcine reproductive and respiratory virus in swine herd participating in an area regional control and elimination project in Ontario, Canada. Transboud Emerg Dis 64, 89-100.

6. Arruda, A.G., Sanhueza, J., Corzo, C., Vilalta, C., 2018. Assessment of área spread of porcine reproductive and respiratory syndrome (PRRS) virus in three clusters of swine farms. Transboud Emerg Dis 65, 1282-1289.

7. Arruda, A.G. Vilalta, C., Puig, P., Perez, A., Albal, A., 2018. Time-series analysis for porcine reproductive and respiratory syndrome in the United States. PLoS One 13, e0195282.

8. Rotolo, M.L., Sun, Y., Wang, C., Gimenez-Lirola, L., Baum, D.H., Gauger, P.C., Harmon, K.M., Hoogland, M., Rain, R., Zimmerman, J.J., 2017. Sampling guidelines for oral fluid-based surveys of group-housed animals. Vet Microbio 209, 20-29.

9. Holtkamp, D., Polson, D., Torremorell, M., Morrison, R., Classen, D., Becton, L., Henry, S., Rodibaugh, M.T., Rowland, R.R., Snelson, H., Straw, B., Yeske, P., Zimmerman, J., 2011. Terminology for classifying swine herds by porcine reproductive and respiratory syndrome virus status. J Swine Health Prod 19, 44-56.

10. Linhares, D.C., Cano, J.P., Torremorell, M., Morrison, R.B., 2014. Comparison of time to PRRSV-stability and production losses between two exposure programs to control PRRSV in sow herds. Prev Vet Med 116, 111-119.

11. Cano, J.P., Dee, S.A., Rovira, A., Muñoz-Zanzi, C., Anil, S.S., Morrison, R.B., 2008. PRRSV vertical transmission dynamics in an endemically infected sow-herd. AASV annual meeting. San Diego, California, USA, 105-108.

12. Graham, J., Rademacher, C., Swalla, R., 2013. Use of oral fluid sampling in suckling pigs for PRRSV monitoring. AASV annual meeting. San Diego, California, USA, 83.

13. Vilalta, C., Sanhueza, J., Alvarez, J., Murray, D., Rorremorell, M., Corzo, C., Morrison, R., 2018. Use of processing fluids and serum samples to characterize porcine reproductive and respiratory syndrome vírus dynamics in 3 day-old pigs. Vet Microbiol 225, 149-156.

Determining the source of *Serratia* and other bacteria in boar semen

Darwin L. Reicks, BS, DVM

Introduction

The presence of certain bacteria such as *Serratia* in the extended semen can cause loss of sperm viability and fertility.[1,2] This can have a significant effect on litter size and fertility.[3] Although the barn is believed to be a primary source of bacteria, the specifics have not been thoroughly defined. Two experiments were conducted to determine the specific source of certain bacteria in the barn.

Materials and methods

Experiment 1

Fifteen boars were collected in 6 different boar studs. Half of the boar studs had a history of *Serratia marcescens* within the prior year. On each boar, prepuce fluid was collected mid-stream and raw semen was collected 5 seconds after the start of ejaculation. All samples were cultured on Tryptic Soy Agar with 5% sheep blood. Plates were read at 48 hours and identification of positive isolates was made at the University of Minnesota Veterinary Diagnostic Lab. Quantification of bacteria in each sample was also determined.

Experiment 2

Four of the same six boar studs were sampled, with three having a history of *Serratia* within 6 months prior to sampling, and one stud having never identified *Serratia* in extended semen. Prepuce fluid and semen were collected as in experiment 1. In addition, a fecal sample was taken directly from the rectum. An environmental sample (Swiffer) was also obtained, by adding approximately 30 mL of Phosphate Buffered Saline to a Swiffer pad, wiping the pad on the slat at the mid-point of the stall after the boar left, and squeezing the Swiffer pad into a 50 mL tube. 16S rRNA gene sequencing was done on all samples in experiment 2 at the University of Minnesota Veterinary Diagnostic Lab. 16S sequencing utilizes PCR to target and amplify portions of the hypervariable regions of the bacterial 16S rRNA gene. Amplicons from separate samples are then given molecular barcodes, pooled together, and sequenced. After sequencing, raw data is analyzed with a bioinformatics pipeline which includes trimming, error correction, and comparison to a 16S reference database.

Results and discussion

Experiment 1

The quantity of bacteria in semen ranged from 300 cfu/ml to 12,000 cfu/mL in positive samples. Overall, 73% of the raw ejaculates had no detectable growth (minimum detection level of 150 cfu/mL). For prepuce fluid, 98% of the samples had growth and overall an average of 500,000 cfu/mL. The frequency of the top 6 most frequently isolated bacteria in prepuce compared to raw semen is shown in Table 1.

Experiment 2

Results were available from three of the four studs at the time of this paper submission. Porphyromonas endodontalis is the most common bacteria isolated out of the prepuce fluids and is present in the highest concentrations at all three studs (see table 2). In the feces, *Prevotella copri* and *Faecalibacterium prausnitzii* were the dominant bacteria found in all three studs.

Results were more variable for the environmental Swiffer, but in general, *Acinetobacter lwoffii*, a common skin inhabitant, was the most common isolate. Results were also variable in semen, with no consistent bacteria dominant.

Serratia marcescens was not a dominant isolate in any of the samples from the three different boar studs, but it was found in lower concentrations in most samples (102 out of 120).

The percentage of isolates in prepuce fluid identified as *Serratia marcescens* varied from 0.03% in stud 1 to 2.4% in stud 3. For feces, the range was from 0.7% in stud 2 and 3 to 1.7% in stud 1. In the semen, the percentage was from 0.3-0.5% in all three studs, and for the environmental Swiffer, between 0.2-0.5%. Although one of the studs has never cultured *Serratia* from extended semen, *Serratia* was identified in 36 out of 40 samples from that stud.

The three-glove method was used,[4] and an attempt was made to obtain a free-catch collection on all boars. Fifteen of 30 boars had no detectable *Serratia marcescens* in the semen. The concentrations were low in most of the boars with 86% of the *Serratia* found in 7 boars out of 30. It is possible that *Serratia* in those boars was higher due to cross contamination with feces or prepuce fluid.

The correlations between the 4 different sample types and quantity of *Serratia* were not particularly strong. The strongest correlations were between prepuce fluid and environment ($r = 0.22$) and between feces and environment ($r = 0.20$), which seems logical.

Summary

This was a first look at results from multiple boar studs using 16S rRNA gene sequencing to determine what bacteria are prevalent in semen, prepuce fluid, feces, and environmental Swiffer samples from boar studs. *Porphyromonas endodontalis*, a gram-negative anaerobic rod, was a dominant bacterium of the prepuce fluid. *Prevotella copri* and *Faecalibacterium prausnitzii*, also gram-negative anaerobic rods, were dominant in the feces from all three studs. *Acinetobacter lwoffii* was the most common bacteria overall on environmental Swiffer samples. Although the three-glove method was used to obtain a free catch collection, bacteria were still found in the semen, with a variety of isolates identified. In experiment 1, most semen samples had no detectable aerobic growth. The 16s rRNA technique appears to be a much more sensitive technique than Tryptic Soy Agar (with 5% sheep blood) plates in identifying all bacteria present and can help us understand what type of normal background bacterial flora are present. *Serratia marcescens* was not a dominant bacteria present in any of the samples, suggesting that when it presents itself in extended semen, it may be due to a disruption or altering of the normal flora in prepuce fluid, feces, semen, or the environment.

Table 1: Data from 90 boars from 6 boar studs showing the top 6 most common aerobic culture isolates from prepuce and raw semen

Prepuce bacteria	No. of boars out of 90 - prepuce	No. of boars out of 90 – raw semen
Pseudomonas aeruginosa	45	7
Gram positive cocci	34	5
Corynebacterium sp	30	7
Staphylococcus sp	21	4
Proteus mirabilis	20	1
Corynebacterium xerosis	18	5

Table 2: Predominant bacteria found by sample type in three boar studs

Sample type	Stud	Isolate	% of isolates
Prepuce fluid	1	*Porphyromonas endodontalis*	26.4
	1	*Staphylococcus aureus*	12.5
Prepuce fluid	2	*Porphyromonas endodontalis*	31.9
	2	*Propionibacterium acnes*	4.3
Prepuce fluid	3	*Porphyromonas endodontalis*	31.3
	3	*Staphylococcus aureus*	7.1
Feces	1	*Prevotella copri*	16.4
	1	*Faecalibacterium prausnitzii*	13.3
Feces	2	*Faecalibacterium prausnitzii*	19.9
	2	*Prevotella copri*	11.0
Feces	3	*Prevotella copri*	19.0
	3	*Faecalibacterium prausnitzii*	14.6
Environmental Swiffer	1	*Staphylococcus aureus*	11.6
	1	*Clostridium perfringens*	7.3
Environmental Swiffer	2	*Acinetobacter lwoffii*	12.9
	2	*Staphylococcus epidermidis*	5.5
Environmental Swiffer	3	*Acinetobacter lwoffii*	10.5
	3	*Staphylococcus epidermidis*	8.2
Semen	1	*Staphylococcus aureus*	9.9
	1	*Prevotella copri*	6.9
Semen	2	*Porphyromonas endodontalis*	18.1
	2	*Staphylococcus epidermidis*	9.2
Semen	3	*Acinetobacter lwoffii*	10.7
	3	*Prevotella copri*	9.9

References

1. Althouse GC, Kuster CE, Clark SG, Weisiger RM. Field investigations of bacterial contaminants and their effects on extended porcine semen. Theriogenology 53 (2000) 1167-1176.

2. Ubeda JL et al. Adverse effects of members of the Enterobacteriaceae family on boar sperm quality. Theriogenology 80 (2013) 565-70.

3. Maroto Martín LO et al. Bacterial contamination of boar semen affects the litter size. Animal reproduction science 120 (2010) 95-104

4. Best collection techniques to avoid bacterial and viral contamination. Boar Stud Managers Conf. Webinar series. September 2018.

Active environmental surveillance of SARS-CoV-2 in midwestern meatpacking plants

S. Storms, DVM; J. Lowe, DVM, MS
Applied Epidemiology Laboratory, College of Veterinary Medicine, University of Illinois, Urbana, Illinois

Introduction

This pilot project investigated the presence of SARS-CoV-2 in the environment of three midwestern meatpacking plants from May to October 2020. Beginning April 2020, during the COIVD-19 pandemic, more than 50 meatpacking facilities closed or severely limited operations due to SARS-CoV-2 infection in the workforce.[1] Meatpacking plants employ hundreds of workers that are in close contact with each other. Infection rates and a fear of being infected in the workplace led to worker absenteeism—creating a backlog of animals unable to be slaughtered. The interruption created an inventory estimated to be nearly 3.4 million market hogs and 1.3 million cattle backlogged for processing in the USA.[2] Meatpackers implemented a plethora of interventions designed to limit the transmission of SARS-CoV-2 in the workplace and build confidence among the workforce. Due to an incomplete understanding of viral transmission, the effectiveness of the implemented infection control practices, individually and in total, is unknown.

Objective

This study's objective was to document current infection control practices and measure the environmental presence of SARS-CoV-2 on surfaces and in the air of meatpacking facilities to assess the effectiveness of newly implemented practices.

Materials and methods

Hygienic and social distancing practices

We evaluated the worker environments for enhanced hygienic and social distancing practices. Areas of evaluation were break rooms, locker rooms, supply counters, human resources, other plant-specific common areas, and the production floor. We also asked plants how they were communicating the risks of coronavirus infection, and the steps that the plant was taking to mitigate risk to employees, and their involvement with the local health departments to their employees and their families.

We collected 208 surface and 32 air samples from June 2020 through October 2020. We sampled high-touch areas from the locations mentioned previously to assess fomite and surface transmission, including microwaves, timeclocks, breakroom surfaces, restroom surfaces, vending machines, and countertops. We used a new pair of gloves and 67cm^2 swatches of electrostatic cloth for surface sampling. One cloth was wiped across a surface area to be sampled, such as all vending machine buttons in the break room. Cloths were placed into 15 mL tubes with 5 mL of BHI viral transport media. Samples were kept on ice for transport to the lab until processing. The tubes were vortexed for 15 seconds in the lab, and then the viral transport media was decanted and submitted for rtPCR at the University of Illinois Veterinary Diagnostic Laboratory.[3] Plant B used PrimeStore® MTM media (Longhorn Vaccines & Diagnostics, San Antonio, TX, USA) for viral transport medium, which fixes and inactivates virions for shipment without ice.

Air sampling

We sampled the surface of air returns and vents in the break areas and restrooms to access air-based transmission potential. We collected air samples from common areas, such as lunchrooms, locker rooms, and the production floor during periods of high worker activity in those areas. We sampled approximately 6000 L of air through a commercial electrostatic dry filter (Innovaprep, Drexel, MO, USA) at a measured flow rate between 70-200 L per min. Twelve of the samples were of 144,000 L by collecting air over 24 hours in the same location as the 6000 L samples. Determination of the presence of viral RNA in all samples was by rtPCR at the University of Illinois Veterinary Diagnostic Laboratory.[3]

Results

All plants were in close contact with their local health departments. One plant screened the entire workforce for coronavirus, and all positive cases were asked to quarantine for 14 days with paid leave. A difference was seen in employee communication about coronavirus mitigation strategies. Plants A and B were in contact with the different employee communities, many of which are of different ethnicities and languages. These two plants saw a much-reduced level of absenteeism, reflected in almost normal production throughout the observation period. Plant C was not in contact with the employee communities about risks and mitigation strategies. This plant had the highest level of absenteeism and loss of production.

All plants had increased hygienic and social distancing practices. All plants required masks, and two required face shields for all employees. Two of the plants enforced social distancing and mask compliance routinely.

Before entering all facilities, nurses screened workers' temperatures, and two of the plants installed thermal cameras as a secondary temperature check. Any employee with a temperature greater than 38°C was denied entry into the facility and asked to get a coronavirus test. All hired more workers to perform routine cleaning of surfaces in the break areas and locker rooms up to three times per shift. This consisted of wiping down surfaces to remove dirt and applying disinfectant and spraying disinfectant on surfaces, such as lockers and benchtops. Plant A used small droplet (< 10 µm) chlorine dioxide disinfection during the night to sanitize common areas and installed UVC lights in the ventilation systems, expanded their break rooms, and locker rooms to prevent worker contact. Plant B uses ozone disinfection once a week in different areas of the plant.

Steps to prevent worker contact have been implemented. Plant A renovated indoor spaces to double the size of the lunchroom. Plant B built a new building to expand the lunch space, and Plant C provided outdoor areas for workers on breaks. Plants A and B installed three-foot-high plexiglass dividers at lunch tables to prevent workers from sitting directly across or next to someone. After a worker sat there for a meal, a card indicated it was used, and cleaning crews cleaned all lunch stations between workers. Plants A and B installed more timeclocks to prevent congestion of workers at the beginning and end of shifts along with dividers in hallways to facilitate one-way foot traffic past any congestion point. Plants A and B have doors propped open or foot pulls installed to minimize hand contact with surfaces. A and B have installed plastic partitions between workers on the production floor, with plans to install metal partitions in the future.

238 surface and air samples as of 11/12/2020 were negative for SARS-CoV-2 on PCR. Two suspect positives were seen from Plant B the month of 10/2020.

Discussion

For the two plants that provided employee testing data, the number of cases of plant employees was less than or equal to the number of cases in the surrounding communities (Table 1). According to the plants, approximately 80% of the employees live in the surrounding community. These data suggest that the plants are not contributing to the community-based spread of coronavirus. Further work is needed to determine the causality relationship between lower viral plant environmental loads and within plant transmission rates. All employees were screened for coronavirus at Plant C in early May, and many asymptomatic infections were detected in the workforce. This number may be a better reflection of the actual infection rate in the community.

Our sampling throughout the summer and fall of 2020 shows that the newly implemented mitigation strategies have reduced the amount of virus in the plant environment to below detectable limits. Continued environmental sampling in plants and monitoring of community coronavirus cases will help shape sampling strategies as seasonal colds and influenza infections rise this winter. Continued hygienic and social distancing efforts by the plants will help keep employees healthy and plants open.

References

1. McCarthy, R., & Danley, S. (2020, June 23). Map: COVID-19 meat plant closures. Retrieved November 12, 2020, from **https://www.meatpoultry.com/articles/22993-covid-19-meat-plant-map**

2. Lowe, J. F. (2020, June 17). June 15th Pork and Beef Market Summary. Retrieved November 12, 2020, from **https://vetmed.illinois.edu/ael/june-8th-pork-and-beef-market-summary-2/**

3. Kong, W., Li, Y., Peng, M., Kong, D., Yang, X., Wang, L., & Liu, M. (2020). SARS-CoV-2 detection in patients with influenza-like illness. *Nature Microbiology, 5*(5), 675-678. **doi:10.1038/s41564-020-0713-1**

Table 1: Number pf positive SARS-CoV-2 tests in the plant or surrounding community

	Plant A	Plant C
Cases in plant	70 (3.4% of workforce)	89 (14% of workforce)
Cases in community	63	227

Number of PCR positive employee tests in two of the plants sampled during June and July of 2020. Employee information was kept anonymous. Plant A's community is the surrounding county, and Plant C's community is the surrounding zip code. Plant C tested the entire workforce in May 2020.

Experimental intravenous, intratracheal, and intranasal inoculation of swine with SARS-CoV-2

A. Buckley[1]; S. Falkenberg[2]; M. Martins[3]; M. Palmer[4]; K. Lager[1]; D. Diel[3]
[1]Virus and Prion Research Unit, National Animal Disease Center, USDA, Agricultural Research Service, Ames, Iowa; [2]Ruminant Disease and Immunology Research Unit, National Animal Disease Center, USDA, Agricultural Research Service, Ames, Iowa; [3]Department of Population Medicine and Diagnostic Sciences, Animal Health Diagnostic Center, College of Veterinary Medicine, Cornell University, Ithaca, New York; [4]Infectious Bacterial Diseases Research Unit, National Animal Disease Center, USDA, Agricultural Research Service, Ames, Iowa

Introduction

Severe acute respiratory syndrome coronavirus 2 (SARS-CoV-2) is novel coronavirus that causes disease (COVID-19) in humans, and it is hypothesized to have a zoonotic origin. Since swine and other livestock species are susceptible to other coronaviruses, research has been initiated to determine the susceptibility of livestock species to SARS-CoV-2 and if livestock can play a role in spread of the virus. Multiple groups have reported that swine are not susceptible to SARS-CoV-2 via the intranasal inoculation route; therefore, two different routes of inoculation (intravenous and intratracheal) were evaluated to determine if swine were susceptible to SARS-CoV-2.

Materials and methods

Twenty-four 3-week-old pigs were purchased from a commercial swine herd and transported to the National Animal Disease Center in Ames, IA. Control pigs were housed in ABSL-2, while challenged pigs and their contacts were housed in BSL-3AG facility. Three challenge routes were performed: intravenous (IV, n = 4, 344-347), intratracheal (IT, n = 4, 348-351) and intranasal (IN, n = 4, 352-355). Sham cell culture lysate challenge was administered to control pigs (n = 2/route) with 2 control pigs untreated. Finally, one contact pig was added to each challenge group and the control room on day 2 post inoculation. Virus was isolated from a Malayan tiger that developed respiratory signs after infection with SARS-CoV-2 (TGR/NY/20). IV pigs received 2 mL, while both IT and IN groups received 5 mL of 6.8 × 106 TCID50/mL virus solution. Temperatures were recorded daily using a subcutaneous microchip. Nasal/oral swabs, rectal swabs, and group oral fluids were collected on 0-7, 10, 12, 14, 18 and 21 days post inoculation (dpi). Serum and whole blood were obtained at 0, 3, 7, 14, and 21 dpi. Samples were tested by PCR for the presence of virus and serum was utilized in a serum virus neutralization assay to test for neutralizing antibodies. At 21 dpi animals were euthanized and tissues were collected for PCR testing and histology.

Results

Immediately after IV challenge, pigs that received either virus or sham cell culture lysate began to vomit, but quickly recovered. Two IT pigs had increased temperatures 2-3 dpi, but only one pig had a temperature recorded greater than 40.5°C on 13 dpi from the IV group (Figure 1). Otherwise, pigs in this study did not demonstrate any signs.

The IV group had two pigs (344, 347) PCR positive on 3 dpi in the buffy coat and one pig (345) positive in a nasal/oral swab on 4 dpi. The remaining samples were all negative by PCR. All pigs in the IV group developed virus neutralizing antibody (VN) titers of 1:32 and 64 on 7 dpi but a reduction in those titers was observed on 14 dpi (1:8 to 1:32, Table 1). By 21 dpi all pigs had titers of 1:8.

In the IT group, all pigs were positive in nasal/oral swabs on 1 dpi, suggesting likely detection of the inoculum virus. On 2 dpi, SARS-CoV-2 RNA was detected in only two pigs (350, 351). Beyond 2 dpi, one pig (351) was PCR positive on 7 dpi, with RNA being detected on nasal/oral swabs. In rectal swabs, pig 350 was positive on 1, 2, and 7 dpi. Pig 351 had positive rectal swabs on 1 and 3 dpi. In addition, the IT group had positive oral fluids on 6 dpi. Pig 350 developed a titer of 1:32 on 21, while pig 351 had a titer of 1:16 on 14 dpi and dropped back down to 1:8 on 21 dpi (Table 1).

All animals in the IN group were positive in nasal/oral swabs on 1 dpi, which again could be detection of inoculum given the route of inoculation. Two animals (353, 355) were positive on 2 dpi and one animal (354) on 3 dpi. Moving to rectal swabs, pig 355 had positive swabs on 1 and 2 dpi. Of note, pig 352 had positive rectal swabs on 2, 3, 4, 5, 10, and 12. In addition, this group was PCR positive in oral fluids on 4 dpi. Only pig 352 seroconverted with a titer of 1:32 on 14 dpi and 1:16 on 21 dpi (Table 1).

The contact pigs did not present any PCR positive samples and no seroconversion was observed in any of the contact animals.

At necropsy, there was no gross pathology visible in the pigs. PCR results on tissues, virus isolation on PCR positive samples and histology are pending. In addition, RNAScope will be utilized to analyze location and quantity of the ACE2 receptor on kidneys and other select tissues from necropsy.

Conclusion

Our results corroborate earlier studies showing that pigs are likely not susceptible to SARS-CoV-2 infection. Detection of SARS-CoV-2 RNA in secretions at early time points post-infection suggest likely detection of residual inoculum virus. The lack of sustained antibody responses in inoculated animals support these observations. Not surprisingly, contact animals remain negative throughout the experiment, demonstrating lack of transmission between swine.

Groups from China, Germany, and the US have previously reported that pigs challenged intranasally/intratracheally were not susceptible to infection with SARS-CoV-2, although Canada has recently reported that pigs challenged intranasally were susceptible to SARS-CoV-2.1,2,3 Our PCR and VN results were similar that that reported by Canada. We did find PCR positive nasal swabs and rectal swabs and VN activity in some animals from all challenge groups, however, viral nucleic acid detection and antibody titers were not sustained throughout the course of the experiment. While viral RNA detection was likely result of detection of

Figure 1: Pig temperatures

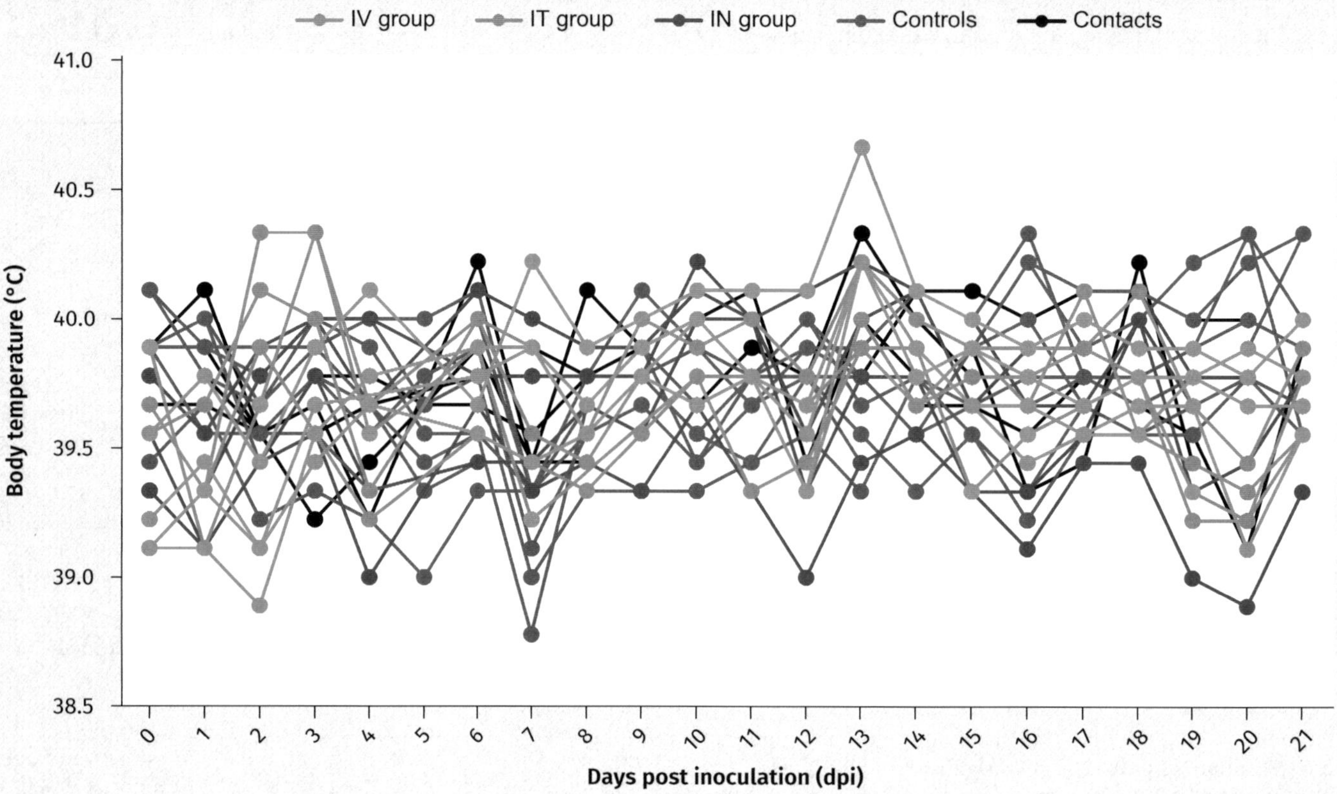

Table 1: Serum virus neutralization titers

Pig number	3 dpi	7 dpi	14 dpi	21 dpi
344	8	64	8	<8
345	<8	32	32	8
346	<8	64	16	8
347	<8	32	8	8
348	<8	<8	<8	<8
349	<8	<8	<8	<8
350	<8	<8	8	32
351	<8	<8	16	8
352	<8	<8	32	16
353	<8	<8	<8	<8
354	<8	<8	8	<8
355	<8	<8	<8	<8
364	<8	<8	<8	<8
365	<8	<8	<8	<8
366	<8	<8	<8	<8

inoculum material, antibody responses are most likely a result of a response to non-replicating virus antigen inoculated in the pigs. Thus, it is unlikely swine are a reservoir and contributing to the epidemiology and spread of SARS-CoV-2 in the human population.

References

1. Meekins, D.A., Morozov, I., Trujillo, J.D., Gaudreault, N.N., Bold, D., Carossino, M., Artiaga, B.L., Indran, S.V., Kwon, T., Balaraman, V., Madden, D.W., Feldmann, H., Henningson, J., Ma, W., Balasuriya, U.B.R., Richt, J.A., 2020. Susceptibility of swine cells and domestic pigs to SARS-CoV-2. Emerg Microbes Infect 9, 2278-2288.

2. Schlottau, K., Rissmann, M., Graaf, A., Schon, J., Sehl, J., Wylezich, C., Hoper, D., Mettenleiter, T.C., Balkema-Buschmann, A., Harder, T., Grund, C., Hoffmann, D., Breithaupt, A., Beer, M., 2020. SARS-CoV-2 in fruit bats, ferrets, pigs, and chickens: an experimental transmission study. Lancet Microbe 1, e218-e225.

3. Shi, J., Wen, Z., Zhong, G., Yang, H., Wang, C., Huang, B., Liu, R., He, X., Shuai, L., Sun, Z., Zhao, Y., Liu, P., Liang, L., Cui, P., Wang, J., Zhang, X., Guan, Y., Tan, W., Wu, G., Chen, H., Bu, Z., 2020. Susceptibility of ferrets, cats, dogs, and other domesticated animals to SARS-coronavirus 2. Science 368, 1016-1020.

National and Iowa impacts of African swine fever in the United States

Miguel Carriquiry[1]; Amani Elobeid[2]; David Swenson[2]; Dermot Hayes[2]
[1]Universidad de la Republica, Uruguay; [2]Iowa State University

Introduction

African swine fever (ASF), a deadly viral animal disease that causes up to 100% fatality in pigs and wild boars, has affected several countries in Asia including China, Viet Nam, Cambodia, South Korea, North Korea, Indonesia, and the Philippines, among others (Figure 1).[1] The first case was discovered in China in August 2018 and, despite the culling of millions of pigs, the disease has spread very quickly across the continent. While no cases have been detected in the United States, the disease is currently also present in nine countries in Europe and 23 countries in Africa, and there is a risk that it will spread to other parts of the world. This outbreak has resulted in the loss of millions of animals both due to the disease and from culling, burning, and burial. Because there is no cure or vaccine for the disease, the economic effects are large and far-reaching. (Note: As of the time of this writing there have been announcements that an effective vaccine has been developed but that it will take time and significant resources in order for it to be widely available and used (**https://www.bloomberg.com/news/articles/2020-01-31/promising-african-swine-fever-vaccine-no-panacea-scientists-say**).

Imports from countries where ASF is present are prohibited and the loss of pigs result in devastating losses to producers. The US is currently free of the disease (eradicated in 1978). The USDA's assessment of the likelihood of an outbreak in the US through legal importation of live pigs and swine products and by-products indicates it is negligible to low with low uncertainty (moderate uncertainty for swine products and by-products). However, the likelihood of the disease entering the US through illegal entry of swine products and by-products is high with low uncertainty (negligible to low with moderate uncertainty for illegal entry of live pigs) (USDA APHIS:VS).[3] Jurado et al. (2019) estimate the risk of entry of ASF into the US if pork is smuggled in air passenger luggage.[2] They find that the risk is high especially from China, Hong Kong and Russia with five US airports accounting for over 90% of the risk. Given the potentially devastating impact of this disease, the USDA has been actively working to safeguard against the entry of ASF into the country including surveillance and testing, monitoring cargo and travelers from affected countries, collaborating with and urging producers to follow strict on-farm protocols and best practices, and restricting pork imports from affected countries.[4] In addition to the adverse effects on producers (loss of export markets and profits) outlined above, an outbreak in the US would results in potentially enormous costs related to slaughtering infected domestic and wild herds and restricting the movement of animals between states. The purpose of this report is to examine the economy-wide impacts of the elimination of export markets due to a hypothetic outbreak of ASF in the United States. We first establish a baseline scenario, which represents the status quo (no ASF disease exists). Then we consider two scenarios, one where we assume that all US exports are eliminated over the ten years of the projection period (2020 to 2029) and another where the export market recovers after two years of zero pork exports. In the first case, a ten-year projection period allows time for the industry to reach a new equilibrium where the industry shrinks (producers making losses exit the market) and pork is sold only in the domestic market. The scenarios are compared to the baseline to estimate the impact of this downsizing on the US economy. Results show that the costs associated with an ASF outbreak are large and require risk mitigation and safeguards in the form of insurance against drastic price reductions.

Figure 1: ASF outbreak in Asia.

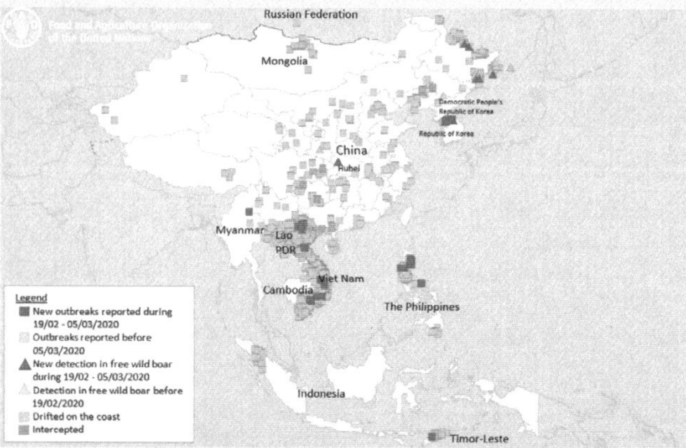

Source: United Nations Food and Agriculture Organization (FAO), (accessed January 31st, 2020). http://www.fao.org/ag/againfo/programmes/en/empres/ASF/**situation_update.html**

Overview of the likely impact of ASF on the US economy

The US is a major exporter of pork. In October 2019, US pork exports accounted for 30% of world pork exports at 3 million metric tons, carcass weight equivalent (second only to the European Union) (USDA, 2020).[5] Additionally, the US saw significant increases in pork exports to China because of the ASF in Asia. Therefore, an outbreak of swine fever in the US would result in a significant reduction is US pork exports and a loss in export markets. The inability of the US to export pork would lead to a flood of pork in the domestic market putting downward pressure on prices and affecting the profits of hog producers.

Loss of pork export markets because of ASF would create an oversupply of meat on the domestic market with significant price reductions throughout the marketing system. Live hog prices would fall to encourage the US consumer to eat more pork. The availability of inexpensive pork in the US domestic market would lead to price reductions in competing proteins such as chicken, eggs and cheese. Consumers in the rest of the world who are suddenly cut off from imported pork will necessarily reduce consumption and turn towards domestic proteins or to meat imports from other exporting countries. The US may lose market shares and its status as a major exporter of pork for a significant amount of time after the first outbreak. As the US meat sectors adjust to lower demand, feed-grain use will fall and employment in the US livestock sector and its affiliated industries will suffer. Second-round impacts will include a reduction in the US trade balance and in rural employment. These results are provided in a separate report on the economic impacts of a US ASF outbreak at the national and state (Iowa) levels.

Methodology

This analysis will quantify some of the impacts on the US agricultural markets as well as the national and state economies using two models: a global agricultural modeling system and an input-output model for the US, namely the Center for Agricultural and Rural Development (CARD) agricultural modeling system and the IMpact Analysis for PLANing (IMPLAN) model, respectively. The agricultural model system provides the impact on US and global agricultural commodities in terms of supply, utilization, and prices. The input-output model provides industry-level impacts in terms of changes in employment, labor income, industrial output, and value added. Both models capture the interlinkages between sectors, which is important given that in the interrelations involved in the supply and demand for agricultural products, factors that disrupt the market for one commodity are likely to affect others.

Swine fever scenarios

This section presents the results for both the elimination of US pork exports for all the years of the projection period (all-years scenario) and for only two years (2-year scenario).

Impact on US agricultural commodities

Pork

As shown in Figure 2 below, hog prices (Barrows and Gilts, National Base) decrease in both the all-years scenario and the 2-year scenario when compared to the baseline. Hog prices fall by approximately 47% in the first year of the outbreak in the all-years scenario and eventually the industry shrinks because of the losses incurred by producers (Figure 3). Prices then stabilize and are 1.8% lower than the baseline by the end of the ten-year projection period because pork exports continue to remain at zero. Figures 2 and 3 show that, in the 2-year scenario when pork exports start to recover after the first two years, hog prices initially decline by 47% and then start to climb back to baseline levels as the industry returns to normal profits. The contraction of the pork industry is evident in Figure 4, which highlights the fact that ASF would have a devastating effect on pork producers especially in the all-years scenario where production declines by almost 30% by the end of the projection period even as margins return to baseline levels. The 2-year scenario also shows a very small contraction in the industry in the long term given that exports eventually return to normal levels. Thus, only a small percentage of hog producers exit the market, and for the most part, the losses to the industry are temporary in the 2-year scenario. Lower prices and quantities sold lead a decline in the pork industries revenue. Figure 5, which compares pork revenues between the baseline and the two scenarios, shows that the pork revenues decline dramatically in the first few years and, while they do start to increase in subsequent years, the revenues reach baseline levels only in the last three years of the projection period in the case of the 2-year scenario. Revenues remain well below the baseline in the all-years scenario and never return to pre-outbreak levels. Over the ten-year period, losses in total revenue add up to $15 billion in the 2-year scenario and a little over $50 billion in the all-years scenario.

Beef

As is evident in Figure 6, cattle prices (Nebraska Direct Steers, 1,100 – 1,300 pounds) in the all-years scenario initially decline (by over 4% in the first year as seen in Figure 7) because meat consumers substitute toward the less expensive pork. Cattle prices start to recover over the projection period after the pork industry adjusts to the outbreak in the long run. As in the case of hog prices, Figure 7 shows that cattle prices eventually return to slightly above baseline levels (less than 1% higher, on average, relative to the baseline). Similar results occur in the 2-year scenario, but the prices increase faster than in the all-years scenario after the initial drop because pork exports start to recover in the third year of the projection period. In the 2-year scenario, cattle prices go back to baseline levels by the end of the projection period. Beef

Figure 2: Impact of ASF on US live hog prices (levels).

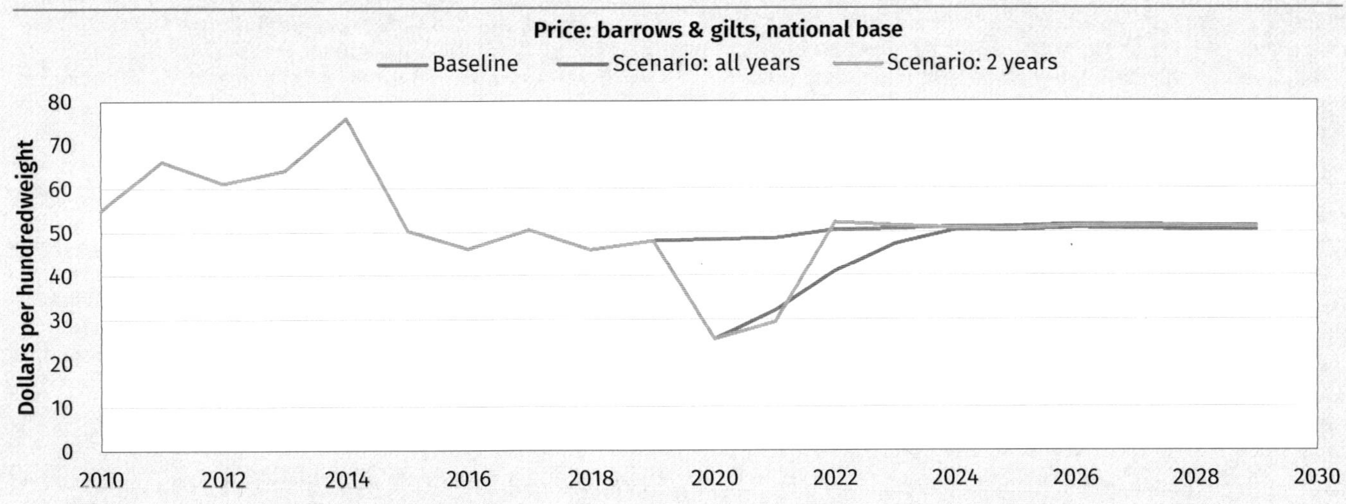

Figure 3: Impact of ASF on US live hog prices (percent change from baseline).

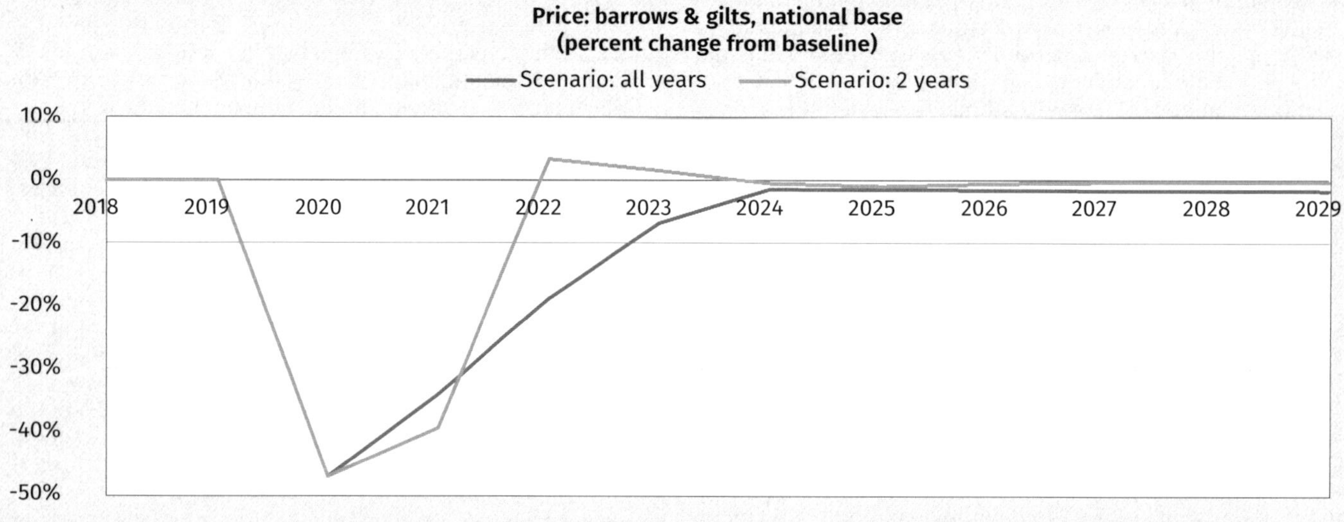

Figure 4: Impact of ASF on US pork production.

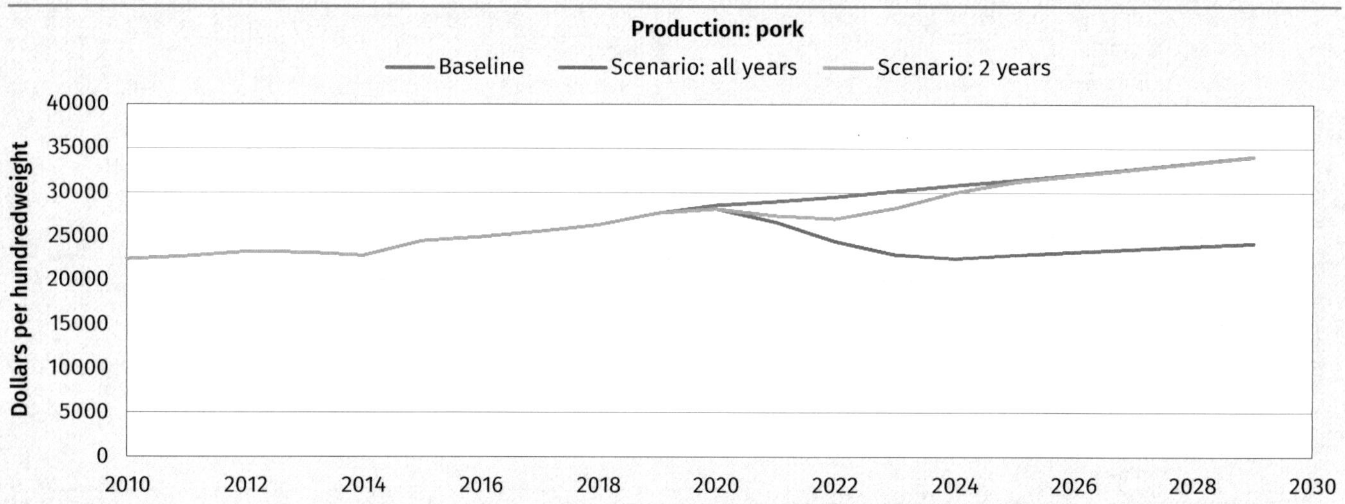

production in both scenarios does not deviate very much from baseline levels (Figure 8). However, production of beef does initially decline slightly relative to the baseline in response to the lower prices resulting from the fall in domestic use before recovering back to close to baseline levels. There is stronger demand for beef from the international market (as pork supply declines) especially in the first few years and this demand helps mitigate the reduction in prices in the earlier years and the faster recovery of prices over the projection period.

Poultry

The impact on poultry prices and production is not large. In both scenarios, broiler prices remain close to baseline levels (Figure 9). Like in the beef sector, the poultry sector loses some domestic market share to the less expensive pork. Domestic use decreases by an average of 1% in the first couple of years in both scenarios relative to the baseline. After that, domestic use for poultry in the all-years scenario is about 0.4% lower than the baseline, on average, while use in the 2-year scenario returns to baseline levels. Although domestic use declines, there is increased demand from the rest of the world. In the first two years, poultry exports are, on average, over 8% higher than the baseline. The higher export demand puts upward pressure on broiler prices. However, the price increase is dampened by the lower domestic demand. In the all-year scenario, prices remain less than 1% higher than the baseline (Figure 10). As with prices for hogs and cattle, broiler prices converge to baseline levels in the long run in the 2-year scenario (by the end of the projection period). Poultry production increases in response to the higher poultry prices and settles at about 1% higher than the baseline in the all-years scenario and at baseline levels in the 2-year scenario by the end of the projection period (Figure 11).

Corn

Figure 12 compares the results for the all-years scenario and the 2-years scenario in terms of corn prices. As the figure shows, corn

Figure 5: Impact of ASF on US pork revenue.

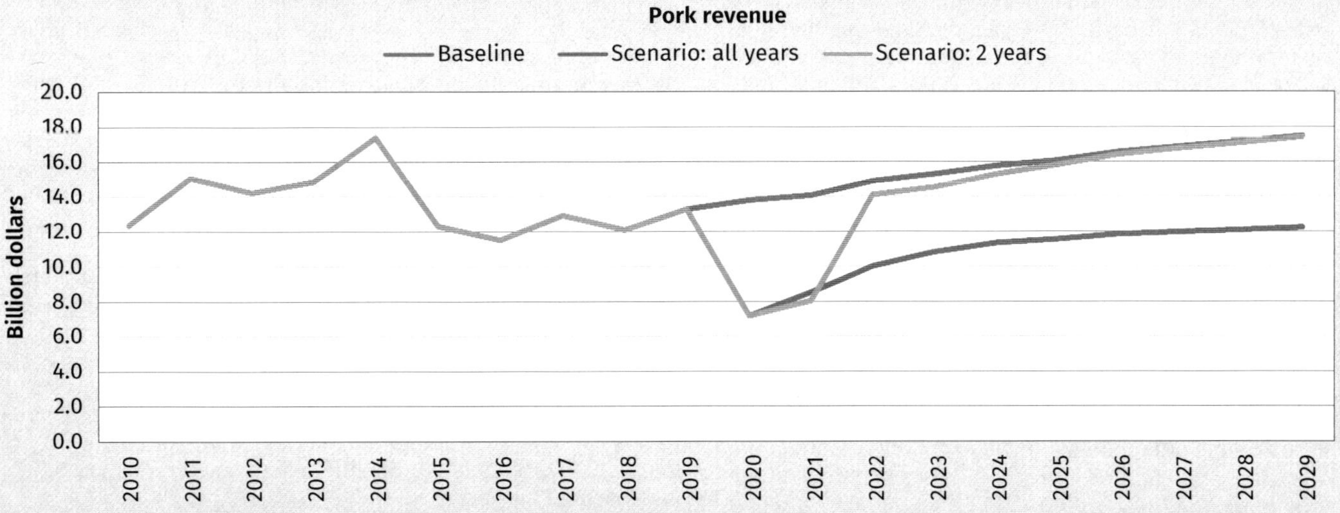

Figure 6: Impact of ASF on US cattle prices (levels).

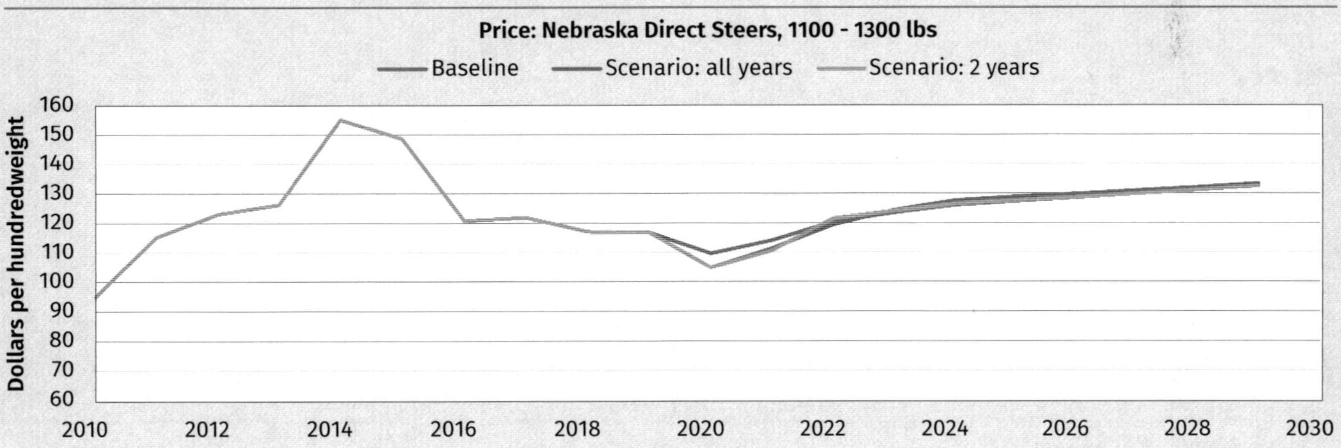

Figure 7: Impact of ASF on US live cattle prices (percent change from baseline).

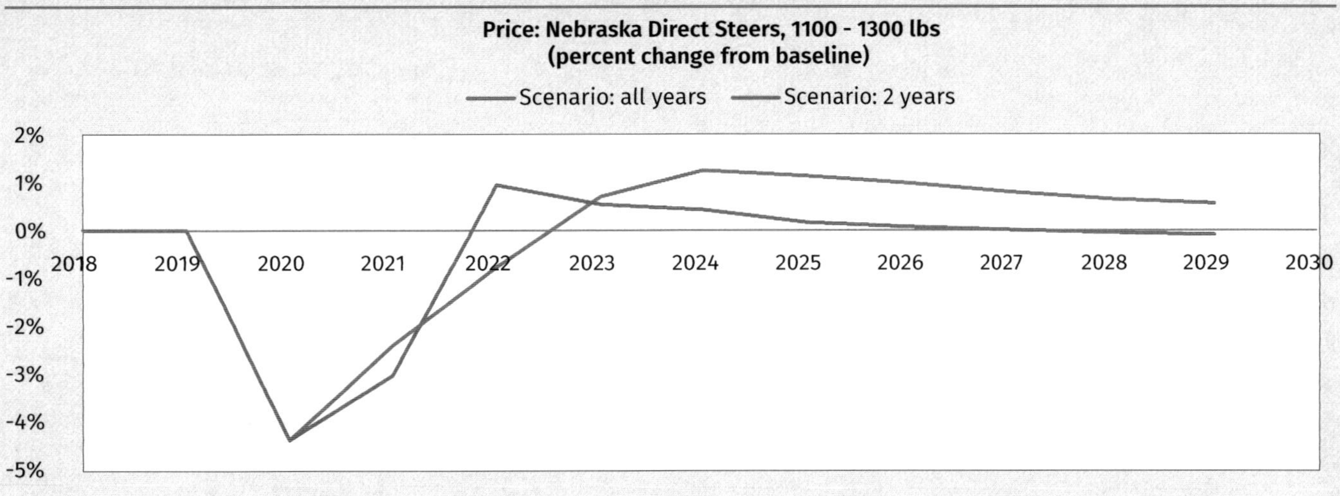

prices do not change very much between the baseline and the two scenarios. However, they are slightly lower than the baseline in both cases. This is a reflection of lower domestic demand. However, the beef industry experiences higher demand from the rest of the world, which mitigates the drop in prices due to lower domestic use, especially in the all-years scenario. Production decreases in response to the lower prices in both scenarios, although the decrease in production is lower in the 2-years scenario (average of -0.2% over the projection period) when compared to the decrease in the all-years scenario (average of -0.4%) (Figure 13).

Soybeans and soybean meal

Figures 14 and 15 present the impacts of an ASF outbreak on soybean prices and production. The outbreak lowers the price of soybeans in the all-years scenario relative to the baseline, thus decreasing soybean production by about 0.5% by the end of the projection period. Production also declines in the 2-years scenario by a smaller percentage than the all-years scenario and remains close to baseline levels. The impact on soybean meal is larger when compared to soybeans. In the all-years scenario, the decline in soybean meal prices ranges between 2% and 7% over the ten-year period as a result of a drop in domestic use (Figure 16). We do not see an even lower drop in soybean meals prices because of increased demand from the rest of the world that dampens the impact from lower domestic use. Results are more muted in the 2-years scenario with prices decreasing by about 2%, on average, in the first two years when the outbreak occurs and then recovering to reach close to baseline levels by the last year of the projection period. Changes in production are small in both scenarios, ranging between -0.2% and -0.8% in the all-years scenario, and by an average of -0.09% in the 2-years scenario (Figure 17).

Figure 8: Impact of ASF on US beef production.

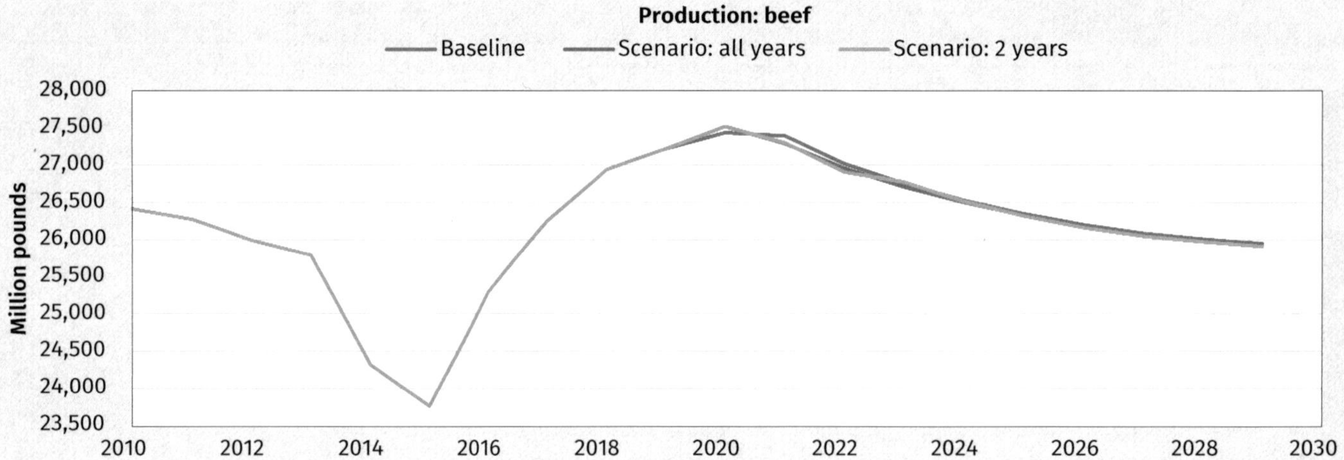

Figure 9: Impact of ASF on US broiler prices (levels).

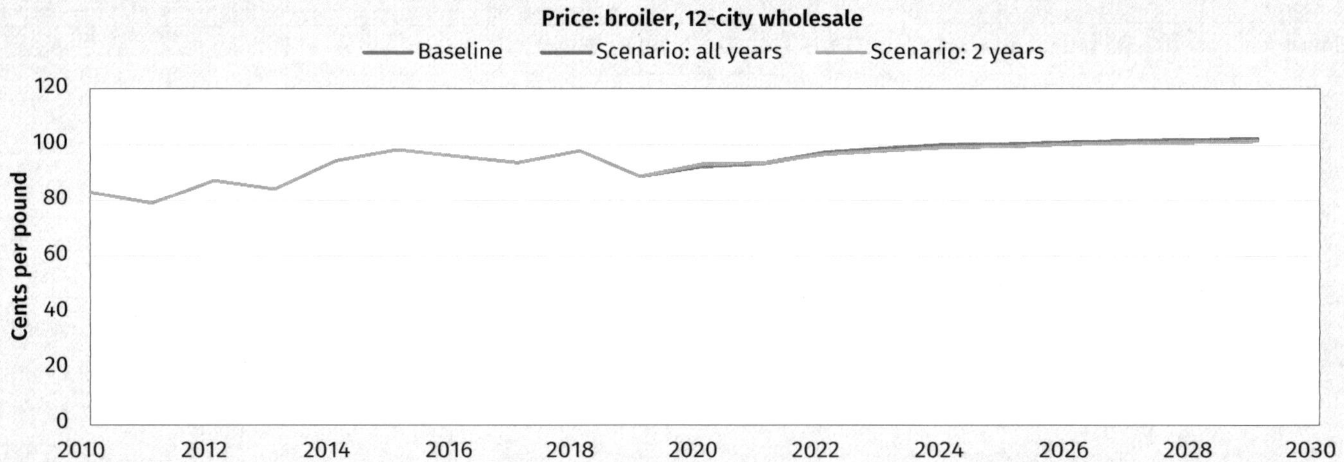

Impact on the US national economy and the state of Iowa economy

To isolate the Iowa (state) results from the US (national) results, two IMPLAN models were built. The percentage allocation of the commodity output changes over time for Iowa are based on Iowa's direct output in each of the commodities analyzed as a share of the US values as contained in the IMPLAN model. An exception, however, is made for the Iowa share of broilers. As the Iowa's poultry industry is overwhelmingly dominated by laying hens and egg production, with comparatively small levels of broiler production, the percentage of farm broiler chicken inventories in Iowa compared to the nation from the 2017 Agricultural Census is used.

For each model, for the US and for Iowa, two distinct impact summaries are generated, one considering the all-years and one considering the 2-year scenario. Separate impacts are compiled for each commodity – corn, soybeans, soybean meal, pork, beef, and broilers – and the unique, non-duplicative effects are then summed.

The following four tables display the annual value of the total economic impacts caused by deviations between the scenario and the forecast baseline for each year through 2028/2029 expressed as if they were reductions or gains in the values for those commodities in the IMPLAN model. The total values reported are the sum of all direct (producer or processor level), indirect (the supplying sectors), and the induced (household spending) amounts. Values are reported for jobs (full and part-time), labor income (the value of all wages and salaries plus employer-provided benefits), value added (includes labor income plus returns to investors and indirect tax payments), and total output (the sales value of production).

Table 1 shows that, for the US, in the all-years scenario, job losses grow from 6,380 in the first year, 2020/2021, to 142,485 by 2028/2029. Value added, which is analogous to gross domestic product, is reduced by $563.6 million in the first year, but declines to $11.1 billion by the last year when all multiplied-through consequences are summed.

In the 2-years scenario (Table 2), job losses are 6,201 in the first year, climb to 39,542 in losses by 2022/2023, and then fall to 220 jobs by the last year. Value added is reduced by $544.45 million in the first year, grows to a $2.99 billion reduction by 2022/2023, and then turns to a positive $25.64 million in 2028/2029.

For Iowa (Table 3), in the all-years scenario, job losses grow from 1,100 in year one to 22,076 by the last year. Value added is reduced by $93.00 million in year one and losses grow to $1.78 billion by 2028/2029.

In the 2-years scenario for Iowa (Table 4), initial year job losses of 1,089 grow to 5,862 by 2022/2023 before recovering to just 63 jobs lost by 2028/2029. Value added reduces by $91.65 million in year one, grows to $470.6 million in the third year, and then recovers to a negative $2.95 million in the last year.

References

1. Pitts, Nathan and Tim Whitnall. 2019. "Impact of African swine fever on global markets." Australian Bureau of Agricultural and Resource Economics and Sciences (ABARES), Department of Agriculture, Australian Government. Released 17 September 2019. **https://www.agriculture.gov.au/abares/research-topics/agricultural-commodities/sep-2019/african-swine-fever**

2. Jurado, C., Mur, L., Pérez Aguirreburualde, M.S. et al. Risk of African swine fever virus introduction into the United States through smuggling of pork in air passenger luggage. Sci Rep 9, 14423 (2019). https://doi.org/10.1038/s41598-019-50403-w

3. USDA APHIS:VS. "Qualitative Assessment of the likelihood of African swine fever virus entry to the United States: Entry Assessment." USDA: APHIS: VS: Center for Epidemiology and Animal Health, Risk Assessment Team. Fort Collins, CO. **https://www.aphis.usda.gov/animal_health/downloads/animal_diseases/swine/asf-entry.pdf** (Accessed January 31st, 2020).

4. USDA. 2019. "USDA Continues to Prevent African Swine Fever from Entering the US: Provides New Resources to Raise Awareness of the Deadly Disease." Press Release Release No. 0034.19. **https://www.usda.gov/media/press-releases/2019/03/19/usda-continues-prevent-african-swine-fever-entering-us.** (Accessed January 31st, 2020).

5. USDA. 2020. "Pork Trade: Selected Countries Summary." Livestock and Poultry: World Markets and Trade, Global Market Analysis, United States Department of Agriculture Foreign Agricultural Services, January 10, 2020.

Figure 10: Impact of ASF on US broiler prices (percent change from baseline).

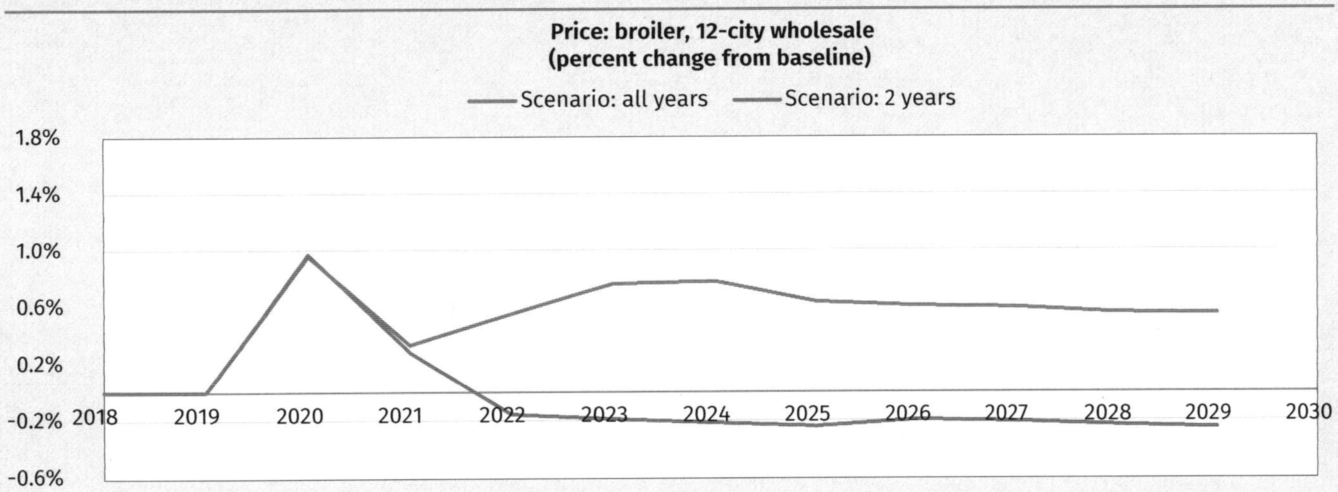

Figure 11: Impact of ASF on US poultry production.

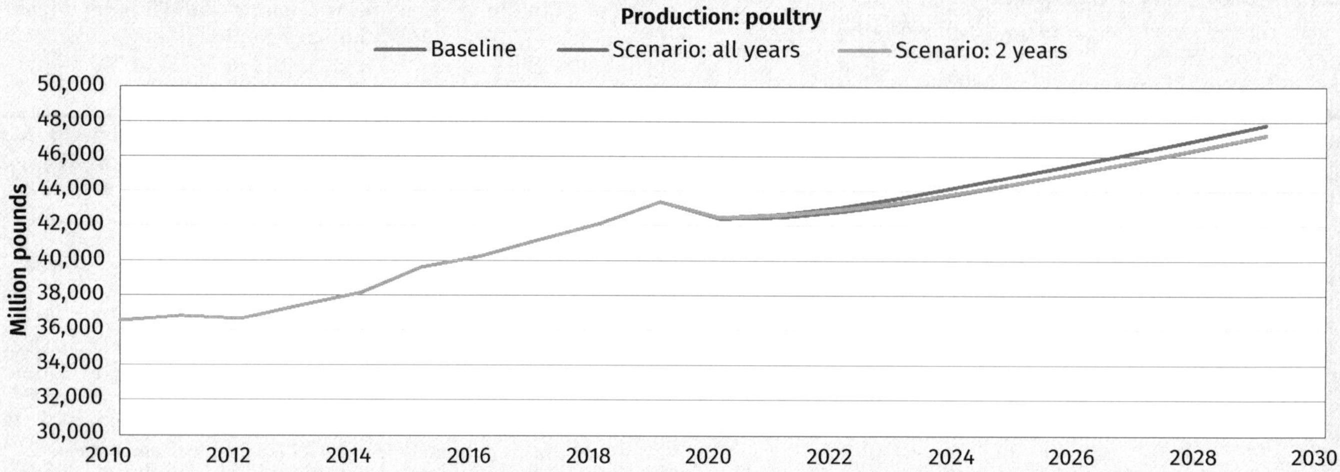

Figure 12: Impact of ASF on farm-level corn prices.

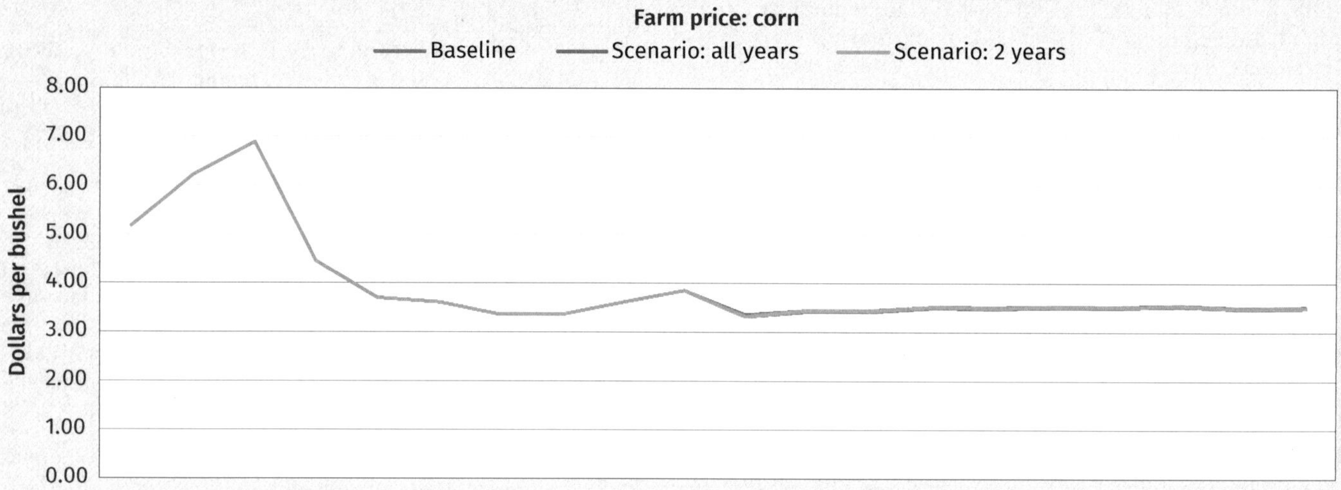

Figure 13: Impact of ASF on US corn production.

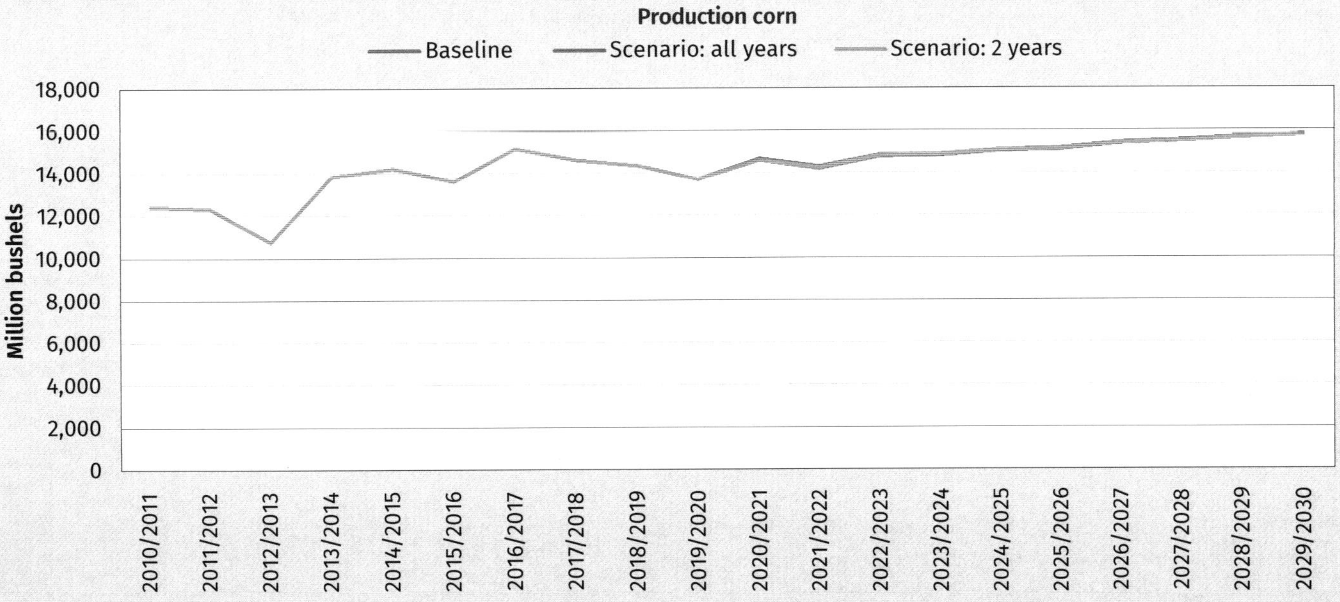

Figure 14: Impact of ASF on farm-level soybean prices.

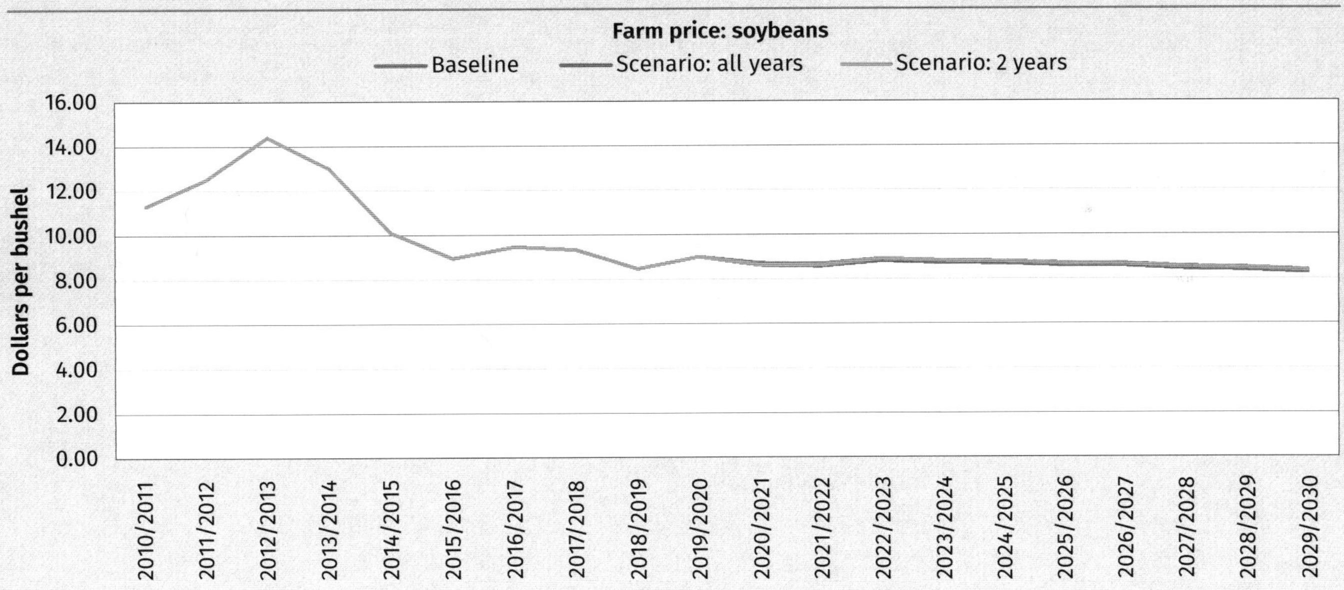

Figure 15: Impact of ASF on US soybean production.

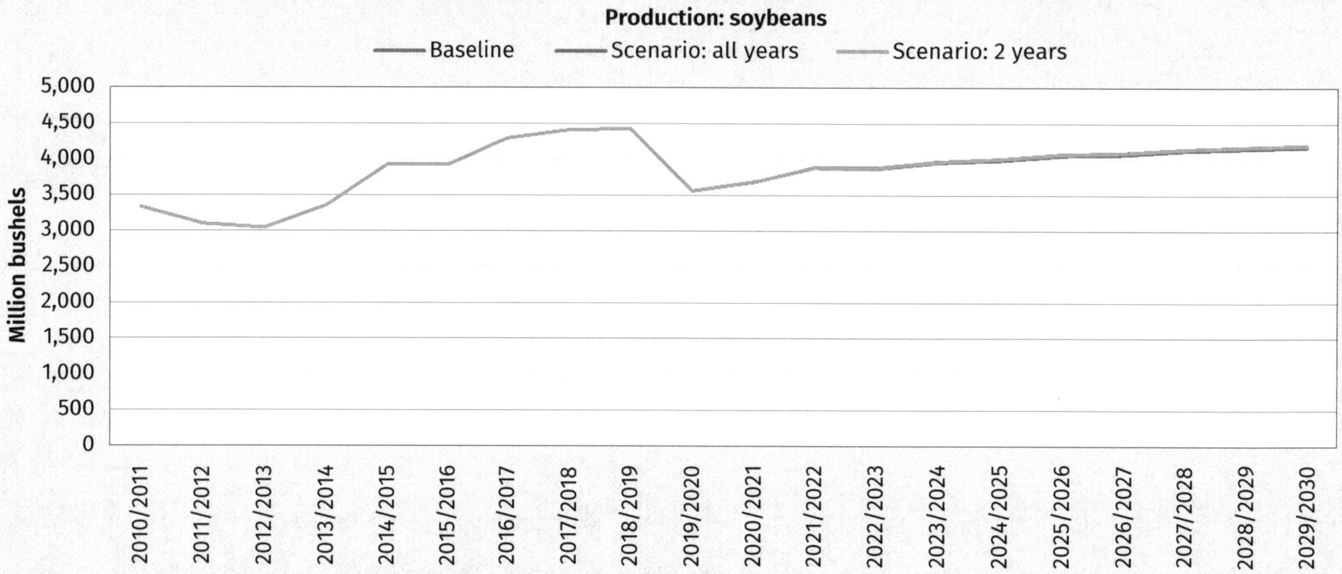

Figure 16: Impact of ASF on soybean-meal prices.

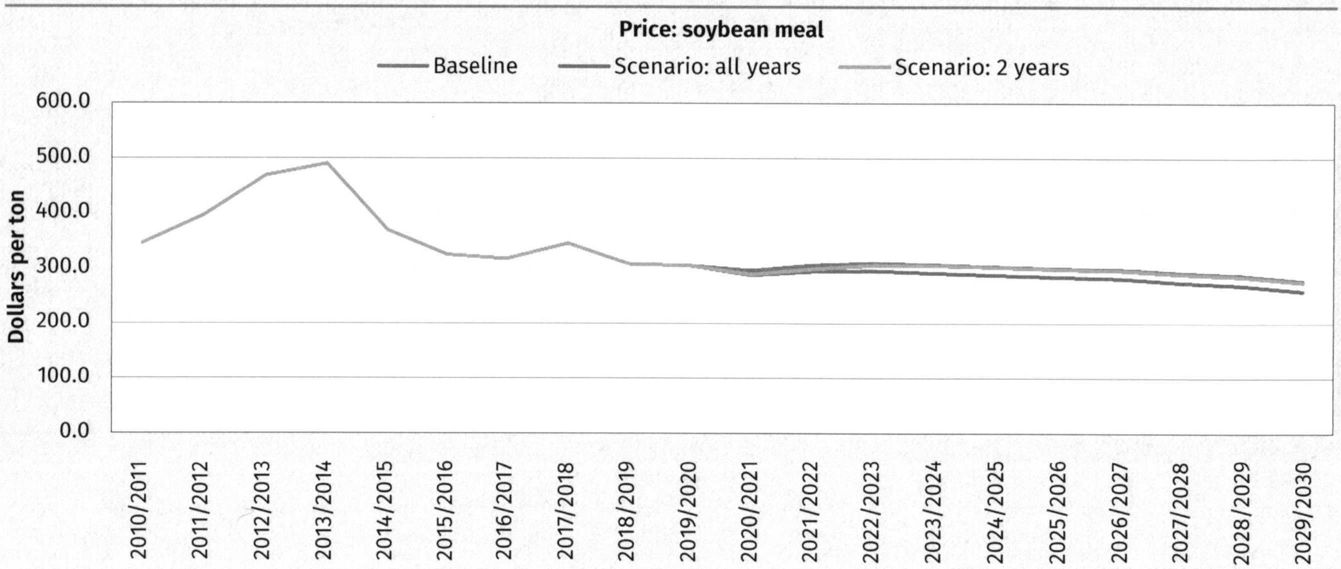

Figure 17: Impact of ASF on US soybean meal production.

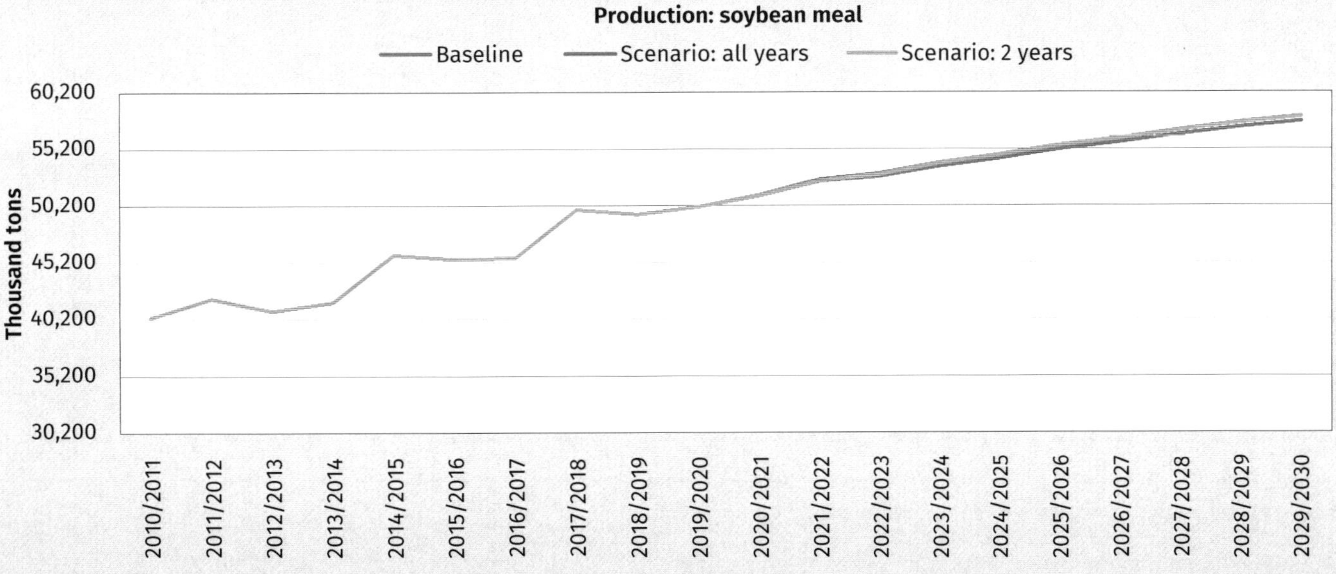

Table 1: Total annual US economic impacts in the all-years scenario.

Impact year	Jobs	Labor income	Value added	Output
20/21	(6,380)	$ (317,794,408)	$ (563,635,854)	$ (1,026,235,887)
21/22	(40,772)	$ (1,742,746,216)	$ (3,133,789,933)	$ (6,099,686,705)
22/23	(79,680)	$ (3,435,000,850)	$ (6,182,397,022)	$ (11,823,032,295)
23/24	(112,398)	$ (4,849,488,047)	$ (8,728,212,009)	$ (16,616,148,840)
24/25	(127,001)	$ (5,478,779,024)	$ (9,864,758,382)	$ (18,677,870,992)
25/26	(129,065)	$ (5,565,280,949)	$ (10,022,754,233)	$ (18,924,035,222)
26/27	(132,925)	$ (5,730,211,848)	$ (10,321,677,252)	$ (19,479,102,710)
27/28	(137,735)	$ (5,940,806,637)	$ (10,702,287,099)	$ (20,181,536,388)
28/29	(142,485)	$ (6,148,788,621)	$ (11,078,484,090)	$ (20,852,837,503)

Table 2: Total annual US economic impacts in the 2-years scenario.

Impact year	Jobs	Labor income	Value added	Output
20/21	(6,201)	$ (307,422,781)	$ (544,447,946)	$ (975,326,706)
21/22	(27,188)	$ (1,146,045,216)	$ (2,066,891,418)	$ (3,981,271,074)
22/23	(39,542)	$ (1,656,555,877)	$ (2,988,808,223)	$ (5,716,747,137)
23/24	(25,954)	$ (1,100,336,336)	$ (1,985,835,379)	$ (3,586,724,544)
24/25	(9,374)	$ (373,612,353)	$ (680,851,796)	$ (1,156,895,385)
25/26	(1,920)	$ (52,308,994)	$ (99,872,474)	$ (110,009,300)
26/27	(754)	$ (2,874,681)	$ (10,973,359)	$ 29,348,507
27/28	(638)	$ 252,133	$ (4,231,916)	$ 33,799,959
28/29	(220)	$ 16,954,371	$ 25,637,803	$ 95,398,369

Table 3: Total annual Iowa economic impacts in the all-years scenario.

Impact year	Jobs	Labor income	Value added	Output
20/21	(1,100)	$ (46,387,658)	$ (93,000,851)	$ (241,968,195)
21/22	(5,603)	$ (224,216,855)	$ (451,204,466)	$ (1,149,383,507)
22/23	(11,768)	$ (471,485,987)	$ (948,009,209)	$ (2,388,420,284)
23/24	(16,894)	$ (676,924,938)	$ (1,360,356,513)	$ (3,420,866,059)
24/25	(19,364)	$ (776,152,451)	$ (1,559,682,132)	$ (3,918,534,069)
25/26	(19,829)	$ (794,959,008)	$ (1,597,284,429)	$ (4,013,991,806)
26/27	(20,452)	$ (820,043,070)	$ (1,647,748,158)	$ (4,142,132,633)
27/28	(21,254)	$ (852,416,811)	$ (1,712,748,409)	$ (4,306,196,093)
28/29	(22,076)	$ (885,606,711)	$ (1,779,440,690)	$ (4,473,264,825)

Table 4: Total annual Iowa economic impacts in the 2-years scenario.

Impact Year	Jobs	Labor Income	Value Added	Output
20/21	(1,089)	$ (45,766,279)	$ (91,650,756)	$ (237,101,882)
21/22	(3,891)	$ (155,830,800)	$ (313,757,499)	$ (801,034,605)
22/23	(5,862)	$ (233,831,028)	$ (470,569,531)	$ (1,190,475,856)
23/24	(4,395)	$ (176,034,922)	$ (353,851,828)	$ (881,819,160)
24/25	(1,699)	$ (67,157,023)	$ (135,126,645)	$ (332,084,382)
25/26	(405)	$ (14,992,549)	$ (30,142,363)	$ (70,485,362)
26/27	(159)	$ (5,165,953)	$ (10,464,708)	$ (22,331,607)
27/28	(117)	$ (3,539,283)	$ (7,147,858)	$ (14,900,764)
28/29	(63)	$ (1,440,193)	$ (2,945,845)	$ (4,306,118)

A comparison of active surveillance protocols to support pre-movement guidelines for African swine fever

M. Farr[1,2], DVM; S. Malladi[3], PhD; O. Beemer[2], DVM, MS; M. Rotolo[2,4], DVM, PhD; A. Ssematimba[3,5], PhD; M. Culhane[3], DVM, PhD; M. Remmenga[2], PhD

[1]Veterinary Diagnostic & Production Animal Medicine, College of Veterinary Medicine, Iowa State University, Ames, Iowa; [2]United States Department of Agriculture, Animal and Plant Health Inspection Service, Veterinary Services, Strategy and Policy, Center for Epidemiology and Animal Health, Surveillance Design and Analysis, Fort Collins, Colorado; [3]Secure Food Systems Team, College of Veterinary Medicine, University of Minnesota, Saint Paul, Minnesota; [4]Pig Improvement Company (PIC), Hendersonville, Tennessee; [5]Department of Mathematics, Gulu University, Uganda

Introduction

African swine fever (ASF) is a devastating swine disease that has resulted in the loss of more than a quarter of the global swine population in an ongoing outbreak since 2016. Emergency preparedness for an ASF outbreak in the United States is critical to reduce economic impacts and minimize disruptions to the food supply. During an outbreak, standard pre-movement surveillance protocols for premises within a control area wanting to move pigs are needed to support swine production business activities within the US. In order to develop effective strategies, we first compared the ability of active surveillance protocol options to detect ASF virus (ASFV) prior to movement. Protocols were evaluated with respect to these key outcomes: the probability of detecting ASFV prior to movement and the predicted number of infected pigs in those herds that were undetected at the time of movement.

Materials and methods

Model parameterization and implementation

Various testing protocol options were simulated using a stochastic within-herd disease transmission model parameterized for moderately virulent ASFV strains (Malta 79, Netherlands 86). Industry partners provided normal production data from 248 finisher herds. Time to clinical signs and disease state durations, including the latently infected period, the infectious period for pigs that die, and the infectious period for pigs that recover were informed by experimental data from published studies[1-3] and estimated using a maximum likelihood estimation approach. The results from this model were then used to simulate the likelihood of detection under various spread scenarios and active surveillance protocols (20,000 iterations each). Spread scenarios varied by adequate contact rate and the percentage of sick pigs due to routine causes unrelated to ASF. Within each spread scenario- three active surveillance protocols were simulated and the likelihood of detection with individual animal testing was compared to the effect of pooling samples in groups of five samples per pool. The time of exposure into a naïve population was randomly assigned for each iteration between 3-30 days prior to movement for a population representing a medium-size finisher swine herd with herd size randomly chosen from 1000 to 3999 swine.

Active surveillance protocols evaluated

We evaluated three different active surveillance protocol options for finisher swine moving to processing. The protocols were based on diagnostic testing via whole blood and tissue samples combined with observational surveillance of clinical signs of disease for 14 days prior to movement. Sample collection was prioritized first from available sick pigs, followed by dead pigs and then live pigs for diagnostic testing. Each surveillance protocol was evaluated on individual and pooled individual animal samples, where whole blood or tissue samples were combined in pools of five respectively. All protocols targeted sampling of pigs 24 and 48 hours prior to movement, with some variation in additional testing rounds and levels of design prevalence. The baseline protocol (protocol one) consisted of testing at 5% disease prevalence (62 samples) split evenly over two days just prior to movement, or 31 samples collected on each day. Protocol two added upon protocol one with an additional testing round of 62 samples split evenly over two days 1 week prior to movement, for a total sample size of 124. We then compared the baseline protocol to a 1% design prevalence (protocol three), increasing the sample size from 62 to 314 samples, split evenly over two days prior to movement.

Results and discussion

Probability of detection

The probability of detecting ASFV in a finisher herd prior to movement ranged from 73-83% across all scenarios. While increasing the frequency of sample collection and sample size from the baseline protocol did show a statistically significant increase in the probability of detection for some scenarios, the magnitude of this increase was minor. One factor contributing to the lower probability of detection is the long latent period for ASF, which results in slow disease transmission during the initial stages of herd infection. When comparing individual testing to the effect of pooling five samples, pooling samples did not significantly alter the probability of detection. Implementing a pooled sampling protocol may reduce the number of tests run by laboratories, conserving resources, while allowing for a greater representation of herd disease status. It is important to mention the detection probability was lower in scenarios with a higher proportion of sick pigs due to routine causes unrelated to ASF. A farm's baseline morbidity negatively impacts the probability of detecting ASFV by reducing the chances of selecting an ASFV positive pig for sampling from amongst many sick pigs. In this situation, increasing the sample size for herds with a high baseline morbidity can help achieve a slightly higher probability of detection.

Predicted number of infectious pigs at the time of movement

The predicted number of infected pigs in those herds that were undetected at the time of movement was also evaluated. The estimated number of ASFV-infected pigs in those herds that were undetected at the time of movement was < 8 pigs. The need to maintain business continuity requires mitigation of the risk of moving ASFV infected pigs to prevent the spread of disease.

Surveillance is only one component of mitigating the spread of ASFV

Based on the results for both probability of detection and predicted number of infected pigs in undetected herds, we can deduce that surveillance alone is not capable of detecting ASFV at an acceptable level that confidently allows for the movement of disease-free pigs. Surveillance must be combined with additional preventive measures established at the farm level, such as implementing heightened biosecurity measures. As the US swine industry has learned from porcine epidemic diarrhea virus (PEDV), these heightened biosecurity measures should include minimizing the introduction of potentially contaminated material or persons through: 1. stringent use of personal protective equipment, 2. disinfection of fomites, 3. disinfection of transportation equipment, and 4. limiting contact between trailers or drivers and the other facilities during unloading.[4] In addition, a strategy should be in place to prepare for the potential of receiving ASFV infected pigs such as isolating any new arrivals for a minimum of one incubation period.

References

1. deCarvalho Ferreira, H.C., et al. Transmission rate of African swine fever virus under experimental conditions. *Veterinary Microbiology* 165 (2013) 296-304

2. deCarvalho Ferreira, H.C., et al. African swine fever virus extraction patters in persistently infected animals: a quantitative approach. *Veterinary Microbiology* 160 (2012) 327-340

3. Guinat, C., et al. Experimental pig-to-pig transmission dynamics for African swine fever virus, Georgia 2007/1 strain. *Epidemiology & Infection* 144 (2013) 25-34

4. Lee, C. Porcine epidemic diarrhea virus: an emerging and re-emerging epizootic swine virus. *Virology Journal* (2015) 12: 193

Active regional surveillance for early detection of exotic/emerging pathogens of swine: a comparison of statistical approaches for selecting farms to be sampled

D. Hu, BS; T.-Y. Cheng, DVM, MS; P. Morris, BS, MS; J. Zimmerman, DVM, PhD; C. Wang, PhD
Iowa State University, Ames, Iowa

Introduction

Farm selection plays a crucial role in livestock disease surveillance. In the past, farm selection was based on conditional random sampling, i.e., selecting farms using predefined criteria (stratified) or according to a random starting point with fixed intervals (systematic). However, the growing recognition of spatial dependence in disease spread (autocorrelation) brings the validity of these approaches into question. In this study, 6 statistically-based sampling methods were compared in terms of the probability of detection when 1 to 5% of farms were positive.

Materials and methods

To conduct the analysis, a hypothetical region (195 × 300 kilometers in size) populated with 6,000 livestock farms (Figure 1) was used to model the spatiotemporal spread of a contagious agent. The model used in the study was derived from a disease spread model proposed by Ju et al. (2020)[1]:

$$\lambda_{f,t)}=\lambda_{f\text{-}1,t}+\alpha\Sigma_{f'\epsilon N_f}\exp(-d_{f,f'})\,p_{f',t-1}$$

where $p_{f,t}$ denoted the disease prevalence within farm f at time t where $f \epsilon \{1, 2, ..., F\}$ and $t \epsilon \{1, 2, ..., T\}$. N_f is the set of farms within 5 km radius of farm f. $\lambda_{f,t}$ indicated the odds of being infected. α overall determined the degree of the spatial dependence from all other farms and $d_{f,f'}$ denoted the distance between farm f and f' ($f \neq f'$). The model was then used to simulate disease spread in the hypothetical region and compare simple random sampling (SRS) to 6 "spatially balanced" sampling methods (Figure 1)[2-7] in terms of the probability of detection at low prevalence (1-5%): generalized random-tessellation stratified (GRTS), local pivotal method (LPM), spatially correlated Poisson sampling (SCPS), local cube method (LCUBE), balanced acceptance sampling (BAS). Notably, this study did not assess the efficiency of on-farm sampling and assumed that disease classification of farms selected for sampling was 100% accurate when on-farm prevalence was ≥ 10%.

Results and discussion

For a fixed sample size and at a regional disease prevalence of 2%, differences in the probability of detection were observed among sample selection methods. In particular, SCPS and LCUBE had the highest probability of detection among the six methods at low prevalence. For example, assuming 60 out of 6,000 farms (1%) were sampled, the probabilities of detecting ≥ 1 positive farm were > 80 % for SCPS and LCUBE, and < 70% for SRS.

This study showed that spatially balanced sampling methods performed better than SRS at low prevalence, i.e., were more likely to select a positive farm. These results can be explained by the fact that disease spread follows a spatially dependent pattern such that neighboring farms are more likely to have similar disease status than distant ones: disease is not random. Unlike SRS, spatially balanced sampling methods used a variety of statistical approaches to adjust for spatial correlation and, thereby, improve detection efficiency. Because they achieve earlier detection with fewer samples, these methods should be considered for surveillance of transboundary, emerging, and foreign animal diseases.

Acknowledgments

This research was supported by grants from the National Pork Board (grant # 17-174; 1776 NW 114th Street, Clive, Iowa 50325) and the Swine Health Information Center (grant #18-198; 809 Wheeler Street, Ames, Iowa 50010).

References

1. Ju J., Wang, C., Rotolo, M., Zimmerman, J., (2020). Modeling regional disease spread over time using a dynamic spatio-temporal model – an application using porcine epidemic diarrhea virus data collected in Iowa, *US Prev Vet Med* 18.

2. Stevens, D.L., Olsen, A.R., 2004. Spatially Balanced Sampling of Natural Resources. *Journal of the American Statistical Association* 99, 262-278.

3. Deville, J.C., Tille, Y., 1998. Unequal probability sampling without replacement through a splitting method. *Biometrika* 85, 89-101.

4. Bondesson, L., Thorburn, D., 2008. A list sequential sampling method suitable for real-time sampling. *Scandinavian Journal of Statistics* 35, 466-483.

5. Grafström, A., 2012. Spatially correlated Poisson sampling. *Journal of Statistical Planning and Inference* 142, 139-147.

6. Grafström, A., Tillé, Y., 2013. Doubly balanced spatial sampling with spreading and restitution of auxiliary totals. *Environmetrics* 24, 120-131.

7. Robertson, B.L., Brown, J.A., McDonald, T., Jaksons, P., 2013. BAS: balanced acceptance sampling of natural resources. *Biometrics* 69, 776-784.

Figure 1: Illustration of farm sampling patterns (n = 200 samples) for each of 6 methods where × represents a farm selected for sampling in a hypothetical region containing 6,000 farms (gray dots). Results based on one simulation for each method. A) Generalized random-tessellation stratified sampling[2]; B) Local pivotal method[3]; C) Spatially correlated Poisson sampling[4,5]; D) Local cube method[6]; E) Balanced acceptance sampling[7]; F) Simple random sampling.

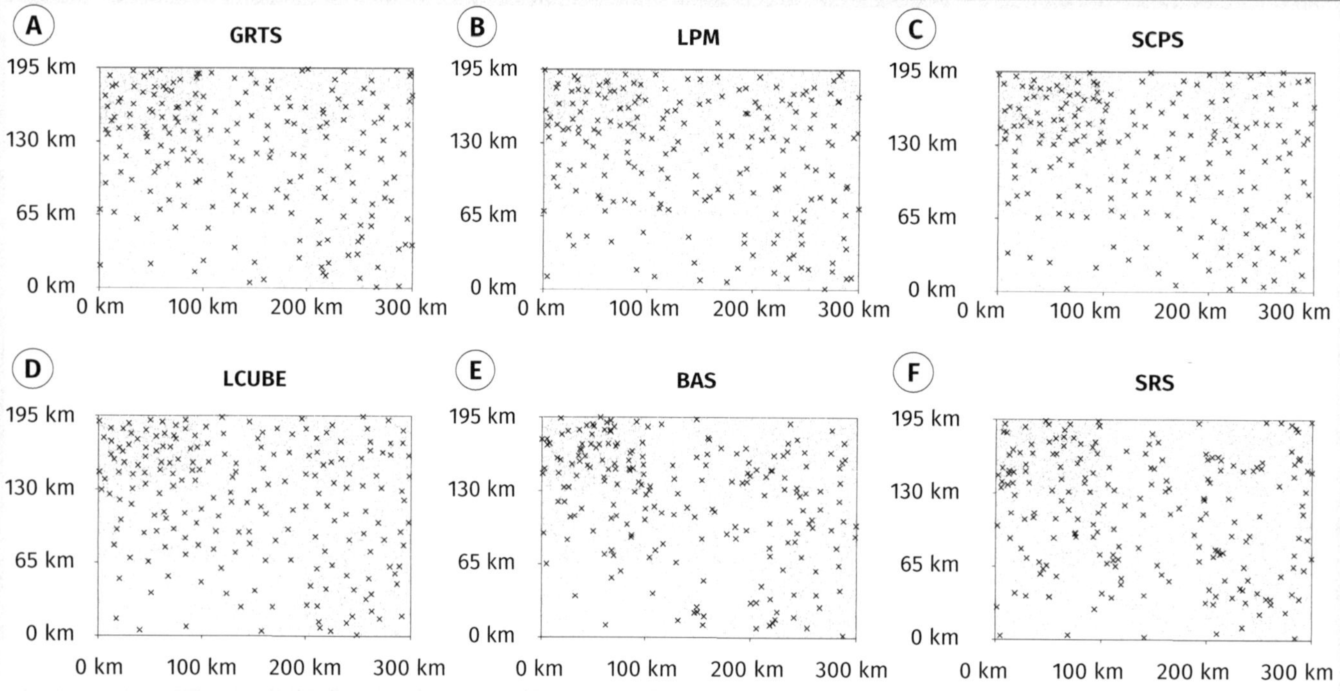

Use of two demonstration projects to evaluate viral survival in feed

S. Dee[1]; A. Shah[2]; R. Cochrane[1]; T. Clement[3]; A. Singrey[3]; R. Edler[1]; G. Spronk[1]; M. Niederwerder[4]; E. Nelson[3]
[1]Pipestone Applied Research, Pipestone Veterinary Services, Pipestone, Minnesota;
[2]SAM Nutrition, Eden Prairie, Minnesota; [3]Department of Veterinary and Biomedical Sciences, South Dakota State University, Brookings, South Dakota; [4]Department of Diagnostic Medicine/Pathobiology, College of Veterinary Medicine, Kansas State University, Manhattan, Kansas

Introduction

The hypothesis that feed ingredients could serve as vehicles for the transport and transmission of viral pathogens has been successfully proven by multiple investigators under laboratory conditions.[1,2] To validate these data, two demonstration projects were conducted to test whether three significant viruses of swine could survive in feed ingredients under real-world shipping conditions.

Materials and methods

In project 1, 30g samples of soybean meal (organic and conventional), lysine, choline, and vitamin A were spiked with a 2 mL mixture of PRRSV 174, PEDV, and SVA and transported for 21 days in the trailer of a commercial transport vehicle, encompassing 14 states and 9741 km. Samples were tested for viral genome by PCR and viability by bioassay at the end of the transit period. In project 2, the same timetable and route was employed; however, 1-ton totes of soybean meal (conventional and organic) and complete feed were used to better represent real world conditions. In addition, each tote was spiked using a 10 mL ice cube containing all 3 viruses at a concentration of 5 logs TCID50/virus ("hot spot model"). Sampling of totes employed a protocol validated by Jones and others[3] and samples were tested by PCR. For assessment of viability, samples from soy-based products were tested by bioassay and the totes of complete feed were consumed by pigs via natural feeding behavior.

Results and discussion

In project 1, viable PRRSV, PEDV and SVA was confirmed in both soy-based products by bioassay. In addition, viable PRRSV and SVA were detected in samples of vitamin A, while viable SVA was detected in samples of lysine and choline. Regarding project 2, it is currently in progress and results are scheduled to be available at the 2021 AASV meeting. The results of project 1 provide further evidence that select viral pathogens of pigs can survive in certain feed ingredients under of actual commercial transit shipments. However, due to acknowledged limitations including small amounts of feed inoculated with relatively large volumes of liquid, it is important to complete project 2 before any final conclusions can be drawn. The goal of a "hot spot" infection model via ice cube inoculation is an attempt to replicate real world settings. Furthermore, the ability to test a method designed to sample large volumes of feed for the presence of viral agents will add additional value to the study. Finally, while the assessment of viability by bioassay is useful, the ability to employ natural feeding behavior adds more reality to the study design.

References

1. Dee SA et al. Survival of viral pathogens in animal feed ingredients under transboundary shipping models. *PLOS ONE* 2018, 13(3): e0194509.

2. Stoian AMM et al. Stability of classical swine fever virus and pseudorabies virus in animal feed ingredients exposed to trans-pacific shipping conditions. *Transbound Emerg Dis* 2020. **doi: 10.1111/tbed.13498**

3. Jones et al. Validation of sampling methods in bulk feed ingredients for detection of swine viruses. *Transbound Emerg Dis* 2019. **doi: 10.1111/tbed.13326**

Use of a demonstration project to test the effect of extended storage on viral survival in feed

S. Dee[1], DVM, MS, PhD; A. Shah[2]; R. Cochrane[1], PhD; F. Wu[1], PhD; T. Clement[3]; A. Singrey[3]; R. Edler[1], MS; G. Spronk[1], DVM; M. Niederwerder[4], DVM, PhD; E. Nelson[3], PhD
[1]Pipestone Applied Research, Pipestone Veterinary Services, Pipestone, Minnesota;
[2]SAM Nutrition, Eden Prairie, Minnesota; [3]Department of Veterinary and Biomedical Sciences, South Dakota State University, Brookings, South Dakota; [4]Department of Diagnostic Medicine/Pathobiology, College of Veterinary Medicine, Kansas State University, Manhattan, Kansas

Introduction

Feed and feed ingredients have been proposed as vehicles for the transboundary transport and transmission of viral pathogens to the US.[1-3] To mitigate this risk, the North American swine industry has attempted to reduce viral viability in feed using a variety of approaches, including periods of extended storage under controlled environmental conditions. However, while widely applied, these protocols have been primarily based on mathematical estimates of half-life, not experimentally derived data. Therefore, the purpose of this study was to design a demonstration project to evaluate the effect of extended storage on the survival of swine viral pathogens in feed ingredients under real-world conditions.

Materials and methods

The study was based on the hypothesis that controlling temperature during storage would enhance the success of the protocol. Samples of soybean meal (organic and conventional), lysine, choline, and vitamin A were spiked with a mixture of PRRSV 174, PEDV, and SVA (total dose = 2×10^5 $TCID_{50}$) for each virus and stored at 20°C for 30 days. As controls, a duplicate set of samples were stored outside during February in Minnesota for the same period. Following 30 days, samples were tested by PCR and swine bioassay.

Results and discussion

Upon completion of the storage period, all outside samples were PCR-positive and contained infectious virus. In contrast, all inside samples were PCR positive but bioassay negative. During internal storage, the storage area remained at a consistently warm temperature (mean = 20.1°C) with a low % RH (mean = 35%) throughout the 30 days. In contrast, the external environment was generally cold (mean = -8.8°C) and moist (% RH = 77%) and varied over time. Under the conditions of the study, the internal storage protocol successfully inactivated PRRSV 174, PEDV and SVA across all five ingredients. In contrast, all three viruses survived in all five ingredients following external storage. This study demonstrated that an extended storage period of 30 days at a temperature of 20°C was effective at reducing the viability of three significant viral pathogens of pigs across multiple feed ingredients.

References

1. Dee SA et al. Survival of viral pathogens in animal feed ingredients under transboundary shipping models. *PLOS ONE* 2018, 13(3): e0194509.

2. Stoian AMM et al. Stability of classical swine fever virus and pseudorabies virus in animal feed ingredients exposed to transpacific shipping conditions. *Transbound Emerg Dis* 2020. **doi: 10.1111/tbed.13498**

3. Stoian AMM et al. Half-Life of African Swine Fever Virus in Shipped Feed. *Emerg Infect Dis.* 2019 Dec 17;25(12). **doi: 10.3201/eid2512.191002**

A novel objective quantifiable measure of lameness in gestating sows

A. Shulman[1], BS; M. Boulanger[1], BS; T. D. Parsons[2], VMD, PhD, DACAW
[1]School of Veterinary Medicine, University of Pennsylvania, Philadelphia, Pennsylvania;
[2]Swine Teaching and Research Center, University of Pennsylvania, Kennett Square, Pennsylvania

Introduction

Sow lameness has been shown to be one of the major causes of premature culling in sow herds, is associated with poor reproductive performance, and is undoubtedly a welfare concern. Group housing provides opportunities for improved welfare through increased social interaction and mobility, but may increase the risk of lameness. Current visual-based lameness scoring systems suffer from being subjective, lack reproducibility, and require active human engagement.[1] Improved methods for the early and successful identification of lameness portend better treatment outcomes and promise to improve sow longevity, productivity, and welfare. In group-housed systems, there is ample opportunity for lameness to be repeatedly assessed while sows walk toward high-value locations (e.g. electronic feeders). The objective of this study was to develop a novel and improved lameness detection paradigm that is both objective and quantifiable based on the rate at which sows walk.

Materials and methods

Pregnant commercial sows (n = 25, DNA line 241) ranging from 0-4th parity were trained to individually run back and forth on a 13 m U-shaped course with a reward feeder at each end (Figure 1). Sows were rewarded with 37 mL of their standard feed each time a sow correctly traversed the course. Sows ran the course for 15 minutes, or less if prohibited by lameness. Sows were video recorded for the entirety of each trial from a fixed angle to capture all instances of ambulation. Thirty-second anonymized video clips of ambulation were shown to a large-animal veterinarian who scored each sow's lameness using a visual assessment scale (VAS) that ranged between 0-150.[2] The video clips were randomly duplicated and mirrored to screen for observer reliability, producing 44 total clips. Sows were considered lame if they were scored as ≥ 40. Ambulation rate (AR; m/min) was calculated for each sow through video analysis (OpenShot Video Editor version 2.5.1). The number of completed runs over the course was counted, multiplied by the length of the course, then divided by the total time spent ambulating. A linear mixed effects model was used to elucidate the ability of AR to predict VAS with VAS as the outcome variable, AR as a fixed effect, and animal ID as a random effect (R Program version 4.0.2). Standard test characteristics were also calculated from the data.

Results

The model demonstrated a significant inverse relationship between AR and VAS scores ($\chi^2 = 23$, df[1], $P < 0.0001$) as every 10 m/min increase in AR predicted a 26 point decrease in VAS. The data suggests that animals with an AR < 75 m/min corresponded to animals identified as lame with an 82% accuracy and yielded 21 true positives, 15 true negatives, 5 false positives and 3 false negatives (Figure 2). This corresponds to a test sensitivity of 87.5%, and specificity of 75%.

Discussion

Standard gait-based grading of lameness on scales ranging from 0 to 5 or less lack both sensitivity and reliability.[1] The data described here suggest that AR can be an objective and useful measure of lameness in sows. In particular, AR offers a high degree of sensitivity which is ideal for a screening test to identify animals in the earliest stages of lameness. VAS scores provide an alternative to common gait scoring scales as it allows for more refined distinctions in lameness scores and was chosen as the standard against which to compare AR. Further work needs to address inter-observer reliability of the VAS scores. AR hopes to be obtained from the daily activity of sows, especially in pen gestation, and captured with video or wearable fit-bit like devices. However, validation of AR as lameness measure under commercial conditions also will require examining how additional on-farm factors contribute. In conclusion, AR is a novel, potentially autonomous, method described here that portends advanced detection of sow lameness and promises to improve sow longevity, productivity, and welfare.

References

1. Nalon et al. Comparison of the inter- and intra-observer repeatability of three gait-scoring scales for sows. *Animal.* 2014;8:4

2. Bos et al. Effect of locomotion score on sows' performances in a feed reward collection test. *Animal.* 2012;9:10

Figure 1: Course diagram. Arrows display the typical pattern of ambulation along the course.

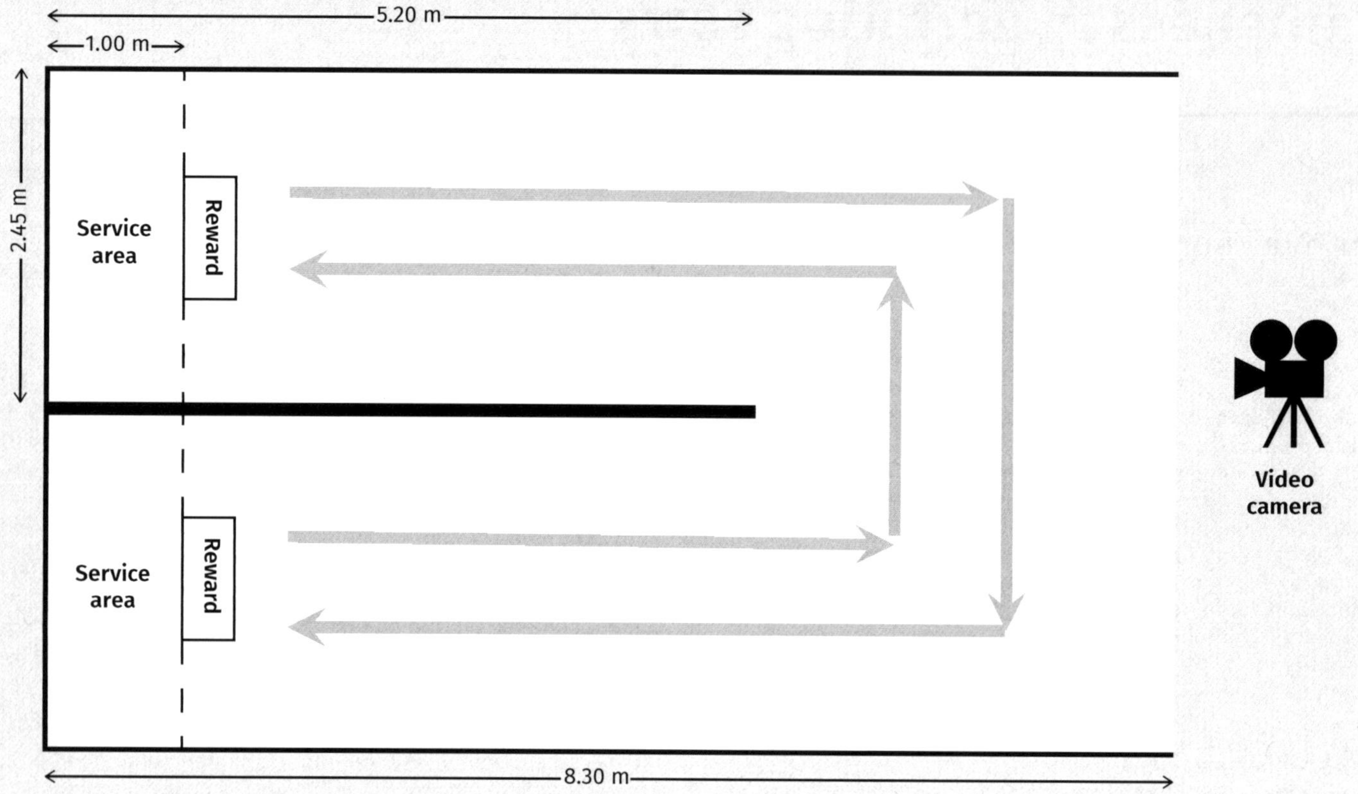

Figure 2: Ambulation rate as a predictor of lameness

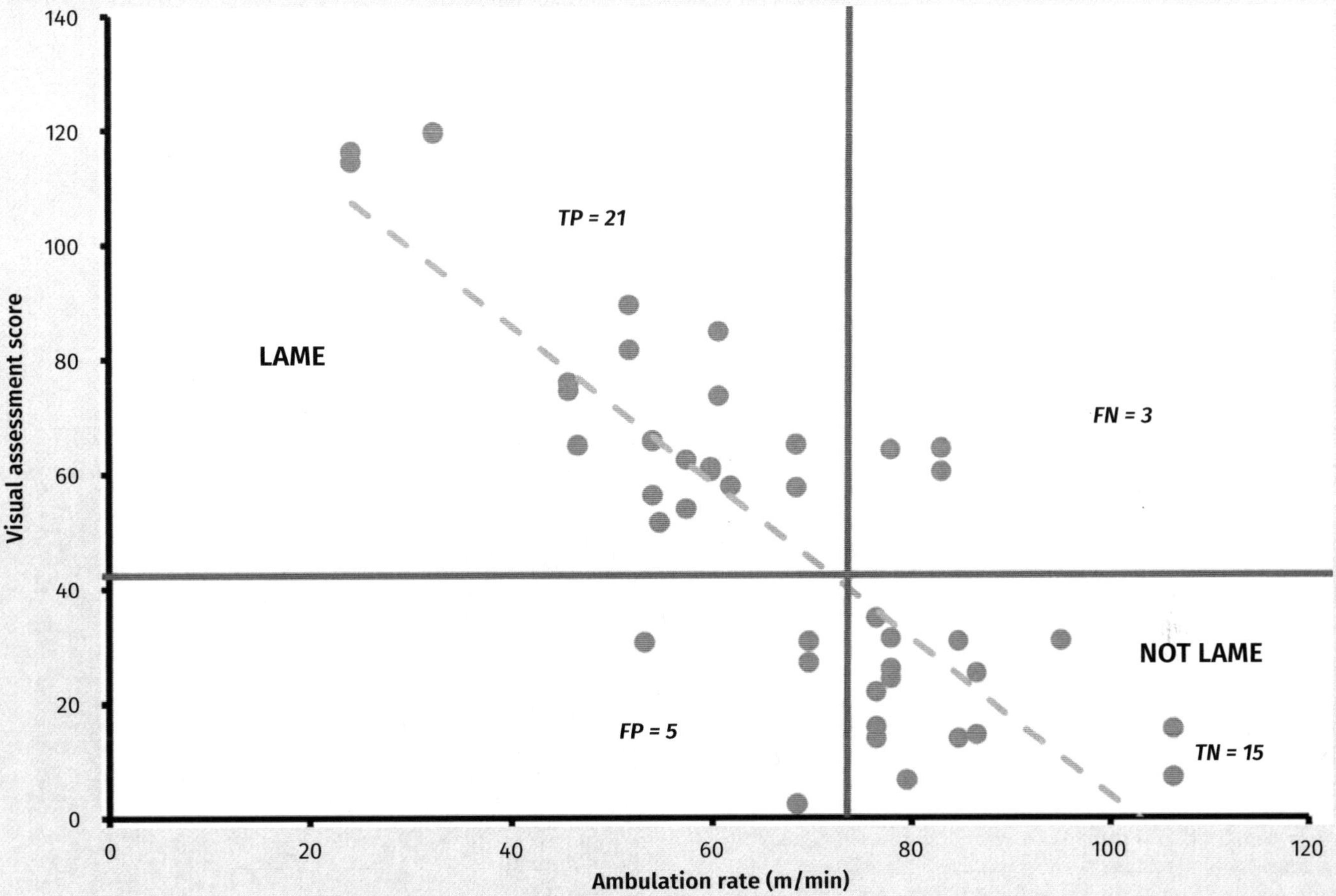

Evaluation of the finishing mortality removal process and factors that contribute to contamination spread

Kris Kovach[1]; Lynn Pavlovic[2], VMD
[1]Iowa State University, Ames, Iowa; [2]Iowa Select Farms, Iowa Falls, Iowa

Introduction

Rendering trucks can pose a biosecurity concern as these trucks have the potential of introducing pathogens onto a finishing site. Without the proper biosecurity protocols in place, cross-contamination during mortality removal can occur, further enhancing the risk of introducing diseases into the herd.[1] The objective of this study was to evaluate the mortality removal process on different style finishing sites using fluorescent powder (Glo Germ™) as a proxy for contamination to determine what associated factors contributed to the greatest amount of contamination spread and whether implementing various biosecurity protocols could reduce potential disease introduction from on-site rendering boxes to pigs in the barns.

Materials and methods

This observational study was performed using four 133 gram bottles of fluorescent powder (Glo Germ™, Glo Germ Company, Moab, UT) that were mixed with 33 grams of fluorescent lotion (Glo Germ™, Glo Germ Company, Moab, UT) and spread within a 90 cm × 70 cm PVC grid placed directly in front of the rendering box.[2] Once the fluorescent powder was applied, an average of 13 hours of time elapsed before the site was re-visited to observe the distribution of the powder. Seventeen different finishing sites, nine sites with a standalone office and eight sites with an attached office, were chosen to measure contamination based on the method of mortality removal being conducted, and specific areas on these sites were categorized as low-risk, moderate-risk, and high-risk based on the likelihood of introducing disease into the herd if a pathogen were to be found in these zones. The dirty side of the office entryway/bench was classified as low-risk, the clean side of the entryway was deemed moderate-risk, and inside the pig space was characterized as high-risk. Caretakers were blinded and unblinded to determine if knowing they were being monitored would alter the spread. A ultraviolet light was used to observe the fluorescent powder. Various mortality removal processes were compared to determine what method and factors contributed to the greatest amount of spread.

Results and discussion

Figures 1, 2, and 3 depict when biosecurity protocols are not followed.

Glo Germ™ was observed on surfaces ranging from rendering truck driver's boots to pigs inside the barn, depending on the dead removal procedures. Maintaining a clean/dirty line with removal equipment, showering after mortality removal, washing clothing daily, and having designated mortality removal boots were critical in preventing it from being tracked into the barn. When caretakers were unblinded, they tended to follow protocols more prudently, which resulted in less spread than blinded caretakers. Sites with a standalone office had less contamination in the moderate and high-risk areas compared to sites with attached offices. Glo Germ™ was observed in the high-risk areas at 50% of the sites with attached offices. In contrast, Glo Germ™ was not detected in these high-risk areas at any of the sites with a standalone office. Glo Germ™ was found in the moderate risk areas at 75% of the sites with an attached office parralled to 67% of the sites with a standalone office. When booties were worn over barn boots by the caretakers, Glo Germ™ was found inside the pig space at 67% of the sites compared to 17% of the sites where designated dead removal boots were worn by the caretakers. The separate mortality boots had more spread within the site entryway, but the booties caused more contamination from the hands and top of the boots. Use of personal vehicles to remove dead pigs

Figure 1: Contamination on barn boots when barn boots with booties were worn, instead of the designated mortality boots, to remove dead pigs.

Figure 2: Contamination on hands, gloves and coveralls after bringing dead pigs to the rendering box. The rendering box is a high-risk area for disease, so a shower is always required after removing deads.

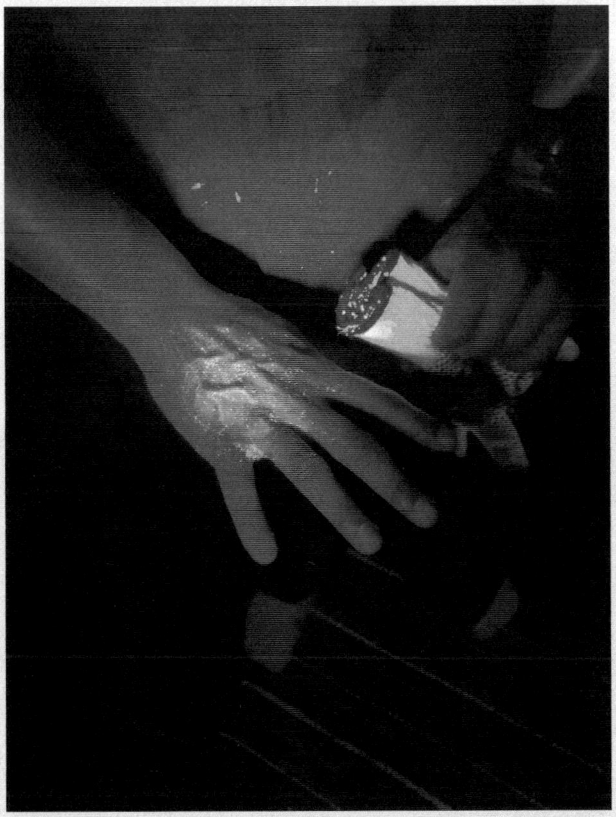

Figure 3: Contaminated coveralls were left overnight and were not directly placed into the washing machine and washed prior to showering out of the site.

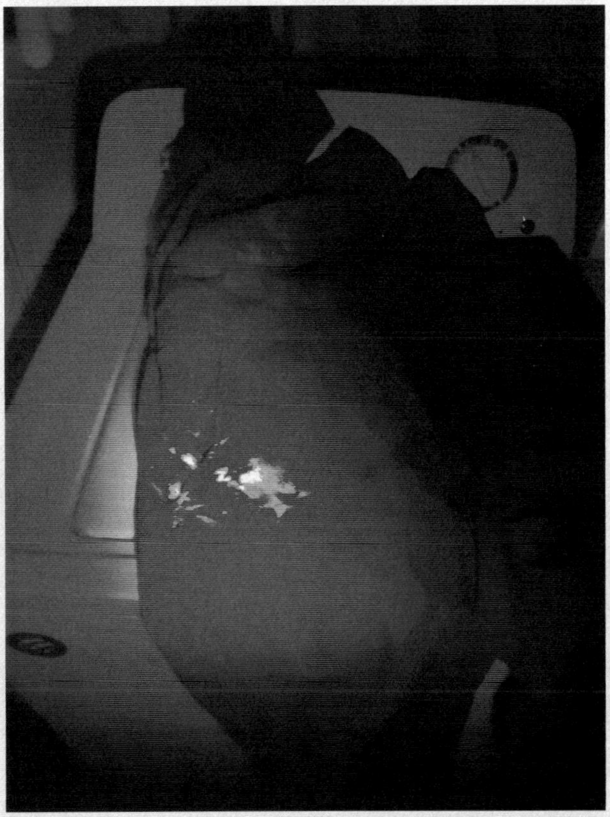

resulted in more contamination being tracked back to site offices, compared with the use of mortality carts and skid loaders to remove dead pigs. Glo Germ™ found in personal vehicles may be transferred to other sites if proper clean/dirty lines are not maintained. Having separate mortality removal equipment, avoiding re-entering barns after removal, and wearing designated mortality boots were the best means of preventing pathogen transfer into high-risk areas. These findings reinforce that if proper protocols are not followed, disease introduction might occur.

References

1. Lowe J. (2012). PRRSV contamination of mortality storage areas prior to disposal by rendering in high and low dense swine production regions. *Allen D. Leman Swine Conference.* 201.

2. Ruston C., Linhares D.C.L., Holtkamp D.J. (2020). Evaluation of a staged loading procedure for the load-out of market pigs to prevent the transfer of swine pathogen-contaminated particles from livestock trailers to the barn. *National Hog Farmer.*

Case study evaluating sow farm natural planned exposure on farms using intensive rotavirus feedback with varying levels of porcine circovirus type 3

Brian Johnson[1], BS; Amy Maschhoff[2], DVM; Melissa Hensch[2], DVM; Corrine Stoffel-Frugé[2], DVM
[1]University of Illinois College of Veterinary Medicine, Urbana, Illinois; [2]The Maschhoffs, LLC, Carlyle, Illinois

Introduction

Sow farms use natural planned exposure (NPE) for farrowing house diarrhea management, specifically against mucosal pathogens such as rotavirus.[1] Typically, a tissue and/or manure matrix targeting pathogen(s) of interest are administered to replacement gilts and gestating dams to stimulate an immune response. There is potential to spread non-targeted pathogens, resulting in iatrogenic disease.[2] However, this impact has not been well documented. Recent industry reports indicate that porcine circovirus type 3 (PCV3) can negatively impact sow farm performance,[3] therefore a more quantifiable understanding of PCV3 levels within NPE material may be useful in managing sow farm health.

Objective

The objective of this study was to compare the execution of a production system's NPE protocols between sow farms and quantify the amount of PCV3 within the NPE material administered on those farms.

Materials and methods

Seventeen commercial sow farms across the Midwest without epidemic disease challenges were enrolled in this study during the summer of 2020. Epidemic diseases were defined as any current challenge that could significantly interfere with farm production, such as category I or II porcine reproductive and respiratory syndrome virus (PRRSV), porcine epidemic diarrhea virus (PEDV), or other diseases. Gestation housing design variation was present across the farms. Farms D, E, H, and I were classified as group housing. Farm F was classified as a combination of both group and individual housing. The remaining farms were classified as individual housing. Each farm follows standard porcine circovirus type 2 (PCV2) vaccination protocols for replacement gilts and weaned piglets. A survey was completed by each farm's manager and veterinarian separately to evaluate farm protocols for NPE. These questions included both specific and open-ended questions created to assess the individual farm's NPE protocols against the system's standard operating procedure (SOP). For replacement gilt NPE, the system's SOP is to include fresh feedback, intensive rotavirus feedback, and mummies in the NPE material. Fresh feedback refers to fresh manure from dams and/or fresh manure/diarrhea from piglets in the farrowing house. Intensive rotavirus feedback refers to an autogenous tissue sample prepared and stored on individual farms for rotavirus control. For pre-farrow NPE, the system's SOP is to include fresh feedback and intensive rotavirus feedback in the NPE material.

One intensive rotavirus feedback sample from each farm was analyzed by polymerase chain reaction (PCR) for PCV2, PCV3, and rotavirus A, B, and C at an accredited veterinary diagnostic laboratory to determine the cycle threshold (Ct) value.

Results and discussion

Farm managers indicated a wide variety of answers when asked about collection frequency for intensive rotavirus feedback. The most consistent answer was quarterly or every three months, or an attempt to do so. However, the indicated collection frequency answered in the survey did not always agree with how long the current intensive rotavirus feedback sample had been used on the farm. The system SOP did not include a specific intensive rotavirus feedback collection frequency for farms to follow. It typically depends on several factors and is based on recommendations from the farm's veterinarian.

Farm NPE protocols are shown in Table 1, along with their compliance to the SOP. Evaluation of the SOP compliance found that 47.1% of farms were following the system's SOP for replacement gilt NPE, and 70.6% of farms were following the system's SOP for pre-farrow NPE. This evaluation also showed that 29.4% of farms were following the system's SOP for both replacement gilt and pre-farrow NPE, while the remaining farms had deviations in one or both NPE protocols. It is important to note, that some of these farms did have acceptable protocol variations due to veterinarian modifications and farm limitations.

Table 2 shows the PCR lab results. All farms had detectable levels of rotavirus C in their intensive rotavirus feedback. Over half of the farms had rotavirus A, but rotavirus B was not detected at any farm. A high Ct for PCV2 was detected in three of the seventeen farms. For PCV3, 23.5% of farms had a Ct less than 28, 29.4% had a Ct between 28 and 36, 41.2% had a Ct greater than 36. Two farms (C and M) submitted two samples. Farm M had a Ct between 28 and 36, and greater than 36. Figure 1 displays the PCV3 PCR Ct range variation across the farms.

This study evaluated the on-farm compliance of NPE protocols and demonstrated the presence of PCV3 within the NPE material in thirteen out of seventeen sow farms. The survey showed that a system's SOP may not be fully implemented on each sow farm. The significance of PCV3 within NPE material is unknown but warrants further investigation to understand the role it plays in disease, farm stability, and overall sow farm health. The potential for NPE protocol farm variation and non-targeted pathogen(s) present in NPE material should both be considered by the veterinarian when managing sow farm health.

Table 1: Farm NPE protocols and compliance to SOP

Farm	Replacement gilt NPE protocol	Replacement gilt NPE protocol compliant to SOP (Y/N)	Pre-farrow NPE protocol	Pre-farrow NPE protocol compliant to SOP (Y/N)
A	FF, IRF, mummies	Y	FF	N
B	FF, IRF, mummies	Y	FF	N
*C	FF, IRF, mummies	Y	FF, IRF, mummies	N
D	FF, IRF, mummies	Y	FF, IRF	Y
E	FF, IRF, mummies	Y	FF, IRF	Y
F	FF, IRF, mummies	Y	FF, IRF	Y
G	FF, IRF	N	FF, IRF	Y
H	FF, IRF	N	FF, IRF	Y
I	FF, IRF	N	FF, IRF	Y
J	FF, IRF	N	FF, IRF	Y
K	FF, mummies	N	FF, IRF	Y
L	FF, mummies	N	FF, IRF	Y
*M	FF, IRF, mummies	Y	FF, IRF	Y
N	FF, mummies	N	FF, IRF	Y
O	FF, IRF, mummies	Y	FF, IRF	Y
P	FF, IRF	N	FF	N
Q	FF, IRF	N	FF	N
Percent of farms compliant to SOP		**47.1%**		**70.6%**

* Farms submitted 2 IRF samples for evaluation
FF: Fresh feedback
IRF: Intensive rotavirus feedback

Table 2: Intensive rotavirus feedback lab results

Farm	Rotavirus serogroups	PCV2 PCR Ct	PCV3 PCR Ct
A	C	> = 37	17.5
B	C	> = 37	36.3
*C	A,C	> = 37, 31.2	33.9, 34.8
D	C	35.9	> = 37
E	C	> = 37	> = 37
F	C	> = 37	22.1
G	A,C	> = 37	28
H	C	> = 37	34.6
I	A,C	> = 37	> = 37
J	C	> = 37	35.8
K	A,C	> = 37	> = 37
L	A,C	> = 37	27.3
*M	A,C	> = 37, 35.1	> = 37, 34.2
N	A,C	> = 37	31.9
O	A,C	> = 37	36.5
P	A,C	> = 37	26.2
Q	A,C	> = 37	36.9

* Farms submitted 2 IRF samples for evaluation

Figure 1: PCV3 PCR Ct range variation across farms

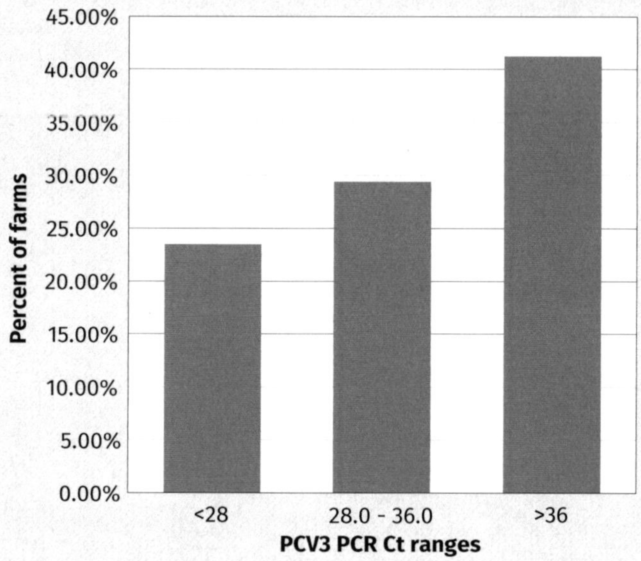

References

1. Schwartz. The Science of Feedback. Twenty-second Annual Swine Disease Conference for Swine Practitioners. 2014;51-66.

2. Robbins and Byers. What do we really know about feedback to gestating dams? AASV. 2013;537-540.

3. Creel et al. Porcine circovirus type 3: update on field experiences with Sequivity™ vaccination for this emerging pathogen. AASV. 2020;107-108.

Evaluation of different fogging methods to disinfect supplies prior to entering into farms using a cardboard box model contaminated with porcine epidemic diarrhea virus and a swine bioassay

Z. Talbert[1], BS; J. Connor[2], DVM, MS; A. Lower[2], DVM; G. Silva[2], DVM, PhD, MS
[1]University of Illinois College of Veterinary Medicine, Urbana, Illinois;
[2]Carthage Veterinary Services LTD, Carthage, Illinois

Introduction

Porcine epidemic diarrhea virus (PEDV) is a highly contagious enteric corona virus that impacts swine populations worldwide. PEDV causes high morbidity and mortality rates, can travel on objects, and can survive several days when in the appropriate environment.[1,2] Between 2013 and 2014, PEDV killed over 8 million pigs in the United States alone, resulting in $900 million-$1.8 billion in losses.[3] Biosecurity continues to be an important part of pathogen bioexclusion on production farms today. To mitigate deadly pathogens like PEDV from entering production facilities, multiple layers of biosecurity practices are put in place. Examples of methods that are often used on farms are UVC chambers and fumigation rooms. Although compliance and proper execution of these biosecurity practices are important, ensuring the methods are efficacious before implementing them are crucial.

Recent studies have shown that full surface coverage of items with a disinfectant passing through fumigation rooms (FM) is not being achieved, meaning pathogen entry through these rooms is likely occurring.[4-8] In order to fully deactivate a pathogen in a timely manner, it is important that a chemical effective against the targeted pathogen is chosen, used at the correct dilution, and given sufficient time to interact with the infectious agent.[9] The objective of this study was to evaluate different fogging methods to disinfect PEDV infected cardboard boxes prior to farm entry using the Biosecurity Box (BB) (JBI Distributors, Red Oak, Iowa), Ultrasonic Fogger (UF) (5 Disc Mist Maker, House of Hydro, Ft. Myers, Florida), and Hurricane Ultra II Fogger (HUF) (Curtis Dyna Fog, Ltd) and compare their efficacy through the use of a swine bioassay.

Material and methods

This study was conducted at a Biosafety Level 2 facility that was "I" shaped, with a central office, four rooms for each treatment, and 5 pens in each room (Figure 1). The fumigation room constructed and used for item entry was 2.43 × 1.52 × 2.28 m in size. The different fogging methods tested were the BB, UF, and HUF (Figure 2).

The BB prototype used was 0.91 × 0.91 × 0.91 m in size. The BB is a fully enclosed disinfectant chamber that has two doors, one on the clean side and one on the dirty side of the farm. The BB has a built-in exhaust fan to eliminate chemical fumes to prevent employee exposure. Items can be placed on the inside wire rack and exposed to disinfectants through the BB 360-degree internal misting system. This system emits water droplets that are approximately 20 microns in size. Cardboard boxes were placed in the center of the wire rack for each treatment in this study. For the duration of the study, the BB was placed in the pre-existing fumigation room. The UF has five large 0.02 m misting heads that contain high-speed vibrating discs to aerosolize fluids. For proper aerosolization of fluids, the UF unit needs to be floated in the fluid that it will be emitting. This system is able to emit fluid particles that are approximately 3-5 microns. To contain the aerosolized fog, the UF was placed inside the BB, with a small fan, for each replication. The HUF is an automated fogger that is currently used on most production farms today. This fogger has an adjustable neck to allow for different fluid emitting angles to be achieved. For the duration of this study, the neck angle was set at 80 degrees and the flow rate was set to the highest setting of 283 mL per minute. This system releases fluid droplets that are approximately 5.5 microns. For consistency, the HUF was placed 1.5 m away and angled towards the pre-existing wire rack for five replicates.

The disinfectant used for each trial is Decon7 (Decon7 Systems LLC, Scottsdale, Arizona). This disinfectant uses the cleaning power of hydrogen peroxide to penetrate and deactivate toxins at a molecular level. The multi-part aqueous solution contains hydrogen peroxide, a four-chain quaternary ammonia formulation, and several other key disinfectant ingredients. This formula helps the disinfectant penetrate biofilms and gives the hydrogen peroxide the ability to produce destructive hydroxyl free radicals to attack membrane lipids, DNA and other cellular components.[9] At the recommended 1:10 dilution, Decon7 has shown to be effective against: bacteria, viruses, fungi, mold, and several other pathogens.[10]

Twenty, four-week old PEDV negative pigs were used as a bioassay. Five pigs were randomly placed in individual pens for acclimation two days before the start of the study. Each room was assigned a treatment through the use of an online random number generator: BB, UF, HUF, and positive control (PC). Cardboard boxes, 0.02 × 0.02 × 0.02 m in size, were used to mimic supplies entering facilities through a fumigation room. To infect the cardboard boxes, a solution was made which consisted of PEDV infected pig intestines and approximately 10 mL of phosphate buffer saline solution (PBS). Each cardboard box was sprayed with 6 mL of the blended PEDV solution, totaling a concentration of inoculation of 9.89×10^9 per box. Boxes were given five minutes to dry on a wire rack after the application of PEDV.

Figure 1: Diagram depicting the floor plan of the research facility and different treatment rooms. The location of each shower is designated with the letter S (PC = Positive Control; HUF = Hurricane Ultra II Fogger; BB = Biosecurity Box; UF = Ultrasonic Fogger).

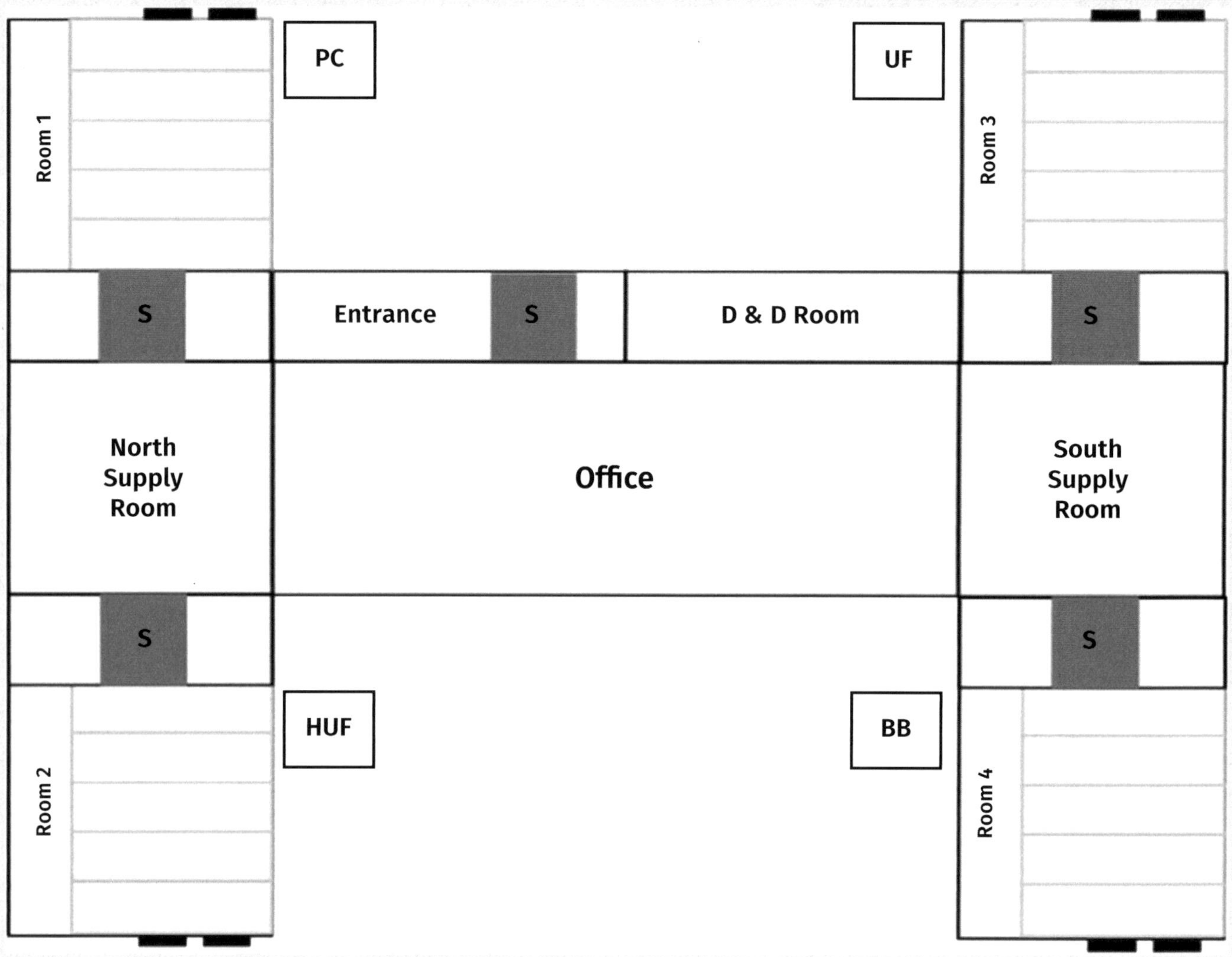

After drying, a single cardboard box was exposed to a 1:10 dilution of Decon7. Tape was placed in the center of the wire racks used to ensure the cardboard boxes were placed in the same spot for each replication. Cardboard boxes in the BB were exposed to chemicals for three minutes, whereas the other treatments ran for five minutes. These run times were randomly chosen and based on past coverage studies in the literature. Following the disinfectant exposure, the cardboard boxes were given an additional five minutes of dry time. After this time, a sterile 4" × 4" gauze pad, held by hemostats, was dipped into PBS and used to swab the surface of the box. The material collected from swabbing was orally inoculated into a pig and the swabbed cardboard box was placed in the pen. Inoculation of the material gathered on a cardboard box post disinfectant exposure and drying was conducted for consistency to ensure all of the pigs interacted with the box in the same way. The cardboard box was then placed in the pen to resemble an infected person to pig interaction when contact with the boxes occurred. Three pigs in the PC group received the same treatment as described above, minus the exposure of the cardboard boxes to Decon7. Two of the 5 pigs in the PC group were inoculated with 8 mL of the PEDV stock solution to ensure the virus was alive. Due to space constraints, no negative control group was used. Each treatment for the BB, HUF, and UF was replicated 5 times. Pigs were assessed twice daily using a grading scale and humanely euthanized according to the AASV/AVMA guidelines if needed. Observations continued for five days. Rectal swabs for PCR analysis were taken on day five and submitted to the Indiana Animal Disease Diagnostic Laboratory.

Results

All pigs in the PC, UF, and HUF treatment groups tested PEDV positive on PCR analysis at the conclusion of the study. In comparison, the pigs in the BB group all tested negative for PEDV on PCR analysis on day 5 (Figure 3). A Fisher's Exact Test was used to compare the average cycle threshold (CT) values from the BB to the UF, HUF, and PC groups. The analysis generated a *P*-value of 0.0003 (Figure 3).

Figure 2: A) The Biosecurity Box prototype B) Hurricane Ultra II Fogger and C) Ultrasonic Fogger used during the study. PEDV-covered cardboard boxes, 6" × 6" × 6" in size, were placed in the center of the wire rack and exposed to the Decon7 mixture for the allotted treatment time for each replicate.

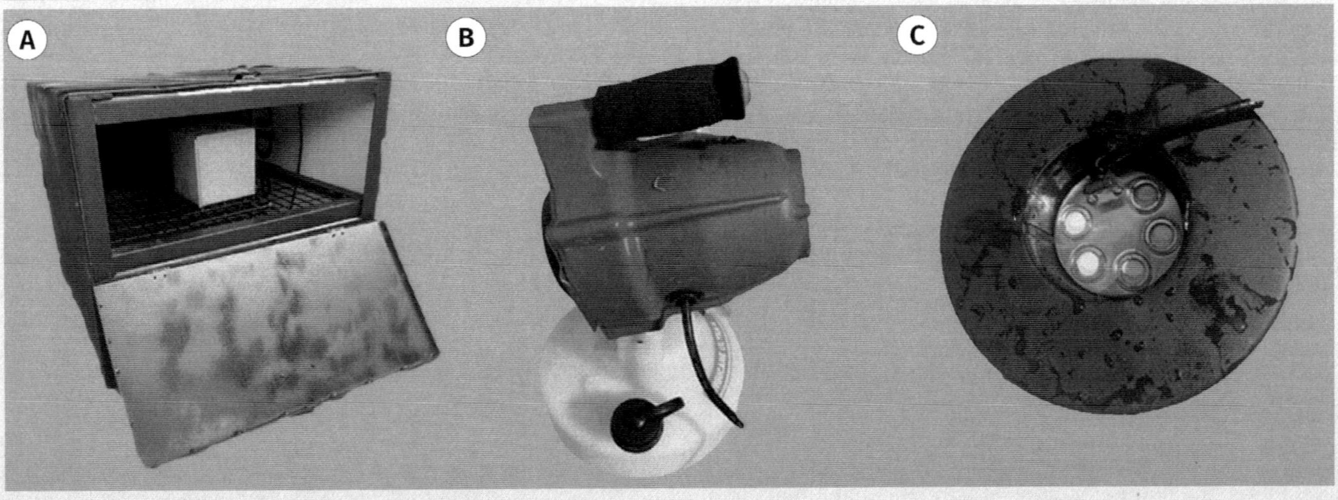

Figure 3: Average PEDV Ct values on day 5 for treatments BB, UF, HUF, and PC by PCR. A *P*-value of 0.0003 was generated when comparison of the BB to the UF, HUF, and PC was conducted using a Fisher's Exact Test.

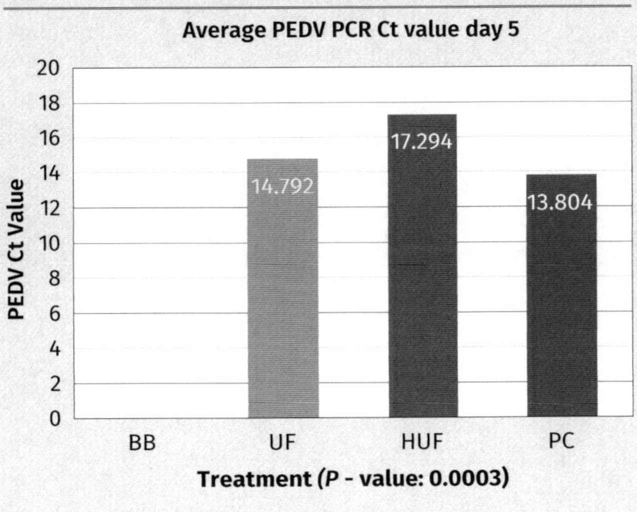

Discussion

To ensure items are sanitized, physical contact of each side with a disinfectant is needed. In this controlled study, the HUF and UF failed to inactivate PEDV from the cardboard boxes, resulting in exposure of PEDV to naïve pigs. In comparison, the BB was able to successfully deliver Decon7 to all sides of the cardboard boxes and therefore inactivate PEDV below its infectious dose range.

To our knowledge, this is the first study to ever test the transmission potential and effect that a cardboard box infected with PEDV has on naïve pigs. Furthermore, this is the first study to demonstrate a fumigation room method that was able to emit a disinfectant that covered all surfaces of a cardboard box and prevent pigs from pathogen exposure. The results from this study demonstrate the importance of investing in advanced biosecurity practices and products, like the Biosecurity Box. If such investments are made, it is equally important to ensure items are properly spaced on the wire rack and correct dilutions suggested by the chemical manufacturers are used. Aside from the ability of the BB to successfully deliver and eliminate a fomite from entering a facility, this product prevents the employee's exposure to harmful chemical fumes and has the potential to improve item entry compliance and safety in all production facilities. In an effort to improve productivity and the health of your herds, it is crucial that the failing fumigation rooms be addressed.

References

1. Pascual-Iglesias, A., Sanchez, C. M., Penzes, Z., Sola, I., Enjuanes, L., & Zuñiga, S. (2019). Recombinant Chimeric Transmissible Gastroenteritis Virus (TGEV) - Porcine Epidemic Diarrhea Virus (PEDV) Virus Provides Protection against Virulent PEDV. *Viruses*, 11(8), 682. **https://doi.org/10.3390/v11080682**

2. Kim, Y., Krishna, V. D., Torremorell, M., Goyal, S. M., & Cheeran, M. C. (2018). Stability of Porcine Epidemic Diarrhea Virus on Fomite Materials at Different Temperatures. *Veterinary sciences*, 5(1), 21. **https://doi.org/10.3390/vetsci5010021**

3. Liu, X., Lin, C., Annamalai, T. *et al.* Determination of the infectious titer and virulence of an original US porcine epidemic diarrhea virus PC22A strain. *Vet Res* 46, 109 (2015). **https://doi.org/10.1186/s13567-015-0249-1**

4. Z. Talbert; *A new "twist" to the currently failing fumigation room,* Leman Conference Annual Meeting; 2019.

5. Z. Talbert; *Boxing pathogens out of your farm through the use of science, innovation, and ingenuity,* Leman Conference Annual Meeting; 2020.

6. Leuck K., LeFevre C., Jansen M., Amodie D. (2020). Evaluation of location and fogger type on disinfectant surface area coverage of supplies entering a commercial sow farm. Student Research Presentations, 51st Am Assoc Swine Vet Annual Meeting., 2020.

7. Madigan J., Brown S., Holtkamp D., Schaefer N. (2020). Evaluation of foggers and stir fans to improve coverage of fogged disinfectants used in supply entry rooms. Student Research Presentations, 51st Am Assoc Swine Vet Annual Meeting., 2020.

8. Schultz R., Blair B., Lowe J. (2020). A fluid dynamics approach for the evaluation of chemical disinfection via fogger. Student Research Presentations, 51st Am Assoc Swine Vet Annual Meeting., 2020.

9. Center for Disease Control. Chemical Disinfectants. 2008. Available at: **https:// www.cdc.gov/infectioncontrol/guidelines/disinfection/disinfection- methods/chemical.html**. Accessed 20 October 2019.

10. Decon7 - JBI Distributors & Services - Disinfectants & PPE Supplies. 2016. Available at: **https://www.jbidistributors.com/decon7/**. Accessed 20 October 2020.

Drinking water flow rates and usage in nursery pigs with various water pressures and nipple orifice sizes

H. Everett[1], BS; S. McCallister[1], BS ; J. Pittman[2], DVM, MS DABVP
[1]College of Veterinary Medicine, North Carolina State University, Raleigh, North Carolina;
[2]Smithfield Hog Production, North Region, Waverly, Virginia

Introduction

Unregulated drinking water in swine facilities has many implications such as, environmental impacts, medication wastage, and excess production costs. While there are many factors for water usage that cannot be controlled at the farm level including health status, diet formulation, and ambient temperature,[1] by controlling flow rate through pressure regulators and nipple orifice size, producers and veterinarians can minimize wastage, more accurately deliver water medications, and reduce cost. Previous literature has shown that approximately 80% of on farm water usage is contributed to animal drinking.[2] It's also been demonstrated that water wastage ranges from 25% to 40% of drinking water usage.[3] A pre study survey of a production system in North Carolina identified 52% of nursery sites lacked pressure regulators which provided the opportunity to install regulators and quantify differences in water usage. The objective of the study was to assess the impact of pressure regulators, nipple orifice settings and drinker type on water usage in nursery pigs.

Materials and methods

The study was conducted June through September on two nursery sites in eastern North Carolina. Rooms within each site were identical in design and water system, but differed between sites. Site B housed 1,200 pigs/room at 20 pigs and one bowl drinker per pen. Site H housed 1,600 pigs/room at 30 pigs and one hanging drinker with two nipples per pen. New pressure regulators and water meters were installed for each room. Waterlines were flushed and nipples cleaned or replaced prior to starting the studies. Treatment settings were based on preliminary observations and measurements. Treatments were a combination of pressure regulation at 15, 25, 30, or unregulated (UR) psi and nipple orifice sizes (diameter); medium 5/64 mm (M), large 7/64 mm (L) or random (R). Random was a combination of small (1/16 mm), medium, large, and extra-large (5/32 mm) settings. Rooms at each site were randomly assigned to one of four treatments; M at 15 psi, R at 15 psi, R at 25 or 30 psi, and L at UR psi. Flow rates were randomly collected within treatments to ensure they met or exceeded industry recommendations (946 mL/min).[4] Measurements were collected with a graduated 1000 mL painters cup. Measurements were collected for 15 seconds and then converted to minutes for analysis. Water usage readings (in gallons) were collected daily during the duration of the nursery phase using water meters. Only days that rooms were fully stocked (35 days) were analyzed. Mortality was collected daily and used for calculations. Data was utilized to calculate total and weekly water usage per pig, and to model weekly 5-day medication stock solution requirements (1:128 rate water medicator) for 1,000 pigs.

Results

Average flow rate for treatments ranged from 1,016 (site B, M at 15 psi) to 2,960 (site H, L at max) mL/min and demonstrated a positive correlation to psi and nipple orifice size. Mortality ranged from 1.1 to 2.1% (site B) and 0.8 to 1.0 % (site H). Site B, room M at 15 psi used 22.0 gal/pig compared to 68.2 gal/pig for room L at UR psi. Site H room M at 15 psi used 58.9 gal/pig compared to 101.7 gal/pig for L at UR psi. Trends in weekly gallons of estimated stock solution increased concurrently with psi and nipple orifice size. For example, week 3 stock solution at site B, room M at 15 psi required 34 gal compared to 108 gal for L at UR psi equating in a 74 gal stock solution savings per 1000 head when based on water usage. Full data comparison including additional treatments are located in Table 1.

Discussion and implications

Under the conditions of this study, drinking water usage was reduced by controlling flow rate through pressure regulators and nipple orifice size, while still meeting daily water recommendations. Savings in water medication costs will depend on the mixing methodology utilized, but savings could be substantial if a standard ppm method is used. It is estimated from these results that waste management costs ($0.004-0.008 per gallon) could result in $0.18-0.36/pig cost savings. Additional savings may be available if the producer pays for the water source. The data allows veterinarians to be more informed and judicious in their prescriptions of water medications. For the producer, this information represents decreased medication cost depending on medication strategy, decreased environmental impact through reduced water usage and waste and overall decreased production cost.

References

1. Almond. Water: Optimizing performance while reducing waste. *46th annual North Carolina Pork Conference.* 2002.

2. Muhlbauer R.V, Moody L.B, Burns R.T, Harmon J, Stalder K. Water Consumption and Conservation Techniques Currently Available for Swine Production. *Pork Checkoff Research Report - Environment.* 2010;10.

3. Li, Y., L. Che´nard, S. Lemay, H. and Gonyou. Water intake and wastage at nipple drinkers by growing-finishing pigs, *J. Anim. Sci.* 2005. 83:1413–1422.

4. Pork Checkoff. Feed/Water Access. *Site Assessment Guide 3.0.* 2016;36.

Table 1: Water flow and usage rates for different pressure settings, nipple orifice sizes and drinker systems in nursery aged pigs

	Site B- bowl drinkers				Site H- hanging nipples			
Treatment	1	2	3	4	1	2	3	4
Water pressure (PSI)	15	15	30	UR[a]	15	15	25	UR[a]
Nipple orifice size[b]	M	R	R	L	M	R	R	L
Average flow rate (mL/min sec)	1017	NC[c]	1850	2815	1227	1818	2299	3709
Standard deviation flow rate	211	NC[c]	665	550	284	562	642	632
Number of days	35	35	35	35	35	33	31	35
Total gallons per pig	22.0	26.3	54.3	68.2	58.9	83.4	81.3	101.7
Weekly total gallons per pig								
Week 1	1.1	1.1	1.1	1.6	6.3	NC[c]	9.0	NC[c]
Week 2	2.6	3.4	3.2	4.0	7.9	9.5	11.8	13.5
Week 3	4.3	6.2	7.0	13.8	9.4	12.5	14.2	15.6
Week 4	7.7	7.5	15.4	24.1	10.4	14.8	15.6	18.5
Week 5	6.8	7.9	27.0	20.1	13.4	18.8	21.1	23.0
5-Day stock solution (per 1000 pigs)								
Week 1	6.3	6.1	5.9	9.2	35.3	NC[c]	50.2	NC[c]
Week 2	14.7	19.0	18.1	22.6	43.9	52.9	65.9	75.2
Week 3	24.3	34.4	39.2	77.2	52.5	69.5	79.3	86.8
Week 4	43.2	41.6	86.1	134.6	58.1	82.6	87.2	103.2
Week 5	38.0	44.3	150.4	112.1	74.5	104.9	117.5	128.5

[a] UR pressure was achieved by not placing a regulator in-line and allowing normal well pressure fluctuations to occur

[b] M = medium, L = large, R = random sizes of Small, M and L orifices

[c] Not Calculated; unrecorded or insufficient data to calculate due to load in or load out of partial groups

Data was calculated from a study where daily water meter readings from two nursery sites with four rooms at each site (site B and H), were set up with pressure regulators, water meters and various nipple orifice size settings to assess drinking water usage through the nursey phase. Site B utilized bowl drinkers and site H used hanging water nipples. Total and weekly gallons per pig were key calculated metrics. An estimation of 5-day stock solution volume (at 1:128 injection rate) for each week for 1,000 pigs was modeled from the data.

The effects of oral antibiotic administration on vaccine-induced H1N1 influenza immune responses in weanling pigs

P. Haenig; S. Storms, DVM; J. Lowe, MS, DVM
Applied Epidemiology Laboratory, College of Veterinary Medicine, University of Illinois, Urbana, Illinois

Introduction

Influenza A virus in swine (IAV-S) has significant economic impacts on swine production and is a threat to human health. Both the humoral and cell-mediated components of the immune system are important in vaccine efficacy. Past studies have used IFNγ to measure T-cell response to antigen stimulation. T-cells, part of the cell-mediated immune response, secrete the pro-inflammatory cytokine IFNγ. IFNγ has been shown to initiate transcription of antiviral factors for protection against viruses, as well as vaccine efficacy.[4,7] IL-10 is an anti-inflammatory cytokine produced by cells of the immune system. One study in mice infected with influenza A virus found that IL-10 leads to decreased viral clearance, antibody production, and subsequent protection.[3]

Recent human and murine studies show that receiving broad-spectrum oral antibiotics immediately prior to administration of the annual trivalent influenza vaccine can decrease IgG production and long-term immunity to H1N1 influenza strains.[2] Understanding the relationship between antibiotic administration and vaccine efficacy will improve production outcomes that benefit the swine industry and promote human health.

Materials and methods

Twenty weanling piglets from a herd free of PRRS, *M hyopneumoniae*, APP, and IAV-S were used for the study. Serum ELISA tests confirmed the negative status of the group. Ten piglets received 22mg/kg chlortetracycline via oral drench for five days, and ten controls were drenched an equal volume of water. On day five, both groups received a commercial monovalent H1N1 swine influenza vaccine per label instructions (FluSure® Pandemic, Zoetis); and were boosted 28 days later.

Serum was collected on days 0, 14, 28, 35, 42, and 56 following the initial vaccine to analyze homologous (CA-2009 pH1N1) antibody titers by hemagglutination inhibition assays. The samples were centrifuged at 3500rpm for 20 minutes, and the serum was collected and heat-treated at 56°C for 30 minutes. The sera were blocked for non-specific binding with kaolin for 20 minutes, centrifuged, and blocked for 20 minutes with 0.5% standardized chicken RBCs.[5] HI assays were performed in 96-well V bottom plates by incubating 50µL of CA-2009 homologous virus at 4 HU and 50µL of two-fold serial dilutions of sera for 30 minutes. After incubation, 50µL of 0.5% standardized RBCs were added and allowed to hemagglutinate or precipitate.[1,6] Wells were read after 30 minutes of incubation.

INFγ and IL-10 antibody-coated plates were used to enumerate influenza-specific memory T-cells (IFNγ) and cytokine response (IL-10) at days 39 and 60 using Single Analyte R&D Systems ELISpot kits. Briefly, peripheral blood mononuclear cells (PBMCs) were collected on days 39 and 60 after the initial vaccination. Cell preparation tubes were used to separate the PBMCs from the RBCs and plasma. PBMCs from each pig were washed, and 5×10^5 cells were added to each well in triplicate. The PBMCs were stimulated with 3×10^5 TCID 50 of homologous virus added to each well and incubated at 37°C and 5% CO_2 for 18 hours. Negative control wells and SEB mitogen control wells were paired with each triplicate sample. After incubation, plates were washed and developed using provided substrates. The plates were dried completely, and spots were counted manually under a stereomicroscope.

One pig in the antibiotic treatment group was removed from the study due to signs consistent with *H parasuis* on day 24; one pig from treatment and two from the control group were removed on day 60 for bacterial pneumonia.

Results and discussion

CA-2009 specific antibody end titers were measured for all serum samples. At the time of booster vaccine (day 32 of trial), 1 of 19 pigs had an HI titer of 80. All others were less with no detectable titers in 11 of 19 the pigs. On day 39, 18 of 19 pigs had a three-fold or greater increase in HI titer with a mean of 942 for treatment and 824 for controls. On day 46, titers were elevated two-fold in all pigs compared to titers at booster with means of 854 for treatment and 1944 for controls. On day 60, 11 of 17 pigs had titers >2650, 1 at 1280, 1 at 640, 2 at 320, and 1 at 80 with means of 3450 for treatment and 4880 for controls (Table 1 and Figure 1).

There were no significant differences in mean ELISpot counts between the antibiotic treatment and control groups at any time point (Table 2 and Figure 1). There were no associations between HI titer and INFy cell counts in individuals at 7 or 28 days following booster.

Interestingly, four of 17 pigs had an antibody response of at least two-fold less (80-640 vs. 2560+) than their peers 14 days after the booster. These low responding pigs had similar memory specific T-cell counts at days 7 and 28 after booster. These pigs also had similar HI titers at booster but lower HI titers at days 14 and 60

Table 1: Hemagglutinin inhibition assay titers

	Day 0	Day 32	Day 39	Day 46	Day 60
HIA treatment	0	5	424.8	549	1810
HIA control	0	7	520	905	2792

Geometric means of HIA titers for antibiotic treated and control sera throughout the trial period. No significant differences were seen between groups.

Figure 1: HIA titers and ELISpot assay results from day 0 to day 60.

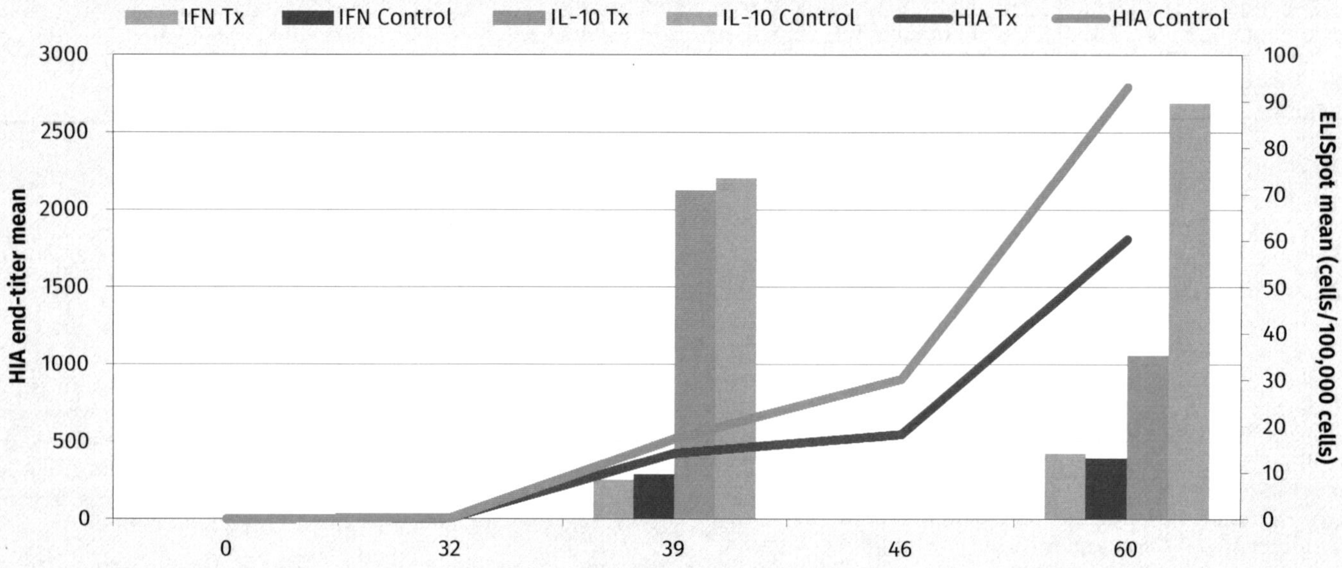

Table 2: Treatment pigs HIA

	IFNγ treated	IFNγ control	IL-10 treated	IL-10 control
Day 39	8.5	9.6	70.7	73.5
Day 60	14.2	13.2	35.2	89.7

The amount of CA-2009 memory specific T-cells per 100,000 PBMCs, displayed as IFNγ responsive cells, and as IL-10 cytokine response to virus stimulation. No significant differences were seen between groups.

after booster, suggesting that only their humoral immune response, but not the cellular immune response to the booster was reduced.

Our study found that concurrent chlortetracycline antibiotic administration at the time of influenza vaccination does not impair cellular or humoral immune responses in weaned pigs. A number of pigs had reduced immune responses to vaccination despite being housed in a low-stress environment with proper vaccine administration and freedom from other major swine pathogens. Further investigation is needed to elucidate this impaired response's cause and determine if the immune response is also impaired with different or multiple vaccines in young pigs.

References

1. Eisfeld, A. J, et al. Influenza A virus isolation, culture and identification. *Nature Protocols* 9, 2663-2681 (2014).

2. Hagan, T., et al. Antibiotics-driven gut microbiome perturbation alters immunity to vaccines in humans. *Cell* 178, 1313-1328 (2019).

3. Sun, K., et al. A detrimental effect of interleukin-10 on protective pulmonary humoral immunity during primary influenza A virus infection. *Journal of Virology* 84, 5007-5014 (2010).

4. Olson, Z. F. Differential measures of cellular IFN-y responses in swine following influenza vaccination. Iowa State University Capstones, Theses and Dissertations. 2014.

5. Spackman, E. Animal Influenza Virus, Second Edition. Springer Protocols. 2014.

6. WHO Global Influenza Surveillance Network. Manual for the laboratory diagnosis and virological surveillance of influenza. *WHO*. 2011.

7. Yung, B. S., et al. Novel IFN-y ELISpot reveals robust T cell responses elicited after influenza nucleoprotein DNA vaccination in New Zealand White rabbits. *Vaccine* 37, 903-909 (2019).

Effect of particle size and H_2O_2/PAA concentration on the efficacy of an aerosol decontamination system for items entering swine farms

Erin Kettelkamp; Suzanna Storms, DVM; Benjamin Blair, DVM; James Lowe, DVM, MS
Applied Epidemiology Laboratory, College of Veterinary Medicine, University of Illinois, Urbana, Illinois

Introduction

The swine industry continues to invest significant effort to heighten biosecurity practices within production systems. Aerosolized chemical disinfection or fumigation of supplies is a common practice to mitigate pathogen entry onto swine farms. However, recent evaluations state the use of commercial foggers results in inconsistent and incomplete disinfectant contact to all sides of objects.[1,2] Current fumigation systems require manipulation of objects, a large volume of disinfectant use, and result in impractical saturation of objects to overcome the lack of empirical disinfection.[3]

Currently used in hospitals and pharmaceutical clean rooms, a "Dry Fog System" (Cantel Medical Inc., DFS) has been implemented as a more effective fumigation process.[6] Dry fogging utilizes a liquid disinfect and compressed air to generate an ultrafine droplet.[6] Unlike a wet fog, the micro-droplets do not settle onto surfaces right away and avoid any excessive condensation or saturation of liquid to the room and objects.[6] Peracetic acid (PAA) and hydrogen peroxide (H_2O_2) are commonly used as oxidizing agents that fully biodegrade to acetic acid (vinegar), O_2, and water through the DFS.[5] As a disinfectant, peracetic acid (PAA) has been widely used in the food processing, dairy, and beverage industries for its non-toxic byproducts, and enhanced lipid-soluble microbicidal effects superior to H_2O_2 alone.[6] Evidence has shown that a DFS using a combination of H_2O_2 and PAA is capable of inactivating microbial agents in the presence of a soil load.[6]

While upholding biosecurity standards is critical to protect the health of swine farms, recent fumigation chamber research would suggest this is a weakness in preventing pathogen spread. Use of dry fog shows promise to overcome the difficulty of current methods through the use of small particles. The objective of the study was to evaluate the effect of particle size and H_2O_2/PAA concentration on aerosol decontamination for supplies entering swine barns.

Materials and methods

An air-tight decontamination room measuring 1.75 × 1.42 × 2.51m (6.27m3) was used for this study. Humidity loggers were placed at 4 locations in the center of the room (labeled A, B, C, D) (Figure 1). A standard household de-humidifier was used in the treatment room to achieve the desired relative humidity (RH) start point < 60% for all experiments.[7]

Commonly utilized Hurricane Ultra II (Curtis Dyna-Fog, HUF) foggers produce a flow rate dependent median particle size of 14 to 26 microns.[4] The DFS produces a controlled median 5.9 um droplet to ensure even dispersion throughout the space comparable to gas or vapor.[5] Two proprietary blends of PAA and hydrogen peroxide (H_2O_2), Actril (A), and Minncare (M) (Cantel Medical) have been approved by the EPA for use as aerosol disinfectants in medical clean rooms with droplets averaging 5.9um.[5]

A (1% H_2O_2/0.08% PAA) and M (22%/4.5%) were used as the sterilant and were compared to Intervention (IVN, 4.25% H2O2) (Virox Animal Health). Three droplet sizes were compared, 5.9um (SD; DFS), 14 um, (MD, HUF@1 gal/min), and 26um (LD, HUF@4.5 gal/min). A dry fog nozzle connected to an air compressor was used to produce a small droplet (SD) size (5.9 um) when operated at a constant 75 psi. A HUF fogger was used at the low setting for a medium droplet size (MD) (14 um) and the high setting for a large droplet size (LD) (26 um).

200 mL of disinfectant were applied at four concentrations: 1% H2O2/0.08% PPA (A), 1% H_2O_2/ 0.21% PAA (MC9.5), 2% H_2O_2/ 0.42% PAA (MC19), and 2% H_2O_2 (IVN) diluted in deionized water just before use. The delivery device was placed on the floor just off-center of the room. Two boxes were placed on a wire shelf and 5log and 6Log Geobacillus spore Biological Indicator (BI) strips were placed on all 6 sides of box A. BI strips are the standard method to assess cleanroom decontamination. They are a robust pathogen more resistant to decontamination than bacteria and viruses.[5]

The room remained sealed for 1 hour where the effective RH was maintained below 85% to prevent condensation and limited aerosol spread.[7] BI strips were incubated overnight at 55 °C in tryptic soy broth. If no growth was observed, it was interpreted as a reduction of a bacterial load equal to the concentration of the biological indicator.

Additionally, 1.0 mL of Ingelvac PRRS® MLV (MLV) was plated onto sterile petri dishes, treated by a direct spray of A and MC to validate that H_2O_2/PAA inactivates PRRSV. MLV inoculated petri dishes were also attached to all 6 sides of box A and subject to MD MC19 treatment. All viral samples were transported to the lab for virus isolation. 1.0 mL of D/E Neutralizing broth was added to each treated plate to inactivate any remaining disinfectant. 250 ul of solution were plated onto MARC-145 cells and incubated for 5 days at 37 °C where no cytopathic effects (CPE) were observed.

Furthermore, 1.0 mL of a manure slurry was plated onto sterile petri dishes attached to box A, treated with MD MC19 to challenge a variation in Log 5 and Log 6 reductions. The fumigated slurry plates subject to bacterial culture to compare BI strip log reductions and the ability of H_2O_2/PAA to decontaminate surfaces with organic material present.

Figure 1: Orientation of box sampling locations relative to room and aerosolized chemical source. Fogger was placed on the floor, BI strips were elevated 12" above the floor surface. The chamber was 1.75 × 1.42 × 2.51m ($6.27m^3$) in size.

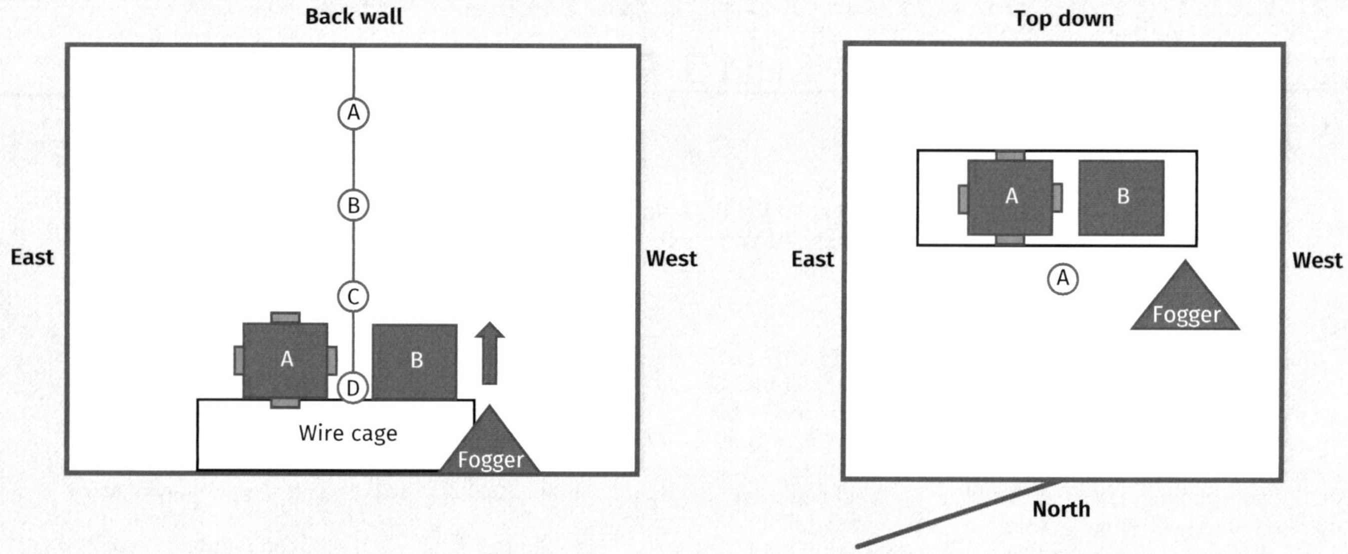

Results and discussion

Decontamination effectiveness was dependent on both particle size and H_2O_2/PAA concentration. The results are summarized in Figure 2. The SDMC19 treatment resulted in a 5-log reduction on 18/18 sides and a 6-log reduction on 17/18 sides. MDMC19 treatment resulted in 12/12 sides having a 5-log but only 8/12 sides having 6-log decontamination. All other combinations had at least 1 side without a 5-log reduction.

Direct contact of A and MC and treatment of MD MC19 to MLV resulted in complete viral inactivation in all orientations within the treatment room. No bacterial growth was observed from MD MC19 treated manure slurry plates from each side of the treatment box.

These results indicate that to achieve decontamination of supplies entering a swine farm droplet sizes averaging less than 14um with H2O2/PAA concentrations greater than 4.2/0.13 ppm is required. Optimal disinfection was achieved with droplet sizes < 6um.

Due to the safe biologic nature of H2O2/PAA, the non-toxic by-products may be safely applied in numerous areas of swine farm supply entry. With no soaking or residue when applied through a DFS, this makes it a suitable decontamination practice for personal items, covered lunches, and various farm supplies.

This study demonstrates the dry fog system's ability to inactivate resistant pathogens with or without the presence of organic material. The DFS technology can be implemented in pre-existing decontamination rooms and offers a safer alternative to traditionally used harmful chemicals. These data support that effective aerosol decontamination can be achieved with droplets smaller than currently used in swine barns when using oxidative disinfectants at high concentrations.

References

1. Kettelkamp, Erin, et al. (2019) Impact of Item Orientation on the Efficacy of Supply Decontamination with Aerosolized Chemical Disinfectants. *AASV 50th Annual Meeting Proceedings,*111–114.

2. Schultz, Rebekah, et al. (2020) A Fluid Dynamics Approach for the Evaluation of Chemical Disinfection via Fogger. *AASV 51st Annual Meeting Proceedings,* 67–70.

3. Leuck, Kristen, et al. (2020). Evaluation of location and fogger type on disinfectant surface area coverage of supplies entering a commercial sow farm. *AASV 51st Annual Meeting Proceedings,* 57-59.

4. Curtis Dyna-Fog. (n.d.). Hurricane Ultra Operation and Maintenance Manual. Retrieved from **https://www.manualslib.com/manual/1338757/Dyna-Fog-Hurricane-Ultra-2794.html**

5. MAR COR Cantel Medical Company. (n.d.). Minncare Dry Fog Technology. Retrieved from **http://www.mcpur.com/main/library/12_brochures/3028187_(Dry_Fog_Brochure).pdf**

6. Krishnan, J., Fey, G., Stansfield, C., Landry, L., Nguy, H., Klassen, S., & Robertson, C. (2012). Evaluation of a Dry Fogging System for Laboratory Decontamination. *Applied Biosafety,* 17(3), 132-141. **doi:10.1177/153567601201700 30**

7. Wood JP, Calfee MW, Clayton , Griffin-Gatchalian N, Touati A, Egler K. (2013, February). Evaluation of peracetic acid fog for the inactivation of Bacillus anthracis spore surrogates in a large decontamination chamber. Retrieved from **https://pubmed.ncbi.nlm.nih.gov/23434480/**

Figure 2: 5 & 6 Log Biological Indicator strip data. Percent decontamination achieved represents the proportion of sides negative for bacterial growth following treatments of MC, A, and IVN at various particle sizes.

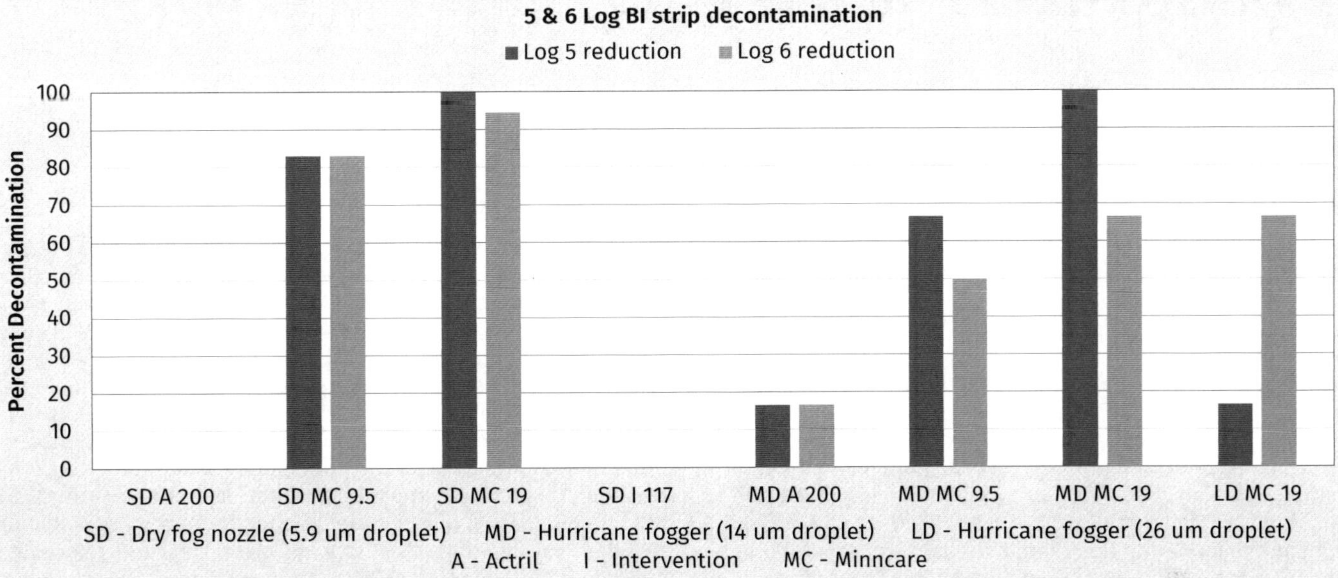

Investigation of *Lawsonia intracellularis* shedding by peri-parturient sows

R. Patton[1], BS; C. LeFevre[2], DVM; F. Leite[3], DVM, MS, PhD; J. Waddell[3], DVM, MBA
[1]The Ohio State University, Columbus, Ohio; [2]Carthage Veterinary Service, Ltd., Carthage, Illinois;
[3]Boehringer Ingelheim Animal Health, Duluth, Georgia

Introduction

Lawsonia intracellularis is a bacterial pathogen with worldwide distribution and the causative agent of porcine proliferative enteropathy, an enteric disease which occurs primarily in grow-finish pigs.[1] The existing data regarding *L intracellularis* shedding from sows in commercial settings is sparse and outdated. Similarly, the role of sows as a source of infection of growing pigs is unclear.[2] The objective of this study was to investigate if *L intracellularis* is shed from peri-parturient sows. Due to recent developments in polymerase chain reaction (RT-PCR) technology, the sensitivity and detection of *L intracellularis* have improved considerably.[3] Sow infection status has been identified as a critical component that must be addressed in control strategies of other common swine pathogens, such as *Mycoplasma hyopneumoniae* in the downstream flow of grow-finish pigs. Understanding of the infection chain of *L intracellularis* and particularly shedding by peri-parturient sows is important for devising strategies to control or potentially eliminate *L intracellularis*.

Materials and methods

This study involved three distinct, large (5,000-6,000 head) commercial sow herds that received replacement gilts from different source farms. Replacement gilts were vaccinated at the source farm with commercial ileitis vaccine at the time of weaning. The herds were selected based on the clinical history of *L intracellularis* and were otherwise of equivalent health status. From each herd, 60 sows in a single weekly farrowing group were selected randomly. Each sow had 5 to 10 grams of feces collected digitally from the rectum within five days of farrowing. These samples were tested individually for *L intracellularis* by quantitative RT-PCR at a public veterinary diagnostic laboratory. Similarly, sixty environmental samples were collected from each herd just prior to the time of weaning (15 – 22 days post-farrowing) within farrowing crates of the same sows from which fecal samples were collected. Each environmental sample was collected from a predetermined area within the farrowing stalls including the creep area. Serum samples from a different sixty random gestating sows in each herd were submitted for *L intracellularis* ELISA serology.

Results and discussion

L intracellularis was found in sow fecal samples in only one of the three herds (Table 1). In all three herds, environmental samples were positive for *L intracellularis*. Sows from each herd were serologic positive via ELISA for *Lawsonia intracellularis,* which may be attributed to both the vaccination history of replacement gilts and continuous exposure to field strains. Positive environmental samples imply that piglets in all three farms were potentially exposed during the suckling phase. While the results of this study indicate a low prevalence of *L intracellularis* in sow feces and environment, this level of exposure could still lead to a clinically significant percentage of piglets being infected, becoming potential shedders, and serving as a source of *L intracellularis* in later stages of production. The results of this study confirmed that sows can shed *L intracellularis* in the peri-parturient period and can be detected by PCR in the suckling pig environment. The implications of these results are that sows are a potential continuous source for vertical transmission of *L intracellularis* to their piglets, potentially leading to transmission among growing post-weaning cohorts in the wean to finish phase.

References

1. E. J. Neumann et al., Ileitis. *Swine Disease Manual,* 2020, pp. 32–34.

2. Guedes R. Update on epidemiology and diagnosis of porcine proliferative enteropathy. *J Swine Health Prod.* 2004;12(3):134-138.

3. Kroll et al., Proliferative enteropathy: a global enteric disease of pigs caused by *Lawsonia intracellularis,* 2005. *Animal Science Publications.* 62.

Table 1: Results of *Lawsonia intracellularis* detection by PCR in fecal and environmental samples and serology

	Herd 1	Herd 2	Herd 3
Fecal	4/68 (5.9%)	0/70 (0%)	0/68 (0%)
Environmental	9/69 (13%)	5/68 (7.4%)	3/67 (4.5%)
Avg gen. copy/mL	254.65	464.60	406.33
Serology	57/60 (95%)	58/60 (96.7%)	60/60 (100%)

Significant gaps exist in the current cull sow surveillance stream for the detection of foreign animal diseases in the US

Selena Clark; James Lowe, DVM, MS; Benjamin Blair, DVM
Applied Epidemiology Laboratory, College of Veterinary Medicine, University of Illinois, Urbana, Illinois

Introduction

Disease, namely foreign animal disease (FAD), remains one of the greatest threats to the US swine industry's profitability. This threat has led to efforts by the USDA-APHIS-VS to surveil for FAD (brucellosis, pseudorabies virus [PRV]) introductions relying on samples collected from within marketing channels. One such channel is the US cull sow marketing network. Previous research has shown that this network acts as a potentially devastating route of disease transmission and dissemination.[1] Because of the potential impact of a FAD introduction, and the risk posed by marketing channels, surveillance to ensure these populations are free of disease is of utmost importance.

The USDA-APHIS-VS utilizes an OIE compliant surveillance plan to ensure the US swine population is free of FAD such as brucellosis.[2] This plan ensures the detection of disease at an assumed prevalence of 1 in 100,000 within the population. To achieve this, a systematic, stratified random sample of blood and tissues is taken from cull sows at the time of harvest. Regulators perform this sampling technique to ensure that all states are adequately sampled, guaranteeing complete freedom of disease within the national swineherd.

A proper systematic, stratified collection of samples should result in a dataset reflecting both the US national herd's geographic and individual herd diversity. Such a representative sample is required if the program serves as a sensitive detection method of a FAD across all reaches of the industry while maintaining a high degree of confidence. A truly representative sample of the US swine population is important, as the timely detection of a FAD is crucial to limit the impact of a disease introduction.

Related work, investigating the potential role of cull sow movements in FAD dissemination in the US, resulted in a data set that facilitated this investigation.[3] This project aims to assess the quality of the US swine herd's current sampling by the federal Brucella/PRV surveillance program.

Materials and methods

Data collection was in partnership with the USDA, APHIS Federal Brucellosis Laboratory (Laboratory). The laboratory collected all premises identification number tags (PIN) affiliated with samples submitted for surveillance. These PINs represent sows sampled from US cull sow processing facilities as part of the national program for Brucella/PRV monitoring.

PINs collection occurred over six weeks, one week per month in May, June, July of 2018; February, March, April of 2019; and three additional weeks during May 2020. These PINs represent the entirety of the national surveillance program's samples during the study timeframe. From each PIN, this study recorded the management/sow ID, Premise ID, state, plant, and kill date. In total, the data set consists of 22,996 unique PINs.

This project utilizes a simulated sampling scenario to determine the number of samples that should be collected from each state by random sampling. A simulated cull sow population was generated by using ~1% of each state's estimated population, or the assumed number of cull sows produced weekly.[4] Then by randomly sampling this population an equal number of times as the USDA-APHIS collected sample within the actual data set, a simulated sample was generated. By iterating this technique 100 times, a unique group of random samplings was formed. The mean number of samples from each state was calculated. These calculated sample sizes were compared to the observed sample size to estimate the relative over or under-sampling within individual states.

The effective prevalence that could be detected in each state based on the actual sampling was then calculated utilizing the following formula:

$$n = (1 - (\alpha)1/D)\ (N - ((D-1)/2))$$

Where n is the observed sample size, α is 1 – confidence, N is the population size for the state, and D is the expected number of diseased animals in the state (prevalence).[5] Using the observed number of samples from a given state, n, the effective assumed prevalence of detection could be calculated. The results were then compared to determine the gaps between the effective prevalence of detection and the desired level of detection.

Results and discussion

Nine of the sixteen states with reported sow inventories were determined to be under-sampled. These states included Iowa, Illinois, and North Carolina, three of the four states with the largest swine herds in the United States. Of all the states represented in the sample, Pennsylvania was the most under-sampled. In this case, only 2.4% of the expected number of samples were collected, resulting in an ability to detect a disease prevalence of greater than or equal to 6.8%, approximately 680 times greater than program standards. The national effective prevalence was also found to be severely under-sampled at greater than or equal to 0.004%. That is two fifths of the standard. This under-sampling could allow such a disease to go undetected and potentially result in a widespread outbreak. Along with major under-sampling, several states were determined to be over-sampled. These included Colorado, Michigan, and Utah (Table 1).

Within plant-day sampling bias was also observed. Of all the sampling events recorded, 15% consisted of less than or equal to three total premise IDs collected for that day. Many slaughter plants have daily capacities in the thousands.[6] Therefore, this sample size does not accurately generate a true representation of the number of sows that go through the plants every day.

Table 1: Average collected weekly and simulated premises ID samples per state used to calculate the effective assumed prevalence of diseased animals

State	Weekly collected sample	Simulated sample	Effective assumed prevalence
CO	59.8	30.3	0.203%
IA	114.6	186.4	0.043%
IL	103.4	108.7	0.044%
IN	101.8	47.6	0.025%
KS	35.1	33.3	0.125%
MI	35.2	22.6	0.262%
MN	104.8	104.0	0.039%
MO	86.7	86.9	0.067%
NC	109.3	164.5	0.044%
NE	72.3	83.7	0.054%
OH	97.7	37.5	0.054%
OK	37.7	86.0	0.174%
Other	85.9	84.2	0.040%
PA	0.7	28.7	6.781%
SD	44.8	53.5	0.085%
TX	25.2	29.8	0.140%
UT	7.9	15.7	0.854%
Total	1120.9	1203.3	0.004%

The severe under-sampling of individual states combined with the apparent lack of stratified sampling within plants results in a surveillance program that does not represent the US industry as a whole. This lack of representativeness allows for low prevalence diseases to go undetected. While the program's purpose is to detect Brucella/PRV, utilization of this sampling method may become a vital surveillance tool during a FAD outbreak. However, these results suggest that current sampling strategies may not detect a pathogen until prevalence within the population is as high as 6.8%, depending on the initial introduction's location. With an increase in the effective prevalence of detection, current strategies may result in catastrophic consequences, as they are likely to delay the detection of disease, allowing for widespread dissemination.

The reason for surveillance is to ensure the swine population is free of disease. However, without the proper sampling protocols in place, this is not possible. While sampling of the market channel is a wise strategy in allowing a widespread cross-sectional sample of the entire industry, care must be taken to ensure a stratified random sample, guaranteeing freedom of disease in all parts of the industry. Unfortunately, while the sampling strategy may perform acceptably at a national level, this analysis suggests that a further evaluation of how samples are stratified is required for existing surveillance streams to be genuinely effective tools in combating FADs in the US.

References

1. Blair, B. & Lowe, J. Describing the cull sow market network in the US: A pilot study. *Preventive Veterinary Medicine.* 2019;162:107-109.

2. USDA-APHIS Swine Brucellosis Control/Eradication. 1998. Available at **www.aphis.usda.gov/animal_health/animal_dis_spec/swine/downloads/sbruumr.pdf**. Accessed on 1 July 2020.

3. Hermes,M., Blair, B., Lowe, J. A spatial and temporal analysis of cull sow movements. 2020 **https://www.aasv.org/library/swineinfo/item.php?AASV/2020/064_Hermes.pdf**

4. USDA Quarterly Hogs and Pigs. 2020. Available at **www.nass.usda.gov/Publications/Todays_Reports/reports/hgpg0920.pdf.** Accessed on 5 August 2020.

5. Fosgate GT. Practical sample size calculations for surveillance and diagnostic investigations. *Journal of Veterinary diagnostic Investigation.* 2009

6. Estimated Daily US Slaughter Capacity by Plant (head per day). 2019. Available at **www.pork.org/facts/stats/u-s-packing-sector/**. Accessed on 13 July 2020.

Examining the relationship between hemoglobin concentration in nursery, growing and finishing gilts to be used for breeding replacement

H. Lathom, BS; G. W. Almond, DVM, PhD
North Carolina State University, Raleigh, North Carolina

Introduction

Stillbirths in piglets is a rising issue in swine production both for economic and welfare reasons. Previous studies showed that anemia in sows is a pathological indication for increased incidence of stillbirths.[1] Anemia is often addressed at processing via Iron Dextran injection but is rarely considered as the pig matures. As a result, the relevant parameters for hemoglobin (Hb) concentration in mature animals is less known. To the authors' knowledge, no other study addressed the prevalence of anemia in breeding replacement gilts before selection and their subsequent first litter performance. The objectives of this study were to gain relevant parameters of hemoglobin (Hb) levels in early and late nursery and finishing stages, and to examine the effects of Hb status upon gilt placement/selection and gilt reproductive data.

Materials and methods

A replacement gilt flow from the farm's multiplier herd was used for gilt enrollment. All enrolled gilts followed the same farm flow and were reared identically. Gilts (n = 600) were identified at 4 weeks of age and two days after placement on the nursery farm. Gilts with obvious lesions, such as umbilical hernias and lameness, were not enrolled. Upon inclusion in the study, gilts were individually ear tagged, and a blood sample was taken from the external pinnae via ear prick. Blood was directly collected into cuvettes and placed in a HemoCue 200+ instrument, which was used for all Hb determinations in the trial. Hemoglobin concentrations were then determined for pigs at 9, 15, and 28 weeks of age using the identical collection protocol. Anemia was defined as a Hb concentration of < 10g/dL. Selection of breeding replacement gilts was determined by standard farm protocol based on conformation, size, and overall health, not on hematological status. Farm records were obtained after gilt placement on sow farms and used to analyze gilt reproductive success. Stillbirths were determined visually by farm staff at parturition. For the Hb concentrations, an analysis of variance included age and selection with means compared with Tukey's test (Statistix, Version 10). Pearson correlations were used to assess the relationship among Hb concentrations and subsequent litter characteristics.

Results and discussion

As shown in Table 1, the Hb concentrations were numerically similar within the herd. The Hb status increased ($P < 0.05$) from early nursery (4 weeks of age) to late finishing stages (28 weeks of age). All numerical values are recorded in Table 1. For the selected gilts, the overall litter characteristics included 12.8 + 0.16 liveborn, 0.74 + 0.16 stillborn, 0.29 + 0.04 mummies per litter. The Pearson correlations indicated that the Hb concentrations at 28 weeks of age were positively correlated (correlation coefficient 0.2332, $P < 0.0003$) to the number of stillborn pigs in the first litter. This indicates that as the Hb concentration in the gilt increases, the total number of stillbirths in parity one increases accordingly. This result was not expected. Previous studies indicate that an increasing Hb concentration in the sow correlates to a decrease in expected stillbirths.[1,2] Further investigation is needed to determine the cause of the relationship discovered in this investigation. Finally, the Hb concentrations collected at each life stage can be used as reference values for diagnostics.

References

1. Bhattarai S, Framstad T, Nielsen JP. Association between sow and piglet blood hemoglobin concentrations and stillbirth risk. Acta Veterinaria Scandinavica, 2019;61:1-4. **doi:http://dx.doi.org.prox.lib.ncsu.edu/10.1186/s13028-019-0496-7**

2. Bhattarai S, Framstad T, Nielsen JP. Stillbirths in relation to sow hematological parameters at farrowing: A cohort study. *J Swine Health Prod.* 2018;26(4):215-222.

Table 1: Hb concentrations in gilts at 4, 9, 15 and 28 weeks of age. The "selects" were gilts identified as suitable replacements in sow herds

	Hemoglobin (Hb) concentration in gilts aged 4 to 28 weeks						
	Age (wks)	**Selects Hb concentration**	**SEM**	**No. gilts**	**Non-selects Hb concentration**	**SEM**	**No. gilts**
1st Hb	4	**11.18**	0.1136	317	**10.97**	0.1215	283
2nd Hb	9	**10.54**	0.0737	315	**10.32**	0.0713	267
3rd Hb	15	**11.12**	0.0732	297	**11.10**	0.094	201
4th Hb	28	**12.38**	0.0759	298	**12.21**	0.0983	184

Comparing anesthetic combinations for vasectomizing boars

E. Pratt[1], BS; E. Spronk[2], DVM; J. Steckelberg[2], DVM; N. Garbes[3], DVM; D. Amodie[3], MS
[1]College of Veterinary Medicine, Iowa State University, Ames, Iowa; [2]Eichelberger Farms Inc., Wayland, Iowa;
[3]Zoetis, Parsippany, New Jersey

Introduction

Boars, used for heat detection or promoting estrus in gilts, are commonly vasectomized to prevent unintended pregnancies when using artificial insemination in commercial operations. Vasectomizing boars is a practice that requires general anesthesia. In swine, a common anesthetic combination is Telazol® (tiletamine HCl and zolazepam HCl, Zoetis, Parsippany, NJ), ketamine, and xylazine (TKX). However, TKX can be expensive and recovery from anesthetization can be slow. The objective of this study was to evaluate an alternative to TKX using acepromazine, ketamine, and xylazine (AKX) as a viable alternative anesthetic combination comparing cost, time to unconsciousness, and time to sternal.

Materials and methods

In this study, 33 boars underwent vasectomies and were assigned randomly to four treatment groups. The treatment groups were T01 (n = 9): 4.4mg/kg, Telazol® 2.2mg/kg ketamine, 4.4mg/kg xylazine; T02 (n = 8): 0.22mg/kg acepromazine, 6.6mg/kg ketamine, 4.4mg/kg xylazine; T03 (n = 8): 0.11mg/kg acepromazine, 3.3mg/kg ketamine, 2.2mg/kg xylazine; and T04 (n = 8): 0.22mg/kg acepromazine, 3.3mg/kg ketamine, 4.4mg/kg xylazine. Prior to drug administration, each boar was weighed, assigned to a treatment group, and then the anesthetic combination was administered intravenously; via the external jugular vein. After administration of the anesthetic combination, time until unconsciousness, and time to sternal were recorded. Vital signs, including heart rate, respiration rate, and rectal temperature were collected using cardiac auscultation, thoracic elevation, and a thermometer respectively, and recorded before and after the vasectomy while under general anesthesia with visual monitoring throughout surgery and recovery to assess anesthetic depth.

Results

T02 and T04 induced general anesthesia for the shortest period. Average time to unconsciousness was 2.2 minutes longer for T02 and T04 (3.8 minutes) compared to T01 (1.6 minutes), and 9.8 minutes longer for T03 (11.6 minutes) compared to T01 (Table 1). However, for time to sternal, T01 was 150 minutes, T02 was 84 minutes (1.8 times faster recovery), T03 was 16 minutes (9.4 time faster recovery than T01), and T04 was 61 minutes (2.5 times faster recovery than T01). At least one pig per treatment group had a vital parameter that was not within the normal range, and all the averages were outside the normal range for all treatment groups.

Discussion

Telazol® is a dissociative anesthetic and a tranquilizer, which also has analgesic effects. Acepromazine is a phenothiazine, which leads to sedation and tranquilization. Ketamine is a NMDA receptor antagonist and leads to an induction of general anesthesia. Xylazine is an alpha-2 agonist, which leads to sedation and analgesia. Each drug has its own mechanism of action but in the combinations used, sedation, anesthesia, and analgesia all occur. The normal vital parameters of swine are heart rate: 80-100 bpm, respiration rate: 25-40 rpm, and rectal temperature: 102.7°F.[1] Vital parameters outside of the normal range were attributed to the stress of anesthesia and handling. The decrease in time to recovery could allow veterinarians to better utilize their time performing other farm visit responsibilities. The decrease in the cost of anesthesia could allow producers a way to vasectomize boars more economically. All combinations using acepromazine cost less per kg of body weight than using Telazol®. To place a 25 kg boar under general anesthesia, T01 would cost $22.64, T02 is $5.63, T03 is $2.81, and T04 is $3.50. However, more studies will need to be done to further evaluate the efficacy of using an AKX anesthetic combination.

Reference

1. Dewey CE, Straw BE. 2006. Herd Examination. In *Diseases of Swine*. 10th Edition. Edited by Zimmerman JJ, Karriker LA. Ramirez A, Schwartz KJ, Stevenson GW. Ames: Wiley and Sons; 2012; 10

® Registered trademark of Zoetis Inc.

Table 1: Comparison of vital parameters for anesthetic treatment groups

Tx group	Drug combination (mg/kg of each)				# Boars	Avg wt (kg)	Avg time to unconscious (min)	Avg time to sternal (min)	Avg HR before/ after surgery (bpm)	Avg RR before/ after surgery (rpm)	Avg rectal temp before/ after surgery (°F)
	Telazol	Acepromazine	Ketamine	Xylazine							
T01	4.4		2.2	4.4	9	17.3	1.6	69.1	172.9/146.7	54.7/55.8	104.0/104.0
T02		0.22	6.6	4.4	8	15.9	3.8	37.6	157.0/153.0	48.5/47.0	104.1/103.9
T03		0.11	3.3	2.2	8	16.2	11.4	16.8	133.3/146.0	57.3/42.0	104.3/103.7
T04		0.22	3.3	4.4	8	15.4	3.8	25.8	168.0/133.7	52.4/53.4	104.5/105.1

Risk factors for umbilical defects in commercial pigs

M. Bracy[1], BS; K. Blake[2], DVM, M.Ag; M. Wensley[3], MS; D. Madson[2], DVM, PhD, DACVP; K. Coble[2], PhD; J. Davenport[2], DVM; K. Baker[2], DVM, MS
[1]Oklahoma State University, Stillwater, Oklahoma; [2]JBS Live Pork, LLC., Greeley, Colorado;
[3]Kansas State University, Department of Animal Science, Manhattan, Kansas

Introduction

Umbilical hernias are abdominal wall defects associated with the navel cord which allow mesentery or intestines to prolapse through the abdominal wall. Umbilical abscesses are often confused with umbilical hernias. Pigs with umbilical hernias or abscesses are considered blemished and if severe enough, unmarketable through traditional channels; euthanasia is a common outcome. The objective of this study was to understand on-farm risk factors that impact the incidence of umbilical hernias and abscesses.

Materials and methods

Historical and prospective incidences of umbilical defects were investigated on eleven batch farrowing farms. Six months of historical data was utilized to select sow farms with both high (n = 6) and low (n = 5) incidences of umbilical defects. All farms had the same genetic make-up; PIC 1050 female × DNA 600 boar with an inventory range of 450-950 sows. Incidence of umbilical defects by 10-12 weeks of age was collected prospectively from 22 batches (2 batches per farm). Factors which were hypothesized to impact umbilical defect rates were collected for each batch during three separate site visits by the same observer who was blinded to the farms historical umbilical defect rates. During visit one, a cleanliness score for farrowing rooms, environmental settings and type of flooring was collected. During visit one on batch two, a luminometer (3M™ Clean-Trace™ ATP Monitoring System) was used to evaluate the presence of bacterial ATP on crates immediately after washing. Visit two focused on farrowing practices including induction, attendance at farrowing, piglet processing and compliance with neonatal piglet treatment protocols. Visit three occurred at the nursery where umbilical defects were counted and all affected pigs were necropsied to determine the cause of the defect. Samples for bacterial isolation and susceptibility testing were collected from pigs with umbilical abscesses.

Data was analyzed with a generalized linear model using the PROC GLIMMIX procedure of SAS (SAS Institute, Cary, NC), and a logit link function, to observe the response of pig proportions having umbilical defects at the end of the nursery phase. Fixed effects were the different observational responses (e.g. disinfectant) and a random effect of farm was included in the model to define farm as the experimental unit. Farm evaluated as an independent predictor variable and not found to be statistically associated with umbilical hernia incidence. Within evaluation of all other predictor variables, farm was included in the model as a random effect to properly account for clustered data structure. The response was assumed to follow a binomial distribution. Initially, each individual predictor variable measured on the sow farm was included in the model as a fixed effect to understand the impact of single predictors on the response. If the predictor variable had a *P*-value of < or = to 0.10, it was considered significant and noted to be used in the complete statistical model. For determining the response predicators used in the complete statistical model, a forward selection process was used to build the statistical model by inserting the predictor with the lowest *P*-value (most significant response), fitting a model, evaluating, and then creating a new model by adding the next predictor variable. This process continued until the last predictor variable added was of no statistical significance.

Results

After analysis of all observed factors (Table 1), only batch and use of disinfectant were found to be significant for predicting umbilical defects and were included in the model. Table 2 reports association of these predictor variables. The odds of umbilical defects were 2.43 times greater for batch 1 compared to batch 2 ($P = 0.088$). The odds of umbilical defects were 6.03 times greater when disinfectant was not used ($P = 0.030$). When the interaction of batch and disinfectant use was analyzed, it was found during batch 2, the odds of umbilical defects were 16.90 times greater when disinfectant was not used compared to when disinfectant was used ($P = 0.033$). Approximately 50% of the animals necropsied between 3-12 weeks of age (early nursery, late nursery, and early finisher stage) were true umbilical hernias and the other 50% were abscesses (data not shown). The most common bacteria isolated from swabbed abscesses included *Trueperella pyogenes, Staphylococcus* sp., *Streptococcus* sp., *and Pasteurella multocida*.

Discussion

After extensive search of the literature, this is the first time that disinfection practices have been associated with umbilical defect rate. If a farm did not disinfect, the proportion of pigs with umbilical defects remained very constant between visits and was higher than the farms that did use disinfectant. For farms that did use disinfectant, the proportion of pigs with umbilical defects was even lower at the second visit (Figure 1). This could be due to the act of visiting the farms the first time caused the farms to pay greater attention to detail during the subsequent batch. It may also have been due to undetected factors, such as change in season, as the visits were one month apart. The fact that half of the defects were true umbilical hernias and thus potentially due to genetics and not infection means that risk factors for these defects were not assessed in this study. Following boar genetics in association with umbilical hernia incidence would be a helpful aspect for future research.

Acknowledgement

Thank you to Merck Animal Health for funding and supporting the research project.

Table 1: Analysis of production data to determine association between predictor variables and proportion of piglets weaned with umbilical defects

Evaluation of individual predictors	***P*-value**
Farm	0.361
Batch	**0.047**
Farm age	0.956
Number of farrowing attendants	0.494
Farrowing attendant experience[1]	---
Average parity[1]	---
Crate age	0.959
Floor age	0.944
Crates per room	0.999
Number of farrowing rooms	0.775
Mat age	0.979
Cooling system[1]	---
Heat lamp height	0.601
Flush type	0.584
Wash time[1]	---
Load time[1]	---
Crates flamed[2]	---
Hot water	0.210
Detergent	0.945
Disinfectant	**0.076**
Mat score[1]	---
Crate score	0.604
Room score[1]	---
Induction score	0.530
Naval cords clipped[3]	---
Drying agent	0.872
Split-suckle	0.947
Norfenicol[1]	---
Average processing age[1]	---
Processing tool	0.276
Iron[4]	---
Behind the sow	0.298
Litter score	0.185
Flush depth	0.480

Table 1: Continued

Evaluation of individual predictors	*P*-value
Luminometer[5]	0.603
Total born[6]	0.715
Preweaning mortality[7]	0.131

[1] Data did not converge in SAS.

[2] No farms flamed, therefore there were no differences to analyze.

[3] No farms clipped naval cords, therefore there were no differences to analyze.

[4] All farms administered an iron injection 24-48 hr after birth, therefore there were no differences to analyze.

[5] The average of three luminometers was taken and categories were made. A low luminometer reading (< 15,000) was categorized as a 0. A high luminometer reading (> 15,001) was categorized as a 1.

[6] Total born numbers were divided by number of crates to determine average total born per litter, then categories were made. A low total born (0-13) was categorized as a 0. A high total born (14+) was categorized as a 1.

[7] Pre-weaning mortality percent categories were made. A low pre-weaning mortality (0-12%) was categorized as a 0. A high pre-weaning mortality (13+) was categorized as a 1.

Table 2: Complete statistical model created using forward selection process to determine the association of predictor variables with proportion of piglets with umbilical defects

Predictor variable	*P*-value	Odds	Interpretation
Batch	0.088	2.43	The odds of umbilical defects are **2.43** times greater for batch 1 compared to batch 2 (*P* = 0.088).
Disinfectant	0.030	6.03	The odds of umbilical defects are **6.03** times greater when disinfectant is not used compared to when disinfectant is used (*P* = 0.030).
Batch × disinfectant	0.051	2.15	For batch 1, there was no evidence that the use of disinfectant was associated with proportion of animals with umbilical defects (*P* = 0.662).
		16.90	For batch 2, the odds of umbilical defects are **16.90** times greater when disinfectant was not used compared to when disinfectant was used (*P* = 0.033).

Figure 1: LS Mean and SEM for percent umbilical defects by disinfectant use and batch.

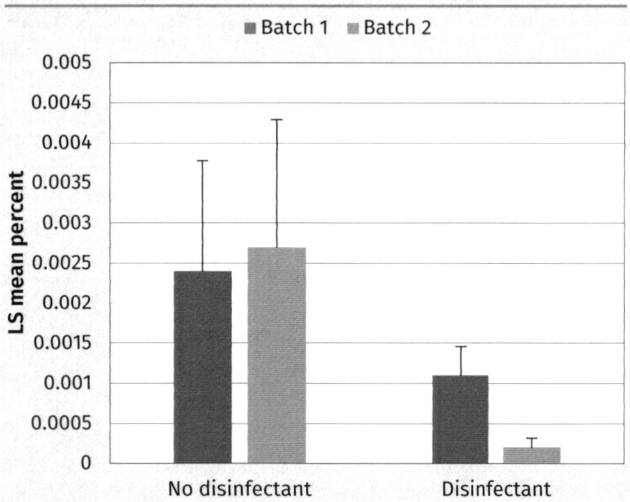

Impact of coccidiosis and treatment with ponazuril on piglet performance

S. McCallister[1], BS; H. Everett[1], BS; L. Britton[2]; C. Phillips[3], PhD; J. Pittman[4], DVM, MS, DABVP
[1]North Carolina State University College of Veterinary Medicine, Raleigh, North Carolina; [2]North Carolina State University College of Agriculture and Life Sciences, Raleigh, North Carolina; [3] Smithfield Hog Production Science and Technology Department, Rose Hill, North Carolina; [4]Smithfield Hog Production North Region, Waverly. Virginia

Statement of the problem

Coccidiosis, caused by the protozoa *Cystoisospora suis*, results in piglet diarrhea.[1] The clinical signs occur as early as five days after birth but are more commonly seen between seven and fourteen days. Morbidity is typically high, while mortality is low. Since coccidiosis fails to respond to antibiotic treatment most farms use different variations of sanitation programs for control.[2] While the efficacy of toltrazuril against *C suis* has been well described,[3] it is not available in the United States. Ponazuril, a similar trazine antiprotozoal, is commonly prescribed off-label by veterinarians to treat and prevent coccidiosis in piglets, however the efficacy has not been widely published. Limited data exists on the impact of *C suis* on piglet performance in commercial herds in this country, making it difficult for veterinarians and producers to establish a return on investment (ROI) for intervention strategies.

Objective(s)

The purpose of this study was to evaluate the effectiveness of ponazuril on piglet coccidiosis and compare performance differences between treated and untreated litters.

Materials and methods

The study was conducted in July and August on a 1500 head multiparous breed-to-wean sow herd, in eastern North Carolina, with a history and recent diagnosis of coccidiosis. Litters from three sequential projected weanings were enrolled into the study across five enrollment days, such that litters were 3 to 6 days of age at treatment. Litters were blocked by farrow date and parity, then randomly assigned as control (CNT) or treated (TX). At enrollment, litters were weighed, pigs were counted and treated per the protocol below. One tube of ponazuril was diluted with 1000mL QS sterile water (19 mg/mL) and each pig in TX litters received 1mL of the dilution per os. CNT litters were not treated. Standard farm piglet movement was allowed prior to, but not after enrollment. Litters were assessed for diarrhea twice weekly. Pooled litter fecal samples were collected at 3-4 days post enrollment, then every seven days and at weaning. Samples were sent to the Iowa State University-Veterinary Diagnostic Laboratory for semi-quantitative fecal floatation. Once a litter tested positive for *C suis*, no additional samples were collected from that litter. At weaning, each litter was weighed, counted and visibly small pigs were weighed individually. Pigs at or less than 10 lbs were recorded as light weight. Average pig weight was calculated from the data collected. Proc GLM of SPSS was used for statistical analysis. Litter was the experimental unit. Main effects of treatment, parity category, enrollment group, and age at enrollment were included in the analysis. For the categorical variable SAS was used for analysis. *C suis* positive litters were compared using odds ratio (OR).

Results

Seventy-six (n = 76) TX and seventy-seven (n = 77) CNT litters were enrolled. There was no difference in start weight or start number between groups. There were no interactions by enrollment day or age. Treatment-parity interaction was observed for number weaned, mortality and light weight pigs, and was included in the final model. Significant increases were observed in TX litter wean weight (+18.9 lbs, $P < 0.01$) and average pig weight (+1.03 lbs, $P = 0.02$) as compared to CNT litters. There was significant reduction in the number of TX litters with diarrhea (-19.8%, $P < 0.01$) versus CNT litters. Trends were noted in litter mortality and number of light-weight pigs between the TX and CNT litters (Table 1). *C suis* was detected in 19 (25.0%) TX and 56 (72.7%) CNT litters; OR 0.125 [0.061, 0.257] ($P < 0.0001$).

Discussion

This study describes the impact of coccidiosis on piglet performance and demonstrates the efficacy of ponazuril against *C suis*. These results are in agreement with the one abstract the authors are aware of on the efficacy of the ponazuril product,[4] and those reported for toltrazuril.[3] Since this study looked at using a preventative treatment method, this likely underestimates the impact of coccidiosis, since no true negative control groups were available to compare. The control of coccidiosis is estimated to have additional positive downstream impact, through increased weight at weaning and immediate post-weaning growth rates, however in this study we only evaluated the impact based on wean weight.[5] Complete effective control of coccidiosis is through increased sanitation methods to reduce environmental contamination and also treatment with an effective anticoccidial to reduce shedding of oocysts into the environment. Sanitation methods described as effective on *C suis* are removal of organic debris, steam cleaning, bleach or ammonium compounds and flaming. Some farms have tried sealing surfaces with paint or water sealant which delays the exposure of piglets to oocyts until they are older.[1,2] Veterinarians and producers can utilize the data herein to estimate a cost of coccidiosis and to calculate a return on investment for additional control measures in their operations. Allied industry can utilize this information to calculate value in further research and testing of potentially effective products, both disinfectants and preventive medications, against *C suis*.

Table 1: Comparison of performance between litters untreated or treated with ponazuril for control of coccidiosis

Treatment	CNT	TX	SE	*P*-value
Number of litters	77	76		
Litter start weight, lbs	56.30	58.37	1.34	0.28
Avg piglet start weight, lbs	4.42	4.59	0.11	0.27
Number pigs per litter at start	12.90	12.82	0.14	0.69
Litter wean weight, lbs	182.99	201.93	4.06	< 0.01
Avg piglet wean weight, lbs	15.22	16.25	0.30	0.02
Number pigs weaned per litter	12.10	12.19	0.17	0.70
Number quality pigs weaned per litter	11.26	11.60	0.21	0.26
Litter mortality, %	6.78	4.60	0.98	0.11
Light weight piglets, %	6.40	4.39	1.03	0.17
Litters with presence of scours, %	97.4	77.6	0.01	< 0.01
Number litters positive for *C suis*	56	19		< 0.001

Comparison of metrics between litters treated (TX) for coccidiosis with ponazuril (19 mg/pig) per os at 3-6 days of age or left untreated (CNT). Litters' dams were blocked by farrow date and parity. Litter weight and pig counts were collected at enrollment and at weaning. Visibly small pigs were individually weighed and classified as no value if less than or equal to 10 lbs. Presence of diarrhea was evaluated twice weekly. Pooled litter fecal samples were collected from each litter weekly or until positive by fecal flotation.

Parity categories (1, 2-3, 4-6)

All variable were tested for trt*parity category, trt*enrollment group, and trt*piglet age at enrollment interactions. There were trt*parity category interactions for # pigs weaned, litter mortality, and # quality pigs weaned so the interaction was included in the model for those variables.

Litter mortality (%) was calculated dividing the number died by the number start and multiplying by 100.

No value piglets at weaning (%) was calculated by dividing the number of pigs < 10 lbs at weaning by the number start and multiplying by 100.

References

1. Lindsay DS, Dubey JP, Santin-Durn M, Fayer R. Coccidia and Other Protozoa. In: Zimmerman JJ, Karriker LA, Ramirez A, Schwartz KJ, Stevenson GW, eds. *Diseases of Swine*. 10th ed. Ames, Iowa: Wiley-Blackwell Publishing. 2012:895-907.

2. Neumann J, Ramirez A, Schwartz K.J. Coccidia. *Swine Disease Manual*. 5th ed. Perry, Iowa. AASV. 2020:125-126

3. Skamparsonis V, Sotiraki S, Kostoulas P, Lentides L. Effect of toltrazuril treatment in nursing piglets naturally infected with Isospora suis. *Vet Parasitol*. 2010;172:46-52.

4. Myers III GH, Kennedy T.J. Efficacy of ponazuril for treatment of coccidiosis caused by experimental infection with *Isospora suis* in neonatal swine. *Am Assoc of Veterinary Parasitologist 46th Annual Meeting*. 2001;32-33.

5. Aliaga-Leyton A, Friendship R, Dewey CE, Todd C, Peregrine AS. *Isospora suis* infection and its association with postweaning performan+ce on three southwestern Ontario swine farms. *J Swine Health Prod*. 2011;19(2):94-99.

The use of an audio-based monitoring technology to detect clinical respiratory distress episodes in growing pigs

A. Witt[1], BS; J. Rudy[2], BS MS; C. Alonso[3], DVM, MS, PhD; A. Fernández-Fontelo[4], MS, PhD; K. Hogan[4], PhD; C. Goodell[5], DVM, MS, PhD, Dipl ACVPM; D. Polson[5], DVM, MS, PhD
[1]Iowa State University College of Veterinary Medicine; [2]Summit Smart Farms; [3]Upnorth Analytics; [4]Humboldt-Universität zu Berlin School of Business and Economics; [5]Boehringer Ingelheim Animal Health USA Inc.

Introduction

Respiratory diseases are a major health threat to growing pigs, including porcine reproductive and respiratory syndrome virus (PRRSV), influenza A virus of swine (IAV-S), and *Mycoplasma hyopneumoniae* (Mhp). Respiratory disease has been reported as a leading cause of mortality in growing pig units.[1] The performance of pigs is often negatively impacted by respiratory pathogens, and can lead to higher production costs, decreases in revenue, and an adverse economic impact on producer businesses. Haden et al[2] evaluated growing pig closeouts and associated diagnostic testing results, estimating the cost of major swine respiratory agents ranging from $0.63 to $10.41 per head marketed.

With the increasing adoption of cloud-based sensor technologies, earlier disease detection, diagnosis and intervention may help improve performance and profitability. In this clinical case, SoundTalks® monitors were used in conjunction with diagnostic testing and the Auto-Regressive and Moving Average (ARMA) models to detect the onset of clinical respiratory episodes, characterize the primary associated infectious agents in groups of growing pigs, and predict future episodes that may help control and prevent possible disease episodes.

Materials and methods

This clinical case study originated from two wean-to-finish sites (two barns, four rooms and 8400 head capacity per site, described here as East and West sites) located one mile apart in the Midwest US. Previous studies done by Polson et al[3] determined the appropriate SOUNDTALKS (ST) monitor placement. Four ST monitors were installed in each room (16 monitors per site). Each monitor was hung above the gate separating two adjacent pens.

The ST system continuously recorded audio data and processed that data into a "Respiratory Distress Index" (RDI). System algorithms established a baseline RDI, and alerts were triggered when RDI values were outside of the expected baseline range. When ST alerts were observed, oral fluids were collected from each ST monitor zone in the affected airspace. Samples were tested by real time polymerase chain reaction (qPCR) for PRRSV, IAV-S and Mhp.

The West site received a new group of pigs in December 2019 sourced from the same company sow farm. The East site was filled in January 2020 from an outside PRRSV stable sow farm. Pigs were marketed from the West and East sites in May and June 2020, respectively. A timeline of diagnostic data results, ST notifications, as well as important dates are demonstrated in Figure 1.

RDI observations from a disease-free period on the West side (December 9th 2019 to March 29th 2020) were modeled through ARMA processes to appropriately consider the data's temporal auto-correlation structure, and estimate the relationship between RDI and relevant covariates such as temperature, relative humidity, wind direction and wind speed. These covariates have been demonstrated to play a role in the survivability and consequently potential transmission of respiratory pathogens between premises.[4-6]

The estimated ARMA model was used to predict point values and their 99% confidence limits for RDI data from April to June. In doing so, the real observations could be directly compared to the model predictions, and thus relevant departures of the real observations to the 99% upper limit were presumably attributable to disease as the model predicted disease-free RDI evolution. As different diseases were proved between April and June, the predicted model outbreaks could be diagnostically confirmed to evaluate the model's prediction capability.

Results and discussion

Several different respiratory disease episodes were characterized at the two sites in both groups of pigs (Figure 2). PCR testing of oral fluid samples confirmed clinical episodes of PRRS, IAV-S and Mhp. ORF5 sequencing confirmed the PRRS virus was a wild-type. IAV-S was detected at the West site after RDI alerts were noted at the end of March (Figure 1 and 2). Within 7 days the East site reported RDI alerts. Prior to the notification of an increase of RDI at the East site, winds were blowing on a daily average of the following: March 31st to the East-Northeast, April 1st to Southeast, April 2nd directly to the East, and on April 3rd wind bearing was to the Southeast. The average wind speed prior to the notification of an increase of RDI at the East site were the following: March 31st had an average speed of 8.5 miles per hour, April 1st had an average speed of 7.4 miles per hour, April 2nd had an average speed of 4.5 miles per hour, and April 3rd had an average speed of 5.6 miles per hour. Average relative humidity prior to the notification of an increase of RDI at the East site were as follows: March 31st had 77% average relative humidity, April 1st had 80%, April 2nd had 72%, and April 3rd had 64% average relative humidity. The IAV-S temporal pattern progression and HA gene sequencing suggest possible aerosol spread of IAV-S from West to East.

Figure 1: Diagnostic and SOUNDTALKS alert timeline.

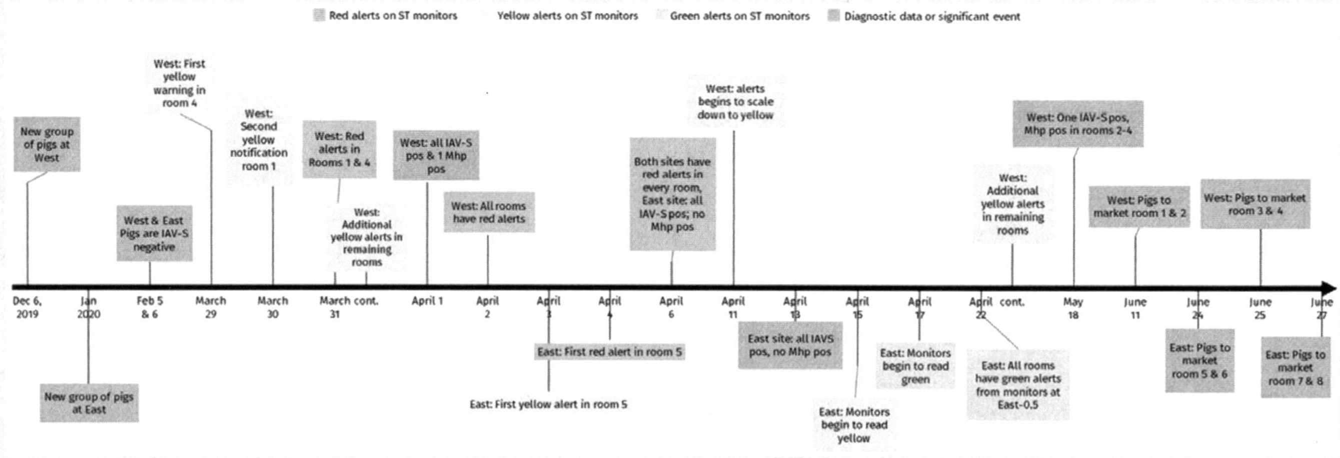

Figure 2: Time series of the Respiratory Distress Index (RDI) showing clinical disease episodes of PRRS, IAV-S and Mhp in pig groups at the West and East finishing sites.

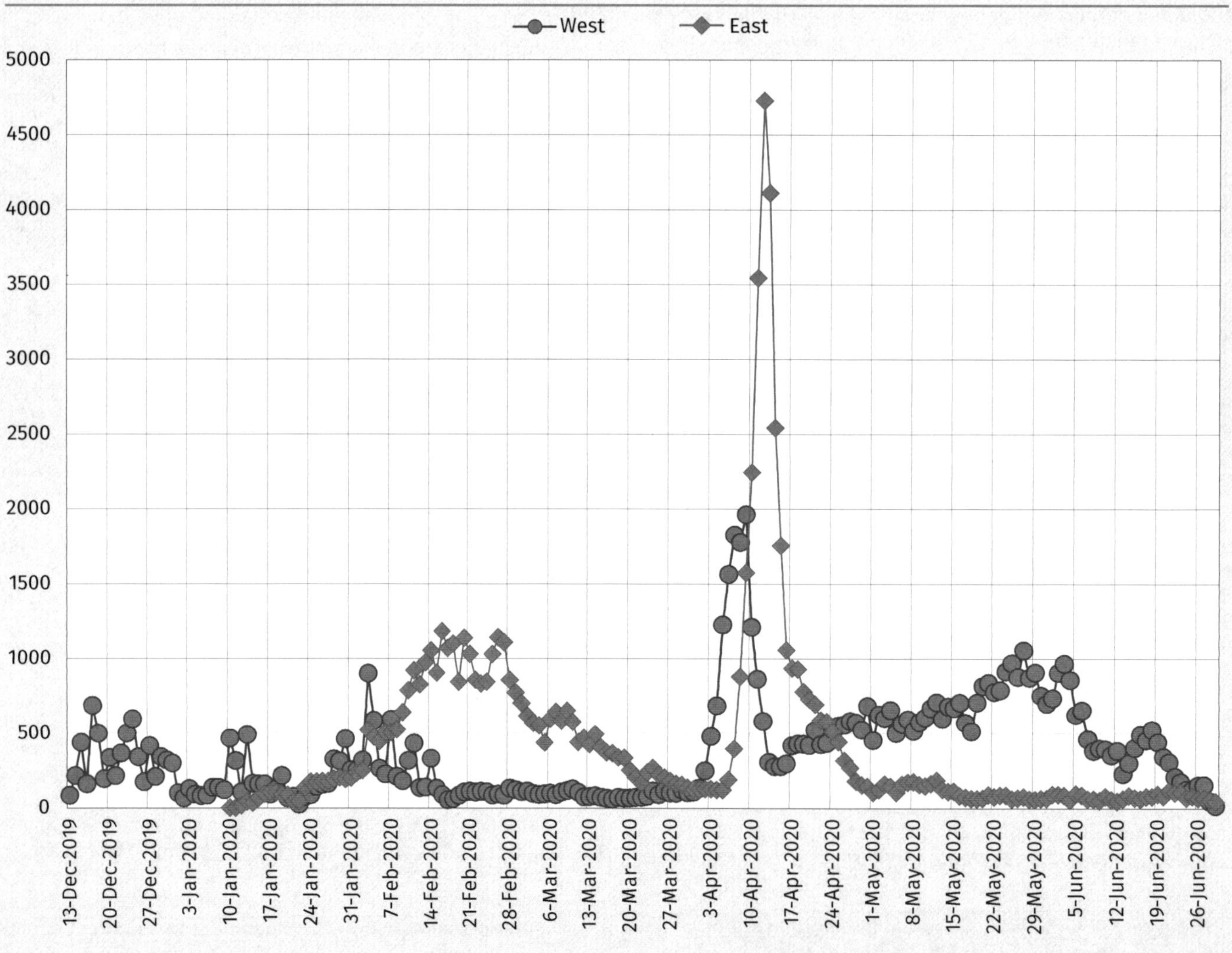

Figure 3 shows how the West side's estimated ARMA model appropriately detected the two outbreaks diagnostically confirmed as Influenza A virus and *Mycoplasma hyopneumoniae* episodes, respectively. SOUNDTALKS technology, diagnostic data and the ARMA modeling method hold promise for predicting expected respiratory behavior and detecting abnormally elevated RDI associated with primary swine infectious disease agents.

With more time series data, ARMA models can be derived in a quasi-real-time surveillance system where new data feed models continuously to readjust to the RDI data evolution. This would provide more robust confidence limits that give options for models to be more accurate of a situation and when outbreak is likely to occur. ARMA models could also include variables including but not limited to seasons, geographic location, biosecurity flows, age, and more. Continuous quasi-real-time data predictions can help producers predict the onset of disease, leading to prevention or earlier intervention. More proactive disease management will increase herd health and lead to positive economic impact for producers.

References

1. National Animal Health Monitoring System. 2017. *Swine 2012: Part III: Changes in the US Swine Industry, 1995-2012,* p. 40, 45. US Department of Agriculture, Fort Collins, Colo.

2. Haden C, Painter T, Fangman T, Holtkamp D. Assessing production parameters and economic impact of swine influenza, PRRS and *Mycoplasma hyopneumoniae* on finishing pigs in a large production system, in Proceedings. 43rd American Assoc Swine Vets Annu Meet 2012;75–76.

3. Polson, D., Playter, S., Berckmans, Z., Qui, Quinn, B. Determining the optimal placement and configuration of an audio-based sensor platform to enable improved detection and characterisation of clinical respiratory episodes in large growing pig airspaces and sites, in Proceedings. 9th European Conference of Precision Livestock Farming 2019; 490-495.

4. Dee S, Otake S, Oliveira S, Deen J. Evidence of long distance airborne transport of porcine reproductive and respiratory syndrome virus and Mycoplasma hyopneumoniae. *Vet Res.* 2009;40(4):39. **doi:10.1051/vetres/2009022**

5. Ypma, R., Jonges, M., Bataille, A., Stegeman, A., Koch, G., van Boven, M., Koopmans, M., Marijn van Ballegooijen, W., Wallinga, J., Genetic Data Provide Evidence for Wind-Mediated Transmission of Highly Pathogenic Avian Influenza, *The Journal of Infectious Diseases*, Volume 207, Issue 5, 1 March 2013, Pages 730–735, **https://doi.org/10.1093/infdis/jis757**

6. Lowen AC, Mubareka S, Steel J, Palese P (2007) Influenza Virus Transmission Is Dependent on Relative Humidity and Temperature. PLoS Pathog 3(10): e151. **https://doi.org/10.1371/journal.ppat.0030151**

Figure 3: ARMA model for the West site showing predicted baseline RDI with 99% confidence intervals.

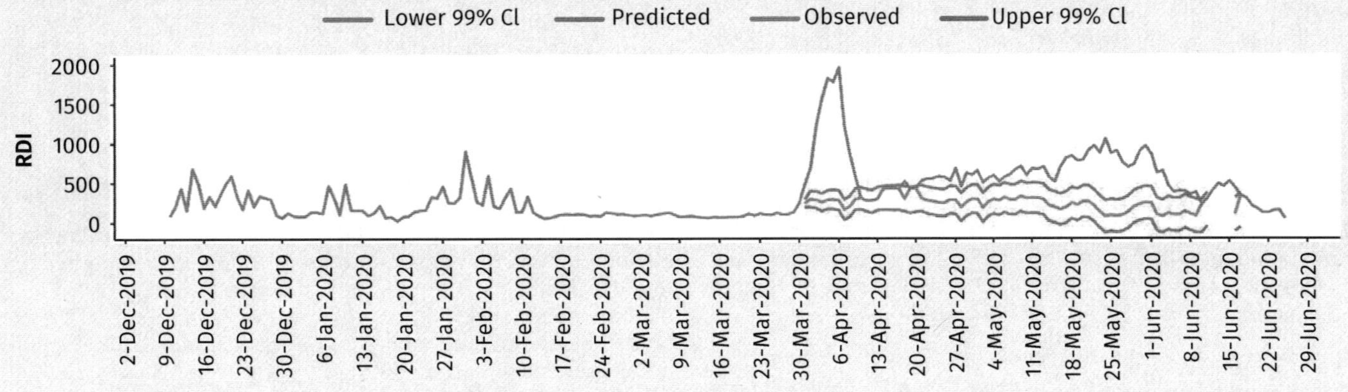

Lack of efficacy of routine and enhanced hygiene practices in removing environmental rotavirus A from rubber piglet comfort mats

Sarah Botkin; James Lowe; MS, DVM; Benjamin W. Blair, DVM
Applied Epidemiology Laboratory, College of Veterinary Medicine, University of Illinois, Urbana, Illinois

Introduction

Porcine rotavirus, a pathogen common in neonatal pigs, is becoming increasingly important in commercial swine breeding herds. Porcine rotavirus type A has been implicated during outbreaks causing enteric disease within piglets both pre and post-weaning.[1,2] Some researchers estimate that rotavirus is endemic within the majority of large commercial swine farms.[3] The clinical manifestation within piglets has led to decreases in both the production and profitability across all segments of the industry.[2] Unfortunately, current mitigation methods have resulted in only a limited impact,[1,4] leaving producers and veterinarians in need of a more complete understanding of such methods.

Because of the fecal-oral route of transmission of rotavirus, persistence within the farrowing environment is suspected to play a key role in overall effect of the disease on a farm.[4] Conventional control methods, such as vaccination, pre-farrowing viral exposure[1] of the dam, and routine hygiene practices have only resulted in limited reductions in the quantity and severity of rotavirus associated disease within commercial swine farms.[1,2,4] This lack of efficacy is the consequence of both the numerous pathogenetic types/strains of the virus[5,6] and its fastidiousness within the environment,[2] allowing viral loads to remain high within the farrowing environment, promoting transmission to piglets. Rotavirus's unique resilience to several forms of disinfection, is in part due to the pathogens resistance to many commonly used chemical disinfectants, as well as the ability to survive at a wide range of temperatures, making effective decontamination of surfaces difficult.[4] This in combination with the lack of clinical efficacy of rotavirus inoculation or vaccination1 allows rotavirus to be a clinical issue in farms.

This study focuses on evaluating the effect various disinfection methods have on viral load on a complex surface. It attempts to determine the best single technique or combination of techniques to use in routine disinfection of the farrowing environment. This project's objective to quantify the impact of each cleaning method on the elimination of environmental rotavirus found within a biofilm on a rubber mats, like those commonly found within farrowing crates.

Materials and methods

Rubber mats, typical of those found within farrowing crates for piglet comfort, were contaminated with a modified live porcine rotavirus vaccine (Pro system Rota, Merck Animal Health). This project performed contamination in a matter to promote the creation of a biofilm for the virus to survive in. To accomplish this, two doses of virus, of approximately 10^4 viral particles per dose, were mixed with 5 grams of feces and 5 mL of water and spread over a 103.2 cm^2 inoculation area (IA). Mats were then reinoculated in the same manner, every other day for a total of three inoculations. Mats were kept in a temperature-controlled room under plastic sheets to create a warm, humid environment.

Forty-eight hours after the final inoculation, the mats were subjected to various decontamination methods. The methods consisted of four arrangements of practices: 1: High-pressure washing alone (HP), 2: HP followed by chemical disinfection with accelerated hydrogen peroxide (Intervention, Virox Animal Health) at a 1:16 dilution with 5 minutes of contact time (DW), 3: Detergent (Barnstorm, Preserve International) applied at low pressure followed by HP (WS) and 4: WS followed by chemical disinfectant (DWS). HP consisted of a flow rate of approximately 17L/min at 310 Bar though a 15° fan nozzle moved laterally back and forth across the surface until visibly clean. Each treatment was replicated 16 times for a total of 64 replicates.

A toothbrush was used to simulate typical pig interaction with the mat during the nursing phase. The mat was briefly placed flat on the floor, and a disposable toothbrush was systematically rubbed across the entire IA. The toothbrush was put into a 50 mL conical tube containing 15 mL of D/E neutralizing broth to ensure inactivation of the disinfectant within the sample. Samples were vortexed, and a 1ml aliquot was removed and sent to the ISU VDL for routine multiplex rotavirus rtPCR for rotavirus A, B, and C. The proportion of samples with rotaviral RNA following treatment was compared with a Z-test.

Results and discussion

All treatments had RNA detected via rtPCR following treatment; HP:11/16, DW:9/16, WS:6/16, and DWS:7/16. However, the proportion of samples with rotavirus RNA following treatment were not found to be statistically different between treatment groups. These data suggest that conventional (high-pressure washing and chemical disinfection) and enhanced (high-pressure washing, detergent, and chemical disinfection) cleaning methods were ineffective in lowering the prevalence of rotavirus A from farrowing comfort mats. The overall prevalence of rotavirus A on mats following cleaning with detergent, disinfectant, or both was 44% (95% CI 31%;61%), demonstrating that rubber comfort mats can serve as a significant environmental reservoir for rotavirus in clinically diseased herds despite proper implementation of cleaning practices.

While it is expected that results would vary depending on the surface being contaminated, this study focuses on the efficacy of disinfection methods on a common complex surface encountered within a farrowing room. The results of this study reveal that current disinfection methods or combination of methods fail to be an effective method for the complete elimination of rotavirus from the environment. These results suggest that further research is necessary to determine if alternative chemical disinfectants, methods or materials compositions could be used to improve decontamination within farrowing, leading to a reduction in clinical disease, losses experienced by the producer.

References

1. Pittman JS, Anderson A, Boyd W, et al. Update on Rotavirus Control from the Field. Published online 2019:87-96.

2. Dewey CE, Carman S, Pasma T, Josephson G, McEwen B. Relationship between group A porcine rotavirus and management practices in swine herds in Ontario. Can Vet J. 2003;44(8):649-653.

3. Kim HJ, Park SI, Ha TPM, et al. Detection and genotyping of Korean porcine rotaviruses. Vet Microbiol. 2010;144(3-4):274-286. **doi:10.1016/j.vetmic.2010.01.019**

4. Chandler-Bostock R, Mellits KH. Efficacy of disinfectants against porcine rotavirus in the presence and absence of organic matter. Lett Appl Microbiol. 2015;61(6):538-543. **doi:10.1111/lam.12502**

5. Parra GI, Vidales G, Gomez JA, Fernandez FM, Parreño V, Bok K. Phylogenetic analysis of porcine rotavirus in Argentina: Increasing diversity of G4 strains and evidence of interspecies transmission. Vet Microbiol. 2008;126(1-3):243-250. **doi:10.1016/j.vetmic.2007.06.006**

6. Lorenzetti E, Stipp DT, Possatti F, Campanha JET, Alfieri AF, Alfieri AA. Diarrhea outbreaks in suckling piglets due to rotavirus group C single and mixed (rotavirus groups A and B) infections. Pesqui Vet Bras. 2014;34(5):391-397. **doi:10.1590/S0100-736X2014000500001**

Development, benefits, and uses of flow cytometry in swine genetics

K. L. Willenburg[1]; B. Belstra[1]; D. Grossi[2]
[1]Fast Genetics, Middleton, Wisconsin; [2]Fast Genetic, Saskatoon, SK

Introduction

The use of flow cytometric sex sorted sperm (SSS) has become a staple in the dairy industry with dairymen electing to combine genomics with sexed sorted sperm for maximizing genetic potential for replacement breeding stock. Inguran Sexing Technologies, Inc has been the driver in SSS and since its inception has produced over 110 million straws of sexed sorted sperm via its 26 sorting labs worldwide and is responsible for more than 50 million replacement heifers. Sexed sorted sperm has been used in other livestock and hobby farming industries such as in pigs, beef cattle, sheep, goats, horses and deer as well as in zoos and marine animals for species preservation. Interest for SSS in the swine industry has been overwhelming and shows the insight of producers for this technology at all levels of the genetic pyramid. Currently, SSS is being used in multiple farms throughout the genetic pyramid at Fast production systems. As this technology continues to improve with increased sorting speeds, sorting efficiencies and post sort sperm survival other applications with flow cytometry are being developed and leveraged. Karyoflow™ as described in detail is an extension of the vast capabilities of flow cytometry enabling Fast Genetics to identify boars with chromosomal aberrations (CB) at a fraction of time compared to traditional methodologies as cytogenetic analysis. Furthermore, multidimensional assessments to develop a mechanistic understanding of cell function is also being explored in multiple species at ST Genetics and Fast Genetics. That said, the following manuscript traces the history of flow cytometry with focus on the implementation of a sexed sorting breeding program within Fast Genetics in addition to the expansion of flow cytometric based technologies.

History of flow cytometry

Flow cytometry (FC) (cyto = cell, metry = measure) in its most basic premise is a measurement of individual cells in a flowing stream of fluid. This technology has grown by leaps and bounds since its early induction in the 1960's – yet it has changed very little in its fundamentals. The number of simultaneous measurements has increased from 1-2 to over 30 parameters today, which has turned into a billion-dollar industry.[1] Today, numerous biomarkers exist for various cells types, which include but are not limited to stem cells, reactive oxygen species, cancer cells, yeast, bacteria, nucleic acids, red and white blood cells, and sperm cells.

The year 2015 was the 50th anniversary of flow cytometry when Mark Fulwyler and Richard Sweet collaborated to build the first sorter that was published in Science.[2] Followers quickly pursued with a group from Stanford that received a patent in 1972 for the first Fluorescence Activated Cell Sorter (FACS) while in Germany, Wolfgang Göhde developed a fluorescence-based flow cytometry device (ICP-11) with Partec in 1968.[3] Cell sorters were marketed shortly thereafter from Becton Dickinson (FACS-II), Partec (ICP-22) and Coulter (Epics). Using a modified Epics sorter, Larry Johnson from the USDA separated chinchilla sperm into two live population[4], which has been the basis for modern day sorting protocols and US Patent #692958. Briefly, for sperm sorting, sperm are incubated with the membrane permeant bisbenzimidazole fluorescent dye, Hoechst 33342 and separated into two live sperm populations based on DNA content with a flow cytometer.[4]

Flow cytometric sorting for gender selection

Flow cytometric sorting is well established in other livestock industries, particularly dairy where in 2016, around 5 million straws of sexed semen were processed in the US[5] and today sexed sorted sperm is used in ~ 3% of the dairy industry.[6] In the swine industry, sexed sorted has not been available for commercial use, yet. However, within Fast Genetics, the application of sexed sorted sperm has shown to create economic value at the genetic nucleus and multiplier farms. At the genetic nucleus level, a faster rate of genetic improvement can be made within specific lines by increasing the selection intensity, whereas at the multiplier level, there are many efficiencies to be realized but doubling the output of the preferred gender without impacting infrastructure cost is highly attractive. Commercially, the value of sexed sorted sperm has been estimated at $5-7 per market hog but will vary depending on the production system and desires of the consumer.

Flow cytometric sperm sorting speeds have been a limitation to the use of sexed sperm in pigs due to the high numbers of sperm required to maximize fecundity (2.5 to 3.0 billion sperm in 75 to 100 mL of extender).[7] Sorting speeds have improved but sorting for a multi-billion cell dose is cost prohibitive and would require a significant amount of time.[8] In order to overcome this obstacle, laparoscopic insemination (LAI) and deep uterine insemination (DUI) have been used to a) reduce the number of sperm and b) deposit the spermatozoa closer to the site of fertilization. From a sorting perspective, these insemination strategies could be a realistic model but would require a cost-effective ovulation induction method and personnel capable of executing these insemination strategies. In a field trial performed by Sexing Technologies, Inc. comparing a single fixed time DUI (600×10^6 non-sorted sperm) and LAI (10×10^6 sorted sperm) versus an estrus based AI control ($> 1.5 \times 10^9$ non-sorted sperm) there were no statistical differences ($P > 0.05$) in farrowing rate (Control = 86.67 ± 0.06, DUI = 82.40 ± 9.82, LAI = 77.78 ± 8.82) or total born (Control = 16.51 ± 0.77, DUI = 15.36 ± 1.28, LAI = 15.00 ± 1.28). Furthermore, in a field trial with hormonally synchronized sows, there were no differences ($P > 0.5$) in pregnancy rates from a semen titration study using a single fixed timed AI with either DUI and PCAI (post cervical artificial insemination) with non-sorted sperm levels ranging from 1.2×10^9 billion to 75 million, but there was a gradual decline in total number of embryos collected as the sperm dose numbers decreased (16.1 ± 0.7 to 11.5 ± 0.7), which was statistically different ($P < 0.05$).[9]

Flow cytometric analysis

Flow cytometry is composed of four main systems: fluidics, optics, electronics and software.[7] Flow cytometry can objectively and accurately determine sperm concentration, viability, mitochondrial function, acrosome integrity, ROS levels, DNA fragmentation, intracellular Ca^{++}, membrane fluidity, apoptotic markers as well as other sperm related parameters. As cells flow through the system, properties like fluorescence, absorbance and light scattering can be detected[7] allowing simultaneous measurement of multiple cell characteristics to be analyzed as well as to acquire data from different subpopulations.

Cell analyzers and sorters continue to become more affordable and user friendly enabling these multi-functional platforms to be used across the swine industry particularly as a boar selection tool that complement phenotypic and genotypic traits, but also in boar studs improving dose quality and ensuring dose concentration accuracy. The industry is pushing for lower sperm numbers in AI doses, which could reduce boar inventory, allow producers to use superior boars and improve piglet uniformity. Counting chambers are accurate but time consuming and flow cytometry offers a good alternative to determine cell concentration since it can discriminate cells from debris that are commonly found in ejaculates.[10]

Flow cytometric diagnostics

Flow cytometric diagnostics is usually associated with human medicine where it has been indispensable for diagnosing immune deficiencies, newborn screening tests and hematopoietic cancers to name a few, but a new venture with this technology was developed by Fast Genetics called Karyoflow™ for identification of males with chromosomal aberrations. It has been reported that ~ 50% of all hypoprolific boars have a chromosomal rearrangement.[11] Pigs with chromosomal abnormalities usually express normal phenotypic traits and growth rates, which allows them to go undetected in the herd.[12] The consequences of CB, as an example of reciprocal translocations (RT) have been associated with decreases in prolificacy with an average reduction of 4 pigs born per litter[12] for the carrier males or from the mates of carrier animals[11] and to date, more than 200 unique chromosome arrangements have been characterized[13] that can have serious economic consequences.[14,15] Depending on the frequency rate, which has ranged from 0.5 to 3.4% depending on the country[15], an average of 4 fewer pigs per litter in the US market of 6.38 million sows[16] could result in losses ranging from $11 to $73 million dollars. It should be emphasized that low fertility based on sow performance could go undetected depending on the number of parities needed for record analysis.[15]

Traditionally, CB, including RT have been diagnosed by chromosome banding and FISH techniques, which is quite laborious, time consuming and requires specialized personnel. Karyoflow has been shown to accurately detect boars and bulls with abnormal karyotypes in a fraction of the time compared to previously mentioned methodology, which will allow all boars and bulls within Fast Genetics and ST Genetics to be screened.

Conclusions

Flow cytometry is a billion-dollar industry and has become a staple in biological research and clinical diagnosis. It has been revolutionary for advances in immunology and cancers. The industry will only continue to flourish as newer fluorochromes are developed and narrower emission spectra and additional lasers are introduced. Flow cytometry offers the opportunity to measure several sperm attributes on every single sperm on several thousands of sperm in a sample, which could be used for sensitive quality control tests.[10]

In terms of flow cytometric sperm sorting, animal agriculture is in its infancy stage as just under 5% of the dairy industry uses sexed sorted sperm, but it is quickly garnering attention with beef on dairy matings in addition to the deer industry for selection of rack size. The swine industry is another market where flow cytometry could be implemented whether it is through multi-fluorochrome analysis for an objective sperm function test or with sexed sorted sperm optimized for each individual production system. Regardless, if the last 50 years of this technology is an indication of the next 50 years then we will continue to see improvements, affordable instrumentation and a revolutionary understanding of sperm physiology.

References

1. Robinson JP and Roederer M. Flow cytometry strikes gold. Science. 2015;350 (6262):739-740.

2. Fulwyler MJ. Electronic separation of biological cells by volume. Science. 1965;150(3698):910-911.

3. Clark DL. A brief history of flow cytometry. **https://fcslaboratory.com/a-brief-history-of-flow-cytometry.**

4. Johnson, L et al. Sex preselection in rabbits: Live births from X and Y sperm separated by DNA and cell sorting. Bio Reprod. 1989;41:199.

5. Moore G and Hasler JF. J.Dairy Sci. 2017;100(12):10314-10331.

6. Mulvany L and Decker S. Sexed-semen is big business in dairy farming and focus of legal battles. **https://www.agupdate.com/news/dairy/sexed-semen-is-big-business-in-dairy-farming-and-focus-of-legal-battles/article_9afde65e-f025-11e7-a519-a73ffa45af03.html.**

7. Knox, R. Artificial insemination in pigs today. Theriogenology. 2016;85:83.

8. Spinaci, M. et al. Storage of sexed boar sperm: Limits and perspectives. Theriogenology. 2016;85:65.

9. Belstra BA et al., J. Anim. Sci. 2020;98(9):1-10.

10. Hossain S et al., Flow cytometry for the assessment of animal sperm integrity and functionality:state of the art. Asian J. Androl. 2011;13:406-419.

11. Pinton A et al., Chromosomal abnormalities in hypoprolific boars. Hereditas. 2000;132:55-62.

12. Ducos A et al., Cytogenic screening of livestock populations in Europe: an overview. Cytogenet Genome Res. 2008;120:26-41.

13. Donaldson B et al., Non-random distribution of reciprocal translocation breakpoints in the pig genome. Genes. 2019;10(10):769.

14. Popescu CP et al., J. Reciprocal translocations in pigs: Their detection and consequences on animal performance and economic losses. J. Heredity. 1984;75(6):448-452.

15. Quach AT et al., Genet Sel Evol. 2016;48:66.

16. Hess A. United states hog inventory up 4%. National hog farmer. Mar 26, 2020. **https://www.nationalhogfarmer.com/marketing/united-states-hog-inventory-4.**

Recent trends in vitamin D serum status in swine

S. Hough, DVM; J. Bergstrom, PhD; J. Hahn, PhD
DSM Nutritional Products

Introduction

Concerns over lameness, fractures, and ill thrift within various phases of production of the swine industry have increased in the past decade. Vitamin D deficiency is rightfully included in differential diagnosis of potential etiologies and risk factors. Vitamin D has long been recognized as essential to the health and function of the skeletal system by maintaining calcium and phosphorus homeostasis.[1] A deficiency can result in rickets, lameness and generalized unthriftiness in younger pigs, while bone fractures can occur in market weight or mature animals.[2] There is growing evidence to support the role of vitamin D in many other metabolic processes, including immune functionality, however little is known about extent and impact of hypovitaminosis D on immune response or animal vigor.[3] Vitamin D can be acquired by oral and parenteral administration, as well as endogenously through exposure to type B ultraviolent (UVB) light, which triggers hydroxylation of cholecalciferol to become vitamin D in the body. Variation in bioactive vitamin D is dependent on a number of factors, including but not limited to exposure to sunlight, feed consumption, and dietary vitamin D stability. Sustained hypovitaminosis D evidenced by reduced circulating levels of 25-hydroxyvitamin D [25(OH)D], can serve as a contributor to metabolic bone disease, which is a broad term used to describe multiple nutritional diseases related to bone growth or remodeling.[4] Swine are susceptible to rickets development because of rapid growth rate and limited exposure to sunlight in modern confinement facilities.[5] Serum 25(OH)D is considered the best biomarker of vitamin D status.[6] Madson et al postulated that, prior to 2010, swine rickets received little attention as historical corn soy nutritional programs provided adequate calcium and phosphorus, and were supplemented with sufficient vitamin D for normal bone growth and homeostasis. The utilization of phytase and alternative feedstuffs due to input costs associated with more economical ingredients may increase potential for variation in quality and quantity of specific nutrients, leading to the potential for clinical or subclinical bone disease.[7] Also, diets, particularly for market swine, are tailored for lean muscle mass accretion and rapid growth with less consideration for bone formation.[8]

Background

In a previous study, published in 2015 by Arnold et.al, serum 25(OH)D concentrations in pigs of different ages and during different seasons were evaluated (Table 1).[4] Vitamin D supplementation within the feed was also evaluated. A total of 1200 samples were included. This serum survey provided evidence that 25(OH) D concentrations may be highly variable across all ages of confined swine. The serum data also highlighted that individual swine within a population may be functioning on suboptimal serum vitamin D concentrations, yet not showing overt clinical signs of deficiency. The implications of subclinical hypovitaminosis D in swine, as well as normal, reference, expected levels of Vitamin D, are not fully understood with current prolific genetics and nutritional factors found in modern swine production.

Recent findings at ISU VDL

Data were extracted from the ISU VDL LIMS system expanding from 2016-2019, while also including the first nine months of 2020 (Table 2). Of the 5,900 accessions (cases), more than 1,700 submission lacked age related information. Data were organized by serum 25(OH)D results and age in weeks, using the same age categories as Arnold et al. 2015. Unlike their study, the lack of case history and context, including but not limited to dietary information, environmental (indoor/outdoor) data, health status, and nutritional supplementation, precluded evaluation of those factors which may have influenced 25(OH)D concentrations in tested individuals. This exercise emphasized the importance and necessity of a comprehensive and complete history in order to provide context to the results. The samples were submitted with the request for vitamin D analysis, while the samples included in the Arnold et al. study were submitted for a primary purpose other than vitamin D surveillance. The average and median serum 25(OH)D concentrations, and relative ranges in association with age are also listed in Table 1. The wide range of values for each age group showcases the variation potential within a population of pigs. The majority of concentrations were found to be below reference ranges established in Puls Vitamin Levels in Animal

Table 1: Arnold et al[4] data

Arnold et al[4] data	Reference Intervals (ng/mL)	Serum (25OH)D (ng/mL) Combined overall mean	Combined overall range
Nursery	18-30	11.4	2.5-62.4
Grower	18-30	19.5	3.4-54.1
Finishing	18-30	26.4	3.7-77.9
Mature sow	35-70	36	4.7-94.5
Mature boar	35-70	38.5	8.9-93.8

Table 2: ISU VDL data for 57 months (January 2016-September 2020)

		Serum (25OH)D (ng/mL)			
Recent ISU VDL data	**No. of submissions**	**Average**	**Median**	**Standard deviation**	**Range**
Nursery	1488	15.3	7.4	21.9	1.5-200
Grower	151	14.6	12.4	9.0	1.5-51.2
Finishing	365	31.5	26.6	21.4	3.5-200
Adult	963	40.2	33.5	30.0	2.5-235

Health.[9] In comparison to a previous study, mature animals similarly possessed greater serum 25(OH)D levels than younger animals. The nursery serum averages for the current data set were higher, but the grower phase samples, if age captured was truly accurate, were lower than in the previous publication. The goal of this exercise was to uncover trends in serum status for the various age groups, but that proved to be difficult due to the incomplete data set.

Forms of vitamin D

Vitamin D can be acquired through injection, diet or endogenously by exposure to UVB light, but typically all routes are biologically inert and require multiple phases of hydroxylation in the body for activation. Hy-D®, a source of 25-OH-D3 for feed or water supplementation, is a DSM product that provides an efficient form of the vitamin that does not require the initial liver hydroxylation step. Also, 25(OH)D3 is more bioavailable than Vitamin D3, due to its higher polarity and 500 times stronger affinity for the vitamin D-binding protein.[10] Looking at the Figures 1-3, you see the difference in supplementation and subsequent serum status. Figure 1 illustrates the difference in serum levels in gilts fed the same equivalents of D3 and Hy-D® from breeding until 28 days of gestation.[11] Even at equivalent levels for D3 and Hy-D®, serum levels were different, indicating improved bioavailability with Hy-D®. This was also seen in recent research by Thayer et al., where higher serum levels were seen in sows, grower and finishing pigs. Figures 2 and 3 show the difference in the grower and finishing phase, where either 1000 IU/kg equivalent or 800 IU/kg equivalent was fed.[12] What does improved serum status translate into for producers? In swine, the use of Hy-D® has been associated with improved structural soundness in gilts.[13] An increase in number of liveborn pigs and weaned piglets in both sows and gilts were also observed when compared to controls.[14,15] In growing animals, Hy-D® has been shown to reduce osteochondrosis lesions.[16] We are still learning about how improving serum status can improve performance parameters but have been able to prove the benefit of a more bioavailable form of vitamin D in different ages, genetics and geographies.

Vitamin D toxicosis in swine is very rare, manifests histologically as mineralization of the kidneys, myocardium, lungs, gastrointestinal tract and blood vessels. Affected pigs can demonstrate lethargy, enteric or respiratory distress and deaths can occur, especially when due to rodenticide exposure.[17] When higher levels are seen in commercial swine, it is likely due to recent, or slightly overzealous supplementation.

Discussion

Vitamin D serum status is not only a consideration for food producing animals who depend on feed supplementation, but also for human health and diet fortification. In recent human literature, observational studies have shown a negative correlation between low levels of mean vitamin D and number of cases of Covid-19 in various European countries.[18] There was also a meta-analysis done in the last year that concluded that vitamin D supplementation was safe and somewhat supportive of respiratory tract health.[19] Vitamin D promotes differentiation of monocytes to macrophages, involved in the production of inflammatory cytokines and support antigen presentation.[20] Vitamin D is a versatile vitamin important for bone metabolism, but also should be considered to have other properties that potentially mitigate impact of everyday challenges. Hence, the evidence suggests that serum status is important for not only bone health, but also for overall health and productivity. Serum evaluation for micronutrients such as vitamin D is useful to evaluate metabolic state. Could it be useful not only for individual animal status, but perhaps also for a population medicine approach similar to other diagnostic tools? A holistic approach must be considered when working up issues that are associated with metabolic bone disease. Not only are serum analyte concentrations important, but assessing bone strength, bone mineralization, and bone histopathology are essential for definitively diagnosing metabolic bone disease conditions in swine.[7] Both infectious and toxic agents that may be causing lameness issues need to be ruled out, so joints, spinal cord and skeletal muscle should also be considered as part of the full examination. Lameness is multifactorial in cause, with infections, nutrition, genetics, conformation, facility, environment, and trauma all having impact, and many clinical expressions of lameness appear to be the result of cumulative insults over time. Astute, accurate and specific case definitions are invaluable in a diagnostic investigation.

Extremely low levels of 25(OH)D3 in serum strengthen a clinical diagnosis of hypovitaminosis D; however, gross or microscopic bone lesions are needed to confirm a diagnosis of vitamin D–dependent rickets.[7] The problem with measuring serum vitamin D in swine is that normal age-based reference intervals for serum 25(OH)D_3 have not been clearly defined in recent years. We must establish what "normal" levels should be and that can only be determined from properly executed prospective studies. We must also consider the sow and how her levels of serum 25(OH)D could impact piglet levels during the critical weaning period. The post-weaning phase is a critical developmental period that is key to securing health and optimum performance in the future. Enteric and respiratory challenges are more likely during the early nursery period. Feed intake may be inconsistent in the first few days post weaning. Most pigs are at a disadvantage with respect to their vitamin D status, as many pigs are born with inadequate serum levels of vitamin D, and tend to have lower serum levels during that early nursery phase.[21] There are likely a number of pigs

Figure 1: Serum levels for gilts after 28 day feeding regimen.

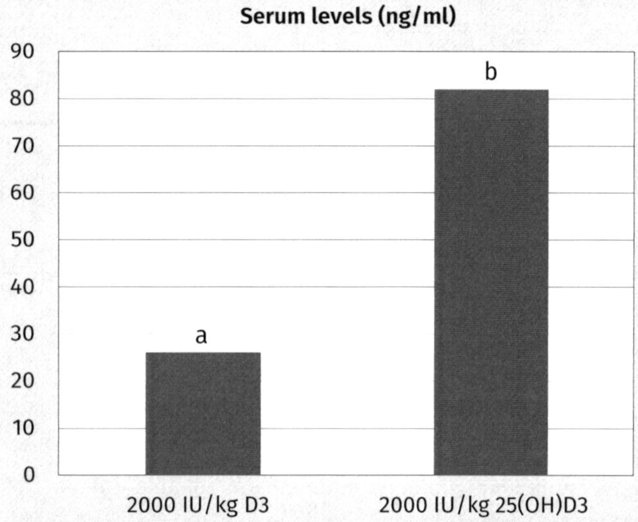

[a,b] Different superscripts are statistically different ($P < 0.05$)

Figure 2: Serum levels for grower pigs after 45 day feeding regimen.

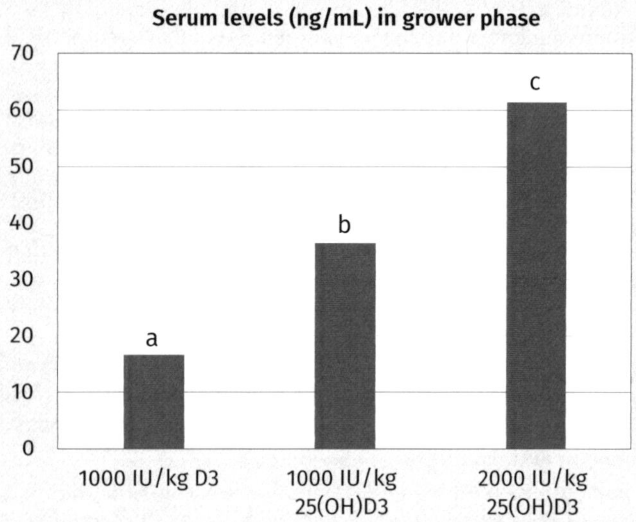

[a,bc] Different superscripts are statistically different ($P < 0.05$)

who never achieve "normal" levels of Vitamin D in serum, yet they go to market. Extrapolation would suggest that pigs can develop rickets if serum 25(OH)D3 is less than 15 ng/mL.[7,22] Calcium and phosphorus ratios should also be considered when evaluating marginal serum vitamin D levels.

Conclusions

A complete history with pertinent information is paramount in working through cases. The inclusion of a full history, clinical observations and signalment helps determine an appropriate diagnostic approach as well as a more definitive diagnosis. This, in turn, may potentially benefit the practitioner or producer with their clinical diagnoses in the future, which could save time,

Figure 3: Serum levels for finisher pigs after 45 day feeding regimen.

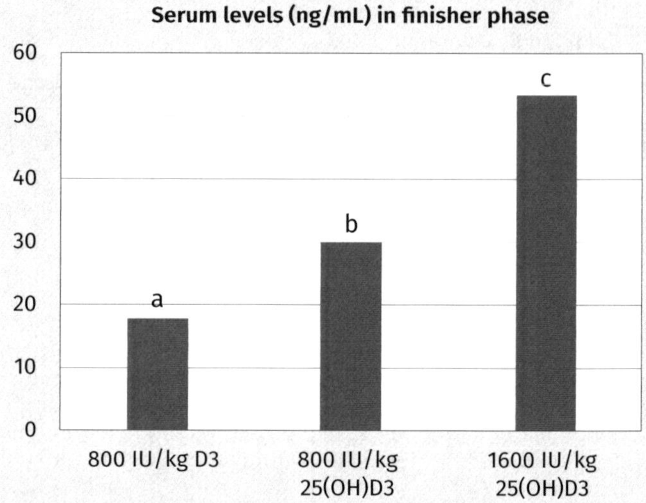

[a,bc] Different superscripts are statistically different ($P < 0.05$)

money and animals. Sample submission is an investment, and should be treated like one.

There is a heightened awareness of nutritional support of immune function. With a recent worldwide Covid-19 pandemic, the relevance of vitamin D status, along with other nutritional factors have been explored and considered in non-healthy states. This too should be considered in veterinary medicine contributions to population health, as nutrients are vital to productivity, welfare and overall wellbeing. Nutrition can often be taken for granted, but mineral and vitamin nutrition are foundational for animal health for all phases of production. Moreover, low cost, marginal nutritional programs can potentially be detrimental, especially with the health and production challenges experienced in the swine industry in current times.

Acknowledgements

Authors would like to thank Dr. Scott Radke and Dr. Kent Schwartz at the ISU VDL for their technical expertise and willingness to help with the interpretation of the data.

References

1. DeLuca, H.F. Overview of general physiologic features and functions of vitamin D. *American Journal of Clinical Nutrition*. 2004; 80, 1689-1696.

2. Meisinger, D. (Ed). The National Swine Nutrition Guide. US Pork Center of Excellence. 2010. Pg 41.

3. Baeke, F., T. Takiishi, H. Korf, C. Gysemans, and C. Mathieu. Vitamin D: modulator of the immune system. *Current Opinion in Pharmacology*.2010; *10*(4), 482–496.

4. Arnold, J., D. Madson, S. Ensley, J. Goff, C. Sparks, G. Stevenson, T. Crenshaw, C. Wang and R. Horst. Survey of serum vitamin D status across stages of swine production and evaluation of supplemental bulk vitamin D premixes used in swine diets. *Journal of Swine Health and Production*.2015; 23 (1), 28-34.

5. Dittmer K.E. and K.G. Thompson. Vitamin D metabolism and rickets in domestic animals: a review. *Vet Pathol*. 2011; 48, 389–407.

6. Jones, G. Metabolism and biomarkers of vitamin D. *Sc and J Clin Lab Invest.* 2012; 72, 7–13.

7. Madson, D.M., S. M. Ensley, P. C. Gauger, K. J. Schwartz, G. W. Stevenson, V. L. Cooper, B. H. Janke, E. R. Burrough, J. P. Goff, R. L. Horst. Rickets: case series and diagnostic review of hypovitaminosis D in swine. *Journal of Veterinary Diagnostic Investigation.* 2012; 24(6) 1137–1144.

8. Maxson P.F., Mahan D.C.: Dietary calcium and phosphorus levels for growing swine from 18 to 57 kilograms body weight. *Journal of Animal Science.* 1983; 56,1124–1134.

9. Puls, R. Vitamin Levels in Animal Health: Diagnostic Data and Bibliographies. Sherpa International. 1994.

10. Burild, A., C. Lauridsen, N. Faqir, H. M. Sommer and J. Jakobsen. Vitamin D3 and 25-hydroxyvitamin D3 in pork and their relationship to vitamin D status in pigs. *Journal of Nutritional Science.* 2016; 5(3), 1-6.

11. Laurisden, C., U. Halekoh, T Larsen and S.K. Jensen. Reproductive performance and bone status markers of gilts and lactating sows supplemented with two different forms of vitamin D. *American Society of Animal Science.* 2010; 88,202-213.

12. Thayer, M., J. L. Nelssen, A. J. Langemeier, J. M. Morton, J. M. Gonzalez, S.R. Kruger, Z. Ou, A.J. Makowski, |and J.R. Bergstrom. The effects of maternal dietary supplementation of cholecalciferol (vitamin D3) and 25(OH)D3 on sow and progeny performance. *Transl. Anim. Science.* 2019; 3, 692-708.

13. Brana, D., J. Gabriel-Landon, J. Cervantes-Lopez and J.A. Cuaron. Nonruminant nutrition: Vitamins and Minerals. *Journal of Animal Science.* 2012; 98(Supp 1), 114.

14. Planchenault, D., B. Renouf, A. Meuter and J.M. Thoby. Effect of dietary 25-hydroxycholecalciferol and cholecalciferol on the performance of gestation and lactating sows and their litter. *Journees Recherche Porcine. 2018;* 50, 143-144.

15. Coffey, J.D., E.A. Hines, J.D. Starkey, C.W. Starkey and T.K. Chung. Feeding 25-hydroxycholecalciferol improves gilt reproductive performance and fetal vitamin D status. *Journal of Animal Science.* 2012; 90, 3783-3788.

16. Sugiyama, T., T.K Chung, H. Yonekura, E. Azem and T. Hayakawa. Effects of 25-hydroxycholecalciferol on the development of osteochondrosis in swine. *Journal of Animal Science.* 2013; 84, 341-349.

17. Straw, B., J. Zimmerman, S. D'Allaire and D. Taylor (Eds). Diseases of Swine, 9th ed. Blackwell Publishing. 2006. Pg 214.

18. Ilie, P.C., S. Steganescu, and L. Smith. The role of vitamin D in the prevention of coronavirus disease 2019 infection and mortality. *Aging Clinical and Experimental Research.* 2020; 32(7), 1195-1198.

19. Martineau A.R., D.A. Jolliffe, L. Greenberg, J.F. Aloia, P. Bergman, G. Dubnov-Raz, S. Esposito, D. Ganmaa,A.A. Ginde, E.C.Goodall, C.C. Grant, W. Janssens, M.E. Jensen, C.P.Kerley, I. Laaksi, S. Manaseki-Holland, D. Mauger, D.R. Murdoch, R. Neale, J.R Rees, S. Simpson, I. Stelmach, G. Trilok Kumar, M. Urashima, C.A. Camargo, C.J. Griffiths, R.L. Hooper. Vitamin D supplementation to prevent acute respiratory infections: individual participant data meta-analysis. Health Technol Assess. 2019;23(2):1-44.

20. Calder, P., A. Carr , A. Gombart and M. Eggersdorfer. Optimal Nutritional Status for a Well-Functioning Immune System Is an Important Factor to Protect against Viral Infections. *Nutrients.* 2020; 12, 1181.

21. Goff, J.P., R.L Horst and E.T. Littledike. Effect of sow vitamin D status at parturition on the vitamin D status of neonatal pigs. *Journal of Nutrition.* 1984;114, 163-169.

22. Heaney R. P., Horst R. L., Cullen D. M., and L. A. Armas. Vitamin D3 distribution and status in the body. *J Am Coll Nutrition.* 2009; 28:252–256.

Digital transformation in the swine industry

M. Hansen
BarnTools LLC, West Des Moines, Iowa

What it means for the producer

As a pork producer, you might think digital transformation has no relevance or immediate value to raising pigs and increasing production performance. Based on our initial survey, we found most producers do not understand the possibilities and impact of digitally transforming a production system. Digital transformation is the act of using technology to create new business processes that were not possible before, which means producers can make advances in all aspects of their system from health to logistics. Digital transformation is impacting every industry and no industry or government is immune.

Our world today is digitization; the future is digital transformation

First, we need to examine the difference between how producers typically operate in a digitized world versus how producers can digitally transform parts of their business processes virtually overnight and realize immediate efficiency gains in all areas of swine production.

Digitization

Today, producers typically manage farm access and site traffic records manually – visitors sign into a farm on a paper sign-in sheet with minimal or, at best. inconsistent oversight and accountability. The "more advanced" production systems take this information and convert it to some type of electronic record, usually an Excel spreadsheet. This would be an example of digitization – doing a manual process the exact same way, but more efficiently with technology (computer and spreadsheets). Many producers are managing data via Excel and email and performing what was a manual process in a digitized format.

Producers conduct operations with countless spreadsheets (and various versions), use file storage such as SharePoint or Microsoft Access to store data in silos, while using email as the main method for sharing that data. Although digitized, the process is siloed, not always current and not accessible or easy to share.

Digital transformation

Digital transformation happens when new processes and outcomes are created using technology advances that enable systems and data to be integrated into one, multi-application platform. To demonstrate the immediate value digital transformation can deliver, let us take a look at biosecurity systems, specifically, farm access management, nights down rules and traceability.

Digital transformation brings accountability to processes that were once built on the honor system. Using geofencing technology and location services on a smartphone, producers can now capture employee or contractor site visits when they cross a site's geofence inside their system. This technology allows users to understand real-time site health status and the associated nights down requirements. With all those integrated pieces of data talking to one another, a mobile app communicates to the visitor and the barn manager if access is approved. A new process has been born.

For a more detailed explanation, let us look at what is occurring simultaneously. When a visitor or employee approaches a farm property with a geofence, the technology now in place can incorporate prior site information, such as site health and site type, and either permit or reject farm access to that individual. Let us take it a step farther. The maintenance man crosses the geofence, which then prompts the system to search the record of his last pig exposure. This information is checked against the health status and nights down rules of those two sites. Using that employee's visit record, the system returns a warning if the maintenance man is about to breach biosecurity. All of this happens in less than a second and automatically. Now that record can be stored in a database with all the other records of site visitors across the system: feed trucks, maintenance, rendering, live haul, etc. These records are relational and build a network which allows veterinarians to have real-time epidemiology and foreign animal disease (FAD) readiness with the click of a button. This is the power of digital transformation technology and connected data intelligence working together.

Implementation and cost

Digital transformation does not mean total disruption of your day-to-day operations or an expensive investment. It is about starting with the right foundation and adding different processes naturally as the data and technology permit. Producers can start by cleaning up the chaos – which simply means consolidating your site information, employee names and roles, along with site-specific health status and associated matrix rules on one integrated digital production platform. So, right away your systems are connected and learning from each other – allowing you to connect departments, people and processes for more informed decision-making across your operation.

It is ironic the world is shifting to a more remote working environment when pork producers have been doing it for decades, given the geographic spread of our farms. Shouldn't we be leaders in remote-first business operations and communications given the head start we have had?

Preliminary findings of digital farm entry logs

J. Ryken
BarnTools LLC, West Des Moines, Iowa

Introduction

In 2019, there was a renewed focus on biosecurity due to the African swine fever (ASF) outbreak in China. In fact, several states conducted ASF outbreak simulations to pressure-test their ability to respond in the case of an actual foreign animal disease (FAD). These simulations, along with a very real FAD threat from China, have exposed and/or reconfirmed gaps in existing biosecurity procedures. BarnTools™ has developed technology to help close those gaps. While developing and implementing the BarnTools solutions, some interesting and important findings around the farm entry visitor logs, a key element of any biosecurity program, were discovered.

Review of requirements

The Secure Pork Supply Business Continuity Plan clearly defines how to develop an advanced biosecurity plan at the Premise ID level. Producers are responsible for developing and maintaining these plans. And, in the case of an outbreak, each state executes their own response. In Iowa, the Department of Agriculture and Land Stewardship will conduct epidemiological investigations to examine any and all pig, people, and equipment movements that could act as a disease transmitting vector.1 The state veterinarian will work with producers to collect 30 days of records based on the guidelines in Table 1.

These records will be used to identify both the potential source of the outbreak and, more importantly, additional sites downstream that may have been exposed. Therefore, compiling accurate records in a timely manner is in the best interest of everyone involved in preventing the spread.

Gaps identified

The simulations were designed to test the states' readiness to respond in the face of an FAD, but also highlighted gaps for producers as well. Some elements in Table 1 were relatively easy for producers to collect from their records. However, many items were either (1) not being recorded, (2) not being stored or not easily retrievable, or (3) not stored in a method that allowed for timely and accurate reporting and sharing.

To better understand these risks, BarnTools researchers conducted interviews with producers and swine veterinarians to collect additional insights into the gaps surrounding biosecurity practices and perceived risks in the industry. We interviewed stakeholders from 10 companies regarding biosecurity. The primary biosecurity concerns identified included vehicle traffic (live haul, feed, rendering, etc.), load crews, maintenance crews (internal and third-party contractors), and other vendors. Additionally, there was no tracking of tools and equipment being brought onto/into farms.

Through the interviews, several existing access management systems were discussed. The most common system used was paper sign-in sheets or logbooks that are filled out upon entry; relying on the honor system to accurately report their last visit and adherence to nights down rules. Most systems rely on farm-wide notifications (e.g. weekly email) with various health statuses and expect employees and visitors to interpret these rules to make their own decisions about entering a farm. Electronic door access codes are used at some sites to limit access. GPS on company vehicles is also used to track traffic in and out of a farm. Finally, there are producers who maintain an approval process and records for all visitors. This process requires approval prior to a person driving onto a site.

These systems represent historical processes that create potential gaps in biosecurity protocol and make tracing a disease outbreak very difficult. First, without real-time health status information, people can unknowingly expose a site because they are unaware of a disease outbreak that had occurred on the site previously visited. Second, it is difficult to hold people accountable using the honor system. This risk is greatest for third-party contractors coming on farm who may have visited a different production system without downtime. Also, with these systems, no electronic log is being created. This means, in the event of an FAD, it is impossible to provide a list of all "traffic" on and off a site.

Comparing requirements to reality

Based on 10 interviews and implementation projects with several swine producers, we have a clearer understanding of the gaps between the prioritized list in Table 1 and the current state of the industry. There are, of course, variations from producer to producer, both in their biosecurity practices and in the perceived risks associated with each element of the table.

Table 1 evaluates the readiness level and the perceived risk associated with the elements. Readiness refers to a producer's ability to easily create the required records. Level 1 is for items where electronic records are generally available and can be retrieved as needed because they are required today as parts of other compliance programs, such as transportation permits. Level 2 are cases where records are being kept in a paper logbook or some ticketing system. From these records, an entry/exit list could be manually stitched together in the case of an emergency, but it would take time and effort. Additionally, the completeness of the data would be questionable. Level 3 is for items where no records exist today because these items are not being tracked. The risk level is ranked as High, Medium or Low, based on what the producers and vets feel is the most likely source of exposure for a disease.

Digital biosecurity

BarnTools has developed technology to effortlessly capture all transactional data occurring on farm. This data, captured in realtime, provides producers both the visibility and accountability for disease vectors traveling within their system. Using an integrated database of sites, people, and disease status information, BarnTools can bring reliable structure around what today is an ad-hoc effort within most producers' operations. Users can be checked into a site in advance or in real-time. Real-time check-ins can happen either via a farm computer or a mobile app.

Each check-in location and timestamp is stored in the database, which means a user's last visit is always known. With this information, any subsequent check-in can be verified against real-time health statuses and provide accountability to the farm access management process. This information can also be used to identify breaches in biosecurity protocol, affording the producer a measure of biosecurity compliance which they can address with staff to reduce the number of breaches.

Digital farm entry logs not only reduce office administration efforts, but also consolidate disparate types of transactional activities into a usable form. Turning data into information can prove invaluable during an emergency FAD event, when state animal health officials require this information to prove a site disease-free (negative status) prior to resuming operations.

General findings

BarnTools is interested in the barriers, real and perceived, for implementing a digital biosecurity system. The findings are quite diverse in nature because of the complexity of the problems related to farm entry management. On the other hand, several benefits have also been identified by our trial users.

Technology readiness

There is a natural resistance to using technology. However, this resistance continues to diminish as smart phones are providing value in all aspects of life. Concerns around technology fall into three arenas: functionality, usability, and privacy.

Functionality, with respect to farm biosecurity, means accuracy is a must. Relying solely on cellular data is insufficient in rural areas. Our testing shows GPS can lead to 99% accuracy. Advances in technology have led to breakthroughs in other industries, such as tracking parcel deliveries. The BarnTools mobile app utilizes such algorithms to provide location reliability.

Usability concerns center around user friendliness. The user interface must be simple and in multiple languages. Usability also means battery life. Any app that creates significant drain on a battery will not be successful. BarnTools uses advanced algorithms to optimize battery life and the user interface is straight forward and user friendly.

Finally, people are concerned about privacy with respect to their location. This requirement is not unique to the farming industry. BarnTools addresses privacy by only tracking the device location once inside the farm's geofence. No location information leaves the phone or is stored unless it is part of the farm's database of sites.

Biosecurity made easy

In working with company veterinarians, it is clear there is tension between wanting to implement the best set of biosecurity rules and needing those rules to be properly interpreted by everyone. BarnTools allows biosecurity rules to be more sophisticated, while giving the ability to easily query the system and find the proper nights down for any movement. This can be done in advance of a site visit preapproval and at the time of check in.

Another challenge solved with digital biosecurity is real-time information and communication. Today, health updates are pushed via email. Companies are always weighing the trade-offs between timely communication and too much communication. This leads to once-a-week updates or blasting health updates to everyone, every day, which creates communication fatigue and, in some cases, does not reach external visitors at all. BarnTools can send real-time, targeted information to those who need to know.

One flexible platform

There are too many siloed systems and apps available today. Each one solves a niche problem. This creates app overload for users who are just trying to do their jobs. A GPS tracking solution can be used to log vehicle activity, but if it is not integrated with health status, the value is diminished. A feed delivery tracker can consolidate information about feed deliveries but does not allow for communication to sites once the feed truck delivers and does not optimize routes based on biosecurity.

Building a generic solution which supports easy implementation across all producers requires forethought and flexibility. Today's home-grown systems are generally ad hoc and living in spreadsheets. A consequence is that high customization has occurred. Implementing a new platform for each producer demands a balanced approach to find a solution that drives data consistency and allows for special requirements.

Turning data into information

Digital entry logs automatically reduce the amount of admin time required to store records. It also provides a systemized method for storing important biosecurity data. Once in the database, it is easy to study trends and turn the data into information that can drive decision-making, coaching behaviors, or investigating problems and finding a smoking gun. This leads to a greater understanding and, ultimately, improved herd health.

Conclusion

Advances in geolocation technology provide opportunity for rapid advancements in real-time biosecurity. BarnTools has developed technology to effortlessly capture transactional data from daily operations. This data, captured in real-time, provides producers both visibility and accountability for disease vectors traveling within their system. Using an integrated database of sites, people, and disease status information, BarnTools can bring reliable structure around what today is an ad-hoc effort within most producers' operations. Digital farm entry logs not only reduce office administration efforts but also consolidate disparate types of transactional activities into a usable form. Turning data into information can prove invaluable during an emergency FAD event, when state animal health officials request and require the information spelled out in the Secure Pork Supply Plan protocol.

There are some challenges to implementing a digital system for tracking entries. These include technology readiness and user behavior. However, the information gained from such a system can be used to prevent disease spread and create improved operational efficiencies. The potential benefits from digital records include real-time notification that can prevent disease introductions, as well as a database of information that can be used for epidemiological tracing efforts and to drive overall improvements in biosecurity compliance. These benefits certainly outweigh the challenges. The looming threat of foreign animal disease and the ever-improving technology for tracking disease vectors will soon make systems like this common place in the pork industry.

Reference

1. Iowa Department of Agriculture. Premises Trace-In and Trace-Out Priority Ranking for Swine Facilities during Foreign Animal Disease Outbreaks by Iowa Department of Agriculture & Land Stewardship. 2019. Available at: **https://iowaagriculture.gov/sites/default/files/animal-industry/pdf/How%20to%20Organize%20Farm%20Records%20for%20an%20Outbreak%20Response.pdf**

Table 1: Premises trace-in and trace-out priority ranking for swine facilities during foreign animal disease outbreaks by Iowa Department of Agriculture & Land Stewardship[1]

Level	No*	Element	Readiness**	Risk***
High-priority	1	Entry and exit of live animals (i.e., any swine moving onto or off the premises, including transportation trucks)	1	L
	2	Entry of animal products (i.e., semen)	1	L
	3	Collection of mortalities ("deads") by rendering	2	M
	4	Entry of regular employees (i.e., those working on the farm on a daily or weekly basis)	2	L
	5	History of any international travel (both employees and visitors)	2	M
Medium-priority	6	Entry of visitors (tours, students, extended family, i.e., anyone)	2	M
	7	Entry of animal health professionals (i.e., veterinarians, veterinary technician, etc.)	2	L
	8	Entry of equipment directly related to daily operations and swine production (i.e., equipment shared between sites, etc.)	3	H
	9	Removal of manure/waste	1	M
	10	Entry of repair/service personnel (i.e., electricians, plumbers, etc.)	2	H
	11	Delivery of feed	2	M
Low-priority	12	Delivery of fuel (i.e., propane, diesel, etc.)	2	L
	13	Removal of garbage	2	L
	14	Delivery of mail and packages	3	M
	15	History of other animals on the site (i.e., pets, wild animals, etc.)	3	L
	16	Entry of pest management personnel (i.e., rodent control, exterminators, etc.)	2	L

* added by author
** Readiness
1 = electronic records are generally available and can be retrieved
2 = records are being kept in a paper logbook or some ticketing system
3 = no records exist today
*** Risk
L = low risk for causing outbreak
M = medium risk
H = high risk

One-year commercial demonstration of methionine hydroxy analogue chelated zinc, copper, and manganese (MINTREX® trace minerals) on sow productivity and longevity

B. V. Lawrence[1], PhD; A. Hintz[1], DVM; B. Buresh[1], PhD; W. Mills[1]; M. Ritter[2], PhD C. Pilcher[2], PhD; M. Warsco[2]
[1]NOVUS International Inc, St. Charles, Missouri; [2]Cargill Animal Nutrition, Brookville, Ohio

Introduction

Pork producers strive to produce more pigs per sow per year, but this should be balanced with pigs per sow lifetime for reducing weaned pig cost. The optimal parity for economic return has been demonstrated to be parity 6.[1] Increasing productivity has come with increasing sow mortality and with fewer bred gilts reaching a second parity. Sow mortality and early gilt loss limits pigs per sow lifetime opportunity. Trace minerals play important roles in hundreds of enzyme systems important for physiological demand including managing free radicals resulting from metabolic pathways and stressors structural development and reproductive function. Trace mineral availability from ingredients and inorganic trace mineral sources is generally low. Highly prolific sows may have increased demand for available minerals due to demand by the developing offspring. New research has revealed the importance of the sow's trace mineral source on her lifetime productivity[2,3] as well as the impact of the dam's mineral source on the offspring[4] and the offspring's lifetime productivity.[3] The objective of this field evaluation was to evaluate the effect of substituting 50% of the zinc (Zn), copper (Cu) and manganese (Mn) in the sulfate form with a similar level of mineral provided as the methionine hydroxy analogue chelate (MINTREX® trace minerals) on sow productivity and longevity.

Materials and methods

Two 3,500 sow farrow-wean units within the same production system were used in the evaluation. One farm remained on their current trace mineral program in which all Zn, Cu and Mn was supplied in the sulfate form of inorganic trace minerals (ITM farm). For the other farm, 50% of the Zn, Cu and Mn in the sulfate form was replaced with a similar level of Zn, Cu and Mn from methionine hydroxy analogue chelate (MINTREX® trace mineral farm). MINTREX® provided 60 ppm Zn, 10 ppm Cu and 20 ppm Mn with the diet for both farms having total mineral levels of 120 ppm Zn, 20 ppm Cu and 40 ppm Mn. Selection of the farms used in this evaluation was based on the system's knowledge of the farms, geographic proximity, genetic make-up of the sow herd, stage of a genetic roll-over, and evaluation of previous 2 years of monthly performance monitor data. To determine historical differences between farms and the likelihood of detecting performance differences, the monthly historical performance monitor data for the previous 24 months was exported from the record keeping system and imported into Excel. The descriptive statistics tool within Excel was then used to create descriptive statistics and confidence intervals for each performance parameter. For the evaluation period, data from the performance monitor report by parity was used for the period beginning May 26, 2019 through May 30, 2020. To correct for historical differences between farms by parity, parity performance monitor data for a historical period from April 30, 2017 to April 27, 2019 was used. The system was in the process of a maternal line genetic change during the evaluation period. Therefore, performance monitor data for both farms was evaluated based on sow genetic line and parity with Genetic Line A representing the existing sow line and Genetic Line B being the newly introduced sow line.

Results and discussion

Historical Genetic Line A herd level performance monitor analysis

Over the historical period there were 16,973 served-to-farrow events at the ITM farm and 17,154 served-to-farrow events at the MINTREX® trace mineral farm with 99% of services to farrow included from gilt breeding through parity 8 services. The farm level historical Genetic Line A performance data for the ITM farm and MINTREX® trace mineral farm are presented in Table 1. In the historical period, sows at the farm assigned to MINTREX® trace minerals were culled one parity earlier than the sows at the ITM farm (2.7 vs. 3.7). This resulted in 8.1 fewer pigs per sow lifetime for the MINTREX® trace mineral farm (30.3 vs. 38.4). The earlier parity at culling for sows from the MINTREX® trace mineral farm during the historical period was largely due to removal of gilts before their first parity. Otherwise, across parities there was little difference in percent removals by parity. The difference in retention rate and pigs per sow lifetime would not have been detected by looking at average parity farrowed; which was 3.1 for the ITM farm and 3.3 for the MINTREX® trace mineral farm with the farms having similar parity distributions. This discrepancy highlights the need to look not only at average parity and parity distribution, but also parity at removal to gain a better understanding of factors impacting sow lifetime productivity. The MINTREX® trace mineral farm had a lower farrowing rate than the ITM farm (79.4% vs. 88.3%). The lower farrowing rate for the MINTREX® trace mineral farm (79.4% vs. 88.3%) was consistent across parities with the degree of difference being greatest for gilts (86.9 vs. 71.9 for ITM farm vs. MINTREX® trace mineral farm). The wean-estrus interval was also higher for the MINTREX® trace mineral farm during the historical period than the ITM farm (8.3-d vs. 6.5-d) with the MINTREX® trace mineral farm sows having a longer wean-estrus interval in all parities. The longer wean-estrus interval across parities for the MINTREX® trace mineral farm also correlated with a lower percentage of sows bred within 7-d post-weaning (81.3% vs. 92.3%) compared to the ITM farm. The lower percent bred within 7-d post-weaning for the MINTREX® trace mineral farm was also consistent across parities. The MINTREX® trace mineral farm also had a higher percent of repeats

Table 1: Herd level performance averages by period and net change for trial period vs. historical period

Variable	ITM Farm			MINTREX® trace mineral farm			
	Historical period	Trial period	Net change	Historical period	Trial period	Net change	MINTREX® trace mineral farm adjusted change
Reported pigs weaned per sow lifetime	38.4	39.8	1.4	30.3	46.0	15.7	14.3
Average parity culled	3.7	3.8	0.1	2.7	4.4	1.7	1.6
Average parity farrowed	3.1	4.2	1.1	3.3	4.1	0.8	---
Farrowing rate, %	88.3	89.7	1.4	79.4	86.8	7.4	6.0
Total born	13.7	14.5	0.8	13.9	14.7	0.8	0.0
Live born	12.6	13.4	0.8	12.7	13.4	0.7	-0.1
Pre-weaning mortality, %	16.1	17.1	1.0	13.8	13.6	-0.2	-1.2
Weaned/sow farrowed	10.4	11.0	0.6	10.4	10.9	0.5	-0.1
Sow mortality, %	8.3	6.8	-1.5	11.2	9.1	-2.1	-0.6
Wean-to-estrus, d	6.5	5.3	-1.2	8.3	6.3	-2.0	-0.8
Bred within 7 d, %	92.4	94.3	1.9	81.3	89.1	7.8	5.9
Percent repeats, %	3.0	1.4	-1.6	6.1	2.3	-3.8	-2.2

after breeding (6.1% vs 3.0%) with the percent repeats also being consistently higher across parities. Taken together, the farrowing rate, wean-estrus interval, percent bred within 7-d after weaning, and percent repeats, indicated reproductive challenges at the herd that was subsequently assigned to the MINTREX® trace mineral farm. The MINTREX® trace mineral farm had a lower pre-weaning mortality (13.8 vs. 16.1) with the MINTREX® trace mineral farm having a lower pre-weaning mortality in all parities compared to the ITM farm.

Rate of genetic roll-over

Rate of sow genetic roll-over was another consideration that could have impacted rate of removal, and thus retention rate or longevity measures for Genetic Line A at the ITM and MINTREX® trace mineral farms. Rate of genetic line roll-over was evaluated by comparing the number of Genetic Line A females removed during the experimental period with the total number of Genetic Line A services during the same period for both mineral sources. There were no Genetic Line A additions during the experimental period with all new gilts entering the herd being from Genetic Line B. The results indicate that Genetic Line A removals accounted for 36.5% of Genetic Line A services at the ITM farm and 35.2% at the MINTREX® trace mineral farm and was considered equal between farms and thus no factor affecting retention to subsequent parity.

Evaluation period sow productivity results

During the MINTREX® trace mineral evaluation period there were 5,510 served-to-farrow events at the ITM farm and 5,114 served-to-farrow events at the MINTREX® trace mineral farm. Because of the genetic roll-over, the average parity farrowed during the experimental period for the Genetic Line A sows increased to 4.2 for the ITM farm and 4.1 for the MINTREX® trace mineral farm compared to 3.1 and 3.3 respectively during the historical period (Table 1). The parity distributions for the farms were similar during the evaluation period. During the MINTREX® trace mineral evaluation period the MINTREX® trace mineral farm had 6.2 more pigs per sow lifetime at 46 pigs per sow lifetime compared to 39.8 pigs per sow lifetime for the ITM farm. The primary driver of the higher pigs per sow lifetime was improved retention. The MINTREX® trace mineral farm had 6.5 percentage point more sows farrowing a 3rd litter, 6.7 percentage points more sows a 4th litter, and 11.8 and 13.7 percentage point more sows farrowed a 5th and 6th litter respectively (Figure 1) based on percentage of total culls within each parity. The increased retention for the MINTREX® trace mineral farm resulted in an average parity at removal of 4.4 compared to average parity at removal of 3.8 for the ITM farm. Compared to the historical data, during the MINTREX® trace mineral evaluation period the MINTREX® trace mineral farm had 15.7 more pigs per sow lifetime compared to a net increase for the ITM farm of 1.4 pigs per sow lifetime. The importance of sow productive lifetime on economic return has been recently reported by Gruhot et al.[1] Improvements in farrowing rate at the MINTREX® trace mineral farm were another significant contributor to the pigs per sow lifetime increase. Average farrowing rate at the herd level increased by 1.4 percentage points to 89.7% for the ITM farm but increased 7.4 percentage points to 86.8% for the MINTREX® trace mineral farm. Across parities, the large improvement in farrowing rate at the MINTREX® trace mineral farm was driven partially by the absence of Genetic Line A gilts during the evaluation period with improvements in farrowing rate also observed in parities 1 through 4. Total born and live born were increased similarly at the ITM and MINTREX® trace mineral farm on a herd average basis for the evaluation period compared to the historical period. The increase of 0.8 more total born at each farm and 0.7 to 0.8 more liveborn during the evaluation period may have been partially driven by a shift in the average parity farrowed from about 3 in the historical period to about 4 in the evaluation period. However, for parity 1 females, total born at the MINTREX® trace mineral farm increased 0.8 pigs per litter more than the parity 1 increase at the ITM farm for the evaluation period compared to the historical period. This 0.8 pig

Figure 1: Percent of bred gilts reaching each parity

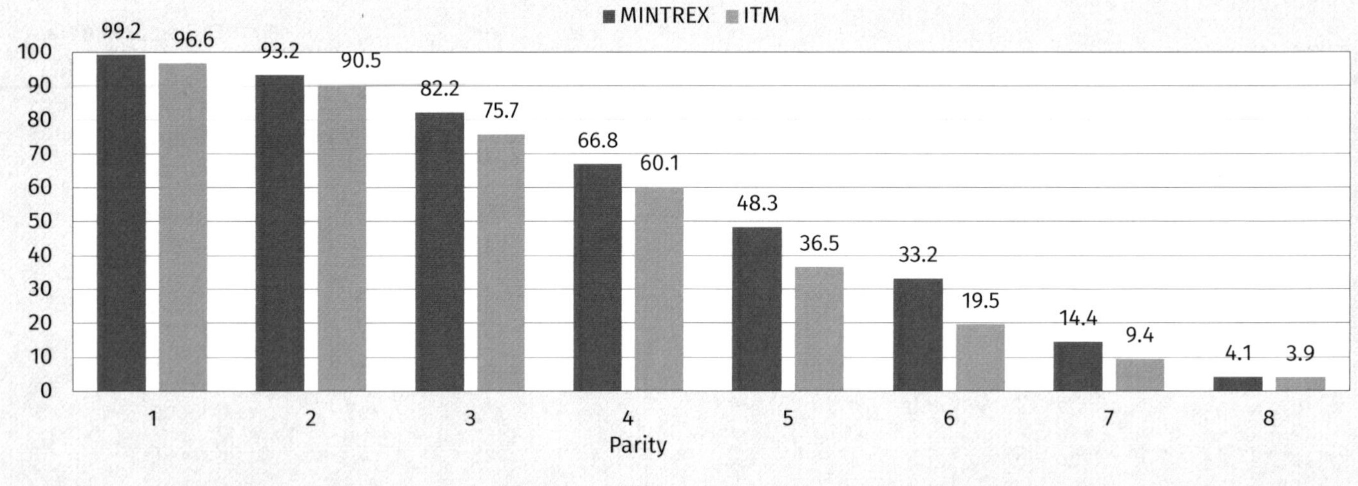

increase in parity 1 total born at the MINTREX® trace mineral farm is important considering the percentage of gilt litters in a typical farrowing group. Similar to total born, liveborn was also only different adjusted for historical differences in first litter females with an increase of 0.7 more liveborn for the ITM farm and 1.0 more liveborn for the MINTREX® trace mineral farm an increase of 0.3 more pigs for the MINTREX® trace mineral farm. At the herd level, pre-weaning mortality increased at the ITM farm from 16.1% to 17.1%, mostly driven by increased pre-weaning mortality in parities 2 and 3. In contrast, pre-weaning mortality at the MINTREX® trace mineral farm declined by 0.2% points from 13.8% to 13.6% for the evaluation period compared to the historical period. Pigs weaned per sow farrowed increased 0.6 pigs at the ITM farm and 0.5 pigs for the MINTREX® trace mineral farm for the evaluation period compared to the historical period. The lack of difference in total born or liveborn further highlights the impact of retention to subsequent parity on pigs per sow lifetime. However, the lower pre-weaning mortality that remained constant at the MINTREX® trace mineral farm while increasing at the ITM farm would have contributed to more pigs weaned per sow farrowed. These results demonstrate the importance of maintaining productivity and retention to subsequent parities as demonstrated for the MINTREX® trace mineral farm with 6.2 more pigs per sow lifetime than the ITM farm during the evaluation period.

Evaluation period wean-estrus and rebreeding efficiency results

As discussed previously, the MINTREX® trace mineral farm had a longer wean-estrus interval, lower percent bred within 7-d, and a greater percentage of repeats during the historical period (Table 1) both on a herd level and across parities. During the evaluation period, the MINTREX® trace mineral farm continued to have a lower farrowing rate (86.8% vs. 89.7%), longer wean-estrus interval (6.3-d vs. 5.3-d), lower percent bred within 7-d post-weaning (89.1 vs. 94.3) and a greater percentage of repeats (2.3% vs 1.4) compared to the ITM farm. Both farms improved during the evaluation period compared to the historical period with the degree of improvement being greater for the MINTREX® trace mineral farm. Corrected for historical differences the MINTREX® trace mineral farm had a net improvement in wean-estrus interval of 0.8 days with 5.9 percentage points more females being re-bred within 7-d of weaning. Corrected for historical differences parity 5 through 8 MINTREX® trace mineral farm sows had a 1-day reduction in wean-estrus interval compared to the ITM farm. Percent bred within 7-d was improved an average of 4.7 percentage points for the MINTREX® trace mineral farm during parties 5 through 8 compared to the ITM farm corrected for historical differences. These results suggest a role of trace mineral status or supply on reproductive efficiency of older parity sows. Further evidence of improved rebreeding efficiency was an improvement of 2.2 percentage points in percent repeats for the MINTREX® trace mineral farm versus the ITM farm when the evaluation period for both were compared to the historical period. The improvement in percent repeats for the MINTREX® trace mineral farm corrected for historical differences for the ITM farm was seen across parities with the improvement ranging from 3.0 percentage points in parity 3 to 6.4 percentage points in parity 6.

Although a field evaluation rather than a controlled study, and a comparison of evaluation period versus historical data, the observations here are in agreement with those of Barea et al.[2] and Zhao et al.[3] These results indicate the importance of increasing sow retention to older parities and its impact on sow lifetime productivity in determining the economic return in the sow herd as demonstrated by Gruhot et al.[1] Evaluating production records for the parity distribution at removal is important for evaluating the impact of retention on lifetime productivity and giving further insight into the reported pigs per sow lifetime. Table 1 provides a summary of the net change difference in performance indicators for Genetic Line A at the MINTREX® trace mineral farm for the evaluation period compared to the historical performance and the evaluation period compared to historical difference for the ITM farm. During the evaluation period the MINTREX® trace mineral farm weaned 6.2 more pigs per sow lifetime in Genetic Line A due to an increase in sows reaching parties 3 through 6. The increased retention resulted in a sows reaching parity 4.4 for the MINTREX® trace mineral farm and 3.8 for the ITM farm while maintaining a similar replacement rate. Other factors contributing to improved pigs per sow lifetime for the MINTREX® trace mineral farm were as follows: greater parity 1 total born, greater improvements in farrowing rate, reductions in pre-weaning mortality, and improvements in rebreeding efficiency including reduced wean-estrus interval and percent repeats. These results indicate that although total trace mineral levels were constant, mineral supplied as the methionine hydroxy-analogue chelate MINTREX® may improve sow lifetime productivity.

References

1. Gruhot et al., An economic analysis of retention in a United States breed-to-wean system. *J. Swine Health Prod.* 2017; 25.

2. Barea et al., Utilisation d'oligo-éléments organiques dans l'alimentation des truies: effets sur la longévité et les performances des portées. *Journées Recherche Porcine,* 2019; 51:167

3. Zhao et al., Chelated trace minerals support sow reproduction. 2012; *Feedstuffs.* June 4

4. Jang et al., Effects of mineral methionine hydroxy analogue chelate in sow diets on epigenetic modification and growth of progeny. 2020; *J. Anim. Sci 98:1*

Equisul-SDT® dose variance study in nursery pigs

N. Winkelman[1], DVM; A. Mueller[1], DVM; R. Winscher[1], CVT; G. Weaver[2], DVM
[1]Swine Services Unlimited Inc., Rice, Minnesota; [2]Aurora Pharmaceutical, Inc., Northfield, Minnesota

Overview

Potentiated sulfa (SMZ-TMP) was first made available for oral use in humans in the US in 1974. Sulfa products by themselves were discovered and made available to humans in the 1930s with sulfadiazine and sulfamethoxazole being first used in 1941. Since then, both sulfas by themselves and potentiated sulfas, have seen widespread use in both animals and humans. Europe has a commercially labeled potentiated sulfa for use in animals. Any animal use of potentiated sulfas in the US is considered extra-label and can only legally be done under AMDUCA provisions. The purpose of this study was to determine the concentration of sulfadiazine/trimethoprim (SDZ-TMP) seen in plasma when pigs are given varying doses of the drug orally via the drinking water.

The purpose of the study was threefold:

1. Measure plasma levels achieved in pigs given 3 different dosages of Equisul-SDT orally via the drinking water;
2. Provide information that shows the dosage necessary to achieve plasma levels equal to or above the MIC for different swine pathogens;
3. Monitor the time period post-treatment at which the drug was no longer present in plasma.

Trial design

Thirty mixed breed, commercial animals, weighing 28-36 lbs. (mean B.W. 31.5 lbs.) and sourced from a PRRS and *Mycoplasma hyopneumoniae* negative herd, were enrolled in the study and divided into three groups with each group receiving a different dose of Equisul-SDT orally via the drinking water (Table 1). The SSUI research facility used was a total mechanical ventilated, confined nursery typical of what is seen in the swine production industry.

The animals were randomly allocated into three treatment groups. One group (2× treatment) received the equivalent dose of 1 bottle (900 mL) into 3 gallons of stock solution with the water line medicator set at a 1:128 dilution rate, one group (1× treatment) received 1 bottle (900 mL) into 6 gallons of stock solution, and one group (0.5× treatment) received 1 bottle (900 mL) into 12 gallons of stock solution. The 1× dose is most commonly used by veterinarians to treat swine pathogens in the field.

Individual blood samples were collected at various time points and the plasma was separated and analyzed at an independent laboratory for the amount of both sulfadiazine and trimethoprim that the samples contained.

Fresh medicated Equisul-SDT water was mixed daily for a seven-day treatment period. Individual PVC tube waterers with a nipple cup assembly were used to mitigate water waste and calculate pen daily water consumption and average drug dose per pig per day.

Trial results

- In Figure(s) 1-5, the various treatments are labeled with the respective treatments being 2×, 1×, and 0.5×.
- Plasma levels of both antibiotics did increase in step with increasing oral dose and is shown in Figures 1 and 2.
- The average plasma TMP level was 70 ng/mL for the 0.5× group, 130 ng/mL for the 1× group, and 350 ng/mL for the 2× group. Average SDZ levels were 1090 ng/mL for the 0.5× group, 1948 ng/mL for the 1x group and 4692 ng/mL for the 2× group (Figures 3 and 4).
- Plasma levels were found to about double in the 1× group compared to 0.5× and somewhat more than double in the 2× group compared to 1× for both antibiotics.
- There was a drop in both plasma SDZ and TMP on Day 3 and Day 7 in treatment 2×. This is difficult to explain since there wasn't a simultaneous drop in water consumption.

Trial conclusions and discussion

- On average, the maximum plasma levels of both sulfadiazine and trimethoprim were achieved within 24 hours of the initiation of treatment at the first time point which samples were collected. The drug is readily consumed, and bioavailability is excellent leading to steady state levels being quickly achieved in the blood, which leads to a rapid onset of action for the animals being treated.

Table 1: Trial design

Treatment	Gallons of stock solution per 900 mL EQUISUL SDT®	Target concentration (mg/#)	Pen #s	# of pigs/pen
2×	3	11.5	1 & 4	5
1×	6	5.6	2 & 5	5
.5×	12	3.0	3 & 6	5
Blood samples from each animal were collected into plasma EDTA tubes on Days 0, 1, 3, 5, 7, 10 14 and 21 and submitted to KCAS Laboratory for HPLC testing.		Total	6	30

- The levels found in the blood were what would be expected for each dosage. Sulfadiazine levels averaged 1090 ng/mL for the 0.5× treatment group, 1948 ng/mL for the 1× treatment group, and 4692 ng/mL for the 2× treatment group over the seven-day treatment period. Blood levels were approximately double for the 1× treatment group compared to the 0.5 group and more than double in the 2× treatment group compared to the 1× group. Trimethoprim levels were very similar to what was seen with sulfadiazine.
- Average plasma levels found with varying doses in the study help to explain the clinical response seen in the field. The 0.5× treatment level didn't achieve a level close to MIC for respiratory organisms. The 1× treatment level achieved a level that was very close to the expected MIC for respiratory pathogens (2 ug/mL) and at some time points was above it. The 2× treatment level achieved an average level that would target more resistant organisms along with providing a full 5-7 days above the expected MIC of most respiratory pathogens.
- The clinician may need to provide a slightly higher dose than 1 bottle (900 mL) per 6 gallons of stock or prescribe the 2× dose for 2-3 days followed by a maintenance dose for 3-4 more days to achieve the best clinical results.
- Palatability of the product is good, as water consumption was on par with body weights and was mostly consistent throughout the trial. Daily water consumption increased gradually during the 7 days of treatment as would be expected.
- The product enters the blood stream quickly and also is eliminated quickly. All samples were below detectable limits (20 ng/mL) within 3 days of removal of the drug from the drinking water (not shown). This was the first time period after removal of treatment at which testing was done. It is probable that many of the samples were negative prior to that.
- Residue concerns are extremely minimal especially if the FARAD recommended withdrawal time of 30 days prior to slaughter is followed.

Figure 1: SDZ average per pig by treatment.

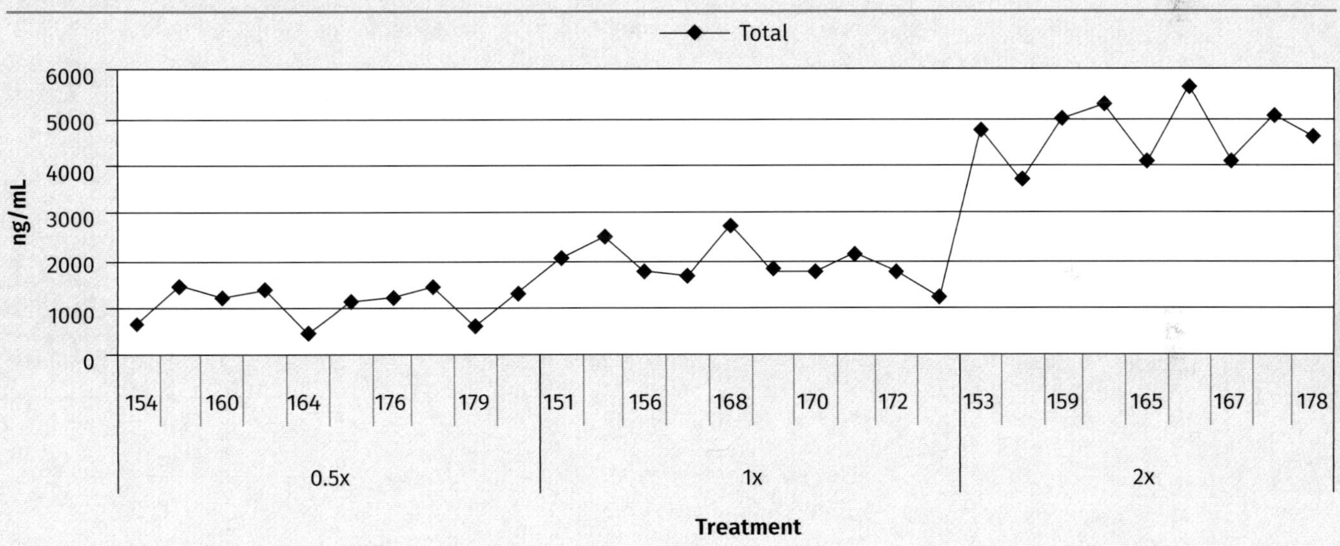

Figure 2: TMP average per pig by treatment.

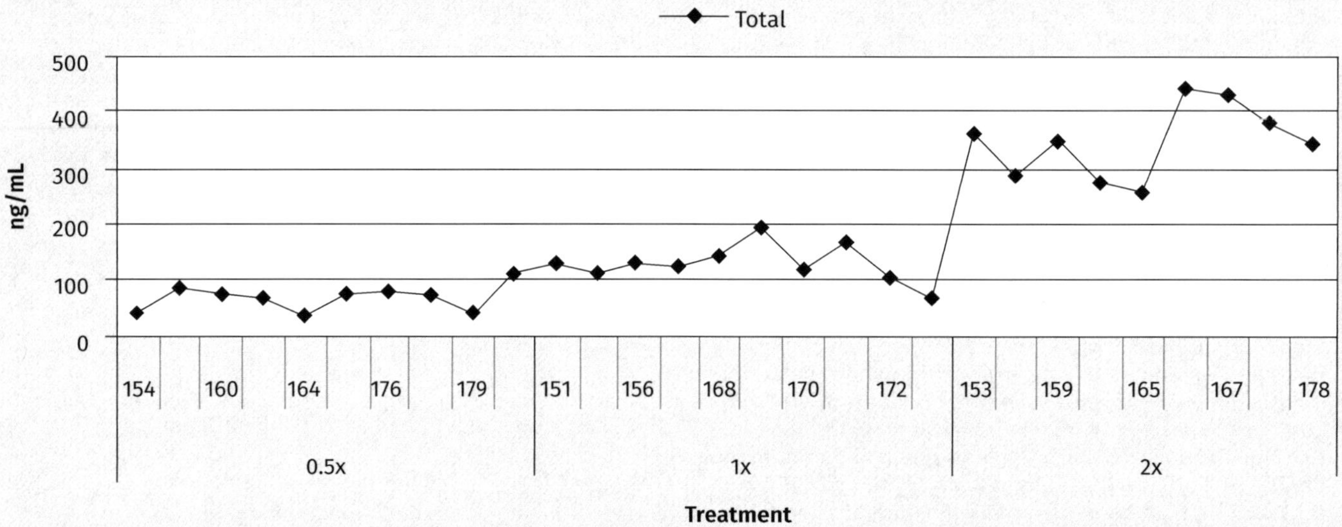

Figure 3: SDZ by treatment and day.

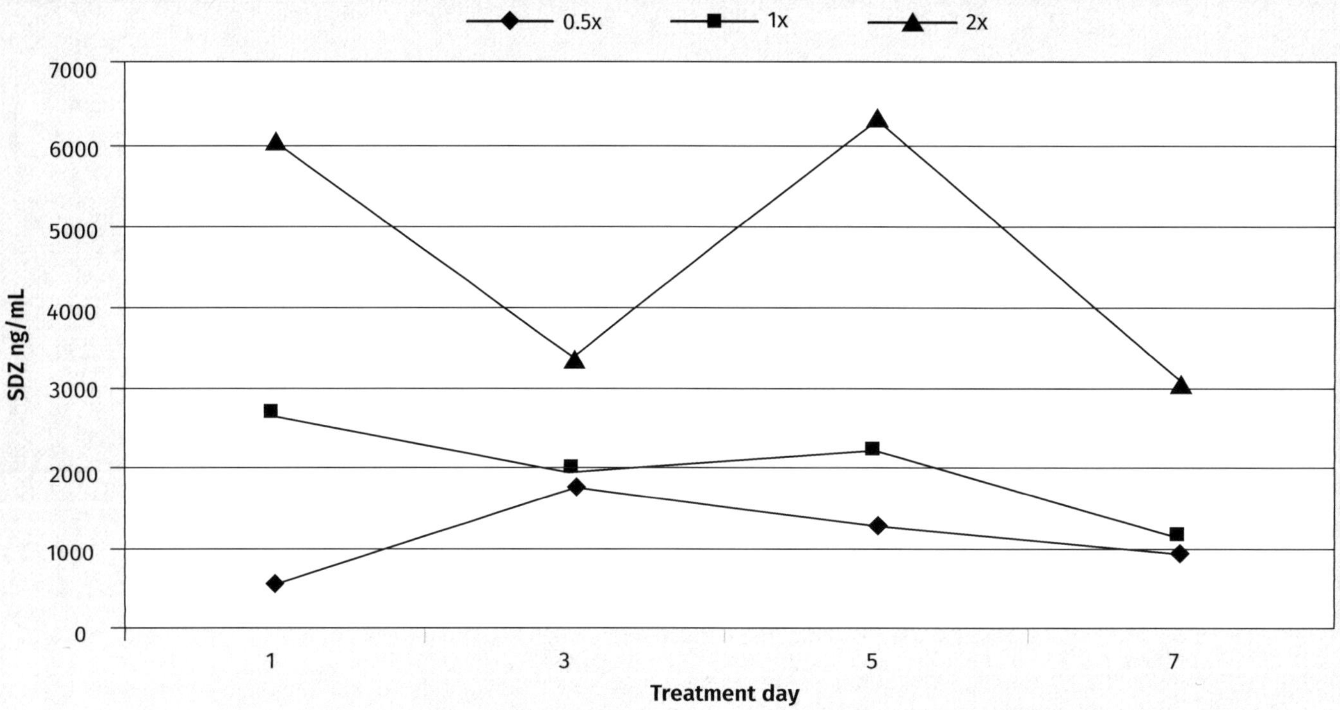

Figure 4: TMP by treatment and day.

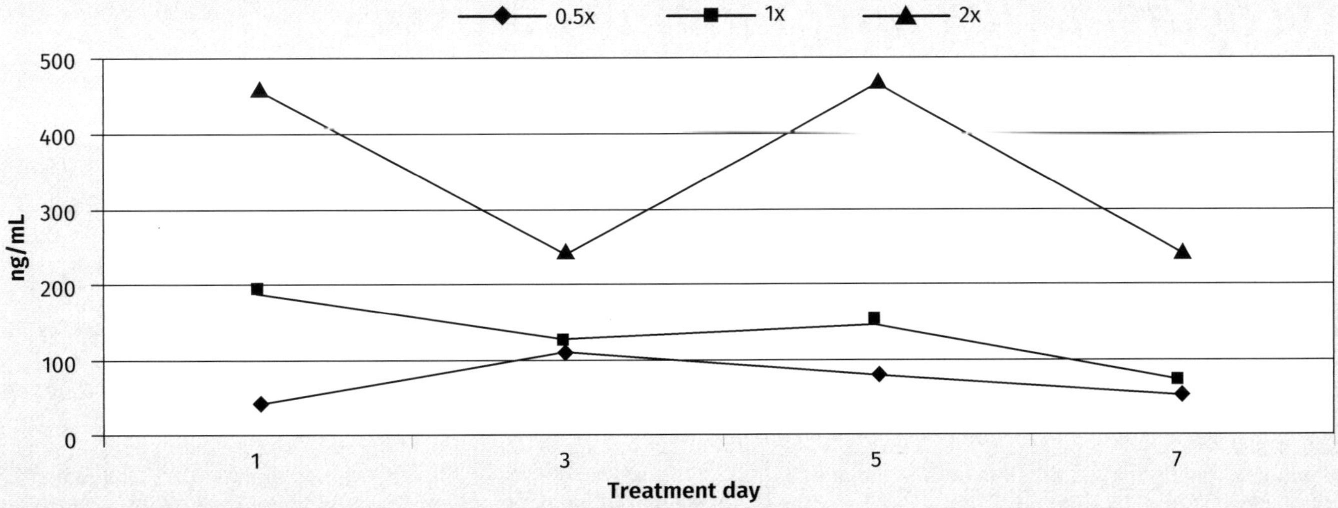

Figure 5: Water consumption per pig by treatment.

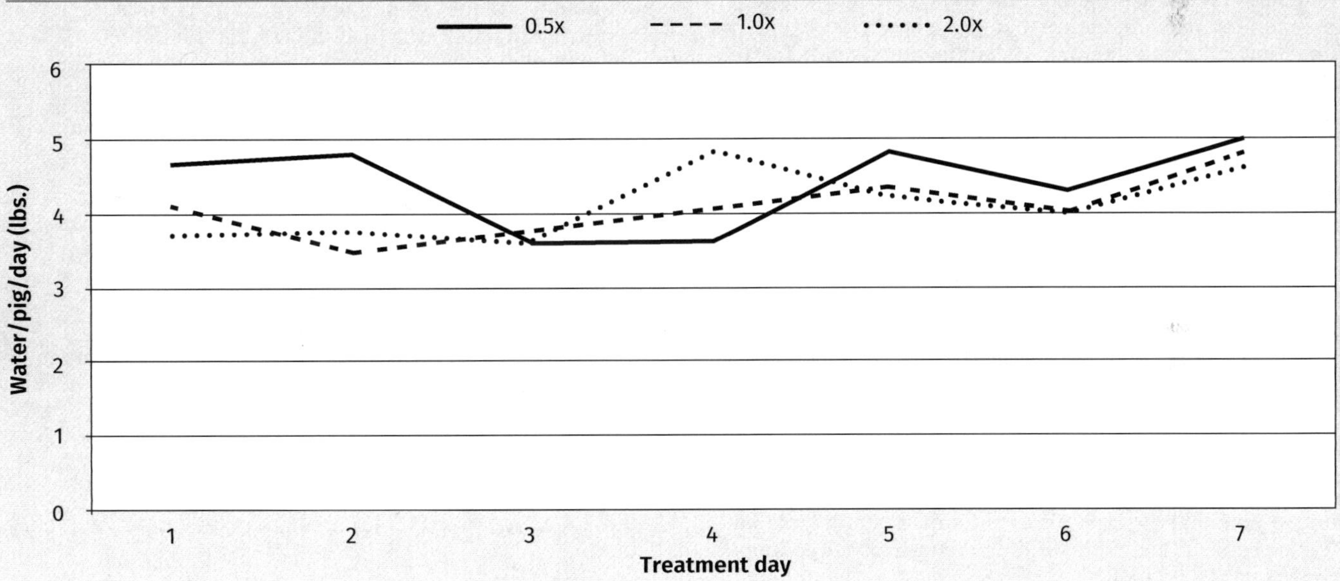

Tiamulin (Vetmulin®): The proven standard in *Mycoplasma hyopneumoniae* control

Lieven Claerhout[1]; Thomas Fangman[2]; David Nolan[2]
[1]Huvepharma® NV Belgium; [2]Huvepharma® Inc. Peachtree City, Georgia

Introduction

A farm-specific *Mycoplasma hyopneumoniae* (M hyo) control program was implemented in a 700-sow breed to wean site in Ireland without depopulating the site. Weaned pigs (27 days old) were moved weekly to a nursery site and at 11 weeks of age moved to all-in all-out finishing barns (all on different sites). This production system was confirmed positive for M hyo based on clinical signs and positive diagnosis of *Mycoplasma hyopneumoniae* via laboratory testing (ELISA serology, PCR on oral fluids and nasal swabs). Samples were obtained from sows, gilts and finishing pigs.

Materials and methods

A control program based on partial depopulation and strategic medication was initiated. The distance to neighboring herds was more than 2 miles. The program was initiated in the summertime. Several actions were taken to improve internal and external biosecurity. Stringent cleaning and disinfection protocols were established. All pigs younger than 10 months of age were removed from the sow herd in order to reduce active shedding and transmission of M hyo. Only animals older than 10 months and suckling piglets remained on the breeding site. Sick animals were removed from the farm or humanely euthanized before the start of the program. The breeding herd was closed to the introduction of gilts and boars five weeks before the start of the medication period. After removing animals under 10 months of age, the remaining sows and gilts were treated via feed with tiamulin 100g/kg microgranulated premix (Vetmulin®,Huvepharma®) at a daily dose of 8-10mg tiamulin hydrogen fumarate/kg bodyweight for two consecutive weeks. Beginning one week prior to the sow treatment protocol and through the sow treatment period (3 weeks total) the suckling piglets received a dose of tulathromycin (2.5mg/kg bodyweight, intramuscularly). Suckling pigs and piglets born during this 3-week period were treated at 5-day intervals with tulathromycin. Piglets originating from sows that were treated for 2 weeks prior to farrowing were considered to be free from Mycoplasma. Mycoplasma monitoring was carried out in sows, nursery and finishing pigs at 14, 23, 30 and 59 weeks after the strategic medication period to evaluate the control program success. All serology tests (ELISA, n = 195) and antigen tests on oral fluids and nasal swabs (PCR, n = 30) were negative for M hyo over a time period of 59 weeks. Lungs at slaughter check did not show typical mycoplasma lesions.

Results

Following this control program clinical signs of *Mycoplasma hyopneumoniae* were not observed. There was an overall improvement of the general health status and performance of suckling, weaned and finishing pigs born after the control. In addition, the administration of Vetmulin® to the breeding herd also supported the control of concurrent infections like ileitis, mycoplasma arthritis and pneumonia caused by other bacteria in the breeding herd. A performance boost was noted immediately after the control program (Table 1, 2 & 3). Improvements were noted in all parameters measured in suckling pigs, nursery and finishing pigs. With the most notable improvement of 12.5 more pounds liveweight in finishing pigs.

Conclusions and discussion

In this multi-site production system, M hyo was successfully controlled by a strategic medication program with Vetmulin® in combination with extensive biosecurity and hygiene measures. This program does not require a total depopulation of the breeding herd or an interruption in farrowing. After treatment protocol, only gilts certified free from enzootic pneumonia were introduced and biosecurity measures were continued. Furthermore, the repeated use of antimicrobials and preventive vaccination against M hyo was discontinued. Continued negative M hyo monitoring results and the absence of clinical signs would suggest that this control program resulted in the elimination of M hyo in this production system.

Note: EU will allow higher feed inclusion levels to obtain the 8-10mg tiamulin hydrogen fumarate/kg body weight. In the USA administering Vetmulin® in the water can be utilized to obtain this dosing recommendation.

Table 1: Suckling pig mortality and weaning weight 16 weeks prior to and 16 weeks after treatment

	16 wk before	16 wk after	Difference
Pre-weaning mortality (%)	13.6	12.3	-1.3
Weaning weight (lb)	13.66	15.97	2.31

Table 2: Nursery pig (*11 WOA*) weight gain, exit weight and mortality 16 weeks prior to and 16 weeks after treatment

	16 wk Before	16 wk After	Difference
	9589 pigs	8171 pigs	1418 pigs
Daily weight gain (lb/day)	0.88	1.05	0.17
Exit weight (lb)	60.52	70.55	10.03
Mortality (%)	3.98	2.22	-1.76

Table 3: Finishing pig weight gain, FCR, days to slaughter, final live weight and carcass weight comparing the last finishing batch before and the first batch after the treatment (n = 650)

	Before	After	Difference
Daily weight gain (lb/day)	1.79	2.06	0.27
Feed conversion rate	2.73	2.54	-0.19
Days to slaughter	120	113	-7.0
Liveweight at slaughter (lb)	267.96	280.5	12.54
Carcass weight (lb)	207.11	213.49	6.38

The *in vivo* and *in vitro* effectiveness of tylvalosin (Aivlosin® Water Soluble Granules 62.5% w/w tylvalosin) at 50 ppm in water for the control of *Mycoplasma hyopneumoniae* following experimental challenge and under field conditions

C. Eliopoulos[1], DVM; E. M. Abbott[2], BVMS, PhD; P. Arruda[3], DVM, MS, PhD; R. Saltzman[3], DVM; K. Rooney[4], MS, DVM; R. Domangue[5], PhD; C. Smith[6], DVM

[1]ECO LLC, Princeton, New Jersey; [2]ECO Animal Health, London, United Kingdon; [3]VRI, Ames, Iowa; [4]The Veterinary Consultancy, LLC, Woodcliff Lake, New Jersey; [5]Statistical Consulting Services, Broadway, Virginia; [6]Pharmgate Animal Health, Wilmington, North Carolina

Introduction

Respiratory disease remains one of the most significant health challenges facing the swine industry and is a leading cause of mortality.[1] Throughout nursery and grow/finish operations, a combination of infectious agents and environmental stressors often leads to swine respiratory disease (SRD). In Guidance for Industry 178,[2] FDA defines SRD as the component of acute respiratory disease in swine associated with bacterial pathogens such as *Actinobacillus pleuropneumoniae, Bordetella bronchiseptica, Haemophilus parasuis, Pasteurella multocida,* or *Streptococcus suis* and is not considered synonymous with "Enzootic Pneumonia," which is a common term used to describe *Mycoplasma hyopneumoniae* co-infection with other bacterial respiratory pathogens, "Porcine Respiratory Disease Complex" (PRDC), which is mixed viral and bacterial infection of two or more respiratory pathogens, or any other complex or syndrome.

Mycoplasma hyopneumoniae (Mhp) is frequently present in affected swine and may potentiate the severity and duration of respiratory infections. Mhp predisposes the pig to secondary bacterial invaders by suppressing and modulating the innate and adaptive immune response of the host.[3] When the immune response is compromised, these bacteria often become secondary pathogens and multiply resulting in more severe clinical signs and lesions [3]

To obtain United States Food and Drug Administration (FDA) approval sponsor companies are strongly encouraged to follow *Guidance for Industry 178*. Section 2 of the guidance describes study design recommendations and criteria CVM intends to use to determine the effectiveness of a new animal drug intended for treatment or control of SRD associated with *Mycoplasma hyopneumoniae* and other bacterial pathogens. These criteria include a multisite field study that should demonstrate clinical effectiveness in pigs with mixed bacterial respiratory infections (see above). At least 30 bacterial isolates from a minimum of 30 pigs across all sites are required for a pathogen to be considered for the label claim. Additionally, to specifically add Mhp to the SRD label claim, the lungs of at least 5-10% of the field study pigs must test positive by PCR for Mhp. A separate induced Mhp infection model study mimicking natural infection must show a clinical effect such as decreased lung lesions in the treated animals. It is also necessary to provide data to support that the inoculating Mhp strain used to develop and validate the challenge model is representative of current North American field strains.

A pivotal, multisite masked, randomized field study comprising 980 pigs from commercial farms with a history of naturally acquired SRD was conducted to demonstrate the effectiveness of tylvalosin (as Aivlosin® Soluble Granules) for the control of SRD when administered at an inclusion rate of 50ppm tylvalosin in drinking water for five consecutive days. Prior investigations at the source farms had confirmed that the desired bacterial respiratory pathogens including *M hyopneumoniae* were present. The studies were run at three locations in the USA and one in Canada. This study was previously summarized in the Freedom of Information (FOI) Summary for the original approval of NADA 141 336 (19 July 2017).[4]

Lung samples from pigs in this study were screened with a quantitative polymerase chain reaction test (qPCR) for the presence of *M hyopneumoniae*. Of 976 samples tested, 185 (19%) were PCR positive for *M hyopneumoniae*. Nine isolates were successfully grown in culture.

To complete the requirements for the addition of Mhp to the current SRD label, a final *in vivo* Mhp challenge study using the validated challenge model was conducted to demonstrate the specific efficacy of tylvalosin against Mhp alone, in the absence of other respiratory bacterial pathogens. As stated above, the efficacy of Aivlosin® Water Soluble Granules against other respiratory bacterial pathogens had already been proven and accepted. Additionally, an *in vitro* MIC susceptibility study was completed comparing the MIC of the ECO Mhp strain used in the induced infection model challenge study against 10 other recent North American field strains.

Aivlosin® is a Registered trademark of ECO Animal Health, Ltd., London, UK.

In vivo challenge study materials and methods

Lung samples from naturally infected pigs displaying clinical signs of mycoplasmal pneumonia (i.e., dry cough) with confirmed exposure (Mhp ELISA+) were obtained from three separate commercial farms. The Mhp strain (DLP4 strain) used in the challenge model development and validation studies originated from one of these farms. It was selected based on the gross appearance of the lung lesions together with a confirmatory Mhp PCR+. Lung tissue from the source pig was homogenized with Friis medium and used to challenge pigs in two developmental induced infection studies. These studies confirmed DLP4 was representative of current North American field strains based on the lung lesions seen in the challenged pigs and that the model was reproducible.

The confirmatory, pivotal challenge study with DLP4 was then initiated. 240 healthy crossbred pigs sourced from a high health vendor were enrolled. Pigs were assigned to weight blocks and pigs within each block were randomly assigned to two treatment (TX) pens (T1 = 50 ppm, T2 = 0 ppm). Every pair of blocks contained one pen of nontreated, challenged pigs (NTX). The NTX pigs were used to determine when the treatment period could begin based upon their lung lesion scores post challenge. Each TX and NTX pen contained six pigs resulting in 16 weight blocks and 8 NTX pens. Pigs were housed in a commercial farm environment at the CRO and had *ad libitum* access to an age appropriate non-medicated, commercial ration from arrival to the end of the study. Strict biosecurity practices were followed throughout the study.

The general health of the pigs, including deaths, was recorded twice daily and any health abnormalities were evaluated by the Clinical Investigator. Scores for coughing, depression, and respiration were recorded once daily. Each morning, the Clinical Investigator or appropriately trained designee stimulated the pigs in each pen to rise and move and counted the number of pigs coughing in each pen over a 2-minute time period. The number of different pigs coughing per pen was divided by the total number of pigs in that pen and multiplied by 100. Any pig that coughed more than once during the 2-minute period was only counted once. Depression and respiration scores were scored as normal, mild, moderate, or severe and any pigs showing clinical signs of secondary bacterial or viral respiratory infection would have been removed from the study.

Nasal swab samples and blood samples were collected from each pig seven days before the challenge procedure (Day -7) and tested by Mhp specific PCR and serum ELISA, respectively. The results of these tests were all negative, proving these pigs had not been previously exposed to Mhp.

Pigs were challenged endotracheally on Day 0 and Day 1 with 10 mL lung homogenate containing DLP4 strain. The Mhp in the challenge material was subsequently quantified to approximately 10^6 CFU Mhp/mL.

Medication began on Study Day 5 (Day of Medication or DM) based upon the lung lesion scores of eight NTX pigs necropsied that day. Disease following the Mhp challenge had progressed sufficiently in these pigs and met the protocol requirements to start treatment. All remaining NTX pigs were removed from the study and 50 ppm TVN and 0 ppm TVN non-medicated control water was dispensed to T1 and T2 pens, respectively, for five consecutive 24-hour periods. After that time, all pigs received non-medicated tap water for a further five days when all were euthanized and necropsied. Gross pneumonic lung lesions were assessed, scored and recorded. Bronchoalveolar lavage (BAL) samples were collected and tested by validated qPCR to determine if Mhp was present in the lung.

Lung lesions consistent with Mhp infection (i.e., pink to dark red, plum or purple areas of consolidation in the cranio-ventral part of the lung and/or cranial part of the diaphragmatic lobe, a.k.a., right and left caudal lobes) were used to assess the areas of each affected lung lobe. The proportional mass of each lobe relative to the total lung mass was calculated as follows: right cranial lobe = 10% of the lung, left cranial lobe = 10% of the lung, left middle lobe = 10% of the lung, right middle lobe = 10% of the lung, accessory lobe = 10% of the lung, right caudal lobe = 25% of the lung, left caudal lobe = 25% of the lung.

Pen feed consumption was measured from Day 0 through the end of the study on DM+10. Individual body weights of all study animals were measured on arrival (Day -8), DM, DM+5 and DM+10.

In vivo challenge study results and discussion

Analysis of water samples collected during the 5-day treatment phase indicated that the overall average inclusion rate in the medicated group was 47.1 parts per million (ppm) tylvalosin (TVN). Tylvalosin was not detected in any of the non-medicated group water samples.

Coughing began three days after the challenge procedure. Bronchoalveolar lavage fluid (BALF) collected at necropsy on DM+10 was positive by PCR for Mhp for all pigs in the 0 ppm and 50 ppm groups. The average coughing score was 5.4% for the 50 ppm TVN group and 16.4% for the 0 ppm TVN group ($P < 0.0001$).Secondary respiratory infection (SRI) did not confound the results, as none of the pigs showed SRI clinical signs antemortem and only four of the 192 TX pigs (~2%) had gross lung lesions indicative of SRI post-mortem. The lung lesion scores of these four pigs were removed from the statistical analysis. Lung lesion score at DM+10 was significantly reduced in the TVN group compared with the control group. Mean lung lesion scores were 5.1% for the 50 ppm TVN group and 10.9% for the 0 ppm TVN group ($P < 0.0001$). These results are presented in Table 1. There were no deaths in either treatment group.

Average daily gain was 0.66 and 0.57 kg/ day for the 50 ppm TVN and the 0 ppm TVN groups, respectively ($P < 0.0001$). Feed consumption, feed efficiency and feed conversion ratio in the period DM to DM+10 were all significantly improved in the treated pigs. Average feed consumption was 1.01 kg/ day for the 50 ppm TVN group and 0.95 kg/ day for the 0ppm group ($P = 0.0048$). Average feed conversion was 1.55 kg feed intake/ kg BW gain for the 50 ppm TVN group and 1.67 kg feed intake/ kg BW gain for the 0 ppm TVN group ($P < 0.0001$).

Adverse events after the start of treatment were reported for five pigs. SRI lung lesions at DM+10 necropsies were reported in one pig in the 50 ppm TVN group and three pigs in the 0 ppm TVN group. Lameness was also reported in one pig in the 0 ppm TVN group. The masked Clinical Investigator concluded these adverse events were not related to investigational veterinary product.

Significant improvement was seen in the medicated pigs compared to the non-medicated control pigs across all study variables. The results of this study provide definitive evidence of the effectiveness of Aivlosin® Water Soluble Granules administered in drinking water at 50 ppm TVN for five consecutive days to control

Table 1: Pen mean lung lesions scores at necropsy and average coughing scores

	Tylvalosin treatment (nominal ppm in water)	
Study variable	**0**	**50**
Pen mean lung lesion score[a]	10.9	5.1 ($P < 0.0001$)
Pen average coughing score[b] (%) in DM+1 to DM+10	16.4	5.4 ($P < 0.0001$)

[a] All 192 TX pigs enrolled in the study survived until the end of study on DM+10, but four pigs (three 0 ppm TVN and one 50 ppm TVN) were removed from the lesion score analysis because gross lesions consistent with secondary respiratory infection were present at necropsy.

[b] Pen coughing score represents the mean of the daily percentages of pigs in a pen coughing over a two-minute period during DM+1 to DM+10.

swine respiratory disease (SRD) associated with *Mycoplasma hyopneumoniae* infection. This is in addition to the label already granted for Aivlosin® Water Soluble Granules for the control of SRD associated with *Bordetella bronchiseptica, Haemophilus parasuis, Pasteurella multocida,* and *Streptococcus suis* in groups of swine in buildings experiencing an outbreak of SRD.

In vitro susceptibility study materials and methods

The following Mhp strains were evaluated against six antibiotics including tylvalosin: the ATCC reference strain 25934, ECO DLP4 strain, the well characterized ISU -232 strain and ten (10) recent strains (2017-2019) obtained from lung tissues of naturally occurring porcine pleuropneumonia clinical field cases. The antibiotic plates used were custom made by Thermo Scientific Sensititer™.

Current guidelines and recommendations for testing veterinary mycoplasma species were followed.[5] All Mhp strains were propagated and grown in Friis broth until a color change from red to yellow/orange was noted. Each was estimated to contain 10^7 color changing units (CCU) / milliliter (mL). One hundred microliters (100 µl) of each grown culture was then diluted in 9.9 mL of fresh Friis broth to provide a Mhp culture with a concentration of 10^5 CCU/mL Fifty microliters (50 µL) of this diluted culture was added to each well of the minimum inhibitory concentration (MIC) plate. The plates were sealed with a plastic film and incubated at 37°C for 7-14 days depending on the strain under test. The lowest concentration of antibiotic inhibiting color change from red to yellow/orange of the medium was considered the MIC of the respective antibiotic.

In vitro susceptibility study results and discussion

The MIC distribution, MIC_{50} and MIC_{90} values for the six antimicrobials against 13 Mhp strains are presented in Table 2. The MIC of the quality control reference ATCC strain 25934 for tylvalosin tartrate was within the expected range (i.e., ≤ 0.015 – 0.06 µg/mL) required in the study protocol.

Tylvalosin tartrate exhibited the highest inhibitory activity on all the Mhp strains tested with MIC values ranging from ≤ 0.03 µg/mL - 0.06 µg/mL and MIC 90 value of 0.06 µg/ml.

Acknowledgements

ECO acknowledges the contribution of Microbial Research, Inc. and Central States Research Centre, Inc.

References

1. National Animal Health Monitoring Service (NAHMS) *Swine 2012 Part III: Changes in the US Swine Industry, 1995–2012*. p. 40, 45

2. *Guidance for Industry178: Recommended Design and Evaluation of Effectiveness Studies for Swine Respiratory Disease Claims*

3. *Diseases of Swine,* Eleventh Edition (Electronic). Edited by Zimmerman, et al. 2019. pp. 863-870

4. NADA 141 336 FDA Freedom of Information Summary

5. Hannan, P.C. Guidelines and recommendations for antimicrobial minimum inhibitory concentration (MIC) testing against veterinary mycoplasma species. *Veterinary Research*. 2000. 31:373-395

Table 2: Summary of *in vitro* susceptibility study results: MIC distribution, MIC_{50} and MIC_{90} values for six antimicrobials against 13 Mhp strains including 10 collected between 2017 and 2019

Antibiotic	MIC (µg/mL)														MIC_{50} (µg/mL)	MIC_{90} (µg/mL)
	≤ 0.03	0.06	0.12	0.25	0.5	1	2	4	8	16	32	64	128	256		
Tylvalosin tartrate	6	7													0.06	0.06
Tiamulin		1	8	4											0.12	0.25
Tilmicosin			1				5	1	3		3				4	32
Lincomycin			1	2	7	3									0.5	1
Chlortetracycline				1		1			8	3					8	16
Tylosin tartrate	1	1			8	3									0.5	1

Development and validation of a reverse transcription real-time PCR assay for detection of PRRSGard®-like vaccine virus

Gaurav Rawal[1], DVM, MS; Wannarat Yim-im[1], MS; Chad Smith[2], DVM; Fabian Chamba[2], DVM, MS, PhD; Charles Francisco[2], DVM, MS; Jeff Okones[2], DVM; Jianqiang Zhang[1], MD, MS, PhD
[1]Iowa State University, College of Veterinary Medicine, Ames, Iowa
[2]Pharmgate Animal Health, Wilmington, North Carolina

Introduction

Porcine reproductive and respiratory syndrome (PRRS) is one of the most economically important diseases costing over $600 million annually to the US swine industry.[1] There are many tools to prevent and control PRRS at the herd level though vaccination is one of the most prominent and efficient strategies to control PRRSV. Reverse transcription real-time PCR (RT-rtPCR) is commonly used for the detection of PRRSV RNA. However, with an increasing use of modified live virus (MLV) vaccines and doubling of PRRSV cases by PCR in last ten years, there comes challenges to interpret PCR-positive results. ORF5 sequencing or whole-genome sequencing can be conducted to distinguish vaccine-like from wild-type PRRSVs although sequencing is relatively expensive and time-consuming. Vaccine-specific PCR for detection of PRRSV vaccine strains is a convenient tool to inform vaccine protocols and differentiate strains. PRRSGard® is a chimeric PRRSV marker MLV vaccine recently developed by Pharmgate Animal Health whose safety and efficacy have been evaluated in both experimental and field trials.[2-5] Here we report the development and validation of a PCR for specific detection of PRRSGard-like vaccine virus.

Materials and methods

PRRSGard® vaccine virus has a unique 23-nucleotide insertion between ORF1b and ORF2 as a genetic marker. Primers and probe spanning this region were designed. Analytical specificity, sensitivity and diagnostic performance of PRRSGard-like PCR were evaluated and compared with a commercial PCR (VetMAX NA&EU PRRSV PCR, Thermo Fisher Scientific) that targets the ORF6 and ORF7 genomic regions. The ORF1b-ORF2 region covering the PRRSGard-like PCR primers and probe, and full-length ORF5 gene were sequenced on some isolates and clinical samples to confirm the PCR results.

Results and discussion

PRRSGard-like and commercial PCR did not cross-react with any of the 27 non-PRRSV swine pathogens. The Ingelvac PRRS MLV, Ingelvac PRRS ATP, Fostera PRRS, Prime Pac PRRS, and Prevacent PRRS vaccine viruses, and 33 PRRSV-2 laboratory and field isolates representing different genetic lineages, tested negative by PRRSGard-like PCR and positive by commercial PCR. Sequencing of the ORF1b/2 target region revealed that none of these PRRSV vaccine viruses (except PRRSGard vaccine) and the 33 laboratory and field PRRSV isolates contained the unique genetic marker of PRRSGard® vaccine virus. This confirmed the accuracy of the PRRSGard-like PCR results. The undiluted PRRSGard® vaccine virus was positive by PRRSGard-like and commercial PCR with CT values of ~18 and ~17, respectively. To determine the analytical sensitivity, 10-fold serial dilutions of PRRSGard® vaccine virus (10^-1 to 10^-8 dilutions, triplicate per dilution) were tested by the two RT-rtPCRs simultaneously. PRRSGard-like PCR was able to consistently detect up to 10^-6 dilution whereas the commercial PCR consistently detected up to 10^-5 dilution. Based on testing serial dilutions of in vitro transcribed RNA, the limit of detection of PRRSGard-like PCR was 16.7 genomic copies/reaction with the Ct cut-off value of 36 when auto baseline and threshold of 0.1 were used. Diagnostic performance of PRRSGard-like PCR was evaluated using 717 serum samples collected from an experimental PRRSGard® efficacy study and two field vaccination trials done with PRRSV naïve pigs. Compared to the commercial PCR, diagnostic sensitivity, specificity, and accuracy of the PRRSGard-like PCR were 96.7%, 99.8%, and 98.7%, respectively. Nine clinical samples with discrepant results between the two PCRs all had high Ct values (34.4-36.5). These nine clinical samples were subjected to Sanger sequencing for the ORF1b/2 and ORF5 regions but sequencing was unsuccessful possibly due to low concentrations (high Ct values) of the virus in these samples. In contrast, the ORF1b/2 and ORF5 regions were successfully sequenced from nine randomly selected samples with Ct ranges of 18-31. All of these nine randomly selected samples had > 99.5% nucleotide identity to PRRSGard® vaccine virus at ORF1b/2 and ORF5 regions, confirming that these nine samples positive by PRRSGard-like PCR truly contained PRRSGard® vaccine virus.

Conclusion

In contrast to a commercial PCR that detects all PRRSV strains, the PRRSGard-like PCR specifically detects PRRSGard-like vaccine virus with comparable analytical sensitivity, specificity and diagnostic performances to the commercial PCR. PRRSGard-like PCR is a convenient tool to differentiate PRRSGard-like vaccine virus from other PRRSV strains and vaccines as well as to inform PRRSV vaccination protocols.

Keywords

Vaccinology, Pigs, Diagnostic Test – Development, Diagnostic Test – Evaluation, Porcine reproductive and respiratory syndrome virus

References

1. Holtkamp et al. Assessment of the economic impact of porcine reproductive and respiratory syndrome on United States pork producers. J. Swine Health Prod. 2013, 21(2), 72–84.

2. Chamba et al. Experimental safety and efficacy of a unique MLV PRRSV vaccine: PRRSGard®. PROC AASV. 2019, 170-174.

3. Smith et al. Evaluation of the response to PRRSGard® adminis-tration in weaned pigs. PROC AASV. 2020, 189-191.

4. Smith et al. PRRSGard® was efficacious against a virulent PRRSV174 challenge. Leman Swine Research Abstracts Sep 19-22, 2020.

5. Smith et al. PRRSGard® had limited performance drag and spread in PRRSV naïve pigs. Leman Swine Research Abstracts Sep 19-22, 2020.

Swine employee survey shows flies and gnats negatively affect employee performance

Gene Spellman, BS
Central Life Sciences, Schaumburg, Illinois

Introduction

Labor management is a challenging factor in profitable swine production. Turnover rates (employee loss and replacement) can be as high as 35%.[1] Costs (direct and indirect) of employee turnover range from 20% of yearly position salary to 150% of a yearly salary.[2] A negative working environment increases employee turnover.[3] Flies and gnats reproduce and live in swine production environments. Flies vomit on, liquify, consume and regurgitate their food. Flies defecate in the swine working environment. Flies are known vectors of swine diseases, such as porcine respiratory and reproduction syndrome (PRRS),[4] porcine endemic diarrhea (PED),[5] and African swine fever (ASF).[6] Flies are also known to carry 65 other diseases, including human illnesses.[7,8] Can flies negatively affect a swine employee's working environment? Searching for the answer to this question was the goal of this survey.

Materials and methods

Central Life Sciences (Central) offers up to 99% fly control, through its in-feed product ClariFly 267.[9] Central commissioned two survey questions to be a part of Summit Precision Production (SPP) Swine Labor Intelligence Study,[10] conducted by Swine Marketing Services (SMS). The questions were: Question 1) Are flies and gnats an annoyance in the barn? Question 2) Do flies and gnats negatively impact my working conditions? These questions were asked of 17 employees of three client swine operations of SMS' Farm Analysis program. The employee responses to these two survey questions are reported in Table 1.

Results

Operation A: 4 out of 5 respondents agreed or strongly agreed to Question 1 AND agreed or slightly agreed to question 2.

Operation B: 1 out of 5 respondents agreed to question 1 AND agreed to Question 2.

Operation C: 7 out of 7 respondents either slightly agreed, agreed, or strongly agreed to question 1. Six out of 7 respondents agreed or slightly agreed to question 2, with one respondent slightly disagreeing to question 2.

Discussion

Operations A and C, based on the respondents' answers, generally felt that flies and gnats were both an annoyance and negatively impacted their working conditions (10 of 12 respondents answered 5 or higher to both questions). These operations, in general, could be assumed to have a fly pressure problem, and it is negatively affecting their employees. Operation B, in contrast, only had 1 of 5 respondents answer that flies and gnats were an annoyance and affected their working environment. It could be assumed that Operation B did not have a fly pressure problem and the employees answered the questions accordingly. In Central's quest to "put anecdotes into analytics", based on these survey question results, it is possible to show that fly and gnat pressure in swine working environments negatively affect employees. This leads to a discussion about the expanding importance of controlling flies and gnats in swine operations. Should fly control be a part of a swine operation's Human Resources strategy, as it relates to employee working environments, retention, and operational reputation and recruitment, in addition to fly control's importance in biosecurity strategy? Would a better working environment, without flies and gnats, reduce absenteeism, reduce employee sickness and health care costs, and improve productivity? Should the cost of controlling flies be shared across more internal functions of swine operations, instead of just the Production function? More research in these areas is needed. Fly control may need to be considered for more reasons across swine operations. This survey shows that

Table 1: Survey results

Farm	A	A	A	A	A	B	B	B	B	B	C	C	C	C	C	C	C
Flies and gnats negatively impact my working conditions	6	4	6	5	5	2	4	2	6	2	5	6	5	3	5	5	5
Flies and gnats are an annoyance in the barns	6	4	6	7	7	2	4	2	6	1	5	6	5	6	7	5	7

Scale
1 Strongly disagree
2 Disagree
3 Slightly disagree
4 Neutral
5 Slightly agree
6 Agree
7 Strongly agree

swine production contaminated with flies and gnats negatively affects employees' working environments.

References

1. National Pork Board. *Employee Compensation and HR Practices in Pork Production 2017.*

2. Moore, S. (November 29, 2012). Employee turnover on dairy farms and what it's costing producers. Michigan State University Extension. Retrieved from **Msue.anr.msu.edu**.

3. Anjum, A. et al. An Empirical Study Analyzing Job Productivity in Toxic Workplace Environments. *Int J Environ Res Public Health.* 2018 May; 15(5): *1035.*

4. Otake S, Dee SA et al. Transmission of porcine reproductive and respiratory syndrome virus by houseflies, (Musca domestica Linnaeus). *Vet Rec* 2003. 152: 73-76

5. Allison G, et al. PEDV positive bioassay reveals houseflies (Musca domestica) can transmit infectious PEDV to pigs. *Proceedings 50th AASV 2019.*

6. Olesen, A, et al. Infection of pigs with African swine fever virus via ingestion of stable flies (Stomoxys calcitrans) *Transboundary Emerging Diseases* 2018, 65: 1152-1157

7. Khamesipour F, et al. A systematic review of human pathogens carried by the housefly (Musca domestica L.). *BMC Public Health.* 2018, 18:1049

8. Zurek L, et al. Insects Represent a Link between Food Animal Farms and the Urban Environment for Antibiotic Resistance Traits. *Appl Environ Microbiol. 2014 Jun;* 80(12): 3562–3567.

9. Spellman, G, et al. New labor-free fly control ClariFly 267 (diflubenzuron) achieves 99% reduction in Housefly (*Musca domestica)* populations in swine facilities. *Proceedings 51st AASV 2020.* 175-176

10. Hoek, J. et al. Humatec™ People-pigs-process-performance: Creating the irresistible workplace. *Proceedings 50th AASV, Orlando, Florida.* 2019; *pp.20-23*

LeeO™ individual animal data system implementation in a commercial sow farm

Brandi Burton[1], DVM; Amber Stricker[1], MS, DVM; Jeff Ley[2], BA; Tyler Holck[3], DVM, MS, MBA
[1]Suidae Health and Production, Algona, Iowa; [2]Prairie Systems, Spencer, Iowa; [3]Feed His People LLC, Gilbert, Iowa

Introduction

Precision pork is a term used in our US swine industry and yet most, if not all, of our information platforms are based upon group data and assessing group means. Individual weights are rarely collected in the farrowing house let alone in the grow-finish stage. Individual animal data has been used in Europe to allow for better decision making on gilt selection and sow culling, piglet viability criteria, weaned pig feeding, market pig selection to maximize income, and traceable attributes for finished pork products.[1]

LeeO™ is an individual animal tracking system designed for paperless data collection in "real time" in the commercial pork supply chain. Utilizing a radio frequency identification (RFID), LeeO™ scans tags and records individual animal weights quickly using a proprietary LeeO™ O Reader and blue tooth technology. This paper will describe the implementation of LeeO™ in a commercial pork production system and the initial lessons learned in the process toward precision pork at the individual pig level.

Implementation

In February 2020, a commercial 2400 sow, multi-site, farrow-to-finish system in the upper Midwest began utilizing the LeeO™ system. The initial implementation involved placing a LeeO™ RFID sow tag in the ear of each due-to-farrow sow in farrowing. When the sows farrowed each live piglet was also given a LeeO™ UHF ear tag and registered into the LeeO™ system including its gender and birth weight using a LeeO™ piglet scale. Mortality and cross-fostering (adoptions in LeeO™) were then recorded using the LeeO™ O Reader and mobile app.

Once the LeeO™ process was started in farrowing, the decision was made to ear tag the entire sow herd with LeeO™ RFID sow tags over a one-week period. Going forward, all piglets are tagged and registered in LeeO™ with gender and birth weight recorded. Pigs are weighed at weaning again and moved to commercial wean-to-finish barns, all traced by location through LeeO™. Mortalities are collected in LeeO™ during the grow-finish phase and individual pig weights are also recorded at some sites using a LeeO™ finishing scale. LeeO™ is also used to identify pigs on the first market loads at some sites prior to shipment to a commercial packing plant for harvest.

Scope of LeeO™ in a commercial setting

Over 60,000+ piglet births, 55,000 birth weights, 45,000 wean weights, 13,000 mortalities, and 12,000 sow services have been recorded and registered into LeeO™ in the first 9 months since implementation on farm. Individual data collection is possible due to the high LeeO™ tag retention which is greater than 99% from birth to wean and greater than 98% through late finishing. From a process standpoint, the sow farm is in the process of converting to paperless data collection, including at breeding, farrowing, and weaning utilizing the LeeO™ system. While the process of tagging and weighing all piglets at birth has added some additional labor in the farrowing house, that labor is compensated by the elimination of data recording and transcription labor that previously occurred at the farm and in the records office.

Immediate on-farm implications

One of the added values of using a system such as LeeO™ is the ability to objectively make decisions based on data collected and available on farm in real-time.

The data has provided key learnings regarding the impact of birth weight on the subsequent performance of the pigs (Table 1). While intuitive, this data had not previously been collected at this farm or in this system and it provoked further investigations and action. For instance, the pre-wean mortality of pigs born less than 2lbs is much higher (63.4%) than the other weight categories of 2-3 lbs (17.4%), 3-4 lbs (8.4%) and > 4 lbs (6.1%). A further investigation birth weight impact on mortality (Figure 1) confirms that a large percentage of piglets born under 2lbs never make it to weaning and those that do survive have a higher post-weaning mortality. Based upon this finding the farm has initiated humane euthanasia of all piglets born under 2lbs with the intent to give the remaining bigger piglets, who are more likely to survive, a better start early in life while on the sow. As a result, wean weights have increased greater than 0.5 lbs and nursery mortality rate has been reduced.

LeeO™ has also been utilized to conduct on-farm research.[2] In effort to further reduce piglet deaths in farrowing, differences between two day-1 treatment options were evaluated. Using LeeO™, sows were randomly allocated and blocked for parity in the farrowing house and assigned to either treatment group 1 or 2. Piglet births, birth weights, mortalities and wean weights were collected and recorded with LeeO™. After 14 weeks and over 18,000 piglets enrolled, the data was analyzed to determine whether there were any performance differences in preweaning mortality or wean weights between the two treatment groups. Results in Table 2 verify successful allocation to treatment with similar parities and birth weights between treatment groups. Evaluation of the data suggested no difference between treatment group allowing the farm to select the most economic treatment option.

Use of LeeO™ going forward

Every day there is more data being collected and more information to be learned on farm with the use of LeeO™. Having this data allows for real-time as well as retrospective analysis as questions arise in the future. For example, one of the sow farm staff noticed a variation in inlet settings in the farrowing rooms and questioned if it may be impacting preweaning mortality. Using LeeO™ to assess piglet births and deaths at specific locations differences were objectively measured in performance which resulted in appropriate adjustments in ventilation.

Further use of LeeO™ is planned for the grow-finish production. Currently all mortalities are tracked through market and an investigation has been conducted looking at the impact of umbilical hernias on mortality including traceback to the farm to assess farm processes and other potential associations. The use of LeeO™ is being expanded at marketing as well where the composition of first market loads is providing valuable economic insight. Ultimately there are many more questions yet to be asked and LeeO™ is a tool that can help to efficiently and objectively provide the answers.

Conclusions

The LeeO™ individual animal data system is easy to implement in a commercial setting and very effective in gathering data from birth to harvest. Additional applications for LeeO™ include the ability to more precisely budget feed in early nursery, provide more precision in marketing at harvest, and provide for better decision making for sow and progeny productivity.

References

1. ter Linde P et al. The value of LeeO, an individual animal identification system adding value to the pork supply chain the Netherlands. AASV, 2020; 337

2. Yeske P et al. Utilizing LeeO individual animal identification system in commercial research. AASV, 2020; 139

Table 1: Impact of birth weight (BW) on performance

Parameter	BW < 2.0 lbs	BW 2.0-2.99 lbs	BW 3.0-3.99 lbs	BW > 4.0 lbs	Total
Number of pigs born	2351	15,506	25,111	10,082	53,050
Number of pigs weaned	786	11,957	21,437	8,850	43,030
Pre-wean mortality (%)	63.4	17.4	8.4	6.1	13.0
Average parity of dam	3.77	3.67	3.65	3.78	3.69
Pig birth weight (lb)	1.72	2.60	3.49	4.42	3.33
Pig wean weight (lb)	10.16	11.82	14.13	16.44	13.9
Wean age (days)	19.5	18.9	18.8	18.8	18.8
Pre-wean ADG (lbs/day)	0.46	0.51	0.59	0.67	0.58

Figure 1: Mortality by birth weight.

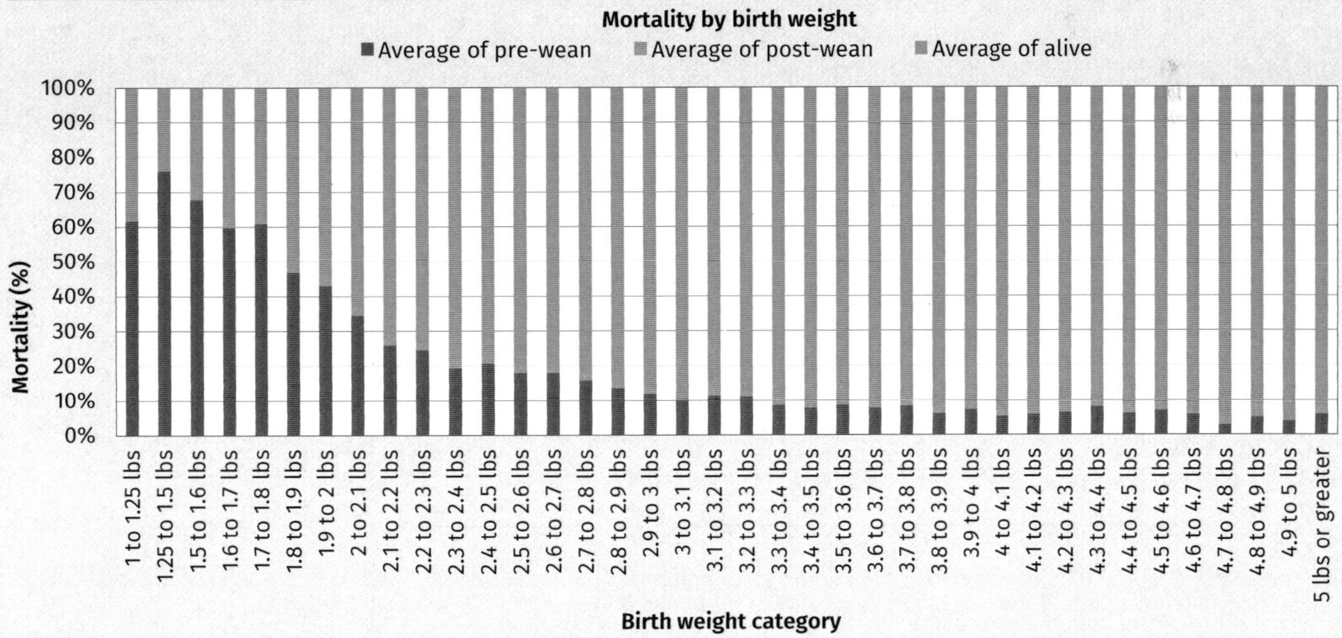

Table 2: Day 1 treatment impact on pre-wean mortality and wean weight

Group	Average of birth weight (lbs)	Average of weaning weight (lbs)	Average of mother's parity	Average pre-wean mortality (%)	Number of piglets enrolled
Treatment 1	3.28	14.25	3.45	20.44%	9644
Treatment 2	3.31	14.25	3.48	19.50%	9065
Grand total	**3.30**	**14.25**	**3.47**	**19.99%**	**18709**

Trends of mycotoxin contamination and mitigation strategies in the US swine industry

L. Zheng[1], PhD; V. Wilson[2], S. Ramirez[1], PhD; U. Hofstetter[3]; G. R. Murugesan[1], DVM, PhD, B. J. Kerr[4], PhD
[1]BIOMIN America Inc., Overland Park, Kansas; [2]Iowa State University, Ames, Iowa;
[3]BIOMIN Holding GmbH, Getzersdorf, Austria; [4]USDA-ARS, Ames, Iowa

Introduction

Mycotoxins are naturally occurring harmful secondary fungal metabolites, which are ubiquitous in agricultural crops. These toxins are detrimental to animal health and can compromise animal performance, making them an essential part of managing food and feed safety. The swine industry frequently utilizes mycotoxin risk mitigation because swine are exquisitely sensitive to multiple mycotoxin types such as deoxynivalenol and zearalenone. Classic signs such as reduced feed intake and vomiting are often used as indicators for exposure, but overlook other costs of mycotoxicosis to production systems. Even at low levels, mycotoxins act as predisposing factors for enteric[1] and respiratory diseases by causing inflammation,[2] modulating the gastrointestinal environment,[2] and immune dysfunction[3] which can contribute to vaccine failure.[4] Additionally, mycotoxins impact reproductive organs[5,6] resulting in reductions in semen quality,[6,7] sow reproductive performance, and may contribute to sow pelvic organ prolapse.

BIOMIN's annual surveys indicate increasing trends of prevalence and co-occurrence of toxins produced by *Fusarium* spp. fungi in US corn in prior years.[8] Historically, the mycotoxins deoxynivalenol, fumonisin, and zearalenone have been the main mycotoxins detected in the US corn-based feed ingredients. The data herein present preliminary information regarding mycotoxin contamination in samples of the 2020 corn harvest in comparison to the previous decade of the BIOMIN survey and recent research in mycotoxin mitigation.[9]

Materials and methods

Corn and corn DDGS used in multiple livestock industries (poultry, ruminant, and swine) were submitted for mycotoxin analysis each harvest year. Twenty-seven new crop corn and 21 corn DDGS samples were submitted as of October 2020. Samples were analyzed for five major mycotoxin groups: aflatoxins (Afla), type A trichothecenes (A-Trich), type B trichothecenes (B-Trich), fumonisins (FUM), and zearalenone (ZEN) utilizing liquid chromatography and tandem mass spectrometry (LC-MS/MS). Data were analyzed using GLIMMIX procedure of SAS with fixed effect of harvest year.

A live evaluation of a feed additive for mycotoxin risk mitigation was conducted in nursery pigs. Biofix® Plus (BPL; BIOMIN America, Inc., Overland Park, KS), which utilizes a multi-pronged approach, was incorporated at an inclusion rate of 1.5 kg/MT during a natural mixed mycotoxin (MTX) challenge. A 2 × 2 factorial design allotted newly weaned pigs (BW = 5.5 ± 0.2 kg) into one of four treatments (105 pigs, 3 pigs/pen, 8 to 9 pens per treatment) with two MTX contamination levels [low (L-MTX) and high (H-MTX); Table 1] and two BPL levels [without (Control) and with (BPL)]. Data were analyzed using the GLIMMIX procedure of SAS to evaluate the main effect of MTX and BPL and their interactions.

Results and discussion

Trend data of prevalence (Table 2) and contamination level (Table 3) show how national mycotoxin challenge varies from year to year. Mycotoxins such as B-Trich and FUM significantly vary in contamination levels in corn from year to year ($P < 0.05$). Regardless of this variation, the last five years have had consistent B-Trich contamination levels, while the preliminary sample pool shows a decrease in FUM contamination from 2019 to 2020. Fumonisin is highly impacted by geographic location, so regional evaluation of risk is paramount. Contamination levels in corn for Afla, A-Trich, and ZEN have remained consistent ($P > 0.05$) over the survey period, and thus should be part of an annual mitigation plan. Harvest year significantly influences contamination levels of all three *Fusarium* spp. produced toxins (B-Trich, FUM, and ZEN, $P < 0.05$) in Corn DDGS. The B-Trich contamination level was decreased ($P < 0.05$) from 2019 to 2020, whereas FUM and ZEN contamination levels remained similar ($P > 0.05$). The 2020 crop risk profile is likely to change drastically as the sample pool expands. Several weather related events of the 2020 growing season – including heat stress, drought, and storm events resulted in crop stress and damage, and regardless of preliminary data, grain quality and mycotoxin contamination risk are still essential for preventative herd health management.

The preliminary results from the 2020 corn harvest suggest a continued risk due to mycotoxins produced by *Fusarium* fungal species, emphasizing the importance of risk mitigation strategies that are effective for multiple mycotoxins.

Performance data of the live study are summarized in Figure 1. Overall, H-MTX exposure resulted in reduced ($P < 0.05$) ADG (0.28 vs. 0.35 kg) and ADFI (0.49 vs. 0.59 kg) during the 5-week period compared with L-MTX diets, and resulting in a numeric reduction ($P = 0.05$) of feed efficiency (0.59 vs. 0.60). Although no interaction of MTX and BPL were observed on cumulative 5-week performance metrics, BPL fed pigs had increased ($P < 0.05$) feed efficiency compared with pigs not fed BPL. The increased feed efficiency regardless of MTX level with BPL treatment, and numerical recovery of 47% of the body weight losses during a high MTX challenge indicates that BPL can be considered a mitigation strategy to address both unpredictable and mixed MTX challenges.

In-vitro adsorption studies have suggested that strategy of feeding clay binders is insufficient for addressing mycotoxins such as deoxynivalenol[10] and zearalenone.[11] In the mixed mycotoxin challenge presented, performance improvements were observed with the addition of an in-feed mitigant with a multi-pronged approach. This specific in-feed mitigant was also able to show full recovery of weight losses seen when high levels of FUM were fed[12] (Figure 2), when other mitigation strategies such as clay and combination of clay with sodium metabisulfite were unable to recover performance losses.[12]

Table 1: Mycotoxin level (μg/kg) in dietary phases by low mycotoxin (L-MTX) and high mycotoxin (H-MTX) diets

Phase	1		2		3	
MTX	L-MTX	H-MTX	L-MTX	H-MTX	L-MTX	H-MTX
Deoxynivalenol	237	2,352	452	4,461	1,083	3,770
Zearalenone	141	358	745	989	293	486
Fumonisins	< LOD*	< LOD	500	100	200	200

* LOD = limit of detection

Table 2: Sample number (N) and toxin prevalence (%) by year

Ingredient	Corn					
Toxin	N	Afla	B-Trich	A-Trich	FUM	ZEN
2011	16	25	75	0	69	38
2012	49	40	24	0	88	8
2013	51	8	25	0	54	8
2014	320	14	59	3	54	18
2015	275	1	74	0	53	13
2016	288	7	79	1	81	40
2017	374	4	70	1	52	24
2018	447	10	76	3	79	45
2019	337	3	81	2	77	46
2020	27	4	44	0	74	11
Ingredient	**Corn DDGS**					
Toxin	**N**	**Afla**	**B-Trich**	**A-Trich**	**FUM**	**ZEN**
2015	6	0	100	0	100	50
2016	56	2	100	0	88	69
2017	81	14	100	0	82	85
2018	43	5	95	0	88	70
2019	40	3	100	8	98	95
2020	21	10	100	0	95	100

Due to the continued risk of mycotoxin co-occurrence, expanding mitigation strategies beyond adsorption such as mycotoxin deactivation with biotransformation and additional support of immune and liver function is essential. Recent research suggests that although not all in-feed mitigation solutions are effective for mycotoxicosis, there are mitigants that are able to address multiple mycotoxin contamination.

References

1. Oswald, I.P., et al. 2003. Mycotoxin Fumonisin B1 increases intestinal colonization by pathogenic Escherichia coli in pigs. Applied and Environmental Microbiology. (2003). 69:10. Pp. 5870-5874.

2. Pinton et al. Toxicity of Deoxynivalinol and its acetylated derivatives on the intestine: differential effects on morphology, barrier function, tight junction proteins, and mitogen-activated protein kinases. *Toxilogical Sciences*. 2012; 130(1): 180-190

3. Pierron et al. Impact of mycotoxin on immune response and consequences for pig health. *Animal Nutrition*. 2016; 2: 63-68.

4. Savard, C. et al. Deoxynivalenol (DON) naturally contaminated feed impairs the immune response induced by porcine reproductive and respiratory syndrome virus (PRRSV) live attenuated vaccine. *Vaccine*. 2015; V. 33, Issue 32, pp. 3881-3886.

5. Cheng, Y.H. et al. Toxicity of different Fusarium mycotoxins on growth performance, immune responses and efficacy of a mycotoxin degrading enzyme in pigs. Animal Research. 55 (2006) 579-590.

6. Acda, Sonia P., et al. "Effects of mycotoxin-deactivating agent on the growth performance of pigs fed ochratoxin and zearalenone-contaminated diets." Philippine Journal of Veterinary Medicine 45.1 (2009).

7. Gbore, F. A., and G. N. Egbunike. "Testicular and epididymal sperm reserves and sperm production of pubertal boars fed dietary fumonisin B1." *Animal Reproduction Science* 105.3-4 (2008): 392-397.

8. Savard, Christian, et al. "Prevention of deoxynivalenol-and zearalenone-associated oxidative stress does not restore MA-10 Leydig cell functions." *Toxicology* 341 (2016): 17-27.

9. E.G. Hendel, S. Curry, P.N. Gott, U. Hofstetter, G.R. Murugesan. Trends in mycotoxin contamination in US Corn. AASV 2019, pp. 177

10. Murugesan, G.R., et al. Prevalence and effects of mycotoxins on poultry health and performance, and recent development in mycotoxin counteractin strategies. Poultry Sciences (2015) 94:1298-1315.

11. Fruhaf, S. et al. Yeast cell based feed additives: studies on aflatoxin B1 and zearalenone. Food Additives & Contaminatns: Part A: Chamistry, Analysis, Control, Exposure & Risk Assessment. (2011) 29:2, 217-231.

12. Rao, Z.X. et al. "Efficacy of Commercial Products on Growth Performance of Nursery Pigs Fed Diets with Fumonisin-Contaminatied Corn." Kansas Agricultural Experiment Station Research Reports. V. 5, Issue 8 (2019).

Table 3: of contamination (ppb) by year

Ingredient			Corn		
Toxin	**Afla**	**B-Trich**	**A-Trich**	**FUM**	**ZEN**
2011	38	4873^{a}	0	3250^{ab}	232
2012	20	932^{bc}	0	2004^{ab}	71
2013	99	629^{bc}	0	2461^{ab}	168
2014	29	1109^{bc}	62	21652^{a}	555
2015	16	551^{c}	0	1745^{ab}	356
2016	28	1245^{b}	314	3893^{a}	485
2017	15	571^{bc}	104	2404^{ab}	266
2018	133	1162b	1366	3305^{a}	310
2019	13	881^{bc}	330	3371^{a}	418
2020	196	773^{bc}	0	1265^{bc}	172
Pooled SEM	72.3	378.6	677.2	815	359.3
P-Value	0.417	< 0.001	0.378	< 0.001	0.978
Ingredient			**Corn DDGS**		
Toxin	**Afla**	**B-Trich**	**A-Trich**	**FUM**	**ZEN**
2015	0	1452^{b}	0	1650^{ab}	95^{abc}
2016	3	2479^{b}	0	3356^{ab}	199^{c}
2017	4	2479^{b}	0	2092^{a}	230^{cb}
2018	3	2369^{b}	0	1231^{b}	353^{c}
2019	36	3779^{a}	181	2614^{ab}	618^{a}
2020	6	2699^{b}	0	3545^{ab}	464^{ab}
Pooled SEM	4.76	412.7	97.8	1041.2	94.8
P-Value	0.378	<0.001	-	0.018	<0.001

$^{a-c}$ Means with differing letters indicate significant difference ($P < 0.05$).

Figure 1: Performance parameter responses during a mixed mycotoxin challenge

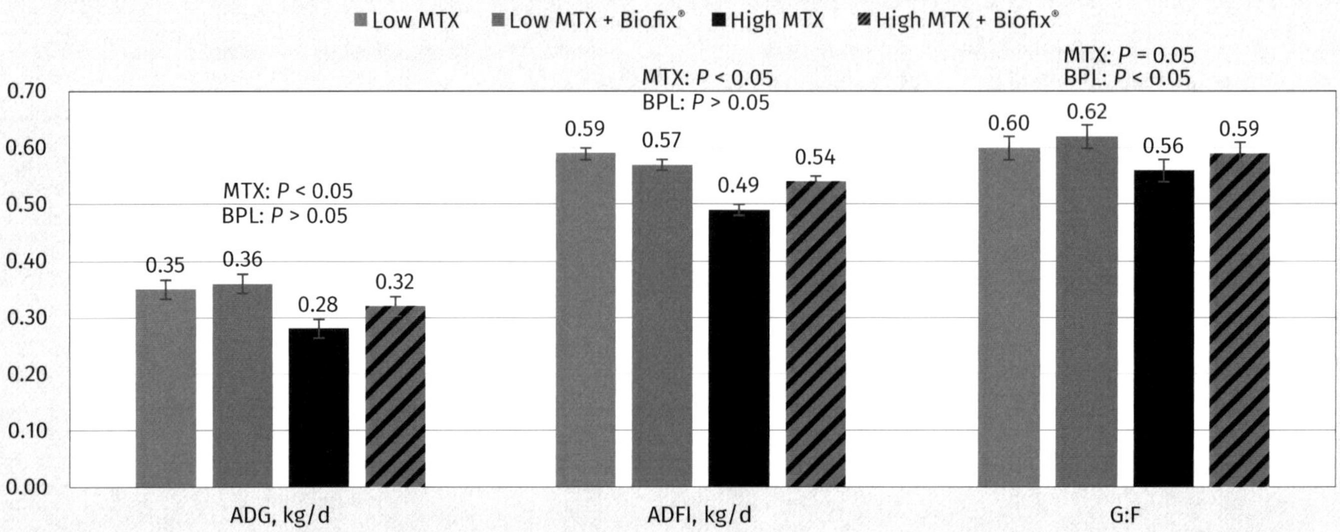

Evaluating effect of a multipronged mitigation approach on performance during a mixed mycotoxin challenge (Low MTX: 0.6 ppm DON, 0.4 ppm ZEN, 0.2 ppm FUM; High MTX: 3.5 ppm DON, 0.6 ppm ZEN, 0.18 ppm FUM). Average daily gain (ADG, kg/d), average daily feed intake (ADFI, kg/d) and feed efficiency (G:F).

Figure 2: Effect of commercially available mycotoxin mitigants on ADG (kg) during high FUM challenges

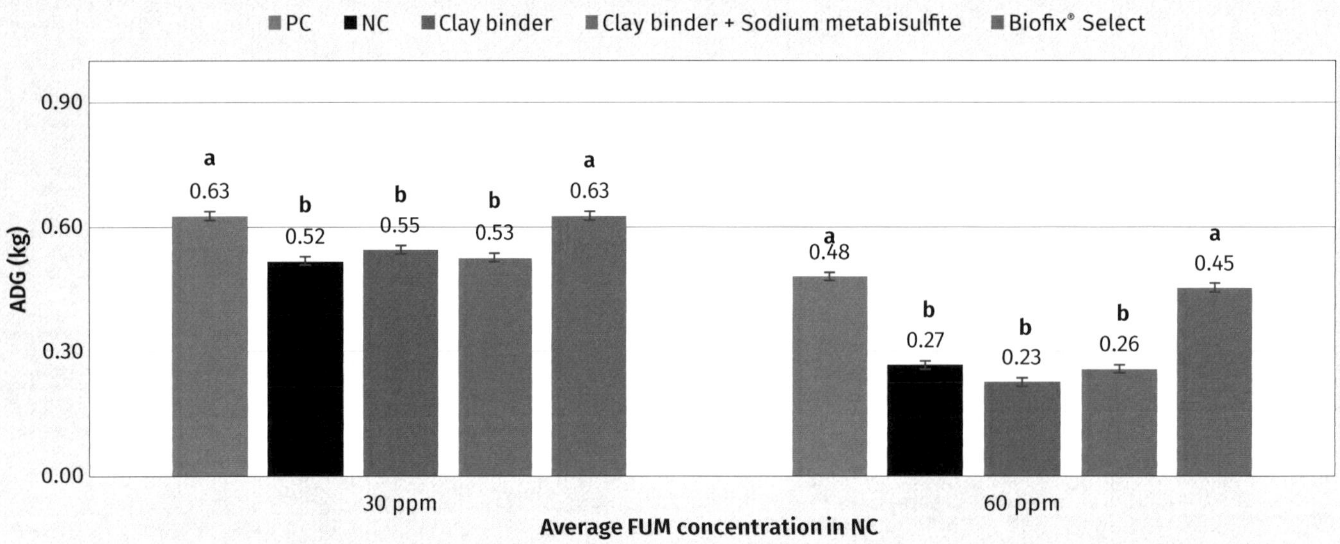

Summary of data from Rao et. al, 2019 from Kansas State University demonstrating effects of commercially available mitigation strategies (clay binder, clay binder with sodium metabisulfite, and a multipronged strategy) on ADG during two high FUM challenge studies (PC: positive control, low FUM; NC: negative control, 30 or 60 ppm of FUM).

Utilization of a fiber technology to agglutinate pathogenic bacteria associated with post weaning diarrhea in nursery pigs

Stacie Crowder[1], PhD candidate; Dr. Kobe Lannoo[2]; Dr. Dari Brown[1]
[1]Land O'Lakes, Arden Hills, Minnesota; [2]AgriImprove, Ghent, Belgium

Introduction

The economic losses associated with post weaning diarrhea are two-fold, with treatment costs and loss of animal performance.[1] This combined with the incidence of drug resistant bacteria increasing over the past few years has led researchers to seek out alternative approaches to support gut health. At weaning the gut is and under stress from dietary changes as well as enteric challenges.[2] This stress leads to an increase in Gr-bacteria, which in turn increases the presence of endotoxins that can lead to inflammation of the intestines. While dietary fiber has been used to increase stool consistency and lower inflammation pressure in nursery pigs, this increased fiber takes up valuable space in nursery formulation. Therefore, the objective of this experiment was to evaluate the effectiveness of a fiber technology to agglutinate pathogenic bacteria commonly associate with post weaning diarrhea in nursery pigs.

Materials and methods

Agglutination was estimated by observing the fiber technology mixed with pathogenic bacteria under a microscope. While this assay did not provide quantification of the agglutination it provided visual representation of the clumping action indicative of agglutination. The next step in the agglutination assay was to quantify the results seen under the microscope with agglutination by filtration. Two identical samples of *E coli* (K88) were put into a 100 mL solution and the fiber technology was added to one of the 100 mL samples. Samples were mixed for 15 minutes and filtered through a Whatmann filter with a pore size of 4-7 µm. Both filtrates were then plated and *E coli* colonies counted to determine amount recovered. *E coli* amounts were determined on the start culture, on filtered *E coli,* and on the residue and filtrate of *E coli* + fiber technology, and calculated to the total CFU amounts. HEK-Blue™-4 cells were used to evaluate the anti-inflammatory properties of the fiber technology. HEK-Blue™-4 cells were maintained in a Dulbecco modified Eagle medium (DMEM) containing Quanti-Blue 1:10. NF-κB activation leads to the expression of a protein soluble embryonic alkaline phosphatase which colors a Quanti-Blue solution from pink to blue signaling a proinflammatory response.

Results

Results from this experiment demonstrate that a fiber technology can agglutinate 99% of the *E coli* and inhibit 78% of the agglutinated *E coli* from growing when plated. LPS challenged HEK-Blue™-4 cells exhibited a 14.9-fold increase in the activation of NF-κB. Treating HEK-Blue™-4 cells with a fiber technology lowered the activation of NF-κB by 12 fold signaling the anti-inflammatory properties of the fiber technology. Further research is needed to determine the effectiveness of agglutination in an in-vivo challenge model as well as to determine the effects on animal performance.

References

1. Peng Ji, Xunde Li and Yanhong Liu (February 10th 2020). Dietary Intervention to Reduce *E coli* Infectious Diarrhea in Young Pigs, *E Coli* Infections - Importance of Early Diagnosis and Efficient Treatment . IntechOpen, **doi: 10.5772/intechopen.91219**

2. Luppi A. Swine enteric colibacillosis: diagnosis, therapy and antimicrobial resistance. *Porcine Health Manag.* 2017;3:16. Published 2017 Aug 8. **doi:10.1186/s40813-017-0063-4**

Robustness advantage of TN Tempo progeny validated using a natural PRRS challenge model

J. R. Dunkelberger[1], PhD; D. Boler[1], PhD; J. M. Eggert[1], PhD; R. Edler[2], MSc; E. Little[2], MSc; D. Hanson[2], BA; S. Dee[2], DVM, PhD; E. F. Knol[3], PhD

[1]Topigs Norsvin USA, Burnsville, Minnesota; [2]Pipestone Applied Research, Pipestone, Minnesota; [3]Topigs Norsvin Research Center B.V., Beuningen, The Netherlands

Introduction

Traditional disease control strategies, such as vaccination and herd management practices, are often used in an attempt to prevent or mitigate the impact of PRRS virus (PRRSV)-infection. However, the efficacy of these methods for PRRS control remains limited. Therefore, the need for alternative or complementary PRRS control strategies is crucial. With the development of high-density genotyping panels and genomic prediction methodologies, the possibility of identifying and selecting livestock for enhanced, natural robustness to disease has emerged as a promising technique for disease control. Ultimately, the success of implementing this approach for the production of livestock with enhanced, natural robustness to disease relies on the existence of naturally occurring genetic variation in robustness to challenge. Results from a number of previous studies provide evidence of naturally occurring genetic variation in host response to PRRSV-infection, supporting the use of genetic selection for this purpose.[1-12] Results include those obtained from a previous study conducted by our group, which showed differences in mortality rate between genetic groups following experimental PRRS challenge.[12] Genetic groups were produced using a commercial crossbred female line bred to one of two sire lines: the Topigs Norsvin (TN) Tempo line or a leading competitor line in the US market.[12] Results showed that, following experimental inoculation with PRRSV 1-7-4, overall mortality rates were 13% and 22% for the TN Tempo-sired group and competitor-sired group, respectively.12 The aim of the current study was to validate the observed natural, between-line genetic variation in robustness to challenge detected in this previous study for TN Tempo-sired finishing pigs: 1.) vs. pigs sired by two additional major competitor boar lines; and 2.) using an alternative (i.e. natural) PRRSV-challenge model.

Materials and methods

Commercial F1 parent females originating from a single genetic supplier were bred to either TN Tempo, competitor Line A, or competitor Line B to produce three groups of crossbred finishing pigs used for this study. At weaning, 702 pigs per genetic group were transported to a commercial research facility. Pigs were sorted into pens by genetic group with an initial stocking density of 27 pigs/pen. The research barn was filled in two wean groups over four days, where each pig had a minimum five-day acclimation period prior to vaccination with a PRRS modified live virus vaccine. Four weeks later, three pigs per pen were randomly selected to serve as seeder pigs and inoculated with $2 \times 10^{3.5}$ TCID50 of the 1-7-4 PRRSV isolate via intramuscular injection.

Analysis of oral fluid samples confirmed the presence of PRRSV 1-7-4 post-challenge. Secondary pathogens detected throughout the study included, but were not limited to: rotavirus, *Pasteurella multocida*, *Actinobacillus suis*, porcine circovirus type 3, and swine influenza virus. Phenotypes recorded throughout the study included body weight and feed intake (FI) at the start and end of test (recorded on a pen basis), used to calculate wean-to-finish average daily gain (ADG) and feed conversion ratio (FCR). Individual (0/1) phenotypes were recorded for full-value status, wean-to-finish mortality (due to any reason), and percent of wean-to-finish mortalities due to PRRSV-infection, specifically, assessed based on exhibiting clinical signs of respiratory distress.

Statistical analyses of FI, ADG, and FCR were conducted by estimating the effect of genetic group (TN Tempo- vs. Line A- vs. Line B-sired) on each trait using a fixed effects model in SAS. Additional effects fitted in the model included: sex (gilt vs. barrow), barn (east vs. west), and date on test (first wean day vs. second wean day) fitted as fixed class effects and age at the end of test as a fixed covariate. This same model (with the exception of age at the end of test) was used to analyze full-value status, mortality (due to any reason), and mortality due to PRRSV-infection as binary traits using a PROBIT model.

Results and discussion

Results from analyses of wean-to-finish performance show that TN Tempo-sired pigs had numerically greater FI (4.26 lbs/day) than Line B-sired pigs (4.15 lbs/day) ($P = 0.05$) and significantly greater feed intake than Line A-sired pigs (4.04 lbs/day) ($P < 0.0001$). Similarly, TN Tempo-sired pigs had numerically greater ADG (1.69 lbs/day) than Line B-sired pigs (1.66 lbs/day) ($P = 0.07$) and significantly greater ADG than Line A-sired pigs (1.65 lbs/day) ($P = 0.007$). Pigs sired by Line A had significantly lower FCR (2.44) than pigs sired by Line B (2.50) ($P = 0.005$) or the TN Tempo line (2.52) ($P < 0.0001$). However, FCR did not significantly differ between Line B- and TN Tempo-sired pigs ($P = 0.26$).

Results also showed that the TN Tempo-sired group had numerically more full-value pigs (90.5%) than the Line A-sired group (87.7%) ($P = 0.09$) and significantly more full-value pigs than the Line B-sired group (82.8%) ($P < 0.0001$). Significant differences in overall mortality rate and mortality rate due to PRRSV-infection were detected among all three genetic groups, where the TN Tempo-sired group had a significantly lower overall mortality rate (7.6%) than pigs sired by either Line A (10.7%) ($P = 0.04$) or Line B (16.2%) ($P < 0.0001$) and significantly fewer mortalities due to PRRSV-infection (29.2%) than pigs sired by Line A (75.3%) ($P = 0.006$) or Line B (76.1%) ($P < 0.0001$).

In conclusion, results from this study show that TN Tempo-sired pigs had numerically, if not significantly, greater FI and ADG than pigs sired by the competitor lines, but significantly greater FCR than pigs sired by Line A. The TN Tempo-sired group had numerically more full-value pigs than the Line A group and significantly more full-value pigs than the Line B group, in addition to a significantly lower mortality rate than either line, whether assessed as mortality due to any reason or to PRRSV-infection, specifically.

Taken together, results from the current study and the previous PRRS study conducted by our group, show that TN Tempo-sired pigs are naturally more robust to a multifactorial PRRS challenge than pigs sired by top competitor sire lines in the US market, whether evaluated using an experimental or natural PRRSV-challenge model.

Furthermore, the differences in robustness observed between the genetic groups used in this study confirm the presence of naturally occurring genetic variation in robustness to disease challenge, providing additional support for the use of genetic selection to breed pigs for enhanced, natural robustness to challenge as a viable PRRS control strategy.

References

1. Lewis CRG et al. Genetic parameters for performance traits in commercial sows estimated before and after an outbreak of porcine reproductive and respiratory syndrome. *Journal of Animal Science*. 2009;87.

2.) Rashidi H et al. Variation among sows in response to porcine reproductive and respiratory syndrome. *Journal of Animal Science*. 2014;92:1.

3. Serão NVL et al. Genetic analysis of reproductive traits and antibody response in a PRRS outbreak herd. *Journal of Animal Science*. 2014;92:7.

4. Boddicker NJ et al. Evidence for a major QTL associated with host response to porcine reproductive and respiratory syndrome virus challenge. *Journal of Animal Science*. 2012;90:6.

5. Boddicker NJ et al. Validation and further characterization of a major quantitative trait locus associated with host response to experimental infection with porcine reproductive and respiratory syndrome virus. *Animal Genetics*. 2013;45:1.

6. Boddicker NJ et al. Genome-wide association and genomic prediction for host response to porcine reproductive and respiratory syndrome virus infection. *Genetics Selection Evolution*. 2014;46:1.

7. Hess AS et al. Comparison of host genetic factors influencing pig response to infection with two North American isolates of porcine reproductive and respiratory syndrome virus. *Genetics Selection Evolution*. 2016;48:43.

8. Dunkelberger JR et al. Effect of a major quantitative trait locus for porcine reproductive and respiratory syndrome (PRRS) resistance on response to coinfection with prrs virus and porcine circovirus type 2b (PCV2b) in commercial pigs, with or without prior vaccination for PRRS. *Journal of Animal Science*. 2017;95.

9. Dunkelberger JR et al. Genomic regions associated with host response to porcine reproductive and respiratory syndrome vaccination and co-infection in nursery pigs. *BMC Genomics*. 2017;18:865.

10. Waide EH et al. Genomic prediction of piglet response to infection with one of two porcine reproductive and respiratory syndrome virus isolates. *Genetics Selection Evolution*. 2018;50:3.

11. Putz AM et al. Novel Resilience Phenotypes Using Feed Intake Data From a Natural Disease Challenge Model in Wean-to-Finish Pigs. *Frontiers in Genetics*. 2019;9.

12. Dunkelberger JR et al. Genetic selection can be used as a viable tool to improve natural robustness to PRRS challenge. AASV. 2020.

Used filter test results and influencing factors across 10 farms

Darwin L. Reicks
Reicks Veterinary Research & Consulting, St. Peter, Minnesota

Introduction

The use of MERV 14, 15, and 16 filters on sow farms and boar studs has become common over the last 15 years. Initially, the thought was filters would need to be changed annually, since that was common practice in other industries. Farms have continued to use these filters much longer however, as the cost per animal unit for the system reduces dramatically with each additional year of use.

The efficiency of particle capture has been previously shown to correlate with the capture of PRRS virus particles.[1] Further, use of filters in the field has been shown to reduce the frequency of new PRRS virus introductions.[2-4]

One limitation on the longevity is whether the filter is still capturing particles with the same efficiency as when new. The second limitation is whether the air flow is too restricted to properly ventilate the farm, which could lead to exaggeration of leak points, production problems related to inadequate ventilation rates, or increased stress on fans and other equipment.

The purpose of this field study was to take a portable filter testing laboratory to farms that had filters in use for long periods of time. By testing at the farm, any questions about filter damage, degradation, or particles shifting in the media could be avoided, and filter selection was done randomly rather than by farm staff.

Materials and methods

10 farms were selected which had MERV14, MERV15, or MERV16 rated filters. Synthetic filters were not considered for this study since they lose their efficiency with use. Filters were in use from 9 to 106 months. Seven sow farms, two finishing sites, and one boar stud were included in the study. 10 filters were tested per farm for a total of 100 filters tested.

A Camfil Clean Air Solutions CamTester PRO air filter testing system was used for the study. Each filter was tested 3 times for particle efficiency in 12 channels, and in accordance with ISO 29462. Filters were not tested to an actual MERV rating. Rather, data were evaluated at 0.4 microns to maximize separation in performance between filters. Pressure drop was tested three times per filter and an average determined. Equipment was recalibrated between each filter to account for wind and other external factors. All filters were tested at an airflow of 1200 cubic feet per minute of air flow.

Data collected on each filter included: months of service (winter months were excluded for seasonally used filters), MERV rating, position of cool cell in relation to filters, whether filter was damp or dry at time of testing, type of pre-filter, type of foliage and distance from foliage around the filter location, type of road and distance from road, proximity of filter to the ground, and whether the filter had observable damage. The filter manufacturer was also recorded. A total of three different filter manufacturers were included in the study and a total of 14 different filter models.

Ammonia testing was also done using a using a Dräger® glass tube method to observe differences in ammonia present outside each farm.

Seven different filter models were tested from three different manufacturers. Four of the 10 farms had 2 different filter types on the same farm. The purpose of different filters on those farms was to have a higher efficiency filter during the colder months and allow air to enter through a lower efficiency filter in the warmer months. There were 54 MERV14 filters, 18 MERV15 filters, and 28 MERV16 filters tested.

Results and discussion

MERV rating

MERV rating stands for minimum efficiency rating value and is a categorization of filter performance based on the percentage of particles captured by the filter across 12 channels of particle size. Filters generally tested into their original categories, regardless of filter age. There were no individual filters that "advanced" to the next MERV rating with increased age and/or plugging. There were significant differences in pressure drop ($P < 0.005$) between MERV14 and MERV15 filters, between MERV15 and MERV16 filters, and between MERV14 and MERV16 filters. There were also significant differences ($P < 0.005$) in particle capture efficiency at 0.4 micron between MERV14 and MERV15 filters, between MERV15 and MERV16 filters, and between MERV14 and MERV16 filters.

Age of filter

There was a small numerical reduction in efficiency of 3-4% for two farms that had MERV14 in place for 7 years (farms RB and RR, figure 1). This may have been to age and deterioration of the media, although there was no visible difference when testing these filters. Those same two farms did not have a significant difference in pressure drop compared to other farms with MERV14 filters (farms RB and RR, figure 2).

There were no apparent differences in efficiency for two brands of MERV15 filters across three different farms. For the MERV16 filters that were tested, farm RS had the lowest efficiency, but this was likely due to having two damaged filters test with lower efficiency. Farm SH, which had a different brand than the others, also tested lower. There did not appear to be an effect of age for the MERV16 filters, even though some of the filters were approaching 8 years of age. Like the MERV14 filters, there was no visible deterioration on any of the MERV15 or MERV16 filters.

Filter from farm RH, while testing in expected ranges for efficiency, had dramatically higher pressure drop compared to any of the other farms. There was significant dust load on this farm due to high traffic, nearby grain handling equipment and feed mill, and filters were visually dirtier on all the prefilters and main filters.

Cool cell position and dampness

Whether a cool cell was positioned in front of the filters made no difference in pressure drop (0.29 inches [W.G.] cool cell first vs. 0.33 inches W.G. filter first). It has been speculated that the wetness of a cool cell pad would result in more resistance through the

Figure 1: Average efficiency by farm and filter MERV rating.

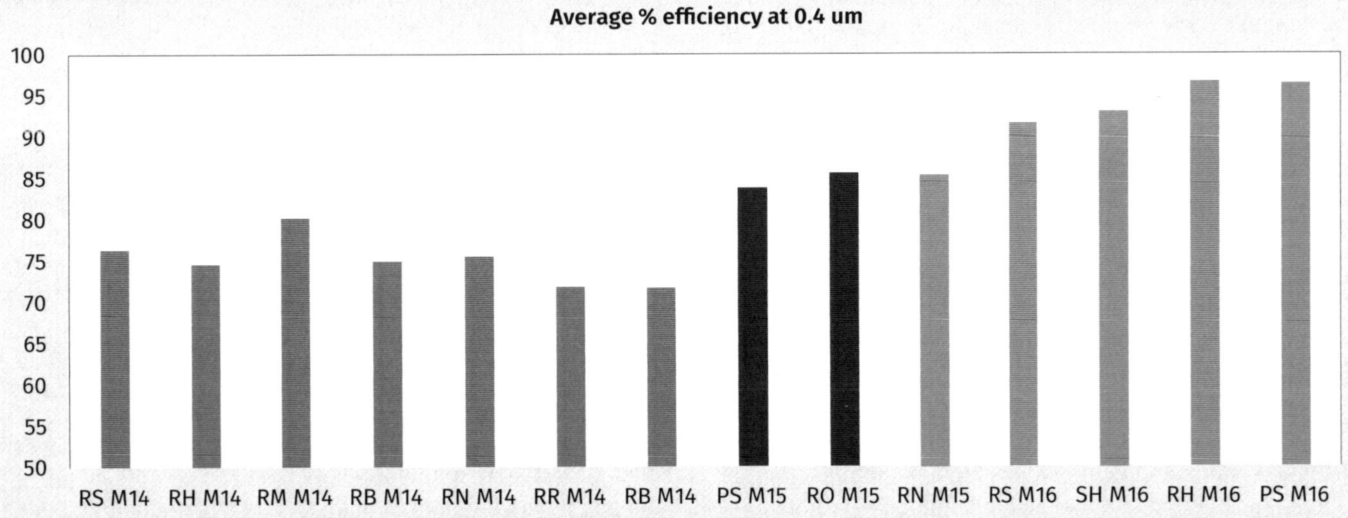

Figure 2: Average pressure drop by farm and filter MERV rating.

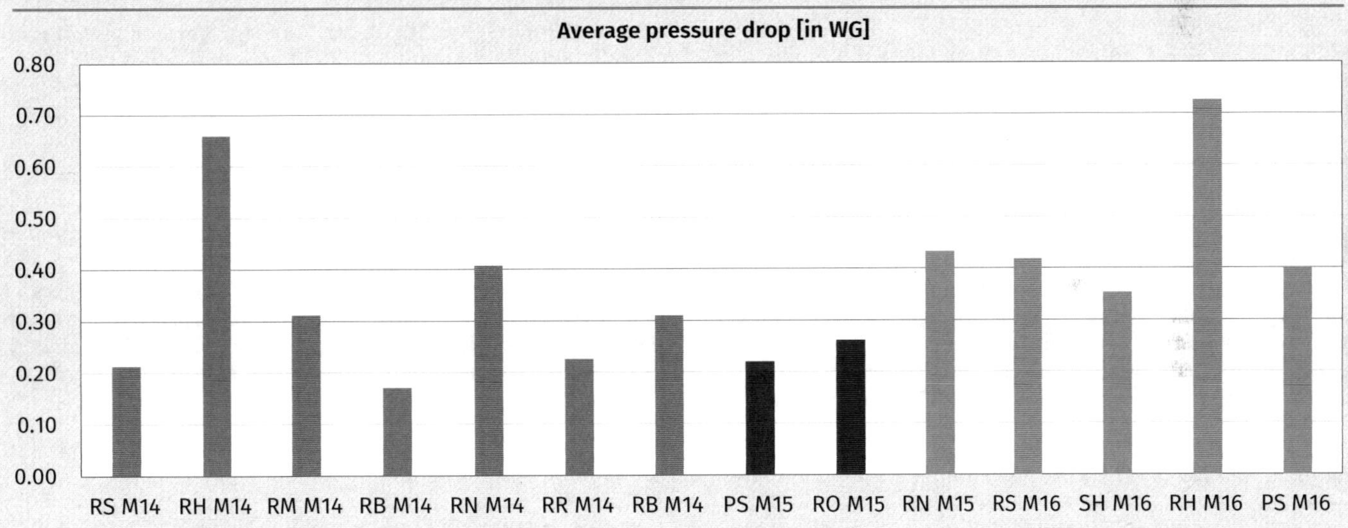

filters and compromised ventilation, but this did not seem to matter with the filters tested in this study. There was also minimal difference whether the filters were damp or dry, 0.29 inches W.G. for the dry and 0.32 inches W.G. for the damp filters.

It should be noted that none of the filters tested were wet to the point of saturation. We only observed filters that were damp, in cases where the cool cell was operating, and the cool cell was in front of the filters. One could speculate that there could be differences in cases where the filters are saturated.

Type of prefilter

Two of the farms using MERV16 filters in the attic went significant periods of time with no prefilter. While filters were noticeably dirtier, there was not much difference in pressure drop compared to farms using a MERV16 filter with a prefilter on banks (0.37 vs. 0.40 inches W.G.). This would suggest that in the attic, it may not be necessary to keep prefilters in place year round for some farms, which would improve air flow and reduce static pressure, particularly in the summer when ventilation is at maximum rates. There was not enough diversity in prefilter options to make any other comparisons.

Type of foliage, road surface, and separation distance

Regarding the type of foliage surrounding the location of the filters in the farm, the road surface, and the separation distance of the buildings from the road surface, there were some differences regarding pressure drop.

The effect of foliage on the area surrounding the filters (figure 3) had a significant impact on the pressure drop for MERV14 and MERV15 filters in filter banks ($P < 0.05$). Filters in banks that were surrounded by grass around the buildings and premise had lower pressure drop than filters that were on a farm surrounded by a farm field or gravel. For MERV14 filters, the pressure drop average across 24 filters surrounded by grass was 0.24 inches W.G.

compared to a pressure drop of 0.40 inches W.G. for 30 filters surrounded by dirt fields or gravel. There was a numerical, but not significant difference for MERV16 filters, which were primarily in attics. There was no effect of separation distance of the filters from field or grass.

The differences in pressure drop suggest that farms where the filters are not surrounded by grass lawns may need to change filters around twice the frequency as those that are surrounded by grass surfaces. While this has been obvious to farm staff when changing prefilters, there also appears to be a big difference with the MERV14, 15, and 16 filters.

There was not enough data to compare whether the road surface had an impact on pressure drop, mainly because all the farms with MERV14 filters were located on gravel roads. Also, there was not enough diversity in distance from the road to adequately compare the impact of filters close to the road compared to those further from the road. One could speculate that there would be a similar effect as with foliage, that filters near to extremely dusty environments are likely to have a much shorter lifespan.

Proximity to ground

When comparing tested filters that were in a bank within 2 feet of the ground to other filters in a bank, there was numerically more pressure drop in those that were at the ground, although the differences were not dramatic. For example, with MERV14 filters, those on the bottom row (within 2 feet of the ground) had a pressure drop of 0.32 inches W.G. compared to 0.28 for those in middle and upper rows of a bank. In the case of MERV15 filters, there was 0.30 inches W.G. for filters near the ground compared to 0.24 for filters in the upper rows of a bank. There were not enough MERV16 filters in banks to make comparisons.

Summary

Across 10 farms, testing 10 mechanical filters per farm, there were no significant or alarming results in regard to age of filter, when using mechanical filters that had been in place an average of 5.6 years, with a range of 9 to 106 months in use. The main significant different among filter was the MERV rating of the filter, both on efficiency and on pressure drop. For example, old MERV14 filters are not going to "advance" to a MERV15 performance level even when in place for many years. Likewise, higher MERV rated filters are not likely to descend to a lower level even in place for many years.

Filters located on farms that are exposed to high levels of particulate in the air such as grain handling equipment, feed mills, or being surrounded closely by corn or soybean dirt fields, are likely to plug filters much quicker than those that are surrounded by areas of grass lawn. Not only do prefilters plug up quicker, but the fine particulate penetration into the MERV14, 15, or 16 filter may mean those farms may need to replace filters twice as often, adding to significant cost differences in maintaining a filtration system over time.

The data in this study only refers to mechanical filters and does not apply to synthetic or antibacterial filters. It should also be noted that the differences in efficiency were all evaluated at 0.4 micron, and nonvisible larger particles are more likely to support viable virus over longer distances.[5]

References

1. Torremorell M, Alonso C, Olson B, Zarling D, Kuehn T. Standards of filter testing and testing used filters for PRRS virus. Leman Swine Conference 2014.

2. Dee S, Spronk G, Reicks D, Ruen P, Deen J. Further assessment of air filtration in preventing PRRS infection in large breeding herds. Vet Record. 2010. 167:976-977.

3. Dee S, Cano JP, Spronk G, Reicks D, Ruen P Pitkin A, Polson D. Evaluation of the Long-Term Effect of Air Filtration Occurrence of New PRRSV Infections in Large Breeding Herds in Swine-Dense Regions. Viruses 2012, 4, 543-662.

4. Reicks D. Filtration: 10 years and 100 farms later...what have we learned. Leman Swine Conference. 2014

5. Alonso C, Goede D, Morrison R, Davies P, Rovira A, Marthaler D, Torremorell M. Evidence of infectivity of airborne porcine epidemic diarrhea virus and detection of airborne viral RNA at long distances from infected herds. *Veterinary Research* 2014, 45:73.

Figure 3: Average pressure drop for filters surrounded by grass vs. dirt fields.

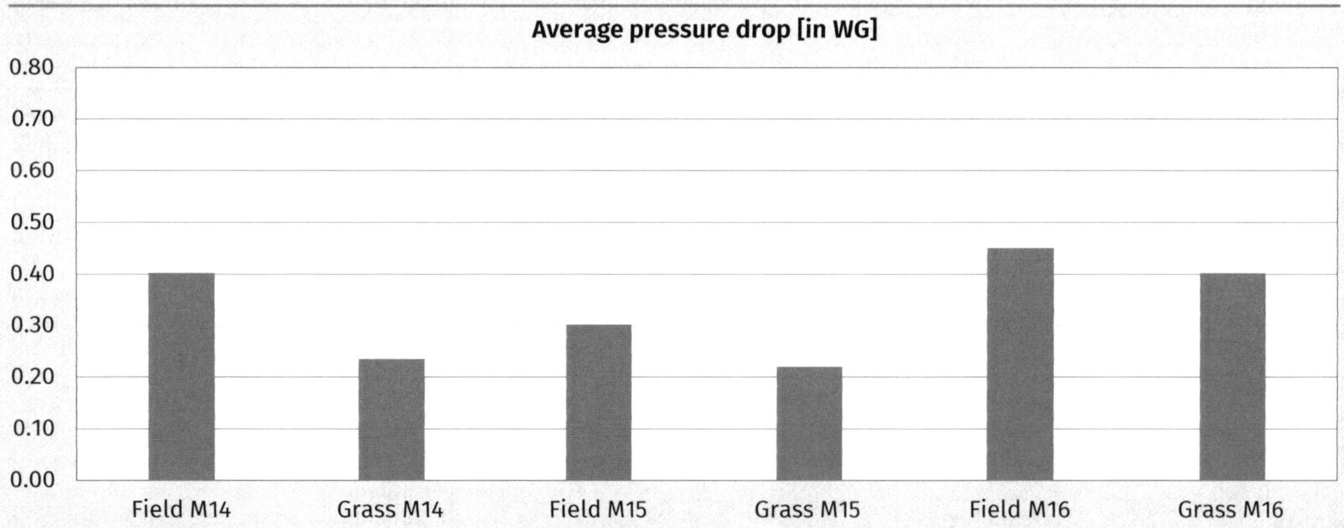

A case study of reproductive failure associated with porcine circovirus type 3 in an endemically infected commercial sow farm

Channing Sebo[1], DVM; Allison Knox[2], DVM; Grant Allison[2], DVM; Bailey Arruda[3], DVM, PhD; Kelly Greiner[2], DVM; Derald Holtkamp[3], DVM, MS; Brad Thacker[1], DVM, PhD; Chong Wang[3], PhD
[1]Merck Animal Health, De Soto, Kansas; [2]Walcott Veterinary Clinic, Walcott, Iowa; [3] Iowa State University, Ames, Iowa

Introduction

Porcine circovirus type 3 (PCV3) was first diagnosed in the US in 2015, with the primary disease manifestation of reproductive failure.[1] The true costs and implications of the disease as well as the role of vaccination for controlling the disease merit further research. PCV3 SEQUIVITY® vaccination of sow herds has been used in the field since 2018.[2] Vaccination impact has been primarily measured through monitoring processing fluids longitudinally over time, with apparent reduction or elimination of PCV3 approximately 16 weeks post-vaccine implementation.[2] The need for field evaluations with non-vaccinated controls was evident in order to understand the full impact of vaccination on sow farms. In addition, there was also a need to evaluate sample types other than processing fluids for monitoring sow herd PCV3 status and to aid in the diagnosis of PCV3 in farms experiencing reproductive failure.

Materials and methods

A commercial 3,000 head sow farm with a history of PCV3 associated reproductive failure was recruited for the study. Previous diagnostic testing had shown processing fluids as well as stillborn and mummified fetuses to be consistently PCR positive for PCV3. The sow farm had previously not used any PCV3 vaccine, was porcine circovirus type 2 (PCV2) stable, *Mycoplasma hyopneumoniae* stable, and PRRSV positive but stable[3] at the start of the trial. One-half of the gilts entering the farm in three monthly shipments between June and August 2019 were randomly assigned to vaccine or control treatment groups with the vaccine group receiving a 1ml, PCV3/PCV2d SEQUIVITY® vaccine twice, 3-4 weeks apart and at least 2 weeks prior to breeding. PCV2d was included in the vaccine as a part of the farms ongoing maintenance program and the impact of the PCV2d portion of the vaccine was not evaluated in the trial due to minimal number of PCV2d positive samples (data not shown). Overall, 418 gilts entered the farm during this time period. A subset of these gilts was randomly selected for trial enrollment, 87 vaccinates and 82 controls. Gilts in both treatment groups went through routine gilt acclimation and vaccinations in accordance with on-farm protocols. The farm staff was blinded to the treatment group. Cross-fostering was not allowed for male piglets while female piglets were cross-fostered as needed. The entry-to-farrow interval was calculated for each gilt based on the farm's production records, and a PRRS status was assigned based on the ongoing processing fluid monitoring. Gilts were bled at arrival (prior to vaccination), at pregnancy check (approximately 35 days gestation) and again 2-3 weeks prior to farrowing. At farrowing, any litter with a combined total of 3 or more mummified or stillborn fetuses was submitted to the Iowa State University Veterinary Diagnostic Laboratory (ISU-VDL) for a complete reproductive failure work-up. From all litters of enrolled gilts, both colostrum and processing fluids were collected by litter. Processing fluids included testicles from all male piglets within the litter. No tails were included in processing fluid samples. At weaning, 3 male piglets per litter were randomly selected for blood collection. Gilt serum, colostrum, processing fluids, and piglet serum were all tested using a duplex qPCR for PCV2 and PCV3. Gilt serum was pooled by 5 within each treatment group and all PCV3 positive pools were retested individually. Processing fluids and colostrum samples were tested individually by litter. Piglet serum was pooled prior to submission so no individual piglet testing was done. The reproductive workup on fetal tissues included testing for the following pathogens: PCV3, PCV2, porcine circovirus type 1 (PCV1), porcine reproductive and respiratory syndrome (PRRSV; both US and European variants), *Leptospirosis*, and porcine parvovirus 1 & 2 (PPV-1, PPV-2). A histologic evaluation was performed on all submitted litters as well. PCV3 *in situ* hybridization (ISH) was performed on tissues at the discretion of the pathologist. The "standard" positive/negative cutoff for the PCV3 portion of the duplex PCR was a Ct value of 37.0. A PCV3 reproductive failure diagnosis was defined by the following: PCV3 PCR positive with Ct values < 20 and histologic lesions consistent with PCV3 infection.

Univariate statistical analysis comparing qPCR results from litters from vaccinated gilts versus control gilts was performed by Kruskal-Wallis One-Way AOV, General One-way AOV, Chi-Square or Fishers Exact Test depending on data type and distribution (Statistix™ 10, Analytic Software, Tallahassee, FL). A multivariate analysis of litter PCR data for each sample type was analyzed using linear mixed logistic regression models with vaccination, entry-to-farrow interval and PRRS status as fixed effects and entry group as a random effect (PROC GLIMMIX, SAS Institute Inc., Cary, NC). Gilt serum PCR data was analyzed using a repeated measures mixed logistic regression model with vaccination, entry-to-farrow interval and PRRS status as fixed effects and entry group as a random effect. (PROC GLIMMIX, SAS Institute Inc., Cary, NC).

Results and discussion

Reproductive diagnostic investigation

A total of 169 gilts and their litters, 87 vaccinates and 82 controls, were evaluated in the trial. Thirty-eight litters met the reproductive workup criteria of a combined total of 3 or more mummified or stillborn fetuses and all but one were submitted to the diagnostic laboratory. Natural challenge of PCV3 occurred on the farm throughout the trial based on finding PCV3 positive samples throughout the study and in all three gilt entry groups. A summary of the results with regard to the final diagnosis and PCV3 PCR results for each reproductive case are presented in Table 1. The denominator for the rates presented in Table 1 is the overall number of litters evaluated in the study and not the number of litters submitted for diagnostic investigation.

Table 1: Results of reproductive failure case workups

			Unit	Treatment group: Vaccine	Treatment group: Control
Litters enrolled			Number of litters	87	82
Litters submitted for reproductive work up			Number of litters	16	21
			Percent	18.40%	25.60%
Laboratory test results			**Case results**		
PCV3 PCR	**PCV3 lesions**	**PRRSV PCR**	**Unit**	**Vaccine**	**Control**
Positive	Positive	Negative	Number of litters	0[a]	6[b]
			Percent	0[a]	7.3%[b]
			Ct average (Ct Range)	Not detected	10.7 (5.6-18.4)
Positive	Negative	Negative	Number of litters	3	3
			Percent	3.45%	3.66%
			Ct average (Ct Range)	33.7 (33.4-34.2)	31.1 (27.9-33.5)
Positive	Negative	Positive	Number of litters	0	4
			Percent	0%	4.88%
			Ct average (Ct Range)	Not detected	34.5 (32.0-36.0)
Negative	Negative	Positive	Number of litters	4	2
			Percent	4.60%	2.44%
Negative	Negative	Negative	Number of litters	9	6
			Percent	10.34%	7.32%

[a,b] Values with different superscripts within a row were significantly different ($P < 0.05$).

All fetal tissues were PCR negative for PCV1, PCV2, European PRRSV, *Leptospirosis,* PPV-1 and PPV-2. Prior to the start of the trial, the farm was PRRSV stable, but became unstable in November 2019 based on PRRSV PCR positive processing fluids through routine on-farm monitoring. The PRRSV instability was also evident as 10 of the 37 litters submitted for a reproductive work-up (4 vaccinates and 6 controls) were PRRSV PCR positive and most had lesions consistent with PRRSV infection.

A PCV3 reproductive failure diagnosis was defined by the following: PCV3 PCR positive with Ct values < 20 and histologic lesions consistent with PCV3 infection. Six litters, all in the control group, met the criteria for PCV3 reproductive failure diagnosis. PCV3 associated lesions observed during the histologic evaluation included perivascular aggregates of predominately lymphocytes in multiple organs including the heart, lung, kidney and liver, locally extensive myocardial fibrosis, and perivascular lymphocytes and plasma cells with rare areas of gliosis in the cerebrum. PCV3 was detected by ISH in all six litters. No reproductive failure due to PCV3 occurred in vaccinated litters.

The cause of reproductive failure was not identified in 21 litters (12 vaccinated and 9 control litters). A subset of these litters had low amounts of PCV3 without lesions present; three vaccinated litters with Cts ranging from 33.4-34.2 and seven control litters with Cts ranging from 27.9-36.0. The significance of detecting PCV3 at these Cts and without the presence of lesions is unknown. Of the 10 cases (3 from vaccinated litters and 7 from control litters) that were PCV3 positive but lacked lesions supporting the diagnosis of PCV3 reproductive failure, 4 of the cases from control litters were diagnosed with PRRSV. The fetal age at death for all three diagnostic outcomes was similar with an average age at death of 100 days and a range of 79-111 days.

Gilt serum

Approximately 10% of both vaccinated and control gilts entering the farm were positive for PCV3 with no statistical difference between the two groups (Table 2). The vaccinated gilts had higher average Cts (lower virus levels) compared to controls at both pregnancy check and pre-farrow as well as lower percent positives at each sampling time point. Combining the pregnancy check and pre-farrow results, 3.4% of vaccinated gilts were PCV3 positive versus 28.0% of the control gilts. The odds that serum was positive at either pregnancy check or pre-farrow for the control gilts was 3.14 times higher than for the vaccinated gilts. No significant findings were found with regard to the PCV2 PCR results (data not shown).

Litter sample results

The results of the litter sample results are presented in Table 3. Processing fluids, colostrum, and serum all had higher average Ct (lower virus levels) in vaccinated litters compared to control litters. Vaccinate litters had lower percent of litters positive vs. control for all three sample types, but piglet serum was the only sample that was statistically significant (3.4% for vaccinates versus 24.4% for controls). The odds that serum from control litters was positive was 9.02 times higher than vaccinated litters serum.

To account for and differentiate PCV3 mild infections and environmental contamination in these litter samples, an alternate, "severe" Ct cutoff was established for each sample type. The cutoffs were established by calculating the mean Ct value, regardless of the treatment group, minus one standard deviation. A separate calculation was made for each sample type. Interestingly, the piglet serum had the greatest correlation between the "standard" percent positive based on the laboratory's cut-off (37.0) and the "severe" percent positive. This finding supports the expected lack of environmental contamination in serum versus the other sample types.

A litter status criteria was created for all three samples types. If any of a litter's samples (colostrum, processing fluids or piglet serum) were positive using the standard cut-off, their status was established as positive. Using the standard cutoffs, the percent positive in vaccinated litters was 85.1% while 97.6% of control litters were positive. Using "severe" cut-off to classify samples, the percent positive in vaccinated litters was 13.8% while 35.4% of control litters were positive. The odds of a control litter having at least one positive PCV3 sample based on the "severe" cutoff was 3.41 times that of vaccinated litters.

Multivariate analyses indicated that vaccination group was the only factor that influenced the Ct values and positivity rates for all three sample types, with the exception of processing fluids. The Ct values and percent positive using the laboratory cut-off ($Ct < 37$) were influenced by the number of days from entry to farrowing and PRRSV status. Increasing days from entry-to-farrowing tended to reduce the rate of positive litters and increased Ct values (less virus). Positive herd PRRSV status tended to reduce the rate of PCV3 positive litters. There was no impact of these two factors on the percent positives for all three sample types using the "severe" cutoffs. No significant findings were found with regards to PCV2 PCR results for any of the litter sample types (data not shown).

A secondary objective of this field study was to guide producers and veterinarians with regard to sample collection for PCV3 diagnostics. Processing fluids are commonly collected on sow farms for monitoring pathogens like PRRSV. Given the ease of collection and how common and routinely these samples are collected, veterinarians can consider using processing fluids for long-term strategic monitoring for PCV3 and PCV2 for sow farm stability as well as when assessing intervention strategies such as vaccine implementation. Based on our findings, using alternate "severe" positive/negative cutoffs versus the laboratory standard cutoff may be indicated although further work is needed. In this study, we took a "mathematical" approach to establishing an alternate cutoff. A potentially better approach would be to use a "biological" approach for determining a suitable cutoff.

Another secondary objective of this study was to measure reproductive performance in relation to the PCV3 associated reproductive failure, but this was impossible due to interference by PRRSV infections documented throughout the study. Unfortunately, the reproductive production record analysis was confounded and is not reported.

Table 2: PCV3 PCR results – gilt serum

Sample type	Unit	Vaccine	Control
Gilt arrival	Average Ct	36.8	36.9
	Lowest Ct	26.0	34.1
	% Positive	10.3%	8.5%
Pregnancy check	Average Ct	37.1^{a}	36.6^{b}
	Lowest Ct	36.1	29.0
	% Positive	2.3%^{a}	13.4%^{b}
Pre-farrow	Average Ct	37.1^{a}	36.5^{b}
	Lowest Ct	36.3	28.8
	% Positive	1.1%^{a}	20.7%^{b}
Pregnancy check or Pre-farrow	% Positive	3.4%^{a}	28.0%^{b}

^{a,b} Values with different superscripts within a row were significantly different ($P < 0.05$).

Table 3: PCV3 PCR results – litter samples

Sample type	Unit	Vaccine	Control
Processing fluids	Average Ct	33.5^{a}	32.3^{b}
	Lowest Ct	22.4	19.1
	% Positive	81.6%	90.2%
	PCT ≤29.1	8.0%^{a}	20.7%^{b}
Colostrum	Average Ct	34.8^{a}	32.8^{b}
	Lowest Ct	25.4	21.0
	% Positive	64.4%	75.6%
	PCT ≤30.2	8.0%^{a}	22.0%^{b}
Piglet serum	Average Ct	36.8^{a}	35.7^{b}
	Lowest Ct	27.6	26.3
	% Positive	3.4%^{a}	24.4%^{b}
	PCT ≤33.9	3.4%^{a}	19.5%^{b}
Litter status	% Positive	85.1%^{a}	97.6%^{b}
	% Severe	13.8%^{a}	35.4%^{b}

^{a,b} Values with different superscripts within a row were significantly different ($P < 0.05$).

Conclusion

PCV3 associated reproductive failure characterized by mummified and stillborn fetuses was diagnosed in this study with documented lesions and significant viral loads as well as ISH staining. All of the six litters diagnosed with PCV3 associated reproductive failure were from control litters.

This field evaluation examined the role of the SEQUIVITY® PCV3 vaccination of gilts and the impact of vaccination on reproductive failure in a field setting. None of the six litters diagnosed with PCV3 associated reproductive failure were from vaccinated litters. There was also significant reduction of viremia in vaccinated gilts compared to controls at both the pregnancy check and prior to farrowing, based on Ct values and positive/negative assessments.

Samples collected from vaccinated litters also demonstrated the value of PCV3 SEQUIVITY® vaccination with lower amounts of virus (higher Cts) and fewer vaccinated gilt litters positive compared to control gilt litters.

Overall PCV3 reproductive disease deserves further investigation on farms experiencing increased rates of mummified and stillborn fetuses. The impact and value of vaccination needs further investigation. However, the results of this study suggest that PCV3 vaccination warrants consideration within swine operations.

References

1. Byers, E. 2018. Porcine circovirus type 3: Practitioner perspective. Proc. 49th AASV Annual Meeting, 445-447.

2. Creel, J et. al. 2020 Porcine circovirus type 3: update on field experiences with Sequivity™ vaccination for this emerging pathogen. Proc. 51st AASV Annual Meeting, 107-108.

3. Holtkamp D, Polson D, Torremorell M, and committee members. Terminology for classifying swine herds by porcine reproductive and respiratory syndrome virus status. J Swine Health Prod. 2011;19:44-56.

Efficacy of commercial PCV2 vaccine using a PCV2d and PRRSV co-infection model

Brett O'Brien[1], VMD, MS; Keith DeDonder[2], DVM; Kelly Lechtenberg[2], DVM, PhD; Kimberly Crawford[1], DVM; Brad Thacker[1], DVM, PhD, Dipl AVBP; Eric Burrough[3], DVM, PhD, DACVP
[1]Merck Animal Health, De Soto, Kansas; [2]Midwest Veterinary Services, Manhattan, Kansas;
[3]Iowa State Veterinary Diagnostic Laboratory, Ames, Iowa

Introduction

Porcine circovirus type 2 (PCV2) remains an endemic disease within the US swine population and thus vaccination for this disease is a critical component of current swine herd vaccination programs. In the past decade, we have seen a shift in the prevalence of PCV2 subtypes with PCV2d now being the dominant subtype circulating in US swine herds. Therefore, it is necessary for producers and veterinarians to have confidence that the commercial PCV2 vaccines utilized in their operations offer protection against PCV2d. Experimental studies conducted previously have demonstrated the efficacy of Circumvent® PCV G2 against PCV2d.[1-4] The objective of this study was to demonstrate the efficacy of Circumvent® PCV G2 against unvaccinated controls in a PCV2d/PRRSV co-infection challenge model.

Materials and methods

Sixty pigs were sourced from a commercial sow farm that was PRRSV naïve and IAV-S and *Mycoplasma hyopneumoniae* stable. Three pigs per litter were enrolled at weaning from twenty litters in which no cross-fostering had been permitted and had been pre-screened to verify litters were PCV2 PCR negative and had low PCV2 maternal antibodies (PCV2 IFA titer ≤ 320). Blocking by litter, pigs were randomly assigned to one of three treatment groups (T01: unvaccinated control; T02: Circumvent® PCV G2- 2 dose program at 3 and 6 weeks of age (WOA); T03: Circumvent® PCV G2- 1 dose program at 3 WOA). See Table 1 for treatment group schematic. Pigs were weaned into one room at the Midwest Veterinary Services research facility and were housed in 20 pens by individual litter for the entire study. All pigs were given three days to acclimate to their new environment prior to receiving their first vaccination. All pigs were co-challenged intranasally with a PCV2d field isolate (5 logs of virus per pig) and PRRSV (6 logs of virus per pig) at 9 WOA. All pigs were treated with Excede® per label directions one day prior to challenge to help control the level of secondary bacterial infections. Clinical observations were recorded daily throughout the study. All pigs were bled immediately prior to vaccination at 3 and 6 WOA, immediately prior to challenge, and then weekly until necropsy. All pigs were necropsied four weeks post-challenge (13 WOA) and fresh and fixed tissues (lung, liver, spleen, kidney, tonsil, ileum, and mesenteric and bronchial lymph nodes) were collected and transported to the Iowa State University Veterinary Diagnostic Laboratory (ISU VDL) within 24 hours of collection. Pigs were weighed immediately prior to vaccination at 3 WOA, immediately prior to challenge, and one day prior to necropsy so that average daily weight gains during the vaccination and challenge phases of the study could be calculated.

All serum was tested at ISU VDL by PCV2 qPCR and PCV2 ELISA and at the Kansas State University Veterinary Diagnostic Laboratory (KSU VDL) to determine PCV2a, PCV2b, and PCV2d IFA titers. Serum collected one week post-challenge was also tested at ISU VDL by PRRSV qPCR and sequencing to confirm each pig was appropriately challenged.

Lung, spleen, tonsil, and mesenteric and bronchial lymph nodes were tested by PCV2 qPCR to assess tissue viral loads, histopathology to determine the severity of microscopic lesions characteristic of PCV2 disease (specifically lymphoid depletion and granulomatous inflammation) and PCV2 IHC to determine the extent of tissue colonization. Lung tissues were tested by PRRSV qPCR to further confirm the effectiveness of the PRRSV co-challenge. Bacterial cultures were conducted on fresh tissues with visible gross lesions at necropsy. Lymphoid tissues (tonsil, spleen, Peyer's patches, bronchial lymph node (BLN), and mesenteric lymph node (MLN)) were scored by a pathologist at ISU VDL who was blinded to treatment group. Tissues were scored based on the level of lymphoid depletion (0-3; 0 = none, 1 = mild lymphoid depletion (LD) with loss of overall cellularity, 2 = moderate LD, 3 = severe LD with loss of follicle structure), level of inflammation (0-3; 0 = none, 1 = mild histiocytic to granulomatous inflammation, 2 = moderate histiocytic to granulomatous inflammation, 3 = severe histiocytic to granulomatous inflammation with replacement of follicles), and level of IHC staining (0-3; 0 = none, 1 = sparse staining, 2 = moderate staining, 3 = extreme staining). Percent positives per group, mean and total tissue scores were calculated and analyzed for each scored category to assess the level of PCV2 disease in each pig and to determine the level of efficacy against the experimental challenge provided by each vaccinated group compared to the unvaccinated control group.

Table 1: Treatment group schematic

Treatment group	Vaccine	Abbreviation	Dose (mL)	Timing
T01	Control (sterile saline)	CON	2	3 WOA
T02	Circumvent® PCV G2	CVENT2	1	3 & 6 WOA
T03	Circumvent® PCV G2	CVENT1	2	3 WOA

Statistical analysis was performed by statisticians who were blinded to treatment group. Block (litter) was included as a random intercept term in the models used. For the analysis of PCV2 tissue scores, ADG, and body weights (BW), general and generalized linear mixed models were used in PROC GLIMMIX. Fixed effects included treatment group (+/- sex and the interaction). Nonparametric methods and/or exact confidence intervals were calculated when GLM did not converge or make valid pairwise comparisons. For PCV2 qPCR, PCV2 ELISA, and PCV2 IFA's a repeated measures analysis was performed with fixed effects for treatment, time, and their interaction and random effects to account for covariance among observations within pigs and clustering within pens. Pairwise comparisons were adjusted for multiple interactions using Tukey methods. Statistical significance was assigned in this study at $P \leq 0.05$.

Results

All study pigs were confirmed PCV2 PCR negative until the time of challenge and the unvaccinated control group remained PCV2 IFA negative until the time of challenge. Pig health aside from the PCV2 and PRRSV challenge was good throughout the entire study as no significant secondary diseases were observed.

The challenge model was successful as evidenced by 100% of pigs being confirmed viremic by PRRSV PCR 7 days post-challenge (virus sequence was 100% homologous to the challenge strain given), 100% of pigs regardless of the treatment group being confirmed PCV2 PCR positive in lung tissues, and severe PCV2 lesions were observed in the positive control group.

The level of PCV2 protection conferred by vaccination was assessed in the tissues collected at necropsy for the presence and severity of lymphoid depletion, lymphoid inflammation, and IHC staining intensity as well as viral tissue loads. For lymphoid depletion, both Circumvent PCV G2 treatment groups showed a significantly lower proportion of pigs positive for lymphoid depletion than controls ($P < 0.05$, see Table 2). When comparing mean and total LD scores, both CVENT2 and CVENT1 had significantly lower scores than controls ($P < 0.01$) but were not different from one another. The proportion of pigs positive for granulomatous inflammation and mean inflammation scores were lower in Circumvent PCV G2 vaccinated pigs versus controls ($P < 0.05$, see Table 3). Total inflammation scores were significantly highest in controls at 5.75 compared to the other treatment groups ($P < 0.01$, CVENT2 = 0.35; CVENT1 = 0.45). Both vaccinated groups had a significantly lower proportion of pigs with PCV2 IHC staining in tissues than controls ($P < 0.05$, see Table 4). Both vaccine groups showed reduced PCV2 viral tissue loads when compared to controls which had a mean qPCR (copies per mL) of 16,058,316,006 which equates to a CT of 12.8 ($P < 0.01$). CVENT2 group had the lowest levels of virus present in the tissues by qPCR at 148,970 copies per mL(CT = 29.7) which was more than a 99.9% reduction in viral loads when compared to

Table 2: Incidence of lymphoid depletion per treatment group and tissue scores

Treatment group	% Positive (95% CI)*	Mean LD score	Total LD score
Control	100 (83, 100)	1.27^{a}	6.56^{a}
Circumvent® PCV G2- 2 dose	10 (1, 32)	0.00^{b}	0.10^{b}
Circumvent® PCV G2- 1 dose	30 (12, 54)	0.15^{b}	0.80^{b}

* CI that do not overlap indicate significant differences at $P < 0.05$

a-b Columns with different superscripts listed denote statistical significance at $P \leq 0.01$

Table 3: Incidence of granulomatous inflammation (GI) per treatment group and tissue scores

Treatment group	% Positive (95% CI)*	Mean GI score	Total GI score
Control	100 (83, 100)	1.11^{a}	5.75^{a}
Circumvent® PCV G2- 2 dose	30 (12, 54)	0.30^{b}	0.35^{b}
Circumvent® PCV G2- 1 dose	25 (9, 49)	0.30^{b}	0.45^{b}

* CI that do not overlap indicate significant differences at $P < 0.05$

a-b Columns with different superscripts listed denote statistical significance at $P \leq 0.01$

Table 4: Percentage of IHC positives per treatment group and tissue scores

Treatment group	% Positive (95% CI)*	Mean IHC score	Median Total IHC score
Control	100 (83, 100)	2.24^{a}	12.5^{a}
Circumvent® PCV G2- 2 dose	10 (1, 32)	0.00^{b}	0.0^{b}
Circumvent® PCV G2- 1 dose	15 (3, 38)	0.24^{b}	0.0^{b}

* CI that do not overlap indicate significant differences at $P < 0.05$

a-b Columns with different superscripts listed denote statistical significance at $P \leq 0.01$

controls and was followed by CVENT1 at 9,736,435. The two dose vaccination program lowered the viral loads in tissues best.

Figure 1 shows the percentage of pigs per treatment group that were ever recorded viremic post-challenge. Consistent with the tissue data, the level of viremia post-challenge was significantly lowest in the two dose vaccine group followed by the one dose ($P < 0.01$, see Table 5).

Serology results during the vaccination phase show profound titer (ELISA and IFA's) increases present in both Circumvent PCV G2 treatment groups following vaccination ($P < 0.01$). All treatment groups showed a marked PCV2 IFA titer increase during the post-challenge time period but the magnitude of this response was largest in the control group. The same effect was noted with PCV2 ELISA titers on all but the CVENT2 group which actually declined slightly post-challenge. Across all PCV2 IFA tests, the same post-vaccination responses previously referenced were similar between tests although the magnitude of the effect differed (PCV2b IFA was highest followed by PCV2d then PCV2a).

There was no difference in ADG (lbs.) between treatment groups during the vaccination phase of this study (CON = 1.36; CVENT2 = 1.44; CVENT1 = 1.43). However, the ADG post-challenge was significantly higher in the Circumvent PCV G2 groups compared to the controls ($P < 0.01$, CON = 1.41; CVENT2 = 1.65; CVENT1 = 1.7).

Discussion

As in previous studies, Circumvent PCV G2 offered a high level of PCV2d protection when compared to controls as evidenced by reduced lymphoid depletion, granulomatous inflammation, tissue colonization, and viral tissue and serum loads as well as improved ADG post-challenge. The two dose vaccination program conferred the strongest level of PCV2 control in this study which is consistent with the Merck Animal Health's dosing regimen recommendations for Circumvent PCV G2's use in the face of a more significant PCV2 challenge. Given the results of the study described herein, producers and veterinarians should remain confident that Circumvent PCV G2 provides solid protection against PCV2d even in the face of significant health challenges such as PRRSV co-infection as demonstrated in this experimental challenge model.

References

1. Strait E. et al. Experimental PCV2d challenge of pigs vaccinated with prototype vaccines containing ORF2 from PCV2a, PCV2b, or PCV2d. AASV. 2017;189-190.

2. Thacker B. et al. Circumvent® PCV-M G2: PCV2d challenge study and field production update. AASV. 2016;151-153.

3. Opriessnig T. et al. A commercial vaccine based on PCV2a and an experimental vaccine based on a variant mPCV2b are both effective in protecting pigs against challenge with a 2013 US variant mPCV2b strain. *Vaccine.* 2014;32:230.

4. Opriessnig T et al. Commercial PCV2a-based vaccines are effective in protecting naturally PCV2b-infected finisher pigs against experimental challenge with a 2012 mutant PCV2. *Vaccine.*2014;32:4342.

5. Opriessnig T et al. Experimental reproduction of postweaning multisystemic wasting syndrome in dual infection with *Mycoplasma hyopneumoniae* and porcine circovirus type 2. *Veterinary Pathology.* 2004;41(6):624-40.

Figure 1: Percentage of PCV2 viremic pigs post-challenge

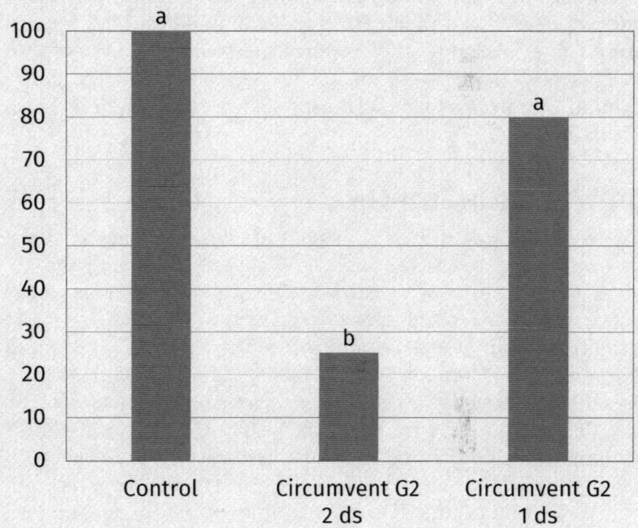

Different superscripts listed denote statistical significance at $P \leq 0.05$

Table 5: Post-challenge PCV2 viremia in serum (genomic copies/mL)

Treatment group	1 wk. post-chall.	2 wks. post-chall.	3 wks. post-chall.	4 wks. post-chall.
Positive Control	389	361,049,977[a]	141,881,404[a]	22,001,359[a]
Circumvent® PCV G2- 2 dose	0	4,564[c]	1,540[c]	1,498[c]
Circumvent® PCV G2- 1 dose	257	328,302[b]	111,009[b]	104,963[b]

Columns with different superscripts listed denote statistical significance at $P \leq 0.01$

Efficacy of a two dose Porcilis™ Ileitis regimen against experimental *Lawsonia intracellularis* challenge

Brad Thacker[1], DVM, PhD; Greg Armbruster[1], DVM, MBA; Brett O'Brien[1], VMD, MS; Robyn Fleck[1], DVM; Kimberly Crawford[1], DVM, MS; Nathan Winkelman[2], DVM; Adam Mueller[2], DVM
[1]Merck Animal Health, Madison, New Jersey; [2]Swine Services Unlimited, Inc., Rice Minnesota

Introduction

In 2015, Merck Animal Health introduced to the US market Porcilis® Ileitis, an inactivated, ready-to-use vaccine for the control of ileitis caused by *Lawsonia intracellularis*.[1] The original label provided one dosing option, a single, 2 mL intramuscular injection given to pigs at 3 weeks of age or older. Now, in addition to the single dose option, the vaccine has been recently approved for administration using two, 1 mL injections starting as early as 3 days of age with a booster 3 weeks later. This added flexibility allows producers to choose a vaccination protocol that best fits their operations. The challenge study reported here evaluated the efficacy of the Porcilis® Ileitis two, 1 mL dosing regimen compared to unvaccinated controls using a well-established experimental challenge model.[2]

Materials and methods

Three-week-old, cross-bred, *Lawsonia* seronegative pigs (n = 64) were used in the study. On study day -1, pigs were tagged, and their gender and litter were recorded at the sow farm. Pigs were then weaned at approximately 3 weeks of age (WOA) and moved to an isolation facility for the remainder of the study. At arrival, pigs were weighed, bled and randomly allocated to treatment within litter with blocking by start weight and gender. The next day (study 0), 32 pigs were vaccinated IM with 1 mL of Porcilis Ileitis (vaccinated group). On study day 20, the same pigs were vaccinated again with 1 mL of Porcilis Ileitis. Thirty-two pigs were not vaccinated with Porcilis Ileitis at any time (control group) and received 1 mL of sterile saline on study days 0 and 20. All pigs were weighed and bled on study days -1, 41 and 63. On study day 41, all pigs were challenged by oral gavage with 40 mL of a gut mucosal tissue homogenate containing approximately 2.04×10^9 of *Lawsonia intracellularis*. Pigs were then observed every second day for general demeanor, abdominal distension and fecal consistency and color (data not shown). Fecal samples for qPCR testing were collected on 0, 7, 14, and 21 days post challenge (DPC). Fecal shedding data is reported as the sum of copies per mL for samples collected on 7, 14 and 21 DPC.

On 23 DPC, pigs were necropsied and the intestinal tracts were removed for macroscopic lesion scoring and sample collection. For macroscopic lesion scoring, jejunum, ileum, cecum and colon were scored for lesion severity (0-3 scale) and lesion length. These values were used to calculate a total lesion score that represented the sum of lesion severity multiplied by the lesion length (cm) for the four segments of the intestinal tract. For histopathology and immunochemistry (IHC) evaluations, two tissue sections were collected. Ileum was collected 1 inch proximal to the ileocecal junction from all pigs. A second tissue section was collected from an area that exhibited macroscopic lesions. If no lesions were observed, the second section was collected 10 cm proximal to the ileocecal junction. Tissues were placed in 10% buffered formalin solution immediately after collection. Both evaluations were scored on a 0-3 scale and the maximum score between the two sections was used for analysis. Because the histopath and IHC scores were the same for each pig, the results of each evaluation are reported as one variable. For tissue PCR, two areas were sampled. Ileal mucosa was scraped two inches proximal to the ileocecal junction from all pigs. A second scraping was performed from an area that exhibited lesions. If no lesions were observed, a second scraping was performed just proximal to where the second tissue section was obtained for histopath/IHC. The average of the two scrapings from each pig was used for analysis.

All laboratory testing except for serology was done at the Iowa State University Diagnostic Laboratory using standard methods and previously reported scoring criteria.[3]For serum antibody testing, the bioScreen Ileitis Antibody ELISA (Svanova, Uppsala Sweden) was performed by Merck Animal Health. Statistical analysis was performed by Kruskal-Wallis One-Way AOV, General One-way AOV, Chi Square or Fishers Exact Test depending on data type and distribution (Statistix™ 10, Analytic Software, Tallahassee, FL).

Results and discussion

The results are presented in Table 1. Fecal samples from all pigs were PCR negative at challenge. One pig in the vaccinated group died prior to challenge due to causes unrelated to the study.

The Porcilis Ileitis vaccinated group had significantly less macroscopic and microscopic lesions and lower levels of tissue infection and fecal shedding compared to the control group. The percent reductions by parameter in the vaccinated group compared to the control group were: total lesion score, 96.6%; histopath/IHC, 94.0%; mucosal tissue infection, 99.7%; feces at necropsy, 99.8%; and total fecal shedding, 92.7%. Weight gain from challenge to necropsy was significantly higher in the vaccinated groups compared to the control group ($P = 0.001$) and the percent improvement compared to controls was 26.9%. As for ELISA results, nearly all pigs were seronegative at the time of the first vaccination. Vaccination with two, 1 mL doses of Porcilis Ileitis induced a high level of serum antibodies by the time of challenge while the control group was seronegative. Following challenge and at necropsy, there was a slight increase in serum antibodies in the vaccinated group while 100% of the pigs in the control group seroconverted and had antibody levels similar to the vaccinated group.

The challenge study reported here clearly demonstrated that two, 1 mL doses of Porcilis Ileitis administered at 3 and 6 weeks of age provided a high level of protection against an experimental challenge with *Lawsonia intracellularis* at 9 weeks of age (3 weeks after

Table 1: Efficacy of Porcilis™ Ileitis (PI) in pigs vaccinated with 1 mL at 3 and 6 weeks of age (WOA) against experimental challenge with *Lawsonia intracellularis* at 9 WOA compared to non-vaccinated control pigs

Parameter	Variable		Treatment group: PI – 1 mL 3 & 6 WOA	Treatment group: Control
Group size	No. pigs		31	32
Macroscopic lesions	Total score		0.6^{a}	3.8^{b}
	Lesion length (cm)		6.2^{a}	74.2^{b}
	Total score x length		7.4a	220.2^{b}
Histopath and IHC	Maximum score		0.13^{a}	2.31^{b}
	% with score	0	93.5	18.8
		1	3.2	3.1
		2	0.0	6.3
		3	3.2	71.9
Tissue infection	Copies per mL		$168{,}883^{a}$	$48{,}399{,}559^{b}$
Feces at necropsy	Copies per mL		$29{,}393^{a}$	$11{,}876{,}150^{b}$
Fecal shedding	Total copies per mL		$1{,}016{,}194^{a}$	$13{,}586{,}018^{b}$
Weight gain	Average daily gain (lb)	Start to challenge	0.839^{a}	0.808^{a}
		Challenge to necropsy	1.598^{a}	1.260^{b}
Lawsonia ELISA	Percent inhibition	At 1st Vac.	0.9^{a}	4.6^{a}
		At challenge	80.5^{a}	4.0^{b}
		At necropsy	89.2^{a}	83.4^{a}
	No. positive/tested	At 1st Vac.	$1/32^{a}$	$3/32^{a}$
		At challenge	$31/31^{a}$	$0/32^{b}$
		At necropsy	$31/31^{a}$	$32/32^{a}$

[a,b] Values with different superscripts within a row were significantly different ($P < 0.05$).

the second vaccination). This dosing regimen provided effectively controlled Ileitis based on both gross and microscopic lesions, reducing the level of tissue infection and fecal shedding of *Lawsonia intracellularis,* and increasing the rate of body weight gain (ADG) along with inducing high levels of *Lawsonia intracellularis* antibodies post vaccination. The severity of macroscopic lesions in control pigs in this study was approximately 67% greater than a previous study we reported in 2018.[4] Accordingly, the results reported here confirm the efficacy of Porcilis Ileitis, administered as two, 1 mL injections at 3 and 6 weeks of age against the relatively severe challenge model used in this study.

References

1. Thacker B et al. Field experiences with Porcilis™ Ileitis: The newest tool for controlling *Lawsonia intracellularis*. AASV. 2017; 191.

2. Winkelman N et al. Validation of an ileitis mucosal homogenate seeder pig model to study the natural transmission of *Lawsonia intracellularis* in grow-finish pigs. AASV. 2017; 170.

3. Roerink F et al. Porcilis™ Ileitis: 20-week duration of immunity against *Lawsonia intracellularis* challenge. AASV. 2016; 144.

4. Thacker B et al. Efficacy of Porcilis™ Ileitis against *Lawsonia intracelluaris* challenge and post-vaccination seroresponse kinetics. AASV. 2018; 400.

Optimizing gut health in the face of enteric endemics

M. Hensch[1], DVM; J. Miller[1], DVM; J. Spencer[2], PhD; J. Lee[2], MS
[1]The Maschhoffs, LLC, Carlyle, Illinois; [2]United Animal Health Inc, Sheridan, Indiana

Introduction

Endemic enteric disease, caused by Rotavirus type A (RVA), Rotavirus type C (RVC), *Escherichia coli*, *Salmonella* and numerous other pathogens, can significantly impact performance in both the pre-wean and post-weaning stages of production. In a large production system, co-mingling piglets sourced from different sow farms of like-genetic and epidemic disease status post-weaning is often common practice. However, ensuring the same endemic disease status across, and even within sow farms is difficult. While vertical transmission from sow to piglet has been well documented, less is understood about lateral transmission of endemic diseases in comingled flows. Connecting the pre-weaning phase to the nursery phase of growth relative to endemic pathogen transmission could be a key to better understanding intervention strategies at both the sow and nursery level. To better understand the diversity and profile evolution of enteric pathogens within a production system, The Maschhoffs, LLC utilized PathKinex™, a comprehensive diagnostic surveillance tool from United Animal Health Inc. This tool was applied across two scenarios, 1) a large comingled flow and 2) a field investigation looking at pathogen evolution in tagged pigs over time. In the field investigation, a water-dispersible *Bacillus*-based product (PROVENT® ECL/1 PLUS, United Animal Health Inc, Sheridan, IN) was utilized in select barns to evaluate impact on mortality and antibiotic usage.

Large comingled flow investigation

The Maschhoffs, LLC production system includes several larger sow flows, each consisting of three to ten sow farms that send weaned pigs to fill nursery sites at the same time. Historically, challenges lay in managing endemic enteric pathogens in select flows and implementing effective intervention strategies. Better understanding the diversity within and between sow units, and the impact this diversity has on the lateral transmission of pathogens between sources of weaned pigs in the nursery, was the goal of this investigation.

Material and methods

A flow was identified with three sow farms consistently comingling weaned piglets into wean to market (WTM) sites with diagnosed endemic enteric concerns. These specific farms were all of the same porcine reproductive and respiratory syndrome virus (PRRSV), *Mycoplasma hyopneumoniae* (MHp), and enteric coronavirus status. Sow farms were also filled from the same gilt source and gilt developer unit (GDU). The clinical picture relative to enteric endemics at the sow farms was highly variable both within farm and between farms. Farm 1 and Farm 2 had 10-20% prevalence of diarrhea in piglets diagnosed as RVA which varied week by week. Farm 3 had a more severe clinical picture of 20-30% diarrhea, diagnosed with a range of pathogens including RVA, RVC, *E coli*, and *Clostridium perfringens*. All farms were utilizing a company protocol for pre-farrow natural planned exposure (NPE) using fresh feces, in addition to intensive Rotavirus NPE with individual farm specific material. As these pigs were placed in the nursery, pigs would have a high prevalence of diarrhea starting 2-3 days post placement and continuing for 2-3 weeks. Endemic pathogens identified in the nursery phase via university diagnostic lab analysis included RVA, RVC, Rotavirus type B (RVB), *E coli*, and various *Salmonella* serovars. Piglets exhibiting clinical disease required treatment following company protocol with both intramuscular antibiotics as well as oral antibiotics.

This investigation used PathKinex™ to evaluate pathogen diversity within and between flows. In the PathKinex™ process, genomic DNA and total RNA was extracted from rectal swabs and tested with a panel of qPCR assays for genus-or species-specific marker or virulence genes from *E coli*, *Clostridium* species. *L intracellularis*, *Salmonella* RVA and RVC. At each sow unit, five scouring litters and three non-scouring litters were identified. Within each of these litters, a rectal swab was collected from the sow and two of her offspring in the litter. Non-scouring litters were of the same age, and cross-fostering had not occurred. The three sow farms weaned into two different nurseries (Nursery 1 and Nursery 2) where pigs from each sow flow were penned separately. Rectal swabs were collected from scouring (approximately 5 pigs/sow unit) and non-scouring pigs (approximately 3 pigs/sow unit). Gene data were statistically analyzed to assess the differences in presence and quantity between sick and healthy pigs, but also unique differences between sow units and the offspring weaned into the two common nursery sites.

Results

Consistent with previous diagnostic lab results, RVC was correlated to each sow unit's sow population and to the downstream piglet populations. When evaluating the results from the three sow farms, RVC was the only pathogen target detected in every sow sampled at each sow farm. However, a variation in gene quantity was observed among sows within a specific farm (Figure 1). For RVA gene quantity, there were sows within each farm that had no detected virus, which shows the challenge of consistency with NPE.

Considerable differences were detected in *E coli* toxin and adhesin genes among sows at the same farm, but also between the different sow farms (Figures 2 and 3). Genes present at the sow farms (*K88, Paa, STa*, and *Stx2*) were also present in piglet rectal swabs in nurseries (Figures 4 and 5), and there was evidence of lateral transmission between flows when comingled in the nursery. However, the detection of F18 and LT toxin genes was only in nurseries, not sow farms, suggesting origin of introduction to these virulotypes during transport or at the nursery and not transferred from the sow units.

Conclusion

The diversity within individual farms could have multiple causes and highlights the difficulty in managing on farm NPE. Factors influencing NPE efficacy may include maturity of the animal, exposure differences, different protocols for GDU NPE, or on farm

Figure 1: PathKinex™ results of pathogen gene quantities detected in individual sows at three separate sow farms whose offspring are co-mingled in the nursery.

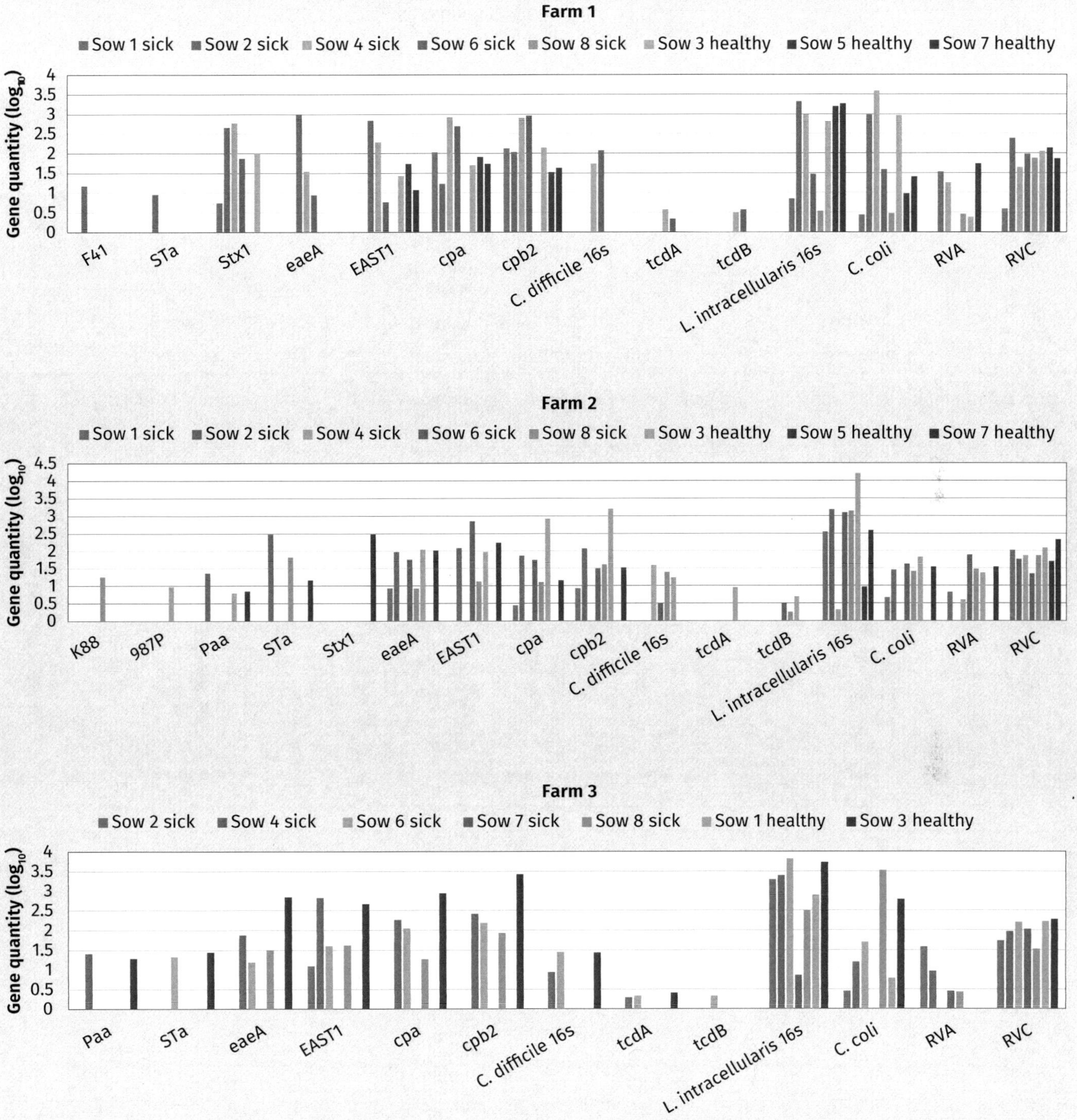

Figure 2: PathKinex™ results of *E coli* adhesin gene quantities detected in sows and suckling piglets from three individual sow farms.

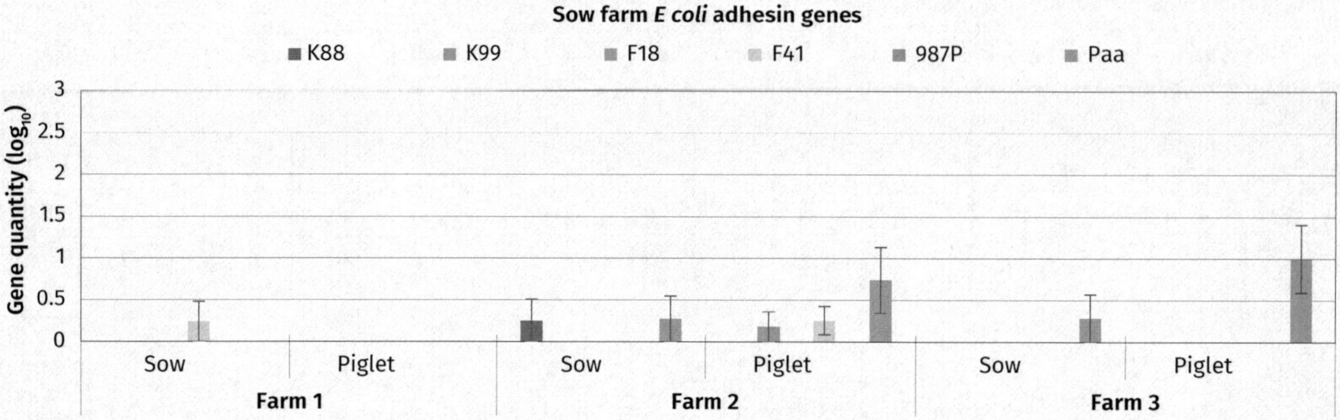

Figure 3: PathKinex™ Results of *E coli* toxin gene quantities detected in sows and suckling piglets from three individual sow farms

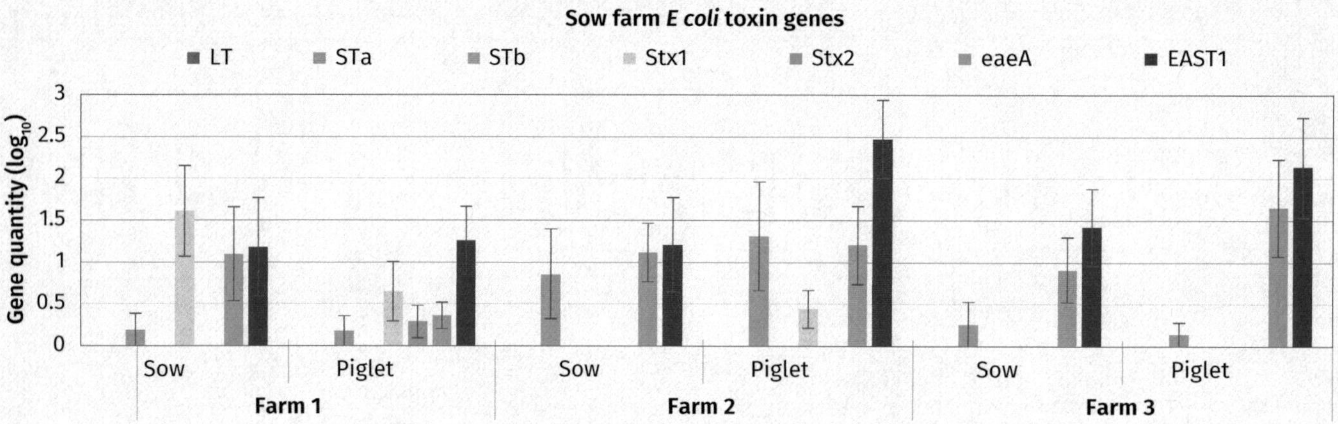

Figure 4: Pathkinex™ results of *E coli* adhesin gene quantities detected in weaned pigs originating from three different sow farms and co-mingled at two different nursery sites.

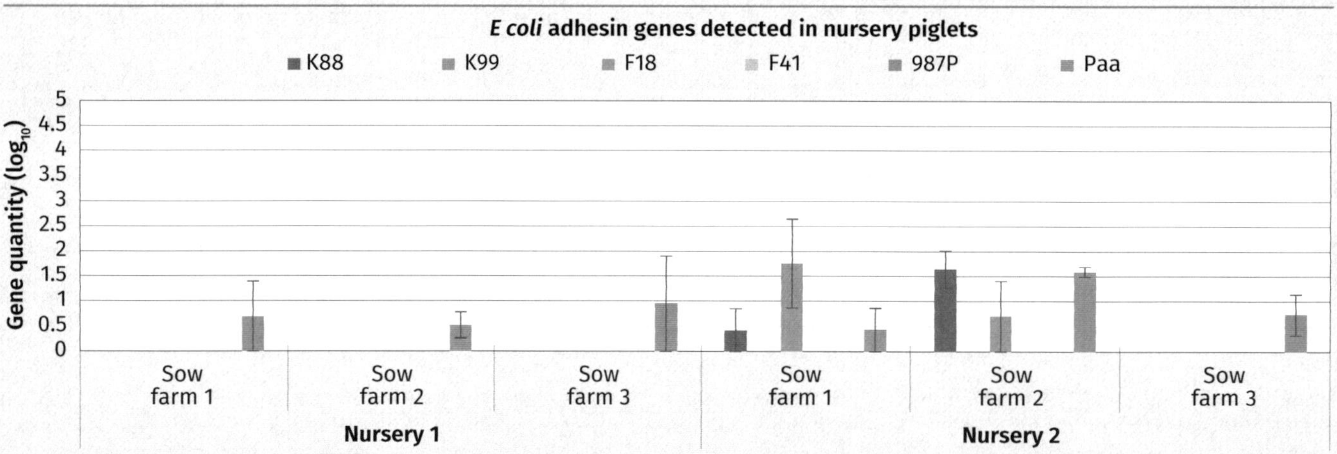

Figure 5: Pathkinex™ results of *E coli* toxin gene quantities detected in weaned pigs originating from three different sow farms and co-mingled at two different nursery sites.

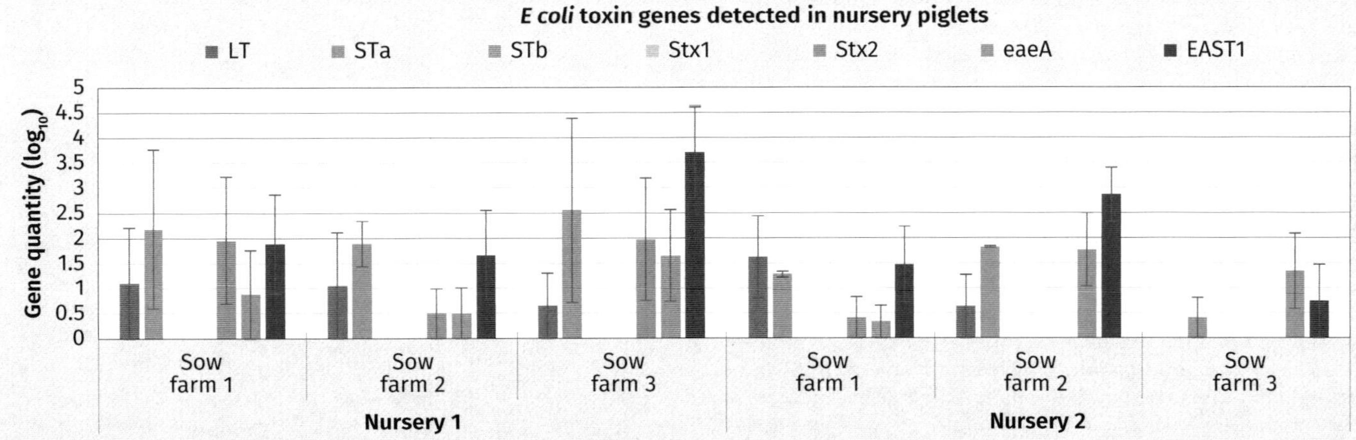

execution. For The Maschhoffs, LLC this identified an opportunity within each farm to firm up NPE protocols, including consistency of timing and quality of execution. NPE was audited and gaps in execution were identified on farm. It also caused consideration of additional exposure within the GDU system. The next difference identified was across different farms. After testing and biosecurity protocols were met, intensive rotavirus NPE containing RVA and RVC was started in the GDU in an effort to improve gilt immunity across farms. The same intensive rotavirus NPE was then used in all sow farms. No known negative effects were seen in the farrowing house or on nursery performance due to this change. There also remained the issue of identifying origin of pathogens specific to the nursery phase. Whether these were opportunistic pathogens from other sow farms causing secondary infection in combination with stressors of weaning, or site-specific issues, this was still unknown. PathKinex™ could be used to help determine site specific issues in comparison to vertical transmissions by looking at sow farm sources, as well as, multiple different nursery sites over time. If the same secondary pathogens can be consistently identified in different nursery sites, this would likely be secondary to the sow farm derived pathogens. By detecting what is coming from the sow herd and identifying changes occurring based on nursery pig age, PathKinex™ can help identify interventions such as sow herd NPE, on farm execution of NPE, or strategies to address site-specific nursery problems. Long term impact of the specific NPE changes are still pending.

Field investigation

Often in comingled flows with endemic enteric pathogens causing clinical disease, oral antibiotics are consistently used as an intervention strategy. This, however, can change gut flora substantially and be of a negative impact to the pig.[1] In this investigation, PathKinex™ was used by The Maschhoffs, LLC to determine pathogen profile of pigs entering a WTM site and evaluate the profile evolution leading up to an expected break at day 28 post wean. PROVENT® ECL/1 PLUS was used in select barns to evaluate the impact of a *Bacillus*-based microbial on nursery mortality and antibiotic treatments needed.

Material and methods

A flow consisting of three sow farms (independent of previously discussed flows) were selected due to a long history of enteric endemic disease. As with the previous case, this flow maintained the same epidemic pathogen status at the sow farm level (PRRSV, MHp, and enteric coronavirus). Clinical diarrhea was minimal at the sow farm, yet post-wean pigs would experience 20-40% diarrhea at placement, followed by 5-10% prevalence of chronic cases of *Salmonella* Typhymurium later in the nursery phase (approximately day 28). Historical nursery diagnostics returned most often RVA and RVC, in addition to *Salmonella* Typhymurium. Multiple oral antibiotics had been used via water administration to treat the pigs including gentamycin, neomycin, and sulfadiazine/trimethoprim with little return on results.

A PathKinex™ strategy was developed to investigate the pathogen profile of pigs determined to be sick or healthy from the selected flow. Pigs were tagged and rectal swabs collected one day prior to weaning and at multiple timepoints throughout the nursery phase to establish a baseline enteric pathogen profile originating from the sow farm and to evaluate the evolution of the profile through the nursery phase. Independent of the diagnostic strategy, an evaluation of PROVENT® ECL/1 PLUS, was also conducted. A four barn, tunnel ventilated commercial WTM site located in central Iowa was utilized for both evaluations. Two barns were selected as Control (no PROVENT® ECL/1 PLUS, total of 7997 pigs) and two barns were selected as Treatment (PROVENT® ECL/1 PLUS, total of 7498 pigs). Barns were filled from the same sow flow over a two-week time period. All barns were treated identical relative to company protocols, and water medications were administered on an as-need basis with veterinary oversight.

PathKinex™ diagnostics were conducted only on pigs from the control group (no PROVENT® ECL/1 PLUS). Five healthy and five sick pigs were first identified at the sow farm, swabbed and ear tagged one day prior to weaning. The sow farm selected represented the most disease concerns at placement into the WTF facility. Sick pigs had clinical signs of diarrhea, poor body condition, and/or poor thriftiness. In addition to these 10 pigs, five additional sick pigs were swabbed and tagged at placement in the WTM site and were serial swabbed again at 14 days post placement and 28 days post placement when a clinical disease break occurred with 100% diarrhea incidence.

PROVENT® ECL/1 PLUS administration started in the treatment barns at placement and was continued until four weeks post-weaning when a disease break typically occurred. PROVENT® ECL/1 PLUS was offered through a water medicator, mixed daily at 50 grams: 1 gallon of water per 500 pigs and then dosed at 1 oz to 128 gallons of drinking water. Barn production data was collected including mortality, number of injectable treatments (morbidity), and water medications ran in both Treatment and Control barns.

Results

Numerous pathogens were identified over time in both sick and healthy populations (Figure 6). PathKinex™ swabs indicated a pre-wean pathogen profile including multiple pathogen genes, most notably *E. coli* toxin genes (Figure 7) and RVC. Swabs collected on Day 1 post wean, detected additional *E coli* toxin genes (*LT, STx1, Stx2*) not found at the sow farm suggesting new exposure and a more diverse profile. Rotavirus also tended ($P = 0.08$) to be higher in sick pigs. Day 14 swabs revealed a similar *E coli* profile (with exception of *Stx1* not detected) with increased gene quantities despite no clinical health break. However, on day 28, a disease event with severe diarrhea and increased mortality was observed. Day 28 swabs indicated a similar profile to day 14, however many of the *E coli* toxin and adhesin genes increased by two $\log_{10}$ in quantity (Figure 7), suggesting a relationship between *E coli* gene quantity and the clinical break. Over the course of the entire study, only one pig was found to have *Salmonella*.

When comparing Treatment (PROVENT® ECL/1 PLUS) barns to Control (no- PROVENT® ECL/1 PLUS) barns, pigs in Treatment barns did not experience a clinical diarrhea break on day 28 and did not require water medications; while pigs in Control barns required gentamycin in the water due to disease event on day 28 post placement. Treatment barns also had 3.42% less pigs treated with an injectable antibiotic, and 0.85% improvement in mortality compared to the control barns. Due to the nature of the evaluation being a field study, statistics were not performed and values are performance means (Table 1).

Conclusions

Rotavirus is known to damage enterocytes and change the luminal environment to one that may favor the colonization of pathogenic *E coli*.[2] Following viral infection, the *E coli* profile appeared to evolve over time until an overwhelming quantity created a clinical disease event. Based on the progression of disease and the early indicators of a change in profile before a clinical event, PathKinex™ may be a tool to predict clinical events by establishing a pathogen profile and understanding the evolution over time. Not finding *Salmonella* was also surprising, but pigs were not observed for purposes of this study past the 28d mark and the pathogen may have changed in timing of presentation in this specific case. PROVENT® ECL/1 PLUS is a multiple strain, *Bacillus*-based, water dispersible microbial product developed for oral delivery in pigs. In this case where pigs had inciting viral factors at placement, later followed by subsequent pathogenic bacterial growth and enteric disease. Results from use of PROVENT® ECL/1 PLUS included reduced mortality, morbidity and decreased water medication usage.

Figure 6: PathKinex™ results of total pathogen gene quantities detected 1 day pre-weaning and at multiple time-points post-weaning, leading up to a clinical enteric disease break.

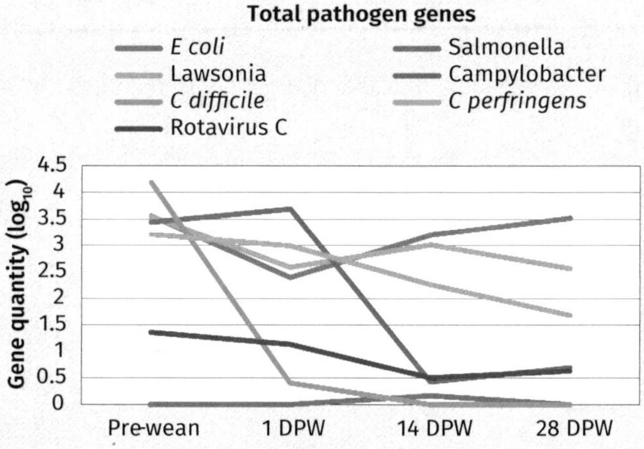

Figure 7: PathKinex™ results of *E coli* toxin and adhesin gene quantities detected 1 day pre-weaning and multiple time-points post-weaning, leading up to a clinical enteric disease break.

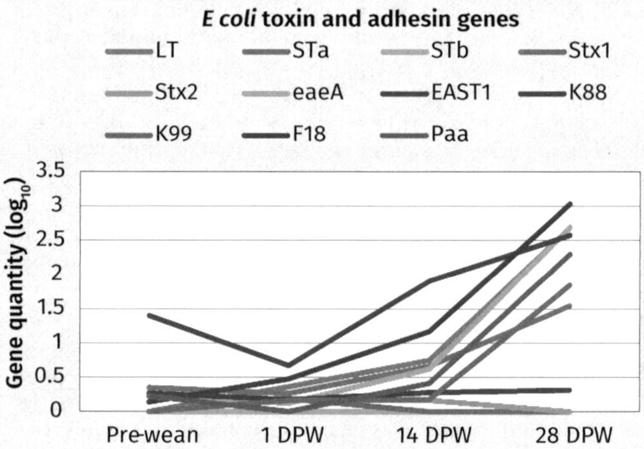

Overall conclusion and implications

Understanding endemic enteric diseases can be a challenge in modern swine production where comingling piglets from multiple sow farms is a common occurrence. Identifying effective intervention strategies for downstream enteric disease starts with understanding origin of pathogens, as well as factors further influencing clinical disease. PathKinex™ was a valuable tool used to understand how homogeneous enteric pathogens were at both an individual sow farm level and across sow farms that make up a commercial WTM flow. In order to understand if the cause of enteric disease was site specific at a nursery, the sow farm information was critical. In this case, results were an indicator of inconsistent NPE practices across farms in addition to lateral transmission after pigs were comingled at placement. This information was used to improve NPE at the GDU, where replacement gilts can have common exposure to Rotavirus types.

In this field investigation, continuing to gain a more proactive understanding of how enteric pathogens change and develop over time could lead to more effective intervention strategies. In this case, the *E coli* profile was developing weeks sooner than when

Table 1: Average total mortality and antibiotic utilization in control barns and treatment (Provent® ECL/1 PLUS) barns

Room/treatment	Total mortality (%)	Total injected (%) (morbidity)	Total water antibiotic pulses
1 Control	3.92	13.47	1
2 Control	2.66	14.54	1
3 Provent® ECL/1 PLUS	2.63	12.79	0
4 Provent® ECL/1 PLUS	2.25	8.37	0
Treatment average			
Control	3.29%	14.01%	1
Provent®	2.44%	10.58%	0

Total mortality, and antibiotic treatments of pigs offered a direct fed microbial Treatment (Provent® ECL/1 PLUS, 7498 pigs); or a Control (no- PROVENT® ECL/1PLUS, 7997 pigs). Provent® ECL/1 PLUS was started at placement into wean to market facility and offered continually for 36 days post placement. Antibiotic injections included Enrofloxacin, Penicillin, Lincomycin, Oxytetracycline while water antibiotic included Gentamicin for 5 days.

actual clinical disease presented itself and proactive treatment, such as PROVENT® ECL/1 PLUS, proved an appropriate intervention. During this evaluation, PROVENT® ECL/1 PLUS was used for 36 consecutive days, a longer duration as compared to traditional use. However, having the above knowledge and intervention steps at the GDU may allow for a much more fine-tuned approach, potentially resulting in further benefits of reduced mortality and morbidity while decreasing the cost of intervention used. Additional considerations for future interventions include, but not limited to crossflow NPE or PROVENT® ECL/1 PLUS at the GDU, in combination with PathKinex™ surveillance to monitor changes in pathogen prevalence and quantity. Producing a replacement gilt with balanced gut health and minimal pathogen profile upon entering the sow farm would further decrease variation of endemic enteric pathogens within the sow farm, simplifying the approach needed in WTM.

References

1. Looft, T. et al. (2012) In-feed antibiotic effects on the swine intestinal microbiome. PNAS. Vol. 109: 1691-1696

2. Lecce, J. G., Balsbaugh, R. K., Clare, D. A., & King, M. W. (1982). Rotavirus and hemolytic enteropathogenic Escherichia coli in weanling diarrhea of pigs. Journal of Clinical Microbiology, 16(4), 715–723.

The power of big data analytics to generate data-driven production decisions

G. Krahn, PhD; C. Eden, BS; B. Altman, BS; J. Spencer, PhD; J. Lee, MS
United Animal Health, Sheridan, Indiana

Introduction

Implementing technologies to capture individualized pig performance data opens the door to deeper understanding of factors driving performance in swine production systems. The result of this is the ability to gain insights that can help make informed, data-driven decisions. United Animal Health's utilization of the LeeO Individual Animal Identification System and Big Dutchman feed system allows for routine collection of precise, individualized data that is not commonly found across the swine industry. The focus of individual pig information is to gather distinct results, gain a deeper understanding into production metrics and deliver insights to help producers make data-driven decisions.

Leveraging the capabilities of LeeO technology and United Animal Health's research farms, big data analytics are used to investigate the effects of body weight impact on lifetime performance, and gain understanding into the importance of lactation feed intake on lactation performance and subsequent reproductive performance. This information lends to identifying appropriate strategies to drive system-wide performance in swine production systems.

Materials and methods

Analyses were performed on individual pig data collected from June 2018-July 2020 at United Animal Health's Research Farms to examine potential factors impacting sow and progeny performance. Radio-Frequency Identification (RFID) tags were applied to all pigs, and the LeeO individual animal identification system collection was used to collect individual pig data. LeeO is a precision livestock farming system that uses RFID technology to track litter information, weights, inseminations, mortalities and other on farm activities.

LeeO was utilized to record weights at multiple time points, including: within 24-hours of birth (N = 72,199 pigs), during weaning (N = 62,875 pigs), at the end of the nursery phase (N = 37,186 pigs) and end of the grow/finish phase (N = 1,033 pigs). During analysis, pigs were distributed into eight birth weight categories and ten wean weight categories. Mortalities were recorded in LeeO and mortality information was used evaluate pre-wean mortality rate and nursery mortality.

Sows were individually weighed pre-farrowing, at weaning and an internal regression model was used to estimate post-farrow weight. Lactation weight change was calculated as the difference between sow weight at weaning and post-farrow weight. All litter information (N = 4,318 litters), wean events, inseminations and pregnancy checks were recorded in LeeO. Wean to first service interval (WTFSI) was based on the first breeding post weaning. Farrowing rooms were equipped with the Big Dutchman feed system which recorded individual daily lactation feed intake and was used to calculate lactation average daily feed intake (ADFI). During analysis, sow data was split into five lactation ADFI increments, based on parity of sow during farrowing.

Results and discussion

Effect of body weight on mortality and subsequent growth performance

Pre-wean mortality was the lowest in pigs weighing ≥ 3.5 lbs. at birth ($P < 0.05$) and mortality increased as birth weight decreased (Table 1). Low birth weight pigs remained lighter weight ($P < 0.05$) at weaning and through the end of the nursery phase (42-days post-weaning).

Pigs with heavier wean weights, had lower nursery mortality and remained heavier weight ($P < 0.05$) through the nursery and grow/finish phases (Table 2). For every additional 1-lb. bodyweight at weaning, an additional 2.4 lbs. bodyweight was achieved at the end of the nursery phase. Furthermore, each additional 1-lb. increase in nursery weight, resulted in a 2.2-lb. improvement in bodyweight at 21-weeks post weaning. During all phases of production, interventions should be implemented to increase weight which will maximize subsequent growth performance.

Lactation feed intake influences litter performance and reproductive performance

Sows with increased lactation ADFI, regardless of parity, weaned heavier pigs ($P < 0.05$) and litters with lower weight variation at weaning (Table 3). Collectively, the two advantages found from higher lactation ADFI assist in reducing the number of small pigs at weaning (< 8 lbs.), whom also, consequently, have higher nursery mortality. Sow lactation weight change improved ($P < 0.05$) in all parities when lactation feed intake increased (Table 3).

Post weaning, a larger percentage of high feed intake parity 1 females returned to heat by day 5, compared to low feed intake parity 1 females. Higher lactation feed intake also improved subsequent conception rates, along with higher total born and born alive in subsequent litters. Managing sows during lactation continues to be an essential component of a sow's productivity. An emphasis should be placed on management practices that help maintain feed intake during lactation.

Impact of Wean to First Service Interval on reproductive performance

Sows serviced for the first time 3-5 days post weaning had increased conception rates, weaned sow utilization and subsequent total born and born alive (Figure 1). Compared to sows serviced on day 3-5, sows serviced on day 6 had a reduction in conception rate (13.38 points), weaned sow utilization (11.46 points), total born (0.42 pig) and born alive (0.61 pig). Results suggest WTFSI is a key metric in overall reproductive efficiency and should be monitored to make informed decisions that improve sow herd performance.

Table 1: Effect of birth weight on pre-wean mortality, wean weight and nursery weight

Birth weight (lbs.)	Number born alive	% of total	Pre-wean mortality	Wean weight (lbs.)[1]	Nursery weight (lbs.)[2]
1.0-1.5	2,549	3.53%	77.99%[f]	7.96[h]	35.8[h]
1.5-1.9	5,908	8.18%	32.30%[e]	9.52[g]	40.4[g]
2.0-2.4	11,741	16.26%	14.90%[d]	11.30[f]	45.2[f]
2.5-2.9	17,518	24.26%	8.68%[c]	12.92[e]	49.2[e]
3.0-3.4	17,001	23.55%	6.11%[b]	14.27[d]	52.6[d]
3.5-3.9	11,418	15.81%	4.86%[a]	15.43[c]	55.6[c]
4.0-4.4	4,623	6.40%	4.20%[a]	16.49[b]	57.6[b]
> 4.5	1,441	2.00%	3.33%[a]	17.71[a]	60.3[a]
All	72,199	100%	12.47%	13.42	50.7

1 Adjusted to 21-day wean age (n = 62,875)

2 Adjusted to 42 days post-weaning (n = 37,186)

Columns with different superscripts are statistically different ($P < 0.05$)

Table 2: Effect of wean weight on nursery mortality, nursery weight and finishing weight

Wean weight (lbs.)	Number of pigs	% of total	Nursery mortality	Nursery weight (lbs.)[1]	Finishing weight (lbs.)[2]
5.0-7.0	601	1.62%	25.91%	35.6[j]	251.3[f]
7.0-8.0	797	2.14%	11.07%	38.2[i]	252.1[f]
8.0-9.0	1,341	3.61%	6.12%	39.2[h]	263e[f]
9.0-10.0	2,200	5.92%	4.23%	42.3[g]	264.6[ef]
10.0-11.0	3,054	8.21%	3.59%	44.5[f]	273.2[de]
11.0-12.0	4,203	11.30%	2.57%	46.8[e]	277.5[d]
12.0-13.0	5,125	13.78%	2.03%	49.1[d]	285.9[c]
13.0-14.0	5,591	15.04%	1.86%	51.4[c]	296.0[b]
14.0-15.0	4,921	13.23%	1.78%	53.7[b]	301.9[ab]
> 15.0	9,353	25.15%	1.42%	57.4[a]	304.3[a]
All	37,186	100.00%		50.7	281.6

1 Adjusted to 42 days post-weaning (n = 37,186)

2 Finishing weight: 21-weeks post-weaning (n = 1,033).

Columns with different superscript are statistically different ($P < 0.05$)

Summary

In summary, utilizing modern technologies to capture individualized pig data can identify, quantify and correlate factors influencing efficiencies in swine production systems. Analysis of data captured at United Animal Health's DEO Research Farm further confirms the importance of strategies to drive sow lactation feed intake, resulting in improved sow reproductive fitness and downstream progeny performance.

Table 3: Impact of lactation average daily feed intake (ADFI) by parity on wean weight, wean weight CV, lactation weight change, percent heat by day 5, conception rate and subsequent total born and born alive

	Lactation ADFI (lbs.)	Wean weight (lbs.)	Wean weight CV[1]	Lactation weight change (lbs.)[2]	Percent heat by day 5	Conception rate	Total born	Born alive
Parity 1	< 8	11.91[d]	19.07[c]	-55.34[e]	85.70%	75.13%	15.00	13.69
	8-10	12.34[d]	18.24[bc]	-37.11[d]	87.43%	81.00%	15.68	14.53
	10-12	12.71[cd]	17.37b	-23.70[c]	87.97%	81.43%	15.88	14.54
	12-14	13.03[b]	16.88[ab]	-11.37[b]	90.56%	83.96%	16.26	14.82
	> 14	13.39[a]	15.88[a]	1.66[a]	90.13%	88.29%	17.07	15.03
Parity 2-5	< 10	12.32[e]	22.00b[c]	-67.44[e]	78.91%	63.93%	14.29	13.84
	10-12	13.02[d]	21.41b[c]	-33.87[d]	99.80%	91.32%	17.05	15.47
	12-14	13.47[c]	20.45[c]	-15.14[c]	96.34%	94.10%	17.42	14.82
	14-16	13.87[b]	19.79[b]	-1.44[b]	97.47%	92.52%	18.12	15.61
	> 16	14.14[a]	19.24[a]	11.90[a]	98.68%	95.54%	18.07	15.43
	All	13.83	19.82	-6.53	92.91%	87.84%	17.46	15.06

1 Wean weight CV: coefficient of variation.
2 Lactation weight change = sow weight at weaning - post-farrow weight.
By parity, columns with different superscripts are statistically different ($P < 0.05$).

Figure 1: Impact of wean-to-first service interval on reproductive performance.

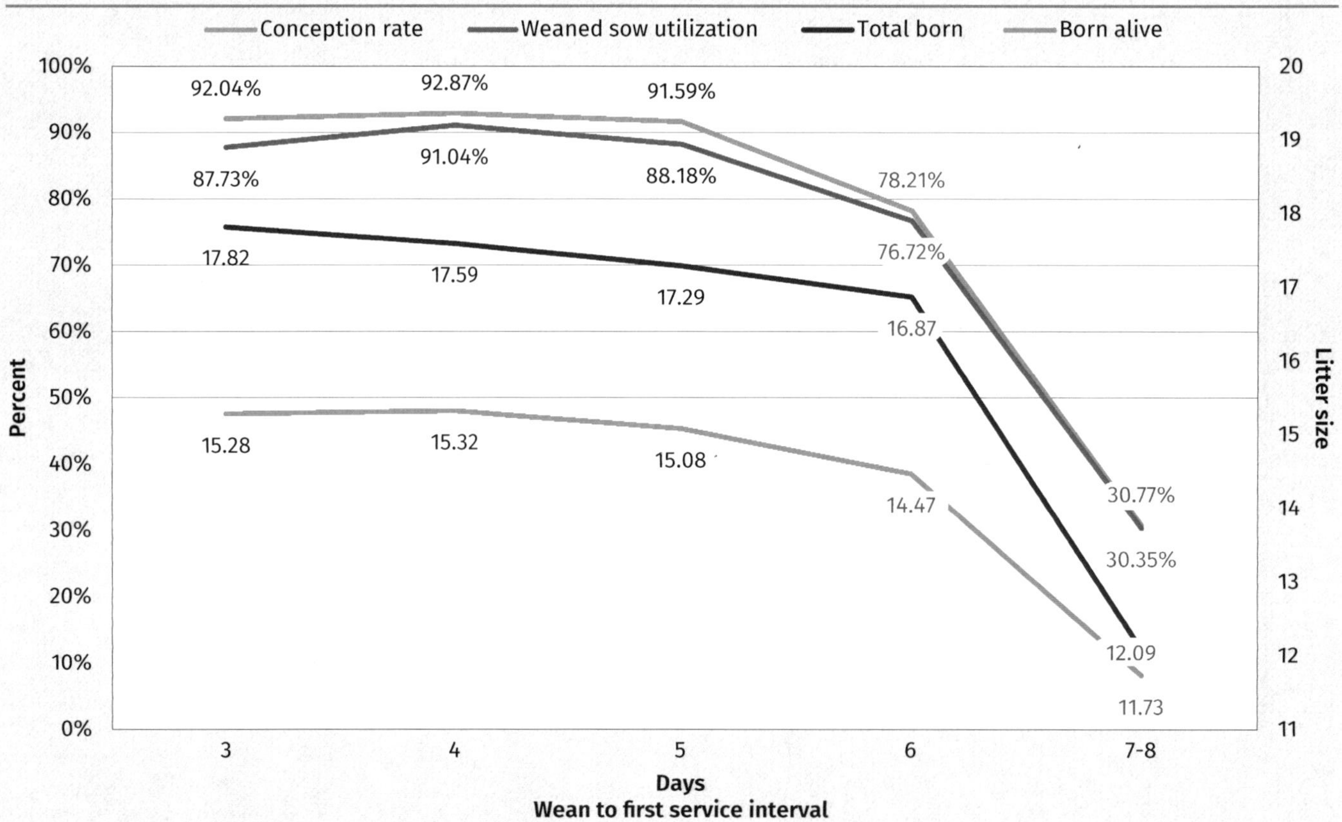

Same-day surveillance diagnostics for critical same-day decisions

C. Kunas[1]; T. Goode[2]
[1]BioChek NL, Reeuwijk, The Netherlands; [2]BioChek USA, Scarborough, Maine

Introduction

Rapid, reliable and efficient laboratory diagnostics is key for veterinarians and producers to take critical decisions in pig production. Facing threats through domestic and foreign swine diseases, such as PRRSV and ASFV, results need to be available as quickly and comprehensive as possible. In order to optimize this process, BioChek (The Netherlands) in cooperation with Biotecon Diagnostics (Germany) develop an integrated approach for viral and bacterial disease diagnostic and surveillance: This includes new sampling techniques, sample preparation, pathogen detection and AI-based evaluation of results.

Most room for improvement in time-to-result is to shorten shipping routes by establishing easy to use, well proven diagnostic systems at the place where the results are required: the veterinary practice.

The objective of this part of the validation study was to assess the practicality of a newly developed extraction kit for RNA and DNA viruses in combination with an easy-to use extraction robot in a non-laboratory environment.

Materials and methods

A broad variety of relevant virus species, including PRRSV and ASFV, and different sample materials, such as saliva, whole blood, feces and tissue, were used in the validation study. All samples were processed using Biotecon's Magnetic Preparation Kit VI in combination with the Roboprep® 32 extraction machine (see Figure 1), which automatically performs all lysis and purification steps for 1 to 32 samples in less than 45 minutes. RT-PCR detection was done with BioChek and other commercially available kits.

Results and discussion

The use of Magnetic Preparation Kit IV in combination with the automatic Roboprep® 32 extraction system allows employees with little laboratory experience to perform extraction of virus RNA and DNA from complex matrices. A full extraction run takes less than 45 minutes and will result in highly purified RNA/DNA, comparable or superior to other extraction devices (comparison data not shown). Specificity was tested with a range of relevant target viruses, results (see Table 1) show neither false negative nor false positive results. Matrix compatibility was tested by spiking different samples with PRRSV. All tested sample matrices could be processed without inhibition and with comparable high efficiency, showing that the Magnetic Preparation Kit VI extraction is compatible with all tested matrices. Prevention of cross contamination was evaluated by purification of spiked pig serum samples (Adenovirus and Influenza A) together in one run, pipetted alternately with negative controls (water). 96 out of 96 spiked samples were correctly identified as positive with a mean Cq of 26.69 (Influenza) and 30.83 (Adenovirus) respectively, while all non-spiked samples were correctly identified as negative.

Figure 1: Easy to use extraction Robot, 32 samples in less than 45 minutes.

Conclusion

Validation data shows that using the Magnetic Preparation Kit IV in combination with Roboprep® 32 is a safe and rapid method for sample preparation of viral RNA and DNA from a wide variety of matrices. Using this easy to learn and easy to use method with the BioChek integrated approach, including pathogen detection and result evaluation, allows small veterinary practices (2 – 5 employees or larger) to drastically reduce time-to result for laboratory diagnostics.

Table 1: Validation of specificity

Family	Species	Nucleic acid	Result
Adenoviridae	Adenovirus	DNA	+
Arteriviridae	PRRSV-1	RNA	+
	PRRSV-2	RNA	+
	highly pathogenic PRRSV	RNA	+
Asfarviridae	African swine fever virus	DNA	+
Coronaviridae	Porcine epidemic diarrhea virus	RNA	+
	SARS-CoV-2	RNA	+
Hepeviridae	Hepatitis E virus	RNA	+
Orthomyxoviridae	Influenza A virus, H3N8	RNA	+
	Influenza A virus, H1N1	RNA	+
Parvoviridae	Porcine parvovirus	DNA	+
Picornaviridae	Hepatitis A virus	RNA	+
Unspiked	-	-	-

Table 2: Verification of matrix compatibility

Matrix	Rate of recovery	Mean Cq	CV [%]
PBS	4/4 = 100 %	23.32	1.13
Allantoic fluid	4/4 = 100 %	23.96	0.86
Oral fluid	4/4 = 100 %	25.81	1.33
Whole blood	4/4 = 100 %	26.86	1.89
Feces	4/4 = 100 %	23.70	0.49
Serum	4/4 = 100 %	23.51	1.22
Unspiked	0/4 = 0 %	negative	-

PBS = phosphate-buffered saline

Tracking individual pigs for genetic nucleus-multiplication and GDU performance

Tom Stein, DVM, MS, PhD
Maximus Systems & Maximum Ag Technologies, Saint-Bruno, Quebec, Canada and Jenison, Michigan

The Maximus software ecosystem now includes individual pig tracking from birth through lactation, nursery, gilt development unit (GDU) and entry to the breeding herd. Initially developed for one of North America's largest pork producers as a replacement for out-of-date legacy genetic management software, it is now considered for more general individual pig tracking applications.

The software is based on an individual database record for each piglet and each one linked to their birth mother as well as their foster mother (if necessary). Tattoo and radio-frequency identification (RFID) tags are assigned just after birth when piglets are weight and teats are counted. The tattoo and RFID tag numbers can be the same (easiest to administer) but may be different. Stillborn pigs get a tattoo but not an RFID tag. The piglet record includes birth-weight, number of teats (functional, necrotic, inverted), anomalies (either associated with the piglet or the litter). These attributes are used in the logic for culling or selecting gilts, or castrating males. Data are imported from a company's BLUP software before farrowing and linked to each dam which enables the logic or rules e.g. "castrate x number of males."

Rules and logic are based on breed-specific algorithms for birth-weight, number of functional teats, etc. The entire castration process is managed by the software and a list linked to each sow before she farrows.

Litter inventory and reconciliation processes are built into the software. These are based on movement records for each piglet and include fosters on/off, castration, and weaning. The software places any missing pigs in a virtual "cart" which allows barn staff to find the missing piglets and enable the reconciliation process.

Piglet weights can be taken at farrowing, at pre-weaning (5 days before weaning), at entry to a nursery, at entry to finishing, and at selection. Treatments are entered for each piglet when given. All deaths and the weight at death are recorded. Same thing for confirmation scoring.

Before sows are bred after weaning, the software enables geneticists to upload the mating program for each sow. This plan comes up on the mobile data entry software when you scan the sow's RFID tag. Boars to be used are listed in an optimized mating plan or a non-optimized plan and all key inbreeding and Eestimated breeding value (EBV) coefficients are shown on the screen.

Individual gilts and boars are followed into each location and are assigned to groups in the nursery, in finishing, and in the GDU. These groups will be managed as typical nursery, finishing or wean-to-finish groups except that every pig in the group will be identified. The individual gilts are entered automatically into the sow farm from the GDU based on movement records.

Thus, this software fully supports a genetic nucleus-multiplication system as well as tracking individual piglets through lactation, nursery, finishing or wean-to-finish, and GDU coupled with their entry into the breeding herd. It can be used for commercial sow farms; for example, starting individual pigs in GDU groups, enabling closeouts for those groups, and seamlessly adding individual gilts to commercial sow farms, i.e. having an individual sow record that starts with gilts in the GDU phase.

Figure 1: Maximus mobile genetics software on a tablet computer. Data entry and review for recording litter information and individual piglet records.

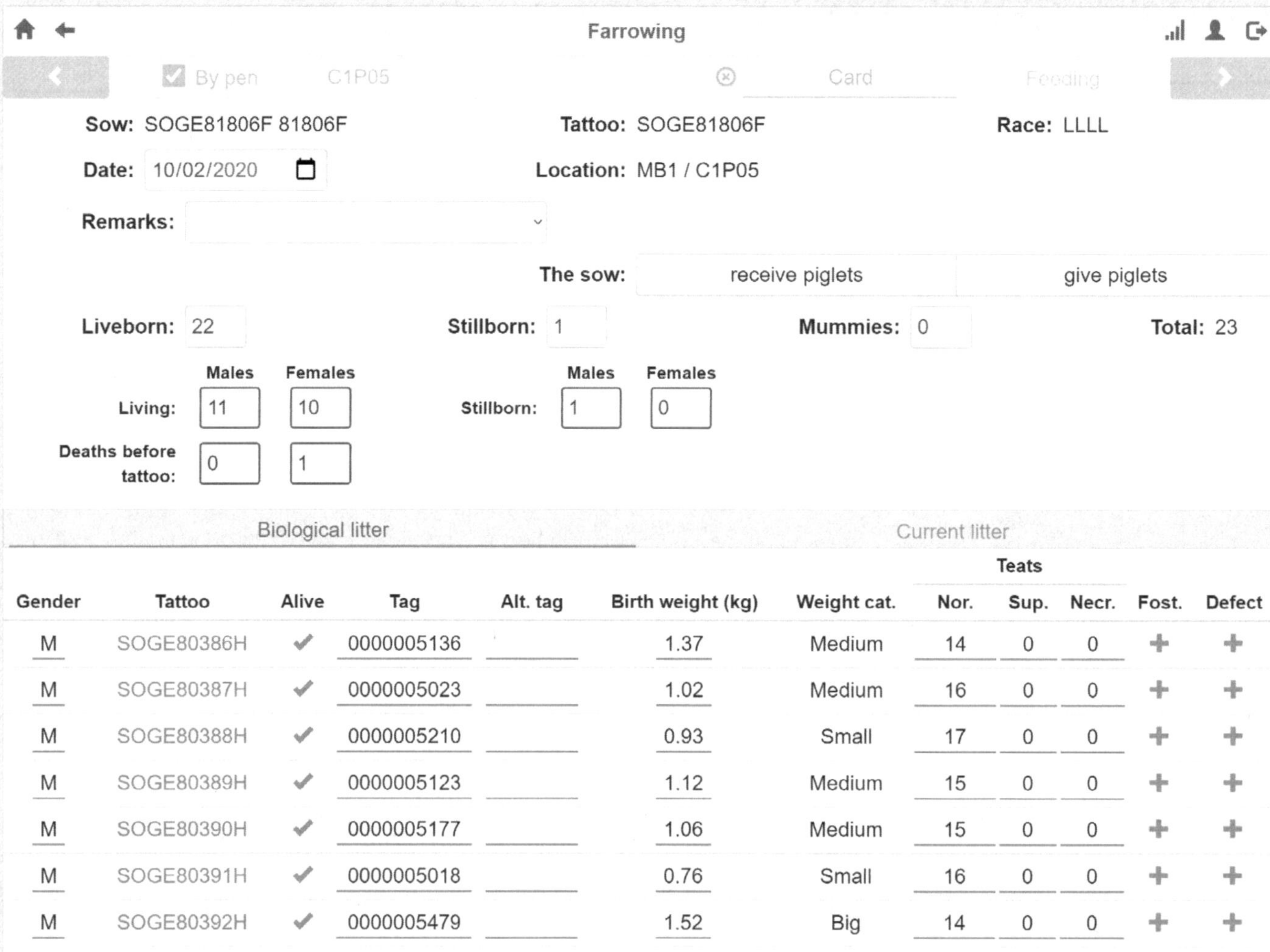

Farrowing

By pen C1P05 Card Feeding

Sow: SOGE81806F 81806F **Tattoo:** SOGE81806F **Race:** LLLL

Date: 10/02/2020 **Location:** MB1 / C1P05

Remarks:

The sow: receive piglets | give piglets

Liveborn: 22 **Stillborn:** 1 **Mummies:** 0 **Total:** 23

	Males	Females
Living:	11	10
Deaths before tattoo:	0	1
Stillborn:	1	0

Biological litter | Current litter

Gender	Tattoo	Alive	Tag	Alt. tag	Birth weight (kg)	Weight cat.	Teats Nor.	Teats Sup.	Teats Necr.	Fost.	Defect
M	SOGE80386H	✔	0000005136		1.37	Medium	14	0	0	+	+
M	SOGE80387H	✔	0000005023		1.02	Medium	16	0	0	+	+
M	SOGE80388H	✔	0000005210		0.93	Small	17	0	0	+	+
M	SOGE80389H	✔	0000005123		1.12	Medium	15	0	0	+	+
M	SOGE80390H	✔	0000005177		1.06	Medium	15	0	0	+	+
M	SOGE80391H	✔	0000005018		0.76	Small	16	0	0	+	+
M	SOGE80392H	✔	0000005479		1.52	Big	14	0	0	+	+

Figure 2: Maximus mobile genetics software on a tablet computer. Data entry and review for recording individual piglet weaning information.

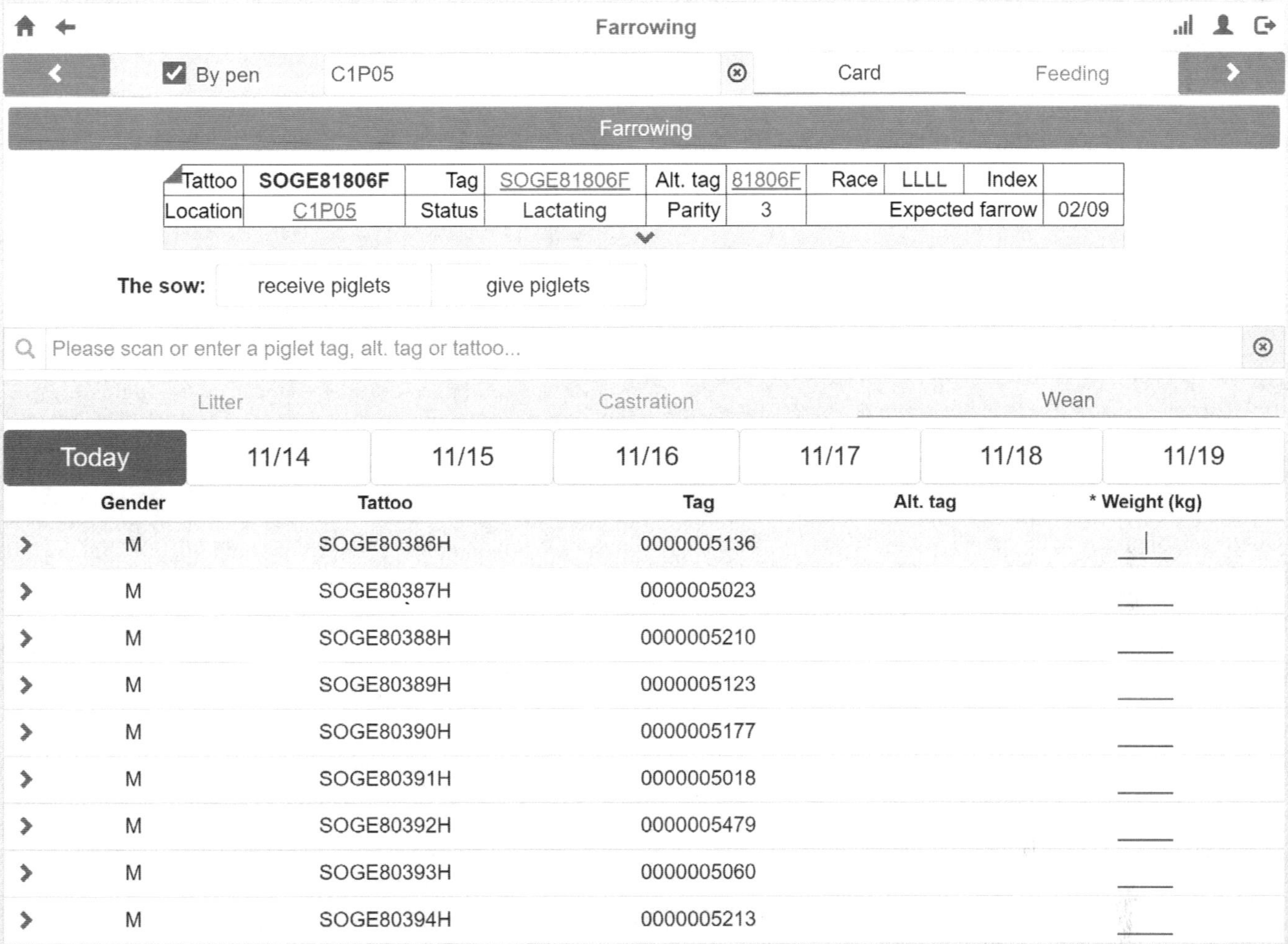

Comparative efficacy evaluation of two modified-live PRRS vaccines against a heterologous PRRSV 1-7-4 challenge

R. Philips, DVM; G. Haiwick, BS; D. Whiteman, MS; J. Hermann, DVM, PhD; B. Fergen, MS, PhD
Boehringer Ingelheim Animal Health USA Inc., Duluth, Georgia

Introduction

The use of Ingelvac PRRS® vaccines can significantly reduce lung lesions following challenge with heterologous PRRSV isolates (86-94% ORF5 nucleotide similarity) in the three-week-old pig respiratory challenge model.[1] In addition, it has been demonstrated that the implementation of a systematic methodology for PRRS control that utilizes modified-live vaccine for the control of wild type-PRRSV infections can mitigate the consequences of infection on health and performance.[2,3] It is necessary for modified-live vaccines to be effective against heterologous challenge with current PRRSV field isolates. Ingelvac PRRS® MLV demonstrated heterologous protection against challenge with current PRRSV lineage-1 RFLP 1-7-4 and 1-3-4 isolates.[4] The objective of this study was to evaluate the efficacy of Ingelvac PRRS® MLV (BIAH-Boehringer Ingelheim Animal Health, Duluth, Georgia) and Prevacent® PRRS (Elanco Animal Health, Greenfield, Indiana) in a three-week-old pig respiratory challenge model, using a heterologous PRRSV RFLP 1-7-4 field isolate that was isolated from North Carolina in 2016.

Materials and methods

A randomized, blinded vaccination-challenge study was performed in 147, three week old pigs from a PRRS naïve source. Pigs were confirmed to be PCR negative for PRRS prior to the initiation of the study. The study consisted of 5 treatment groups of pigs. Group 1; n = 36, was vaccinated with Ingelvac PRRS® MLV (2 mL intramuscular (I.M.) per label recommendations), Group 2; n = 36, was vaccinated with Prevacent® PRRS (1 mL I.M. per label recommendations) and Group 3; n = 34 was vaccinated with a novel experimental PRRSV type-2, lineage-1 modified-live vaccine (WS-96; 2 mL I.M.). Groups 1, 2 and 3 were vaccinated on day 0. Group 4; n = 36, was a non-vaccinated challenge control treatment group that received a 2 mL placebo I.M. injection on day 0. Group 5; n = 5, was a non-vaccinated and non-challenged strict control group that served as an environmental control. Pigs were housed in rooms by group during the vaccination period. Group 5 pigs were removed and euthanized prior to challenge and examined for lung lesions and confounding bacterial pathogens. At day 27 of the study (D27), all pigs were comingled and challenged with 2.0 mL I.M. and 2.0 mL intranasally (1 mL per nostril) of inoculum containing $10^{4.6}$ $TCID_{50}$/mL of virulent PRRSV RFLP 1-7-4. Ingelvac PRRS® MLV and Prevacent® PRRS have 87.9% and 88.7% similarity respectively to the PRRSV 1-7-4 challenge isolate based on ORF 5 sequence. Blood was collected from all pigs on days 0, 7, 14, 21, 27, 29, 31, 33, 35 and 42. Serum samples were tested by RT-PCR for the presence of viremia and by ELISA for the presence of anti-PRRSV antibody. Viremia testing for PRRSV was done by the Iowa State University Veterinary Diagnostic Lab, Ames, Iowa, using Quantitative PCR and results reported as "positive or negative" in addition to PRRSV quantity in units of genomic copies per ml serum. Pigs were weighed at day 0, day of challenge (D27), and termination of the study (D42) to assess average daily weight gain (ADWG). On D42 (14 days post-challenge), all pigs were necropsied and lungs were scored for the presence of macroscopic lesions. Data were analyzed using SAS® Generalized Linear Mixed Models. Pairwise comparisons between groups were conducted as appropriate using a level of confidence of 0.05 with multiplicity-adjusted p-values to indicate statistical significance. The multiplicity adjustment was accomplished using ADJUST=SIMULATE in the LSMESTIMATE option in SAS® GLIMMIX.

The study design is summarized in Table 1.

Results

Lung lesions. The day 42 (14 days post-challenge) lung lesion assessment found the vaccinated treatment groups to have significantly lower mean gross lung lesion scores compared to the non-vaccinated (Placebo) challenged treatment group. The mean percent lung lesion scores for Ingelvac PRRS® MLV, Prevacent® PRRS and WS-96 were 5.67%, 8.11% and 7.15% respectively, compared to a mean lung lesion score of 16.11% for the non-vaccinated challenged treatment group. No significant differences were noted between the vaccine groups; however, pigs vaccinated with Ingelvac PRRS® MLV had the lowest numerical mean lung lesion score.

Post-challenge lung lesion results are summarized in Table 2.

Average daily weight gain. ADWG was evaluated during the pre-challenge and post-challenge phases of the study. During the post-vaccination/pre-challenge phase of the study (day 0-day 27), there was no significant difference in ADWG between treatment groups. The pre-challenge ADWG was 1.16, 1.06, 1.10 and 1.14 lbs per day for the Ingelvac PRRS® MLV, Prevacent® PRRS, WS-96 and Placebo groups respectively. In the post-challenge phase of the study (day 27-day 42) the vaccinated treatment groups demonstrated a significant increase in ADWG compared to the non-vaccinated (Placebo) challenged treatment group. ADWG for Ingelvac PRRS® MLV, Prevacent® PRRS and WS-96 was 1.41, 1.10 and 1.35 lbs per day respectively, compared to a 0.84 lbs per day ADWG for the non-vaccinated challenged group. The Ingelvac PRRS® MLV vaccinated group demonstrated a significant increase in post-challenge ADWG compared to the Prevacent® PRRS vaccinated group.

Table 1: Study design

Group	N	Treatment	Day 0	Day 27	Day 42
1	36	Ingelvac PRRS® MLV	Vaccinated	PRRSV 1-7-4 challenge	Necropsy – lung lesion evaluation
2	36	Prevacent® PRRS	Vaccinated		
3	34	Novel WS-96 vaccine	Vaccinated		
4	36	Placebo	Sterile diluent Non-vaccinated		
5	5	Strict control	Non-vaccinated	NA	NA

The pre and post-challenge ADWG results are summarized in Tables 3 and 4.

Viremia. Following challenge, the non-vaccinated challenge control pigs became viremic within 24 hrs of infection (D29), and all animals remained viremic during the post-challenge period. In comparison to the challenge control group, the Ingelvac PRRS® MLV and WS-96 vaccinated groups had significantly lower levels of viremia as measured and analyzed by $\log_{10}$ genomic copies in the area under the curve (AUC) of the post-challenge period of the study - D29 through D42; ($P = 0.0008$ and $P = 0.007$ respectively). The post-challenge level of viremia in the Prevacent® PRRS vaccinated group was not significantly different from the non-vaccinated challenge control group during the post-challenge period of the study ($P = 0.11$).

Post-challenge viremia results are summarized in Figure 1.

Discussion and conclusions

The primary objective of this study was to evaluate the efficacy of Ingelvac PRRS® MLV and Prevacent® PRRS against a current and virulent heterologous PRRSV 1-7-4 isolate in a respiratory challenge model. As evidenced in this study, PRRSV infection can negatively impact pig performance. Mitigating this negative biological impact can have a significant economic benefit. In this study, the Ingelvac PRRS® MLV group demonstrated a significant improvement in ADWG compared to both the Prevacent® PRRS and non-vaccinated control groups during the post-challenge period. Ingelvac PRRS® MLV had a 22% and 40% higher ADWG when compared to the Prevacent® PRRS group and the challenge control group respectively. In addition, Ingelvac PRRS® MLV demonstrated a significant reduction in the level of post-challenge viremia compared to challenge controls where the Prevacent® PRRS group was not significantly different from challenge controls in the level of post-challenge viremia.

This study is another example demonstrating the ability of Ingelvac PRRSV® MLV to protect against a relevant PRRSV challenge. Implementing vaccine in a systematic methodology for PRRS control can mitigate the consequences of infection and subsequently improve the health and performance of pigs.

Table 2: Day 42 lung lesions (mean %) in vaccinated and non-vaccinated pigs challenged with PRRSV 1-7-4

Group	Treatment	Mean percent lung score (95% CI)
1	Ingelvac PRRS® MLV	5.67[a] (2.60, 9.82)
2	Prevacent® PRRS	8.11[a] (4.36, 12.88)
3	Novel WS-96 vaccine	7.15[a] (3.53, 11.92)
4	Placebo - challenge control (non-vaccinated)	16.11[b] (10.79, 22.26)

* Different letters indicate significant ($P \leq 0.05$) differences

Table 3: Post-vaccination/pre-challenge ADWG (day 0-27)

Group	Treatment	ADWG in lbs. (95% CI)
1	Ingelvac PRRS® MLV	1.16 (0.98, 1.34)
2	Prevacent® PRRS	1.06 (0.88, 1.24)
3	Novel WS-96 vaccine	1.10 (0.92, 1.28)
4	Placebo - challenge control (non-vaccinated)	1.14 (0.96, 1.32)

* No significant difference in pre-challenge ADWG between groups

References

1. Patterson, A., et al. 2013. Modified live PRRSV vaccination is efficacious following challenge with eight genetically diverse PRRSV isolates. *A.D. Leman Swine Conference Proceedings.*

2. Oropeza, A., et al. 2012. A methodical approach to PRRS management, a success story of production improvements. *IPVS Proceedings.*

3. Kolb, J., et al. 2015. A summary of four large scale system-based PRRS control projects. *AASV Proceedings.*

4. Patterson, A, et al. 2019. Efficacy of Ingelvac PRRS® modified live virus vaccine against heterologous Porcine Reproductive and Respiratory Syndrome virus challenges. *J Vaccines Vaccin 10:407.*

Table 4: Post-challenge ADWG (day 27-42)

Group	Treatment	ADWG in lbs. (95% CI)
1	Ingelvac PRRS® MLV	1.41[a] (1.28, 1.54)
2	Prevacent® PRRS	1.10[b] (0.97, 1.24)
3	Novel WS-96 vaccine	1.35[a,b] (1.20, 1.49)
4	Placebo - challenge control (non-vaccinated)	0.84[c] (0.70, 0.97)

* Different letters indicate significant ($P \leq 0.05$) differences

Figure 1: Post-challenge viremia – least square means of genomic copies on $\log_{10}$ scale by group and day.

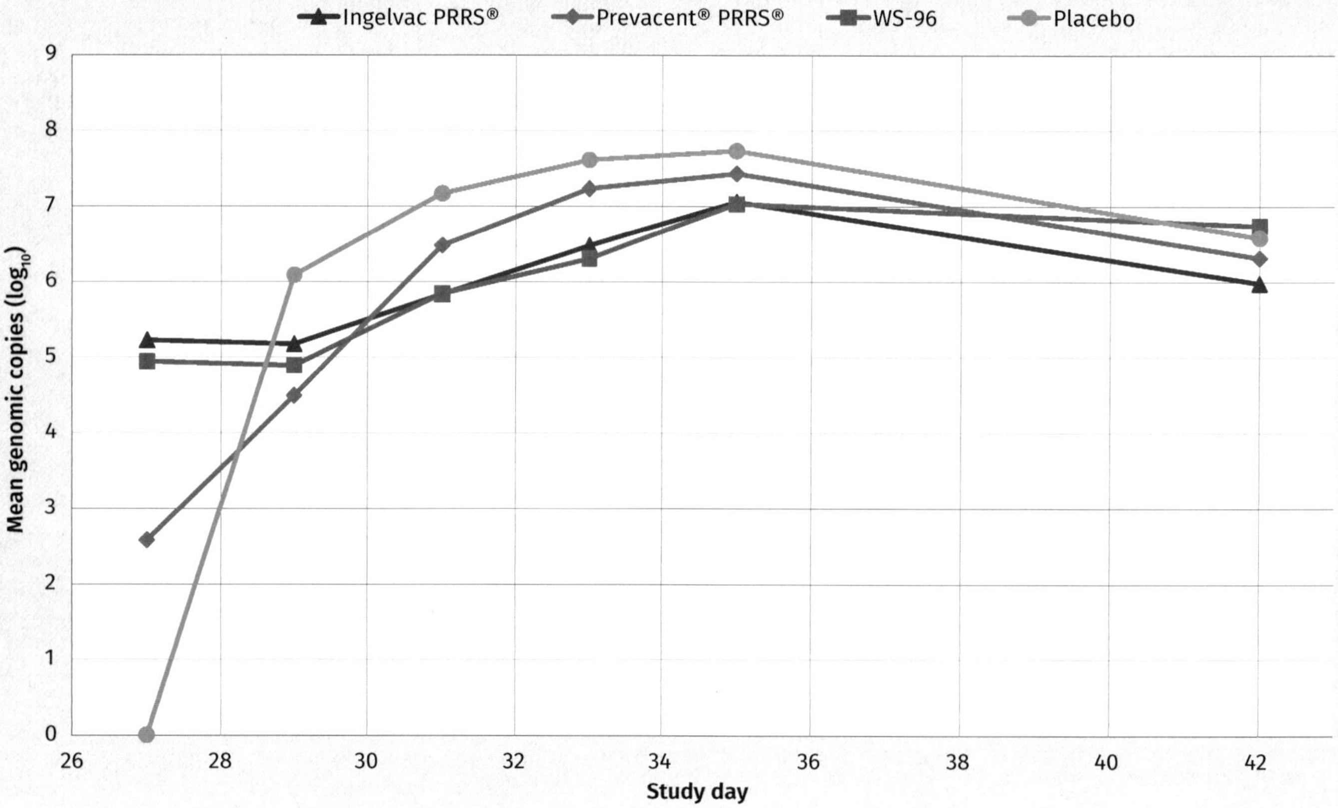

Enterisol® Ileitis, what we have learned in twenty years controlling *Lawsonia intracellularis*

Fernando Leite, DVM, MS, PhD; Greg Cline, DVM; John Waddell, DVM, MBA; Jeremy Kroll, PhD; Jeff Husa, DVM; Edgar Diaz, DVM; Oliver Gomez-Duran, DVM, PhD
Boehringer Ingelheim Animal Health USA Inc.

Introduction

The goal of every swine veterinarian is to solve health challenges with effective tools to promote wellbeing and production performance. Porcine proliferative enteropathy (PPE) or "ileitis", is one such health challenge. Thankfully, the swine industry has had an effective tool to help in controlling this disease for twenty years.

PPE was first described by Biester and Schwarte in the 1930s who reported on field cases and experimental reproduction of disease.[1] They described epithelial proliferative lesions in the intestine of swine that were reproducible by feeding intestinal contents and mucosal scrapings to susceptible pigs. The etiology could not be determined at the time and they commented that animals in field cases were infected with *Salmonella,* which they thought could potentially contribute to the disease. In the 1970s the research group of Dr. Gordon Lawson began investigating this disease making crucial findings.[2] Gebhart et al. (1993) found that the bacterium responsible for proliferative enteropathy was a new species that was named in 1995, *Lawsonia intracellularis.* This was after much effort to achieve pure cultures of the bacterium and to fulfill Koch's postulates.[3-5] Six years later, in 2001, the first vaccine against the disease, Enterisol® Ileitis, was launched in the United States.

Enterisol® Ileitis use in controlling PPE

Enterisol® Ileitis is an oral live attenuated vaccine, containing a non-virulent strain of *L intracellularis* to mimic natural infection and induce mucosal immunity with one dose. The efficacy of Enterisol® Ileitis against PPE was first demonstrated by Kroll et al. at the 31st AASV Annual Meeting in 2001.[6] The study showed that providing vaccine in the drinking water to pigs conferred immunity and significantly reduced the colonization of *L intracellularis* as well as gross and microscopic lesions.[6] These findings were reproduced and peer-reviewed in 2004, when Kroll et al. proved the vaccine could be provided both by oral drench and in the drinking water of pigs to confer significant protection against PPE.[7]

Field studies followed; at the 33rd AASV Annual Meeting, Kolb & Sick (2003) presented a summary of 5 field trials evaluating the efficacy of Enterisol® Ileitis in the United States.[8] This evaluation included a total of 55 grow/finish groups, with over 46,900 vaccinated pigs. Kolb found that average daily gain (ADG) was significantly improved in all 5 trials ($P < 0.05$), and mortality and cull rates were also improved in the herds that were experiencing clinical signs due to PPE.[8] McOrist & Smits (2007) conducted a randomized, blinded, negative-controlled clinical field efficacy study to evaluate the effect of Enterisol® Ileitis in three herds in Australia. The authors found that vaccination improved ADG and concluded it was efficacious in conditions of both clinical and subclinical presentations of PPE.[9] Since then, several studies have evaluated Enterisol® Ileitis under different field conditions and in different countries including Mexico, Germany, Philippines, Japan, Netherlands, Finland, Denmark and South Korea. All of these studies demonstrated improved performance metrics with the use of Enterisol® Ileitis.[10-19]

It is important to consider as well that Enterisol® Ileitis has not only been shown to be an important tool in grow/finish herds but also in protecting gilts and sows against PPE and the hemorrhagic form of PPE, porcine hemorrhagic enteropathy (PHE). Soon after the launch of the vaccine, Waddell et al. (2003) described a vaccination protocol in a newly constructed 1250-sow breeding farm (gilt multiplier) and its downstream gilt offspring wean-finish site, in which immunization against *L intracellularis* rather than routine feed or water antibiotic medication was used to prevent and control PPE.[20] They observed no outbreaks of any form of PPE at either site in spite of the complete absence of any feed or water antibiotic use when immunizing animals. They also noted that non-vaccinated gilts from this same original gilt source that were shipped to at least three different non-vaccinated herds during this same stocking timeframe did experience PHE outbreaks. Cantor et al., (2008) found similar results in which administering one dose of Enterisol® Ileitis to gilts prior to their selection and shipment led to the cessation of deaths and clinical signs from PHE.[21] Sanford (2006) performed an evaluation of clinical cases of PPE in gilts after vaccination and found field observations that would suggest the duration of immunity for Enterisol® Ileitis vaccinated gilts is at least 2 years and probably greater than 3½ years.[22] A reduction in cases of PHE has also been observed in grow-finish pigs when implementing vaccination in piglets.[23]

These studies and results demonstrate that one dose of Enterisol® Ileitis, given by oral drench prior to weaning or through the water in later ages is effective in promoting production performance and controlling disease under various field conditions. Maternal immunity interference has not been found to be an issue with this vaccine when used as recommneded.[24]

Vaccination as a tool to reduce antimicrobial use

Several of the studies that implemented Enterisol® Ileitis vaccination noted a beneficial reduction in the use of antimicrobials on the farm. Voets & Hardge (2007) evaluated this specific effect on three different systems.[25] They found it was possible to remove a substantial amount of antimicrobials with the use of vaccination and maintain, if not in some cases improve performance.[25] Nerem (2009) investigated performance differences between continuous tylosin in feed during finishing versus a protocol of receiving one dose of Enterisol® Ileitis prior to entering the finisher. This study also concluded that it is possible to change from routine use of antimicrobials in the finisher to prevention of PE by vaccination without sacrificing performance. It was also noted that vaccination was the most economical strategy.[26]

As it pertains to feed additives, it is interesting to consider that some feed additives have been found to promote or act in synergy with Enterisol® Ileitis. Bourgot et al. (2017) investigated feeding sows a short-chain fructo-oligosaccharide (scFOS) prebiotic during the last third of gestation and throughout lactation. They found that maternal supplementation of scFOS led to an increase of *L intracellularis* specific IgA levels in piglets that received Enterisol® Ileitis.[27] Muller et. al, (2018) investigated the supplementation of pigs with a zinc amino acid complex while receiving Enterisol® Ileitis and subsequent *L intracellularis* challenge. They found that in comparison to vaccination alone, the combined use of Enterisol® Ileitis and zinc amino acid complex led to a greater reduction in mortality than vaccination alone.[28] The effects of zinc on the immune system have been known for some time and have also been found to help the pig respond to *L intracellularis* challenge.[29]

While a medication free window of three days prior, and three days following vaccination are needed when implementing Enterisol® Ileitis vaccination, not all antimicrobials affect the vaccine. Husa et al. (2010) showed that administration of ceftiofur or tulathromycin did not alter vaccine efficacy.[30] High levels of zinc oxide as well as acidic conditions have also been previously found to not affect Enterisol® Ileitis.[9]

The cost of PPE and return on investment of vaccination

Reduction in the use of antimicrobials is also a factor that contributes to the return on investment of Enterisol® Ileitis. Voets & Hardge (2007) reported a 53% reduction of antibiotic use and an $8.60 higher economic benefit per pig in one system that began the use of Enterisol® Ileitis.[25] The cost of PPE has recently been updated by Dr. Holtkamp to range from $5.98 to $17.34 per marketed pig.[31] According to Holtkamp, the major source of economic losses associated with ileitis arises from productivity losses caused by the disease. The disease may also result in an increase in the percentage of culled pigs, and in some cases it may cause mortality.[31] Kolb & Sick (2003) found a benefit cost ratio greater than 7:1 in a summary of 5 field trials as compared to maintaining a conventional continuous feed medication program for ileitis.[8] More recent trials have found increased revenue of $4.80, $4.10 and $5.93 per marketed pig based on production performance with the implementation of Enterisol® Ileitis vaccination.[16,18,19]

Immunity to *L intracellularis* infection and vaccination

Recently it has been shown that *L intracellularis* can replicate inside of macrophages.[32] Macrophages are an important antigen presenting cell and can be recruited to many sites of the body, including the intestine.[33] It has also been recently shown that when given time to replicate, within three days, *L intracellularis* is capable of inducing the expression of inflammatory cytokines such as TNF and IL-8 by activation of the innate immune system.[34] This combination of replication not only in enterocytes but also in macrophages, and the ability to induce inflammatory cytokines is likely significant for the induction of a robust immune response. This is especially true for the induction of a cellular mediated immune response and activation of CD8 cytotoxic T cells. The latter recognize intracellular antigens and have been suggested to be the main drivers of protection against PPE.[33,35]

Studies have also measured the humoral immune response to *L intracellularis* and vaccination. While Enterisol® Ileitis does not generally lead to seroconversion in animals, a commercial killed bacterin vaccine does.[7,36] Serum antibody levels are not correlated with protection to PPE.[7,36] Interestingly, it has been found that oral vaccination with Enterisol® Ileitis not only leads to the generation of *L intracellularis* specific IgA in the ileum of animals, but also *L intracellularis* specific IgG can be found in intestinal mucosal tissue. Another interesting observation is that not only oral administration but also intramuscular administration of Enterisol® Ileitis can lead to *L intracellularis* specific IgG levels in the gut.[37] Intramuscular administration of Enterisol® Ileitis is off-label and cannot be recommended.

The future

As the field of veterinary medicine progresses, so does our understanding of what leads to health, disease and production performance. The gut microbiome, the community of all microorganisms found in the intestine, has been a recent research focus since a strong association has been found among gut microbiome composition and overall health and production performance.[38,39] We have found that *L intracellularis* not only can lead to lesions, diarrhea and decreased performance but also to a significant change of the gut microbiome community.[40,41] Interestingly we have also found that Enterisol® Ileitis in addition to the afore mentioned effects in reducing lesions, clinical signs, improving performance and inducing an immune response, also leads to changes in the composition of the gut microbiome. One of the immediate consequences found of this effect in gut microbiome composition was reduced *Salmonella* shedding in co-infected animals.[42,43] This is important as it could have consequences for the transmission of *Salmonella* as well as the promotion of food safety. A change in microbiome composition induced by Enterisol® Ileitis has also been associated with a significant reduction in *L intracellularis* shedding in vaccinated animals.[44] These results suggest that Enterisol® Ileitis prevents not only lesions and traditional measures of disease, but also prevents negative impacts that *L intracellularis* infection can have on the gut microbiome, which are likely to cause increased susceptibility to other pathogens and decreased intestinal function. The other impacts this could have, as we better understand the relationship of gut health and the gut microbiome on overall health remain to be seen.

While the afore mentioned studies have set the standard for vaccination protocols for PPE, there is still room for more investigation and questions to be answered. Interestingly, successful but temporary eradication of PPE has been achieved when focusing on achieving a negative sow and gilt population.[45] Recently, Patton et al., (2021) have described that peri-parturient sows do shed *L intracellularis* and in fact, the pathogen can be detected in the farrowing crate environment.[46] This raises many questions as to the significance of dams as a source for *L intracellularis* infection in later stages of production and consequently if there is any value in changing current vaccination protocols. It must also be considered that one of the hypotheses for the occurrence of PHE versus the chronic and non-hemorrhagic form of PPE, is exposure of older naïve animals to *L intracellularis*. Thus eradication attempts may also benefit from the immunization of animals to prevent this costly form of the disease.[47]

As it pertains to vaccination protocols, recent research has also found that Enterisol® Ileitis can confer immunity when administered intramuscularly along with 3FLEX® vaccine.[48] Although this vaccine combination and route of administration remain off label, and cannot be recommended, it may also provide direction for future development efforts.

Conclusion

Few swine vaccines have had the long track record, efficacy, and supporting data that Enterisol® Ileitis has. The future is bright as we better understand all of the benefits of Enterisol® Ileitis and refine ways to use this significant tool to promote swine health and production.

References

1. Biester & Schwarte. 1930. Intestinal Adenoma in Swine. Am J Pathol. 1931 Mar;7(2):175-185.6.

2. Gebhart & Guedes. 2010. *Lawsonia Intracellularis*. In Pathogenesis of Bacterial Infections in Animals, 10th ed., 2020.; 03–12. Wiley-Blackwell.

3. Gebhart et al. Ileal Symbiont Intracellularis, an Obligate Intracellular Bacterium of Porcine Intestines Showing a Relationship to Desulfovibrio Species. Int J Syst Bacteriol. 1993 Jul;43(3):533-8.

4. Lawson & Gebhart. Proliferative Enteropathy. J Comp Pathol. 2000 Feb-Apr;122(2-3):77-100.

5. McOrist, et al. Characterization of *Lawsonia intracellularis* Gen. Nov., Sp. Nov., the Obligately Intracellular Bacterium of Porcine Proliferative Enteropathy. Int J Syst Bacteriol. 1995 Oct;45(4):820-5.

6. Kroll et al. Efficacy of an avirulent *Lawsonia intracellularis* (Enterisol® Ileitis) vaccine in swine. AASV; 2001; 149-156.

7. Kroll et al. Evaluation of Protective Immunity in Pigs Following Oral Administration of an Avirulent Live Vaccine of *Lawsonia intracellularis*. Am J Vet Res. 2004 May;65(5):559-65.

8. Kolb J. & Sick F. Summary of field trials implementing Enterisol® Ileitis against ileitis. AASV; 2003; 243-244.

9. McOrist & Smits. Field evaluation of an oral attenuated *Lawsonia intracellularis* vaccine for porcine proliferative enteropathy (ileitis). Vet Rec. 2007 Jul 7;161 (1):26-8.

10. Diaz & Chevez. Evaluation of a protocol to control *Lawsonia intracellularis* using live vaccine in a farm in Mexico. IPVS, 2006; 198.

11. Viekilde et al. Effect of Enterisol® Ileitis Vet. In a Danish Pig Herd. IPVS, 2006; 244

12. Bulay et al. Field Efficacy of Enterisol® Ileitis Vaccine in a Philippine Commerical Farm. IPVS 2006;200.

13. Yamaguchi et al. A Field Safety And Efficacy Trial Of Enterisol® Ileitis Vaccine In Japan. IPVS 2006; 199.

14. Voets & Hardge. A Meta-Analysis on the Global Efficacy and Economics of Enterisol® Ileitis. IPVS, 2006; 201.

15. Klien et al., Longitudinal field case study on the impact of ileitis vaccination on performance. IPVS, 2010; 710.

16. Park et al. Efficacy of a commercial live attenuated *Lawsonia intracellularis* vaccine in a large scale field trial in Korea. Clin Exp Vaccine Res. 2013;2:135-139.

17. Schuttert & Seenaert, Field Observation: Successful control of an ileitis outbreak in a Dutch sow herd by oral vaccination. IPVS, 2014; 39.

18. Peiponen et al., Effect of a live attenuated vaccine against *Lawsonia intracellularis* in weaned and finishing pig settings in Finland. Acta Vet Scand (2018) 60:18.

19. Ochoa et al., Effect of an oral live vaccine against *Lawsonia intracellularis* on the performance of pigs at a production system in Mexico. AASV; 2020: 319.

20. Waddell et al. Ileitis vaccination controls ileitis in the complete absence of oral antibiotics in a start-up breeding farm. AASV; 2003:245-246.

21. Cantor et al., Case control of acute ileitis or proliferative haemorrhagic enteropathy in gilts. IPVS; 2008; 54.

22. Sanford S. Field Observations of Duration of Immunity in Enterisol® Ileitis Vaccinated Replacement Gilts. IPVS; 2006, 195.

23. Seo & Yoon. Effects of Enterisol Ileitis against PHE in Finishing Pigs. IPVS; 2016, 669.

24. Kroll et al., Maternal Immunity Associated with *Lawsonia intracellularis* Exposure and Vaccination. IPVS; 2004, 255.

25. Voets & Hardge. Reducing antimicrobial use in pig production by vaccination. Leman Swine Conference; 2007; 42.

26. Nerem J. *Lawsonia intracellularis* vaccination as an alternative to dietary antimicrobial medication in finishing swine. Safe Pork; 2009;1851.

27. Bourgot et al. Maternal short-chain fructo-oligosaccharide supplementation increases intestinal cytokine secretion, goblet cell number, butyrate concentration and *Lawsonia intracellularis* humoral vaccine response in weaned pigs. Br J Nutr. 2017 Jan;117(1):83-92.

28. Muller et. al, Effect of feeding Availa-Zn and/or vaccine on growing pig performance and mortality during *Lawsonia intracellularis* challenge. Leman Swine Conference. 2018; 25.

29. Leite et al., The effects of zinc amino acid complex supplementation on the porcine host response to *Lawsonia intracellularis* infection. Vet Res. 2018 Sep 10;49(1):88.

30. Husa et al. Compatibility of Enterisol® Ileitis oral *Lawsonia intracellularis* vaccine for swine when administered concurrently with parenteral ceftiofur sodium or tulathromycin. AASV; 2010. 283-284.

31. Holtkamp D., Economic losses associated with ileitis. 2019.

32. Pereira et al., Survival of *Lawsonia intracellularis* in porcine peripheral blood monocyte-derived macrophages. PLoS One. 2020 Jul 31;15(7):e0236887.

33. Tizard. Vaccines for Veterinarians. Elsevier, 2020.

34. Leite et al., The effects of *Lawsonia intracellularis, Salmonella enterica* serovar Typhimurium and co-infection on IL-8 and TNFalpha expression in IPEC-J2 cells. Vet Microbiol. 2019 Apr;231:76-79.

35. Cordes et al., Cell-mediated and humoral immune responses in pigs following primary and challenge-exposure to *Lawsonia intracellularis*. Vet Res. 2012 Feb 8;43(1):9.

36. Roerink et al., A novel inactivated vaccine against *Lawsonia intracellularis* induces rapid induction of humoral immunity, reduction of bacterial shedding and provides robust gut barrier function.Vaccine 2018 Mar 7;36(11):1500-1508.

37. Nogueira et al., Effect of the route of administration on the mucosal and systemic immune responses to Lawsonia intracellularis vaccine in pigs. Aust Vet J. 2015 Apr;93(4):124-6.

38. Kim & Isaacson, The intestinal microbiome of the pig. Anim Health Res Rev. 2012 Jun;13(1):100-9.

39. Neiderwerder, Role of the microbiome in swine respiratory disease. Vet Microbiol. 2017 Sep;209:97-106.

40. Borewicz et al., Changes in the Porcine Intestinal Microbiome in Response to Infection with *Salmonella enterica* and *Lawsonia intracellularis*. PLoS One. 2015 Oct 13;10(10):e0139106.

41. Leite et al., Evaluation of the porcine gut microbiome response to *Lawsonia intracellularis* infection. AASV; 2020; 333.

42. Leite et al. Vaccination Against *Lawsonia intracellularis* Decreases Shedding of *Salmonella enterica* serovar Typhimurium in Co-Infected Pigs and Alters the Gut Microbiome. Sci Rep. 2018 Feb 12;8(1):2857.

43. Visscher et al. Spread of an Experimental *Salmonella* Derby Infection in Antibiotic-Treated or *Lawsonia intracellularis* Vaccinated Piglets. Animals (Basel). 2018 Nov 12;8(11):206.

44. Leite et al., Vaccination Reduces the Shedding of *Lawsonia intracellularis* and Alters the Small and Large Intestine Microbiome. Leman Swine Conference. 2020; 37

45. Pia C. Lessons Learned in *Lawsonia intracellularis* Eradication Effort. James Mckean Swine Disease Conference; 2005; 112-114.

46. Patton et al., Investigation of *Lawsonia intracellularis* shedding by peri-parturient sows. AASV; 2021.

47. Collins A. Advances in Ileitis Control, Diagnosis, Epidemiology and the Economic Impacts of Disease in Commercial Pig Herds. Agriculture 2013, 3, 536-555.

48. Beckler, Investigation of Enterisol® Ileitis Given Intramuscularly and with the Presence of Feed Grade Antimicrobials. Leman Swine Conference; 2020; 43.

Salmonella enterica serovar I 4,[5],12:i:- update on risk factors and control

Fernando Leite, DVM, MS, PhD
Boehringer Ingelheim Animal Health USA Inc.

Introduction

Salmonella are Gram-negative bacteria that belong to the family Enterobacteriaceae. They are facultative anaerobes, facultative intracellular pathogens, non-spore-forming and are almost all motile via flagella.[1] The genus *Salmonella* was named after the American veterinarian Daniel Salmon in 1900 for his work to characterize the bacteria, which was first isolated from swine and originally called *Bacillus choleraesuis*. There are only two species of *Salmonella* which are *S enterica* and *S bongori*. *S enterica* is further subdivided in to six subspecies and most serovars of clinical importance to animals and humans are of the *S enterica* subspecies enterica (I).[1] Serovars are based on the White-Kauffmann-Le Minor scheme, which classifies bacteria based upon immunologic characterization of two surface structures, lipopolysaccharide (LPS) and flagella. Currently, *Salmonella enterica* has more than 2600 serovars.[1]

Where do the numbers I 4,[5],12:i:- come from?

To better understand the I 4,[5],12:i:- serovar, it is important to know how it got its name. The "I" is in reference to the subspecies enterica, the following three numbers "4,[5],12" are in reference to the O antigen or LPS. Finally, the last two designations "i" and "-" are in reference to the flagella or H antigen. The reason why a "-" negative symbol is present is because the I 4,[5],12:i:- serovar only has one flagellar antigen and not two. Most *Salmonella* pursue the ability of "phase variation" or changing the composition of their flagella between two antigens or "phases".[1] This is also why the I 4,[5],12:i:- serovar is considered "monophasic", because it only has one phase and antigen to make up its flagella, thus the negative sign denotes a missing flagella antigen. This characteristic is what sets it apart from the Typhimurium serovar which has a very similar antigenic formula of I 4,[5],12:i:1,2.[1] *Salmonella* serogroups are based on the O antigen, both Typhimurium and I 4,[5],12:i:- are of the same serogroup B.[1] A full serovar name is only assigned to isolates of *Salmonella enterica* subsp. enterica which meet the full antigenic definition for a serovar. Since *S* I 4,[5],12:i:- strains don't express the phase-2 flagella antigen, they lack a full antigenic formula and cannot be given a full name following the standard convention.[1]

How is *Salmonella* I 4,[5],12:i:- any different from other serovars?

Clinical salmonellosis in swine can be broadly separated in to septicemia caused by *S* Choleraesuis and enterocolitis associated with *S* Typhimurium or other serovars.[2] Experimental challenge studies have shown that the clinical presentation of serovar I 4,[5],12:i:- infection is of enterocolitis similar to the clinical presentation of Typhimurium.[3,4] *Salmonella* 4,[5],12:i:- also possesses similar ability as serovar Typhimurium to colonize the tonsils and ileocecal lymph nodes, and to be shed in the feces.[4] *S* 4,[5],12:i:- has been considered a variant of the Typhimurium serovar, due to its similarity. Strains of *S* I 4,[5],12:i:- have also been found to be frequently multidrug resistant.[5]

How has the prevalence of *Salmonella* I 4,[5],12:i:- changed over time?

Changes in the prevalence of different *Salmonella* serovars are known to occur, although the reason for the changes are not always clear. The incidence of disease and diagnosis of serovar Choleraesuis, has gone down drastically in North America. This has been attributable largely to the use of vaccination.[2] While the incidence of Choleraesuis has gone down, the incidence of Typhimurium has remained high but a stark increase in the prevalence of I 4,[5],12:i:- has been observed. According to Naberhaus et al., (2019), since 2014, *S.* I 4,[5],12:i:- has been the dominant serovar identified from swine clinical samples at the ISU-VDL, with *S* Typhimurium becoming the second most common serovar identified.[6] A similar trend in the rise of the prevalence of *S* I 4,[5],12:i:- has also been identified by researchers at the University of Minnesota Diagnostic laboratory.[7]

What are some potential risk factors that could promote *Salmonella* I 4,[5],12:i:-?

Several studies have found that pelleted feed is a risk factor for increased *Salmonella* infection.[8] One of the proposed reasons for this is because pellets have to be made from very finely ground ingredients which reduces the transit through the digestive tract and therefore does not achieve a protective low intestinal pH that could impact *Salmonella*.[8] As it regards to feed, high levels of zinc oxide (2,500 mg Zn/kg of diet) have been found to not be protective against *S* Typhimurium and were associated with longer and higher shedding levels of the bacteria in feces when compared to pigs fed lower levels of zinc oxide (50 to 150 mg Zn/kg of diet).[9] Interestingly, *S* I 4,[5],12:i:- may have a competitive advantage when copper is present in feed. Some strains of *S.* I 4,[5],12:i:- possess a mobile genetic element (Salmonella Genomic Island-4) that contains metal tolerance genes for copper, silver and arsenic. It has been shown that the presence of SGI-4 does allow for *S* I 4,[5],12:i:- to grow significantly better in the presence of copper, especially under anaerobic conditions, such as those found in the gut.[10] In accordance with these findings, the combination of elevated concentrations of zinc oxide (2000 mg/kg) and copper sulfate (200 mg/kg) has been shown to maintain higher levels of *S* I 4,[5],12:i:- shedding in comparison to pigs fed a basal diet without elevated levels of Zn and Cu (165 mg/kg Zn and 16.5 mg/kg Cu).[11] The presence of metal tolerance genes in *S.* I 4,[5],12:i:- may provide a competitive advantage to these strains along with their increased resistance to antimicrobials.

Other research has indicated that both PRRSV infection and *Lawsonia intracellularis* infection are risk factors for increased *Salmonella* shedding.[12] This led to research to investigate if *L intracellularis* control strategies could impact *Salmonella*. Two independent trials have now found that vaccination against *L intracellularis* with Enterisol® Ileitis can reduce the shedding of *S* Typhimurium and *S* Derby, respectively in co-infected animals.[13,14] Visccher et al. (2018) also found that when comparing Enterisol® Ileitis to tylosin use, vaccinated animals had a significant decrease in the *Salmonella* colonization of lymph nodes.[14] Enterisol® Ileitis is not labeled for controlling *Salmonella* and cannot be recommended for this use. It is likely that controlling for other diseases may also disfavor *Salmonella* on the farm.

It is also important to consider that farm conditions can vary and risk factors may not always be evident or the same between farms. Recently, Nair et al., (2020)[15] evaluated potential risk factors associated with *Salmonella* shedding in 50 cohorts of weaned pigs in Canada and did not identify a significant risk factor associated with *Salmonella* infection status.[15]

Do vaccines protect against *Salmonella* I 4,[5],12:i:-?

While some protection can be conferred across serogroups, protection is partially serogroup specific. Foss et al. (2013)[16] evaluated the effect of oral live Choleraesuis (serogroup C1) and Typhmurium (serogroup B) vaccines against matched and unmatched serogroup challenge. They found that serogroup matched vaccination and challenge (ie serogroup B vaccine and B challenge) led to greater protection than serogroup un-matched challenge (ie serogroup C vaccine and B challenge).[16] Neubauer & Roof (2015)[17] evaluated the effect of the Choleraesuis vaccine strain (SC 54) and the Typhimurium vaccine strain (421/125) present in Enterisol Salmonella T/C® separately, for their ability to protect against a virulent Typhimurium challenge. They found that the Typhimurium vaccine strain present in Enterisol Salmonella T/C® conferred significantly greater protection than the Choleraesuis vaccine strain as measured by clinical signs of diarrhea, and average daily weight gain post challenge.[17] Supporting the findings of Foss et al.[16] (2013). Enterisol Salmonella T/C® is the only vaccine commercially available in the US to contain a live Typhimurium vaccine strain. Theuß et al., (2017) [18] evaluated an oral live Typhimurium vaccine for its capacity to protect against a monophasic *S* I 4,[5],12:i:- challenge. They found that the vaccine did lead to a significant reduction in the incidence and magnitude of clinical signs as well as a reduction of the colonization of the intestine and lymph nodes.[18] Previous research has also shown that when compared to another vaccine, Enterisol Salmonella T/C® vaccination significantly reduced the colonization of carcass lymph nodes by *Salmonella*.[19]

As it pertains to vaccination protocols, recent research has highlighted sow vaccination as a strategy against *S* I 4,[5],12:i:-. Both Smith et al. (2017) and Davies et al. (2016) have found that a prefarrow vaccination protocol with a Typhimurium vaccine did lead to a decrease of clinical signs and fecal shedding in downstream pigs challenged with *S* Typhimurium and *S* I 4,[5], 12:i:-.[20,21] Recently, Casanova-Higes et al. (2019) highlighted the effect that sows can have in transmitting *Salmonella* to piglets, and the importance colostrum intake could have on disease control.[22] While passive immunity could play a role in protecting piglets, Hur et al. (2011) when comparing different vaccination protocols to protect piglets, found that there is a significant effect of active immunity and vaccinating piglets even when they are born to an immunized dam when evaluating protection against a Typhimurium challenge.[23] Similar findings have been observed with an autogenous vaccine against *S* I 4,[5],12:i:-, in which the combination of sow and piglet vaccination conferred significant protection and improved weight gain.[24]

References

1. Ryan, O'Dwyer, Adley. Evaluation of the Complex Nomenclature of the Clinically and Veterinary Significant Pathogen Salmonella. 2017. Biomed Res Int. 2017. 2017:3782182.

2. Stevens M., Gray J. T. 2013. "Salmonella Infections in Pigs." In Salmonella in Domestics Animals, 263–94. **doi:10.1079/9781845939021.0263.**

3. Shippy DC, et al. Porcine Response to a Multidrug-Resistant *Salmonella enterica* serovar I 4,[5],12:i:- Outbreak Isolate. Foodborne Pathog Dis. 2018. 15(5):253-261.

4. Naberhaus SA et al. Pathogenicity and Competitive Fitness of *Salmonella enterica* serovar 4,[5],12:i:- Compared to Salmonella Typhimurium and Salmonella Derby in Swine. Front Vet Sci. 2020. 30;6:502.

5. Elnekave et al., *Salmonella enterica* serotype 4,[5],12:i:- in Swine in the United States Midwest: An Emerging Multidrug-Resistant Clade. Clin Infect Dis. 2018. 5;66(6):877-885.

6. Naberhaus et. al, 2019. Emergence of Salmonella enterica serovar 4,[5],12:i:- as the primary serovar identified from swine clinical samples and development of a multiplex real-time PCR for improved Salmonella serovar-level identification. J Vet Diagn Invest. 2019. 31(6):818-827.

7. Hong et al., Serotypes and Antimicrobial Resistance in Salmonella enterica Recovered from Clinical Samples from Cattle and Swine in Minnesota, 2006 to 2015. PLoS One. 2016 Dec 9;11(12):e0168016.

8. Andres V.M., Davies R.H, Biosecurity Measures to Control Salmonella and Other Infectious Agents in Pig Farms: A Review. Comprehensive Reviews in Food Science and Food Safety Volume 14, Issue 4. 2015.

9. Janczyk P. et al., 2013 No protective effects of high-dosage dietary zinc oxide on weaned pigs infected with Salmonella enterica serovar typhimurium DT104. Appl Environ Microbiol. 2013. 79(9):2914-21.

10. Banchu P. et al, SGI-4 in Monophasic Salmonella Typhimurium ST34 Is a Novel ICE That Enhances Resistance to Copper. Front Microbiol. 2019. 24;10:1118.

11. Bearson B. et al, The Role of Salmonella Genomic Island 4 in Metal Tolerance of Salmonella enterica Serovar I 4,[5],12:i:- Pork Outbreak Isolate USDA15WA-1. Genes (Basel). 2020. 30;11(11):E1291.

12. Beloeil P.A. et al., Risk factors for Salmonella enterica subsp. enterica shedding by market-age pigs in French farrow-to-finish herds. Prev Vet Med. 2004. 30;63(1-2):103-20.

13. Leite F. et al. Vaccination Against Lawsonia intracellularis Decreases Shedding of Salmonella enterica serovar Typhimurium in Co-Infected Pigs and Alters the Gut Microbiome. Sci Rep. 2018.12;8(1):2857.

14. Visccher C. et al, Spread of an Experimental Salmonella Derby Infection in Antibiotic-Treated or *Lawsonia intracellularis* Vaccinated Piglets. Animals (Basel). 2018. 12;8(11):206.

15. Nair S. et al., Identifying Active Salmonella Infections in Swine Nurseries Using Serology and Bacterial Culture and Evaluating Associated Risk Factors. Animals (Basel). 2020. 27;10(9):1517.

16. Foss D. et al Protective immunity to Salmonella enterica is partially serogroup specific. Vet Immunol Immunopathol. 2013. 1;155(1-2):76-86.

17. Neubauer A., Roof M. Evaluation of the Enterisol® Salmonella T/C vaccine strains against challenge with a virulent *Salmonella* serovar Typhimurium isolate. AASV, 2015: 171-174.

18. Theuß et al., Immunogenic potential of a Salmonella Typhimurium live vaccine for pigs against monophasic Salmonella Typhimurium DT 193. BMC Vet Res. 2017.17;13(1):343.

19. Kolb et al., Vaccination with Enterisol® Salmonella T/C reduces Salmonella enterica colonization of ileocecal lymph nodes in growing pigs. IPVS, 2016:808.

20. Smith et al. Maternal vaccination as a Salmonella Typhimurium reduction strategy on pig farms. J Appl Microbiol. 2018. Jan;124(1):274-285.

21. Davies et al., Use of an attenuated live Salmonella Typhimurium vaccine on three breeding pig units: A longitudinal observational field study. Comp Immunol Microbiol Infect Dis. 2016. Jun;46:7-15.

22. Casanova-Higes et al., Weaned piglets: another factor to be considered for the control of Salmonella infection in breeding pig farms. Vet Res. 2019. Jun 18;50(1):45.

23. Hur et al. Efficacy of a novel virulence gene-deleted Salmonella Typhimurium vaccine for protection against Salmonella infections in growing piglets. Vet Immunol Immunopathol. 2011. Feb 15;139(2-4):250-6.

24. Ruggeri et al., Inactivated Salmonella enterica serovar Typhimurium monophasic variant (*S. Typhimurium* 1,4,[5],12:i-) in sows is effective to control infection in piglets under field condition. Vet Microbiol. 2015. Oct 22;180(1-2):82-9.

Field applicable technology in the swine industry: SoundTalks® cough monitoring and Remote Insights© movement analysis in a PRRS and IAV-S co-challenge study

K. Hogan[1], PhD; A. Sponheim[1], DVM; E. Fano[1], DVM, MS, PhD; Dale Polson[1], DVM, MS, PhD, J. Cairns[2]; Michael C. Keyes[2]; Pipestone Applied Research[3]
[1]Boehringer Ingelheim Animal Health USA Inc., Duluth, Georgia; [2]Remote Insights, Minneapolis, Minnesota; [3]Pipestone Applied Research, Pipestone, Minnesota

Introduction

Until somewhat recently, assessing pig movement and cough has been largely subjective, as it was often manually collected and dependent on individuals in the barns listening for cough. With the advent of auditory technology and sophisticated algorithms, cough can now be assessed automatically and precisely. Using recent technology developed to provide objective insights into swine respiratory distress and activity of pigs in the barn, we can shed new light on traditional study observations. In this study, vaccine impact was assessed in a PRRS and IAV-S co-challenge study, while cough and pig movement technology was used to quantify the impact on clinical signs and pig activity. The following case study describes how various technologies can be combined in the barn. Additionally, we highlight examples of the types of new information that can be gathered and used to assess animal health.

What is SOUNDTALKS?

SOUNDTALKS is sound monitoring technology that can detect evidence of respiratory distress in swine. Using proprietary algorithms developed by SOUNDTALKS audio research engineers, monitors containing microphones continuously listen for cough. The basic measurement used by SOUNDTALKS is called the Respiratory Distress Index (RDI). The SOUNDTALKS algorithms filter out non-relevant sounds (squealing, grunting, screaming, sounds from fans, sounds from feedlines and feeders, other mechanical sounds, etc.). Sounds relevant to respiratory distress (cough) are then aggregated continuously and the RDI daily value is generated. Additionally, an LED alerting system is present on the bottom of each monitor to notify those in the barn of changes in the level of cough. Using a green-yellow-red alerting scheme, caregivers are provided at-a-glance information on the levels of cough in different parts of the barn. Previous research has consistently found that SOUNDTALKS can provide early warnings (yellow alerts) to coughing 2-5 days before the caregivers in the barn reported hearing cough.[1]

What is REMOTE INSIGHTS activity monitoring?

Remote Insights© (RI) uses the Internet of Things (IoT) and artificial intelligence (AI) to continuously monitor the activity of pigs detected through wireless ear tags. The ear tags capture the motion of the pigs and machine learning models translate those motions into activity categories of dormant vs. active. These categories are trended based on an established baseline for each pig, and deviations are identified that have been correlated to events of interest, like disease challenge, estrus, nesting and/or labor.[2]

Materials and methods

An IAV-S and PRRSV co-challenge study assessing the impact of a Live Attenuated Influenza Vaccine (LAIV) and Ingelvac® PRRS MLV was performed at Pipestone Applied Research in Spring 2020. In brief, three treatment groups, 1 treatment group/room at a BSL2 facility, were included in the study. Treatment group one (T1, Room 3) included non-vaccinate controls. Treatment group two (T2, Room 5) included vaccinated with 2 mL of INGELVAC PRRS MLV intramuscular at weaning. Treatment group three (T3, Room 6) included piglets vaccinated with 1 mL of LAIV intranasal at 5 days of age and 2 mL of INGELVAC PRRS MLV intramuscular at weaning.

Acoustic SOUNDTALKS sensors were placed in each trial room and an initial baseline for each room was gathered over 48 hours upon placement of pigs at weaning. After the initial baseline was calculated, the algorithm continuously assessed the processed audio data for significant deviations from the baseline, which were translated into warnings (yellow LED lights) and alerts (red LED lights) by the monitors. As the pigs grew, the baseline was continually reassessed to ensure that it was appropriate for that group of pigs and the environment at that specific time. Ninety-one pigs/treatment group were fitted with REMOTE INSIGHTS ear tags one week prior to the IAV-S challenge (day 14). Like SOUNDTALKS, a baseline was established over the course of one week and deviations from the baseline were reported to indicate changes in behavior and movement.

Three weeks post-wean (day 21), all pigs were challenged with a heterologous H3N2 IAV-S strain A/swine/Ohio/A01354299/2017/H3N2. One week later (day 28), all pigs were challenged with a wild type heterologous PRRS 1-7-4 isolate.

Results and discussion

The information reported in this proceeding is exclusively the information from the SOUNDTALKS and REMOTE INSIGHTS portion of the trial. Movement and cough data were exported from their respective databases periodically throughout the trial, as well as a final data pull at the conclusion. Due to a logistics issue, RI baseline data in Room 3 was not initiated until the day of the IAV-S challenge and is thus the baseline represented as the average data of all animals.

Figures 1, 2 and 3 show the RI activity level in each room as a rolling 24 average for all pigs in the room. The y-axis is a converted value of the pig movements that are smoothed and zeroed. Green, yellow and red bars are shown in the background of all Figures indicating the aggregated level of cough in the room for a given day. The baseline, or normal level of noise and cough in the room, is shown by a green shading; yellow shading indicates a significant change in the amount of noise in the room, optimized to detect cough; and red shading indicates that a respiratory event is underway in the room (and coughing is likely audible to the human ear). These colored alerts are displayed based on the RDI values that are displayed in Figure 4.

SOUNDTALKS alerts in the NVC room showed extended levels of cough from shortly after the IAV-S challenge to close to the end of the trial (Figure 1). The alert pattern in the PRRS MLV room (Room 5; Figure 3) indicate slightly few days at red level coughing than the NVC, but the extended number of days of yellow alerts after the red coughing episode indicate an extended recovery. The alerts in the LAIV + PRRS MLV room (Room 6; Figure 2) show a considerably shortened duration of red and yellow alert days compared to the other two rooms. This is supported by Figure 4, which shows that levels of respiratory distress were highest in the NVC pigs, followed by the PRRS MLV vaccinated pigs, and the LAIV + PRRS MLV vaccinated pigs.

An interesting finding was that activity as measure by RI showed a decrease in activity in all three rooms when SOUNDTALKS was alerting red or yellow. There appears to be an inverse relationship between RDI and RI activity levels, indicating that when the pigs are experiencing a respiratory event that is detected by the SOUNDTALKS monitors, the overall activity of the animals decreases. Despite the distinct differences between the two technologies, there appears to be an agreement in the data that warrants further investigation.

Various technologies are becoming increasingly prevalent on farm. Research like the trial presented here allow for confidence to be built in devices that may be less familiar to veterinarians or producers for in-barn use. Additionally, information such as this helps to validate new technologies with one another.

References

1. Polson, D. et al., 2018. Precision livestock farming (PLF) for pig health and production: Sound as a diagnostic sample. AASV, 2018.
2. Remote Insights. Personal communication

Figure 1: RI activity data and SOUNDTALKS alert data for non-vaccinated controls (T1, Room3).

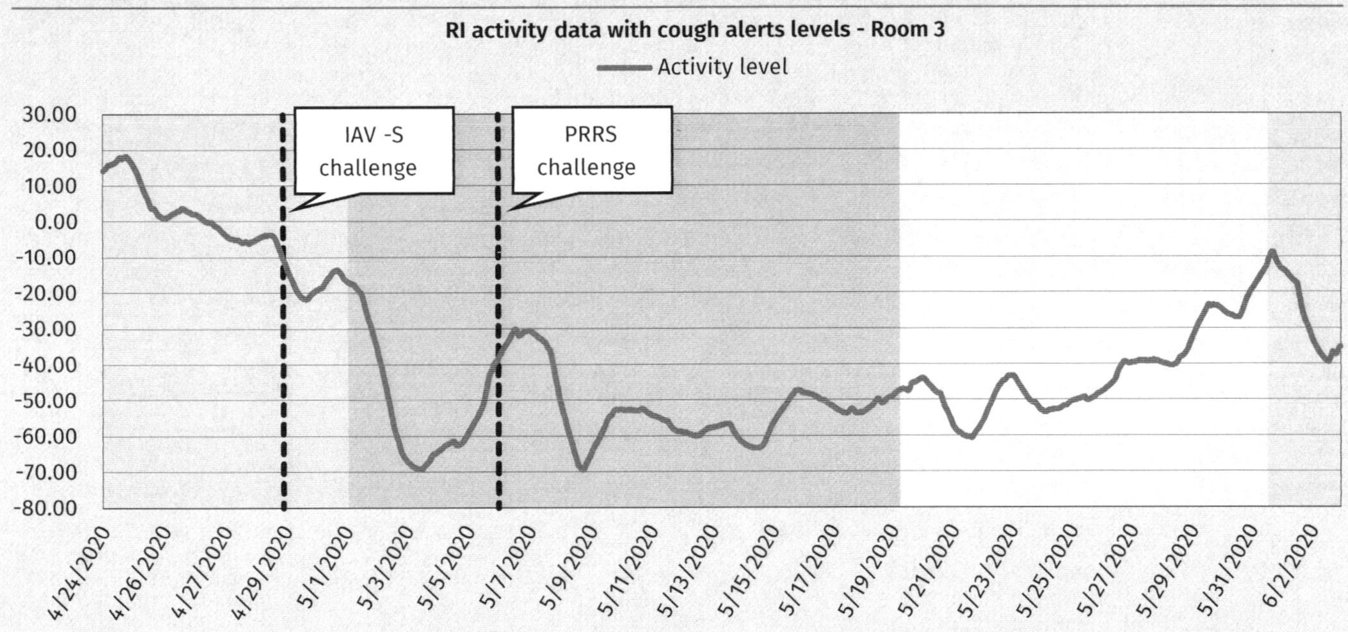

Figure 2: RI activity data and SOUNDTALKS alert data for INGELVAC PRRS MLV treatment group (T3, Room 5).

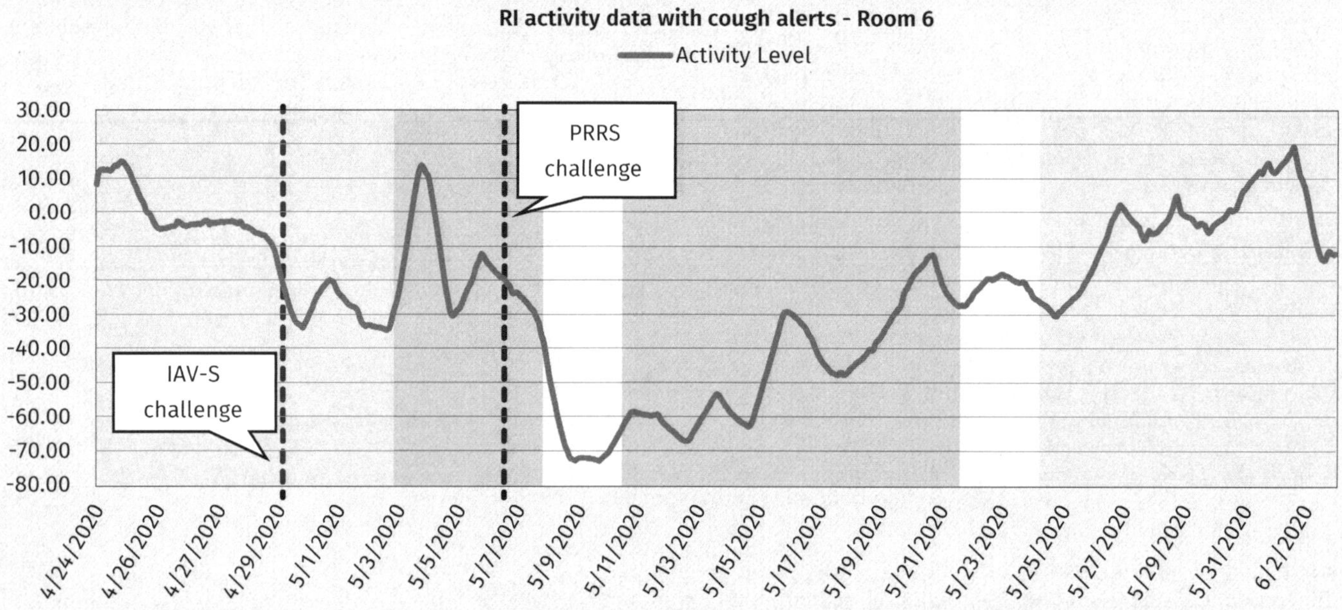

Figure 3: RI activity data and SOUNDTALKS alert data for LAIV + INGELVAC PRRS MLV treatment group (T3, Room 6).

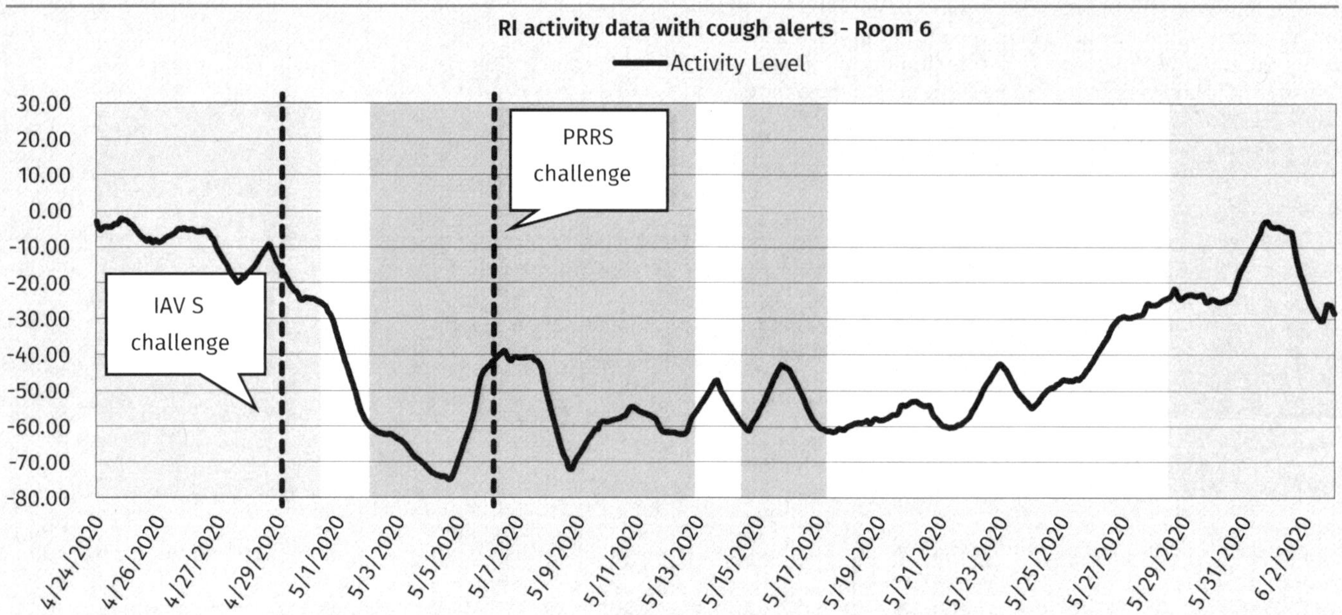

* background color is SoundTalks alert color

Figure 4: SOUNDTALKS Respiratory Distress Index (RDI) data, adjusted for daily mortality by room in each 24 hour period. IAV-S and PRRS challenges are denoted by red bars.

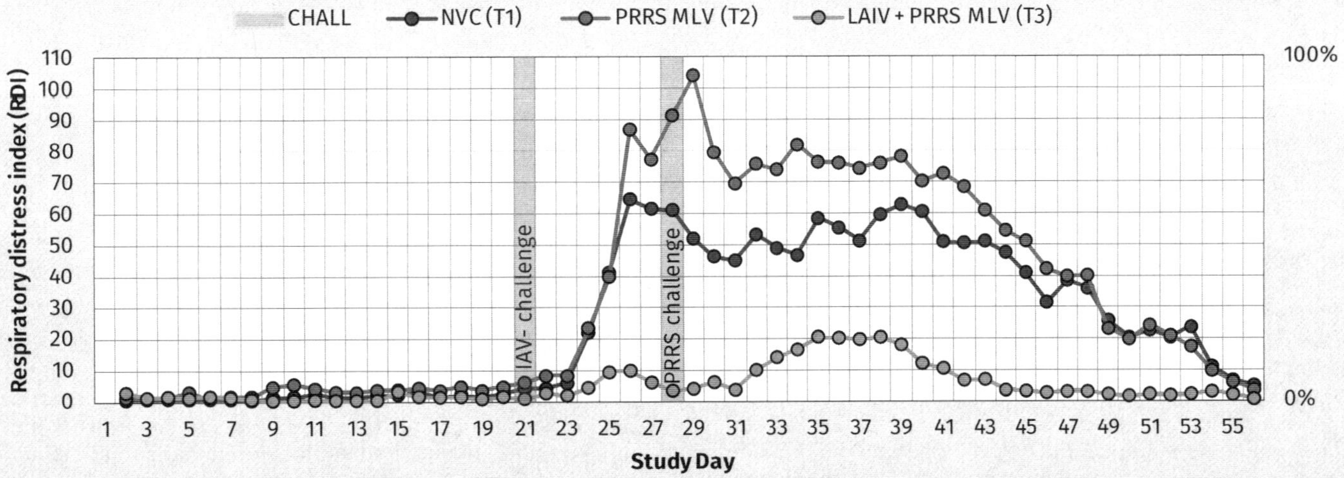

Easier autogenous vaccine management through GVL

Paul Yeske, DVM
Swine Vet Center

Introduction

Managing and tracking custom-made vaccine isolates can be challenging with all the responsibilities of practicing veterinarians. The Newport Laboratories Inc. HerdPoint software tool, now part of the GVL platform, offers a complete solution to easily track existing isolates on file, view expiration dates, access product numbers and specific product isolates, and manage non-adjacent herd approval applications. GlobalVetLink (GVL) and Newport Laboratories recently worked together to rebuild HerdPoint from the ground up, resulting in streamlined workflows and a more intuitive, user friendly tool. This paper will provide an overview of the updated platform and demonstrate how veterinarians are using this tool to manage their client's isolates and autogenous vaccines.

About HerdPoint

HerdPoint is a custom-made vaccine management tool that is exclusive to customers of Newport Laboratories. HerdPoint gives veterinarians greater insight into their ongoing isolate management and improves the efficiency of the nonadjacent-herd approval process. Through HerdPoint, veterinarians can:

- Track existing isolates on file, view expiration dates and submit requests for extensions
- Immediately access product numbers and specific product isolates
- View and track nonadjacent-herd approval applications
- View, submit and sign documents digitally

HerdPoint transition to GVL

HerdPoint is now integrated into GVL's online platform, making it more user friendly for veterinarians to view and manage their clients' isolates. As part of the transition to GVL, all features and functionalities of HerdPoint are now fully optimized for use on the most modern and up-to-date internet browsers, including Chrome, Firefox and Edge. GVL rebuilt the software behind HerdPoint from the ground up to streamline workflows and make the tool more intuitive and user friendly, giving veterinarians greater visibility into the status of their clients' isolates and non-adjacent herd requests. One of the key enhancements made to HerdPoint as part of the transition to GVL's online platform is a focus on prioritizing the information veterinarians need most, such as expiring isolates. This allows veterinarians to quickly view and manage their clients' isolates.

Increased efficiency

With a more intuitive interface, veterinarians can spend less time managing isolates and more time helping their clients. Many veterinarians already use GVL to manage animal health and movement documents. Now they will also be able to manage their clients' custom-made vaccine isolates with Newport Laboratories in the same platform. Customer support for HerdPoint has also been enhanced, with experts from both Newport Laboratories and GVL ready to support users depending on the nature of the question and follow up needed.

HerdPoint features

HerdPoint provides veterinarians and veterinary technicians with easy access to isolates and products, as well as notifications to assist with management of isolates and requests. Under the Custom-made vaccines tab in the GVL platform, users can easily access the Isolate and Product pages (Figure 1). From the isolates page, a veterinarian or veterinary technician has access to the Isolate table (Figure 2). They can toggle between the Active and Expired tables to view isolates depending on their status. The search feature provides easy access to a specific isolate. The table can be sorted by the individual columns to provide a streamlined view and isolate actions are easily accessible from the Action column. By hovering over Actions next to a specific isolate, the user will be presented with a list of actions depending on the status of the isolate. These actions may include view isolate, request or view extension, or request or view release. By clicking on the isolate number under the isolate details, or by selecting view isolate from the actions dropdown menu, a veterinarian or veterinary

Figure 1: Easy navigation to isolates and products.

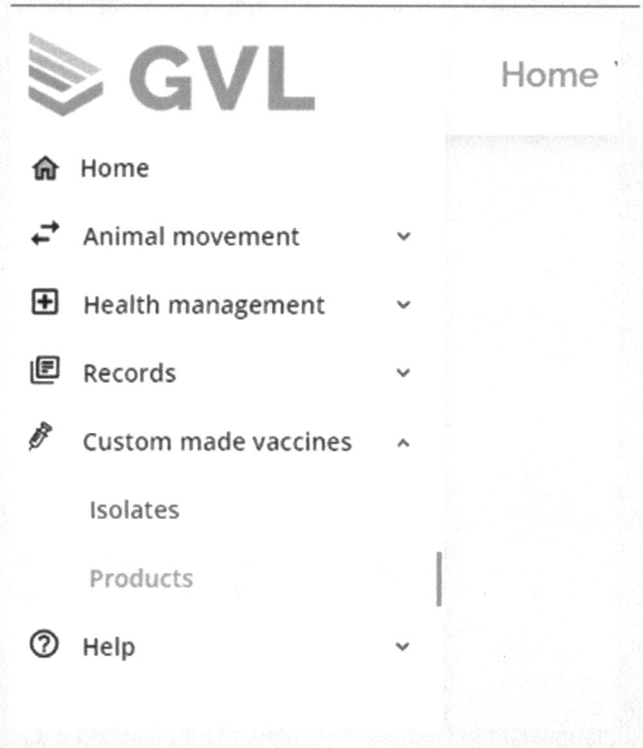

Figure 2: HerdPoint isolate table in GVL platform.

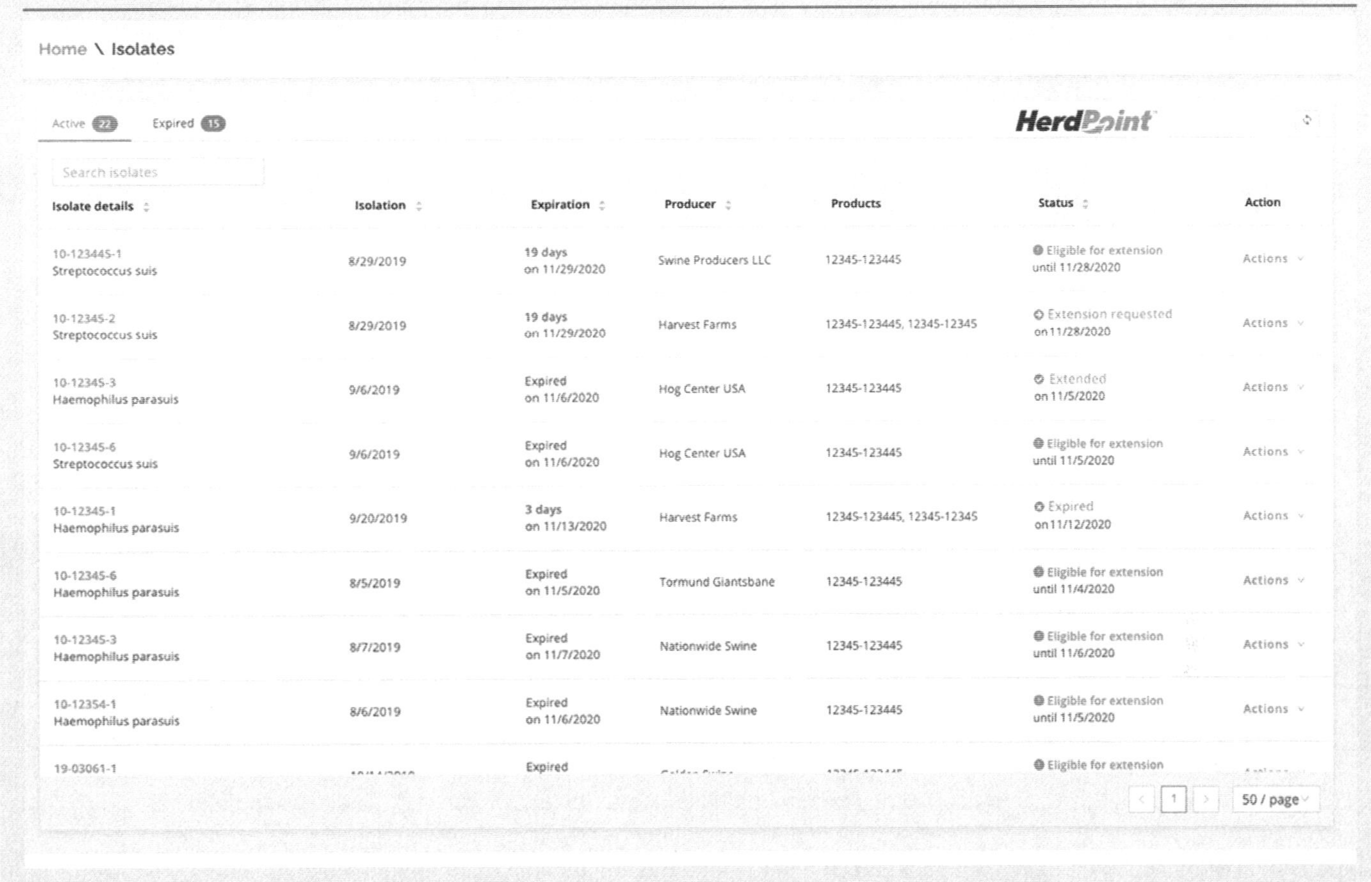

technician can view the isolate information including the status, expiration date and products associated with the selected isolate. From the products page, a user has access to the Products table which has multiple tabs to easily view non-adjacent herd requests and where a specific request is in the process.

Non-adjacent herd request process

The non-adjacent herd request process is initiated by Newport Laboratories. Once a non-adjacent herd request has been initiated, the veterinarian or veterinary technician can access it from the Open tab on the Products page. They will then need to create the herd list for this specific request, which can be uploaded via a CSV file. Once uploaded, the herd list will be attached to the request. A veterinary technician can create and upload the herd list and then save the non-adjacent herd request for the veterinarian to complete. The veterinarian will need to enter their diagnosis or clinic signs of the disease(s) observed and the summary of the epidemiological pattern, including the cause and/or effect of the disease(s) situation that links the nonadjacent herd(s) to the herd(s) of origin. They can select statements from the lists provided, create new statements for the specific request or a combination of both (Figure 3). Once they have selected their statements, they can submit the non-adjacent herd request for review. After submission, they can view the progress of approval of their non-adjacent herd request from the Products page.

HerdPoint notifications

Email notifications for the extension and non-adjacent herd request processes are built into HerdPoint. For isolates in the extension request process, an email is sent when isolates are available for extension. If an extension is requested, the clinic will receive email notification when the extension is approved. If an extension request is not approved, the clinic will receive email notification of the extension return with a message from Newport Laboratories. If an isolate is not extended, an additional email notification will be sent five days before expiration letting the clinic know they still have time to request an extension. When extended isolates are within 60 days of expiration the clinic will also receive email notification of those isolates. For the non-adjacent herd request process, when a request form has been created by Newport laboratories the clinic will receive email notification to complete the form. The clinic will also receive an email notification when the non-adjacent herd request has been approved. If a non-adjacent herd request has been sent back requiring additional information, the clinic will receive an email notification with the request and what additional information they need to include on the request. State veterinarians also receive notification of non-adjacent herd requests that need their approval as well as notification when they have approved a request. These email notifications assist veterinarians and clinic staff in managing the extension and non-adjacent herd approval request processes.

Figure 3: Non-adjacent herd request attending veterinarian assessment statements.

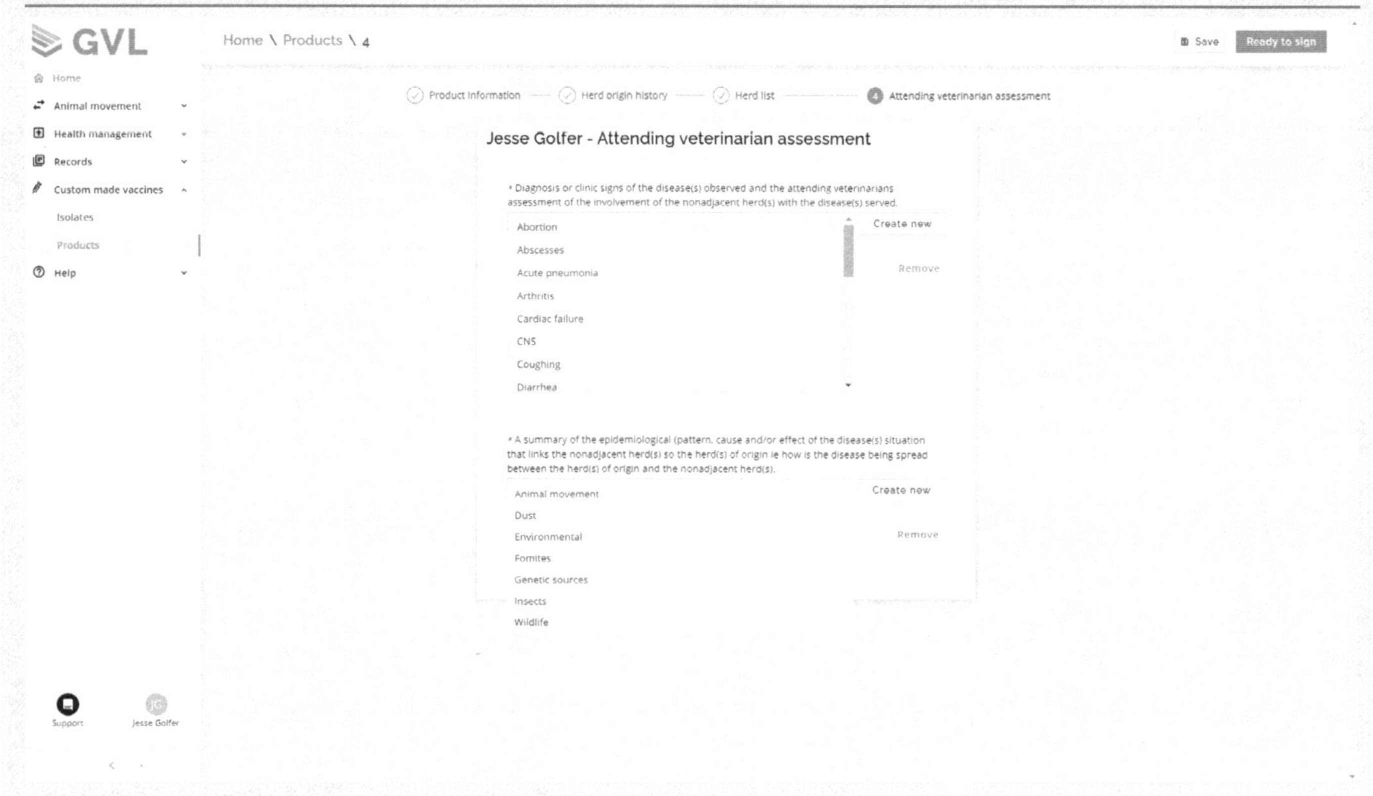

Improvement for veterinarians and veterinary staff

The rebuilt HerdPoint provides easy access to information and automation of processes. It is now easier for veterinarians and veterinary technicians to understand isolate and product information at a glance. This greater transparency of data provides the opportunity for easy decision making. Veterinarians and veterinary technicians can also easily view the status of isolates and request extensions and releases. The built-in sorting features and email notifications provide easier management of isolates and non-adjacent herd requests and the simplified non-adjacent herd request process allows for quick processing of applications.

Summary

It is now simpler for veterinarians to manage and track custom-made vaccine isolates and product information with HerdPoint, now part of the GVL platform. Customer support for HerdPoint has been enhanced, with experts from both Newport Laboratories and GVL ready to support users. GVL and Newport Laboratories worked together to rebuild HerdPoint from the ground up and the result is streamlined workflows and a more intuitive, user friendly tool for veterinarians.

GVL

GVL is the nation's leader in providing easy-to-use, cloud-based animal health solutions for food and companion animal health practitioners. GVL's services provide intuitive online certification solutions that enable users to create professional animal health records quickly and accurately. All GVL solutions are easy to use, efficient, secure, accessible, and compliant. GVL works closely with the USDA and regulatory bodies, and also partner with practice management, production management and feed management software providers to keep all stakeholders connected.

Newport Laboratories Inc.

Maintaining the health of your animals is essential to the success of any livestock operation. Newport Laboratories works to ensure that veterinarians and livestock producers are equipped with the preventative, diagnostic, and educational tools necessary to ensure peak performance. The livestock industry changes quickly, and these changes often bring unique disease challenges. Discover an easier way to manage isolates through HerdPoint. This exclusive software tool tailored to veterinarians lets you:

- Track existing isolates on file, and view expiration dates
- Immediately access product numbers and specific product isolates
- View, request and track non-adjacent herd approval applications
- View, submit and sign documents digitally

Efficacy of *Bacillus subtilis* and *Bacillus licheniformis* (BioPlus 2B®) on sow lactation performance, litter growth, nursery performance and health of pigs challenged with PRRS virus post-wean

Steven P. Lerner[1]; Sage R. Becker[2]; Scott A. Dee[3]; Steve J. Kitt[4]; Mark J. Bertram[4]; Todd E. Williams[3]; Brent J. Pepin[3]; Dan B. Hanson[3]; Erin A. Little[3], Paul F. Frahm[3]; Roy Edler[3]
[1]Chr Hansen, Inc., Athens, Georgia; [2]Iowa State University, Ames, Iowa;
[3]Pipestone Veterinary Services, Pipestone, Minnesota; [4]First Choice Livestock, LLC

Background

Daily feeding of effective probiotics has been repeatedly demonstrated to support normal gastrointestinal (GI) functions and a beneficially balanced gut microbiome in sows, piglets, and growing pigs. This subject was recently reviewed by Liao and Nyachoti, 2017.[1] Aside from the typical array of GI functions that are positively affected by probiotics, namely digestion, absorption, barrier, and immune functions, there is a growing body of evidence that other organ systems, e.g. the lungs, may derive benefits from the actions of probiotics and a healthy gut microbiome.

Purpose

The purpose of this study was to determine the effect of daily feeding of BIOPLUS 2B® to sows from 31 days pre-farrowing through weaning on their performance and that of their litters and to determined the effect of continued daily feeding of BIOPLUS 2B to piglets through 28 days post-weaning in the presence of a severe challenge with PRRSV.

Materials and methods

(Figure 1) Animals and variables measured
(Table 1) Treatment groups
(Table 2) Arrangement of treatments at nursery

Results and discussion

Results of BioPlus 2B vs Control in Sows/Litter and Nursery
Figure 2A Sows and litter
Figure 2B Nursery
Figure 2C Nursery
Figure 2D Nursery

For sows fed BioPlus 2B® for the last 31 days of gestation and throughout lactation, there was no effect on litter size, but there was a significant increase in post-birth weight and a tendency for an increase in weaning weight. Daily feeding of BioPlus 2B to pigs from dams that were fed BioPlus 2B were significantly better able to survive a PRRSV challenge, to grow at a normal rate, and

Figure 1: Animals and variables measured.

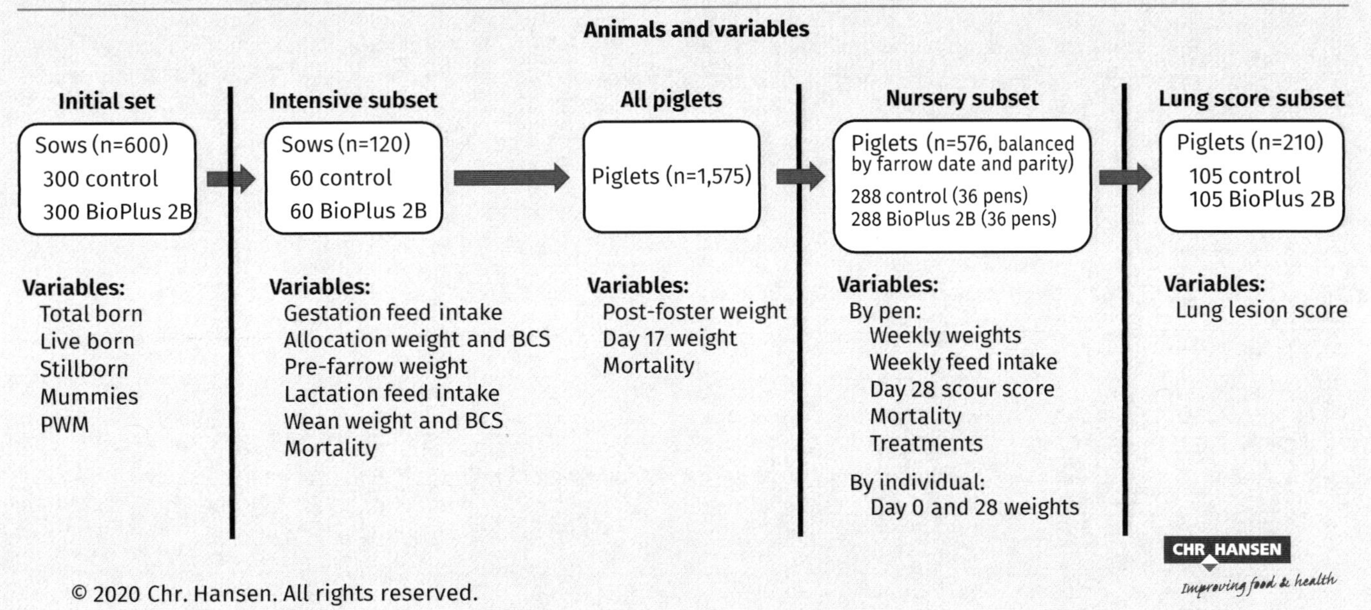

Table 1: Treatment groups

Treatment group	Sows	Pigs	PRRS challenge 7 d post-wean
Control (1, 3, 5)	No probiotic	No probiotic	Yes
BioPlus 2B® (2, 4, 6)	BioPlus 2B	BioPlus 2B	Yes

Table 2: Arrangement of treatments at nursery

Room	1	2	3	4	5	6
Treatment	1	2	1	2	1	2
BioPlus 2B on sow farm	No	Yes	No	Yes	No	Yes
BioPlus 2B in nursery	No	Yes	No	Yes	No	yes
PRRSV challenge	Yes	Yes	Yes	Yes	Yes	Yes
Feed bins	A	B	C	D	E	F

PRRSV challenge: 1-7-4 variant, administered IM on d7 post-wean, $2 \times 10^{3.5}$ $TCID_{50}$
Feeding rate of Bacilli: 1.6×10^9 CFU/kg, piglet = 1.1×10^9 CFU/kg
Statistical analyses: SAS Proc GLM and Proc GLIMMIX

to convert feed to weight at a normal rate compared to pigs that didn't receive the effective probiotic. Among the surviving pigs, there was no difference in morbidity at 28 days post-weaning, as no effect of treatment on scouring or lung scores were noted.

Daily feeding of an effective probiotic improves capacity of pigs to survive and to perform normally in the presence of a PRRSV challenge.

Reference

1. Liao SF and Nyachoti M 2017 Using probiotics to improve swine gut health and nutrient utilization Animal Nutrition 3:331-343

Figure 2: Results of BioPlus 2B vs control in A) sows and litter; B) nursery - wean wt vs 28d post wean wt; C) nursery - ADG and ADFI; D) nursery - scour score, lung lesions and mortality.

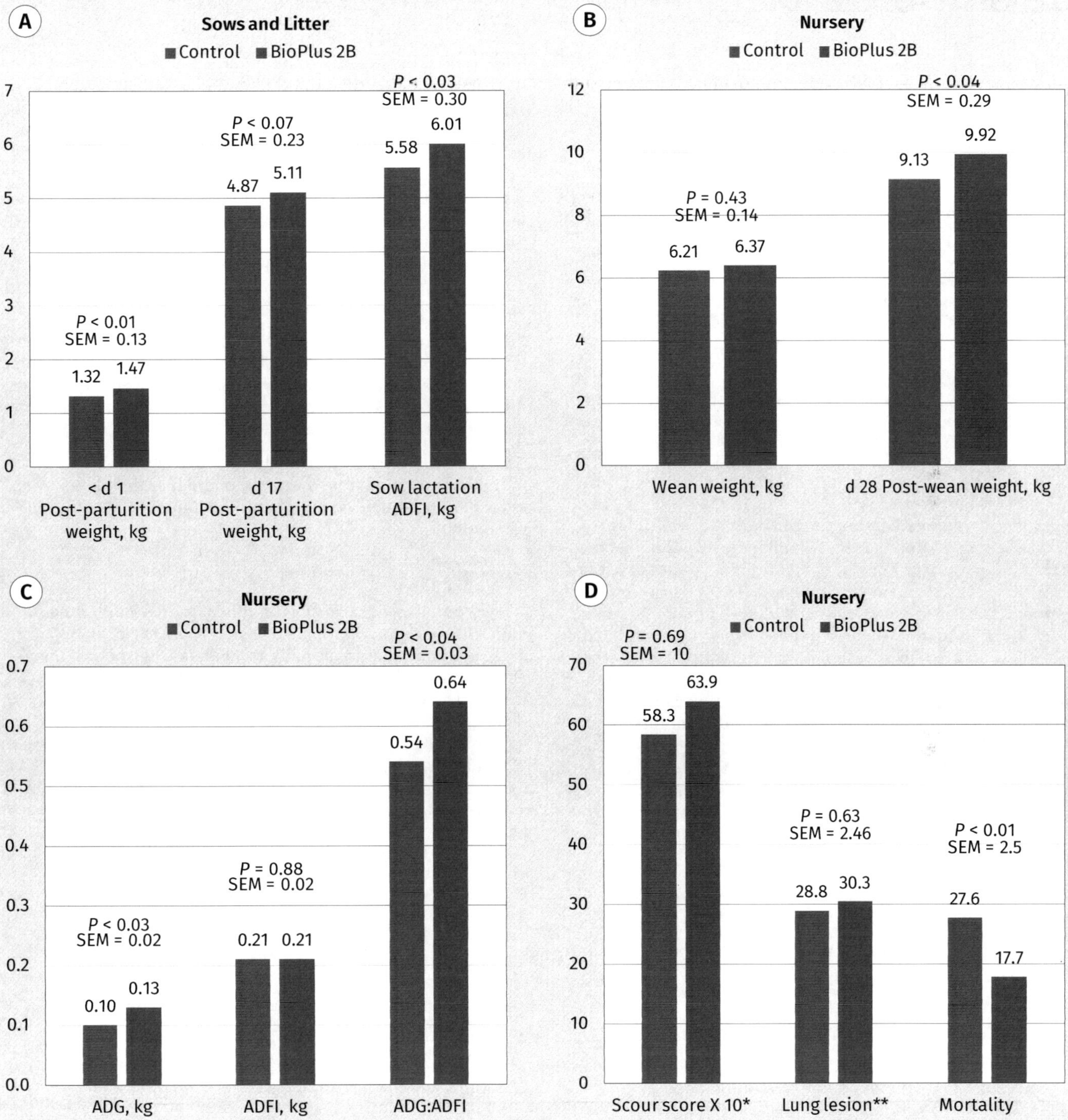

* Pen average: 0 = no scour; 1= moderate scour; 2 = severe scour

** infected % X factor

How to enhance biosecurity and save money at the same time!

Jeff Odle
LANXESS Corporation, Actives and Disinfectants, Pittsburgh, Pennsylvania

Introduction

The swine industry in the United States was poised to have an excellent 2020 with new trade agreements in place with China, Japan, Korea, Mexico and Canada. Unfortunately, the COVID 19 pandemic hit the USA and swine producers struggled to survive, literally and financially. Pathogenic challenges did not go away for the animals and in fact biosecurity became even more important to workers, for their own personal safety. Some organizations look at cleaning and disinfection (C&D) as added expense but it can be an opportunity to save money and contribute to the bottom line.

Purchasing a C&D product that does not deliver, whether it be superior soil removal for a detergent or lack of efficacy against the intended pathogens for a disinfectant, is a waste of money. This paper will look at the cost benefit of an effective C&D program.

Start with a detergent

Even though formulated detergents add cost, the overall return on investment can be profound. Cleaning not only prepares the surface for disinfection but also decreases the number of pathogens by physically removing them. It is obviously easier for an antimicrobial to be effective when there are fewer microorganisms present. Organic matter, such as dried on feces, hard water scale, and biofilms all provide a protective barrier, which prevent the disinfectant from coming into contact with the pathogen. Biofilm also reacts with the disinfectant thus reducing concentration of available antimicrobial agent. Bacteria within a biofilm are 1,000 times more resistant to antimicrobials than bacteria in suspension. A study[1] conducted by the University of Prince Edward Island, shows presoaking the surface reduced the amount of time needed for cleaning over no presoaking. Presoaking followed by a commercial detergent achieved a 46% reduction in the time required to clean each pen (Table 1).

One could argue, cleaning is actually more beneficial than disinfecting by reducing the population of microorganisms by 80-90%. It is not chemically possible to open biofilms by power washing with water alone. Detergents lower the surface tension from 72 dynes per centimeter to 30 dynes per centimeter making soils easier to remove. Organic soils are better removed by using alkaline cleaners whereas hard water scale responds more favorably to acidic pH cleaners. Foaming cleaners allow longer contact time, especially on vertical surfaces, for penetration and removal of soils.

This same study also evaluated the impact on animal productivity by tracking days to market with a target finish weight of 110kg (243lbs). Pens where a presoak was followed by a commercial detergent and a peroxygen disinfectant resulted in a 5.2% reduction in number of days to market.

Time Is Money is an aphorism attributed to Benjamin Franklin. In this case, time saved equals less cost in labor, less power and water used, reduced downtime and improved feed to gain productivity.

Cost in-use

Disinfectants are used at varying concentrations and come in many different size containers. It is important to calculate cost in-use in order to get an accurate comparison. Most liquid disinfectants are used at a 1:256 dilution rate meaning a 1-gallon jug of concentrate will dilute to 256 gallons in-use. Price divided by number of gallons equals cost in-use. Powder based products, such as Virkon™ S, are used at a 1:100 or 1:600 concentration. Therefore a 10lb pail of powder concentrate at 1:100 will make 123 gallons and at 1:600 a total of 738 gallons of in- use solution. If larger volume containers are available, these are generally sold at even lower cost per pound or gallon resulting in additional savings.

Table 1: Average wash time per pen with varied washing protocols[1]

	Detergent	Presoak	Wash time per pen (minutes)	Difference (minutes)	Time savings %
Cold water	-	-	68.03	0	0
	+	-	59.80	-8.23	12.1
	-	+	41.39	-26.64	39.1
	+	+	36.38	-31.65	46.52
Hot water	-	-	52.61	-15.42	22.6
	+	-	46.24	-21.79	32.0
	-	+	41.88	-26.15	38.43
	+	+	36.81	-31.22	45.9

A disinfectant must be effective against the pathogens of concern otherwise it is wasted money and effort. Does the disinfectant have the proper label claims proving efficacy? Does it work under the environmental conditions in which it is being used, such as cold temperatures in winter or fast contact times for boot baths?

Most disinfectants are used at a single label recommended concentration for a broad spectrum kill of the listed pathogens on the label. Where lower concentrations are specified for known and specific diseases, it allows the end user to tailor their use to the application. Broad spectrum kill is important where the most susceptible animals are present or where a disease challenge has been identified. In applications like transportation, viral transfer of PRRS and PEDV from farm to farm is the greatest concern. If lower disinfectant concentrations proving efficacy against known pathogens, like PRRS and PEDV, are listed on the label, then there is an opportunity to significantly reduce cost in-use for certain applications.

Conclusion

Providing animals a clean and healthy environment, allows them to grow and thrive at their utmost potential. Good biosecurity starts with cleaning. The standard use dilution test used for testing disinfectant efficacy calls for only 5% organic challenge. Cleaning reduces the number of pathogens present and prepares the surface for an effective disinfection. Understanding the importance of cleaning, calculating cost in-use, and using a proven disinfectant chemistry which provides efficacy over a wide range of conditions and applications will allow for enhanced biosecurity and cost savings. The return on investment will be measured in healthier animals with greater productivity, less therapeutic costs and lower mortality.

Reference

1. Hurnik, D. 2005. Investigations into optimal washing and disinfection techniques for pig pens. London Swine Conference – Production at the Leading Edge 6-7

Targeted development of rotavirus vaccines using Medgene's experimental prescription platform (RxP) against current circulating strains

A. Young, PhD; A. Petersen, MS

Introduction

Swine rotaviruses are a prevalent and economically devastating disease that continue to plague the swine industry.[1] Current vaccines necessarily rely on outdated strains that may not cover the full breadth of rotaviruses impacting producers today. Rotaviruses exist in multiple genotypes (generally denoted P and G types) due to reassortment of viral genes, a more flexible and temporally responsive vaccine platform is required to match the needs of producers with modern vaccine technology. We have shown that a baculovirus-produced, rotavirus specific bivalent vaccine elicited high titers of neutralizing antibodies against the target Rotavirus A. Mice were immunized with several putative subunit vaccines, and blood collected to analyze immunoreactivity. Mice immunized with a bivalent vaccine developed neutralizing titers exceeding 1:640 whereas control mice failed to develop any titers. Interesting, the inclusion of additional protein targets did not increase vaccine effectiveness, and in fact a trivalent vaccine reduced apparent neutralization titers to 1:320. Based upon this encouraging data, we have continued to examine the specificity of the immune response in vaccination trials. Specifically, mice were vaccinated with baculovirus-produced proteins whose sequences were derived from circulating strains of protein Rotavirus. Sera obtained from these vaccinated animals was then used in standard serum neutralization (SN) assays as a correlate of expected protection. While these experiments are ongoing, we have established critical epitopes within the target proteins as evidenced by effective cross-neutralization in vitro, combined with in silico analysis of predicted protein structure. Based upon the data obtained, it is clear that cross-protection against significantly divergent Rotavirus A isolates can be generated, and these predictive algorithms are currently being tested directly in swine. Given new regulatory pathways, this work opens the door to the availability of targeted vaccines to protect against currently circulating strains of Rotavirus A, and the potential for development of vaccines against other strains of Rotaviruses (B and C) that can be easily produced using analogous technology.

Materials and methods

Sequences of target Rotavirus isolates were obtained from routine diagnostic samples submitted to AAVLD Accredited Diagnostic Laboratories. Recombinant baculoviruses were produced in order to generate proteins corresponding to the obtained sequences, according to our USDA filed Outline of Production. Experimental prescription platform vaccines were produced according to the filed Outline of Production containing either one or two relevant proteins from each isolate. Mice were vaccinated twice at 3-week intervals with an appropriately reduced dose of the swine vaccine, and serum harvested 28-35 days post second vaccination. Sera was then tested in standard virus neutralization assays using either homologous or heterologous Rotavirus isolates.[2,3] In addition, viral sequences corresponding to the native virus, or the adapted sequence used in recombinant baculovirus were analyzed for predicted secondary structure and immune epitopes. In a final series of experiments underway at time of writing, test vaccines corresponding to a subset of those used in mouse studies were administered to feeder pigs, and sera collected for analysis of cross-neutralization with homologous and heterologous virus.

Results and discussion

Selection of appropriate viral proteins for test vaccines

In initial studies, a field isolate was identified and successfully cultured in our laboratory (termed Rotavirus A1). Based upon viral sequences derived from the genome, baculoviruses were produced independently, each of which expressed a single target viral protein from the homologous virus Rotavirus A1. These proteins were prepared according to our filed Outline of Production, inactivated, and formulated either individually or in combination vaccine products that were then administered to separate groups of mice. Serum harvested from these mice was then tested for neutralization of homologous virus, and the results presented in Table 1.

Cross-neutralization of mouse sera against homologous and heterologous virus

Based upon the results obtained in Study 1, a second study was planned to examine the relative importance of sequence conservation of the target proteins in providing protection in mice to either heterologous or homologous virus. Briefly, mice were vaccinated with each of two proteins individually or with a

Table 1: Serum neutralization titers to rotavirus A1 following vaccination with various protein targets

Protein target	SN Titer
Protein 1	1280
	160
Protein 2	40
	40
Proteins 1,2	1280
	640
Proteins 1,2,3	320
No vaccine	< 20

combination of both proteins, and the result sera analyzed for cross-neutralization. The original virus designation, as well as the recombinant baculovirus construct designation are presented in Table 2 for one experiment. Based upon the serological results obtained in vitro, the relative importance of each protein and the importance of sequence homology to the target virus was compared, with the results expressed in Table 3. This is representative data of one such experiment, with further experimental data being acquired at time of writing. Importantly, it was evident that one of the two selected viral proteins appeared to provide the majority of protection observed in these studies, and that the sequence homology between the target viral protein and the vaccine construct was not highly predictable. However, further analysis of individual epitopes conserved between the virus and vaccine construct was predictive, and several putative neutralizing epitopes were identified based on this work.

Overall discussion

Rotavirus infection remains a major concern for both human and animal health.[4] Due to the highly unstable nature of the viral genome, and the large number of circulating strains, development of universal vaccines has been difficult. Subunit vaccines, representing one or more proteins from the rotavirus genome have shown success against homologous viral challenge, but less effective against heterologous viruses.[5,6] Given new regulations enacted by the USDA-Center for Veterinary Biologics, it is now possible to generate rapid vaccines corresponding to individual circulating strains of immediate need.[7] A current challenge is to investigate the feasibility of a multivalent vaccine approach that would address the majority of circulating strains of Rotavirus A, thereby providing protection against this highly significant pathogen of the swine industry. We have therefore focused on first identifying protective protein combinations from the virus that are expected to provide protection in the field against homologous virus. Based on this work, we have identified 1 to 2 proteins that together appear to provide high levels of neutralizing antibody in mice. The next challenge is to identify specific conserved epitopes within those proteins that are critical to providing neutralizing titers. Prior work has suggested that neutralizing titers of approximately 1:80 appear to correlate with protection in gnotobiotic pig models, and were therefore used as a baseline for our work.[6,8] Interestingly, protein sequence homologies between recombinant protein and native viral sequences as low as 77.9% were found to be unimportant, in that high neutralizing antibody titers predictive of in vivo protection were generated following vaccination with homologous constructs. In contrast, higher sequence conservation was not a guarantee of neutralizing titers, given that 3 viruses with higher sequence conservation to the vaccination construct failed to generate detectable neutralizing titers. Further observation of the sequence differences between these virus proteins and the target construct identified several target regions that may correspond to target neutralizing epitopes. Together, these data indicate that the RxP approach can generate highly successful vaccines with the potential to protect against homologous challenge, and greater understanding of the nature of target epitopes within the viral proteins may extend the success of these vaccines against heterologous challenge. Generation of a successful, multivalent rotavirus vaccine formulation should be possible, and greater analysis of virus diversity in the field will be crucial to establishing an appropriate formulation.

References

1. VanderWaal, K., and J. Deen. 2018. Global trends in infectious diseases of swine. *Proc Natl Acad Sci U S A* 115:11495-11500.

2. Feng, N., L. Hu, S. Ding, M. Sanyal, B. Zhao, B. Sankaran, S. Ramani, M. McNeal, L.L. Yasukawa, Y. Song, B.V.V. Prasad, and H.B. Greenberg. 2019. Human VP8* mAbs neutralize rotavirus selectively in human intestinal epithelial cells. *J Clin Invest* 129:3839-3851.

3. Ward, R.L., A.Z. Kapikian, K.M. Goldberg, D.R. Knowlton, M.W. Watson, and R. Rappaport. 1996. Serum rotavirus neutralizing-antibody titers compared by plaque reduction and enzyme-linked immunosorbent assay-based neutralization assays. *J Clin Microbiol* 34:983-985.

Table 2: Protein sequence homology between the native viral protein and test baculovirus

Baculovirus	Rotavirus isolate					
construct	6-091920	3-080820	1-073020	8-100620	7-100620	4-090920
VRx-20087	82.3%	82.0%	77.9%	82.8%	82.3%	77.8%
VRx-20088	78.8%	78.5%	95.4%	81.9%	N/A	NA

Table 3: Serum neutralization titers following vaccination with either a combination or single-protein vaccine

Vaccine	Rotavirus isolate					
construct	6-091920	3-080820	1-073020	8-100620	7-100620	4-090920
Both	Neg	Neg	1:525	Neg	1:390	1:431
VRx-20087	Neg	Neg	1:593	Neg	1:403	1:403
VRx-20088	Neg	Neg	1:80	Neg	Neg	Neg
Sham	Neg	Neg	1:48	Neg	Neg	Neg

4. Wilhelm, B., L. Waddell, J. Greig, A. Rajic, A. Houde, and S.A. McEwen. 2015. A scoping review of the evidence for public health risks of three emerging potentially zoonotic viruses: hepatitis E virus, norovirus, and rotavirus. *Prev Vet Med* 119:61-79.

5. Jiang, B., Y. Wang, and R.I. Glass. 2013. Does a monovalent inactivated human rotavirus vaccine induce heterotypic immunity? Evidence from animal studies. *Hum Vaccin Immunother.* 9:1634-1637. **doi: 1610.4161/hv.24958.** Epub 22013 Jun 24956.

6. Wang, Y., M. Azevedo, L.J. Saif, J.R. Gentsch, R.I. Glass, and B. Jiang. 2010. Inactivated rotavirus vaccine induces protective immunity in gnotobiotic piglets. *Vaccine.* 28:5432-5436. **doi: 5410.1016/j.vaccine.2010.5406.5006.** Epub 2010 Jun 5415.

7. Erdman, M.M., N.E. Clough, and P.J. Hauer. 2020. Review of updated regulations and product license categories for veterinary vaccines in the United States. *Journal of the American Veterinary Medical Association* 257:1142-1147.

8. Jiang, B., J.R. Gentsch, and R.I. Glass. 2002. The role of seerum antibodies in the protection against Rotavirus Disease: An Overview. *Clinical Infectious Diseases* 34:1351-1361.

A new weapon for decontamination of surfaces and premises from African swine fever virus: accelerated hydrogen peroxide

Jose A. Ramirez, MS., PhD, EVP & Lead Science Advisor
Virox Animal Health, Oakville, ON L6H 6R1 Canada

Introduction

The resurgence of African swine fever over the past 18 months in commercial swine herds in Asia and Eastern Europe, and more recently in European wild boar, has resulted in renewed interest in practical and effective options for eradication of the virus from environmental reservoirs. Given an effective vaccine is currently unavailable, effective disinfection of contaminated surfaces remains a key weapon for prevention and control. Although there are a number of research studies documenting the efficacy of diverse disinfection options against the virus, differences in methodology and criteria make it difficult to select a particular disinfection option with certainty. The fact that the virus is fairly unique - it is the only member of the only genus (Asfivirus) in the Asfarviridae family - also makes application of the usual criteria for disinfectant susceptibility based on virus morphology difficult. This has resulted in a variety of different recommended options for disinfection around the world.

In the US, the Department of Agriculture (USDA) has issued a guidance on which US-EPA registered disinfectants "...may be used against ASFV in/on farm premises and related structures and equipment."[1] The USDA guidance advices that the first recourse in response to the virus should be to use a disinfectant which has been subject to the full evaluation prescribed for disinfectants in the federal pesticides law Section 3.[1] It also allows for several other choices that can be used if a Section 3 registered disinfectant is not available - a provision of the federal pesticides law for emergencies known as Section 18. The latest addition to the Section 3-approved disinfectants for use against ASFV is the accelerated hydrogen peroxide (AHP®) technology branded Intervention® and Prevail™ for the swine production sector.[1]

In this paper, we review the results of recent in-vivo studies evidencing the biocidal efficacy of AHP® against ASFV using the two most widely accepted testing methodologies. We discuss the practical implications of this work, with special focus on a comparative assessment of the different options now available to US producers and veterinarians, and how factors other than biocidal efficacy can affect the outcome of the ASFV-disinfection effort.

Methods for evaluation of virucidal efficacy

Traditionally, in-vitro methods for the evaluation of biocidal activity of chemicals have relied on two types of methods, suspension assays and surface carrier tests. In both cases, the criteria for efficacy can be qualitative or quantitative in nature, depending on whether the criteria is a growth/no growth observation on culture media or a specific count for colonies (in the case of bacteria or fungi) or infected cells (in the case of viruses).

In a suspension method, the organism inoculum (sometimes in the presence of an organic load to simulate use under "dirty" conditions) is added to a volume of disinfectant. After the specified contact time, the mixture is neutralized by dilution or chemical neutralization and any survivors are recovered in culture media. In a carrier method, a specified volume of inoculum and added soil is dried on the surface of a coupon or petri dish. After the appropriate drying period, a specified volume of disinfectant at the required concentration is added to cover the dried inoculum. After the specified contact time, the inoculum-disinfectant mixture is diluted sufficiently to arrest the biocidal activity of the disinfectant. The eluate is then transferred to the appropriate culture media or, in the case of viruses, to monolayers of suitable host cells at the appropriate 10-fold dilutions. In a quantitative test, after incubation, the viability titer of the organisms is compared to the titer in the control carriers and a logarithmic reduction is calculated and compared to a pre-established reduction criterion for efficacy.

In general, suspension tests are simpler and easier to perform; however, it is widely accepted that they may not represent use of the disinfectant under real-world conditions.[2] Additionally, it has been noted that they may not present a sufficient challenge to the disinfectant product.[2,3] In contrast, since in a carrier test the test virus is dried on the inanimate surface; it is widely considered to be a better representation of actual in-use conditions. In most practical situations, the infectious agents will be mixed with organic matter (feces or bodily fluids) and in the worst case will have dried unto an environmental surface. Thus, the disinfectant has to be able to penetrate or dissolve the dried contaminants and suspend the organism-soil mixture so that the disinfectant active can be transported through the membrane or capsid walls of the target organisms. Clearly, test products that do not possess a high degree of detergency or emulsive power will generally be less effective in both the carrier test and in field conditions.[4] This is also the reason why many researchers have found that many disinfectants will pass a suspension test at a given concentration but will require a much higher concentration to pass a carrier test.[2-4]

For many of the so-called transboundary animal disease viruses (e.g. foot-and-mouth disease, African swine fever, classical swine fever, etc.); most of the published data for disinfectant efficacy against these viruses has been based on suspension testing.[5,6] The reason for this is that historically, many of these viruses have been more economically relevant in Europe; where traditionally, the prevalent methodology for virucidal testing has been based on suspension methods - a virucidal carrier method has just been recently published less than two years ago.[7] Furthermore, the first study reporting disinfectant action vs. ASFV using a carrier methodology was the relatively recent paper of Krug et al.[5] As a consequence, many of the disinfectants recommended by the USDA as effective against ASFV most likely will not have been evaluated under the more demanding and realistic conditions of the carrier test methodology, but rather under the less stringent suspension testing assay.[5]

A key problem with surface carrier testing is that many viruses lose significant infectivity during the drying step on the test carrier.[2,3] This results in a very small difference between the virus that remains intact after drying and the overall limit of virus detection, which makes measuring the reduction in virus titre due to disinfectant action very difficult. One solution to this problem was described by Krug et al.[5] The technique involves using large inoculum volumes in regards to the volumes of disinfectant and neutralizer in order to minimize the loss of detectable virus due to dilution. This technique enabled the successful performance of the carrier testing described below.

Evaluation of efficacy of AHP® disinfectant technology against ASFV

The test was carried out by Dr. Esther Blanco at the Instituto Nacional de Investigacion y Tecnologia Agraria y Alimentaria (INIA) in Madrid, Spain.[8] The INIA laboratory is the European Community Reference Laboratory for African swine fever for the European Union.[9] Both a suspension test and a carrier test were carried out with the AHP®-based products Intervention® and Prevail™, at the recommended dilution of 2 oz. per gallon (1:64) in hard water (1:40 was used for Prevail™ per label instructions), and specified contact time of 5 min. The suspension testing followed the methodology specified in the European standard test method EN14675 for the evaluation of virucidal activity of disinfectants for use in veterinary applications.[10] The carrier testing methodology was in accordance to the standard test method for virucidal activity of disinfectants published by the American Society for Testing of Materials (ASTM);[11] one of the methods accepted by the USEPA as evidence of virucidal efficacy for a commercial disinfectant product.[12]

The results of the testing are shown on Table 1. With a logarithmic reduction of more than 4-logs, Intervention® meets the required efficacy performance for virucidal activity against the African swine fever virus in both the US and in Europe.[10,12] The same type of efficacy was observed for Prevail™, with a log10 reduction of ≥ 4.1.[8]

A further evaluation was carried out with real time PCR (RT-PCR). In this technique, total DNA is extracted from the VERO cells incubated with both control and neutralized disinfectant solution. The results of the RT-PCR assay for the tested samples under the conditions described above showed a cycle threshold of 36.08; below the threshold of positivity of 35.00 for this particular assay. This result further supports the notion that the virus was fully inactivated by Intervention® at a dilution of 2 oz./gal and 5 min contact time.[8]

Practical considerations

The influence of factors other than intrinsic biocidal efficacy of the chemical disinfectant has been discussed previously.[13] Hygiene outcomes in the field are heavily influenced by the quality of the cleaning and disinfection process and the ability (or willingness) of working crews to comply with it. For example, a disinfectant that cannot be easily used with standard pressurized power washing systems prevalent in farms and trailer washing station runs a higher chance of resulting in a failed hygiene outcome. In particular, in our experience in the field, the following factors can determine whether premises will be effectively disinfected or not:

- erroneous concentration due to manual preparation or dilution,
- missed spots due to lack of visual cues for coverage like foam,
- incomplete or sub-optimal treatment of vertical surfaces due to rapid disinfectant run-off,
- less-than thorough application due to worker discomfort from a respiratory and/or eye/skin irritant.

While these potential failure modes are critical in the day-to-day biosecurity program of a healthy farm looking to keep environmental infectious pressure low, they are crucial in the one-time terminal decontamination of an ASFV-positive facility before restocking, or in farms with heightened biosecurity measures due to proximity to an affected area. It is thus instructive to review the different choices recommended by USDA and approved by USEPA based on biocidal efficacy criteria in a practical context, with particular attention to the factors outlined above. A summary of the disinfection options from USDA is presented in Table 2.[1]

Aside from the newly approved Intervention®,[14] the table includes four other formulated products that have gone through a full EPA review and approval for the specific claims against ASFV. It also includes three products that have received an exemption from the EPA for use in case none of the EPA-approval products are available.[1] We will not focus here on the exemption products (Sodium hypochlorite, Citric acid, Thymol) given these are fallback solutions in case others are unavailable and the user has no other choices at their disposal. We focus our attention on the first-choice, commercial products.

All of the EPA-approved products recommended by USDA are chlorine precursors, that is; they generate chlorine species in solution when dissolved in water. Three of the products are based on the substance sodium dichloro-isocyanurate (DCC), the sodium salt of a chlorinated hydroxytriazine that in the presence of water ultimately decomposes into hypochlorous acid and isocyanurates.[15] The fourth product is a powdered mixture of potassium monopersulfate (PMPS) and sodium chloride that when dissolved in water react to form hypochlorous acid.[16,17]

Because these products are highly concentrated powders (or compacted powder tablets) special care must be taken in their

Table 1: Summary of results of testing of Intervention™ accelerated hydrogen peroxide against African swine fever virus

Test method	Recovered virus log-titre (control)	Recovered log-titre after 5 min exposure to disinfectant	log reduction
Suspension Test	5.5	< 0.5	≥ 4.6
Carrier Test	5.5	0	5.5

handling, particularly in avoiding exposure to skin and eyes during preparation of in-use solutions, as well as avoiding inhalation of their dust/fine powder. It is also critical to avoid contaminating the original packaging with moisture, as this could cause the generation of highly poisonous chlorine gas and compromise the shelf life of the powdered stock-product.

An important operational difficulty with powder/tableted products is the need to prepare a diluted liquid stock of an intermediate concentration that can be used as the source to feed a chemical injection system. With DCC powders or tablets, since the solubility limit in water is quite high (approximately 227 gr/L[18]), it is possible to prepare a liquid stock solution of say 10% concentration by weight (100 gr of DCC powder per Liter of water) which is then diluted through an injector with a dilution tip for 4 oz/gal (1:32). In this way, the resulting solution out of the injector nozzle will be diluted by a factor of 300 with respect to the packaged powdered product (1:300); this delivers the roughly 1,000 ppm chlorine required (see Table 2). Having said this, in our experience, this calculation for a two-step dilution just described is not easily understood by most individuals with operational responsibility, thus the potential for incorrect end-dilution is very high.

The same preparation protocol is typically required with the other chlorine-precursor product based on PMPS.[19] In this case however, the choices of intermediate stock dilutions/volume are more limited due to the low water solubility of the product of 62 gr/L (roughly 9 oz. per gallon).[19] At the limit of solubility, the strongest stock solution that can be prepared is at about 6% by weight. Adding more powder product than this will result in precipitation and settling - the liquid in the stock solution will not "hold" more than 62 gr per Liter (or 9 oz per gallon). Because the required in-use concentration is 1% (10 gr per Liter of water), this means that the dilution occurring at the injector drawing from the 6% stock solution should be 1:6. In other words, the injector should be drawing 21 oz. of stock solution for every gallon of water dispensed by the power washer. In practice, this means that to treat a farrowing room that could require, say 50 gal of diluted disinfectant, the operator would have to prepare slightly more than 8 gal of stock solution. This makes it impossible for crews to use hand-held bottle-foamers (as the bottle in these foamers is typically designed to hold between 48 to 96 oz. of liquid) and this would require repeated refilling or multiple units. Furthermore, with a standard power washer delivering 4.5 gallon per minute of water, a 96 oz. bottle-foamer would be emptied in about 1 minute, clearly impractical. For farms with overhead quick-connects affixed to the ceiling that are spread-out through the barn (where the injector-foaming wand assembly is moved throughout each connection point with the chemical concentrate as the crew progress through the barn), it requires using two 5-gal pails of the stock solution and moving them throughout the different connection drop-points in the barn.

Aside from the operational difficulty and risk for calculation errors intrinsic to the preparation of intermediate stock solutions, there is the risk of blockage of the small orifice in the chemical injector from residual debris from the tablet or powder. The orifices in these injectors range in the order of thousands of an inch and can easily be blocked by small particles remaining from incomplete dissolution of the powder in the stock solution. Furthermore, because of the low flow of stock chemical into the injector (in relation to the diluent water jet) blockages are rarely noticed by the operator until they realize that the liquid level in the stock solution is not changing. Furthermore, DCC or PMPS chlorine precursors do not typically exhibit a clear visible cue for application (e.g. highly visible foam), which further contributes to the operator not realizing chemical is no longer flowing properly due to injector orifice blockage.

The lack of visible cues on application also contribute to the misapplication (both under- and over-application) of the chemical, resulting in missed spots or excess use/waste. A more critical issue, however, is the immediate runoff of solution when applied to vertical surfaces. This is particularly critical in the successful terminal decontamination of a farm due to be restocked after a bout with ASFV. In order for the disinfectant to be effective, the surfaces have to remain wet for a period of 10 minutes. With water-thin solutions lacking a foam phase, runoff of excess liquid from vertical surfaces occurs very quickly. This accelerates drying on these surfaces possibly limiting the extent of the contact time of the disinfectants.

In contrast, Intervention® is a liquid concentrated disinfectant that can be easily dispensed through existing chemical injection/power washing equipment increasingly adopted in the day-to-day cleaning & disinfection programs in swine farms and livestock trailer facilities. The ASFV cleaning & disinfection protocol would be no different than that employed in the day-to-day biosecurity program. This reduces any need for additional training and therefore the potential for errors due to changes in protocol. Crews simply connect the 1-gallon bottle of concentrate to their power-washing system injector or fill the handheld bottle-sprayer foaming wands. Crews using larger volumes (such as trailer washing stations) can simply insert the injector dip-tube into a pail or drum of Intervention® concentrate. The risk of miscalculating or mis-preparing an intermediate stock solution is eliminated. The heightened risk of chemical exposure to a concentrated powder or stock solution is unnecessary. The potential for clogging of the chemical injector from undissolved particles is nonexistent.

When dispensed through a standard foaming nozzle, operators can easily judge application by the presence of the highly visible foam. This reduces product waste and minimizes the risk of missed spots. Intervention® has been formulated to produce a durable, high-cling foam on application. Therefore, gravity-induced run-off from vertical surfaces is reduced (due to the foam's lower specific weight); while at the same time, the foam phase impairs the evaporative diffusion process that drives surface drying. Furthermore, the reduced required contact time of 5 minutes (vs. 10 minutes for the other chemical options) further ensures that vertical surfaces are properly decontaminated.

Conclusion

The suitability of a chemical disinfectant to deliver expected hygiene outcomes goes beyond the biocidal properties of the chemical. Many additional factors that arise from the chemical and physical properties of the disinfectant need to be considered as well. We have discussed some of these elements, with particular emphasis on issues related to preparation and application in field conditions as well as the importance of potential health hazards and chemical exposure. The recent validation of Intervention® efficacy versus the ASFV, and subsequent US EPA approval provide producers and veterinarians with a superior alternative to fighting environmental contamination by ASFV.

Table 2: Summary of FIFRA Section 3 approval products recommended by USDA APHIS for mitigating African swine fever virus. Health hazards information has been taken from the safety data sheets published by the manufacturers. The products marked with a **§** are based on an exemption approval, only to be used in emergencies where the recommended US EPA approved products are not available.[1]

Packaged active	In-use active	Health hazards	Potential issues
48.21% NaDCC (29-31% free available chlorine)	Varies with commercial product but in the range of 958-1076 ppm free available chlorine @ 10 min contact time.	• Causes serious eye irritation • May cause respiratory irritation • Contact with small amounts of moisture release poisonous chlorine gas	• Needs to be freshly prepared and used within 1 day • Stock solution needs to be manually prepared • Potential for insoluble particles to clog chemical injectors • Corrosive to equipment and surfaces
21.41% PMPS, 1.5% NaCl (9.75% FAC)	1% (975 ppm free available chlorine - hypochlorous acid), 10-minute contact time	• Corrosive, irreversible eye damage and skin burns	• Stock solution needs to be manually prepared - Insoluble @ higher concentrations than 62 gr/L (1:16). • Potential for insoluble particles to clog chemical injectors • Corrosive to equipment and surfaces
Sodium Hypochlorite**§** (exemption approval 5.25%, 8.25%, 12.0%, 12.5%)	3%, 15 min or 30 min	• Causes serious eye irritation • May cause respiratory irritation	• Stock solution needs to be manually prepared • Needs to be freshly prepared if diluted with hard water
Citric acid**§** (powder close to 100% active).	3%, 30 minutes	• Corrosive, irreversible eye damage and skin burns	• Stock solution needs to be manually prepared • Potential for insoluble particles to clog chemical injectors
Thymol**§** (exemption approval, only for aircraft and aircraft loading ramps)	0.05% Thymol, 15 minutes	• None of note	• Pressurized can, ready to use spray, impractical for farm and livestock trailers

References

1. APHIS ASF virus disinfectants **https://www.aphis.usda.gov/animal_health/emergency_management/downloads/asf-virus-disinfectants.pdf.** Accessed October 30th 2020.

2. Bellamy, K. A review of the test methods used to establish virucidal activity, J. Hospital Infec., 30(S), 1995 389:396.

3. Sattar, S. Springthorpe, S., Adegbunrin, O., M. AbuZafer A. disc-based quantitative carrier test method to assess the virucidal activity of chemical germicides, Journal of Virological Methods, 112(1/2), Sept. 2003, 3:12.

4. Quinn, P. J. Evaluation of veterinary disinfectants and disinfection processes, in Disinfection in Veterinary and Farm Animal Practice, Blackwell Scientific Publications, Oxford, UK, 1987, 66:116.

5. Krug, P.W., Lee, L.J., Eslami, A.C., Larson, C. R. and L. Rodriguez. Chemical disinfection of high-consequence transboundary animal disease viruses on nonporous surfaces, Biologicals, 39 (2011) 231:235.]

6. De Lorenzi, G., Borella, L., Alborali, G.L., Prodanov-Radulovic, Stukelj, M. and S. Bellini. African swine fever: A review of cleaning and disinfection procedures in commercial pig holdings, Research in Veterinary Science, 132 (2020) 262:267.

7. EN 16777:2018, Chemical disinfectants and antiseptics - quantitative non-porous surface test without mechanical action for the evaluation of virucidal activity of chemical disinfectants used in the medical area - Test method and requirements (phase 2/step 2). Comité Européen de Normalization, Brussels, EU.

8. Blanco, E. Virucidal efficacy of a disinfectant for use on inanimate environmental surfaces. Virus: African swine fever Virus (ASFV), Project Number 18CISA093.2, 12.26.2019., INIA, Madrid, Spain.

9. EEC Council Directive 2002/60/EC of June 27 2002 laying down specific provisions for the control of African swine fever and amending Directive 92/119/EEC as regards Teschen disease and African swine fever, Official Journal of the European Communities, L 192/27.

10. EN14675:2003 Chemical disinfectants and antiseptics - quantitative suspension test for the evaluation of virucidal activity of chemical disinfectants and antiseptics used in veterinary field - Test method and requirements (phase 2, step 1). Comité Européen de Normalization, Brussels, EU.

11. Standard Test Method to assess virucidal activity of chemicals intended for disinfection of inanimate, nonporous environmental surfaces, Designation:E1053-11, ASTM, West Conshohocken, PA, 2013.

12. Product performance test guidelines OCSPP 810.2200 Disinfectants for use on hard surfaces- efficacy data recommendations, EPA 712-C-07-074, USEPA, Washington D.C., March 12, 2012.

13. Ramirez, J.A. Theory and practice of cleaning and disinfection in swine biosecurity, Proceedings of the 51st annual meeting of the AASV, Atlanta, March 7-10, 2020.

14. News Release, Virox Technologies Corp., **https://www.viroxfarmanimal.com/wp-content/uploads/2019/08/News-Release-ASFV-Intervention-and-Prevail-02082019.pdf**, accessed Oct. 30, 2020.

15. Falk, N.A., Macnaughtan, M., Taheri, A. and W. McCormick. Chlorine and Chlorine Compounds, in Block's Disinfection, Sterilization and Preservation, McDonnel, G. and J. Haysen (Eds), Wolters Kluwer, Philadelphia, 2021, 280:305.

16. Auchincloss, T.R. Biocidal, particularly virucidal compositions, US Patent No. 4,822,512, granted. Apr. 18 1989.

17. Ramirez, J. New Technologies in Disinfection and Infection Control, in Disinfection & Decontamination, Principles, applications and related issues, G. Manivannan (ed.), CRC Press, Boca Raton, FL.

18. Ulmann's Encyclopedia of Industrial Chemistry, 5th Edition, W. Gerhartz, G. Schultz (eds), 1996. WIley-VCH, Weinheim, Germany.

19. Virkon™ Protecting against African swine fever virus breaching our US borders, Marketing brochure, Lanxess Corp., Pittsburgh, PA, 2019.

The relationship between early life measures and finishing performance

C. Bruns, PhD; B. McNeil; J. Sonderman, PhD; T. Rathje, PhD
DNA Genetics, Columbus, Nebraska

Introduction

While wean age can impact nursery performance,[1] weaning weight can impact later performance independently of wean age. Early weight measures (e.g. weaning weight) can be indicators of finishing performance in pigs. Heavier weaning weights have been found to be associated with improved growth rate and feed efficiency in nurseries;[2] however, the relationship with finisher growth rate and feed efficiency was not significant[2]. It has also been shown that heavier pigs at weaning reach market with fewer days on feed post-weaning.[3] A better understanding of the impact of weaning weight within a given wean age will guide strategies to manage wean weight that produces optimum lifetime performance. The aim of this study was to estimate the relationship between early weight measures and later performance in pigs. Utilizing data from 3 purebred lines, the effect of weaning weight on nursery and finisher growth rate, backfat, loin depth, and feed efficiency accounting for the wean age of the pig was determined. Additionally, the effect of weaning weight on nursery and finisher mortality was examined.

Materials and methods

Data for this study were obtained from 9 sow farms with similar genetics and management protocols. All pigs included in the study were from purebred Yorkshire, Landrace, and Duroc populations. Pigs were born after January 2017 and weaned before July 2020. All pigs were weaned into a nursery (~3 weeks of age) and moved into a finisher (~11 weeks of age) to be performance tested for 12 weeks. Summary statistics by line are shown in Table 1. Individual feed intake during the finisher phase was recorded at one site on boars only. The boars were sampled from each weekly wean group from all 9 sow farms. The minimum wean age was 16 days and the maximum wean age was 26 days for all 3 breeds. The minimum number of pigs in any age and line combination was 200.

The relationship between weaning weight and nursery average daily gain (ADGn) was estimated within wean age using a model with fixed effects of sex and wean age, random effects of litter and year-herd-week of weaning, and a covariate for weaning weight nested within each wean age. Finisher average daily gain (ADGf), 10th rib backfat (BF), and 10th rib loin depth was fitted with a model containing fixed effects of sex and wean age, random effects of litter and pen, and a covariate for weaning weight nested within each wean age. The model for feed to gain ratio (F:G) was to the same as the other finisher traits, but excluding a sex effect. These models allowed for different regression estimates across weaning weight for each day of wean age. Nursery and finisher mortality were modeled using a fixed effect of sex, random effects of litter and year-herd-week of weaning, and a covariate for weaning weight.

Results and discussion

The estimated regression coefficient and standard errors for weaning weight within wean age on ADGn, ADGf, BF, LED, and F:G are shown in Tables 2 and 3. The effect of weaning weight on nursery and finisher growth rate decreased as age increased for all 3 lines, suggesting that weaning weight has a larger impact on post-weaning performance when pigs are weaned earlier. A

Table 1: Nursery and finisher performance summary statistics[a]

	Yorkshire	Landrace	Duroc
N	59167	63628	99048
Wean age (days)	21.7	21.2	20.1
Weaning weight (kg)	5.64	5.73	4.93
Average daily gain – birth to weaning (g)	200	206	172
Nursery mortality (%)	4.22	3.84	4.97
On-test weight (kg)	27.6	28.2	27.5
On-test age (days)	78.0	76.3	75.3
Average daily gain – nursery (g)	389	406	408
Finisher mortality (%)	3.17	2.14	1.61
Off-test weight (kg)	98.6	105.7	110.0
Off-test age (days)	162.4	161.0	160.2
Average daily gain – finisher (g)	842	915	971

[a] Pigs born after 1/1/2017 and weaned before 7/3/2020

Table 2: Estimated regression coefficient (SE) for weaning weight (kg) within wean age on nursery average daily gain (ADGn, grams) and finisher average daily gain (ADGf, grams)

	Yorkshire		Landrace		Duroc	
Wean age	**ADGn[a]**	**ADGf[b]**	**ADGn**	**ADGf**	**ADGn**	**ADGf**
16	32.9 (4.3)	30.4 (5.8)	36.7 (4.3)	23.5 (6.3)	25.8 (2.1)	32.6 (3.3)
17	27.2 (1.3)	25.4 (1.8)	24.5 (1.1)	31.2 (1.7)	32.7 (0.6)	33.7 (1.1)
18	25.6 (1.1)	25.6 (1.5)	27.1 (0.9)	29.6 (1.4)	33.3 (0.5)	31.6 (0.9)
19	24.3 (1.0)	26.3 (1.4)	27.7 (0.9)	30.0 (1.3)	31.2 (0.5)	28.8 (0.9)
20	26.6 (0.9)	28.9 (1.2)	27.4 (0.9)	30.5 (1.3)	31.3 (0.6)	28.9 (1.0)
21	23.0 (0.8)	24.2 (1.0)	24.7 (0.7)	28.5 (1.1)	29.9 (0.6)	28.8 (0.9)
22	23.3 (0.6)	21.1 (0.8)	25.2 (0.6)	27.1 (0.8)	28.6 (0.5)	25.9 (0.9)
23	20.6 (0.5)	21.3 (0.7)	23.7 (0.5)	25.5 (0.7)	27.1 (0.5)	24.8 (0.8)
24	20.1 (0.5)	19.4 (0.8)	22.5 (0.6)	23.9 (0.9)	25.4 (0.7)	22.7 (1.2)
25	21.4 (0.7)	20.2 (1.0)	20.3 (1.3)	20.4 (2.0)	25.2 (1.4)	22.3 (2.2)
26	19.1 (1.1)	18.7 (1.5)	13.1 (2.9)	16.1 (4.1)	27.2 (3.2)	17.1 (5.1)

[a] Model includes fixed effects of sex and wean age, random effects of birth litter and year-herd-week of weaning, and covariate of weaning weight nested within wean age

[b] Model includes fixed effects of sex and wean age, random effects of birth litter and pen, and covariate of weaning weight nested within wean age

similar trend was found for LED, but not for BF and F:G. While both wean age and weaning weight impact BF and F:G, there was no indication that increasing wean age resulted in a systematic change in the estimated regression coefficient for weaning weight.

The generalized additive model curve and confidence intervals for predicted nursery and finisher mortality rates across weaning weights are shown in Figures 1 and 2. It is clear from the figures that while there is a relationship between weaning weight and nursery mortality, the impact has been eliminated for pigs that survive to enter the finisher. These results indicate that a pig's ability to survive is less impacted by weaning weight later in life compared to immediately post-weaning.

While the effect of wean age and weaning weight on later performance has been established previously, few studies have investigated the effect of weaning weight within wean age. The results of this study indicate the importance of adjusting for weaning age when determining the weaning weight effect on later performance. The impact of weaning weight on growth rate and LED diminishes with increased wean age. Management practices that increase wean age could improve post-wean performance regardless of the average weaning weight.

References

1. Leliveld et al. Effect of weaning age and post weaning feeding programme on the growth performance of pigs to 10 weeks of age. Livest Sci 2013;157:225.

2. Mahan and Lepine. Effect of pig weaning weight and associated nursery feeding programs on subsequent performance to 105 kilograms body weight. J Anim Sci 1991;69:1370.

3. Wolter and Ellis. The effects of weaning weight and rate of growth immediately after weaning on subsequent pig growth performance and carcass characteristics. Can J Anim Sci 2001;81:363.

Table 3: Estimated regression coefficient (SE) for weaning weight (kg) within wean age on backfat (BF, mm), loin eye depth (LED, mm), and feed to gain ratio (F:G)

	Yorkshire			Landrace			Duroc		
Wean age	**BF[a]**	**LED[a]**	**F:G[b]**	**BF**	**LED**	**F:G**	**BF**	**LED**	**F:G**
16	0.62 (0.16)	1.25 (0.35)	0.051 (0.033)	-0.34 (0.19)	1.43 (0.37)	-0.019 (0.025)	0.35 (0.07)	1.36 (0.19)	0.013 (0.011)
17	0.39 (0.05)	1.25 (0.11)	-0.006 (0.010)	0.36 (0.05)	1.28 (0.10)	0.009 (0.010)	0.30 (0.02)	1.63 (0.06)	0.008 (0.005)
18	0.32 (0.04)	1.20 (0.09)	0.019 (0.008)	0.35 (0.04)	1.46 (0.08)	0.007 (0.008)	0.32 (0.02)	1.52 (0.05)	0.010 (0.005)
19	0.39 (0.04)	1.21 (0.08)	0.027 (0.008)	0.32 (0.04)	1.41 (0.08)	-0.000 (0.008)	0.27 (0.02)	1.37 (0.05)	0.011 (0.004)
20	0.40 (0.03)	1.27 (0.07)	0.005 (0.007)	0.29 (0.04)	1.50 (0.08)	0.018 (0.008)	0.29 (0.02)	1.34 (0.06)	0.010 (0.005)
21	0.35 (0.03)	1.26 (0.06)	0.013 (0.006)	0.35 (0.03)	1.36 (0.06)	0.017 (0.006)	0.30 (0.02)	1.36 (0.06)	0.010 (0.004)
22	0.33 (0.02)	1.16 (0.05)	0.015 (0.005)	0.33 (0.03)	1.20 (0.05)	0.010 (0.005)	0.27 (0.02)	1.30 (0.05)	0.014 (0.004)
23	0.30 (0.02)	1.15 (0.04)	0.007 (0.004)	0.32 (0.02)	1.16 (0.04)	0.008 (0.005)	0.23 (0.02)	1.28 (0.05)	0.008 (0.004)
24	0.29 (0.02)	1.01 (0.05)	0.007 (0.005)	0.31 (0.03)	1.08 (0.05)	0.010 (0.006)	0.30 (0.02)	1.20 (0.08)	0.006 (0.005)
25	0.34 (0.03)	1.10 (0.06)	0.017 (0.006)	0.32 (0.06)	0.87 (0.12)	0.009 (0.014)	0.19 (0.05)	1.08 (0.13)	0.011 (0.011)
26	0.30 (0.04)	1.05 (0.09)	-0.006 (0.010)	0.31 (0.12)	0.85 (0.25)	-0.025 (0.042)	0.20 (0.10)	1.04 (0.30)	0.019 (0.034)

a Model includes fixed effects of sex and wean age, random effects of birth litter and pen, and covariate of weaning weight nested within wean age

b Model includes fixed effect of wean age, random effects of birth litter and pen, and covariate of weaning weight nested within wean age; trait measured on boars only

Figure 1: Predicted nursery mortality by weaning weight.

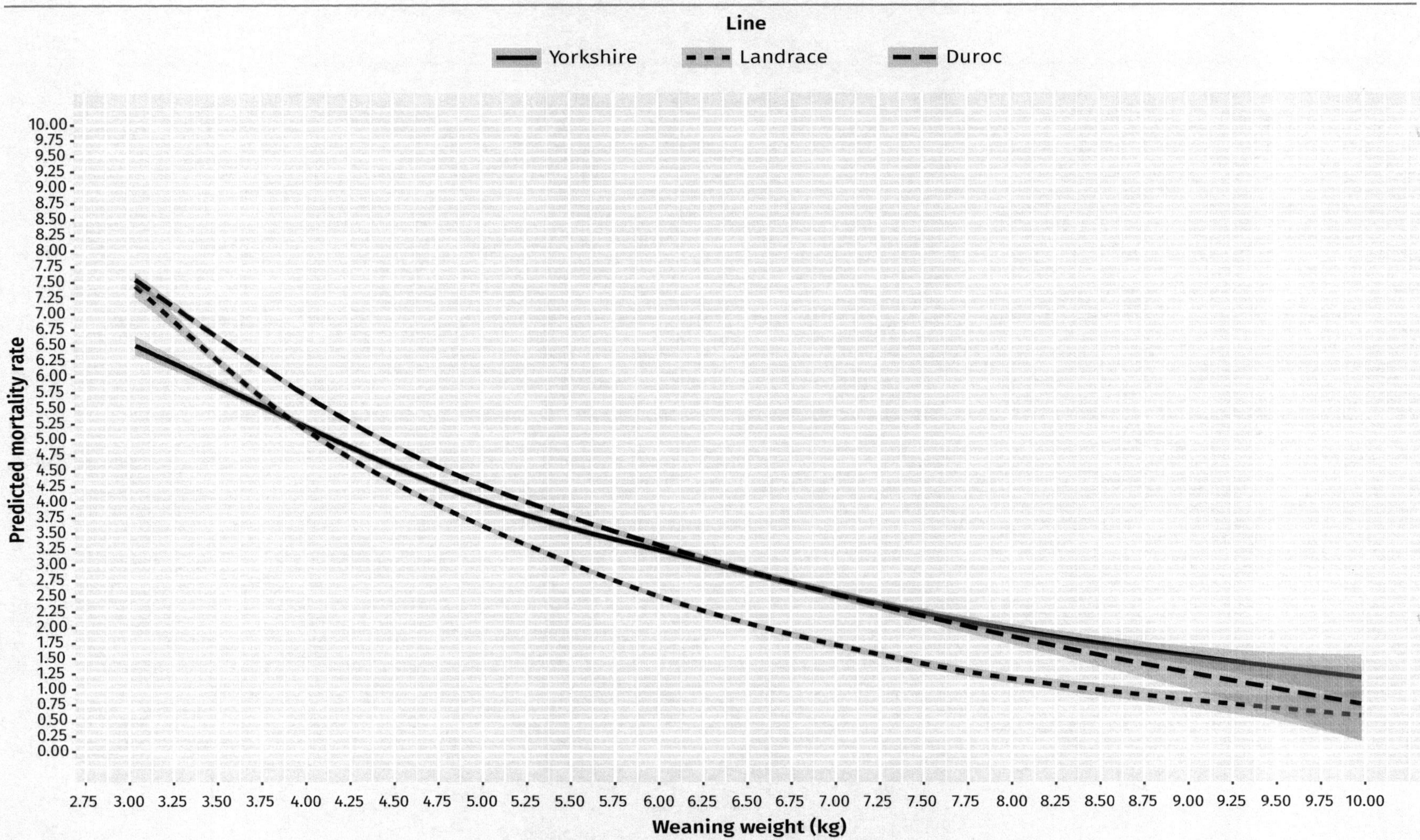

Figure 2: Predicted finisher mortality by weaning weight.

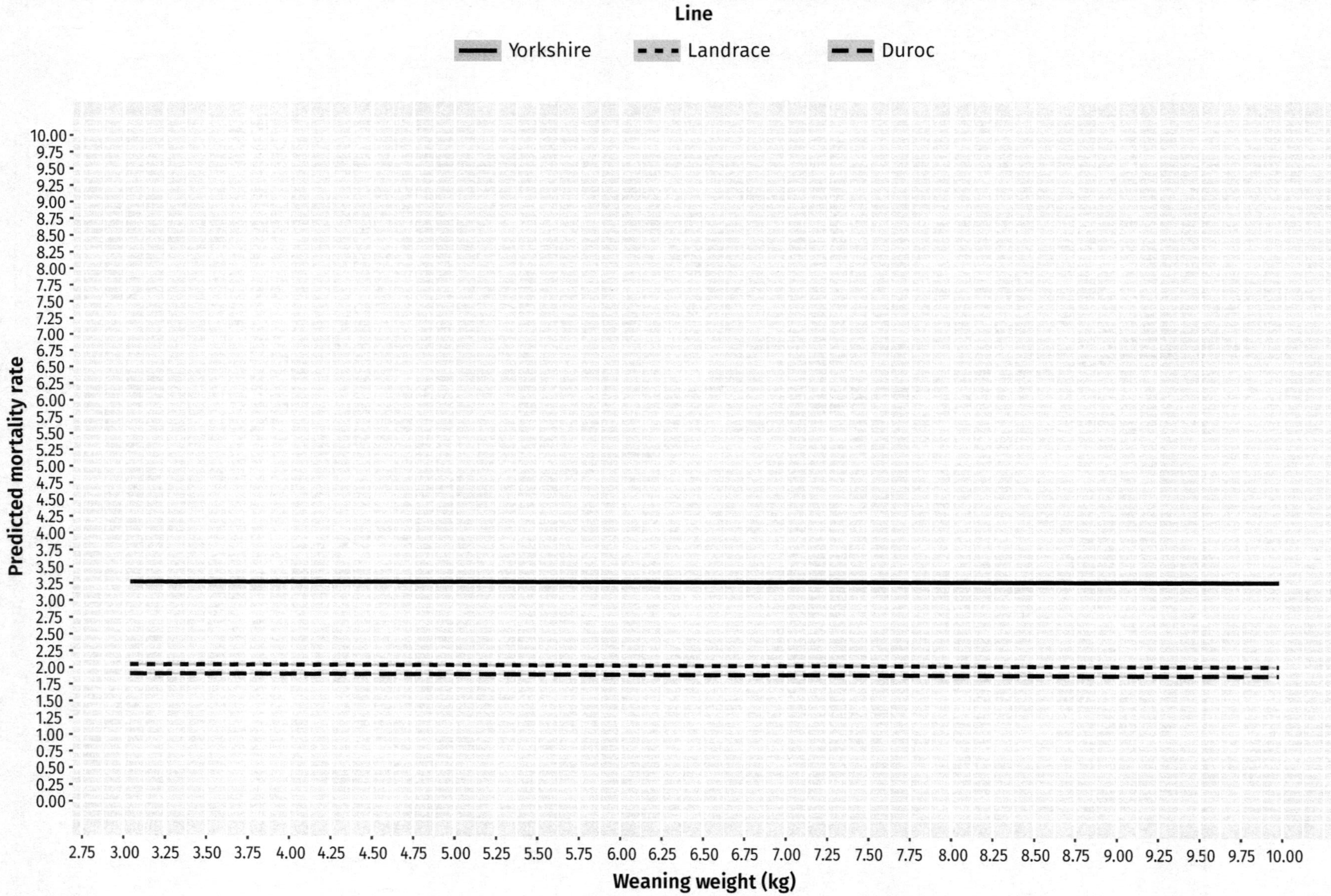

Culture of care longevity measurements post training

Sarah Probst Miller[1,2], DVM; Seth Krantz[3], DVM
[1]AgCreate Solutions, Inc; [2]Pork Avenue Training Portal; [3]Tosh Pork

Introduction

Often it is important to prove that change of behavior has occurred and that the change has had lasting effect. This study looks at the longevity of gained competencies for low pressure handling.

Materials and methods

Baseline competency measurements

To measure human behavioral impact on pig welfare related to animal handling, a 3rd party veterinarian applied observed animal handling events and applied the Common Swine Industry Audit (CSIA)[3] inspection guidelines to every pig and pig housing location on Farm A. The veterinarian Opportunities for immediate actions as well as employee training were identified. The system's production team and veterinarian discussed with the 3rd party veterinarian and agreed on an action and training plan.

Training plan

This targeted response plan included first a Farm A specific re-training plan on animal handling followed by some lesson specific topics that worked to address missing expressed competencies. Farm A's successful completion would be followed by a system-wide re-training plan on low pressure animal handling. This plan was initiated with a live manager training meeting to explain the situation and its ramifications to achieve an understanding of the importance of successful completion. Farm A's training was implemented immediately utilizing Pork Avenue Training Portal (PATP)[7]. The program components per learning experience include online training, graphical SOP's and in-barn verification which employed the *See it. Do it. Teach it.* methodology.[1,4-11] Training delivery occurred via learning experiences delivered on PATP's online learning management software.[7] Simultaneously, an internal and external verification plan was detailed and executed post re-training to measure expressed competency results using the lesson's In-Barn Verification Checklists as guidance documents. All training accomplishments were recorded on PATP's online learning management software and a transcript of learner accomplishment created per learner.[7]

See it.

Farm A chose to administer most videos and quizzes as a group and have a discussion after each lesson on skills they needed to express differently on farm. Groups were formed based on preferred language of English or Spanish. Learners were informed prior to the videos that they needed to learn the topics well enough that they could teach it back to a verifier. The See it portion of the experience was marked as complete on the individual employee's transcripts.

Do it.

Then the learners continued to complete skills which were manager or mentor observed until verified. If desired, pictorial SOP's were available for review in English and Spanish.

Teach it.

Internal verification included having individuals teach what they learned back to an In-barn Verifier. In-barn Verifiers presented themselves to each employee as "their student" and encouraged them to "pretend I am a new employee and teach me how to do this task." While they spent most of the verification experience listening, Verifiers utilized the In-barn Verification Checklists to make sure all key competencies were taught back successfully according to the farm protocol and asked about a key competency if it was not brought up specifically. When individuals successfully passed their In-barn Verifications, Verifiers marked this part of the learning experience complete on their transcript.

Multi-year assessment plan

3rd party remeasurement of competencies occurred at one-year post training and two-years post training.

Results

The baseline expressed competencies for the training curriculum chosen was at a 26% competency expression rate on farm for chosen handling competencies. Post external verification of training, staff averaged a 95% handing competency expression rate on this single farm. This is a 69-percentage point difference and a 265.4% increase in handling competency expression on farm. Retraining throughout the entire system occurred throughout Fall and Winter of 2018/19. External verification of other farms occurred in December 2018 and January 2019, and results averaged 90-some percent with one farm at 69% and the rest in the high 90's. Retraining occurred at the farm with the 69%. Two years post initial training, a 3rd party assessment was repeated. The range of farm scores were between 88% and 100% expressed competency. On the two-year assessment the average expressed competency had moved from a 90% on average to 95.2% on average. The improvement can be credited to most farms maintaining or improving their expressed competency score and the farm with the lower score committing to the training and application of competencies.

Conclusion and discussion

Using the "See it. Do it. Teach it." methodology[1,2,4-6,9] has effectively improved competency expression in other studies from a 59% level of competency expression to an 85% competency expression level.[8] This case study also achieved improved levels of expressed competency across a whole system and has a track record thus far for competencies remaining expressed two years post initial training.

References

1. Bear, Anne, Technology, Learning, and Individual Differences, Journal of Adult Education, 2012

2. Cercone, K. (2008). Characteristics of adult learners with implications for online learning design, AACE Journal, 16(2), 137-159.

3. Common Swine Industry Audit, 2018. Available at **https://lms.pork.org/CommonIndustryAudit**.

4. Dale, Edgar, Audio visual methods in teaching. Holt, Rinehart, and Winston, 1965.

5. Dale, Edgar, Cone of Experience, Education Media, 1969

6. Gaetano, Kristian, 5 ways adults learn differently than children, Learn Kit, 2016

7. Pork Avenue Training Portal. 2018. Available at **www.porkavenuetraining.com**

8. Probst Miller, Sarah, Impact of two training implementation plans on post-training competency expression of personnel on farm, AASV Proceedings, 2018

9. Probst Miller, Sarah, Applying the science and physiology of how adults learn to our on-farm approach as practitioners, AASV Proceedings, 2016

10. Probst Miller, Sarah, The impact of trained and untrained personnel on product implementation and subsequent pig health, AASV Proceedings, 2016

11. Ryland, Bruce, Fostering Critical Reflection in Adulthood, by Jack Mezirow & Associates, San Francisco: Jossey-Bass, 1990

Evaluation of the effect of tracheal sample pooling on *Mycoplasma hyopneumoniae* PCR detection

R. Mugabi[1]; A. P. Poeta Silva[1]; M. L. Rotolo[2]; S. Krantz[3]; P. Harms[2], M. J. Clavijo[1,2]
[1]Iowa State University, Ames, Iowa; [2] PIC®, Hendersonville, Tennessee; [3]Tosh Family Farms, Paris, Tennessee

Introduction

Mycoplasma hyopneumoniae (*M hyopneumoniae*) is the etiological agent of enzootic pneumonia as well as a primary pathogen in porcine respiratory disease complex. *M hyopneumoniae* is found in swine herds across the globe, with economic losses estimated at $0.63 to $10.12 per market pig.[1] Due to the importance and economic effects of this pathogen, multiple research efforts are geared towards biosecurity, vaccination strategies, and antimicrobial therapy in an effort to prevent and control *M hyopneumoniae*. A vital tool in prevention and control of any pathogen is the implementation of a sensitive and cost-effective surveillance program. Due to the slow disease process of *M hyopneumoniae* and accompanying delay in clinical signs, confidence in the ability to detect this pathogen through surveillance is paramount to maintaining health status.

There are two main approaches to *M hyopneumoniae* surveillance, detection of antibody or antigen. Historically, monthly serum samples for enzyme-linked immunosorbent assay (ELISA) testing have been employed in swine herds to monitoring health status. This approach has delayed sensitivity due to the time needed for antibody production (~3 weeks). Research has shown that tracheal samples tested by polymerase chain reaction (PCR) improves time to detection and provides a higher sensitivity compared to serum samples. The drawback to tracheal samples is the cost difference between ELISA and PCR. PCR assays are generally 5-6x the cost of the ELISA test. In an effort to reduce cost, pooling tracheal samples has been suggested as a means to achieve improved sensitivity without being cost-prohibitive. In some cases, pooling can reduce diagnostic sensitivity depending on disease prevalence and size of pools, therefore' **the objective of this study was to evaluate the effect of pooling tracheal samples on detection of *M hyopneumoniae*.**

Materials and methods

A bank of negative and positive *M hyopneumoniae* samples were collected from populations of known status. All samples were tested by PCR to determine status and Ct value. For each Ct value from 27 to 36, five samples were selected from the collected tracheal samples. Three pool sizes were evaluated for each of the selected Cts, pools of 3, 5 and 10. Each of the five positive samples had 3 replicates per pool size. The bank of negative tracheal samples was used to create the pools alongside the positive samples to control for sample diversity. In addition, an individual aliquot of each positive sample was tested to confirm the Ct value at time of pooling. In total, 450 PCRs were performed for the pooled samples and 50 were performed for the individual samples. Each sample was tested using the commercial kit MagMAX™-96 Pathogen RNA/DNA kit (Applied Biosystems, Carlsbad, CA) and amplified using the TaqMan® Fast Virus 1-Step Master Mix with the added by the AmpliTaq® 360DNA Polymerase (5U/uL) (ThermoFisher Scientific), and primer/probe described for Mhp183.[2] PCRs were run to 37 cycles on the automated Applied Biosystems® 7500 Real-Time PCR (ThermoFisher Scientific). A sample was considered positive at Ct < 37.

Ct values of individual samples were 2-units binned, i.e., 25 ≤ Ct < 26.9, 27 ≤ Ct < 28.9, 29 ≤ Ct < 30.9, 31 ≤ Ct < 33.9, 34 ≤ Ct < 35.9. Variations in Ct values were compared among pool sizes and individual samples (bins of 2-units) using a mixed-effect linear regression model. Pooled sample Ct values were considered the dependent variable. Pool size, individual sample Ct values and their interaction as independent variables, and the sample (repetition) were considered a repeated measure. Overall pool sensitivity was calculated using a formula (TP/TP+FN), where TP and FP were the number of true positives and false negatives, respectively. A pooled sample was considered positive if 2 of the 3 replicates were PCR positive.

Results

The Ct values of the positive field samples (n = 48) ranged from 25.3 – 35.9. Two un-pooled samples of high CTs 35.6 and 36.8 were negative upon re-testing and were therefore removed from the pooling analysis. Variation in Ct values of pooled samples by pool size within a group is shown in table 1. Coefficient of variation represents variation of the Ct difference within a group. Ct difference is the difference between the mean Ct value of each sample and the Ct of the un-pooled sample within a group. All pool sizes were positive if the individual sample Ct was in the range of 25.3 to 33. When the individual sample Ct was >33, regardless of pool size, negative samples were more likely although not all were negative.

Comparison of individual sample Ct values across pool size of 3, 5 and 10 showed a significant increase in Ct values ($P \leq 0.05$) regardless of the pool size. The Ct difference between individual samples and pooled Ct groups was not significant except the last high Ct group (35-36.9). The overall mean Ct difference was small compared to other groups (Table 1). The sensitivity was numerically high with a sample pooled 3× (se = 95.8), compared to pool 5× (se = 93.8) and 10× (se = 89.6).

Discussion and conclusion

The results of this study support the use of pooling tracheal samples by pool sizes of 3, 5 or 10 if the positive sample was below Ct 33 (threshold Ct). With this information, surveillance programs can incorporate pooling of tracheal samples to improve detection without implementing cost-prohibitive testing.

Table 1: Comparison of the mean Ct difference and the coefficient of variation of the different pools with a given Ct bin/range of un-pooled samples

CT range	Pool	Mean Ct difference	Coefficient of variation
25 ≤ Ct < 26.9	3	1.8	0.74
	5	2.3	0.6
	10	3.25	0.3
27 ≤ Ct < 28.9	3	1.3	0.59
	5	2	0.4
	10	2.74	0.26
29 ≤ Ct < 30.9	3	1.13	0.37
	5	1.6	0.5
	10	2.7	0.3
31 ≤ Ct < 33.9	3	1.53	0.6
	5	2.12	0.4
	10	2.75	0.37
34 ≤ Ct < 35.9	3	1.17	0.8
	5	0.3	3.4
	10	2.02	0.36

References

1. Haden, C.D., Painter, T., Fangman, T., Holtkamp, D., 2012. Assessing production parameters an economic impact of swine influenza, PRRS, and *Mycoplasma hyopneumoniae* on finishing pigs in a large production system. Proceedings from American Association of Swine Veterinarians Annual Meeting. Denver, Colo. p 75-76.

2. Strait E.L., Madsen M.L., Minion F.C., Christopher-Hennings J., Dammen M., Jones K.R., Thacker E.L., 2008. Real-time PCR assays to address genetic diversity among strains of *Mycoplasma hyopneumoniae*. J. Clin. Microbiol. 46:2491-2498.

QuickTip® semen bag: sperm safety through design

R. Grossfeld, PhD; Monika Esch, DVM; R. Koepf; C. Simmet, DVM
Minitube GmbH, Tiefenbach, Germany

Introduction

Porcine artificial insemination (AI) is an important part in pig production. More than 95% of all sows worldwide are inseminated.[1] Each AI dose needs to be temporarily stored in some sort of semen container until the insemination takes place. In many cases, the semen dose is transported and stored in a semen container for many days. The semen container must therefore absolutely guarantee non-toxicity in terms of semen fertility. Minitube's QuickTip® semen bags meet this essential requirement through compliance with numerous regulations, a rigorous quality control and not least, design.

Safety through design

Minitube's QuickTip® semen bags are designed for sperm friendliness and safety. An innovative plastic foil is used to manufacture the bags. Like many plastic foils, it consists of several layers to combine certain physical properties for the final product. While other manufacturers use a three to four layer structure with at least one adhesive layer to keep the plastic films together, Minitube has opted for an innovative compound foil concept entirely free from any adhesive for the QuickTip® bags.

Figure 1 shows a cross section of a typical foil of a commercially available semen bag. The image was produced with polarization microscopy. Next to the outer layer lies the adhesive or glue layer, which can contain repro-toxic substances. The glue layer is separated from the diluted semen by supporting layers. These layers suffer from technically unavoidable variations in thickness. If the thickness of the layer between glue and semen inside the bag decreases below a certain value, the risk that toxins from the glue layer migrate into the semen solution, increases exponentially. This occurrence has been documented in the past[2] with severe consequences for sow fertility. Variations in the thickness of these layers are system-inherent and realistically, cannot be detected 100% in any quality control system.

The foil used to manufacture Minitube's QuickTip® semen bags is designed to eliminate the risk of toxins being released into the semen. The foil for the QuickTip® semen bags is made solely of PET (polyethylene terephthalate) and PE (polyethylene) and does not use any potentially toxic additives.

Figure 2 shows the layered structure of a QuickTip® semen bag.

Figure 1: Typical layer structure of a plastic foil with adhesive used for semen bags; this one is **not** from Minitube (Fraunhofer IVV).

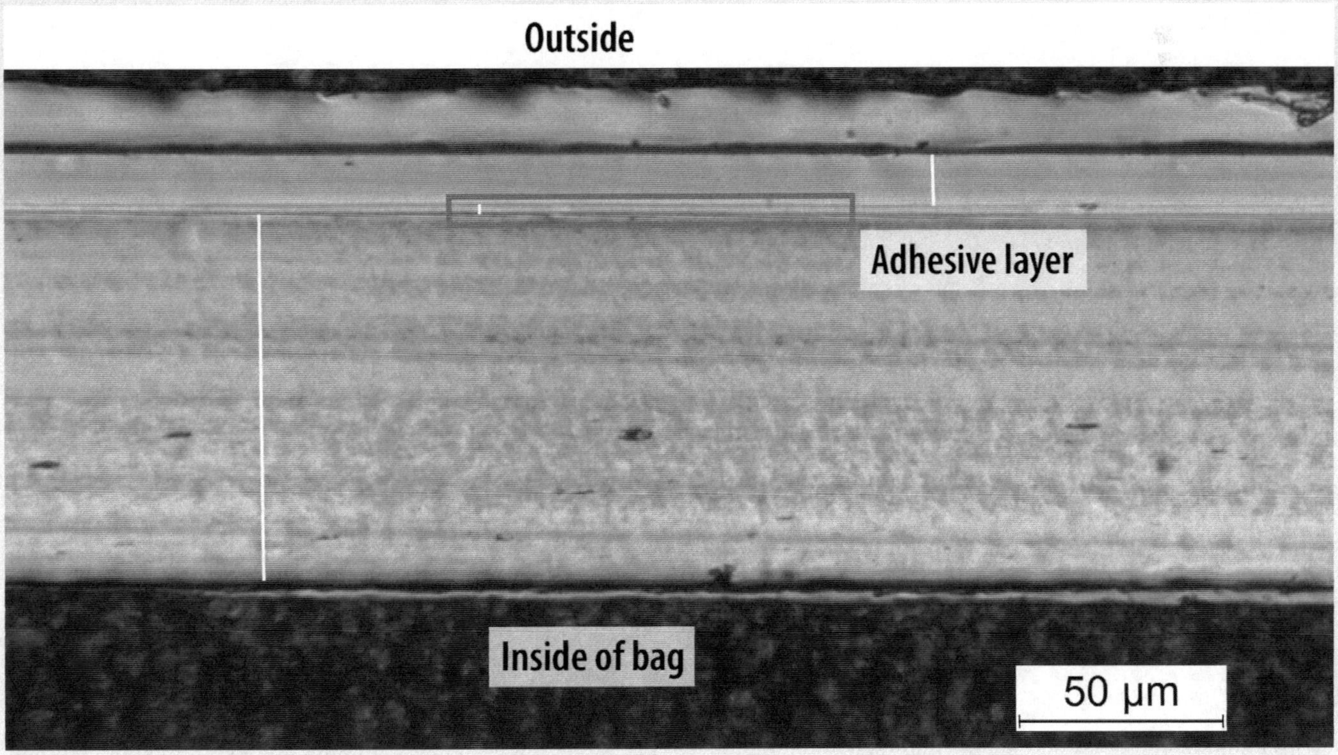

The Minitube foil structure uses PET as outside layer and binds the inner bilaminar PE film to the outer layer with purely physical forces. The QuickTip® semen bag is free from any adhesive chemicals and therefore cannot contain any substances toxic to reproduction. This is an exclusive safety insurance that is only provided by Minitube QuickTip® semen bags. The foil design for Minitube QuickTip® bags is unique in the boar semen bag market and ensures a systemically safe product.

Quality assurance system

All materials used in the manufacture of QuickTip® semen bags must be tested and proven to be biologically inactive and sperm friendly. Minitube as a specialized and ISO 9001:2015 certified producer of semen bags, has a quality control program in place, which consists of a three-level concept for the quality assurance and testing of raw materials and finished products that come into contact with semen.

Sperm tolerance testing

Quality assurance tests are regularly performed at the Unit of Reproductive Medicine of the University of Veterinary Medicine Hannover, Foundation, which operates a DAkkS accredited GLP laboratory in accordance with the international ISO/IEC 17025 norm. Each new raw material must pass the tests before it can enter the production line for boar semen bags. For these tests, boar semen from at least three different boars is extended with Androstar® Plus, divided into split samples and incubated in the QuickTip® boar semen bags to be tested and in glass vials as a control. Semen is evaluated over a semen storage period of at least 6 days. All split samples are stored in a semen storage unit at 17°C. The following tests are performed:

- Total and progressive motility (CASA) after 24, 72 and 144 h
- Acrosome integrity and membrane integrity (flow cytometry, H33342/PNA/PI) after 24 and 144 h
- Thermo-resistance test after 144 h storage at 17°C (incubation at +38°C for 120 min), then analysis of progressive motility (CASA)
- Mitochondria membrane potential (flow cytometry, H33342/PI/JC-1) after 24 and144 h
- Membrane fluidity (flow cytometry, H33342/Yo Pro 1/Merocyanin (M540)) after 24 and 144 h
- Complete morphology (wet mount, phase contrast microscopy 1000×) after 24 and 144 h

Absence of non-intentionally added substances (NIAS)

The QuickTip® semen bag foil does not contain substances like heavy metals, endocrine disruptors and plasticizers like Bisphenol A, Phthalates and Adipates per its formulation and certification. NIAS are any substances that might have been non-intentionally added to the raw material during production or transport and are potentially harmful. These substances can cause adverse biological effects in the event they are added non-intentionally. Most importantly in the context of boar semen preservation, these substances can cause damage to the sperm during storage and diminish fertility and productivity of the inseminated sows.[2] The test for NIAS verifies compliance with these requirements. Therefore, in the third level of the Minitube quality assurance program, the absence of NIAS is controlled with routine testing of the raw materials for endocrine disruptors and plasticizers like Bisphenol A, phthalates and adipates as well as heavy metals. It is a matter of principle for Minitube, as a manufacturer of semen packaging material, to take full responsibility for using only substances and raw materials that are not harmful to animals or sperm.

Figure 2: Layer structure of Minitube's Quicktip® semen bag (Fraunhofer IVV).

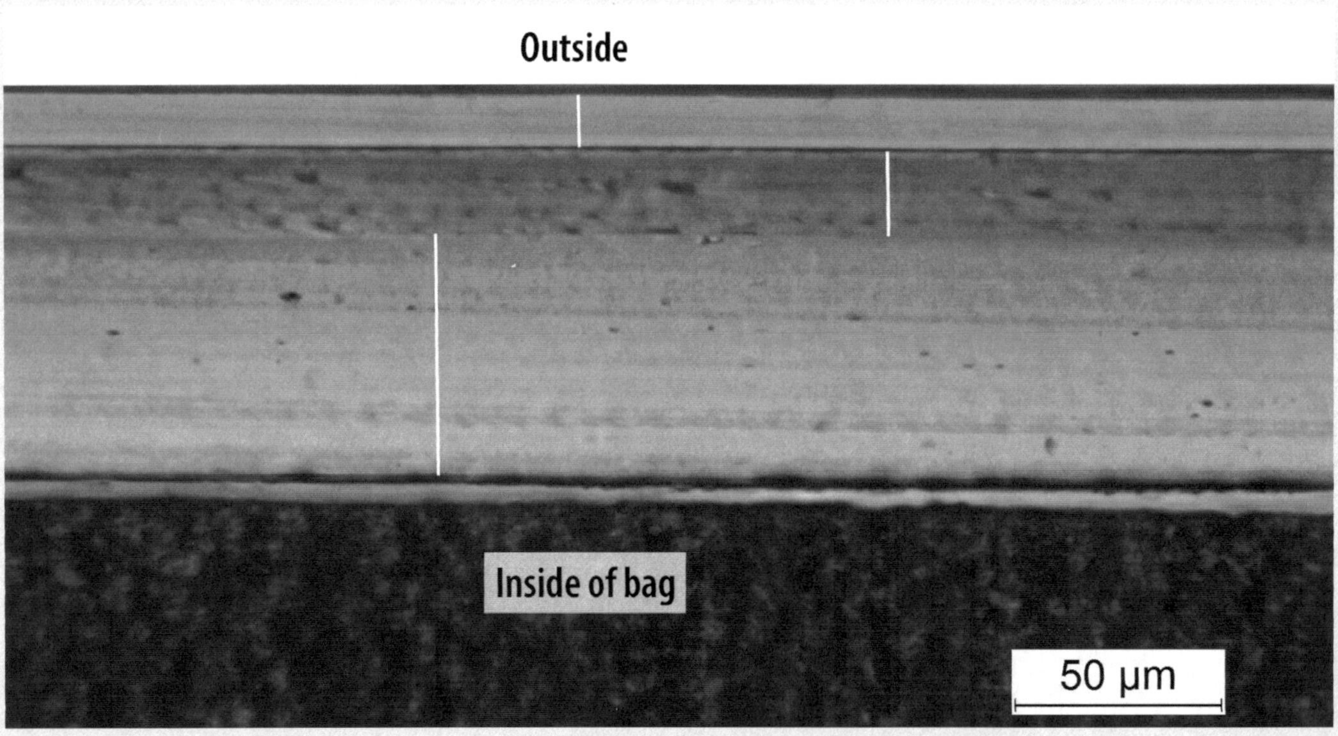

Physical quality control

In addition, all batches of QuickTip® boar semen bags are subjected to physical quality control and are only released for sale after approval by Minitube's quality control management. Tests performed include sealing tests, leak tightness, functionality of the QuickTip® and the accuracy of the dimensions and grid of holes for transporting the bags on the filling machine. The samples used for these tests are randomly drawn from each production batch in accordance with ISO 2859-1-204.

Specifications and compliance

Minitube QuickTip® semen bags are made from polyethylene terephthalate (PET) and polyethylene (PE). PET and PE are known for its inactivity and biocompatibility. They are free from any toxic ingredients like Bisphenol A, heavy metals, phatalates or adipates as these substances would impair the necessary product features. Each batch of the raw material that is assigned for the production of QuickTip® boar semen bags at Minitube's manufacturing plant, must be accompanied by a batch certificate. Unlike other manufacturers' semen bags, Minitube's QuickTip® semen bags are free from the disinfectant Triclosan, which has been demonstrated to have a negative effect on mammalian cells, including porcine spermatozoa.[3]

The raw materials used in Minitube's QuickTip® semen bags meet all of the following standards:

- EC Regulation 1907/2006/EC and amendments (referred to as REACH[4] regulation), and further EC regulation 1272/2008/EC on classification, labelling and packaging of substances and mixtures
- EC Directive (EC) No 2023/2006 on good manufacturing practice for materials and articles intended to come into contact with food
- EC Directive (EC) No 2016/1416 amending and correcting regulation (EU) No 10/2011 on plastic materials and articles intended to come into contact with food
- EC Directive (EC) No 10/2011 on plastic materials and articles intended to come into contact with food
- Commission Regulation (EU) No. 10/2011 of January 14, 2011, including its amendments (EU) No. 1282/2011 of November 28, 2011, (EU) No. 1183/2012 of November 30, 2012, (EU) No. 202/2014 of 3 March 2014, (EU) No. 2015/174 of 5 February 2015, (EU) No. 2016/1416 of 24. August 2016 and (EU) 2017/752 of 28. April 2017
- Empfehlung III „Polyäthylen" von Kunststoffen im Lebensmittelverkehr: Empfehlungen des Bundesinstituts für Risikobewertung (BfR) (former BgVV)., latest amendment February 15, 2016 (BGBl. I S. 198)
- Code of Federal Regulations, issued by the Food and Drug Administration (FDA), paragraph 21 CFR 177.1520 (olefin polymers)

Conclusion

The QuickTip® semen bag is manufactured from foil that does not contain any adhesive chemicals and is therefore sperm safe by design. The raw material for the bags needs to comply to numerus regulations in order to assure sperm safety. The quality assurance procedure of the QuickTip® semen bag is conducted by a certified GLP semen quality laboratory and by excessive internal physical testing. All these tests aim to assure a safe and reliable semen container for boar semen storage.

References

1. Knox RV. Artificial insemination in pigs today. Theriogenology 85, 83-93 (2016) **https://doi.org/10.1016/j.theriogenology.2015.07.009**

2. Nerin, C., Ubeda, J., Alfaro, P. et al. Compounds from multilayer plastic bags cause reproductive failures in artificial insemination. Sci Rep 4, 4913 (2015). **https://doi.org/10.1038/srep04913**

3. Ajao C., Andersson MA., Teplova VV. et al. Mitochondrial toxicity of Triclosan on mammalian cells. Tox. Rep. 2, 624-637 (2015) **https://doi.org/10.1016/j.toxrep.2015.03.012**

4. REACH regulation 1907/2006/EC **https://eur-lex.europa.eu/legal-content/EN/TXT/HTML/?uri=CELEX:02006R1907-20140410&from=DE.**

Boar Better® stimulates gilt puberty with or without live boar exposure

John J. McGlone[1,2], BS, MS, PhD, PAS, ACAWS; Steve Stitzlein[2], BS
[1]Laboratory of Animal Behavior, Physiology and Welfare, Texas Tech University, Lubbock, Texas;
[2]Animal Biotech, Dallas, Texas

Introduction

The United States has over 6 million sows in commercial production. The United States has an average 50% annual replacement rate which means that we need to introduce more than 3 million gilts each year. This is not an easy task as gilts are persnickety to get introduced into the sow herd. While we have published efficacy data using Boar Better (BB) with sows,[1,2] the use of BB to stimulate and mate gilts is a newer application of sexual behavior pheromone technology.

Why do we want so many gilts in the sow herd each year?

First, new gilts should be genetically better than the sows as genetic progress is continuous. Second, sows get injured, die, or get culled for low performance, so they leave the herd. Third, although gilts seem to cost a lot, they really do not. You will get more for a culled sow (during normal a normal marketing times) than a gilt; so, you make a little money trading an old sow for a young gilt. Today's sow price is about 40 cents per pound; so, a 500-pound sow is worth $200. A 280-pound gilt might bring $120, plus you usually pay a negotiated genetic premium, and gilts have fewer pigs than older sows, so it can be a complicated assessment.

The cost to bring a new gilt into the breeding herd is less clear. The animal has a cost, but not all gilts that are intended for breeding will get bred. If a farm brings in 100 gilts, the farm may end up getting only 50-90 gilts into the sow herd. Clearly, if you get only 50% of eligible gilts into the herd as a breed female, the cost of gilt introduction is much higher than if you would get 90% of the gilts bred. Therefore, the goal of the farm is to get as high of a percentage of the eligible gilts into the sow herd as a pregnant female. A new tool can help you get this done.

If you hold gilts in isolation away from adult males and minimally stress them, many of them will not cycle. One must have an actual estrus cycle to be able to breed the young gilt. Many gilts require an adult male exposure or a stressful experience (such as transport) to kick off their first estrus cycle. Ideally, you would like to breed your gilts on their second or third estrus cycle because the first cycle is often less fertile (lower litter size and reduced farrowing rate).

What causes gilts to cycle?

A few gilts cycle with little or no stimulation, but the majority require stimulation. The ovaries of the gilt begin developing while the piglet is in utero (during pregnancy). Eggs begin to develop more rapidly after birth. As time goes on, eggs mature, but they do not ovulate. Something is needed to kick off that first ovulation and essentially this is the initiation of puberty. Those first pubertal eggs are ready to go with the right stimulation. The two key triggers to initiate ovulation are stress and boar exposure. Stress is less reliable or complete when compared to boar exposure. One of the main signals boars provide is a mixture of specific olfactory signals – boar pheromones. But boars are costly and can be dangerous. The goal of each gilt development unit should be to either make the boars more effective or remove them. This is now possible because we have identified the complete boar pheromone, and it is commercially available.

We have been investigating how to use the boar pheromone to assist producers with getting more gilts cycling, bred and into the sow herd. Here we summarize some recent studies. First, we looked at gilt ovary weights to see if we could stimulate the onset of puberty (during puberty the ovary grows larger). Next, we looked at actual ovarian development by collecting and examining ovaries after gilts were harvested. Several people have also been using Boar Better in commercial farms to stimulate gilt puberty.

Boar Better on ropes with no boar: a method for smaller producers

In the first study (Figure 1), cotton ropes were used to expose the gilts to the boar pheromone at Texas Tech University's farm. Each rope was sprayed with 4mL of Boar Better once per day, and there was 1 rope per gilt in each pen on test. Two control groups were evaluated: (1) rope and no pheromone and (2) no rope and no pheromone. While on test gilts were handled in a minimal manner. Once the test was concluded gilts were then harvested and their reproductive tracts were weighed. A gilt reproductive tract is very small before puberty and it grows rapidly once she has started cycling. This gave us confidence that daily exposure of Boar Better to 5 to 6-month-old gilts will cause them to develop puberty without boar exposure. However, when we tried the rope method in large scale commercial settings, the time and cost to maintain ropes was inefficient. This method is best for smaller farms, but not recommended for larger scale farms.

Boar Better sprayed on the snout: with or without boar exposure

In another study we sprayed 170-day old gilts each day on the snout with Boar Better for 4 weeks – with no boar in the barn. The control gilts had no boar and no exposure to Boar Better. Again, at then end of the trial the gilts were harvested and the ovaries were examined for the onset of puberty (Figure 2).

Using Boar Better with a boar: field studies

We have reports of field studies from around the United States and from China where people have sprayed Boar Better either alone (no boar) or while also providing boar exposure. However, these trials are not as controlled as university trials, and the methods vary considerably due to different production systems' flows and needs. This information is considered anecdotal in nature. Each study used a different set of treatments. We observe a range in the field of 30 to 85% of gilts that are identified with a heat no service date (HNS) or are bred and enter the sow herd. Clearly, if a farm gets 90% of gilts bred, this is a large economic advantage compared with getting only 50% of the gilts bred. Table 1 presents data from several field trials showing the percentage of gilts identified in heat when using Boar Better.

Advice on using Boar Better (BB) for gilts

- Daily boar exposure is the best-known way to stimulate gilts onset of puberty
  - Utilizing BB can help increase your boar exposure and stimulation of puberty
- If you do not have access to a live boar to stimulate gilts, then you should be able to get a high percentage of gilts cycling and bred by using BB
- Many methods can be used to apply the BB to induce puberty, including:
  - Spray each gilt on the snout at minimum of once per day with 4 mL of BB (Figure 3)
    (this is the most common and most effective method)
  - Spray a rope daily with BB and let pigs self-administer
  - Other methods of application being studied, including:
    - Fogger systems
    - Automated sprayers (such as those used for insect control)
- Research and development are on-going as we strive to find the most effective application methods
- The 2 main uses for BB within your gilt herd to increase productivity:
  - Developmentally to stimulate onset of puberty
  - Acute use of BB to identity gilts expressing estrus for HNS or breeding purposes

Please reach out to Animal Biotech for technical support in setting up the best protocol for your specific needs: Steve Stitzlein: 614-206-8643; **Steve@animal-biotech.com; Info@animal-biotech.com.**

References

1. McGlone, JJ et al. A novel boar pheromone mixture induces sow estrus behaviors and reproductive success. *Appl Anim Behav Sci* 2019. **https://doi.org/10.1016/j.applanim.2019.104832**

2. McGlone et al. Multi-farm analyses indicate a novel boar pheromone improves sow reproductive performance. Animals 2019. 9:37. **http://dx.doi.org/10.3390/ani9020037**

Figure 1: Gilts were exposed to ropes with BB added on day 170 of age and each day for 30 days before evaluation of the reproductive tracts after slaughter. Paired ovarian weights (left) and a picture of the pen with ropes, BB and gilts (right).

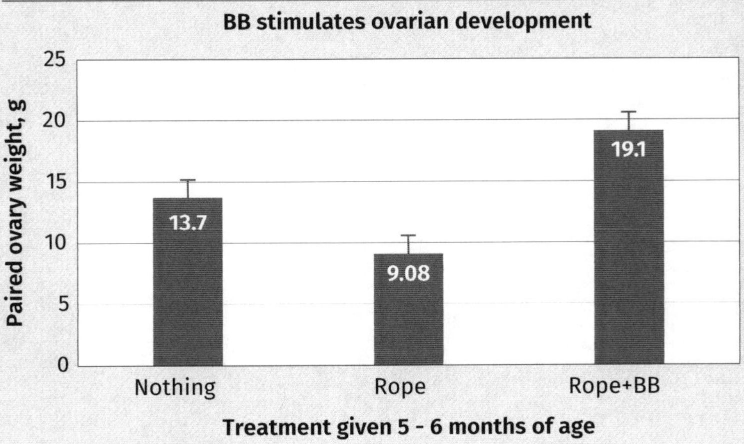

Figure 2: Percentage of 200-day-old gilts that had passed puberty or not after 30 days of exposure to BB or to a control – in both cases with no boar present. Exposing gilts to BB for 30 days resulted in 85.7% of gilts having 1 or 2 cycles within 30 days while control gilts had 64.3% of reproductive tracts that had cycled at least once.

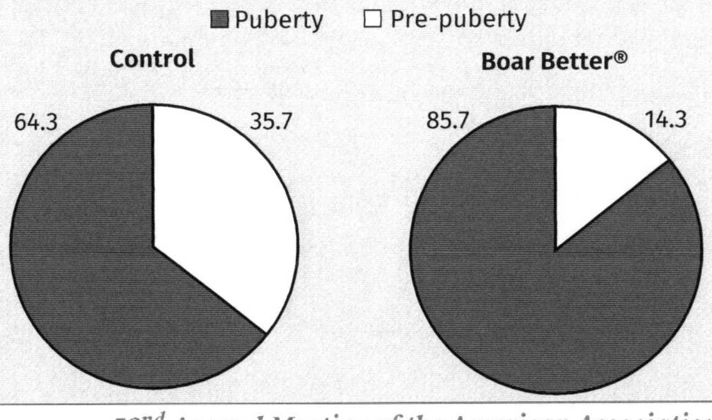

Table 1: Results of field use of Boar Better (BB)

Field data location	Notes	% gilts with an identified heat date when using BB
China	No Boar + BB	89.1%
Illinois	Boar + BB	94.4%

Figure 3: Gilt being sprayed with Boar Better.®

Evaluation of Prevacent® PRRS in a commercial environment using a seeder pig challenge model

S. Dee[1], DVM, MS, PhD, Dipl. ACVM; D. Hanson[1]; R. Edler[1], MS; J. Nerem[1], DVM;
R. Evelsizer[2], DVM; P. Hoffmann[2], DVM; C. L. Puls[2], PhD
[1]Pipestone Applied Research, Pipestone, Minnesota; [2]Elanco Animal Health, Greenfield, Indiana

Introduction

Porcine reproductive and respiratory syndrome virus (PRRSV) is a significant pathogen of pigs that causes detrimental reductions in growth performance, increased medication and treatment costs, and increased mortality. Previous reports suggest that approximately 50% of commercial pigs are naturally infected with wild-type virus during the grow-finish phase.[1] The control of the disease is difficult, as the virus can transfer between populations of pigs via vertical or horizontal transmission, can become airborne, and can travel on fomites.[2] One common method of control and mitigation of disease is the use of modified live vaccines which provide reduction of clinical lesions, improved performance, and reduced mortality.[3] In 2018, Elanco Animal Health (Greenfield, IN, USA) received USDA licensure for a contemporary, lineage 1, modified live PRRSV vaccine, Prevacent® PRRS. This was important as the majority of wild-type strains circulating in the US swine industry presently are classified as lineage 1.[4] At the time of this study, Prevacent PRRS was the only marketed modified live vaccine containing a contemporary lineage 1 strain. Licensing data showed limited shedding and transmission of vaccine virus, as well as a reduction in lung lesions across a number of PRRSV variants and lineages.[5,6] However, to date, there has been no large-scale evaluation of Prevacent PRRS effectiveness under controlled commercial conditions. Therefore, the purpose of this study was to evaluate the effectiveness of Prevacent PRRS under controlled commercial conditions using a seeder pig challenge model.

Materials and methods

All protocols reviewing animal care and use were reviewed by the Pipestone Research IACUC. A total of 2,153 three-week old crossbred maternal barrows were randomly assigned to four treatment groups: 1) non-vaccinated, challenged controls (NVCC, n = 641), 2) vaccinated with Prevacent® PRRS and challenged (PREV, n = 637), 3) vaccinated with Inglevac PRRS® MLV and challenged (ING, n = 635), and 4) non-vaccinated and challenged seeder pigs (n = 240). Pigs were sourced from a herd negative for PRRSV, *Mycoplasma hyopneumoniae, Actinobacillus pleuropneumoniae* (APP), swine influenza virus (IAV), and porcine epidemic diarrhea virus (PEDV). Following allocation at the sow farm (day 0), treatment groups were weaned into two separate locations. Animals allocated to PREV and ING were placed into separate air-tight rooms at one location and administered 1 mL Prevacent PRRS and 2 mL Inglevac PRRS MLV intramuscularly, respectively. Control animals (NVCC and seeder pigs) were placed at a separate off-site location and received no vaccine or placebo. On day 41 of study, control animals (NVCC and seeder pigs) were transported to the study location. All animals were individually weighed and reassigned to treatment pens for the post-challenge study period, such that final pen configuration consisted of eight NVCC, eight PREV, eight ING, and three seeder pigs for a total of 27 pigs/pen. There was a total of 79 pens on study. On day 42, all seeder pigs were challenged IM with 2 mL of a contemporaneous lineage 1, RFLP 1-7-4 strain of PRRSV (titer = $10^{3.5}$ $TCID_{50}$ per mL). The challenge virus was described as highly virulent in commercial farms and was 89.5% homologous to Prevacent PRRS and 87.9% homologous to Inglevac PRRS MLV on ORF-5 sequencing. A total of 30 pigs each from PREV and ING were randomly identified and serially bled on day 0 (prior to vaccination), 7, 14, 28, 41 (prior to challenge), 56, 70, 84, and 112 for determination of viremia using individual quantitative PRRS PCR. Antibody responses (PRRS ×3 ELISA) were evaluated on day 7, 14, 28, and 41. All diagnostic testing was completed at the ISU VDL. Pigs were individually weighed on day 0, 41, and 142 (end of study). On day 142 of study (end), the number of light culls (≤ 172.0 lb live weight), defect culls (physical deformities, i.e. umbilical hernia, tailbites, lameness), and cumulative morbidity and mortality were used to calculate the percentage of full value pigs. Pigs were offered ad libitum access to feed and water throughout the study. Individual pig was used as the experimental unit for diagnostic and growth performance data. Diagnostic data (PCR and ELISA) were analyzed as a repeated measures model using a generalized linear mixed model. The model included the fixed effects of treatment, day, and treatment × day interaction. Body weight and growth rate data were analyzed using the PROC MIXED procedure of SAS (SAS Inst., Cary, NC) with fixed effect of treatment.

Results and discussion

Viremia

Viremia data are presented in Figure 1. Pigs vaccinated with Prevacent had lower ($P < 0.05$) viremia as measured by Ct values on day 7, 14, and 28 of study. Viremia in the post-challenge period was generally similar ($P > 0.05$) between the two vaccinated groups, with the exception of day 56, when PREV had a 2.4 lower Ct value than ING.

Serology

Figure 2 demonstrates the average ELISA response of PREV and ING. Both vaccinated groups had similar ($P > 0.05$) ELISA antibody responses. The mean S:P ratio of both PREV and ING were positive by day 14 post-vaccination and continued to increase through day 41 (day prior to challenge).

Growth performance

The results of the growth performance evaluation are presented in Table 1. There was a difference in weaning weight, with NVCC being heavier than PREV and ING. Consequently, subsequent growth performance data were analyzed using a covariate to account for this difference. Pigs vaccinated with Prevacent were 2.7 lb heavier ($P < 0.05$) than NVCC and 1.5 lb heavier ($P < 0.05$) than ING at day 41 of study, and had 7.4% greater ($P < 0.05$) ADG than NVCC and 4.1% greater ($P < 0.05$) ADG than ING in the

Figure 1: Effects of vaccination on viremia.

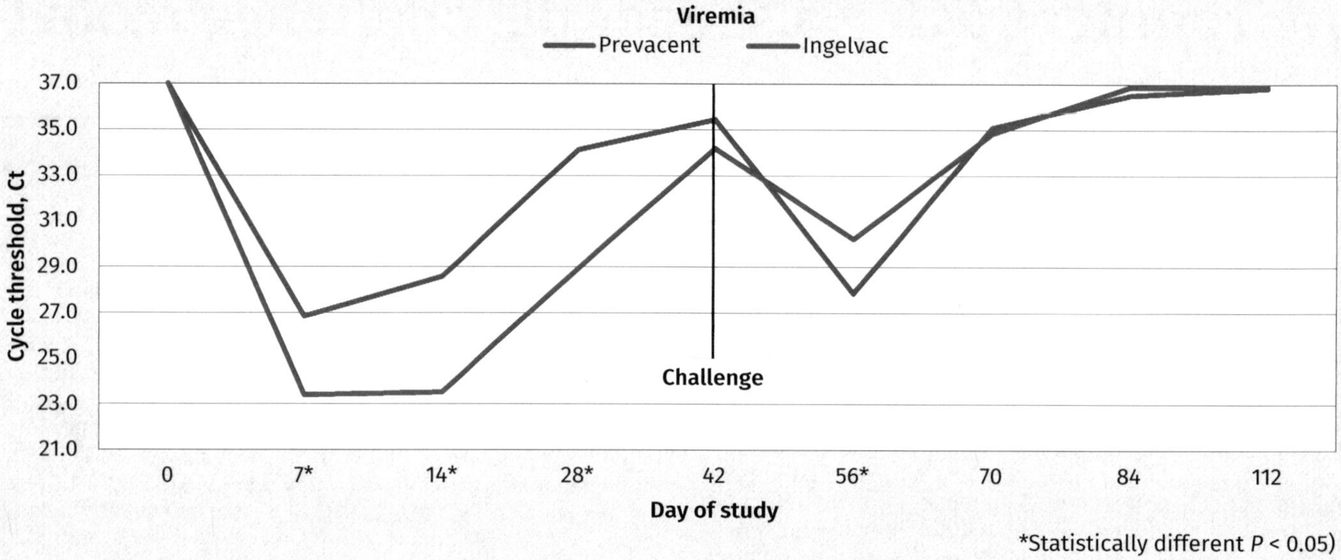

Figure 2: Effects of vaccination on antibody response.

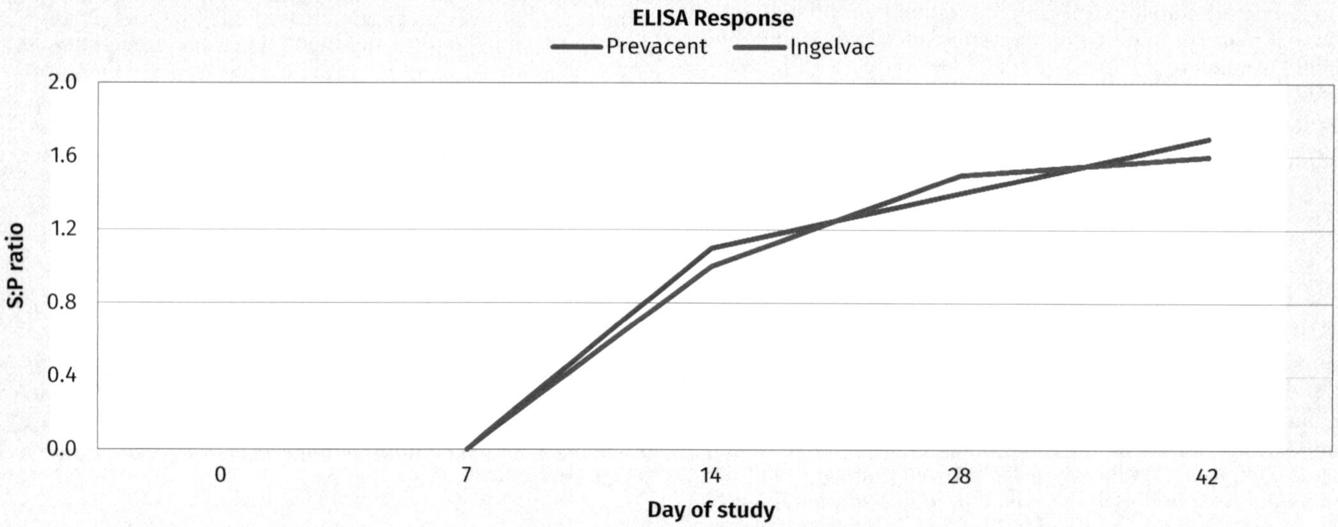

pre-challenge period. Pigs vaccinated with Prevacent were 13.7 lb heavier ($P < 0.05$) at the end of study compared to NVCC and had 5.9% greater ($P < 0.05$) ADG in the post-challenge period. Similarly, pigs vaccinated with Ingelvac were 14.6 lb heavier ($P < 0.05$) at the end of study compared to NVCC and had 7.1% greater ($P < 0.05$) ADG in the post-challenge period. There was no difference ($P > 0.05$) between PREV and ING for post-challenge ADG or final body weight. Overall (pre- and post-challenge period) ADG was 6.1% lower ($P < 0.05$) for NVCC than PREV and ING, which were not different ($P > 0.05$).

Morbidity and mortality

The results of vaccination on the incidence of morbidity and mortality and calculation of full value pigs are presented in Table 2. The incidence of morbidity and mortality was 13 percentage units greater ($P < 0.05$) for NVCC than for PREV and ING, which were not different ($P > 0.05$). There was no difference ($P > 0.05$) between treatments for the incidence of cull lights or for the incidence of defects. Consequently, there was a greater ($P < 0.05$) percentage of full value pigs for PREV and ING, which had similar ($P > 0.05$) levels, than for NVCC.

Table 1: Effects of vaccination on growth performance of pigs

Item	Treatment		
	NVCC	PREV	ING
No. pigs	641	637	635
Body weight, lb			
Weaning (vaccination)	11.7^{a}	11.3^{b}	11.2^{b}
Day 41 (day prior to challenge)	53.5^{c}	56.2^{a}	54.7^{b}
Day 142 (end)	224.0^{b}	237.7^{a}	238.6^{a}
ADG, lb/d			
Pre-challenge period	0.95^{c}	1.02^{a}	0.98^{b}
Post-challenge period	1.69^{b}	1.79^{a}	1.81^{a}
Overall	1.47^{b}	1.56^{a}	1.56^{a}

^{a,b,c} Means with differing superscripts are different ($P < 0.05$)

Table 2: Effects of vaccination on morbidity and mortality of pigs

Item	Treatment		
	NVCC	PREV	ING
Cull light (< 172 lb), %	4.37	3.30	3.15
Cull defect, %	0.57	0.43	0.43
Morbidity and mortality, %	21.2^{b}	8.3^{a}	8.0^{a}
Full value pigs, %	73.8^{b}	87.9^{a}	88.4^{a}

^{a,b} Means with differing superscripts are different ($P < 0.05$)

Discussion

The seeder pig model used in this study has been well established and was effective at inducing clinical signs and mortality in the population of pigs.[7] The study was conducted in a representative commercial facility and used a relatively high health population of pigs. Furthermore, the study procedures were reviewed and approved by IACUC, and a high standard of biosecurity was implemented on the study as vaccine crossover did not occur during the 42-day pre-challenge period. The study also used a contemporary, lineage 1, industry relevant field strain of PRRS for challenge. One potential acknowledged limitation of the study, however, was that treatment groups were housed in separate air spaces during the pre-challenge period. This was implemented by design to minimize the opportunities for vaccine shedding and transmission across the treatment subpopulations.

The finding that PREV had less viremia than ING in the pre-challenge period is in agreement to previous research.[5] It has been shown that as Ct value decreases, viral load increases, with a difference of 3.3 equaling approximately a 10-fold increase in virus.[8,9] Nonetheless, the significance and practical implications of this difference are not clearly defined.

The difference in pre-challenge growth performance between the treatment groups could be, in part, due to treatment groups being housed in separate air spaces in the pre-challenge period. The NVCC were housed at a separate off-site facility than PREV and ING treatment groups, which were housed in the same research location, but in different air-tight rooms within the facility. Nonetheless, these findings are in line with previous research suggesting no negative effect of Prevacent PRRS on growth performance in the post-vaccination period.[9] Both PREV and ING had superior growth performance to NVCC in the post-challenge period, which aligns with previous research.[6,10]

As expected, vaccination with a modified live vaccine significantly reduced morbidity and mortality after introduction of challenge. The results of this study confirm the economic benefit of vaccinating at-risk populations of pigs.

This was the first study to compare PREV and ING in a commercial setting. In general, PREV and ING performed similarly in the post-challenge period. However, this study was conducted in a relatively high health population of pigs, and it is unclear how results may differ in more conventional health populations of animals.

Conclusions

The results of this study confirm previous research demonstrating the efficacy of Prevacent PRRS. Similarly, vaccination with a modified live vaccine proved to be effective at improving growth performance and reducing mortality in the face of a challenge.

References

1. P. Yeske. 2020. Presented at 2020 National Pork Industry Conference (NPIC), Wisconsin Dells, WI.

2. E. Pileri and E. Mateu. 2016. Review on the transmission porcine reproductive and respiratory syndrome virus between pigs and farms and impact on vaccination. Vet. Res. 47:108. **doi:10.1186/s13567-016-0391-4.**

3. A. Patterson, G. Haiwick, J. Hermann, B. Fergen, K. Wakeland, W. Chittick, and R. Philips. 2019. Efficacy of Ingelvac PRRS® modified live virus vaccine against heterologous porcine reproductive and respiratory syndrome virus challenges. J. Vaccines Vaccin. 10:407. **doi:10.35248/2157-7560.19.10.407.**

4. I. Paploski, C. Corzo, A. Rovira, M. Murtaugh, J. Sanhueza, E. Smith, and K. VanderWaal. 2018. Making epidemiological sense out of large datasets of PRRS sequences. Pg. 17 in Proc. Allen D. Leman Swine Conference, St. Paul, MN.

5. C. L. Puls, P. Hoffmann, R. Saltzman, J. Risser, R. Evelsizer, and J. M. Hammer. 2020. Prevacent® PRRS: determining safety. Pg. 134-138 in Proc. 51st AASV Annual Meeting, Atlanta, GA.

6. J. M. Hammer, C. L. Puls, R. Evelsizer. 2019. Prevacent PRRS: a new tool for porcine reproductive and respiratory syndrome virus (PRRSV) control. Pg. 339-340 in Proc. 50th AASV Annual Meeting, Orlando, FL.

7. S. Dee, J. E. Guzman, D. Hanson, N. Garbes, R. Morrison, D. Amodie, and L. G. Pantoja. 2018. A randomized controlled trial to evaluate performance of pigs raised in antibiotic-free or conventional production systems following challenge with porcine reproductive and respiratory syndrome virus. PlosOne. 13e0208430. **doi:10.1371/journal.pone.0208430**

8. D. Polson, E. Mondaca, K. Doolittle, G. Hartsook, and E. Vaughn. 2010. Evaluation of the Ct values of non-quantitative real time RT-PCR as a proxy for determining PRRS viral load and its potential for monitoring swine populations, areas, and regions. Pg. 475 in Proc. 21st IPVS Congress, Vancouver, Canada.

9. P. Gauger. Personal communication.

10. P. Yeske, A. Betlach, R. W. Evelsizer, and J. M. Hammer. 2020. Evaluation of shedding and effect on pig performance of Prevacent® PRRS vaccine. Pg. 49-50 in Proc. Allen D. Leman Swine Conference, St. Paul, MN.

Evaluation of shedding and effect on pig performance of Prevacent® PRRS vaccine in commercial conditions

Paul E. Yeske[1]; Alyssa Betlach[1]; Robert W. Evelsizer[2]; James M. Hammer[2]
[1]Swine Vet Center, St. Peter, Minnesota; [2]Elanco, Greenfield, Indiana

Introduction

Many swine production systems today are vaccinating PRRS negative pigs from at risk populations with PRRS vaccine. Vaccination has been shown to improve performance in the face of field virus infections.[1-3] However, vaccination has reduced post-weaning growth performance in absence of PRRS exposure.[4] Using a modified live PRRS vaccine along with unidirectional pig flow has been shown to help protect PRRS negative weaned pigs placed in high PRRS risk areas as well as reduce shedding of field virus if infection occurs and reduce area PRRS spread. Elanco licensing data for Prevacent® PRRS, from a small group of pigs in a research setting, showed limited transmission from vaccinates to unvaccinated sentinels.[5] The purpose of this study was to see how the Prevacent® PRRS vaccine performed under commercial conditions.

Objectives

1) Identify the presence of airborne shedding of Prevacent® PRRS vaccine, **2)** Identifying the level of viremia and shedding in pigs following vaccination with Prevacent® PRRS vaccine in a commercial setting, and **3)** Measuring any impact of vaccination on performance of the pigs.

Materials and methods

Pigs used in the study were sourced from two PRRS negative (AASV category IV) farms with negative diagnostic history (of monthly statistical sample 95/10 testing both PCR and antibody).

Pigs were placed into a typical commercial wean to finish barn with large pen (150 head per pen) finishing design and vaccinated in the first 1-2 weeks post-placement.

- Deep pit barn for manure storage
- Tunnel ventilated building
- Pit fans for minimum ventilation
- Good history of remaining negative so less likely that field virus contamination would be a complicating factor
- Building had 2 rooms that were separate air spaces and had independent entryways from a common entry area and loading chute area
- Figure 1: Pen layout of barn

Day 0: Prevacent® PRRS vaccine administered at labeled dose (1 mL/pig). Controls were administered 1 mL saline.

Body weights were collected on days 0, 50, and 125 of the study.

Thirty sentinel pigs in each treatment were identified and tagged and serially blood tested on days 4, 6, 8, 15, 22, 27, 36, 43, 50, 64, 81, 92, 106, 125 and 140 during the study.

Oral fluids were collected from each pen at the same time blood samples were collected.

A Cyclonic Air collector was used to collect air samples daily for 28 days; samples were collected from inside the barn, at pit fan, and 1 mile downwind from the site for the day.

Results

All air samples collected were negative from all locations demonstrating that there was no detectable shedding by this method of sample collection (Table 1).

Pigs serologically converted to vaccine showing an immune response to the vaccine (Figure 2 & 3).

The levels of virus dropped quickly following exposure which may explain that there was limited shedding detected from the population. Towards the end of the trial, pigs broke with clinical influenza. This caused the previously vaccinated pigs to shed Prevacent® PRRS throughout the facility. Viremia levels were higher than initial levels following vaccination. Wild type PRRS was ruled out through submitted diagnostics (Figure 4 & 5).

There was no difference in growth or performance between control and vaccinates up to d 125 of study when the first pigs were marketed. There was also no difference in mortality (Table 2).

Conclusions and discussion

In this study, there was no detection of the vaccine virus in the aerosol testing for the first 28 days following vaccination, including testing in the barn, outside the nearest located pit fan, and downwind 1 mile. Other vaccines have been reported to have virus detected at the pit fan and downwind for up to 14 days following vaccination.[6] This is promising for using vaccine next to negative sites with less concern of spread in adjacent areas, while protecting the site from effects of field virus spread into the site. Pigs vaccinated with Prevacent® PRRS developed viremia by 4 days post-vaccination and quickly decreased the level of viremia by day 15 with very low levels by day 30. However, vaccine was still able to infect the non-vaccinated controls in the same room and adjacent room at the end of the study following the clinical outbreak of H1N1 influenza virus at day 120.

In contrast to other vaccines, there was no significant reduction in growth at the end of the nursery phase, finishing phase, or overall, in vaccinated pigs versus sham vaccinated controls.[4]

It is interesting to see the interaction with a field infection of an H1N1 influenza infection, resulting in much higher PRRS viremia levels when compared to animals immediately following vaccination. Since the group broke at the very end of the trial, it was difficult to assess the impact of the influenza A H1N1 on performance and gain.

This study demonstrates Prevacent® PRRS can be a useful tool in the control of PRRS infection while allowing for lower risk of area vaccine virus spread and no negative effect on pig performance.

Figure 1: Pen layout of barn.

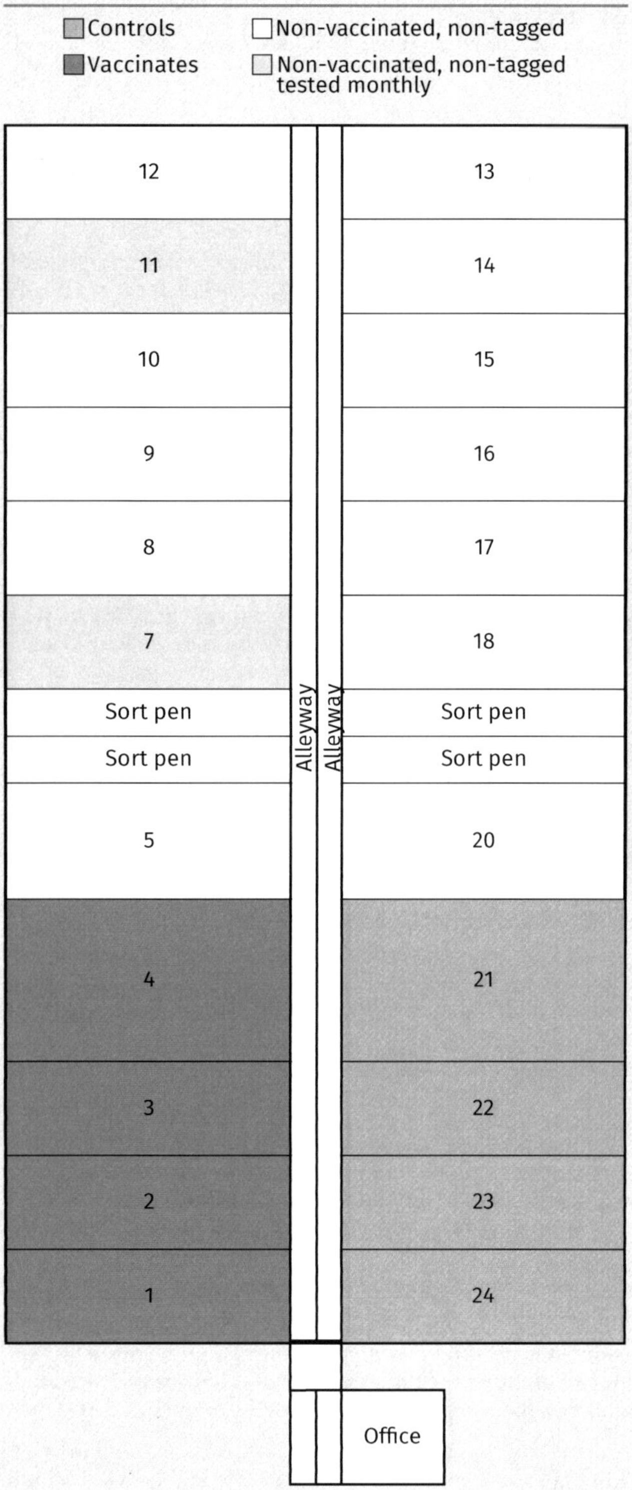

Value of research

There was no evidence of shedding of Prevacent vaccine virus by air sampling in the first 28 days post-vaccination, within the barn, outside the pit fan, or downwind 1 mile.

Following the outbreak of influenza H1N1, there was circulation of the Prevacent® PRRS vaccine both within the room and the adjacent control room, resulting in higher viremia at this time in controls verses original vaccinates.

Prevacent® PRRS did not cause a reduction in growth performance or final body weight compared to controls.

Mortalities were similar in vaccinates as controls.

References

1. Oppressnig T. et al.; Demonstration of Protection Against Heterologous Challenge; J Swine Health and Production 13:246-253.

2. Roof M.; 2007 Cross Protection in 16 Independent Trials; Data presented at a meeting in Poland.

3. Polson D., et.al, 2000. Evaluating the Beneficial Performance and Financial Effects of Pig Vaccination with Ingelvac® PRRS MLV in a PRRS-Positive Multi-Site Production System. Proceedings of the 16th IPVS. Pg: 152.

4. Pretzer S., et al., 1996. The Effects of Porcine Reproductive and Respiratory System (PRRS) vaccination on postweaning growth performance. KSU Swine Day: 83-86

5. Elanco. 2016. Data on File.

6. Yeske P., 2011. A study to evaluate airborne shedding of PRRS virus in pigs following vaccination with Ingelvac PRRS in a commercial production system. Proceedings AASV 43rd Annual meeting.

Table 1: Number of positive airborne samples

	Day of study																											
	0	1	2	3	4	5	6	7	8	9	10	11	12	13	14	15	16	17	18	19	20	21	22	23	24	25	26	27
Downwind	0	0	0	0	0	0	0	0	0	0	0	0	0	0	0	0	0	0	0	0	0	0	0	0	0	0	0	0
Outside pit fan	0	0	0	0	0	0	0	0	0	0	0	0	0	0	0	0	0	0	0	0	0	0	0	0	0	0	0	0
Within room	0	0	0	0	0	0	0	0	0	0	0	0	0	0	0	0	0	0	0	0	0	0	0	0	0	0	0	0

Figure 2: Serologic response of Prevacent® PRRS.

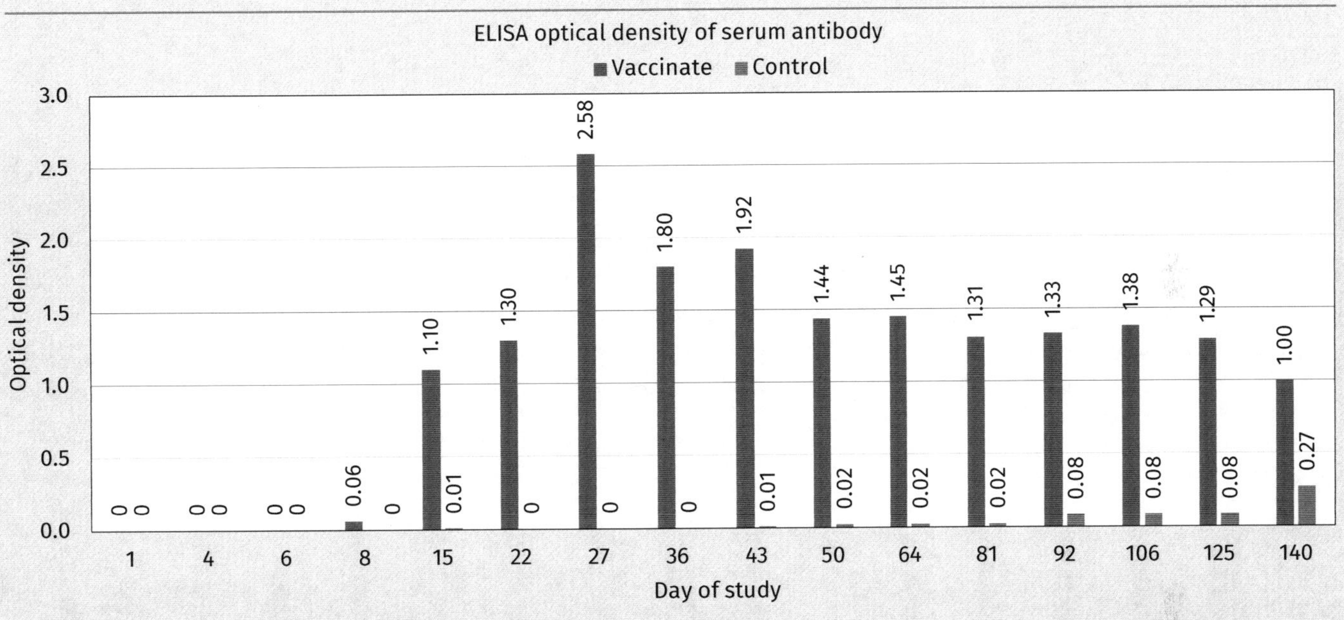

Figure 3: Antibody response of oral fluids of Prevacent® PRRS.

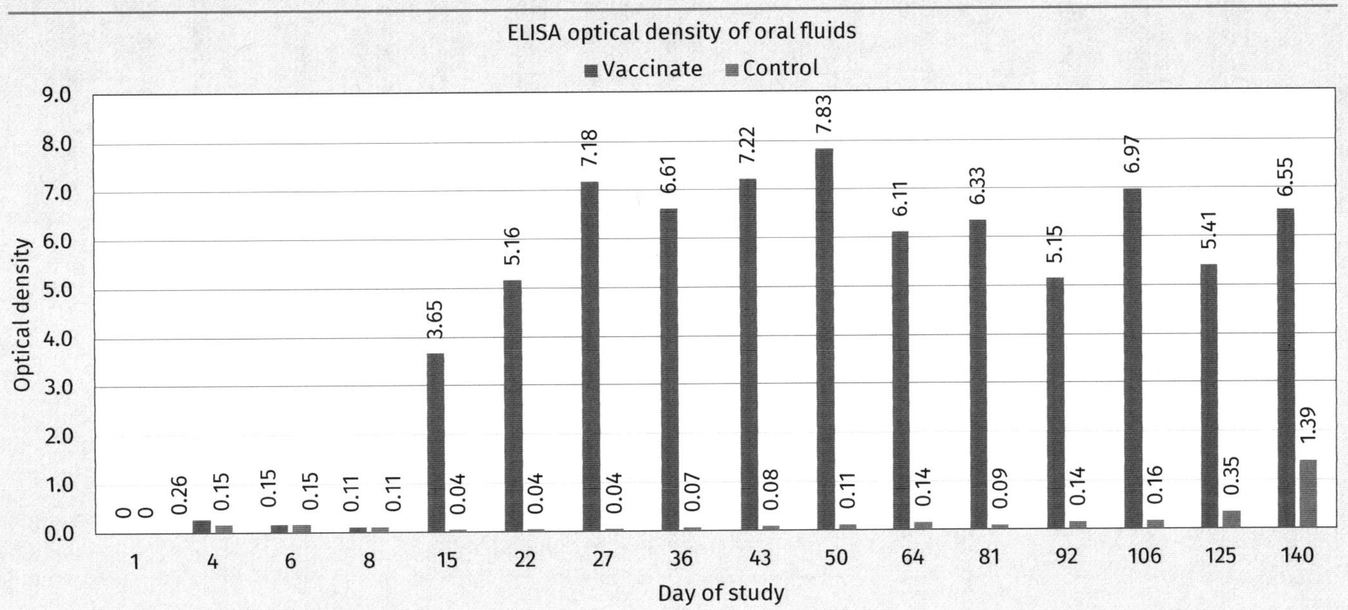

Figure 4: Levels of viremia at various days of study.

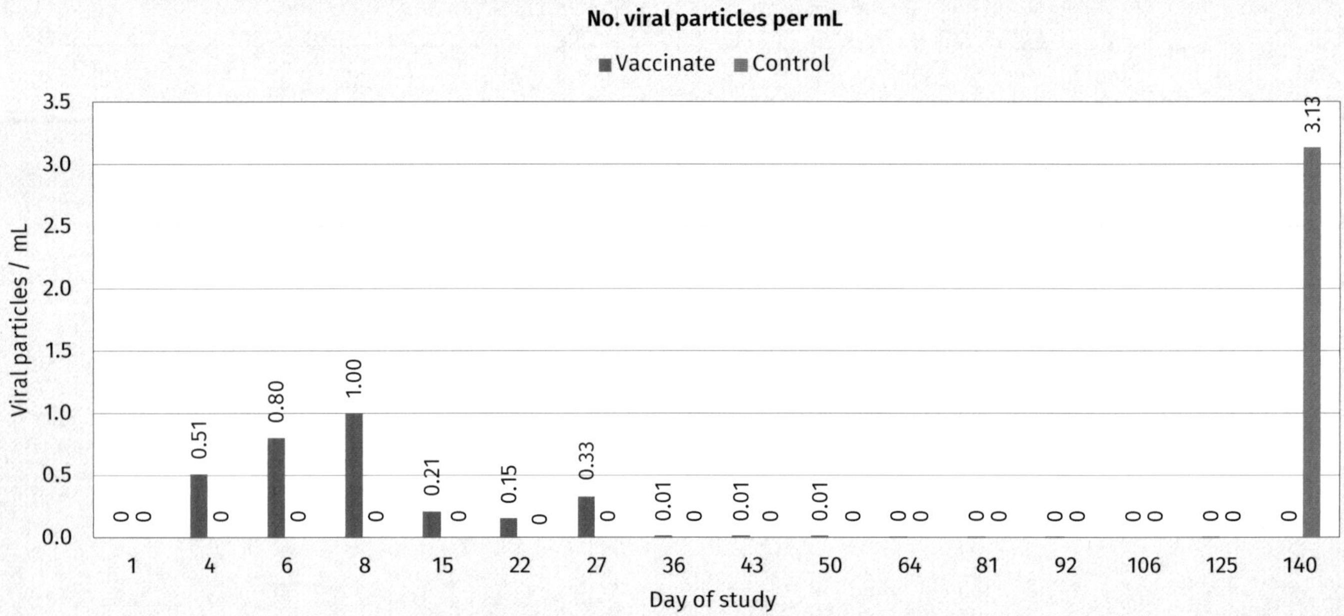

Figure 5: Levels of viremia at various days of study by oral fluids.

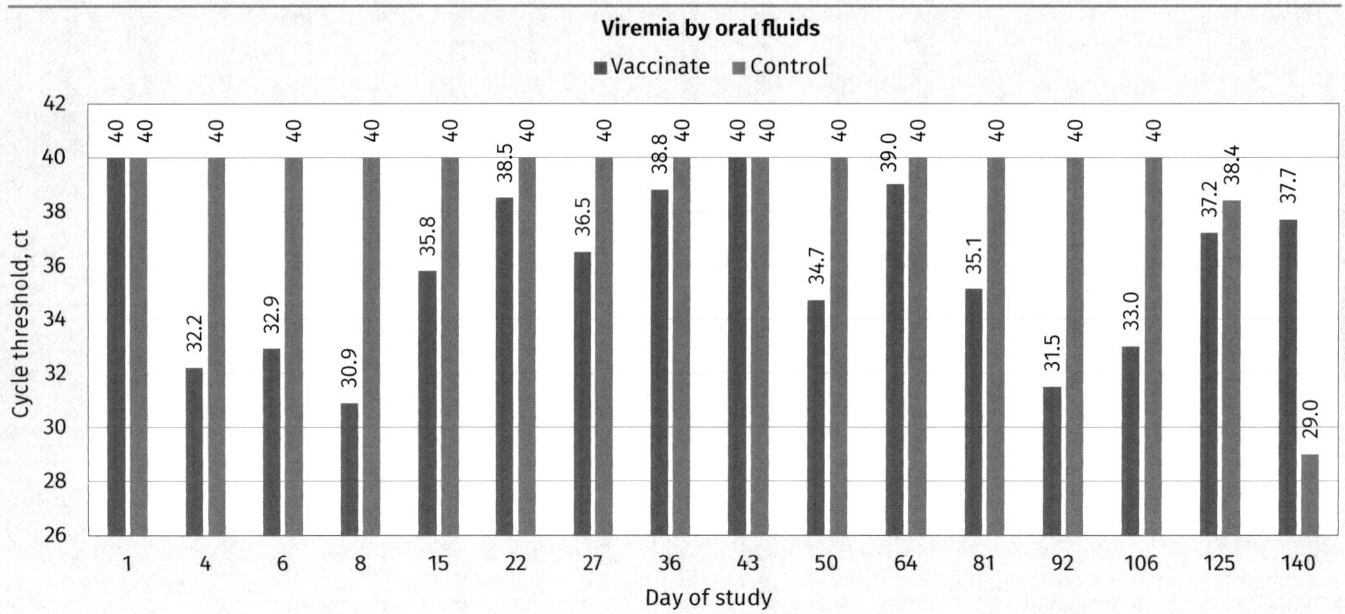

Table 2: Growth performance and mortality

	Treatment			
	Controls	**Vaccinated**	**SEM**	***P*-value**
Weight, lb				
d 0	13.9	13.9	0.15	0.99
d 50	75.1	75.8	0.63	0.27
d 125	232.4	233.1	1.39	0.64
ADG, lb				
d 0 to 50	1.22	1.23	0.01	0.19
d 50 to 125	2.10	2.10	0.01	0.96
d 0 to 125	1.74	1.75	0.01	0.63
Removals, %				
d 0 to 50	2.9	2.8	0.37	0.91
d 50 to 125	5.2	4.8	0.29	0.79

Effects of a single-phase feeding program on growth performance of nursery pigs

H. Cemin, PhD; L. Swalla, MS; J. Pietig, MS; S. Hansen, MS; E. Hansen, MS
Hubbard Feeds, Mankato, Minnesota

Introduction

Nursery diets are typically provided in multiple phases from weaning until approximately 25 kg of body weight (BW) with the purpose of closely meeting the nutrient requirements and physiological capabilities.[1] In the United States, one of the most common feeding programs consists of 3 phases following a feed budget of approximately 1 to 2 kg for phase 1, 5 to 7 kg for phase 2, and 20 to 23 kg for phase 3. Phases 1 and 2 are complex diets formulated with specialty ingredients such as lactose, fermented or enzyme-treated soybean meal, and fish meal, among others, whereas the phase 3 diet is commonly corn and soybean meal-based. While the practice of feeding multiple phases has merit, many production systems have several fill events in the nursery, which in practice makes it challenging to accurately provide the planned feed budgets for all pigs, especially those placed later. Therefore, a simpler feeding program could provide practical benefits to producers. The objective of the current study was to compare a single-phase feeding program, targeted to replace phases 1 and 2, with a traditional feeding program.

Material and methods

A total of 5,090 pigs with an average initial body weight (BW) of 6.0 kg were used in five trials conducted to evaluate the effects of a single-phase feeding program compared to a traditional 2-phase feeding program from weaning until 12 kg of BW. Two treatments consisting of a single-phase program or a 2-phase program were applied for 22 days postweaning. The single-phase program provided a feed budget of 6.8 kg of a diet with 1.38% standardized ileal digestible (SID) Lys. The 2-phase program consisted of a 1.4 kg budget of a phase 1 diet with 1.45% SID Lys and 5.4 kg budget of a phase 2 diet with 1.36% SID Lys. All other nutrients met or exceeded the estimated requirements.[2] The experimental diets were corn, soybean meal, and whey permeate-based. After the experimental period, pigs were fed a common corn and soybean meal-based diet for 23 days. Pens of pigs were weighed on days 0, 22, and 45 to calculate average daily gain (ADG), average daily feed intake (ADFI), and gain-to-feed ratio (G:F). Data was analyzed using SAS software. The initial analysis evaluated trial × treatment interactions; as there was no evidence for differences ($P > 0.10$) for any of the response variables, data from the five trials was combined for further statistical analysis. There was a total of 63 replicates for the 2-phase program and 69 replicates for the single-phase program.

Results and discussion

From day 0 to 22, period during which the experimental diets were provided, pigs fed the single-phase program had higher ($P = 0.041$) ADFI and tended ($P = 0.075$) to be heavier on day 22 (Table 1). There was no evidence for differences ($P > 0.10$) for ADG and G:F. From day 22 to 45, when pigs were fed a common diet, those previously fed the single-phase program had a tendency ($P = 0.057$) for improved ADG. Overall (day 0 to 45), pigs fed the single-phase program had improved ($P = 0.037$) ADG and a tendency ($P = 0.077$) for higher final BW. In summary, the results of the current study demonstrate that a single-phase feeding program fed during the first 22 days postweaning resulted in higher overall ADG and similar ADFI and G:F compared to a 2-phase program. In practice, the main benefit of a single-phase program would be a greater chance of accurately providing the planned feed budget, especially for production systems that have multiple fill days, while maintaining optimal performance and economics.

References

1. Kansas State University. Swine Nutrition Guide. 2019. Available at: **https://www.asi.k-state.edu/research-and-extension/swine/swinenutritionguide/index.html**. Accessed 13 October 2020.

2. National Research Council. Nutrient Requirements of Swine. 11th ed. Washington, DC: National Academies Press. 2012.

Table 1: Effects of phase-feeding program on growth performance of nursery pigs[1]

Item[2]	Nursery program		SEM	Probability, *P*
	2-phase	Single-phase		
BW, kg				
d 0	6.0	6.0	0.03	0.612
d 22	12.6	12.7	0.08	0.075
d 45	24.6	25.0	0.20	0.077
d 0 to 22				
ADG, g	291	296	3.05	0.141
ADFI, g	333	341	2.90	0.041
G:F, g/kg	869	865	5.82	0.611
d 22 to 45				
ADG, g	496	516	7.06	0.057
ADFI, g	756	774	12.98	0.255
G:F, g/kg	655	662	7.35	0.428
d 0 to 45				
ADG, g	395	405	3.99	0.037
ADFI, g	547	559	6.70	0.156
G:F, g/kg	721	725	5.88	0.583

1 A total of 5,090 pigs were used in 5 experiments with 63 (2-phase program) or 69 (single-phase program) replicates per treatment. Experimental diets were fed from d 0 to 22 followed by a common diet from d 22 to 45.

2 BW = body weight. ADG = average daily gain. ADFI = average daily feed intake. G:F = gain-to-feed ratio.

Tetracore ASFV indirect ELISA

Luis G. Gimenez Lirola
Innoceleris, LLC

African swine fever (ASF) is a highly contagious and lethal hemorrhagic viral disease. In absence of vaccine for prevention, ASF control and eradication relies on passive and active surveillance, quarantine of affected farms and slaughter, making this disease economically devastating to swine producers. Due to its varied and unspecific clinical presentation (i.e., from peracute, acute or subacute to chronic or asymptomatic form), and the emergence and co-circulation of less virulent isolates that can cause a milder, nonspecific illness, make this disease difficult to recognize early during an outbreak. Only a combined approach based on early detection and prevention will lead to control of ASF worldwide. In this sense, any effective ASF surveillance plan must be supported by a combination of direct (e.g., Tetracore ASFV real-time PCR) and indirect (e.g., Tetracore ASFV ELISA) laboratory methods for the detection of the infection in acute-subacute or subacute-chronic stages, respectively. Here we present an indirect ELISA, complementary to our real-time PCR, designed to detect specific antibodies against ASF virus in different specimens, including pig serum, plasma, tissue exudate and oral fluids. Key test components (antigen and controls) are recombinant and non-infectious, and all reagents are supplied ready-to-use. The diagnostic performance of this test has been evaluated on a large panel of experimental and field samples (~1000) of known ASFV positive/negative status, including positive samples collected from ASF affected farms in China, and the results will be presented at the meeting.

Impact of a diformate-monolaurate agglomerate in the lactation diet of sows on the performance of suckling piglets

C. Lückstädt[1]; J. Paulsrud[2]
[1]ADDCON GmbH, Bitterfeld-Wolfen, Germany; [2]Apiam Solutions LLC, Sioux City, Iowa

Introduction

It is widely agreed that establishing and maintaining good gut health is an effective strategy against intestinal pathogens and is the cornerstone to competitive animal production. In the past, this strategy was only made possible through the routine use of antibiotic growth promoters (AGPs) in the feed. However, the realization that creating and maintaining a healthy intestinal environment can produce comparable, if not better performance without using AGPs, means that understanding and achieving gut health is essential to productivity and food safety programs alike.

Gut health requires not only the absence of intestinal pathogens, as can be achieved using AGPs, but encompasses questions of effective digestibility and absorption of nutrients, a normal, stable microbiota and a healthy, functional gut mucosa, without inflammation. To this end, it is important to see the gut environment as a living system, influenced by environmental factors, especially those delivered via the feed.

Physiologists have found that maintaining a healthy gut requires up to 25% of the daily protein and 20% of the dietary energy supplied with the feed. This strategy should be carefully planned into the dietary program, in order to not waste resources.[1] Furthermore, the judicious use of feed additives can also support gut health through dietary means.

The application of organic acids and their salts to diets for pigs has been studied extensively for more than 50 years. They have proved especially effective in maintaining growth performance since the ban on antibiotic growth promoters came into effect in Europe. Numerous trials have demonstrated their mode and magnitude of action and established effective doses for piglets, fattening pigs and sows. The use of formic acid and of diformate in particular has been the subject of intense investigation, with the result that we now understand its dose-dependent effect on growth performance and feed conversion in pigs under a range of different environmental conditions and feed formulations.[2] The main mode of action is its antimicrobial effect, which makes it comparable with antibiotic growth promoters; but organic acids also reduce pH in the stomach, which optimizes conditions for pepsin activity; and increases the digestibility of nitrogen, phosphorus and several minerals. This is not only beneficial in sparing nutrients, but it also prevents losses that might otherwise contribute to environmental pollution.[3]

While biosecurity and hygiene in the feed mill and on farm are essential, the acidification of feed ingredients or finished feeds with organic acids also offers considerable benefits to bacterial control. Feed acidification is not only effective within the feed; possibly its biggest benefit occurs within the pig itself.[4]

However, while the antimicrobial impact of organic acids and their salts, including potassium or sodium diformate, is mainly directed against Gram-negative bacteria like Salmonella or *E coli*, medium chain fatty acids (C-6 to C-12) have also been shown to have an antibacterial impact against various Gram-positive bacteria.[5] This is especially true for lauric acid (C-12) and its monoglyceride ester, monolaurate. Lauric acid has the greatest antibacterial activity of all medium chain fatty acids. This effect is magnified if monolaurate is used,[6] making it a promising candidate as an additive or as an alternative to antibiotics for treatment of different diseases.[7]

Despite the well documented impacts of both additives, data on the combined impact of these additives on health and performance in sows and suckling piglets are scarce – and currently only available from Europe. One of such studies[8] reported significant beneficial effects on suckling piglets from a semi-commercial research farm in Germany (Table 1).

The current study therefore tested the effect of a diformate-monolaurate agglomerate in the lactation diet of sows on health and performance of suckling piglets under commercial conditions in the United States.

Materials and methods

The study tested the effect of a dietary agglomerate of sodium diformate with the monoglyceride of lauric acid – monolaurate, in multiparous sows (average parity 1.8) at a commercial farm in South Dakota. Overall, 839 sows took part in the study, lasting from November 2019 until April 2020. The trial was run back-to-back in the farm – the results from November 2019 till January

Table 1: Performance data of piglets from sows fed a lactation diet with or without a sodium diformate - monolaurate agglomerate (Formi® 3G) – modified after Lückstädt[8]

	Control	1.0% Formi 3G	*P*-values
Litter weight at birth [kg]	19.56 ± 3.26	23.67 ± 3.85	0.001
Weight of weaned piglet [kg]	6.91 ± 0.62	7.41 ± 0.68	0.011
Litter weight at weaning [kg]	82.44 ± 7.58	90.78 ± 10.40	0.004

2020 were taken as the negative control, including 414 sows, while data collected between February and April 2020 consisted of 425 sows, which received a commercial lactation diet from one week before farrowing until weaning containing 1.0% of the diformate – monolaurate agglomerate (traded as Formi® 3G, ADDCON). The lactation diet was fed according to a feeding curve within the last days before farrowing, whereas after farrowing *ad libitum* feeding was encouraged. All relevant data for a performance trend analysis (liveborn piglets, losses of liveborn, pre-wean mortality, piglets weaned) according to standards from the National Pork Producers Council (NPPC) have been monitored. Data were analyzed using the t-test and a confidence level of 95% was defined for these analyses.

Results and discussion

Results of the additive usage revealed a strong and significant impact on the performance of the sows (Table 2), as well as the survival rates of the suckling piglets (Table 3).

The number of liveborn piglets per sow and year increased significantly ($P = 0.008$), by almost 9 percent against the control group – while the achieved number of piglets (30.9), was almost 17% higher than the target set by NPPC. The number of weaned piglets in the acidifier group was by 1.7 piglets bigger than the control and was furthermore 12.6% or 3 piglets above the NPPC-target. Additionally, the non-productive days per sow and year were numerically reduced by 3.2 days.

The beneficial impact of the additive was also visible at the survival rates of suckling piglets. Mortality rates, especially between the 2nd day and the 8th day after farrowing, as well as in the piglets above 8 days of age were drastically reduced. Specifically, the mortality in piglets older than 8 days was significantly reduced ($P = 0.02$) by more than 54% compared to the control – and was even 10% better than the target set by the Pork Producers Council, showing again the capability of the sodium diformate-monolaurate additive. Overall, the pre-wean mortality was reduced by 8%, thus leading to higher numbers of weaned piglets, as mentioned above.

Based on the reported data, feeding the acidifier to sows during late gestation and lactation is an efficient tool to improve the profitability of piglet rearing: conditioning the sows; and reducing the impact of pathogenic bacteria on the sow and thus indirectly also to their suckling piglets, as reported previously, at the AASV 2019 meeting. This leads naturally to lower mortalities before weaning, as described in the current study.

We conclude therefore that the usage of Formi 3G can help sow farmers to raise piglets in a sustainable and economic way under mid-west conditions in the United States.

References

1. Hittel J and Lückstädt C (2017): Dietary potassium diformate and medium chain fatty acids – support for a healthy gut in sows during late gestation. Tagungsband 16. BOKU-Symposium Tierernährung 2017, 91-94.

2. Lückstädt C and Mellor S (2011): The use of organic acids in animal nutrition, with special focus on dietary potassium diformate under European and Austral-Asian conditions. Recent Advances in Animal Nutrition – Australia, Vol 18, 123-130.

3. Lückstädt C, Hutter C and Petrovic S (2019): Dietary diformate and monolaurate – support for a healthy gut in sows during lactation – a short review. Proceeding 13th SafePork 2019, 137-138.

4. Lückstädt C, Theobald P and Auerbach H (2010): A review on Salmonella control in fattening pigs through the use of potassium diformate (KDF) – European case studies. Proceedings of the 21st IPVS Congress 2010, 174.

5. Preuss HG, Echard B, Enig M, Brook I and Elliott TB (2005): Minimum inhibitory concentrations of herbal essential oils and monolaurin for gram-positive and gram-negative bacteria. Molecular and cellular biochemistry 272 (1-2), 29-34.

6. Batovska DI, Todorova IT, Tsvetkova IV and Najdenski HM (2009): Antibacterial study of the medium chain fatty acids and their 1-monoglycerides: Individual effects and synergistic relationships. Polish Journal of Microbiology 58 (1), 43-47.

Table 2: Performance data of sows fed a lactation diet with or without a sodium diformate – monolaurate agglomerate (Formi® 3G)

	Control	1% Formi 3G	Difference [%]	Target NPPC
Lactation length [d]	19.5	19.4	n.d.*	n.d.
Liveborn piglets [sow/y]	28.4	30.9	+8.8 ($P = 0.008$)	26.5
Piglets weaned [sow/y]	25.1	26.8	+7.0 ($P = 0.07$)	23.8
Non-productive days [sow/y]	59.1	55.9	-3.2 days	n.d.

* n.d. = not determined

Table 3: Mortality data of suckling piglets from sows fed a lactation diet with or without a sodium diformate – monolaurate agglomerate (Formi® 3G)

	Control	1% Formi 3G	Difference [%]	Target NPPC
Piglet losses, % of liveborn	12.1	9.7	-19.3	10.0
Losses < 2 d, % of liveborn	5.0	5.7	+12.3	6.0
Losses 2-8 d, % of liveborn	5.0	3.1	-36.9	3.0
Losses > 8 d, % of liveborn	2.0	0.9	-54.1 ($P = 0.02$)	1.0
Pre-mean mortality, %	11.7	10.8	-8.0	10.0

7. Rouse M, Rotger M, Piper K, Steckelberg J, Scholz M, Andrews J and Patel R (2005): *In vitro* and *in vivo* evaluations of the activities of lauric acid monoester formulations against *Staphylococcus aureus*. Antimicrobial Agents Chemother. 49, 3187-3191.

8. Lückstädt C (2019): Effects of a blend of sodium diformate and glycerine monolaurate in sow feed during late gestation and lactation on piglet performance until weaning. AASV - 50th Annual Meeting Proceedings 2019, 334.

Effects of vaccination protocols on influenza A virus genetic diversity in co-challenged pigs

C. Li[1], DVM, MS; M. Culhane[1], DVM, PhD; M. Cheeran[1], MVSc, PhD; D. Schroeder[1], PhD; L. Galina-Pantoja[2], DVM, PhD; M. Jansen[2], DVM; D. Amodie[2], MS; M. Mellencamp[2], PhD; M. Torremorell[1], DVM, PhD
[1]College of Veterinary Medicine, University of Minnesota, St. Paul, Minnesota; [2]Zoetis, Parsippany, New Jersey

Introduction

Influenza A virus (IAV) can cause respiratory disease in pigs of all ages, and infected pigs can also promote interspecies transmission as hosts for not only swine-lineage IAV but also avian, and human-lineage IAV strains.[1] Because of its segmented genome, IAV can quickly change its genetic makeup, which often allows it to evade the immune system even in individuals that have been previously exposed to vaccine or to field virus. This process of exchanging gene segments is referred to as reassortment and it can contribute significantly to antigenic shift of IAV.[2]

While reassortment requires host cell infection with more than one IAV, this circumstance is encountered frequently in US swine populations. In fact, a previous study found that 96% of the herds sampled had simultaneous co-circulation of more than one strain.[3]

Very little work has been done to quantify or characterize reassortment of IAV in pigs, especially in the short period of time following dual challenge. This study evaluated the reassortment behavior of IAV in vaccinated and non-vaccinated pigs following a co-challenge model with two different IAV strains.

Materials and methods

The pigs described here were a part of a larger study evaluating different vaccination protocols by using a dual influenza challenge model. Three-week old pigs from a porcine reproductive and respiratory syndrome virus (PRRSV), *Mycoplasma hyopneumoniae,* and IAV-negative sow farm were weighed, randomly allocated to treatment, and vaccinated with live attenuated influenza vaccine (LAIV), commercial whole inactivated influenza vaccine (COM), autogenous whole inactivated influenza vaccine (AUT), or saline (Table 1). On day 36, pigs were challenged by the introduction of two IAV-infected seeder pigs into each room of 10 vaccinated pigs. Seeder pig #1 was infected with an H1N1 Clade 1A3.3.3 gamma IAV and seeder pig #2 was infected with an H3N2 Clade 3.2010.1 human-like IAV. After 5 days of exposure to the seeder pigs, all pigs were humanely euthanized for necropsy and bronchoalveolar lavage fluid (BALF) collection.

Laboratory analyses

Plaque purification to obtain individual viruses[4,5] was performed on BALF samples from up to 4 pigs/treatment group from all treatment groups that had ≥ 2 samples with a viral titer of ≥ 1.75 $TCID^{50}$/mL. A total of 4 treatment groups met this criteria (Table 1). Up to 24 plaques were randomly selected from each sample; a total of two hundred and two plaques were obtained from 13 different BALF samples. After whole genome amplification by SuperScript III One-Step RT-PCR, whole genome sequencing (WGS; Nextseq Illumina sequencing) was performed. Sequences were then manipulated for interpretation through trimming, assembling, annotation, and phylogenetic analysis. Once the sequences were interpreted, each plaque was classified by genotype. The genotypes were classified as non-reassortant (i.e., having all 8 genes from either the original H1N1 or the H3N2 IAV), reassortant (i.e., having individual genes from both the H1N1 and the H3N2 IAV), or mixed genotype (i.e., having full-length gene sequences of the same gene segment from both the H1N1 and H3N2 IAV; e.g., having the full-length H1 and H3 hemagglutinin gene segments sequenced from the purified plaque).

Statistical analysis

The statistical analysis was conducted on the proportion of reassortants by treatment group and did not include plaques classified as mixed genotype. Utilizing R version 3.6.2 (R core team, 2019), the reassortant proportion from different groups was fit into a binomial logistic regression model, allowing for overdispersion.

Results

Of the 202 plaques submitted from 4 treatment groups for WGS, a total of 150 were non-reassortants and contained only the original gene segments of either the H1N1 or the H3N2 challenge strain. Thirty-six plaques were found to be reassortants and further classified into twenty-one distinct genotypes. Additionally, sixteen plaques were found to have mixed gene segments that fell into a total of fifteen different mixed genotypes.

Distribution of genotypes was also compared across treatment groups (Figure 1). When evaluating the percentage of reassortants across treatment groups, it was noted that both groups using a prime-boost regimen (either homologous or heterologous prime boost) saw a significantly lower percentage of reassortants than non-vaccinated, challenged pigs ($P = 0.015$; Figure 2).

Discussion

This pilot study, which consisted of a small number of samples, demonstrated that in a short period of time, six days after a co-challenge with 2 strains of IAV, multiple reassortants with several different genotypes were detected. This substantiates the belief that reassortment of IAV can happen rapidly in swine populations infected with more than one strain of IAV.[6] Additionally, while not all vaccinated groups were able to show a statistical reduction in the proportion of reassortants detected, both prime-boost groups did have a reduced amount of reassortants. This suggests that appropriate vaccination of pigs may aid in the reduction of reassortment in pigs infected with more than one influenza A virus strain. More studies are needed to better evaluate the effects of various vaccination protocols on reassortment to determine under which protocols the likelihood to for reassortment of IAV in pigs will be reduced.

Table 1: Study design and treatment allocations

Treatment group[1]	Vaccination[2] (3 wk /6 wk)	Number of pigs with BALF samples with TCID50/mL = > 1.75	Met criteria for plaque purification? (≥ 2 pigs with BALF samples with TCID50/mL = > 1.75)
T01	COM[3]/COM	3	Yes
T02	AUT[4]/AUT	1	No
T03	AUT/COM	0	No
T04	COM/AUT	0	No
T05	LAIV[5]/None	4	Yes
T06	LAIV/COM	2	Yes
T07	No vac/chal	4	Yes
T08	No vac/no chal	0	No

1 Treatments 1, 2 ,3 ,4, and 7 were in Rooms 3, 5, 7, 9, and 11; Treatments 5 and 6 were in Rooms 8 and 10; Treatment 8 was housed in Room 1.

2 Treatments were applied at 3 and/or 6 weeks; according to label.

3 COM = Commercial vaccine: FluSure XP® a whole, inactivated commercial multivalent vaccine which includes two H3N2 strains (Cluster IV-A and IV-B), one H1N1 strain (Clade 1A.3.2 gamma-2) and one H1N2 strain (Clade 1B.2.2.2 delta-1b).

4 AUT = Autogenous vaccine; a whole, inactivated autogenous multivalent heterologous vaccine which includes one H3N2 strain (Cluster human-like), one H1N2 strain (Clade 1B.2.2.1 delta-1a) and one H1N1 strain (Clade 1A3.3.3 gamma).

5 LAIV = Ingelvac Provenza™, a live-attenuated vaccine with one H3N2 strain (Cluster I) and one H1N1 strain (beta-gamma2-like Clade 1A.2.3-like) that have truncated non-structural influenza (NS-1) proteins.

Figure 1: Number and genotype of plaques by treatment group. R = reassortant. M = mixed.

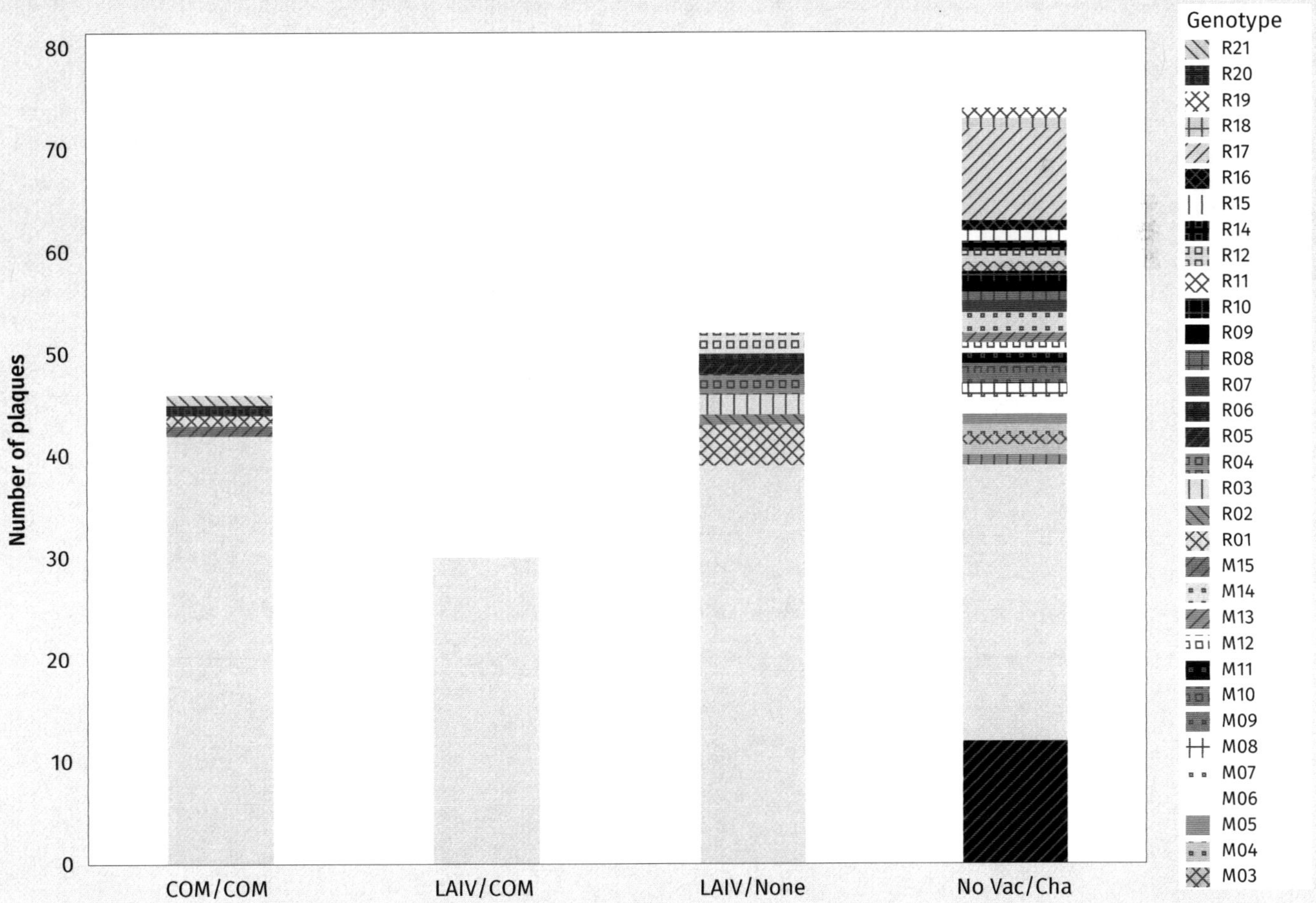

Figure 2: Percent of reassortants and non-reassortants by treatment group.

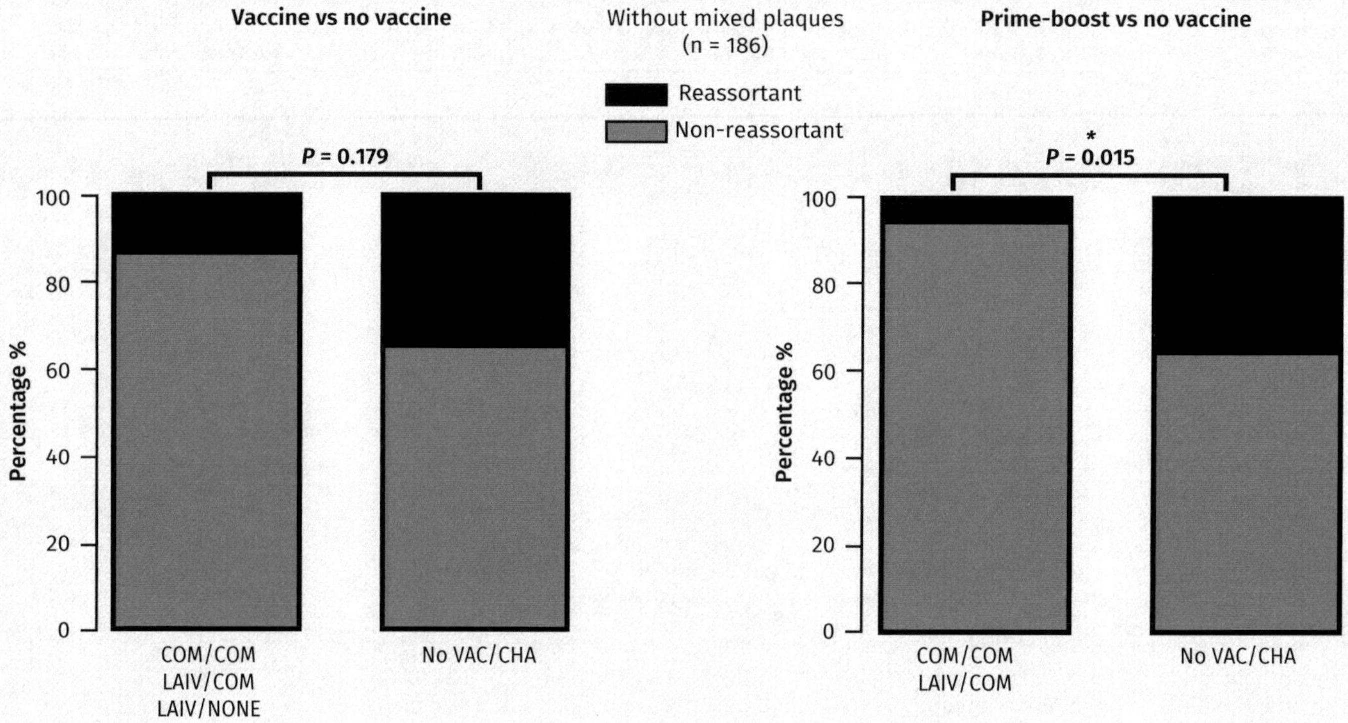

References

1. Zimmerman J, et al. *Diseases of Swine.* Vol. Eleventh edition, Wiley-Blackwell, 2019.

2. Sandbulte,M. R., Spickler, A. R., Zaabel, P. K., and Roth, J. A. (2015). Optimal Use of Vaccines for Control of Influenza A Virus in Swine. *Vaccines (Basel)* 3(1), 22–73. **doi: 10.3390/vaccines3010022**

3. Nirmala J, et al. Is influenza vaccination a key driver of influenza genetic diversity in piglets? 2018 Annual Meeting, American Association of Swine Veterinarians.

4. Murata H, et al. Plaque purification as a method to mitigate the risk of adventitious-agent contamination in influenza vaccine virus seeds *Vaccine.* 2011; 29:3155–3161.

5. Esteban, D. Darwinian Principles Acting on Highly Mutable Viruses. *Virus as Populations: Composition, Complexity, Dynamics, and Biological Implications.* 2016 :73-122.

6. Ma W, et al. The pig as a mixing vessel for influenza viruses: Human and veterinary implications. J Mol Genet Med. 2008;3(1):158-166.

Effects of heterologous prime-boost vaccination against influenza A virus on piglet maternally derived antibodies

C. Mainquist-Whigham[1], DVM, MS; M. Jansen[2], DVM; T. Painter[2], DVM; M. Mellencamp[2], PhD; D. Amodie[2], MS; L. Galina-Pantoja[2], DVM, PhD
[1]Pillen Family Farms, Columbus, Nebraska; [2]Zoetis, Parsippany, New Jersey

Introduction

Influenza A virus (IAV) causes respiratory disease in swine, which can have a significant economic impact on producers. Today, most efforts at controlling influenza focus on the use of vaccination in sows. Both whole herd vaccination and pre-farrow vaccination protocols have been demonstrated to reduce the number of positive piglets detected at weaning.[1] Selection of the appropriate vaccination protocol is often determined by the objective of the influenza control program: reduction or elimination of the virus from the sow herd or mitigation of clinical signs in adult animals is more easily achieved through whole-herd vaccination, while pre-farrow vaccination is a better fit if the primary objective is the protection of piglets post-weaning. Maternally derived antibodies (MDA) against influenza have been previously shown to protect neonatal pigs from clinical disease and reduce virus transmission.[2]

Heterologous prime-boost vaccination is a protocol which involves priming the immune system with one vaccine and boosting with another vaccine.[3] When used in pigs to protect against influenza, heterologous prime-boost has been demonstrated to increase antibody response when compared to conventional (homologous) vaccination.[4] Therefore, utilizing heterologous prime-boost in a pre-farrow vaccination protocol may increase antibody production in sows, resulting in higher MDA in piglets and reduced clinical disease and transmission in piglets consuming their colostrum. The objective of this study was to evaluate two pre-farrow vaccination protocols: heterologous prime-boost vs. conventional, on piglet MDA levels at weaning as determined by hemagglutination inhibition (HI) testing.

Materials and methods

One hundred twenty-five mixed age and parity sows in an IAV positive, *Mycoplasma hyopneumoniae* negative, and porcine respiratory and reproductive syndrome virus (PRRSV) positive commercial sow herd were randomly allocated to one of three treatment groups by parity (Table 1). All sows were part of a single breeding group, and gilts were not enrolled in the study. An RNA particle vaccine (RPV) had been administered to all adult animals in the site on day -71 of the study. Diagnostic tests of post-mortem deep tracheal swabs collected on day -15 detected an H3N2 Cluster IV-A influenza virus. On study day 0, vaccines (RPV or FluSure XP®) were administered to sows according to their respective treatment groups. After the sows farrowed and lactated for approximately 21 days, serum was collected from the three largest pigs in each litter regardless of sex. Serum was submitted for HI testing against two different IAV strains: a homologous HI test against the resident H3N2 cluster IV-A strain, as well as an alternative strain not contained in either vaccine used for the study (an H1N1 beta clade). Seroprotection in piglets was defined as an HI titer of ≥ 40.[5]

Statistical analyses

HI titer data were transformed by the $\log_{10}(\text{titer}+1)$ transformation prior to statistical analysis to stabilize the variance and normalize the data. These data were analyzed by a general linear mixed-model approach using the Proc Mixed Procedure (SAS 9.4, Cary, NC). Transformed titers were analyzed with a model that considered the fixed effect of treatment and the random effects of parity, block(parity), sow(parity block treatment) and the residual error. Least Squares Means (LSMeans) were calculated for each treatment group and compared by the two-sided Student's t-test at the 5% level of significance. HI titer LSMeans were then back-transformed and presented as geometric means.

Seroprotection was defined as a binary variable (yes = 1, no = 0) and analyzed by a linear mixed model approach. Using the SAS Proc Glimmix procedure, seroprotection was analyzed with a model that considered treatment as a fixed effect and parity, block(parity), sow(parity block treatment) and the residual error as random effects. The analysis utilized a binomial error and logit link. The 5% level of significance was used to assess statistical differences.

Results

The heterologous prime-boost (RPV/FSXP) and homologous prime-boost (RPV/RPV) groups had significantly higher HI titers compared with the controls (RPV/None; $P < 0.0001$ and < 0.0043, respectively; Figure 1). The heterologous group also had significantly higher titers compared with the homologous prime-boost group ($P < 0.0001$). This was true not only for the HI titers against the resident H3N2 Cluster IV-A strain, but also the H1N1 beta strain (Figure 1). The percentage of seroprotected animals (HI titer ≥ 40) was also evaluated (Figure 2). Again, RPV/FSXP had significantly higher percentage of protected pigs compared to either RPV/RPV or RPV/None against both IAV strains evaluated ($P = 0.0001$). RPV/RPV was significantly higher than RPV/None against H1N1 ($P = 0.0001$) but not against H3N2 ($P = 0.0722$).

Discussion

In the context of this study, heterologous prime-boost vaccination of sows pre-farrow resulted in higher antibody levels in their piglets at the time of weaning than homologous prime-boost vaccination or controls. This is consistent with results from other studies evaluating antibody production following heterologous prime-boost vaccination compared with more conventional homologous prime-boost vaccination.[4,6,7] However, previous studies have not always shown a clear advantage of heterologous prime-boost vaccination for every influenza strain evaluated. One difference between the vaccine protocols used in this study and the previously mentioned studies is the time interval between the prime and boost doses. Additionally, the two vaccines used in this study

Table 1: Study design and treatment allocations

Treatment group	Vaccination[1] (Whole herd /4 weeks pre-farrow)	Vaccination protocol type	Number of pigs in each treatment group/room
T01	RPV[2]/RPV	Homologous prime boost	50
T02	RPV/FSXP[3]	Heterologous prime boost	50
T03	RPV/None[4]	Control	25

1 The first vaccination was applied to all animals as a part of a whole herd vaccination on study day -71; the second vaccination, when applicable, was applied 4 weeks pre-farrow. All animals received 2 mL of FarrowSure Gold® (Zoetis) at 4 weeks pre-farrow, regardless of treatment group, per normal herd vaccination protocol.

2 RPV = RNA particle vaccine (Merck); a prescription vaccine selected for this herd and containing IAV Cluster 4 H3, Human-like H3, Pandemic H1, Alpha H1, and delta H1.

3 FSXP = FluSure XP® (Zoetis) a whole, inactivated commercial multivalent vaccine which includes two H3N2 strains (Cluster IV-A and IV-B), one H1N1 strain (Clade 1A.3.2 gamma-2) and one H1N2 strain (Clade 1B.2.2.2 delta-1b).

4 Control animals in T03 did not receive any influenza vaccine pre-farrow but did receive 2 mL FarrowSure Gold®.

Figure 1: Geometric mean hemagglutination inhibition (HI) titer of piglets at weaning.

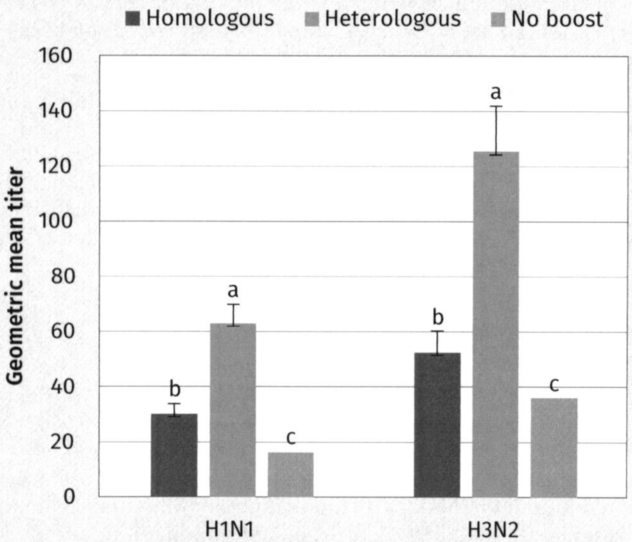

a,b,c Values with different letters are significantly different at $P \leq 0.05$

Figure 2: Percent of seroprotected piglets (hemagglutination inhibition titer ≥ 40) at weaning.

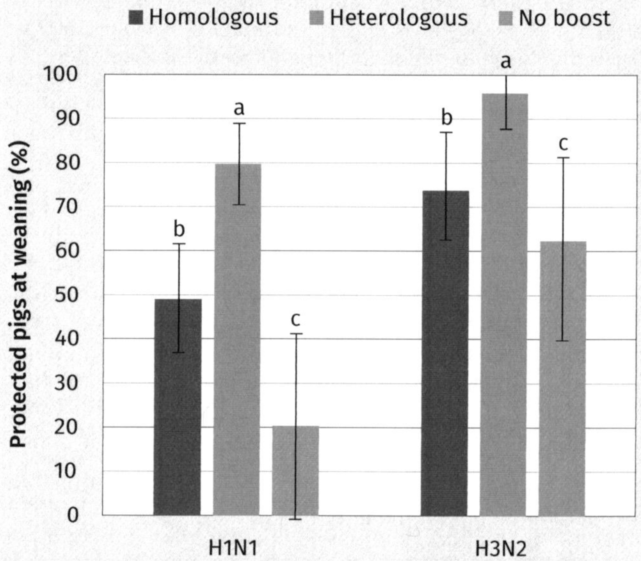

a,b,c Values with different letters are significantly different at $P \leq 0.05$

differed more from each other than in the other studies. For example, one vaccine is an adjuvanted, whole inactivated virus vaccine, and the other is a non-adjuvanted, RNA particle vaccine. We cannot be sure if either, or both, of these differences from previous protocols might contribute to the clear demonstration of the benefit of heterologous prime-boost. It is clear that heterologous prime-boost vaccination has promise for eliciting a higher antibody response against influenza in pigs. Further research is warranted to determine optimal timing and products to utilize.

References

1. Chamba Pardo FO, et al. Effect of influenza A virus sow vaccination on infection in pigs at weaning. *Transbound Emerg Dis.* 2020; 00:1.

2. Torremorell M et al. Influenza prevalence in piglets at weaning and its effect on infection dynamics and clinical disease after weaning. 2017; James D. McKean Swine Disease Conference.

3. Woodland D. Jump-starting the immune system: prime-boosting comes of age. *Trends Immunol.* 2004; 25:98.

4. Van Reeth K, et al. Heterologous prime-boost vaccination with H3N2 influenza viruses of swine favors cross-clade antibody responses and protection. *NPJ Vacc.* 2017; 2:11.

5. Hiroi S et al. Trivalent influenza vaccine-induced antibody response to circulating influenza a (H3N2) viruses in 2010/11 and 2011/12 seasons. *Hum Vaccines Immunother.* 2015; 11:386.

6. Li C, et al. Exploring heterologous prime-boost vaccination approaches to enhance influenza control in pigs. *Vet Res.* 2020; 51:89.

7. Jansen M, et al. Effects of heterologous prime-boost vaccination against influenza A on antibody response in sows. 2020; Allen D. Leman Conference.

The efficacy of two detergents for sanitizing swine farrowing stalls

K. Nisley[1], BS; S. Tousignant[2], DVM, PhD; M. Schleper[3], DVM
[1]Iowa State University, College of Veterinary Medicine, Ames, Iowa;
[2]Boehringer Ingelheim Animal Health USA Inc., Duluth, Georgia; [3]Christensen Farms, Sleepy Eye, Minnesota

Introduction

Sanitation in farrowing stalls is critically important for the health of piglets. The objective of sanitizing rooms is to reduce pathogen load in addition to establishing a hygienic environment for piglets. Unsanitary conditions can limit growth performance through an induction of the immune response. The use of chemical degreasers is important to breakdown bacterial biofilm, a naturally occurring collection of microbial cells on a surface that are enclosed in an extracellular polymeric substance (EPS) consisting of polysaccharides, proteins, phospholipids, and other polymeric substances. Eliminating biofilm allows access to pathogens to effectively inactivate them with disinfectants. The objective of this study was to utilize an adenosine triphosphate (ATP) bioluminescence meter (Hygiena™, Camarillo, California), conventional bacterial culturing, and polymerase chain reaction (PCR) analysis to compare the effectiveness of two degreasers for sanitizing swine farrowing stalls.

Materials and methods

The study was conducted at a single 2,500 sow farm. Six all-in all-out farrowing rooms were used, three were treated with a sodium-hydroxide based degreaser (MS TopFoam Power, MS Schippers, Bladel, Netherlands) and the other three were treated with an acidic degreaser (Barnstorm™, Preserve International, Reno, Nevada) which had previously been used on the farm. In each room, five different stalls were randomly assigned to be sampled by utilizing an excel random-number generator. Samples were collected prior to washing, after detergent application and washing, after disinfection, and the next morning. Each stall was sampled at the back-right corner of the creep area on the floor and also sampled on the vertical wall in the same area. Two 4-inch diameter rings were utilized for sampling, one for ATP bioluminescence and one for culture and a rotavirus panel RT-PCR. Samples were collected with swabs (Stuart's Media swabs, Starplex Scientific, Etobicoke, Ontario) using a systematic swabbing pattern to ensure complete coverage. The culture swabs were scored based on microbial growth, scored 0-4. The ATP meter used determined residual organic matter, expressed in relative light units (RLU). Each phase of the sanitation process was timed, and the amount of degreaser used was measured.

Results and discussion

Both treatments reduced bacteria and ATP levels after disinfection, while both had increased rotavirus C levels (Table 1). TopFoam decreased ATP levels by 7.1% more than Barnstorm while Barnstorm decreased bacterial levels by 8.9% more than TopFoam from before application to after disinfection. There was variance in RLUs across rooms, possibly indicating different levels of bacterial biofilm breakdown, subsequently exposing more ATP and increasing RLUs. Rotavirus C PCR results were similar between both Barnstorm and TopFoam, with an increase of positive results by 43% and 47% respectively. These results do not necessarily indicate viable virus, rather an incomplete removal of rotavirus C components from stalls. On average, TopFoam decreased wash time by 26 minutes and decreased product usage by 597 mL. In order to fully appreciate the trends, more replicates are needed. Further investigation into the successive use of each degreaser each turn could provide valuable information on each degreaser's ability to breakdown the biofilm and more effectively sanitize surfaces. However, breaking down biofilm creates anomalies, such as increased rotavirus C PCR levels. Further studies into parameters such as number of scouring piglets, wean weights, and pre-wean mortality could clarify results. Overall, the degreasers performed similarly with short-term use, however, could perform differently if used on a long-term basis.

Table 1: Mean culture scores, mean percent PCR positive results across all three rooms (suspect on diagnostics was treated as positive), and mean relative light units (RLU) for Barnstorm and TopFoam

	Barnstorm™	TopFoam
	Mean culture scores	
Dirty	3.4	3.2
After power-wash	2.1	2.5
After disinfectant	1.3	1.6
Next morning	0.8	0.5
	Mean PCR % Positives	
Dirty	47%	53%
After power-wash	80%	80%
After disinfectant	90%	90%
	Mean RLU	
Dirty	684.9	494.0
After power-wash	533.5	809.6
After disinfectant	355.8	221.7

Quantifying sow lameness with a time-of-flight camera

Matthew Boulanger; Thomas Parsons, VMD, PhD
University of Pennsylvania School of Veterinary Medicine

Introduction

Public concern for the welfare of gestating sows is driving changes in husbandry practices, pushing producers from individual to group housing of sows. However, increased incidence of lameness is often reported with group housed sows. While group housing can make lameness more noticeable, inter-animal aggression during group formation can also lead to increased leg injuries. Lameness is both a welfare and production problem, as lameness is the second most common cause of premarket culling.[1] However, early detection remains challenging as existing metrics can often lack inter-observer reliability.[2] Thus, any new approach that offers reliable early detection of lameness promises better therapeutic outcomes and improved sow longevity, productivity, and welfare.

Objective

To develop and test approaches to automated lameness detection in sows utilizing topography above the spine detected via infrared cameras and computerized image analysis.

Materials and methods

Twenty-five sows (DNA Line 241) ranging from 0 to 4th parity were trained with feed rewards to run an obstacle course. The course consisted of two adjacent feeders separated by a concrete wall – forming a 13 [m] "U" shape.[3] A Time of Flight (ToF) infrared camera was hung 2.4 [m] above the course. By pointing the camera straight down, a topography of the sow's back was generated, as depicted in Figure 1. The camera was controlled through Matlab to collect data at 20 Hz. A visual camera also recorded trials, wherein pigs ran automatically between feeders for 15 minutes. Trial videos were then presented to a clinician in random order. The clinician graded each video with a tagged visual analog score (tVAS),[4] unaware of the numeric grade. Instead, the tVAS scale presented a spectrum of clinical signs to describe lameness, and the clinician placed a slider along the spectrum. Later, the slider distance was quantified on a 0 to 150 scale. Furthermore, half the videos were randomly selected to be duplicated and horizontally mirrored – testing clinician consistency. The infrared video topography was processed to automatically identify sow spines using curvedness. The elevation of the spine was then fit to a second order polynomial.

Results

The standard deviation of tVAS scoring on mirrored videos ± 4.7 or less than a 3% error. Spine topography was successfully fit to second order polynomials with a mean R^2 of 0.72 ± 0.17. The "a" and "b" coefficients of each polynomial fit were extracted from five separate frames, and the average was compared to tVAS scores. Using this, an "a" coefficient threshold of -0.0258 [mm^2 $pixel^{-1}$] was identified- below which a pig could be considered lame. Using this diagnostic, the "a" coefficient had a sensitivity of 0.785 and a specificity of 0.909. Similarly, a "b" coefficient threshold of 4.12 [mm $pixel^{-1}$] was determined, above which a pig was considered lame. The "b" coefficient had a sensitivity of .785 and a specificity of 0.818.

Discussion

ToF camera development has accelerated in recent history,[5] and integration into mass marketed facial recognition systems has driven down its price. This technology is no longer cost-prohibited for on-farm use and automated detection of lameness promises to be a boon for the management of group housed sows, especially in farms utilizing electronic sow feeding where the sows could be easily monitored by the camera as they exit the feeder each day. The algorithms employed here for lameness detection on ToF images yielded good specificity but were not very sensitive. This is the opposite to what is needed for a screening tool that intends to identify the early onset of lameness where a sensitivity is needed. Future work will explore additional image analysis techniques to see if they yield a more sensitive approach to automated lameness detection with a ToF camera.

References

1. Zhao, Yunxiang, et al. "Analysis of reasons for sow culling and seasonal effects on reproductive disorders in Southern China." *Animal reproduction science* 159 (2015): 191-197.

2. MK Pierdon, TD Parsons. "Reliability of a four point sow lameness score across multiple observers." 2015 Allen D. Leman Swine Conference.

3. Bos, E-J., et al. "Effect of locomotion score on sows' performances in a feed reward collection test." *Animal* 9.10 (2015): 1698-1703.

4. Nalon, Elena, et al. "Comparison of the inter-and intra-observer repeatability of three gait-scoring scales for sows." *Animal: an International Journal of Animal Bioscience* 8.4 (2014): 650.

5. He, Ying, et al. "Depth errors analysis and correction for time-of-flight (ToF) cameras." *Sensors* 17.1 (2017): 92.

Figure 1: Depiction of experimental methods and resulting data. The left subfigure is a picture of the arena used to train pigs in a feed-reward collection test. Trained sows traveled between feedboxes, depicted by red and blue crosses. The corresponding arrows represent the path of the sow around the concrete barrier – outlining the "U" shape of the obstacle course. Furthermore, the red cone to the left of the left subfigure depicts the Time of Flight camera used to record pigs during trials. It was hung vertically – pointed straight down to generate the topography of sow's backs. The right subfigure contains a sample frame of data, collected with the Time of Flight camera. The right subfigure has two different perspectives of the same dataset: (top) a top down perspective, and (bottom) an angled perspective. It should be noted that the spine, which was identified automatically using curvedness, was highlighted in red within both images. Furthermore, a pig icon in the center designates anatomic elements of the datasets. In both the top and bottom figures, the pig's head (cranial) is facing towards the left. Similarly, the tail (caudal) is to the right, with corresponding arrows to the pig icon.

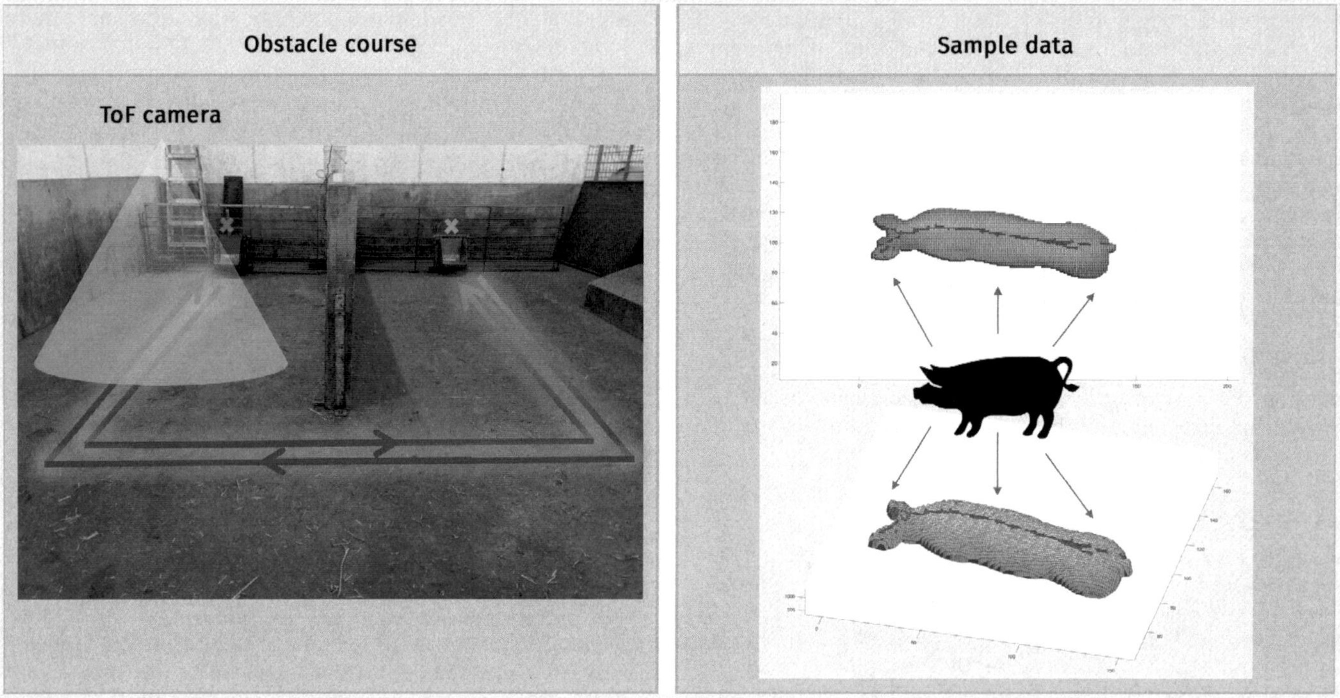

Pilot study to validate passive transfer in piglets

Lucas Buehler[1]; Roy Edler[1], MS; Scott Dee[2], DVM, PhD, MS

[1]The Ohio State University, Columbus, Ohio; [2]Pipestone Applied Research, Pipestone, Minnesota Introduction

Identifying failure of passive transfer is common in other species, but has not been well-documented in swine. Multiple factors can influence the prevalence of neonatal diarrhea including failure of passive transfer, pathogen load, and inadequate environments. The goal of this pilot study was to collect preliminary data to support the potential use of immunoglobulin levels as measured by an immunocrit ratio to evaluate failure of passive transfer in piglets to create a reference value for future work. An immunocrit ratio is defined as the ratio of the precipitate length divided by the total length of diluted serum, and has been previously used to assess failure of passive transfer in other species, but not validated in swine.[1] Therefore, this study was designed to provide reference value to predict neonatal diarrhea using immunocrit levels.

Materials and methods

The study was conducted on a commercial sow farm. Four piglets were selected from 50 different litters. Two piglets were selected from the first six piglets born and the other two piglets were selected from the last six piglets born. Ten piglets were selected for a control group from different litters prior to suckling and places in a separate pen on a 50% dextrose solution. This group was established to document the levels of immunoglobulins in colostrum deprived piglets. All piglets were tagged and bled at 0, 12, 24, 36, and 48 hours of life. Each tube of blood was spun down with a centrifuge and serum was collected. The serum was mixed with 40% ammonium sulfate solution and mixed at a 50:50 ratio with serum from each piglet. The mixture was aspirated into microhematocrit tube and centrifuged. The immunocrit ratio was then recorded from microhematocrit tube. Clinical scoring of piglets with neonatal diarrhea were recorded for five days post parturition, to statistically look for a trend in neonatal diarrhea piglet with certain immunocrit levels.

Results and discussion

The mean immunocrit levels were higher in piglets with evidence of neonatal diarrhea versus piglets with no evidence of neonatal diarrhea (Figure 1). At 24 hours, colostrum-deprived piglets had immunocrit levels averaged 0.0035, while piglets that received colostrum averaged at 0.1215 (P value = 0.6527). While failing to provide reference values, this study demonstrated the use of the immunocrit technology to identify colostrum-deprived pigs. However, due to a limited sample size statistical differences could not be detected. Strengths of this study include having a colostrum-deprived control group to have a baseline for time zero immunocrit levels for colostrum-deprived piglets. Also, the immunocrit testing is rapid, inexpensive, and is able to be performed onsite. A limitation for this study was a lack of taking a time zero baseline for the piglets that received colostrum. In future studies, conducting this research on a larger scale with more data points would be advised.

Reference

1. Peters et al. Reference values for immunocrit ratios to assess maternal antibody uptake in 1-day-old piglets. AASV. 2016;24:36-41

Figure 1: Average Iimmunocrit ratio in neonatal diarrhea and non-neonatal diarrhea piglets over time.

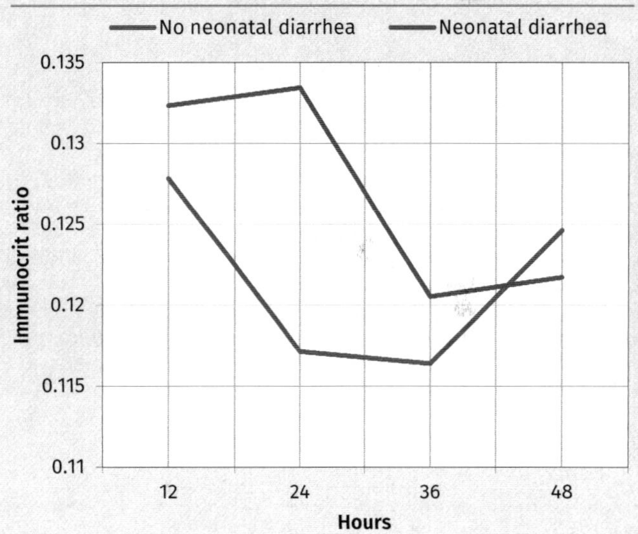

A comparison of two different trailer styles on stress and meat quality in commercial finishing hog transport

M. Boucher[1]; C. Templeton[1], DVM; C. Werth[1], MS; L. Wormsbecher[2], MS
[1]Southwest Veterinary Services (Synergy Services), Listowel, Ontario; [2]Conestoga Meat Packers, Breslau, Ontario

Introduction

In the past decade, livestock transportation in North America has gained increased attention among stakeholders in the industry and from the public due to perceived and real effects on animal welfare, food safety, meat and carcass quality.[1] The objective of this study was to evaluate the effect of on-farm loading techniques between the newly designed Berdex™ trailer and the conventional potbelly using an observational behavioural and handling scoring system during loading in conjunction with pre-slaughter cortisol sampling and post-slaughter meat quality measures (Minolta colour indication and presence of bruising/blood splash/two-tone/PSE). It was hypothesized that the Berdex™ trailer would reduce stress during loading and transportation to slaughter compared to a conventional potbelly trailer, leading to improved animal welfare and meat quality.

Material and methods

The study was conducted on 6 commercial finishing farms and at Conestoga Meat Packers, Breslau, ON. Each farm shipped 4 loads of market hogs to slaughter, 2 on the Berdex™ trailer and 2 on the potbelly. For each shipment, loading time was recorded as well as rate of prod use and pig turnaround (0 = none, 1 = low, 2 = medium, 3 = high) to quantify handling and loading behaviour. These two rates were then combined to assign each trailer a loading score deemed to reflect how well the pigs moved overall onto the trailer. Upon arrival at the slaughter plant, hogs were unloaded and placed in holding pens. At this point, ropes were hung to collect a pooled sample for salivary cortisol measurement. After slaughter, meat quality testing was conducted on randomly selected loins from the carcasses of both trailers to assess Minolta, bruising, blood splash, two-tone meat and PSE measures.

Results and discussion

The Berdex™ trailer held an average of 230 pigs per load compared to the conventional trailer, which held an average of 177 pigs per load. Although the conventional trailer held fewer pigs, the average on farm loading time was 4 minutes longer compared to the Berdex™ trailers. The Berdex™ trailer had a significantly higher load score of 1.25 compared to the conventional trailers load score of 0.975 ($P < 0.05$), meaning prod use and pig turnaround was greater during Berdex™ loading (see Table 1 for values). Mean cortisol, Minolta levels and the average count of PSE variables were not significantly different between the conventional and Berdex™ trailers. The Berdex™ trailer represents important advantages to the swine industry, with faster loading times and more hogs per shipment. Understanding production and handling efficiencies are important to swine practitioners, so they can assist producers with animal welfare and production decisions.[2] The higher loading score achieved by the Berdex™ is likely due to producer novelty of loading these trailers, therefore future research should be geared at addressing this.

References

1. Schwartzkopf-Genswein KS et al. Road transport of cattle, swine and poultry in North America and its impact on animal welfare, carcass and meat quality: a review. *Meat Sci*. 2012;92:3.

2. Geverink NA et al. Effects of regular moving and handling on the behavioural and physiological responses of pigs to preslaughter treatment and consequences of subsequent meat quality. *J Anim Sci*. 1998;76:8.

Table 1: Comparison of loading characteristics between conventional and Berdex™ trailers

	Conventional trailer	Berdex™ trailer
Average number of pigs/load	177	230
Compartment 0	36.1	77.3
Compartment 1	55.9	64.3
Compartment 2	85.6	88.4
Average load time on farm (mm:ss)	39:12	35:12
Average prod use	1.99	2.24
Average pig turns	0.12	0.27
Average load score**	0.975	1.25

** significant $P < 0.05$

Shiga toxin-producing *Escherichia coli* (STEC) detection and characterization among US swine herds

K. Hewitt[1], BS; J. Brown[1], DVM; K. Skoland[1], BS; M. Nickel[1], DVM; C. Ruston[1], DVM; M. Breuer[1], BFA; E. Burrough[2], DVM, PhD, DACVP; D. Sperling[3], DVM, PhD; L. Karriker[1], DVM, MS, DACVPM
[1]Swine Medicine Education Center, Iowa State University College of Veterinary Medicine, Ames, Iowa; [2]Iowa State University Veterinary Diagnostic Laboratory, Ames, Iowa; [3]Ceva Sante Animale, Libourne, France

Introduction

Shiga toxin (Stx2e)-producing *Escherichia coli* (STEC) is one of the most important *E coli* pathotypes in animals and is the etiologic agent of edema disease (ED) in pigs.[1] ED has been reported to have a mortality rate from 50- 90% in herds and can often reoccur.[2] ED still occurs, despite vaccination attempts.[3] A review of historical diagnostic data from a veterinary diagnostic laboratory, revealed that approximately 19% of *E coli* isolates from pigs genotyped from 2009 to 2019 were positive for the Stx2e gene. This study investigated the presence of fimbrial and toxin genes from hemolytic *E coli* isolates of swine to estimate the frequency of STEC in US swine herds. Isolates were further characterized by colony morphology on blood agar (BA) plates and antimicrobial susceptibility.

Materials and methods

Rectal swabs were collected from commercial pigs suspected to be clinically affected by ED (neurologic deficits). Pigs ranged from three to nine weeks of age and were located at 47 different sites; each site was strategically sourced from 13 production companies across the United States. Five clinical pigs were sampled from three, non-adjacent pens for a total of 15 samples per site. Individual pig care criteria was used to identify acute and subacute pigs displaying the following clinical signs: depression, anorexia, central nervous system deficits and/or diarrhea. The producer captured ancillary information including site information, *E coli* history, clinical signs and individual fecal consistency and color. All samples were submitted to a veterinary diagnostic laboratory for bacterial isolation. Once *E coli* was isolated and identified, hemolytic isolates with different colony morphologies (rough, smooth, or smooth-mucoid) were selected to represent all positive pens from each site. This subset of isolates were further characterized by genotyping via polymerase chain reaction (PCR) and antimicrobial susceptibility testing (AST) via broth microdilution.

Results and discussion

Two hundred and thirty-eight of the 690 isolates (34%) were hemolytic on BA, as expected for STEC isolates. One hundred and ten of the 238 hemolytic isolates (46%) were selected systematically by pen to have a mixture of colony morphology for PCR and AST. Twenty-one of the 110 isolates (19%) were characterized as STEC, 16 of the 21 isolates (76%) were PCR positive for the Stx2e toxin gene and the F18 fimbrial gene. The remaining five isolates carried the Stx2e toxin gene, but no fimbrial genes for attachment. The 16 isolates originated from 9 sites (19% of sites); none which vaccinated with a F18 *E. coli* vaccine. As indicated in Table 1, all 16 STEC (F18 + Stx2e) isolates had smooth or smooth-mucoid colony morphologies. AST results were similar to historic diagnostic susceptibility data with significant resistance to several labeled antimicrobials. If a producer has reoccurring *E coli* breaks with high mortality and neurological clinical signs considerations should include, utilizing a vaccine that contains a F18 component and incorporating AST results to guide antimicrobial selection. In congruence with historical diagnostic data, disease-capable STEC is present in approximately 15% of the herds sampled in this study. US veterinarians should focus on *E coli* isolates that are hemolytic on BA with smooth and smooth-mucoid colony morphologies when considering further diagnostic evaluation of suspected STEC cases. Resistance patterns continue to support the need for additional management strategies.

References

1. Gyles, C. L. Pathogenesis of Bacterial Infections in Animals Edited by Carlton L. Gyles ... [et Al.]. 4th ed. Ames, IA: Wiley-Blackwell, 2010, 267-279.

2. Casanova, N. A., Redondo, L. M., Dailoff, G. C., Arenas, D., & Fernández Miyakawa, M. E. (2018). Overview of the role of Shiga toxins in porcine edema disease pathogenesis. Toxicon: official journal of the International Society on Toxinology, 148, 149–154. **https://doi.org/10.1016/j.toxicon.2018.04.019**

3. Zimmerman, Jeffrey J., Locke A. Karriker, and Alejandro Ramírez. Diseases of Swine / Edited by Jeffrey J. Zimmerman, Locke A. Karriker, Alejandro Ramirez, Kent J. Schwartz, Gregory W. Stevenson, Jianqiang Zhang. Eleventh ed. 2019:807.

Table 1: Morphology, genotyping and antimicrobial susceptibility results of the STEC (F18 + Stx2E) isolates

	Number of STEC isolates (F18 + Stx2E)		Number of susceptible STEC isolates (F18 + Stx2E)
Morphology		**Antimicrobial**	
Smooth	7/16	Ampicillin	9/16
Mucoid	0/16	Ceftiofur	16/16
Smooth-mucoid	9/16	Clindamycin	0/16
Gene		Danofloxacin	NI*
STb (toxin)	14/16	Enrofloxacin	3/16
STa (toxin)	9/16	Florfenicol	6/16
K99 (pilus)	0/16	Gamithromycin	NI*
LT (toxin)	14/16	Gentamicin	5/16
F18 (pilus)	**16/16**	Neomycin	4/16
987P (pilus)	0/16	Penicillin	0/16
K88 (pilus)	0/16	Sulfadimethoxine	2/16
F41 (pilus)	0/16	Spectinomycin	NI*
Stx2e (toxin)	**16/16**	Tetracycline	0/16
EAST1 (toxin)	7/16	Tiamulin	0/16
Stx1 (toxin)	0/16	Tildpirosin	NI*
Stx2 (toxin)	16/16	Tilmicosin	0/16
AIDA (adhesin)	0/16	Trimethoprin/Sulphametoxazole	0/16
EAEA (adhesin)	0/16	Tulathromycin	NI*
PAA (adhesin)	2/16	Tylosin	NI*

* NI = no interpretation available based on antimicrobial, organism, species, and tissue combination

Evaluation of initial pig weight and weaning age as indicators of nursery mortality and growth

B. Kitting[1], BS; M. Valles[2]; J. Flohr[2], PhD; R. Robbins[2], DVM, PhD; D. Drebes[2], DVM; M. de Souza[2], MS; F. Zarate[2], MS; R. Schmitt[2], PhD; R. Moreno[2], PhD
[1]University of Pennsylvania School of Veterinary Medicine, Philadelphia, Pennsylvania;
[2]Seaboard Foods, Guymon, Oklahoma

Introduction

Emphasis on increasing the percentage of full value pigs that market through a commercial swine production system continues to be a driving factor for producer profitability; especially given recent over supply of secondary cull markets. The presence of suboptimal pig populations in a production system can divert resources that could be allocated towards more likely full value animals. Many swine producing companies have established stringent weight, age, or quality criteria for a wean pig to be placed in a nursery. Other producers are more conservative and only maintain euthanasia criteria based on welfare and quality criteria. The objective of this study was to characterize performance of wean pigs in the nursery based on weight at placement and weaning age.

Materials and methods

Approximately 1,200 pigs, across 100 litters within a single farrowing room at a PRRSV unstable sow farm, were individually tagged and information associated with their litter environment and dam were recorded. Within the following 24 hours, pigs were weaned to the nursery. On the same day as placement in the nursery, each pig was weighed and characterized individually based on clinical signs and observations. All small, sick or at-risk pigs were separated into ICU pens and treated according to the producer's specific SOP's. During the nursery turn, each pig received diets and feed budgets, along with vaccinations and medications according to current producer's SOP's. Pigs were monitored and mortality was recorded by recovering tag number, weight, sex, and cause of death as determined by necropsy. After 48 days in the nursery, each pig was weighed within twenty-four hours of scheduled placement into the finisher. Data were analyzed as a cohort study with mixed effect models in the lme4 package of R. Models evaluating the fixed effects of initial weight and weaning age. Pen within the nursery room was included as a random effect. Non normally distributed responses (mortality, clinical observations) were analyzed as binomial distributions. Contrast statements for linear and quadratic polynomials for increasing initial weight and wean age were tested. Results were considered significant at $P < 0.05$ and a tendency at $P < 0.10$

Results and discussion

Two-way interactions associated with the fixed effects were not statistically significant ($P > 0.15$). Initial weight in the nursery ranged from 2 to 18 pounds. Figure 1 illustrates predicted growth rate and mortality based on initial weight in the nursery. Heavier initial weight significantly increased nursery average daily gain (ADG) and shipment weight to the finisher (Quadratic, $P < 0.01$), and reduced mortality ($P < 0.01$). Weaning ages observed within the cohort study ranged from 16 to 21 days. Increasing wean age resulted in heavier initial weights at placement (Linear, $P < 0.02$), increased nursery ADG (Linear, $P < 0.01$) and shipment weight to the finisher (Linear, $P < 0.01$): however, it did not impact nursery mortality ($P = 0.65$). Overall, probability of mortality and the direct correlation of initial placement weight and weaning age on nursery growth performance need to be assessed by producers to determine the probability of pigs reaching full value or secondary cull market statuses, and to establish meaningful profitability expectation.

References

1. Roberts, J. Commercial nursery mortality and weaning weight. *Proc Allen D. Leman Swine Conf.* 2000: 194-195

2. Larriestra, A. J., Wattanaphansak, S., Neumann, E., Bradford, J., Morrison, R., & Deen, J. Pre-existing conditions as predictors of mortality and slow growth in nursery pigs. *Proc Allen D. Leman Swine Conf.* 2002: 231-233

Figure 1: Average daily gain and mortality predicted by initial placement weight

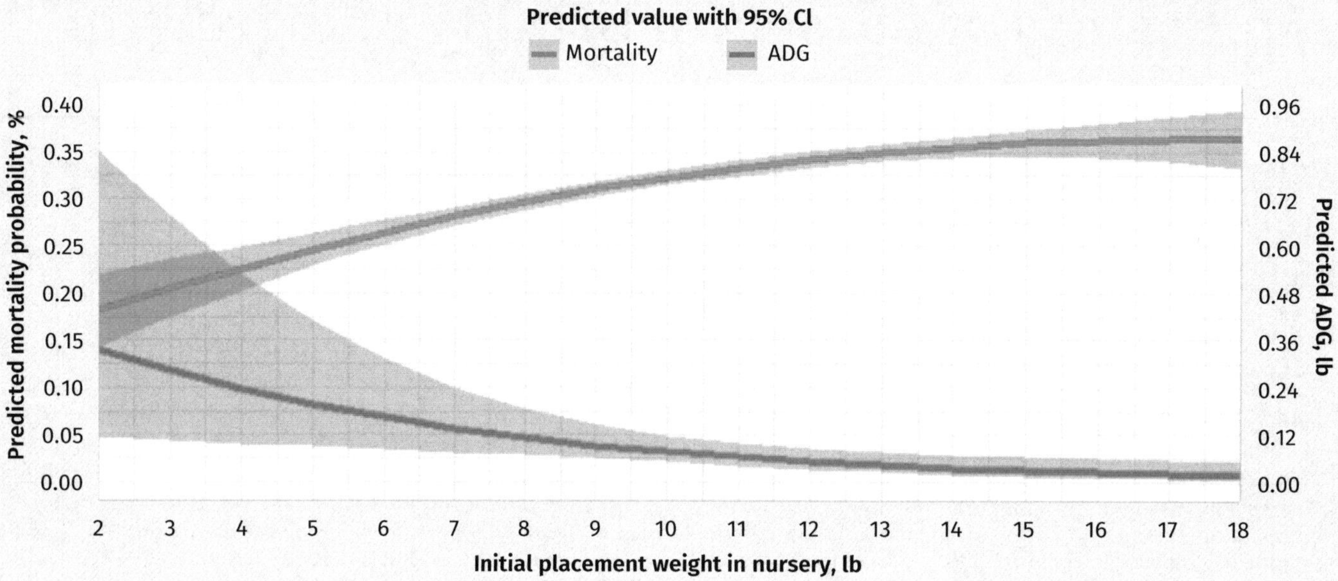

Developing a disease outbreak strategy: a survey of swine slaughterhouse location, capacity and limitations

Ashley Hallowell, MS; Meghann Pierdon VMD, DACAW
University of Pennsylvania School of Veterinary Medicine

Introduction

Depopulation and disposal methods are used as a rapid response for an animal health emergency. Despite planning for depopulation and disposal methods as they relate to African swine fever and other foreign animal diseases, COVID-19 supply chain disruptions have made clear the need for more practical strategies. One solution could involve the recruitment of small slaughter facilities near farms; however, little is known about these processors. Veterinarians and farmers are in need of more information. Therefore, an investigation of slaughterhouse location, capacity and limitations will guide producers and veterinarians in identifying establishments that could help with depopulation and disposal during an emergency situation.

Objective

To assess the location, capacity and limitations of swine slaughterhouses in order to develop potential strategies for use of these facilities when faced with a market disruption or disease outbreak requiring depopulation and disposal of swine.

Materials and methods

Data from the Meat, Poultry and Egg Product Inspection Directory was used to identify establishments conducting swine slaughter within Pennsylvania, New Jersey, New York, Ohio, Delaware, Maryland and West Virginia. A survey was designed to capture management perspectives regarding average and maximum capacity of swine slaughter and potential limitations for reaching maximum capacity. Establishments were called from the beginning of July to the end of August 2020 and were asked if they were willing to participate in a telephone interview. For those that agreed to participate in the telephone interview, information was collected in an online survey (Google Forms) and compiled in an electronic spreadsheet (Excel 2016; Microsoft, Redmond, Washington). Farms and their nearest slaughterhouse were mapped in order to determine distance between farms and slaughterhouses.

Results

Survey results of swine slaughter capacity and limitations are reported in Table 1. Out of 154 establishments across five states, 63 responses (40.9% response rate) were received. Ninety seven percent of swine slaughterhouses that participated in the survey were small establishments, killing less than 100 hogs a week and on average 20.3 hogs a week. The average reported maximum kill capacity at small swine slaughterhouses was 38.5 hogs a week. When asked if hours of swine slaughter could be increased, 33.9% said yes. Limitations for reaching maximum capacity include USDA inspection hours, staffing, facility space (i.e., holding pen and freezer capacity) and high customer demand. The most common suggestion from 24.3% of respondents was to increase the number of inspected swine slaughterhouses in the area. When assessing the distance between farms to their nearest slaughterhouse, farms were located on average 17.4 miles away. The average farm capacity was 2,073 hogs in comparison to the maximum weekly kill capacity for small and mid-large size establishments at 253.9 hogs.

Discussion

The distance between swine farmers and slaughter facilities is not a barrier to using these facilities during an emergency. There is however a mismatch between the number of pigs and the slaughter capacity of the facilities. The nearby slaughterhouses may have difficulty upholding depopulation efforts of these farms during a market disruption or disease outbreak. This survey identified a few areas for increasing slaughterhouse capacity including increasing the inspection hours and taking temporary measures to increase staffing and holding capacity for live animals and freezer space for these facilities. In conclusion, these establishments could be a critical part of the infrastructure needed in planning for depopulation and disposal of swine.

Table 1: Survey questions and responses regarding slaughter capacity and limitations

	No. reporting	% reporting
What is the average weekly kill capacity for pigs at your establishment? (n = 62)		
1-10	21	33.9%
11-100	39	62.9%
101-1,000	0	0%
> 1,000	2	3.2%
What is the maximum weekly kill capacity for pigs at your establishment? (n = 63)		
1-10	14	22.2%
11-100	44	69.8%
101-1,000	3	4.8%
> 1,000	2	3.2%
How many hours a week does the establishment kill pigs? (n = 63)		
1-8	46	73.0%
9-18	12	19.1%
19-40	4	6.4%
> 40	1	1.6%
How many hours a week does the establishment kill other livestock or poultry? (n = 61)		
1-8	33	54.1%
9-18	9	14.8%
19-40	19	31.2%
> 40	0	0%
Could these hours be increased? (n = 62)		
Yes	21	33.9%
No	41	66.1%
How many pigs can the establishment hold prior to slaughter? (n = 58)		
1-50	42	72.4%
51-100	11	19.0%
101-1,000	4	6.9%
> 1,000	1	1.7%
What stunning strategies are used by the establishment? (n = 59)		
Electric	28	47.5%
Captive Bolt	19	32.2%
Gun Shot	10	17.0%
CO_2	2	3.4%
Do you have any comments you would like to add that you think might be important to our understanding of pig slaughter capacity in the region? (n = 37)		
Limited number of slaughterhouses	9	24.3%
Limited staffing	16	43.2%
Limited space	6	16.2%
Government oversight	7	18.9%
High customer demand[1]	8	21.6%

[1] Survey was conducted during COVID-19 pandemic

Phylogenetic analysis of spillover events between swine and human influenza A virus subtype H3N2 between 2014 and 2019 in the United States

D. Kinkade; Suzanna Storms, DVM; Benjamin Blair, DVM; James Lowe, DVM, MS
Applied Epidemiology Laboratory, University of Illinois College of Veterinary Medicine, Urbana, Illinois

Introduction

The surveillance of influenza A virus (IAV) has become an extremely vital aspect in the human health sector and the swine health industry. New IAV strains can be achieved by inter-species transmission, antigenic mutations (drift), and genetic reassortment due to the exchange of genetic segments between two or more IAVs.[1] These characteristics create an extremely genetically variable virus. While these novel strains may have multiple outcomes, there is always the risk of a strain gaining pandemic potential that would have significant public health and economic implications. Previous research has shown that the H3N2 subtype has a high potential for cross-species transmission given its more recent triple reassortant nature and therefore calls for close monitoring.[2] Gaining a better understanding of what these IAV strains are, how they are changing, and what their genetic stability is within our swine herds is imperative to protecting our industry from future outbreaks and economic loss.

Materials and methods

H3N2 sequences collected in human and swine hosts from July 2014 – July 2019 were collected from NCBI's Influenza Virus Database. The HA RNA segment sequences from these strains were then aligned using a MAFFT alignment program. The aligned sequences were then run through IQ-TREE 1.6.12 to produce maximum likelihood trees with an SH-aLRT test and an ultrafast bootstrap with 10,000 replicates for each influenza season.[3] Trees were then viewed and annotated by lineage and origin using known reference strains via FigTree 1.4.4. All swine-origin strains from each year and three human reference strains were then selected, aligned by MAFFT, and ran through the program RDP4 to find and remove any recombinant strains. That file was then used to create another tree using the same methods described above in IQ-TREE and FigTree. The FASTA file was then put into BEAST 1.10.4 and separated into its respective clades and ran under strict clock, uncorrelated relaxed clock, random local clock, and fixed local clock models and under constant size, exponential growth, and Bayesian skyline plot estimations. After selecting the best fitting model based on marginal likelihoods of these models, evolutionary rates were determined using the Bayesian MCMC method implemented in BEAST 1.10.4.[4]

Results and discussion

The 2014-2015 influenza season showed the 2010 human-like swine lineage with no single spillover events. The 2015-2016 influenza season also showed the 2010 human-like swine lineage with no single spillover events. The 2016-2017 influenza season shows the 2010 human-like lineage, the newly emerged 2016 human-like lineage, and one single human to swine spillover event. The 2017-2018 influenza season shows the 2010 human-like lineage, 2016 human-like lineage, and two single human to swine spillover events. The 2018-2019 influenza season showed the 2010 human-like lineage, 2016 human-like lineage, and two single human to swine spillover events. Further research for past and future influenza strains would be beneficial for a better understanding of how these strains are changing. Continued genetic surveillance and analysis of IAV is crucial to ensure the health and safety of swine and human populations.

References

1. Chiapponi C, Ebranati E, Pariani E, et al. Genetic analysis of human and swine influenza A viruses isolated in Northern Italy during 2010-2015. Zoonoses Public Health. 2018;65(1):114-123. **doi:10.1111/zph.12378**

2. Ma W, Lager KM, Lekcharoensuk P, et al. Viral reassortment and transmission after co-infection of pigs with classical H1N1 and triple-reassortant H3N2 swine influenza viruses. J Gen Virol. 2010;91(9):2314-2321. **doi:10.1099/vir.0.021402-0**

3. B.Q. Minh, H.A. Schmidt, O. Chernomor, D. Schrempf, M.D. Woodhams, A. von Haeseler, R. Lanfear (2020) IQ-TREE 2: New models and efficient methods for phylogenetic inference in the genomic era. Mol. Biol. Evol., 37:1530-1534.

4. A. J. Drummond, S. Y. W. Ho, M. J. Phillips and A. Rambaut (2006) Relaxed phylogenetics and dating with confidence. PLoS Biology 4(5): e88.

A comparison of time, product usage and dose validation features between automed® and conventional dosing syringes during wean pig shots

N. Benge[1]; L. Glowzenski[2], MsED, VMD; J. Morgan[3], DVM; L. Magyar[4]; Thad Hinkle[3]; D. Polson[3], DVM, PhD
[1]Iowa State University College of Veterinary Medicine, Ames, Iowa; [2]TriOak Foods, Oakville, Iowa;
[3]Boehringer Ingelheim Animal Health USA Inc., Duluth, Georgia; [4]automed®, Huxley, Iowa

Introduction

The automed® syringe (AMS), a battery-powered delivery device, differs from conventional dosing syringes (CDS) for its ergonomics, ability to log treatments with or without RFID tags, and following this study, ability to detect air in the syringe and track injection location, in-or-out of a pig. Currently, the manual CDS is regularly used in the industry and presents challenges recording treatments and tracking compliance. The objective of this study was to compare the dosing accuracy and injection time between the AMS and CDS during wean pig shots, and develop a function for the AMS to detect injection location. Knowledge gained from exploring these objectives will help practitioners and producers determine if the AMS is a valuable alternative to the CDS.

Materials and methods

To investigate dosing accuracy, product for injections was divided into 2 groups (1-AMS, 2-CDS), and weighed before and after injection. Using product density, the actual product used (APU) was determined from the change in volume. The percentage of projected product usage (%PPU) was calculated using the equation: APU (mL)/projected product usage (mL) = %PPU.

To study the time required to give a group of wean pig injections, the injections were divided into 3 groups (1-AMS with RFID, 2-AMS without RFID, 3-CDS). For group 1, a Full Duplex (FDX), low-frequency RFID tag (Allflex USA Inc., Texas) was scanned before each crate. The time for two people to give a group of wean pig injections was recorded for all three groups, and converted to time per 100 pigs.

To develop the injection location function of the AMS, treatments were divided into 2 groups (1-In Pig, 2-In Bottle). An algorithm was then used to "train" the AMS on injection location. The AMS predictions were then compared to written data of injection location.

Microsoft Excel® (Microsoft Corporation, Redmond, WA) was used for statistical analysis.

Results and conclusions

The results show conventional syringes consistently administer above the projected dose, while a 1mL dose from the AMS administered below the projection, and a 2mL dose administered at almost the exact projection (Figure 1). An explanation for a low 1mL dose volume in the AMS may be in the capacity of the AMS syringe. The plunger location is determined based on reference points, and resolution is lower for a 1mL compared to 2mL dose. Both the AMS and CDS were found to be within ISO 7886-1 tolerance for syringes.[1]

This study has shown that the time for the AMS with RFID, AMS without RFID, and CDS to inject 100 pigs, on average, takes 12 min 00 sec (n = 22,SD = 1 min 41 sec), 10 min 34 sec (n = 29, SD = 1 min27 sec), and 10 min 25 sec (n = 26, SD = 1 min 26 sec), respectively. When using RFID, administration time increased, but time saved in recording treatments should be considered. Notably, there is a shorter read-range of the RFID tag used in this study compared to other low-frequency tags and ultra-high frequency tags, and future studies should explore other types of tags with longer read-range.

For individual injections, the AMS correctly identified injection location on 82.3% of attempts (n = 989) and reported if any air was detected (± 10% administered). Across a session, the device was able to estimate the group coverage within 8.6% of the actual coverage.

Overall, when comparing AMS to CDS, practitioners can expect benefits of automated record keeping and compliance, with comparable dosing accuracy and administration time. The AMS may be a valuable alternative to the more commonly used CDS.

Reference

1. Phillips P. Syringe accuracy (TOGC) estimates for ISO 7886-1 1, 5, & 10 mL. Surgical Materials Testing Laboratory. 2016.

Figure 1: Comparison of product usage between automed® syringes (AMS) and conventional dosing syringes (CDS).

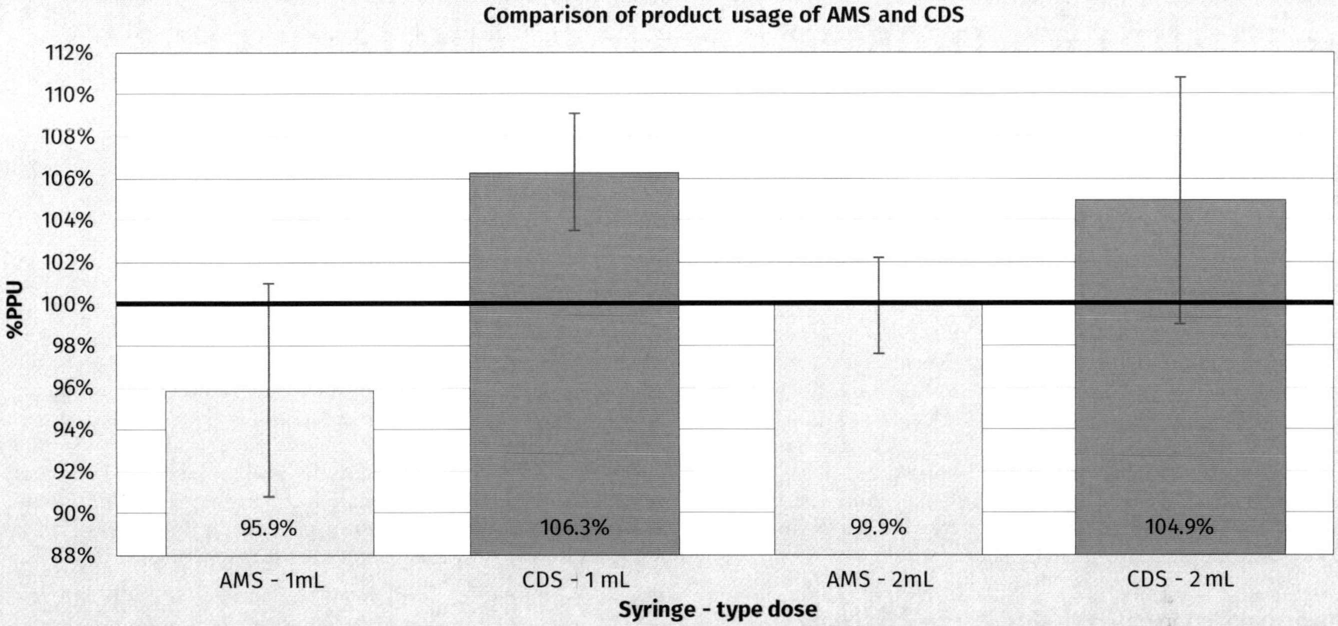

Syringe type	Dose	n	Minimum	Average	Maximum	SD
AMS	1	7	88.0%	95.9%	103.0%	5.1%
CDS	1	5	102.5%	106.3%	110.0%	2.8%
AMS	2	7	96.9%	99.9%	104.5%	2.3%
CDS	2	3	99.0%	104.9%	110.9%	5.9%

Determining incidence and developing a case definition for muscle necrosis and lameness found in high health flow

R. Weidmayer[1]; D. Murray[2]; L. Johnson[2]; A. Sponheim[3]; T. Michels[3]
[1]University of Minnesota, Minneapolis, Minnesota; [2]New Fashion Pork, Jackson, Minnesota;
[3]Boehringer Ingelheim Animal Health USA Inc., Duluth, Georgia

Statement of problem

Lameness in commercial pig production has always been a source of welfare concerns and economic losses. Many infectious diseases such as *Erysipelas rhusiopathiae,* and *Mycoplasma hyosynoviae,* are proven to cause lameness in various ages of pigs in addition to other noninfectious causes. However, despite being naïve to most infectious agents that cause lameness and only minor occurrences of non-infectious lameness, a flow located in Wisconsin has had significant mortality due to lameness.

Description of herd and case history

The flow consists of a 1,630 head sow farm with an on-site nursery and gilt development unit (GDU) as well as six 2,400 head wean-to-finish sites. The sow herd was fully stocked from one source at start-up and has remained closed since. The flow follows Global Animal Partnership (GAP) protocol. The herd is naïve to *Mycoplasma hyorhinis, M hyosynoviae,* and *Glaesserella parasuis. Actinobacillus suis* and *Erysipelothrix rhusipathiae* have not been isolated from joints. Antibiotics administered to wean-to-finish pigs removed from the GAP program has not improved lameness. Growing pigs are from two different sire lines.

To evaluate lameness prevalence, all pigs in the wean-to-finish, sow pens, and gilt development unit were counted and classified as either grade 1, 2, or 3 on three separate occasions. Grade level descriptions were as follows: Grade 1: pigs slowly walked off lameness or slightly limped, Grade 2: obviously lame, Grade 3: non-weight bearing. Swollen joints were also counted. To develop a case definition, samples were submitted from each site, including: whole leg, longissimus muscle, tonsil, lymph nodes, heart, lung, diaphragm, 2nd rib, liver, spleen, kidney, small intestine, large intestine, serum, and joint swabs.

Significant findings

During the observation periods, lameness presentation peaked during the early finisher stage and was elevated in the sow pens and GDU. Figure 1 shows the incidence of lameness based on age of the pig. A greater proportion of male pigs were lame than female pigs. Prevalence of swollen joints did not increase with lameness prevalence. Lameness presentation occurred prior to tail biting at all wean-to-finish sites. There was a substantial increase in the number of lame pigs due to injury while marketing loads.

Interstitial hemorrhage of skeletal muscle was occasionally observed on gross necropsy. Muscle histology observations: scattered swollen muscle bundles, usually mild interstitial hemorrhage to sometimes moderate-severe, muscle necrosis with clusters of macrophages and phagocytosis, inflammation of diaphragm, skeletal muscle, and cardiac muscle. Cartilage persistence in the diaphysis and misaligned columns, low bone density and bone ash was observed on bone samples. Serum analysis show low vitamin A, E, D, low total protein, and high phosphorus. Vitamin D level only approached normal levels during supplementation via water medication. Low levels of Vitamin A, and D were observed in the liver. Porcine parvovirus type 2 was consistently identified. No known infectious cause of lameness was identified on submitted tissues.

Take home messages

Infectious agents known to cause lameness were not identified in any of the samples collected previously or during the study sampling periods. Lameness was observed more frequently in males than females, and tail biting increased the level of lameness, but was not the primary cause. Further work should be done to understand the significance of muscle necrosis, as well as improper bone formation, and the contribution to lameness in both growing animals as well as mature sows and gilts.

Figure 1: Percent of the population classified as lame by pig age (weeks) during observation periods in June, July, and August.

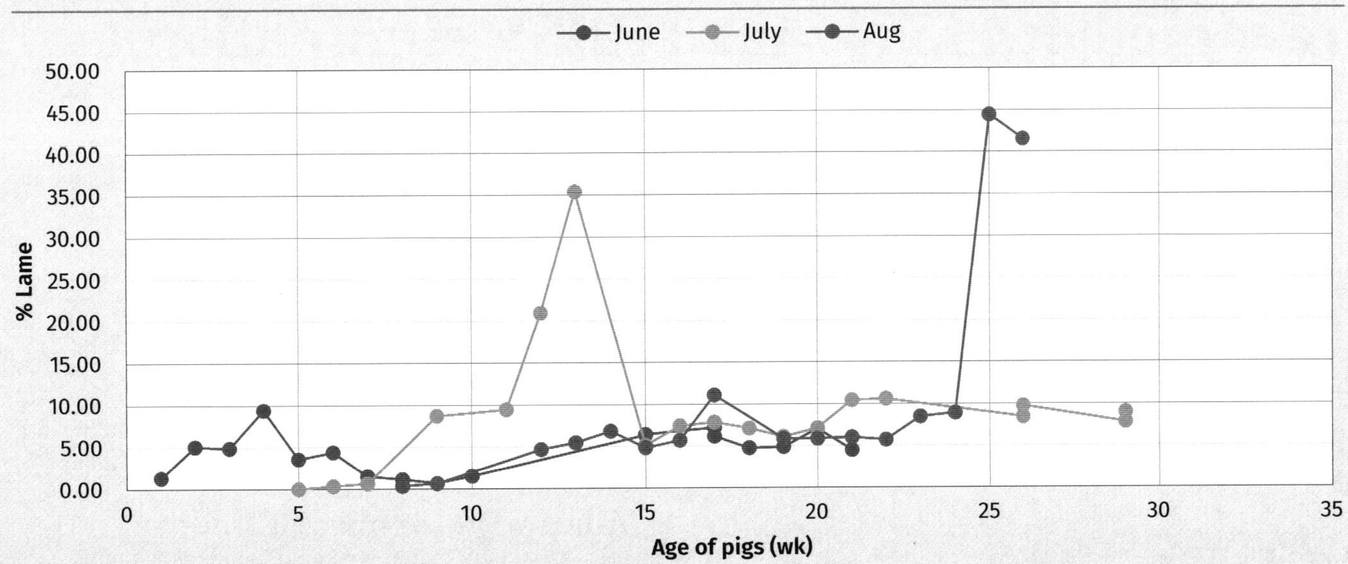

Investigating intermittent PRRSV detection in a breeding herd following PRRSV elimination

Katyann Graham[1]; Kate Dion[2], DVM; Kylie Glisson[2], DVM; Marcelo Nunes-De-Almeida[1], DVM, MS; Amanda Sponheim[3], DVM; Jianqiang Zhang[1], MD, MS, PhD
[1]Iowa State University, Ames, Iowa; [2]The HANOR Company, Enid, Oklahoma;
[3]Boehringer Ingelheim Animal Health USA, Inc., Duluth, Georgia

Statement of the problem

Following a porcine reproductive and respiratory syndrome virus (PRRSV) elimination program, intermittent PRRSV shedding in a sow farm may occur, and sows may produce PRRSV-positive litters over multiple farrowings. Clinical signs are not always observed when PRRSV is detected in biological samples, such as processing fluids (PF), and thus the presence of live infectious virus is unknown.

Case history

On week 35 (2018) a new 2,700 sow farm was stocked with PRRSV positive gilts that were previously administered two doses of PRRSV modified live virus (MLV) vaccine and one dose of live virus inoculation (LVI) and started farrowing on week 51 (2018). The farm finished entering gilts on week 3 (2019) and started a herd closure and PRRSV-elimination program. On week 50 (2019), after 12 consecutive weeks of PRRSV RT-PCR negative PF and 8 weeks of PRRSV RT-PCR negative due-to-wean ear blood swabs, the farm was eligible to receive gilts. On week 14 (2020), the farm began receiving PRRSV MLV vaccinated gilts. The following week, the farm started shipping all weaned sows to another facility and began receiving MLV vaccinated parity one sows. Weekly PF testing was continued from week 50 (2019) and on week 24 and 27 (2020) processing fluids tested PRRSV RT-PCR positive.

Data/diagnostics

Eleven litters (week 24) and 15 litters (week 27) were pooled in the two PRRSV RT-PCR positive PF samples. Individual piglet oral swabs were collected (week 24 and 27) and pooled per litter. One litter from each week tested PRRSV RT-PCR positive. Individual blood, lung and tonsil samples were collected from piglets in positive litters and PRRSV was detected by RT-PCR from 2/11 week 24 piglets (serum and tissue) and 1/6 week 27 piglets (tissue). The two sows from the positive litters were euthanized and serum and all tissue samples were PRRSV RT-PCR negative. Virus isolation using MARC-145 and ZMAC cells from the positive tissue and serum samples were negative.

Sow farrowing histories from week 24 and 27 were paired with weekly PF results and showed that sows contributing to the positive PF from this case could have contributed to previous PRRSV RT-PCR positive pooled PF results.

A bioassay was performed in a BSL-2 facility, inoculating naïve piglets with tissue homogenate and PF from week 24 and week 27 PRRSV positive cases. Results are shown in Table 1.

Significant and applicable findings

The pigs inoculated with tissue homogenate in the bioassay became PRRSV RT-PCR positive. This shows that weekly testing of PF after PRRSV elimination may be used to detect intermittent shedding of infectious PRRSV in the absence of clinical signs. Extensive testing allows identification of sows who are contributing to the positive PF samples and could facilitate a test and remove strategy to prevent further transmission to other pigs. Although the virus in the samples was determined to be infectious, the question remains of how the piglets became positive and why it did not spread.

Lessons learned

Intermittent PRRSV shedding over multiple farrowings in a sow farm does occur. In this case, there was evidence of infectious PRRSV detected in piglets after 24 weeks of RT-PCR-negative results. While only one piglet from each week was found to contribute to the positive PF, there is still a risk that low prevalence could result in reinfection and destabilization of a herd.

Table 1: Serum PRRSV PCR results for the bioassay challenge groups

Challenge group	Route and dose (mL) of administration	Dose (genomic copies/pig)	-1 DPI; Percentage positive for PRRSV PCR	1 DPI; Percentage positive for PRRSV PCR	7 DPI; Percentage positive for PRRSV PCR	14 DPI; Percentage positive for PRRSV PCR
(1) PBS negative control	2 mL IM	0	0% (0 of 5)	0% (0 of 5)	0% (0 of 5)	0% (0 of 5)
(2) PRRSV positive (MN184) control	1 mL IN & 2 mL IM	2,720,000	0% (0 of 5)	100% (5 of 5)	100% (5 of 5)	100% (5 of 5)
(3) PRRSV positive processing fluids	2 mL IM	7,458	0% (0 of 2)	0% (0 of 2)	0% (0 of 2)	0% (0 of 2)
(4) PRRSV positive tissue homogenate	2 mL IN & 4 mL IM	31,962	0% (0 of 8)	0% (0 of 8)	0% (0 of 8)	87.5% (7 of 8)

Determining feasibility of a serum rinse method to detect PRRSV via PCR

K. Edmunds[1], MS; M. Hensch[2], DVM; A. Maschhoff[2], DVM; K. Henness[2], DVM; R. Philips[3], DVM; J. Waddell[3], DVM, MBA
[1]Michigan State University College of Veterinary Medicine, East Lansing, Michigan;
[2]The Maschhoffs, LLC., Carlyle, Illinois; [3]Boehringer Ingelheim Animal Health, Duluth, Georgia

Introduction

As porcine reproductive and respiratory syndrome virus (PRRSV) remains the most economically important disease affecting swine herds in the United States,[1] continued development of highly sensitive and specific detection methods is in producers' best interest. The current system for classifying breeding herds according to their PRRSV status relies on testing serum from piglets for PRRSV by RT-PCR to determine wean pig status and sow herd stability as well as time to negative.[2] Pitfalls of serum sampling include time constraints and animal selection bias. Sow farms frequently administer a vaccine or antibiotic to pigs at weaning. After each injection, needles contain red blood cells and serum residue. A serum rinse method consists of rinsing needles post-injection with a solution which is then tested for PRRSV by RT-PCR. The objective of this study is to determine the diagnostic sensitivity of a serum rinse method for detection of PRRSV by RT-PCR.

Material and methods

At placement, 50 weaned pigs from a PRRSV negative source were vaccinated with Ingelvac PRRS® MLV (Boehringer Ingelheim) vaccine. Five days later, serum from the 50 vaccinated pigs was submitted for testing by RT-PCR, which confirmed PRRSV vaccine-positive status. Unvaccinated pigs from the same source were housed in a separate building on the same site. Immediately prior to the start of study, serum was collected from 60 unvaccinated pigs, which confirmed negative PRRSV status.

At the study's start, 25 PRRSV vaccine-positive pigs with the lowest US PRRSV C_t's were designated as positive controls (PC). Five unvaccinated pigs were randomly selected as negative controls (NC). Individuals from PC and NC were each injected once and serum rinses from each individual were submitted for PCR. Three treatment groups were then established to demonstrate 1:1, 1:10, and 1:25 dilutions. For each group, one vaccine positive pig was injected with 1mL sterile water. The same needle was then used to inject 1 mL sterile water per additional animal within each specific treatment group. The needle was then removed from the syringe in accordance with safety handling protocols and placed directly into a sodium citrate tube containing 0.3 mL sodium citrate and 0.75 mL phosphate buffered saline. The sodium citrate tube was capped and shaken horizontally for 5 seconds. Sealed tubes were submitted to a public veterinary diagnostic lab and tested for PRRSV by RT- PCR.

Results and discussion

One hundred-five serum rinse samples were tested. PRRSV was detected in 8% of samples from the positive control group. PRRSV was not detected in any other treatment group samples. Overall sensitivity and specificity were 2% and 100%, respectively. Positive predictive value was 100%, and negative predictive value was 4.9% (Table 1). Although not the primary objective of the study, these results highlight the validity of previous research on an injection's ability to transfer virus from pig to pig.[3] This can be applied to PRRS elimination projects and to practices for viral transmission reduction, such as changing needles between litters of pigs. Under the conditions of this study, the serum rinse method described lacked the diagnostic sensitivity to detect PRRSV in pigs vaccinated with a modified live PRRSV vaccine in a field setting. While it requires substantial improvement for implementation, serum rinse sampling demonstrates a unique approach to serum testing at the population level.

References

1. Kleiboeker, S. B., et. al. Simultaneous Detection of North American and European Porcine Reproductive and Respiratory Syndrome Virus Using Real-Time Quantitative Reverse Transcriptase–PCR. Journal of Veterinary Diagnostic Investigation, 2005; 17(2), p 165–170.

2. Lopez, W.; Linhares, D. Processing fluids, blood serum, and tail blood swabs to detect PRRSV RNA and PCV2 DNA by PCR-Based assays, ISU James D. McKean Swine Disease Conference, 2017; p 69

3. Madapong, Adthakorn, et. al. Humoral immune responses and viral shedding following vaccination with modified live porcine reproductive and respiratory syndrome virus vaccines. Archives of Virology. 2017; volume 162, p 139 - 146.

Table 1: Sensitivity, specificity, and predictive values for serum rinse samples submitted for PRRSV PCR

	Vaccine-positive	Vaccine-negative	Sample total
PRRSV PCR positive	**2**	**0**	2
PRRSV PCR negative	**98**	**5**	103
Sample Total	100	5	105

* True positives (TP) refer to serum rinse samples from treatment groups that included vaccine-positive pigs that tested positive on PRRSV PCR. False Negatives (FN) are serum rinse samples from treatment groups that included a vaccine-positive pig and tested negative on PRRSV PCR. False positives (FP) are serum rinse samples from vaccine-negative pigs that tested positive on PRRSV PCR. True negatives (TN) are serum rinse samples from vaccine-negative pigs that tested negative on PRRSV PCR.

Effect of *Saccharomyces cerevisiae boulardii* on mummies, stillborns, and wean-to-service interval on a commercial breed-to-wean farm

Isaac N. Goldner[1], BS; Clayton Johnson[2], DVM; Claire Lefevre[2], DVM; Jessica Risser[3], DVM; Gustavo Silva[2], PhD; Michael Shields[3], PhD; Jon Jordon[3], BS
[1]University of Illinois College of Veterinary Medicine, Urbana, Illinois;
[2]Carthage Veterinary Services LTD, Carthage, Illinois; [3]Elanco, Greenfield, Indiana

Introduction

Breed-to-wean (BTW) farm throughput is highly correlated with farrowing house and breeding performance. Increasingly, swine producers are evaluating the impact of alterations to gastrointestinal microflora on BTW throughput. *Saccharomyces cerevisiae boulardii* is a live yeast product with a scientific proven nutritional health history. In the US, it is marketed as Levucell SB. The objective of this study was to measure improvement in production parameters as a result of feeding Levucell SB to gestating and lactating sows. Levucell SB balances the microbiota through creation of an anerobic environment for beneficial bacteria to thrive. It builds natural defenses through modulation of pro and anti-inflammatory responses. Levucell SB also improves gut integrity through increase in villi height and improvement in crypt depth. Through the judicious feeding of Levucell SB, it is expected to see a net positive impact on the production parameters within the treatment group and their respective litters.

Materials and methods

A 5,200 head commercial sow farm, located in the midwestern United States, was selected for the trial. 510 sows and gilts were allocated into either a treatment or control group based on parity and breed date. Each allocation occurred on that sow or gilts 94th day of gestation following her first service. Allocated gilts and sows were randomly placed and housed in gestation crates until gestation day 112, at which time they were transferred to farrowing, all based on normal operational SOP's of the farm. Treatment group was fed 1x106 cfu/gram of feed, approximately 50 grams top dressed per four pounds of feed daily. Control group was fed the standard diet. The administration of Levucell SB began at time of allocation, and then continued through the lactation phase. Various parameters were collected at time of farrowing, through lactation phase, and into breeding. The data was analyzed by PROC GLM procedure in SAS.

Results and discussion

There was a statistical difference ($P = 0.011$) in both number and rate of mummies. Control group had, on average, 0.30 mummies per farrow at a rate of 2.01%; and the Treatment (Levucell SB) group had, on average, 0.18 mummies per farrow at a rate of 1.26%. Levucell SB fed sows demonstrated a numerical improvement in critical BTW farm throughput metrics including stillborns (0.10 piglet improvement per farrow) and wean-to-service interval (0.04 day improvement). The percent bred by 7 days numerically improved by 1.74% for control versus treatment group. Preweaning mortality and average daily gain were collected; however, due to unconventional fall behind movements at the farm this data was not reliable. Fixed costs per sow is a key performance indicator across the US swine industry. One way to accomplish this is to decrease pigs born dead and non-productive days. This study demonstrated that Levucell SB statistically reduced mummies, and numerically reduced stillborns and wean-to-service interval. Given these findings, a sow farm can benefit from feeding Levucell SB because it will help increase viable piglets and decrease rebreeding intervals.

Comparison of disinfectant coverage between traditional and electrostatic fogging methods for supply entry into sow farms

S. Simmons[1], BS; H. Johnson[2], DVM; A. M. Betlach[2], DVM
[1]College of Veterinary Medicine, North Carolina State University, Raleigh, North Carolina;
[2]Swine Vet Center, St. Peter, Minnesota

Introduction

Disinfection and decontamination (D&D) rooms are routinely utilized in sow farms to reduce risk of disease introduction by fogging disinfectant onto incoming supplies. Recent studies have shown that traditional fogging methods do not provide complete disinfectant coverage of supplies, thus posing a biosecurity risk.[1,2] Moreover, modifying fogger numbers, run time, and adding stir fans did not result in complete coverage.[1] Electrostatic fogging utilizes physical principals associated with charge and particle distribution with the goal of providing even coverage. The objective of this study was to compare the disinfectant coverage of cardboard, aluminum, and steel boxes using traditional and electrostatic fogging methods.

Materials and methods

Disinfectant coverage between a traditional (Hurricane Ultra II®, Dyna-Fog Ltd., Jackson, GA) and an electrostatic (Hurricane ES™, Dyna-Fog Ltd, Jackson, GA) fogger was compared. A 1.8 × 2.4 × 1.8 m wooden shed was sealed off to create an airtight mock D&D room. In the room, a fogger was perpendicularly placed four feet from a wire rack. Three 0.3 × 0.3 × 0.3 m boxes of the same material (i.e. cardboard, aluminum, or steel) were equally spaced on the middle shelf of the wire rack. Each box and their sides were labeled with a unique identification number to assess coverage based on orientation and placement. Each side of each box was pre-marked with four 7.6 × 7.6 cm squares, in which two were located in opposing box corners and two were in the center. Foggers were loaded with a solution made from 3.8 L water, 236 mL accelerated hydrogen peroxide disinfectant (Intervention®, Virox Technologies Inc., Oakville, ON, Canada), and 177 mL pink livestock marker (Sprayolo®). Foggers were alternated between replicates for a total of 5 replicates per fogger type using cardboard boxes and 3 replicates per fogger type using steel and aluminum boxes. For each replicate, the experimental conditions included: 10 minutes of fogging time, 15 minutes of drying time, medium particle size setting (20-30 microns), and boxes were grounded using copper wire. Experimental conditions were based on findings from preliminary studies (data not shown). Coverage inside each pre-marked square was measured with the saturation value using the RBG ColorMeter® application (White Marten, Stuttgart, Germany) on an iPhone (Version 13.6.1, Apple, Cupertino, CA). An average of the saturation measurements for each side was used to generate an overall coverage value for each box. Coverage based on fogger type, box position, material type, and their interactions was assessed using a linear regression model. The model considered fixed effects for fogger type, box position, and material. Analysis was performed in R v3.6.1 (R Core Team, 2019). A $P \leq 0.05$ was used for significance.

Results and discussion

Overall coverage differences were not observed based on fogger type and material (Figure 1A). However, differences were evident based on box position (Figure 1B), in which middle boxes had greater coverage than the left and right boxes. Additionally, coverage differences with respect to box side were observed, in which middle boxes had greater coverage than left and right boxes on the side facing the fogger. Aluminum boxes had better coverage than steel and cardboard on top. The results of this study indicated that electrostatic fogging does not provide better coverage than traditional fogging methods under the conditions these experiments were run. Moreover, box position relative to the fogger is the most important factor contributing to coverage.

References

1. Madigan, J., Brown, S., Holtkamp, D., Schaefer, N. Evaluation of foggers and stir fans to improve coverage of fogged disinfectants used in supply entry rooms. Proc AASV. 2020: 83-84.

2. Leuck, K., LeFevre, C., Jansen, M., Amodie, D. Evaluation of location and fogger type on disinfectant surface area coverage of supplies entering a commercial sow farm. Proc AASV 2020: 70-71.

Figure 1: Average saturation based on material (A) or box placement (B) given fogger type.

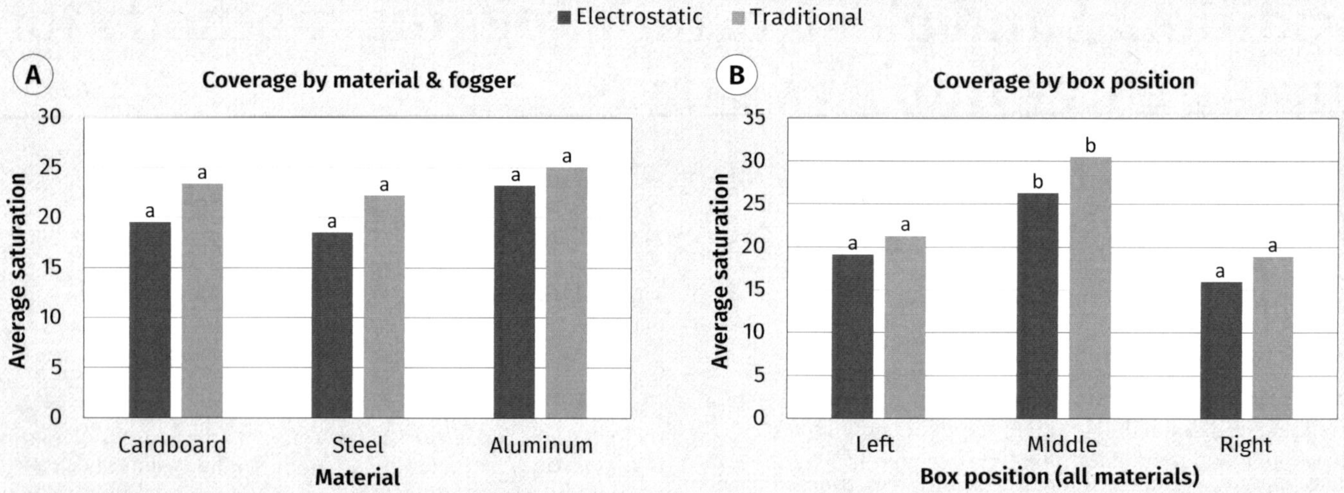

* different superscripts indicate statistical significance

Use of computer-aided design to optimize swine facility airflow

Lucas T. Daniels; Benjamin W. Blair
Applied Epidemiology Laboratory, College of Veterinary Medicine,
University of Illinois at Urbana-Champaign, Urbana, Illinois

Background

Ventilation plays an important role in the productivity, health, and profitability of swine production systems. Methods to systemically access the implications of air movement and corresponding pig comfort include from in barn measurement (IBM) in the build environment and computer-based fluid dynamic models (FDM) for designs prior to construction or in the built environment. FDMs have numerous advantages over IBM including their ability to access air flow and temperate under a wide and dynamic set of conditions with a very high level of resolution, allowing for "what if" analysis to optimize the design and operational settings of the ventilation system. FDM facilitates a robust, systematic approach to ventilation assessment that is not possible with IBM in the built environment. While FDM is used in other industries to assess ventilation system performance, the authors are unaware of its application to livestock building design. This study focused on how FDM can be applied to swine buildings to optimize design and operating conditions during a critical phase of production, the immediate post weaning period.

Objective

Validate a FDM for a single wean-to-finish barn filled with post weaning pigs at optimal and typical but suboptimal ventilation settings.

Materials and methods

Renderings of a single room 2,400 head wean to finish barn measuring 252′ × 80′ were constructed to scale using the computer-aided design program Fusion 360 and assembled using Inventor 2019. The building had 10 – 24″ fans on plenums attached below slat level, 1 - 36″ and 8-54″ fans on the end wall. Inlets consisted of three rows of 10 24 × 48″ 2-sided mechanical inlets on the ceiling and two 32.5′ × 6′ curtains on the end opposite the large fans. All inlets moved a fixed amount based on the number of fans which the controller required to be operating. Two scenarios were tested to investigate air movement at swine level inside the building assuming the building was stocked at twice finisher capacity using the surface area and thermal output equivalent of 4,600 weaned pigs weighing 13 pounds.

The two simulations were designed and executed in Autodesk CFD, to evaluate the effect of the air speed at the inlet with a constant exhaust volume through two fans on the pit of 2500 cfm in a minimum ventilation scenario. In both scenarios both the side and end curtains were completely closed. In simulation 1 (Optimal), the airflow velocity at the inlet was fixed at 735 ft/min through an opening of 0.17 inches. The second simulation (Low speed) a ceiling inlet width of 0.25 inches and an air velocity of 500 ft/min were employed. Once complete, the mean airspeeds at pig level was recorded within 3 pens.

Results and discussion

Airspeed within O were found to be 16.02 ft/min (25.51), 36.06 ft/min (6.38) and 120.04 ft/min (74.76) within the low, medium and high flow pens respectively. Within L airspeed was observed as 3.93 ft/min (3.05), 29.62 ft/min (11.98), and 105.45 ft/min (43.85). These results suggest that significant intra-barn and intra-pen variation exist within current barn design (Figure 1). A decrease in airspeed across the inlet, while decreasing intra-pen variation, increases the amount of total variation in airspeeds within the barn. Thus, decreasing the speed at the inlet may increase the observed variation of the health and production within a barn, However, both scenarios suggest that improvements should be explored to decrease this variation.

Figure 1: Comparisons of airspeeds at pig level within the barns.

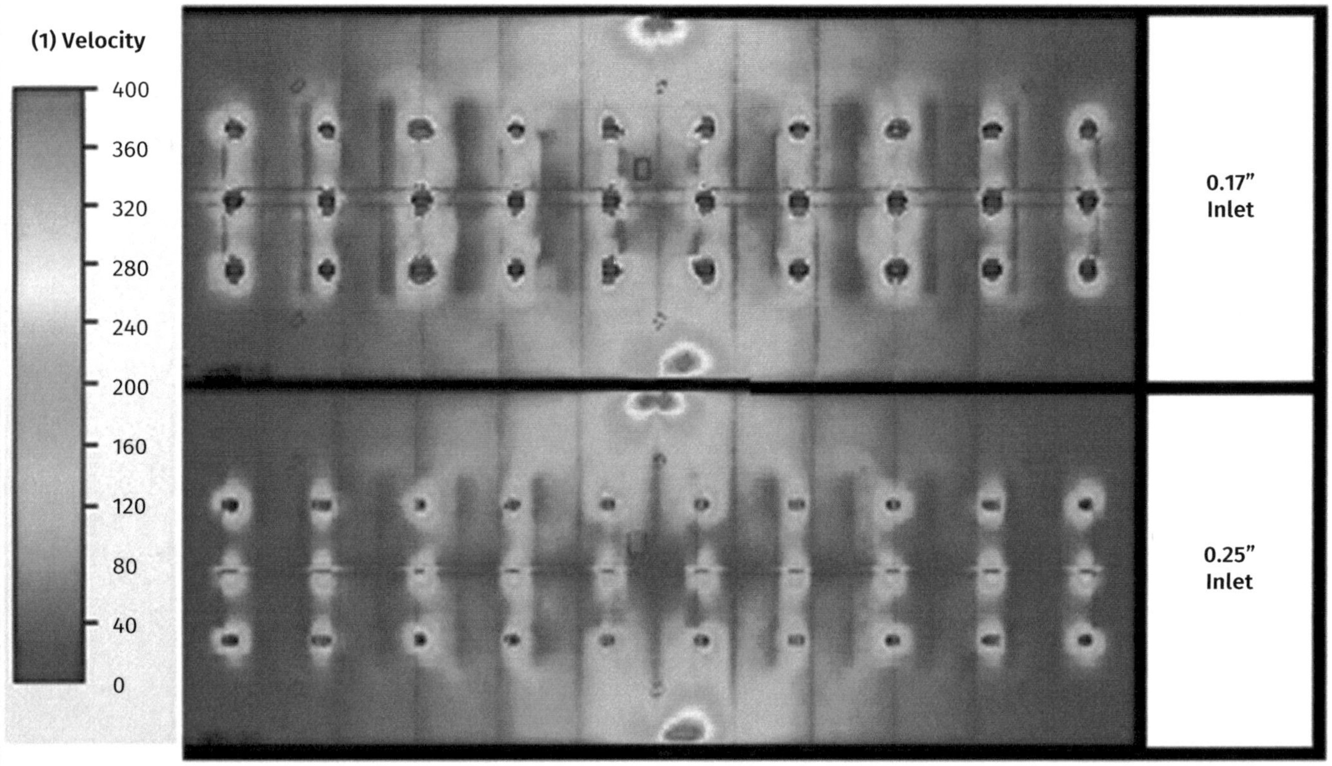

Determination of porcine sapovirus prevalence and clinical correlation in suckling pigs

Amanda Anderson[1], BS; Brandi Burton[2], DVM; Channing Sebo[3], DVM
[1]Iowa State University, Ames, Iowa; [2]Suidae Health and Production, Algona, Iowa; [3]Merck Animal Health, De Soto, Kansas

Introduction

Porcine sapovirus (formerly known as porcine enteric calicivirus) is a non-enveloped RNA virus, closely related to norovirus, in the family caliciviridae. There is minimal literature describing the role of sapovirus in swine enteric disease; however, veterinary diagnostic laboratories and several swine systems in the United States have recently received an increasing number of submissions that test positive for sapovirus by polymerase chain reaction (PCR). Prevalence has been assessed in countries in Europe, Asia, and Africa; however, it has not been extensively studied in the United States. Reported clinical signs and lesions are indistinguishable from other enteric viral pathogens, such as rotaviruses. According to several practicing veterinarians, the undocumented impact of sapovirus may include a slight increase in mortality and reduced weight gain in weaned piglets. The sensitivity and specificity of diagnostic tests, immune response to the pathogen, and prevention and control methods have not been reported. Therefore, the objective of this study was to determine the prevalence and clinical impact of sapoviruses on breeding farms in the United States and determine the demand for further research.

Materials and methods

Five multiparous breeding herds were selected for enrollment into this study. One farm, which had been previously identified as positive for sapovirus, was enrolled to investigate whether sapovirus causes clinical signs independent of other pathogens. On this farm, ten litters of suckling piglets with clinical diarrhea were conveniently selected and enrolled in the trial. Another ten litters that were free of enteric disease were also enrolled. Similar ages of litters were designated on each farm, if possible. From each litter, three pigs were fecal swabbed using individually wrapped, synthetic swabs. Swabs were pooled by litter and will be tested for sapovirus and rotavirus by PCR.

The other four farms were enrolled to assess prevalence. Twenty litters with clinical diarrhea and twenty litters without were assigned to their respective groups each farm. Again, similar ages were selected for enrollment when available. Three pigs per litter were swabbed, and five litters worth of swabs were pooled. PCR testing will be conducted to determine the sapovirus positive samples and used to calculate the percentage of positive samples per farm.

All samples that have been collected to date are being stored in a -80-degree Celsius freezer until sampling is complete. Upon completion of testing, the PCR results, sow parity, clinical signs, and piglet age will be analyzed for trends. Statistical analysis will be completed after all results are received.

Results

Due to the seasonality of enteric diseases on swine farms, it was not possible to complete sampling this summer. Additional samples are being collected, and all swabs will be submitted for testing at the same time. This method was chosen to reduce the variation that can result from testing at multiple times.

Discussion

It should be noted that the system in which this research is being conducted has previously identified sapovirus positive samples on routine diagnostics and rotavirus natural planned exposure preparation. Specimens that have been collected from pigs with enteric clinical signs are sapovirus positive both with and without a diagnosis of other significant pathogens. As an industry, it is crucial to be proactive in researching new pathogens that have the potential to cause clinical disease. This study will help the swine medicine community determine whether this pathogen deserves attention moving forward.

Alternative therapies and their impact on productivity of PRRSV challenged pigs

S. Baker[1]; M. Wagner[2], DVM
[1]Iowa State University College of Veterinary Medicine, Ames, Iowa; [2]Fairmont Veterinary Clinic, Fairmont, Minnesota

Introduction

Porcine reproductive and respiratory syndrome virus (PRRSV) is a significant swine pathogen estimated to cost the US swine industry $664 M annually. Post-weaning PRRSV infections significantly reduce the productivity and performance of growing pigs. While a current industry practice is to vaccinate, stabilize herd immunity, and treat secondary infections due to PRRSV infection, there may be alternative therapies that may improve the productivity of PRRSV challenged pigs. Consumer demand has also caused a shift away from the use of antimicrobials, which presents an opportunity for alternative therapies to gain traction. The objective of this study was to investigate the impact of electrolyte supplements and essential oils, added to the water and feed, on the productivity of PRRSV challenged pigs.

Materials and methods

Weaned pigs (129) from a PRRSV negative (Category 4) sow farm were enrolled in this study. The pigs were divided into 10 groups of 12 to 14 pigs per group with each group having a similar average starting weight. A positive (negative) control received a control diet with no additives in the feed or water and was challenged (not challenged) with PRRSV. The other eight groups were challenged with PRRSV and given different combinations of additives in the feed, water or both (Figure 1). Pigs were individually weighed 1-week post placement. On the same day, 5 pigs per pen in the challenged groups were inoculated intramuscularly with live PRRSV (RFLP type 1-4-3). The non-challenged control group received no virus. Serum samples were collected 7 days post-inoculation to ensure pigs were successfully challenged. Pigs were weighed again on days 14, 33, and 48 post-placement and serum samples were collected from every pig on day 48 to be tested for PRRSV by RT-PCR.

Results

The comparison of alternative feed and water additives and their impact on average daily gain (ADG) and percent of full value pigs (%FVP) is reported in Figure 1. The non-challenged control group out-performed all the challenged groups, as expected. Among the challenged groups, the pigs that received the electrolyte water-soluble solution (VM Mix) and control feed had the best ADG (0.94) and %FVP (100). It was also worth noting that groups that received essential oils in both feed and water did not perform differently from the challenged control group, when considering ADG, but did result in greater %FVP. Additionally, serology results showed all PRRS PCR negative pigs did demonstrate PRRS ELISA titers. Lastly, when comparing RT-PCR results at day 48 post-placement, the group of pigs that received the VM Mix had a lower percentage of positive pigs when, compared to the rest of the groups.

Discussion

Based on the data presented in this study, supplementing pigs challenged with PRRSV challenged with an electrolyte solution (VM Mix) in the water may improve performance and productivity, specifically average daily gain and percentage of full value pigs. Additionally, this electrolyte product is similar in cost to common disease interventions veterinarians may use. The results also suggest that pigs may be more responsive to alternative therapies provided via the water, but the feed application may be a less costly method to improve the number of full-value pigs. The results of this study suggest a larger scale, more costly study with replication is warranted.

Figure 1: Alternative therapies to PRRSV infections and their impact on average daily gain and percentage of full value pigs.[1,2]

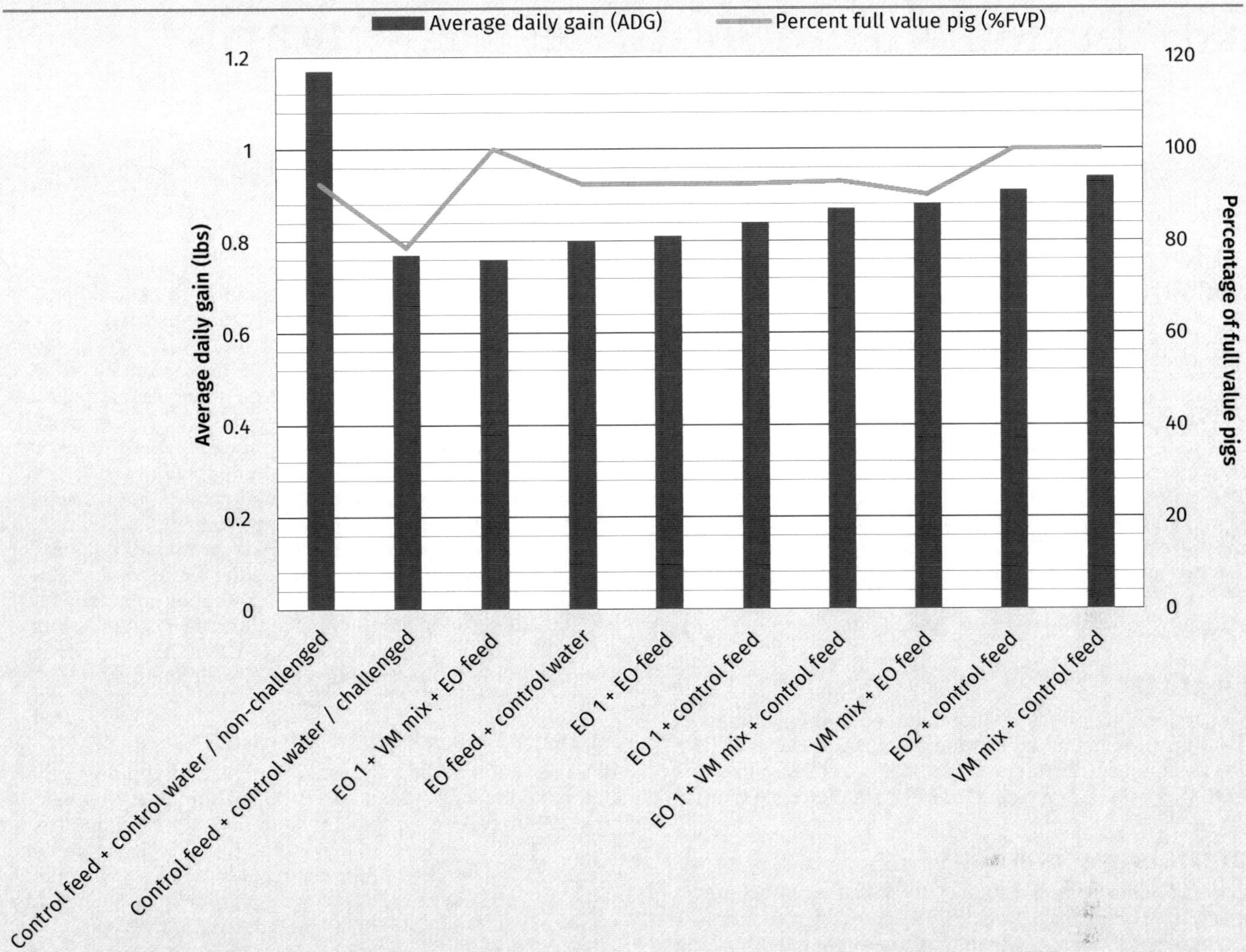

1 N=1 for all treatment groups. Differences in the outcome variables between treatment groups were numerically different, but statistics were not applied to the data due to lack of replication

2 EO1 = essential oil 1 in water, EO2 = essential oil 2 in water, EO feed = essential oil in feed supplement, VM Mix = vitamin and mineral mix in water

Evaluation of porcine rotavirus prevalence and distribution within southwestern Ontario

J. Buchan[1], BSc; H. Jansen[1], BNSc; A. Surmak[2], BSc; K. Vilaca[2], DVM MSc; G. Moser[2], BSc; K. Marenger[2], Dip.Ag.; L. Moser[2], BSc
[1]Ontario Veterinary College, University of Guelph, Guelph, ON, Canada;
[2]South West Ontario Veterinary Services, Stratford, ON, Canada

Problem

Rotaviruses (RVs) are part of the *Reoviridae* viral family and can infect a large variety of species worldwide.[1] Porcine RVs are swine-specific, and the A (RVA), B (RVB), and C (RVC) groups have been established as significant pathogens in pigs.[1] RVs are ubiquitous agents causing diarrhea in neonatal and young pigs, with infections commonly being of high morbidity and low mortality.[2] This virus is a risk to herds due to its highly infectious nature and environmental stability.[2] Furthermore, piglets with poor immunity are more susceptible to RV infection.[1] This puts high importance on establishing strong passive maternal antibody transfer to minimize disease. Yet, current commercial vaccine products are limited to protection against RVA only. Further understanding of the RV prevalence is required to know if commercial products are providing sufficient protection in herds.

Objectives

The objective of this study is to establish a clinical picture of RV within Ontario to better understand the disease's regional distribution and on-farm prevalence. This analysis will also identify potential areas of improvement for RV prevention and control.

Materials and methods

To gather diagnostic information on RV cases, results from RV sample submissions from 3 local clinics (n = 628; PCR testing of A, B, C types) from March 2017 to July 2020 were compiled and analyzed. Results were then amalgamated by case (n = 248) to eliminate bias by the number of samples submitted. Descriptive statistics were calculated to analyze the case results, as well as the distribution of positive cases per RV type by age and geography. From these results, samples from cases with clear clinical impact for groups A, B, and C were selected for sequencing. Sequencing data will be analyzed to determine which genotypes are observed in Ontario (from a small sampling), the genetic diversity (or lack thereof) of strains, and if there are any geographic trends.

Results

Initial results showed that 79% of cases were RVA positive, 57% were RVC positive, and 34% were RVB positive. When considering the age of pigs infected, 69% of suckling pig cases were RVA positive, with only 37% and 13% testing positive for RVC and RVB, respectively. In comparison, 94% of weaner pig cases were RVA positive, 85% were RVC positive, and 62% were RVB positive. Thus, weaner pig cases had similar rates of infection for RVA and RVC, and an overall higher prevalence of RVB infections compared to cases from suckling pigs. Furthermore, suckling pig cases were predominantly single RV type infections (64% positive for one RV type), whereas the majority of weaner cases were mixed infections (85% positive for more than one type of RV). Distribution of positive cases geographically followed the trend of pig density in Southwestern Ontario, with generally a higher prevalence of RVA and RVC in each county compared to RVB. Sequencing results are pending.

Application by practitioners

These results provide practitioners with a more comprehensive understanding of RV cases, specifically within Southwestern Ontario. All RV types are present to some degree within this area, with a higher prevalence of RVA and RVC. The lack of protection against RVC is a shortcoming in current RV prevention methods, and is an important consideration moving forward. Sequencing results will provide more detail in understanding RV cases, and will provide practitioners with more region-specific information. Practitioners are encouraged to continue to test for RV in practice to continue to refine the clinical picture of RV in herds.

References

1. Will AL et al. Evaluation of rotavirus infection and diarrhea in Iowa commercial pigs based on an epidemiologic study of a population represented by diagnostic laboratory cases. *J Vet Diagn Invest.* 1994; 6:416-422.

2. Shepherd FK et al. Porcine rotavirus. *Diseases of Swine,* 10th ed. 2019; 715-722.

The effects of ivermectin on commercial wean pig performance when challenged with porcine epidemic diarrhea virus

C. Burgart[1], MS; M. Hoogland[2], DVM; W. Lincoln[2], DVM; N. Lauterbach[2], DVM
[1]Iowa State University College of Veterinary Medicine, Ames, Iowa;
[2]Smithfield Foods Hog Production Division, Smithfield, Virginia

Introduction

With its detrimental impact on health, porcine epidemic diarrhea virus (PEDV) has a significant negative correlation on swine performance, especially in the United States since its introduction in 2013. Additional strategies to lower the impact of infection against the Coronaviridae family virus is vital. In a recent study,[1] Ivermectin (IVM) showed an ability to inhibit the replication of the SARS-COV-2 virus in-vitro. IVM has been tested in-vitro as an antiviral therapy against multiple viruses that impact human health. The objective of this study was to determine if injectable IVM impacted growth performance or livability in commercial pigs when challenged with PEDV.

Materials and methods

Noromectin® (Norbrook, Inc. USA, Overland Park, Kansas), a broad-spectrum injectable IVM approved for use in swine, was used in this study. Treatment groups were blocked by litter, parity, sex, and weight. There were 779 animals weighed at weaning, identified with a radio frequency identification tag, and placed into one of three treatment groups on a PEDV negative sow farm. The three treatment groups included a control group not injected with IVM, a one dose treatment group that received IVM (0.2 mL) prior to weaning, and a two dose treatment group that received the same dose prior to weaning followed by a second dose of IVM (0.4 mL) ten days later.[2] Groups were placed into a naturally ventilated, wean-to-finish barn with an equal number of replicates from each treatment group in a pen. Animals were confirmed PEDV negative by polymerase chain reaction (PCR) from fecal swabs at placement. On day 3 all animals on the site were exposed to PEDV orally via water medicator. Infection was confirmed using subpopulation fecal swabs on day 7. Individual weights were taken on days 0, 11, and 50. The experimental unit was individual pigs. Proc GLM of SPPS was used for statistical analysis. The main effect of treatment and gender were included in the analysis.

Results and discussion

Results can be found in Table 1. There were no significant differences between treatment groups for average weight at Day 0, Day 11 or, Day 50. There were also no significant differences in average daily gain from day 0 to 50 or from day 11 to 50 following PEDV challenge. No significant difference in mortality between the three treatment groups. In this study, IVM did not have a positive or negative impact on growth performance in commercial wean pigs before or after challenge with PEDV. Further research should be conducted to confirm these results, as well as exploring additional IVM products, dosages, and routes of administration and their potential role in reducing the impact of Coronavirus disease in commercial pigs.

References

1. Leon, Druce, Catton, Jans, Wagstaff. (2020). The FDA-approved drug ivermectin inhibits the replication of SARS-CoV-2 in vitro. Antiviral research, 178, 104787.

2. Lifschitz, Pis, Alvarez, Virkel, Sanchez, Sallovitz, Kujanek, Lanusse. (1999). Bioequivalence of ivermectin formulations in pigs and cattle. Journal of Veterinary Pharmacology and Therapeutics, 22(1), 27-34.

Table 1: Statistical analysis results of individual weights across treatment groups and sex

	Treatment group				*P*-values		
	Two IVM doses	**One IVM dose**	**No IVM**	**Standard error**	**Treatment**	**Sex**	**TRT*sex**
Number of pigs	**291**	**296**	**291**				
Average Day 0 weight (lbs)	15.71	15.65	16.76	0.193	0.913	0.105	0.871
Average Day 11 weight (lbs)	20.12	19.68	19.97	0.249	0.442	0.021	0.986
Average Day 50 weight (lbs)	71.09	70.03	70.78	0.796	0.622	0.009	0.770
ADG (Day 0-50)	1.11	1.09	1.10	0.014	0.511	0.011	0.925
ADG (Day 11-50)	1.31	1.29	1.30	0.016	0.692	0.009	0.707
% mortality	1.67	1.17	2.17	0.898	0.681	0.997	0.388

Efficacy of a surface-bonded quaternary ammonium chloride on the inactivation of porcine reproductive and respiratory syndrome virus (PRRSV) when organic matter is present

Madison Durflinger, BS: Derald Holtkamp, DVM, MS

Introduction

Pigs infected with porcine reproductive and respiratory syndrome virus (PRRSV) take two weeks longer to reach the same market weight as healthy control pigs. The ability to have a residual disinfectant product that can provide continuous anti-microbial protection could be beneficial in reducing the overall pathogen load for PRRSV throughout the pre-weaning stage. Therefore, the objective of this study was to test the efficacy of a surface-bonded silane quaternary ammonium salt for inactivating PRRSV at various application times in the presence of organic matter.

Materials and methods

The surface-bonded quaternary ammonium chloride (Microbe Shield®, Zoono Group Limited) disinfectant or a sham disinfectant of phosphate buffer solution (PBS) was applied 40 days before, 20 days before, or immediately after contamination of virus (PRRSV) onto the coupons. Each application time had three treatment groups, one was a negative control, and two treatment groups received PRRSV inoculum. Either Zoono® or sham disinfectant was applied to each positive control group.

Coupons were constructed by cutting plastic flooring (ChessBlue slate, Hog Slat, Inc.) into 8.9 cm × 9.7 cm squares. This flooring is commonly used in the creep area of farrowing crates. For disinfection, 0.05 mL of Zoono® or PBS was applied to each coupon with an ultrafine mist sprayer. The Zoono® was applied at a rate of 24.5 square meters per liter, which was lower than the label recommended rate of 4.9 to 19.64 square meters per liter.

After disinfection, the coupons were kept at 20°C in plastic storage bins until contamination. During the contamination, 1 mL of fetal bovine serum was spiked with 2 mL of PRRSV MN184 at a titer of 1×10^6 TCID/mL in a 100 mm diameter petri dish. The negative control groups were spiked with PBS. For the 0 minute application times, treatment groups were first contaminated and after a 90-minute drying time, disinfectant was applied. During contamination, the mixture was applied to the top surface of the coupons, and for the 40 day and 20 day application times, virus recovery occurred after a 90-minute contact time. For the 0 minute group a 60-minute contact time occurred before viral recovery. For viral recovery, the surface of the coupon was swabbed with a sterile cotton swab, and the swab was stored in a 5 mL snap cap tube filled with 3 mL of Minimal Essential Medium. Immediately after recovery, the snap cap tubes were placed on dry ice before being stored in a -80 degrees Celsius freezer. The samples were submitted for virus isolation (VI) and quantitation in MARC-145 cells for PRRSV.

Results and discussion

PRRSV was inactivated on the 0 minute treatment groups where viral contamination occurred before disinfectant application. Our negative controls groups came back negative for active viral particles. For the 0 minute application times, the surface bonded quaternary ammonium chloride inactivated PRRSV when organic material was present. Given the application rate used in this study was lower than the label rate, further studies evaluating higher application rates are needed. Results for the -20 day and -40 day group are pending completion of the virus isolation and quantitation.

Effects of fecal microbiota transplant on performance and fecal microbiota in weaned pigs when administered through a water medicator

A. Holowka[1], BS; B. Burton[2], DVM; R. Merica[2], PhD; B. Evelsizer[3], DVM
[1]Lincoln Memorial University; [2]Suidae Health and Production; [3]Elanco Animal HealthIntroduction

Introduction

Fecal Microbiota Transplant (FMT) involves introducing the microbiota found in feces of a healthy animal to the gut of one that is sick with the goal of transferring disease resistance. Notably, FMT is used successfully in humans to treat severe antibiotic-associated diarrhea caused by infection with *Clostridiodes difficile*.[1] A similar process to FMT is rumen transfaunation in cattle, which veterinarians use to treat ketosis, anorexia and several causes of indigestion.[2] In a recent PRRS-PCV2 co-infection study (Niederwerder et. al 2018), a 70% and 60% reduction in mortality and antimicrobial use, respectively, was seen in pigs that received FMT.[3] Additionally, this study found that FMT treatment reduced clinical signs of disease and increased gut microbial diversity, suggesting that the performance of an animal is influenced by the types of microbes that resides in its gut. There have been multiple studies using FMT in experimental environments using swine models, but none have been performed on a large-scale in a commercial grow-finish hog barn. The objective of this study was to assess performance, as measured by average daily gain and mortality, and diversity of gut microbiota in weaned pigs after the administration of FMT for seven days through a water medicator.

Materials and methods

Criteria like that used by Niederwerder et. al,[3] 2018 was employed to select high performing donor sows. Feces that met the criteria were homogenized in a 10% glycerol saline solution at a ratio of 1:5 (w/v). The homogenate was strained first through a medium and then fine cheese cloth before being pooled and thoroughly mixed in a large container. The final FMT solution was aliquoted into 1-liter bottles and stored at 4°C until administration. This study was conducted in a double-stocked 1200-head wean-to-finish barn with two sides. The FMT-treated pigs (n = 1200) were given 5mL/pig/day of FMT solution through a water medicator (set at a dilution rate of 1:128) for seven days. The control group received fresh water continuously during this time. Weights of a randomly selected subset of control (n = 268) and FMT-treated pigs (n = 255) were taken at the start of FMT administration, the end of FMT administration, and 14, 28, 42, 71 and 122 days post-FMT administration.

Results and discussion

While there was no difference in mortalities found between treatment groups during the nursery phase, 0.8% fewer pigs died in the FMT group during the feeder-to-finish phase of the study. FMT pigs were also the first two loads of pigs to market.

Due to the nature of commercial pig production during the summer of 2020, the average weight of the pigs in each group (two controls and two FMT-treated) at the start of FMT was not matched due to differences in age and having to double stock the barn. Notably, the pigs in the FMT 2, which were the last to load in the barn, had an average weight of 16.08 lbs at the start of the FMT as compared to the Control 1 that weighed on average 28.37 lbs. It was not surprising that the average daily gain (ADG) for the FMT 2 group, which had also experienced some health challenges in the nursery, lagged behind that of the other three groups for most of the study (Figure 1). After 42 days post-FMT administration, the pigs were moved to another site where they were single stocked. After this point, the FMT 1 pigs start to have a higher ADG at the time intervals in Figure 1, and at the last time point, FMT 2 pigs continue to surge in their ADG and not only catch up, but surpass Control 2 pigs. Another important factor to note is that these pigs were diagnosed with subclinical ileitis between day 71 post-FMT - pre-market.

Preliminary results from this study as well as those by Niederwerder et. al suggest that FMT may benefit performance of commercial swine herds, especially in the face of an infectious disease challenge.

Acknowledgment

Thank you to the Swine Veterinary Internship Program at Iowa State University.

References

1. Kelly, C. R et al. Fecal microbiota transplant for treatment of *Clostridium difficile* infection in immunocompromised patients. The American Journal of Gastroenterology, 109(7), 1065–1071. **doi.org/10.1038/ajg.2014.133**

2. DePeters E. J et al. Rumen transfaunation. Immunol Lett. 2014;162(2 Pt A):69-76. **doi:10.1016/j.imlet.2014.05.009**

3. Niederwerder M. C et al. Fecal Microbiota Transplantation Is Associated with Reduced Morbidity and Mortality in Porcine Circovirus Associated Disease. Front. Microbiol. 9:1631. **doi: 10.3389/fmicb.2018.01631**

Figure 1: Average daily gain (ADG) at various time intervals per treatment group following FMT administration.

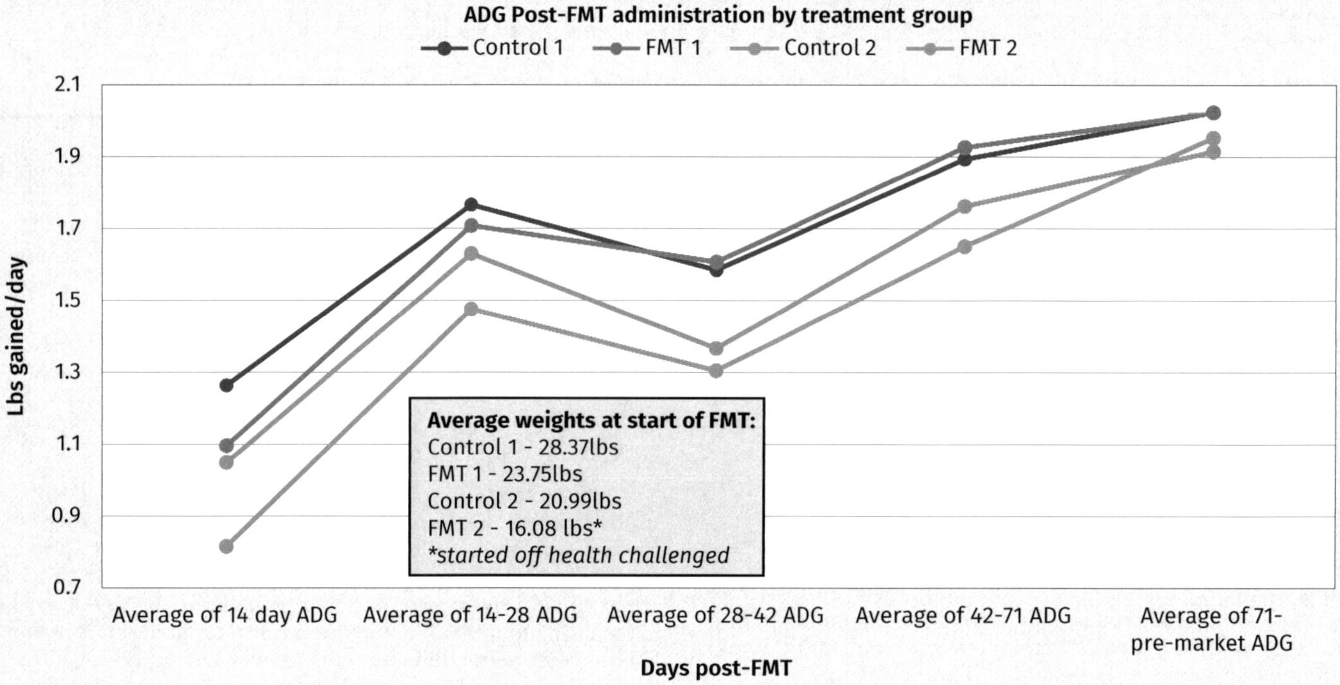

A pilot study to measure failure of passive transfer in suckling piglets

S. Melson[1], BS; S. Dee[2], DVM, MS, PhD ; Roy Edler[2], MS
[1]University of Minnesota College of Veterinary Medicine, St. Paul, Minnesota;
[2]Pipestone Veterinary Services, Pipestone, Minnesota

Introduction

Neonatal diarrhea is a detriment to the performance of piglets in the farrowing house. Multiple factors can influence the prevalence of neonatal diarrhea including failure of passive transfer, increased pathogen load, and inadequate environmental conditions. Colostrum uptake and transfer of maternal antibodies plays an important role in protection against neonatal diarrhea.[1] A refractometer has been used to evaluate failure of passive transfer in calves, but not in piglets.[2] Therefore, the goal of this pilot study was to determine if a refractometer could be a useful tool to identify piglets at risk of neonatal diarrhea due to failure of passive transfer of maternal antibodies

Materials and methods

Based on published data, failure of passive transfer would have taken place in piglets with a serum total protein (STP) level of < 4.0g/100 mL.[3] Using a commercial sow farm, 200 piglets were selected from 50 litters and organized into two groups. Jugular venipuncture was performed on each piglet at 12, 24, 36, and 48 hours post-partum. Diarrhea scores (yes/no) were assigned to each piglet over the first five days of life. STP was measured using a standard refractometer (Fisherbrand Analog Clinical Refractometer). As a control litter exhibiting failure of passive transfer, 10 piglets from five different litters were placed in an empty farrowing crate prior to suckling. These 10 piglets were not allowed to suckle and were fed a 25% dextrose solution. The blood samples from the control litter were collected at 0, 12, 24, 36, and 48 hours post-partum and samples were processed in the same manner as the other groups.

Results and discussion

Mean serum total protein values at 12, 24, 36, and 48 hours postpartum for piglets that received colostrum and experienced neonatal diarrhea (group 1) were 5.61 ± 0.55, 5.44 ± 0.57, 5.25 ± 0.54, 5.35 ± 0.52 g/100 mL, respectively. Mean values for piglets that received colostrum and did not experience neonatal diarrhea (group 2) were 5.62 ± 0.29, 5.53 ± 0.30, 5.43 ± 0.29, 5.35 ± 0.29g/100 mL, respectively. There were no significant differences in STP values between groups 1 and 2 ($P = 0.82, 0.44, 0.25, 0.92$, respectively). In contrast, Mean STP values from the piglets that received no colostrum (control group) at 0, 12, 24, 36, and 48 hours post-partum were 2.56 ± 0.56, 2.70 ± 0.82, 2.66 ± 0.50, 2.66 ± 1.63, 2.73 ± 3.16g/100 mL, respectively. Under the conditions of this study, it is suggested that the use a refractometer could distinguish STP levels for piglets that were truly colostrum-deprived from piglets that received colostrum, although P-values were not calculated for this comparison due to the small sample size of the control group (Figure 1). The refractometer could not, however, distinguish STP values between piglets that received colostrum and exhibited neonatal diarrhea versus those with no evidence of neonatal diarrhea. This pilot study, while small in scope, provided preliminary evidence for the use of a refractometer to identify piglets at risk for neonatal diarrhea.

Figure 1: Graph of serum total protein values for piglets that received colostrum and experienced diarrhea (group 1, n = 57), did not experience diarrhea (group 2, n = 143), and piglets that did not receive colostrum (control, n = 10) over time.

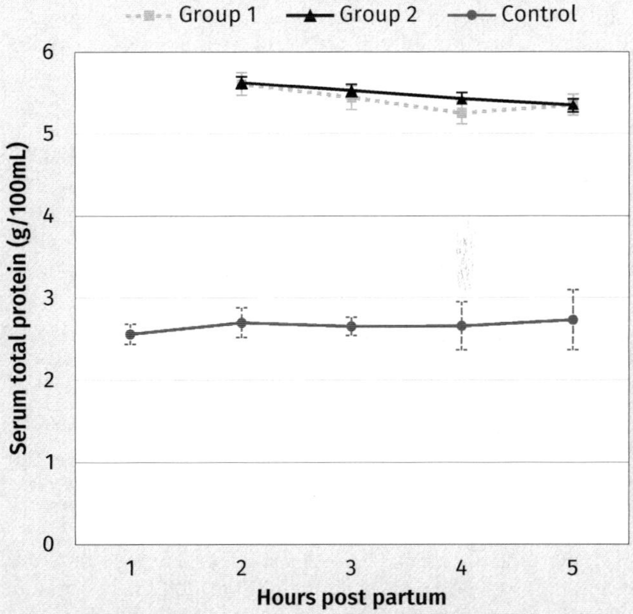

References

1. Quesnel, Hélène, et al. Colostrum Intake: Influence on Piglet Performance and Factors of Variation. *Livestock Science*, 2012.

2. Elsohaby I, McClure JT, Keefe GP. Evaluation of digital and optical refractometers for assessing failure of transfer of passive immunity in dairy calves. *J Vet Intern Med*. 2015. 721-726.

3. Kabalin, A. et al. Serum biochemical parameters in suckling piglets with low and average birth mass. *Veterinarski arhiv* 2019. 171-184.

Using oral fluids to determine the prevalence of porcine circovirus type 2 in growing pigs

T. Pieper[1], BS; D. Boykin[2], DVM; C. Smith[3], DVM; J. Okones[3], DVM
[1]College of Veterinary Medicine, University of Illinois, Urbana, Illinois; [2]Cactus Family Farms, Osceola, Iowa; [3]Pharmgate, Wilmington, North Carolina

Introduction

Porcine circovirus type 2 (PCV2) is ubiquitous throughout US commercial swine herds. Porcine circovirus associated disease (PCVAD) often presents as a postweaning multisystemic wasting syndrome (PMWS), and can compromise the immune system allowing for a number of secondary infections. Despite ongoing vaccination of growing pigs for PCV2, there is evidence of an uptick in PCVAD in the industry, one that can result in significant economic loss.[1] On average, producers have lost 3 to 4 dollars per pig due to PCVAD, with losses reaching as high as 20 dollars.[2] If the timing and prevalence of PCVAD can be established, the information may be used to improve intervention strategies. Therefore, the objective of this study is to determine the relative prevalence of PCV2 within growing pigs using oral fluids and to characterize its manifestation as PCVAD.

Materials and methods

Two separate sow flows were selected for surveillance. Sow Farm 1 was hyopneumoniae (Mhp) negative and porcine reproductive and respiratory syndrome virus (PRRSV) positive stable. The sows there were vaccinated with a modified live vaccine. Sow Farm 2 was PRRSV positive and Mhp positive. Both sow farms had 3,750 head with a moderate risk for PCVAD in the down-stream flow and located in the Midwest. Sow Farm 1 was chosen to represent a typical commercial flow. Sow Farm 2 was chosen as a PRRS positive commercial flow model. Six oral fluid samples were collected by hanging ropes in pens throughout 20 wean to finish barns (2500 head) using random spatial sampling. 10 of these wean to finish barns were sourced from Sow Farm 1, and the other 10 were sourced from Sow Farm 2. Samples were each collected from pigs at six, eight, and 15 weeks of age. The oral fluid samples were tested by RT-PCR for PRRSV and PCV2. The resulting Ct values were then compared in a mean basis to mortality, average daily gain, feed efficiency, and % head sold to secondary market on an age basis.

Results

PCV2 was consistently detected by RT-PCR of the oral fluid samples from both Sow Farm 1 and Sow Farm 2, regardless of the age group sampled. Decreasing PCV2 Ct values, indicating larger viral loads, were detected in each flow as the pigs aged (figure 1). In addition, Sow Farm 2 had noticeably lower PCV2 Ct values than Sow Farm 1. The difference in Ct value increased from 5.8% at six weeks to 7.8% at eight weeks to 17.4% at 15 weeks. PRRSv was also consistently detected on both Sow Farms 1 & 2, with both experiencing an increase in Ct value as the pigs aged. However, Sow Farm 2 experienced a drop in Ct value at eight weeks before quickly rising to meet Sow Farm 1 at 15 weeks.

The sampling of Sow Farms 1 & 2 will continue until the week of November 23, 2020. After this point, the performance values of each site will be compared to their Ct values of PCV2 and PRRSV, to check for correlations.

Discussion

The PCV2 viral load increase over time could be attributed to increasing levels of infection within the population, or to the ease of sample collection as the groups aged. However, the latter possibility is contradicted by relatively level detection of PRRSV across age groups, especially in Sow Farm 1.

These results prove that PCV2 is present within the population measured and that a relative amount can be measured. PCVAD manifestation will be characterized once performance data for each finishing group is fully available. If a correlation between performance and measured PCV2 Ct can be established, an intervention strategy can be developed and implemented. A PCV2 Ct value standard would be established to indicate early treatment of growing pigs, most likely with an antimicrobial or anti-inflammatory agent.

References

1. Kim, W., Simer, R., Prickett, J., Johnson, J., Zimmerman, J., and Yoon, K. Application of oral fluid sampling to population diagnostics for porcine circovirus type 2. American Association of Swine Veterinarians 123-124 (2007).

2. Gillespie, J., Opriessnig, T., Meng, X., Pelzer, K., and Buechner-Maxwell, V. Journal of Veterinary Internal Medicine. Porcine circovirus type 2 and porcine circovirus-associated disease. 23:6, 1151-1163 (2009).

Figure 1: Average Ct values from Sow Farms 1 & 2 at 6, 8, and 15 weeks respectively.

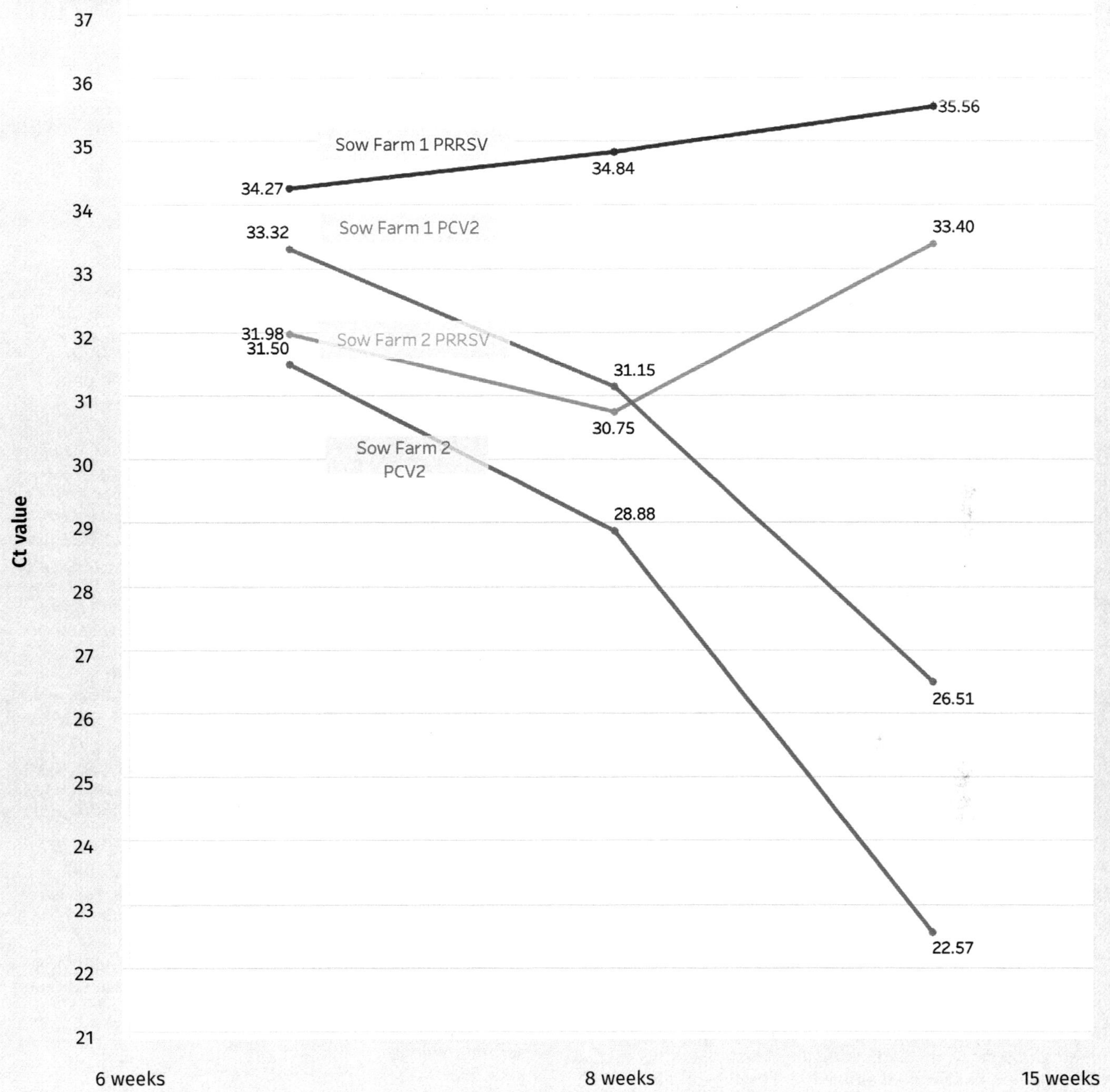

The effect of isotonic protein supplementation on pre-weaning pig growth and mortality within a commercial sow farm

J. Ramirez[1], BS; A. Ornelas[2]; F. Zarate[2], MS; R. Robbins[2], DVM, PhD; D. Drebes[2], DVM; J. Flohr[2], PhD; R. Schmitt[2], PhD
[1]University of Illinois College of Veterinary Medicine, Urbana-Champaign, Urbana, Illinois;
[2]Seaboard Foods, Guymon, Oklahoma

Introduction

Proper development of the neonatal pig's small intestine can lead to improved immune function and better nutrient absorption. Maximizing both functions can reduce pre-weaning mortality, increase weaning weight, and improve postweaning growth and profitability for commercial swine producers. A supplement targeted to improve intestinal development is Tonisity Px, a water-soluble, 3% isotonic protein drink. Previous studies have concluded that this isotonic protein supplement can improve individual pig weights, average daily gain[1] and pre-weaning mortality.[2] The objective of this study was to evaluate the impact of isotonic protein supplementation on growth performance and pre-wean mortality of neonatal pigs in a commercial sow farm.

Materials and methods

For this study, 4 consecutive farrowing rooms within a 13,000 head commercial sow farm were used to establish a total of 480 litters for the trial. Within each farrowing room (120 crates), rows (20 crates each) were randomly assigned to receive 1 of two experimental treatments: Either no isotonic supplement (Control) or 500 mL/ litter of 3% isotonic protein supplement per day from day 2-8 post-farrow (Isotonic supplement). There was a total of 240 litters assigned to the Control and Isotonic protein supplement treatments.

One day post-farrow, all pigs in each test room were tagged with either a yellow (Control) or green (Isotonic supplement) individually numbered button tag and weighed. Throughout lactation, individual pig mortality was recorded, and litters were monitored for clinical signs of rotavirus scours—which had been observed within this farm. Prior to weaning, all remaining pigs were individually weighed to determine post-farrow weight gain.

All data were analyzed as a generalized randomized complete block design using the lme4 package of R. Row within farrowing room was the experimental unit, individual pig was the observation unit, and treatment was the fixed effect. Litter, row within farrowing room, and farrowing room were used as random effects in model. Non normally distributed responses (pre-wean mortality) were analyzed using a binomial distribution. Results were considered significant at $P < 0.05$ and a tendency at $P < 0.10$.

Results and discussion

Average daily gain (ADG) was higher for control pigs ($P < 0.01$) compared to Isotonic supplemented pigs (Table 1). Pre-wean mortality ($P = 0.80$) and weaning age ($P = 0.66$) were not statistically different regardless of experimental treatment. During the trial, clinical signs associated with scours were observed within both treatment groups (25.11% of Control litters and 30.64% of Isotonic supplemented litters), but the prevalence across treatments was not statistically different ($P = 0.21$). To remove any treatment response confounded by the presence of scours, a second analyses was performed which excluded litters with clinical signs. Average daily gain was greater ($P < 0.01$) for control pigs compared to Isotonic supplemented pigs. Weaning weight of control pigs was heavier ($P < 0.04$) than Isotonic supplemented pigs. The results from this study show better growth performance in control pigs over those receiving the isotonic protein supplement. It is important to take into consideration the higher incidence of clinical scours observed within the treatment rooms which could offset neonatal intestinal development expected from isotonic protein supplementation. Additionally, weaning age (17-18 days) of pigs from this experiment were younger than previous trials evaluating isotonic protein supplementation. Future research testing isotonic protein supplementation in the neonatal pig and its impact followed through postweaning performance could help distinguish additional impact above what was evaluated in the current study.

References

1. Betbeze, S., Firth, A., Sievers, C., Johnson, H. 2018. Evaluating the effects of an isotonic protein solution in nursery pigs. 49th Annual Meeting of the American Association of Swine Veterinarians.

2. Carlson, A., Eisenhart, M., Bretey, K., Buzoianu, S., Firth, A. 2019. Early-life administration of Tonisity Px™ isotonic protein drink to pigs improves farrowing livability and growth to end-nursery. 50th Annual Meeting of the American Association of Swine Veterinarians.

Table 1: Individual pig performance and litter scour prevalence

	All test litters			Scour litters removed		
	None	**Tonisity**	***P* =**	**None**	**Tonisity**	***P* =**
Number of pigs started	3151	3167	___	2317	2146	___
Number of pigs started/litter	13.13	13.20	___	13.09	13.09	___
Days on test	16.67	16.31	**0.30**	16.59	16.60	**0.94**
Lactation length (days)	17.83	17.78	**0.66**	17.83	17.84	**0.94**
Start weight (lb)	3.18	3.14	0.61	3.20	3.19	**0.85**
Wean weight (lb)	10.33	9.74	**0.02**	10.60	10.04	**0.04**
Survival rate, %	90.60	90.92	**0.80**	91.47	91.90	**0.71**
Dead rate, %	9.40	9.08	**0.80**	8.53	8.10	**0.71**
Age of death (days)	5.08	5.20	**0.77**	4.87	4.82	**0.93**
Wean pig ADG (lb/day)	0.43	0.39	**0.01**	0.44	0.41	**0.01**
Adjusted ADG (lb/day)	0.38	0.35	**0.02**	0.40	0.37	**0.02**
Scour litters (% Litters)	25.11	30.64	**0.21**			
Age at scour (days)	3.93	3.77	**0.66**			
Mild scours (%)	52.94	52.73	**0.98**			
Severe scours (%)	47.06	47.27	**0.98**			

Modeling avian influenza virus transmission dynamics in migratory waterfowl to assess risk to pig populations in Ohio

H. L. Walker[1]; A. S. Bowman[1], MS, DVM, PhD, DACVPM; A. G. Arruda[1], DVM, MS, PhD; L. W. Pomeroy[2,3], PhD
[1]Department of Veterinary Preventative Medicine, College of Veterinary Medicine, The Ohio State University, Columbus, Ohio; [2]Division of Environmental Health Sciences, College of Public Health, Ohio State University, Columbus, Ohio; [3]Translational Data Analytics Institute, Ohio State University, Columbus, Ohio

Introduction

Infectious avian influenza has the potential to influence more than just bird health; this disease can have rippling effects on biodiversity, livestock, public health, and create significant economic losses worldwide.[1] Migratory waterfowl (MWF) are the wildlife reservoir for avian influenza virus, and usually contract the virus in their northern breeding grounds when an influx of susceptible juveniles are introduced to the population; the juvenile birds become infected individuals capable of shedding the virus during their southward fall migration.[2] These shedding events are the source of spillover that can infect other wildlife and susceptible livestock. Pig populations are especially susceptible to avian influenza viruses because of the ease by which these pathogens cross this species barrier; swine can act as intermediate hosts in which avian influenza viruses can undergo genetic reassortment to create a virus with the potential to infect humans. Due to the potential risks and limited knowledge available, there is a definite need for more research to quantify the dynamics of the virus in MWF and how it could be transmitted to swine. The objective of this study is to quantify influenza transmission in MWF, with an emphasis on the disease dynamics immediately prior to fall southward migration, using disease modeling. From this, we will be able to quantify the probability of influenza transmission from wildlife to swine in Ohio to determine if MWF contribute to swine outbreaks in the fall.

Materials and methods

To model influenza transmission, we categorized the bird population based on their disease status and age (mature or immature) using susceptible-infected-recovered (SIR) models.[3] Several models were calibrated using a dataset with 20,529 entries that describes MWF seroprevelance in various counties throughout Ohio from 1986 to 2019 to estimate the transmission rate and other disease parameters for MWF. From this, we simulated a model to determine probability of swine influenza infection from MWF during the fall and validated our final model with data that describes swine influenza infections from 2015 to 2019.

Results and discussion

The models quantified avian transmission rates that varied monthly, peaking during the fall season in Ohio (Aug-Oct). This is consistent each year and correlates with waterfowl migration patterns, demonstrating the potential for shedding events to occur from juveniles and other MWF to swine particularly in the fall season. Figure 1 shows the estimated force of infection per year with each black dot representing one year.

With the creation of computational tools utilizing time series prevalence data, we can better understand reservoir dynamics of influenza and how migration contributes to cross-species transmission to pig populations. Estimating disease transmission

Figure 1: Transmission rate of avian influenza by year. Each black dot represents one year.

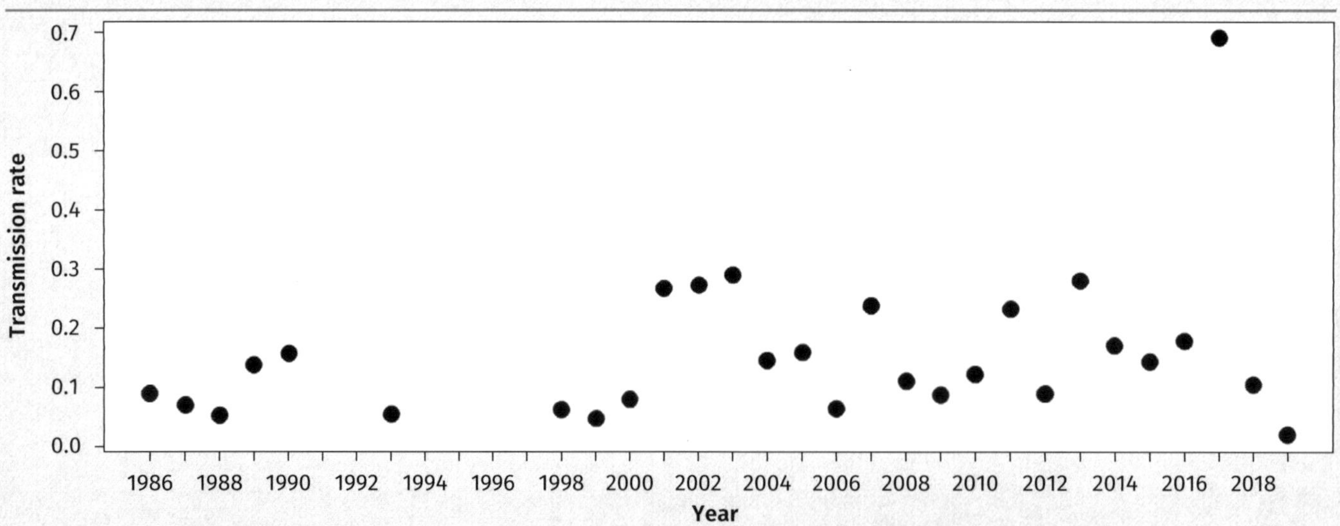

within the MWF populations will provide more information about the avian-pig dynamics of this epizootic virus; this will allow for improved prioritization of disease surveillance in both bird and swine populations and more accurate disease risk assessment among swine production.

References

1. Hénaux V et al. Estimating transmission of avian influenza in wild birds from incomplete epizootic data: implications for surveillance and disease spread. *Journal of Applied Ecology*. 2013; *50*(1), 223-231.

2. Krauss S et al. Influenza A viruses of migrating wild aquatic birds in North America. *Vector-Borne & Zoonotic Diseases*. 2004; 4(3), 177-189.

3. Keeling MJ, Rohani P. *Modeling infectious diseases in humans and animals*. Princeton University Press. 2008

Understanding endemic pathogen and mortality patterns in a commercial wean-to-finish swine production system

Siyu Xiao[1]; Samantha Mehling[1]; Alexandra Henao-Diaz[1]; Jeremy Maurer[2]; Ed Kluber[2]; Rachel Stika[1]; Christopher Rademacher[1]; Jeff Zimmerman[1]; Luis Giménez-Lirola[1]; Chong Wang[1]; Derald J. Holtkamp[1]; Rodger Main[1]; Locke Karriker[1]; Daniel C. L. Linhares[1]; Mary Breuer[1]; Christa Goodell[3]; David Baum[1]
[1]Department of Veterinary Diagnostic and Production Animal Medicine, College of Veterinary Medicine; Iowa State University, Ames, Iowa; [2]Smithfield Foods, Milford, Utah; [3]IDEXX, Westbrook, Maine

Introduction

A commercial wean-to-finish (WTF) swine production system suffered acute late-finishing mortality. Further investigation showed three mortality clusters: cluster I averaged 27.6% mortality, peaking at week 16 on feed; cluster II averaged 8.1% with no peak; cluster III averaged 14.2% mortality, peaking at week 21.[1] *Actinobacillus pleuropneumoniae* (APP) was suspected as the major cause of sudden death, but oral fluid (OF) ELISA testing was performed to determine whether other endemic pathogens were associated with differences in mortality. The objective of this study is to investigate if associations exist between the mortality clusters and antibody levels of selected endemic pathogens.

Material and methods

This study was a continuation of an observational study[1] involving 60 lots with approximately 2500 pigs per lot. OF samples (n = 6) were collected from each lot every two weeks for 22 weeks in a fixed-spatial manner (6 ropes were evenly spaced over the length of the barn).[1] Oral fluids were tested for antibodies against APP, influenza A virus (IAV), porcine reproductive and respiratory virus (PRRSV), porcine epidemic diarrhea virus (PEDV). All ELISA results were reported as sample/positive (S/P) ratios. SAS software (SAS® Institute, Inc., Cary, NC) was used for statistical analyses of ELISA S/P values in a linear mixed model with repeated measures (SAS® PROC GLIMMIX). Tukey-Kramer test was used to detect statistically significant differences in antibody levels between clusters.

Results

ELISA antibody test results of each pathogen are shown in Figure 1: A) APP, B) IAV, C) PRRSV, and D) PEDV. Oral fluid antibody testing at placement revealed antibody (presumably maternal) against APP, IAV, PRRSV, and PEDV, followed by various antibody patterns during the sampling period. APP S/P values declined to zero, except a slight increase occurred in cluster I around week 18 on feed. Cluster II had the lowest APP S/P values. IAV values declined to the lowest point in the first 6 weeks in all clusters then increased variably in clusters. Statistical differences in IAV S/P values occurred mostly in late finishing. PRRSV S/P values declined in the first 4 weeks, then rose in all clusters. During weeks 4 to 8, PRRSV values were lower in cluster III than Cluster II. No significant differences were observed among clusters in PRRSV S/P values after week 12. PEDV S/P values showed the greatest difference among clusters, ranging from no evidence of antibody to high S/P values by week 22. Cluster III PEDV S/P values were significantly lower than clusters I or II.

Discussion

Our results indicated that the patterns of endemic infections did not explain the observed differences in mortality[1] in these populations. That is, except for PEDV, there were minimal differences in the pathogen patterns among clusters, including APP (with only slight increases after week 17 on feed). Therefore, we hypothesize that differences in mortality[1] may be the outcome of multiple factors, including polymicrobial infections and animal husbandry (i.e., as have been reported over-crowding in clusters I and III[1]). Our hypothesis is that the interaction between these factors is more important than the specific endemic pathogens, themselves. This study suggests that OF ELISA is an effective tool for pathogen profile surveillance, but that good husbandry remains the foundation of animal health. Surveillance and husbandry must factor into management decisions.

Reference

1. Mehling S et al. Mortality patterns in a commercial wean-to finish swine production system. Vet Sci. 2019;6(2):49

Figure 1: ELISA antibody test results of A) APP, B) IAV, C) PRRSV, and D) PEDV for each cluster. Similar patterns are shown for APP, IAV, and PRRSV antibody. Differences exist in PEDV.

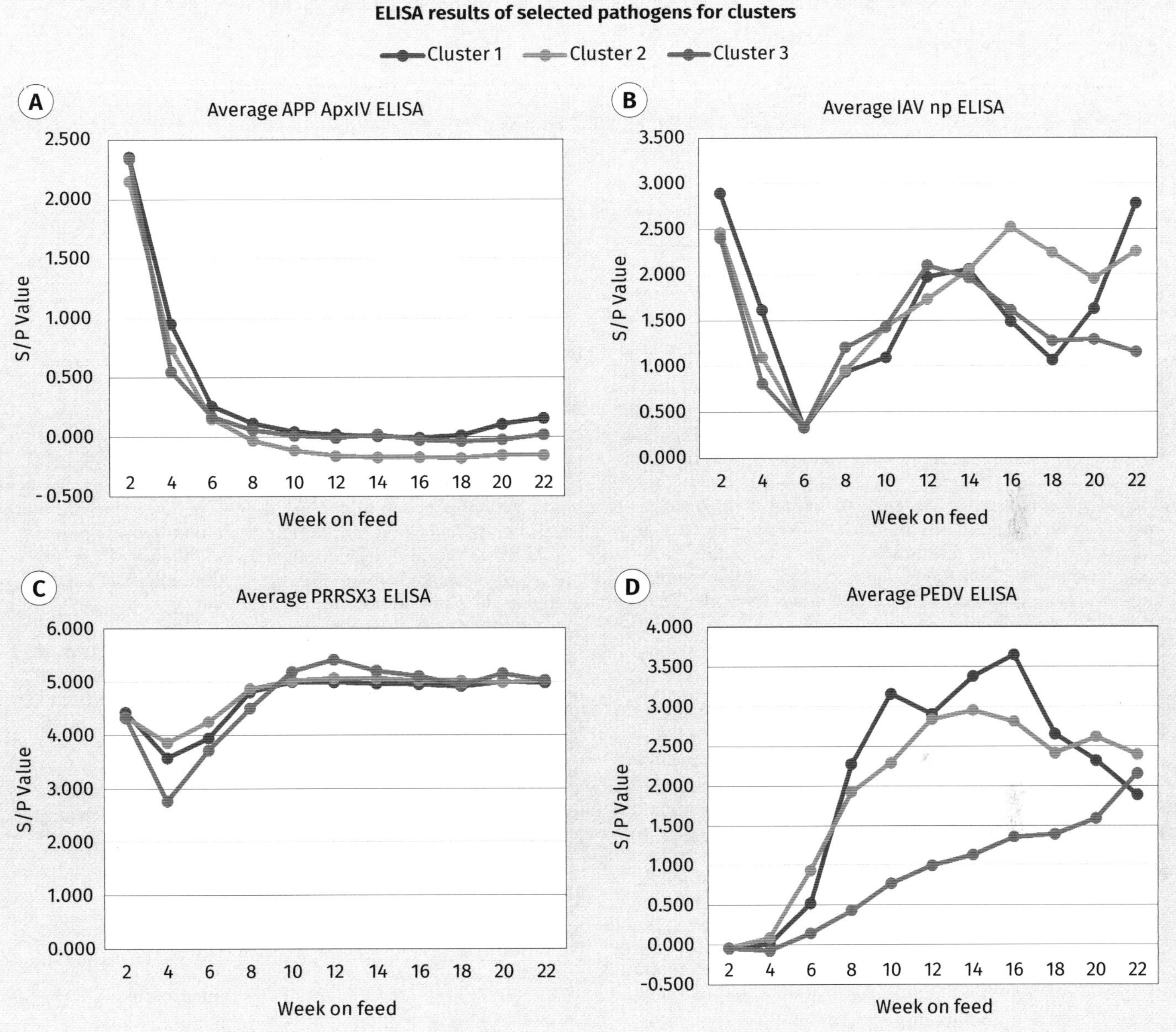

Early detection of lipopolysaccharide-induced sickness behavior in boars utilizing precision agriculture

J. Zimmerman, MS; T. Parsons, VMD, PhD
University of Pennsylvania, School of Veterinary Medicine, Philadelphia, Pennsylvania

Introduction

Behavioral changes in response to illness are well documented and contribute to an animal combating the infection. Most diseases are associated with specific clinical signs; however, some viral diseases become contagious before being identified via clinical signs or diagnostics and have the potential to allow disease spread unnoticed with potentially devastating effects. This is especially problematic for boar studs where an asymptomatic animal could result in the widespread distribution of infected boar semen and result in the widespread dissemination of pathogens throughout the swine industry.[1] Subtle changes in behavior characteristic of sickness behavior are not easily recognized without prolonged direct observation or detailed coding of video data following camera installation. Neither of these approaches are practical for commercial boar studs. The purpose of this study was to develop a practical automated system to detect the onset of sickness behavior in boars based on activity levels monitored by a wearable tri-axial accelerometer.

Materials and methods

A lipopolysaccharide (LPS) injection model was adapted to induce sickness behavior in boars by mimicking a viral infection. Animals were monitored via infrared thermometers, video recorded and a wearable tri-axial accelerometer. Six boars were habituated to individual observation pens, observed for 3 days prior to being injected intramuscular with either LPS (10 µg/kg body weight) or pyrogen-free saline as the control, and observed for an additional 3-days post-injection. Temperature was recorded in the auricular area for the first 5 hours post injection with an infrared thermometer to confirm the effect of LPS. Prior to habituation, auricular temperature taken via infrared thermometer was validated against rectal and environmental temperature. Using a multiple linear regression model, auricular temperature was used to predict rectal temperature during the immune challenge to minimize behavioral perturbations. Behavior data was collected using a 2-minute fixed sampling interval to obtain a frequency for each behavior including standing, sitting, lying, walking, eating, drinking, and exploring the environment. Tri-axial accelerometer data provided postural data on standing, lying, and sitting as well as an overall activity level. Data was analyzed utilizing a linear mixed effects model with temperature as the outcome variable and treatment and time after injection as fixed effect and animal as a random effect.

Results and discussion

Boars injected with LPS had higher body temperatures scores ($c^2 = 5.6$, df(1), $P < 0.05$) compared to the saline injected controls (Figure 1). Preliminary data analysis supports the onset of sickness behavior concurrently with the onset of fever. Ongoing analysis is focused on the sensitivity of the tri-axial accelerometer in the detection of these behavioral changes. Taken together, these preliminary results support the efficacy and validity of using wearable devices with sensors for movement data to assess animal health prior to presenting clinical signs. Artificial insemination is the most commonly used reproductive technology in the swine industry and collected semen is distributed across a wide geographical area. Semen has been shown to harbor different porcine viruses and the window between viremia and clinical signs can span days. Identifying the pre-clinical onset of sickness behavior greatly enhances producers and swine veterinarians ability to control disease and prevent the spread of both endemic diseases such as porcine reproductive and respiratory virus or even trans-border diseases like African swine fever.

Reference

1. Guérin, B., and N. Pozzi. 2005. "Viruses in Boar Semen: Detection and Clinical as Well as Epidemiological Consequences Regarding Disease Transmission by Artificial Insemination." Theriogenology 63 (2 SPEC. ISS.): 556–72. **https://doi.org/10.1016/j.theriogenology.2004.09.030**

Figure 1: Average increase in boar auricular temperature after LPS injection.

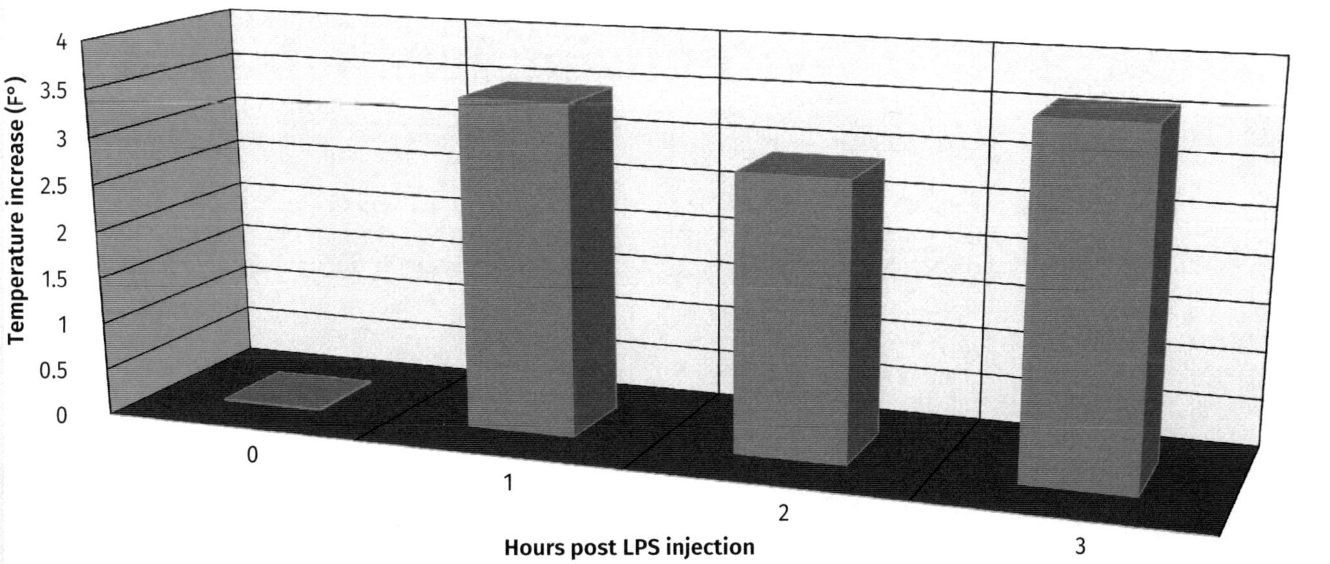

Two-stage sampling and risk-based sampling are more efficient than simple random sampling for detection of PRRSV-viremic piglets in farrowing rooms

M. N. Almeida[1]; M. Zhang[1]; W. Lopez[2]; C. Vilalta[3]; J. Sanzueza[4]; C. Corzo[5]; J. J. Zimmerman[1]; D. C. L. Linhares[1]
[1]Iowa State University, Ames, Iowa; [2]PIC, Hendersonville, Tennessee; [3]UpNorth Analytics, Arbeca, Spain; [4]Catholic University of Temuco, Temuco, Chile; [5]University of Minnesota, Minneapolis, Minnesota

Background

The need for a method to optimally sample for the detection of PRRSV-viremic piglets is well-recognized. The lack of such a method is problematic for assuring the status of PRRSV-negative herds and/or herds undergoing PRRSV elimination. The challenge relies in finding the balance between the cost of surveillance and the cost of failing to detect PRRSV-infected piglets. Traditional farrowing room surveillance has based sample size on hypergeometric distribution (probability of success in random draws from a finite population without replacement) and piglet selection on simple random sampling, but other approaches are possible. Thus, the objective of this field-based study was to compare simple random sampling (SRS), two-stage sampling (2SS), and risk-based sampling (RBS) in terms of the sample size needed to achieve 95% power of detection (POD) under prevalence scenarios observed in commercial sow farms.

Methods

The dataset was based on testing (PRRSV RT-PCR) 4,510 serum samples from nursing piglets in 422 litters in 12 breed-to-wean PRRSV positive sow farms with different prevalence scenarios. The POD for three sampling methods (SRS, 2SS, RBS) was estimated by simulated sampling from these data. For SRS, a given number of piglets (*n.sample*) were randomly selected without replacement within a room. If ≥ 1 positive piglet(s) was detected in the simulation, the trial was regarded as a "success". For each *n.sample*, the proportion of successful trials (10,000 iterations) was defined as the POD. This process was continued until the *n.sample* that achieved 95% POD was determined. The POD for 2SS and RBS was likewise estimated by simulation, with some differences in the procedures. For 2SS, the first stage was random selection of litters and the second stage was random selection of piglets within litters. If *n.sample was* ≤ total number of litters, then *n.sample* litters were selected without replacement in the first stage and one piglet was sampled from within each litter in the second stage. If *n.sample* was > total number of litters, then all litters were included in the first stage, with one piglet randomly sampled from each litter in the second stage. Thereafter, the remaining samples were selected randomly (litter and piglet within litter) until *n.sample* piglets were sampled (without replacement). Likewise, RBS used a two-stage approach, but unlike the random selection in the first stage of 2SS, RBS first targeted high-risk litters (as determined by a risk factor analysis). When no more high-risk litters were available, 2SS was then used to complete the number of piglets to be sampled (*n.sample*).

Results

2SS required 1-to-25 fewer samples than SRS to detect ≥ 1 viremic piglet in 13 of 16 rooms and the same number of samples in the 3 remaining room. RBS required 1-to-7 fewer samples than 2SS to detect ≥ 1 viremic piglet in 7 of 16 rooms, the same number of samples in 6 rooms, and 1 more sample in 3 rooms (Table 1).

Conclusion

SRS, the strategy upon which most sample size tables for disease detection are based, detected viremic piglets with recommended sample sizes for a given prevalence. However, it was less efficient than either 2SS or RBS. RBS and 2SS were roughly equivalent, but RBS targeting first parity sows and litters with ≤ 12 piglets resulted in improved PRRSV detection efficiency. Future work will provide sample size guidelines for each approach.

Table 1: Sample size to detect ≥ 1 viremic piglet using simple random sampling (SRS), two-stage sampling (2SS), or risk-based sampling (RBS) based on observed data

PRRSV viremic piglets (%)		**57.3**	**55.5**	**35.6**	**28.1**	**24.4**	**20.9**	**20.1**	**17.9**	**17.9**	**17**	**13.3**	**6.5**	**4.5**	**3.8**	**3.4**	**1.3**
Farm-room		**A-1**	**J-1**	**J-2**	**E**	**I**	**K-1**	**K-4**	**K-3**	**A-3**	**H**	**C-1**	**A-2**	**K-2**	**A-4**	**C-3**	**G-2**
Population Size		**157**	**211**	**163**	**135**	**271**	**67**	**179**	**39**	**151**	**176**	**60**	**201**	**222**	**53**	**117**	**240**
90%	SRS	3	3	6	8	9	10	10	10	11	13	14	32	44	37	50	127
	2SS	3	3	6	7	6	6	8	4	10	11	7	14	30	33	47	97
	RBS	3	4	6	7	5	5	7	5	10	11	7	14	27	33	46	98
95%	SRS	4	4	7	9	11	12	12	13	13	17	19	41	55	42	58	139
	2SS	4	4	7	8	8	8	10	5	11	13	7	17	37	38	56	114
	RBS	4	5	7	9	6	6	9	5	11	13	7	15	30	36	54	115
99%	SRS	6	6	10	14	14	16	17	18	19	23	26	57	76	48	79	180
	2SS	5	6	11	11	11	11	16	7	15	16	7	19	52	44	67	141
	RBS	6	7	9	11	8	11	13	7	14	16	7	17	46	42	69	135

Influence of number of parturitions on hematological parameters in multiparous sows of intensive production between two consecutive weaning in Chile

Alejandra Jiménez; Joaquín Álvarez-Norambuena; Carolina Baeza; Álvaro Ruiz
Departamento de Patología y Med. Preventiva, Fac. Cs Veterinarias, Universidad de Concepción

Introduction

Hematology is one of the tools that can provide valuable information regarding health of animals. However, in swine production it is rarely used due to the cost and lack of blood refence intervals (RI).[1] Previous hematological studies of sows in different reproductive status are not common and have been carried out with small groups of animals, with emphasis at specific times during pregnancy or lactation.[2,3] Variations between lactating and pregnant sows has been reported.[4] There is an absence of specific information between the different parturition of the sow in Chile. The aim of this study was to determine the influence of farrowing number on the variations in hematological parameters of multiparous sows in Chilean production between two consecutive weaning.

Materials and methods

In order to do this, 307 healthy multiparous sows with 1 to 8 farrowing from two swine farms from Chile where chosen

Sows were classified by parturition P1, P2, P3, P4 and ≥ P5. They were tested six times, T1 at weaning, T2 at almost 24 hours before insemination, T3 at 30 days of gestation, T4 at 60 days of gestation, T5 at 90 days of gestation and T6 at second weaning.

The estimated blood parameters were count of red blood cells (RBC), hemoglobin concentration (HGB), hematocrit (Hct), total plasma protein concentration (TPP), platelets count (Pt), total leukocyte count (LEU) and its differential (eosinophils (EOS), lymphocytes (LIN), monocytes (MON) and neutrophils (NEU)) the results were analyzed using repeated measured ANOVA and Friedman's test.

Results

There was a decrease ($P < 0.05$) in RBC (Figure 1A) and Hct as the number of farrowing of the sow increased as well as LEU (Figure 1B), NEU, and LIN (WBC) parameters. The TPP had higher values ($P < 0.05$) on sows with ≥ 5 parturitions, in all sampling events except T3. EOS, HGB, MON and Pt did not show statistically significant differences ($P > 0.05$) between parities at different sampling times.

Conclusions and discussion

The decrease of red blood cells and white blood cells is concordant with Ježek et al[4]. In humans this phenomenon might be because of the reduction of hematopoietic stem cells, less hormonal stimulus and the decrease of space used by hematopoietic tissue in bone marrow (due to fat infiltration).[5] Given the similarity between humans and pigs, this may be the answer for this reduction.

The concentration of total proteins rises as farrowing increases, this have been described in other species as a physiological increase of globulins in plasma, this may lead this results in sows.

Further investigation is needed to understand the factors that are affecting these parameters in sows, and how to improve animal-welfare, by standardizing hematological parameters.

References

1. Perri, A., et al. 2017. Can Vet J, 58, 371-376.
2. Elbers, A., et al. 1994. Vet Q, 16, 127-130.
3. Žvorc, Z., et al. 2006. Vet Arhiv, 76, 245-253.
4. Ježek, J., et al. 2018. J Swine Health Prod, 26, 72-78.
5. Mahlknecht, U., et al. 2010. Exp Ther Med, 1(6):1019-1025.

Figure 1: Hematological values of multiparous sows according to their number of farrowing of A) RBC and B) LEU. No statistical differences were found in times with red circle (*P* > 0.05).

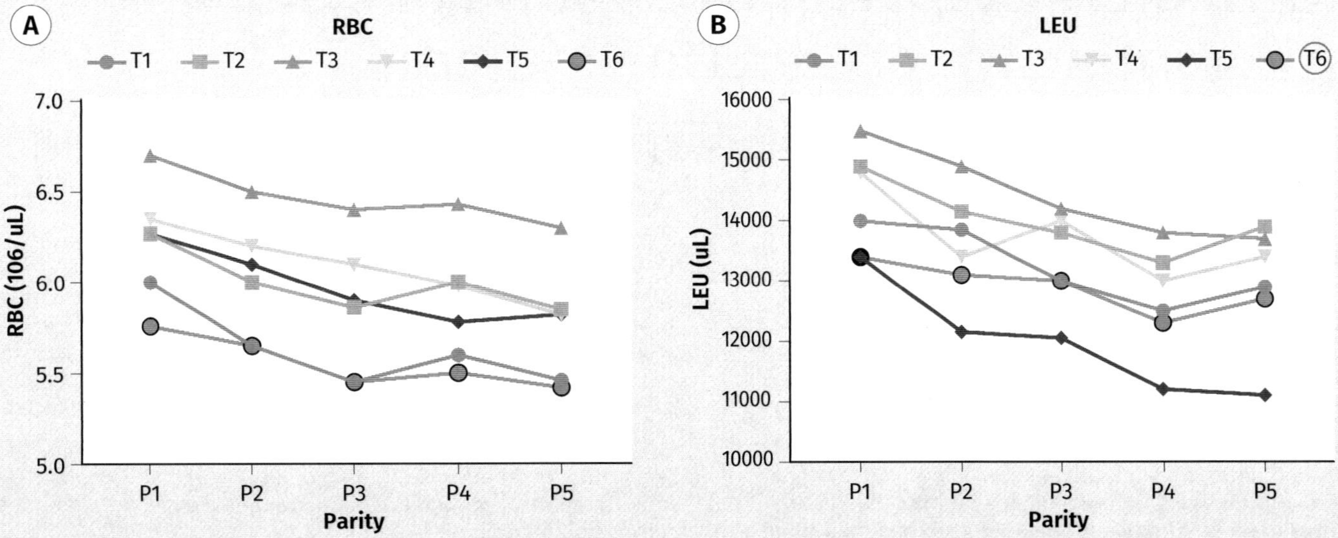

Influence of sampling time on hematological parameters of multiparous sows of Chilean swine intensive production systems

Alejandra Jiménez; Joaquín Álvarez-Norambuena; Carolina Baeza; Álvaro Ruiz
Departamento de Patología y Med. Preventiva, Fac. Cs Veterina\rias, Universidad de Concepción

Introduction

Hematology is one of the tools that can provide valuable information regarding health of animals, however, in swine production it is rarely used due to the cost and lack of blood reference intervals (RI).[1]

Previous hematological studies on sows in different reproductive status are not common and have been carried out with small groups of animals, with emphasis at specific times during pregnancy or lactation.[2,3] Previous reports indicate that physiological changes in sows, specifically those induced by lactation and gestation cause variations in the hemogram.[3,4] The aim of this study was to determine hematological parameters of healthy multiparous sows in different moment of gestation and at weaning.

Materials and methods

The blood samples were taken from 307 healthy multiparous sows from two swine farms from Chile, (a multisite and a wean-to-finish farm). Each sow was tested six times, T1 at weaning, T2 at almost 24 hours before insemination, T3 at 30 days of gestation, T4 at 60 days of gestation, T5 at 90 days of gestation and T6 at second weaning.

The estimated blood parameters were red blood cell count (RBC), hemoglobin concentration (HGB), hematocrit (Hct), total plasma protein concentration (TPP), platelets count (Pt), total leukocyte count (WBC) and its differential (eosinophils [EOS]), lymphocytes (LIN), monocytes (MON) and neutrophils [NEU]). The results were analyzed using repeated measured ANOVA for normal distributed variables and Friedman's test, for non-normal distributed variables.

Results

There was a significant increase ($P < 0.05$) in red blood cells (RBC, HGB and Hct) by T3 and after a trend to decrease ($P < 0.05$) to the second weaning, similar to white blood cells, LEU, LIN and MON. In case of NEU and TPP had higher values ($P < 0.05$) at weaning (T1 and T6). At T5 WBC and Pt show a significant decrease ($P < 0.05$).

Conclusions and discussion

Changes in red and white blood cells in the hemogram during the last trimester of gestation has been described previously in various species included swine. A decrease in hematological variables occurs due to hemodilution caused by an increase in plasma volume in the mother. This is a physiological phenomenon induced by the fetus's growth.[3]

Difference in other studies in the total count of white blood cells associated with the increase of neutrophils induced by stress differs in this study.

Complete blood count in Chile is like what was previously reported, despite environmental factors that may affect the hematological parameters. The difference in WBC count due to the increase of neutrophils may be due to avoiding sampling close to farrowing because of the increased stress that entails.

It is necessary to consider these variations when interpreting blood tests of sows in production, to make an appropriate clinical diagnosis, with the proper reference intervals for the species and its characteristics.

References

1. Perri, A., et al. 2017. Can Vet J, 58, 371-376.
2. Elbers, A., et al. 1994. Vet Q, 16, 127-130.
3. Žvorc, Z., et al. 2006. Vet Arhiv, 76, 245-253.
4. Ježek, J., et al. 2018. J Swine Health Prod, 26, 72-78.

Table 1: Hematological reference interval from samples of multiparous sows according to their sample time

Param	Unit	RI T1	RI T2	RI T3	RI T4	RI T5	RI T6
				Red Cells			
RBC	10^6/µL	4.41 - 6.83^{a}	4.86 - 7.12^{b}	5.07 - 7.85^{c}	4.72 - 7.42^{b}	4.58 - 7.36^{b}	4.24 - 6.87^{a}
HGC	g/dL	10.11 - 14.3^{a}	10.7 - 15.0^{c}	11.3 - 16.1^{d}	10.1 - 15.2b^{c}	9.79 - 15.51^{b}	9.89 - 14.56^{a}
Hct	%	28.13 - 41.8^{b}	29.5 - 42.4^{c}	31.1 - 44.6^{d}	29.14 - 43.1^{c}	27.87 - 42.3^{b}	27.25 - 40.6^{a}
				White Cells			
LEU	µL	7883.14 - 18611.5b^{c}	9209.7 - 18750.4^{de}	9579.29 - 19120.1^{e}	8722.99 - 18658.5^{cd}	6843.21 - 17279.5^{a}	7815.69 - 17949.59^{b}
NEU	µL	4593.7 - 13513.5^{d}	4443.8 - 12496.2^{c}	3753.9 - 11795.01^{b}	3775.36 - 11619.08^{b}	2435.61 - 11034.09^{a}	4311.24 - 13185.38^{cd}
LIN	µL	879.78 - 6118.32^{a}	2061.42 - 6771.96^{b}	2484.99 - 8409.71^{d}	2099.34 - 7994^{c}	1706.73 - 7104.65^{b}	734.07 - 5865.85^{a}
MON	µL	79.43 - 977.24^{a}	123.03 - 1412.54^{bc}	112.2 - 1862.09^{c}	104.71 - 1318.26^{b}	93.33 - 1071.52^{a}	95.5 - 977.24^{a}
EOS	µL	0 - 1595^{a}	0 - 177^{6d}	0 - 1826^{d}	0 - 1771^{bc}	0 - 1548^{d}	0 - 1807^{b}
				Others			
TPP	g/L	66.8 - 95.3^{cd}	66.1 - 89.3^{a}	64.5 - 88.9^{a}	62.34 - 96.5^{b}	67.7 - 91.7^{bc}	69.67 - 94.5^{d}
Pt	10^3/µL	413.93 - 906.33^{e}	389.87 - 843.73^{d}	232.31 - 887.27^{c}	254.76 - 860.52^{c}	87.12 - 723.62^{a}	144.5 - 814.24^{b}

Columns with different letters are significantly different for the same metabolite ($P < 0.05$).

Retrospective study of the serological behavior of *Lawsonia intracellularis* in swine intensive production system from Chile

Joaquín Álvarez-Norambuena; Paula Gädicke; Álvaro Ruiz
Departamento de Patología y Med. Preventiva, Fac. Cs Veterinarias, Universidad de Concepción

Introduction

Lawsonia (L) intracellularis is a disease, producing a condition known as proliferative enteritis.[1,2] This disease generates a lot of economic losses around the world, because it reduces the efficiency of food conversion.[3] Serology can be used to identify infected herds; passive immunity drops and seroconversion time. However, finding high antibody titers only indicates exposure to this bacterium.[4,5] In South America there is little information on the prevalence and serological behavior of this disease. The main objective of this study is to determine the prevalence and serological behavior of *Lawsonia intracellularis.*

Materials and methods

In order to do that the ELISA serological results of *L intracellularis* were analyzed, from samples of 23 swine production farms in Chile (15 multi-site and 8 one-site), carried out during the years 2013 to 2018 in the Swine Diagnostic Laboratory of the University of Concepcion, Chile.

Fifty-two serological profiles in total (n = 1947), corresponding to the commercial line, of different age were analyzed. The samples were classified in A: 3 to 5 weeks; B: 6 to 8 weeks; C: 9 to 11 weeks; D: 12 to 14 weeks; E: 15 to 18 weeks; F: 19 to 22 weeks; and, G: 23 to 25 weeks. Additionally, the production system was classified as multi-site and one-site.

The ELISA test used was Svanovir® *L intracellularis*/Ileitis-Ab, following the standardized protocol by the manufacturer (Svanova®).

The data was analyzed by ANOVA, and if it was not normal by the Kruskal-Wallis test.

Results

A prevalence of 49.46% between 2013 and 2018 was found. A significant difference between one-site and multi-site systems between years was observed. Serological profile behaves different between year and production systems ($P < 0.05$) (Figure 1).

Conclusions and discussion

Different types of behavior can be observed between years. Multi-site system (Figure 1A) lean to have high Pi values due to the passive immunity, then the one-site system. Then the Pi values decrease as it starts to drop the maternal antibodies and after this increase constantly as seroconversion starts to rise. In 2016, the serological profile was a little different, because there is an early increase of the Pi values with a subsequent decrease and then a late seroconversion. In 2017, the Pi values tends to increase constantly after the 6 to 8 weeks.

On the other hand, one-site systems (Figure 1B) there were no samples in 2015 and 2016. In 2013, 2014 and 2017, the serological profiles behave like 2017, but in 2014 and 2017 a little decrease was observed in the last samples age. In the year 2018, there is an irregular serological pattern, where there is not a clear trend in the Pi values.

The data suggest that the passive immunity drops around 8 to 9 weeks of age, and seroconversion starts around the 11 to 15 weeks of age. Seropositivity in one-sites, without regard to the passive immunity, is 58.0% and in multi-sites is 55.5%. In this study, al the farms were positive, so farm prevalence is 100%.

References

1. McOrist et al. 1995. Int J Syst Bacteriol, 45, 820-825.
2. Smith, D., et al. 2001. Vet Microbiol 82, 331-345.
3. Collins, A., 2013. Agriculture 3, 536-555.
4. Knittel, J., 1998. American Journal of Veterinary Research 59, 722-726.
5. Kroll, J., et al. 2005. Clin Diagn Lab Inmunol 12, 693-699.

Figure 1: Line chart of the average Pi values, measured by ELISA, to *L intracellularis* of A) 15 multi-site swine farms and B) 8 one-site swine farms, classified by group of age and year.

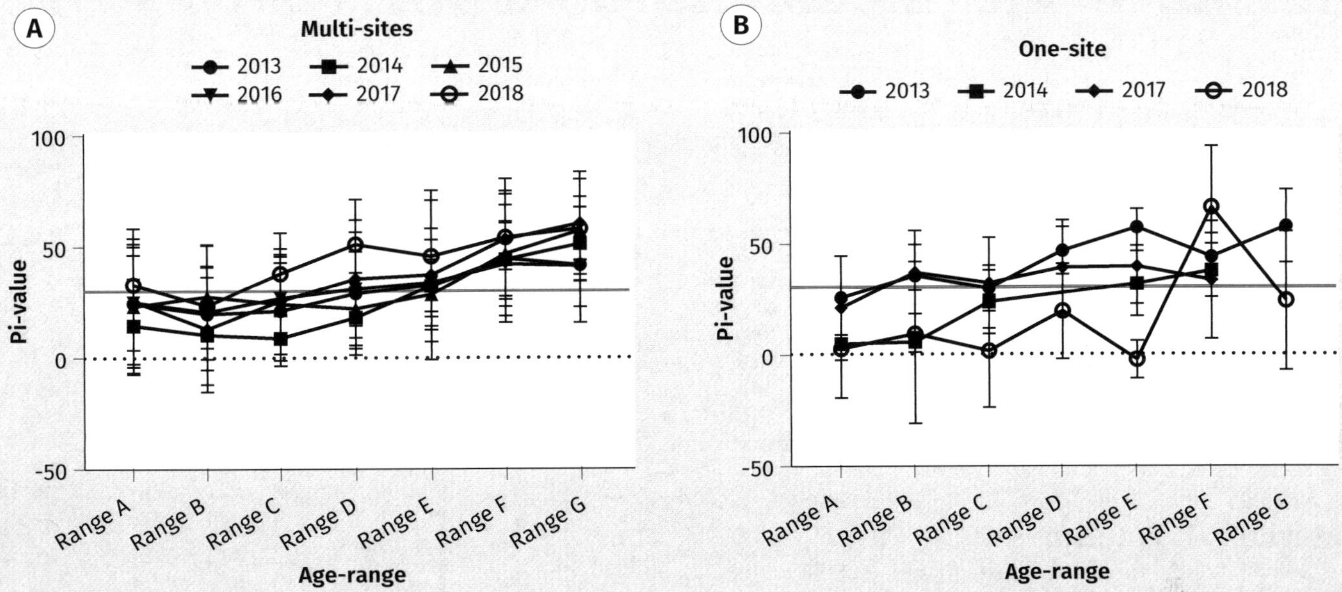

Serological behavior of influenza A virus in Chilean swine intensive production systems

Joaquín Álvarez-Norambuena; Constanza Villarroel; Álvaro Ruiz
Departamento de Patología y Med. Preventiva, Fac. Cs Veterinarias, Universidad de Concepción

Introduction

Influenza A virus (IAV) is endemic in pig populations worldwide. The commonest subtypes are H1N1, H1N2 and H3N2.[1] This disease generates a lot of economic losses globally from US$ 3 to 10.[2] Serology, through ELISA, is widely used in order to understand viral circulation and how the virus affects different population in different geographic areas[3] but did not differentiate immunity originated by vaccination or field exposition.[4] In South America, and specially in Chile, there is little information on the serological behavior of this disease. Therefore, the aim of this study was to determine the serological behavior of Influenza A virus in swine.

Materials and methods

Swine IAV ELISA serological results from samples of 15 swine production farms in Chile, carried out during the years 2014 to the first quarter of 2020 in the Swine Diagnostic Lab of the University of Concepcion, Chile, were compared and analyzed.

Forty-one serological profiles in total (n = 2202 serum samples), corresponding to the commercial line, of different age were analyzed. The samples were classified in A: 3 to 5 weeks; B: 6 to 8 weeks; C: 9 to 10 weeks; D: 11 to 13 weeks; E: 14 to 15 weeks; F: 16 to 18 weeks; G: 19 to 22 weeks; H: 23 to 25 weeks.

Samples were analyzed by commercial IDEXX Influenza A Ab Test (IDEXX Laboratories, Westbrook, ME, USA), following the standardized protocol by the manufacturer.

The data was analyzed by ANOVA with LSD Fisher test and with Kruskal-Wallis when distribution was not normal.

Results

Different behavior by year was reported ($P < 0.05$) (Figure 1A), Nevertheless, the not a clear decay of the passive immunity, with the exception of the years 2018, when a decay of it was observed after range B (6 to 8 weeks) (Figure 1B).

Conclusions and discussion

Average serological differences between years were observed ($P < 0.05$) (Figure 1A) showing that the year 2018 has lest variation and that there is not statically significant difference between years 2014, 2019 and 2020.

The analysis of the results per year considering the ranges of ages (Figure 1B), showed a different behavior among age range between years. In range A lean to have lower S/N ratios due to the passive immunity, showing a trend to increase up to range C. However, there is a tendency to decrease the values after this range without a uniform pattern, suitable for the variation of the presentation of this disease in swine industry.

Figure 1: Line charts of the S/N values, measured by ELISA, to IAV of 15 swine farms A) ELISA results classified by year and B) ELISA average results classified by age-range. Red line shows threshold level (S/N ratio of 0.6) for a positive result.

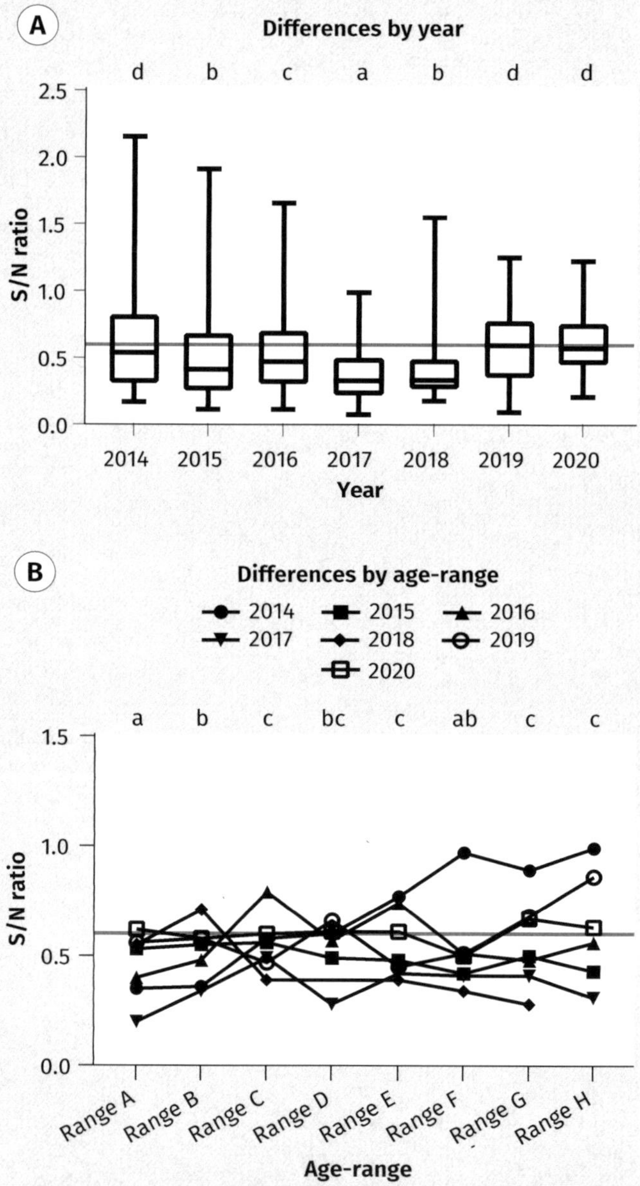

[abcd] Different letters show statistical significance difference among years or average age range, calculated from all the samples of all years in that range ($P < 0.05$).

Different from 2014, where there is a trend to increase S/N values (pigs are getting negative to the test), which could be because to the introduction of SIV vaccines to the country in 2013, that was used in sows. In contrast, in 2018 there is a tendency to decrease the S/N values over the time, and most of the population was always positive.

The data suggest that the passive immunity could drops around 6 to 8 weeks of age, but apparently it overlaps with seroconversion by field infection, with a high percentage of positive animals in the different age ranges.

References

1. Lewis, N., et al. 2016. elife, 5, e12217.
2. Haden, C., et al. 2012. AASV Annual Meeting, 75-76.
3. Tse M., et al. 2012. Clinn Vaccines Immunol 19, 334-337
4. Panyasing, Y., et al. 2016. Transbound emerg dis, e328-e338

Time to first wild-type PRRS virus detection and risk factor analysis in wean to finish pigs

Jose Angulo[1,2]; Samantha Jansen[1]; My Yang[1]; Albert Rovira[1]; Peter Davies[1]; Jeff Zimmerman[3]; Montserrat Torremorell[1]
[1]University of Minnesota; [2]Zoetis Inc; [3]Iowa State University

Introduction

Despite the significant cost of PRRS, we have limited information on the distribution and frequency of PRRSV infections in growing pigs as well as potential risk factors involved in the introduction of wild-type viruses particularly in vaccinated herds. The objective of this study was to determine wild-type PRRS virus (WT-PRRSV) detection time in vaccinated growing pigs housed in wean to finish sites as well as risk factors associated to WT-PRRSV detection.

Material and methods

We collected data from wean-to-finish (WTF) sites (n = 63) across 10 production systems in the Midwest US. One cohort of pigs per site was enrolled and site selection criteria included: 1) pigs sourced from PRRSV negative or stable breeding herds (based on the AASV PRRSV breeding herd classification), 2) pig placement in all-in/all-out sites located in medium to high density areas, 3) willingness to collect monthly oral fluid samples for PRRSV testing, and 4) willingness to share site level information on production and management practices. 60 WTF sites fulfilled the criteria as three WTF sites where PCR positive at placement and were excluded from the study.

Most cohorts studied, 90% (54 of 60), used PRRS modified live virus (MLV) vaccine and 56% (30 out of 54) of the herds used the vaccines either at placement or within two weeks post-placement into the WTF, and 44% (24 out of 54) used the vaccines before weaning either at processing or around weaning age. The PRRSV status from the breeding herds that sourced the pigs to the WTF sites were negative without using PRRSV modified live virus (MLV) vaccine (52%, 31 out of 60), or negative using a PRRSV MLV (48%, 29 out of 60). Oral fluid samples were collected at each WTF site at 3, 8, 12, 16, 20 and 25 weeks post-placement using fixed spatial sampling (8 ropes per site). Oral fluid samples (N = 2,584) were tested individually by PRRSV RT-PCR. Sequencing (ORF5) was performed at each positive sampling event on the sample with the lowest RT-PCR cycle threshold (Ct) value. PRRSV-ORF5 sequences were aligned and classified as wild-type PRRSV using a cut-off of > 2.0% nucleotide difference from vaccine viruses references. In addition, we captured site level information on site characteristics, demographics, management practices, vaccine protocols and biosecurity practices using a survey administered to the production system veterinarians.

Results and discussion

A Kaplan-Meier survival analysis was conducted in all vaccinated sites considering first wild-type PRRSV detection and then survival curves were compared between sites that received pigs vaccinated prior to weaning and sites with pigs vaccinated after weaning into the WTF site. Fifty two percent (95% CI 37, 67) of WTF sites did not have WT-PRRSV detected during the study period. The mean for time to first WT-PRRSV detection in all sites was 20.6 weeks (95% CI 19, 22). When comparing sites vaccinated at the sow farm versus those vaccinated at the WTF, the mean time to detection was 19 weeks and 22 weeks respectively, with no statistical differences for survival curves (*P*-value 0.211). Logistic regression included thirteen variables captured in surveys having detection to WT-PRRSV as response variable. A relevant effect in increased WT-PRRSV detection risk (odd ratios) was observed for using external crew for unloading pigs and using outsourced truck for disposal of dead pigs. A decrease risk was observed for receiving pigs from a single source, having all-in-all-out by site, truck driver helping unloading pigs and having dedicated equipment to the site for moving dead pigs. PRRSV infections were distributed throughout the growing period with WT-PRRSV detection being more frequent towards the end of the growing phase. Our results suggest that timing of PRRSV vaccination does not have an effect on timing of WT-PRRSV detection highlighting the importance of biosecurity. Overall, we identified some practices that appear relevant to improve biosecurity and minimize the risk of WT-PRRSV circulation in wean to finish pigs.

A longitudinal study on PRRSV detection in swine herds with differing demographics and PRRSV management strategies

Tariku J. Beyene; Andreia G. Arruda
Department of Veterinary Preventive Medicine, The Ohio State University, Columbus, Ohio

Introduction

Detection of porcine reproductive and respiratory virus (PRRSV) especially under low prevalence scenarios remain one of the main challenges to understand, prevent and control PRRS in swine herds. The main objective of this longitudinal study was to describe detection of PRRSV in farms of different demographics and under different PRRSV management strategies over a one-year period, and investigate the association between PRRSV PCR positivity and predictors of interest.

Materials and methods

Five swine farms were selected and recruited in a volunteer basis: three breeding herds and two growing pig herds. The farrow-wean herds consisted of a 4,000-sow herd that had a PRRSV outbreaks 6 months prior to enrollment and that was not using vaccination or field exposure (partially immune), a 6,000-sow herd that had a new outbreak at enrollment, and a 2,500-sow herd using a modified-live virus (MLV) vaccine program (all sows twice a year). The two grower herds included a 3,550-head wean-finish herd that was PRRSV naïve with a new PRRS outbreak at enrollment and no vaccine administered; and a 2,800-head finisher for which both sow and piglets in the source herd were vaccinated with an MLV. Sampling was conducted monthly for approximately 12 months during February 2019 - March 2020 using a combination of tonsil scrapings (individually) and processing fluids (composite) for breeding herds; and tonsil scrapings (individually) and oral fluids (composite) for growing pig herds. For each sampling event, sixteen total samples (8 for each sample type) were collected using fixed spatial sampling.[1] Samples were sent to the University of Minnesota and tested using RT-PCR. A Ct value threshold of 35 was used as cut-off.[2] Multivariable mixed effect logistic regression models were built separately for breeding and growing pig herds to investigate whether farm and sampling type were associated with a positive result, accounting for repeated sampling over time. Results were used to estimate the variability of PRRSV detection between farms. $P < 0.05$ was considered significant, and all statistical analyses were conducted using R software (version 3.5.3).

Results and discussion

In this study, 710 samples were collected across all farms. The overall PRRSV detection risk by PCR across all samplings and all herds was 21.1%; with the highest total proportion positive (35.6%) being in the finisher herd that was vaccinated with MLV; and the lowest total proportion positive (4.8%) being in the breeding herd that had an outbreak 6 months prior. PRRSV detection risk in breeding herds using tonsil scrapings and processing fluids were 5.1% [95% CI: 3.6, 6.5%] and 20.7% [95% CI: 16.7, 24.7%], respectively. Overall, processing fluids yielded the largest proportion of PCR positive results compared to tonsil scrapings. In growing pig herds, detection risks using tonsil scrapings and oral fluids were 52.3% [95% CI: 49.1, 55.5%] and 16.2% [95% CI: 13.8, 18.9%], respectively. For this case, tonsil scrapings yielded the largest proportion of positive PRRSV detection when compared to oral fluids. Finally, breeding herd model results showed a significant variation of PRRSV detection between farms, with almost half of the total variation in PRRSV positivity being explained by the farm in which the sample was collected at. This was not true for growing pig herds. The PCR method used so far in this study does not discriminate between vaccine and field strains of the virus, therefore, whole genome sequencing is underway.

References

1. Rotolo ML et al. Sampling guidelines for oral fluid-based surveys of group-housed animals. *Vet. Microbiol.* 2017;209:20-29.

2. Kim WI et al. Effect of genotypic and biotypic differences among PRRS viruses on the serologic assessment of pigs for virus infection. *Vet. Microbiol.* 2007;123:1-14.

Investigation of Enterisol® Ileitis given intramuscularly and with the presence of feed-grade antimicrobials

Dana Beckler, DVM, MS
Gut Bugs Inc., Fergus Falls, Minnesota

Introduction

Enterisol® Ileitis vaccine, is a highly effective, successful product approved for use in healthy post-weaning pigs for the prevention of porcine proliferative enteropathy due to *Lawsonia intracellularis* to be administered via the drinking water, or via oral drench. No data exists on the efficacy of this product administered via intramuscular injection when combined with 3FLEX® vaccine and when administered in conjunction with concomitant antimicrobial therapy. The objectives of this study were to evaluate the efficacy of Enterisol® Ileitis vaccine when combined with 3FLEX® vaccine and administered via intramuscular injection as well as to evaluate the efficacy of this vaccine combination when administered to pigs receiving concomitant in-feed antimicrobial treatment.

Materials and methods

Pigs were obtained post-weaning at 17-21 days of age and randomly divided into 3 treatment groups with 24 animals each. The treatment groups were: non-vaccinated positive challenge control (PC); Enterisol® Ileitis given IM with 3®FLEX vaccine (EIIM); Enterisol® Ileitis given IM with 3®FLEX vaccine with concomitant antimicrobial administration (EIIMATB). Vaccine preparation was done by reconstituting the lyophilized form of Enterisol® Ileitis with 3FLEX® vaccine. This resulted in a final 2 ml dose containing modified live *L intracellularis* antigen along with PCV2, *Mycoplasma pneumonia* and PRRSV MLV vaccine fractions. To investigate if in-feed antimicrobial treatment would inhibit the efficacy of vaccination with the live modified Enterisol® Ileitis vaccine, the EIIMATB group received in-feed tiamulin (35 ppm) and chlortetracycline (400ppm) beginning one week prior to vaccination, and continuing through one week post-vaccination. All animals were challenged with mucosal homogenate containing virulent *L intracellularis* 28 days post vaccination and necropsied 21 days later. Clinical signs, average daily weight gain (ADG), fecal shedding, as well as macroscopic and microscopic lesions were evaluated.

Results

Both vaccinated treatment groups led to a significant reduction in diarrhea clinical score ($P < 0.05$). Both vaccinated groups also led to a significant ($P < 0.5$) increase in ADG following challenge. The PC group had a post challenge ADG of 1.43 lbs while the EIIM and EIIMATB groups had ADGs of 1.65 and 1.71 lbs, respectively. Macroscopic lesions scores were decreased in both the vaccinated groups with values of 0.67 (EIIM, $P < 0.05$) and 1.04 (EIIMATB, $P = 0.2$) compared to the non-vaccinated (PC) group which had an average score of 1.55. Similarly, microscopic lesions were also decreased by vaccination with an average score of 2.05 in the PC group compared to averages of 1.29 ($P < 0.05$) and 1.57 ($P = 0.14$) in the EIIM and EIIMATB groups, respectively. The IM vaccine combination was found to be safe as there were no adverse reactions.

Conclusions

The two injectable vaccine groups compared to the non-vaccinated group following *L intracellularis* challenge improved ADG, led to a reduction of macroscopic and microscopic lesion scores, reduced the fecal shedding of *Lawsonia* and reduced clinical signs. Little or no evidence of antimicrobial interference was observed with the combination of Enterisol® Ileitis and 3FLEX® intramuscular vaccine. This study demonstrates a new tool for swine producers that warrants further investigation.

Turnover events of animal caretakers and productivity in Ohio swine farms

N. J. Black, MPH-VPH; A. G. Arruda, DVM, MS, PhD
Department of Veterinary Preventive Medicine, The Ohio State University, Columbus, Ohio

Introduction

Attracting and maintaining quality animal caretaking personnel has been one of the major issues the US swine industry has faced in recent years. Turnover rates among caretakers in swine farms in the US have been reported to be between 20 and 30% depending on farm size.[1] Furthermore, turnover events can be costly and may have an impact on productivity and overall animal health. The objective of this study was to investigate the association between turnover events and herd productivity parameters during a one-year period in eleven breeding farms within two production systems in the state of Ohio.

Methods

A retrospective study was conducted using eleven farms within two production systems located in Ohio. Human resources and production data for the year of 2019 was obtained at the week level. A linear mixed-effects model was fit in STATA 15 using restricted maximum likelihood estimation. The primary outcome of interest was the number of pigs weaned per sow the week following a turnover event (short-term effect). Fixed-effects included a turnover event, which was defined as when an employee left the farm, whether the employee left voluntarily (employee decided to leave or the company made the decision), the average number of employees on the farm, average herd size, season, and preweaning mortality (which was used as a proxy for disease occurrence). Random effects included farm and production system to account to the effect of clustering. Univariable models were fit to screen for potential variables to be offered for final model selection, using a cutoff of $P < 0.2$. Final model was selected through a backwards stepwise approach. A P-value < 0.05 was used to declare significance and $P < 0.10$ was used to declare tendency.

Results

Systems 1 and 2 contributed 2 and 9 farms, respectively. Average herd size among the farms ranged from 1,225 to 6,180 pigs with a median of 2,500 pigs and average number of full time employees ranged from 5 to 20 employees with median of 12 employees. There were a total of 152 turnover events during the 52 weeks in 2019, with 4 and 148 turnover events in systems 1 and 2, respectively. Of these, 3 and 15 of the turnover events were involuntary. The frequency of turnover events among all the farms over the examined period ranged from one to 32 events, with five of the nine farms in system 2 having more than 20 turnover events in a single year. The mixed-effect model did not show a significant association between an employee turnover event and the number of pigs weaned per sow the following week, after accounting for herd size, number of employees, preweaning mortality, season, system and farm. Herd size was significantly associated with the number of weaned pigs per sow, with an increase of every 1000-head resulting in an increase of one pig weaned for every five sows ($P = 0.019$). Season was also significantly associated with number of pigs weaned per sow, with the highest productivity occurring in the spring and the lowest in the summer. The within-farm intraclass correlation coefficient (ICC) was 0.17, indicating that 17% of the unexplained variability in number of pigs weaned per sow can be attributable to between farm differences within a system. The within-system ICC was negligible, indicating little variability between systems.

Conclusion

This analysis provided insight into the amount of employee turnover that swine producers can experience in a single year. We have found no association between the turnover events and production when examining short-term effects (production impact in the week following turnover). Considering the time period between turnover events and production impacts is unknown, further analysis will explore long-term effects, and incorporate external factors such as county's unemployment rates.

Reference

1. National Pork Board and AgCareers, 2017. Employee compensation & HR practices in pork production. Pork Board, Des Moines, IA, USA. Available at: **http://porkcdn.s3.amazonaws.com/sites/all/files/documents/PorkStore/04937.pdf**.

Inactivating multiple coronavirus interferon antagonists attenuates pathogenesis of porcine epidemic diarrhea virus

A. C. Buckley[1]; X. Deng[2]; A. E. Pillatzki[3]; K. M. Lager[1]; K. S. Faaberg[1]; S. C. Baker[2]
[1]Virus and Prion Research Unit, USDA-ARS-National Animal Disease Center, Ames, Iowa;
[2]Department of Microbiology and Immunology, Loyola University Chicago, Maywood, Illinois;
[3]Animal Disease Research & Diagnostic Laboratory, South Dakota State University, Brookings, South Dakota

Introduction

Porcine epidemic diarrhea virus (PEDV) is a coronavirus that causes diarrhea and mortality in neonatal pigs. Available efficacious vaccines to prevent PEDV infection are limited and many producers continue to use feedback to boost farm immunity, which can carry the risk of unintentional exposure of animals to other agents. The objective of this research was to create an alternative PEDV vaccine candidate by attenuating the virus. Attenuation in this study was focused on inserting mutations into proteins known to block host interferon responses also known as interferon antagonists.

Materials and methods

An infectious clone of wild-type PEDV (icPEDV-WT) was previously generated and used to engineer a recombinant virus with four mutations (icPEDV-mut4). The four mutations were located in 3 proteins: nonstructural proteins 1, 15, and 16. Sequencing was performed to confirm mutations. Growth kinetics of both viruses were compared on Vero and PK1 cell lines and interferon gene expression was measured by RT-qPCR .

Thirty-four piglets were weaned between 3-6 days of age and assigned to 3 treatment groups: icPEDV-WT (n = 13), icPEDVmut4 (n = 13), and controls (n = 8). Piglets were inoculated orally with 2 mL of virus solution with a titer of 500 $TCID_{50}$/pig or sham media. Piglets were rectal swabbed and given a clinical diarrhea score on 0-10, 12, 14, 18, and 21 days post inoculation (dpi). Clinical diarrhea scores were assigned with the following criteria: 0 = normal feces, 1 = soft stool, 2 = semi-liquid stool, 3 = liquid feces, 4 = voluminous watery diarrhea. Serum was collected on 0 and 21 dpi. Two piglets were euthanized on 2 dpi for collection of jejunum and ileum tissue samples and the remaining animals were euthanized on 21 dpi. Swabs were tested by RT-qPCR for the presence of virus and serum tested for IgG and neutralizing antibodies. Structural damage and virus location in tissues were analyzed by H&E and immunohistochemistry (IHC) respectively. Viral sequencing for the regions of site-specific mutation was completed on fecal swab RNA from late in infection.

Results

Growth kinetics on interferon non-responsive Vero cells were similar between icPEDV-mut4 and icPEDV-WT. In contrast, growth on interferon responsive PK1 cells of icPEDV-mut4 was reduced 1000-fold compared with wild-type, which would suggest that reduced interferon antagonism resulted in limited viral replication. This was supported by higher levels of interferons measured in icPEDV-mut4 infected PK1 cells compared to icPEDV-WT infected cells.

All piglets inoculated with icPEDV-WT developed watery diarrhea and achieved a score of at least 3 for one day, while all piglets inoculated with icPEDV-mut4 had scores of 0 or normal feces for the entire study. Viral shedding quantified by RT-qPCR was consistently 100-1000 times lower in rectal swabs of icPEDV-mut4 infected animals. All challenged pigs developed an antibody response to infection, though icPEDV-mut4 piglets had lower IgG and neutralizing antibody levels compared to icPEDV-WT. Histology of intestinal sections from icPEDV-mut4 piglets showed no significant pathologic lesions. In contrast, icPEDV-WT piglets had classic microscopic lesions with villous atrophy. IHC detected viral antigen in icPEDV-WT tissue sections but not in icPEDV-mut4 tissues. Sequencing of the viruses over regions of mutation revealed that the mutations were stable.

Conclusion

Inactivating multiple interferon antagonists created an attenuated PEDV. Further research will need to evaluate the efficacy of this virus as a vaccine candidate.

Determining the minimum infectious dose of *Mycoplasma hyopneumoniae* using diluted pooled tracheal sample fluid as an inoculum administered intratracheally

B. Burton, DVM; R. Merica, PhD
Suidae Health & Production, Algona, Iowa

Introduction

*Mycoplasma hyopneumoniae (*Mhp*)* is a major respiratory pathogen that significantly reduces the growth potential of pigs. Since economic losses due to *M hyopneumoniae* infection can be quite significant,[1] producers have turned to eradicating the pathogen from sow herds. While research has shown that lateral transmission is generally minimal, vertical transmission from sow to piglet is believed to be the greatest contributor to the propagation of the pathogen.[2] Many Mhp elimination strategies are described in the literature, but a common theme among most closure protocols is exposing replacement gilts to the pathogen. Use of Mhp-infected lung homogenate as an inoculum has remained relatively unchanged since its introduction in 1971.[3] Historically, cultures of the pathogen have been used experimentally to infect pigs[4] but due to the fastidious nature of the bacterium, this proved to be a difficult undertaking. Preliminary results from the Suidae Health & Production laboratory have demonstrated that the Mhp inoculum can be obtained through antemortem tracheal sampling and used to infect gilts in a herd closure scenario.

Objective

The main objective of this study was to determine the minimum dose of Mhp required to infect gilts intratracheally using various dilutions of tracheal fluid that had been obtained by pooling samples from Mph-infected pigs.

Materials and methods

There were four treatment groups with 12 pigs in each. The inoculum was prepared from previously tested tracheal fluid diluted with Friis media. Pigs in treatment groups 2, 3, and 4 were inoculated with a 1:3, 1:9, and 1:27 dilution, respectively, of the inoculum used for group 1. All doses of the inoculum were tested by Mhp PCR and Ct values were as followed for group1-4 in that order: 23.5, 24.4, 26.3, 28.0.

48 gilts were randomly selected, tagged for identification, and assigned to one of four treatment groups. Gilts in each group were penned together and treatment pens were evenly spaced throughout the barn with as much distance as possible between them. Once pigs were assigned to a treatment group, they were tracheal sampled and then inoculated intratracheally with the appropriate dose of Mhp. Pre-inoculation tracheal samples from all 48 gilts were sent to ISU-VDL and tested individually for Mhp by qPCR to confirm the negative status of the herd. All subsequent tracheal samples taken throughout the course this study were also individually tested for Mhp by qPCR.

At 7 days post-inoculation (DPI), tracheal samples were taken from all 48inoculated pigs as well as and 48 uninoculated pigs that shared the same treatment pen (12 per group). The 48 uninoculated pigs were tagged for identification if follow up testing was required.

Results

The treatment 1 inoculated group resulted in 100% positive gilts 7 days after inoculation (see Figure 1) with an average Ct of 30Inoculated gilts in treatment groups 2-4 also showed high rates of infection that ranged from 75 to 92%. Treatment 1 and 3 uninoculated gilts were 25% positive7 days PI suggesting there was some transmission of Mph within the pen. All gilts including those unioculated that were found to be negative on day 7 retained that status through 21 days post inoculation.

Discussion

All doses of Mph tested were found to infect between 75-100% the gilts in each group. While the highest dose resulted in 100% Mph positive gilts, no association of dose and infection rate was observed indicating that the minimal dose was not reached. While the majority of inoculated gilts were successfully infected with Mph, no evidence of transmission to unioculated gilts was found after day 7 PI. A confounding factor in the spread of Mph to uninoculated gilts may have been that the herd was infected with the Influenza virus just at the onset of the study.

The goal of this project is to provide the industry a better alternative to using lung homogenate for exposure to Mhp. By using the fluid that is collected during tracheal sampling, a farm reduces the economic impact of losing a replacement gilt, avoids the welfare concern of euthanizing healthy pigs, and reduces the risk of potential adverse health effects on exposed animals.

References

1. Schwartz, M., 2015: Cost of *M hyopneumoniae* in growing pigs. Allen D. Leman Conference.

2. Yeske, P. 2018. *Mycoplasma hyopneumoniae:* Lateral transmission and gilt exposure methods. AASV 2018 Proceedings. 482-484.

3. Kuhn R.G. 1971. The Action of Certain Antibiotics and Ether on Swine Zoonotic Pneumonia. Can. J. Comp. Med. January 1971. Vol. 35, 1-4.

4. Kobisch, M. et al. 1996: Experimental infections of swine. Tully J.G., Razin S, eds. Molecular and Diagnostic Procedures in Mycoplasmology. Vol II. San Diego, CA: Academic Press. 371-376.

Figure 1: Percent Mhp PCR positive 7 days post-inoculation

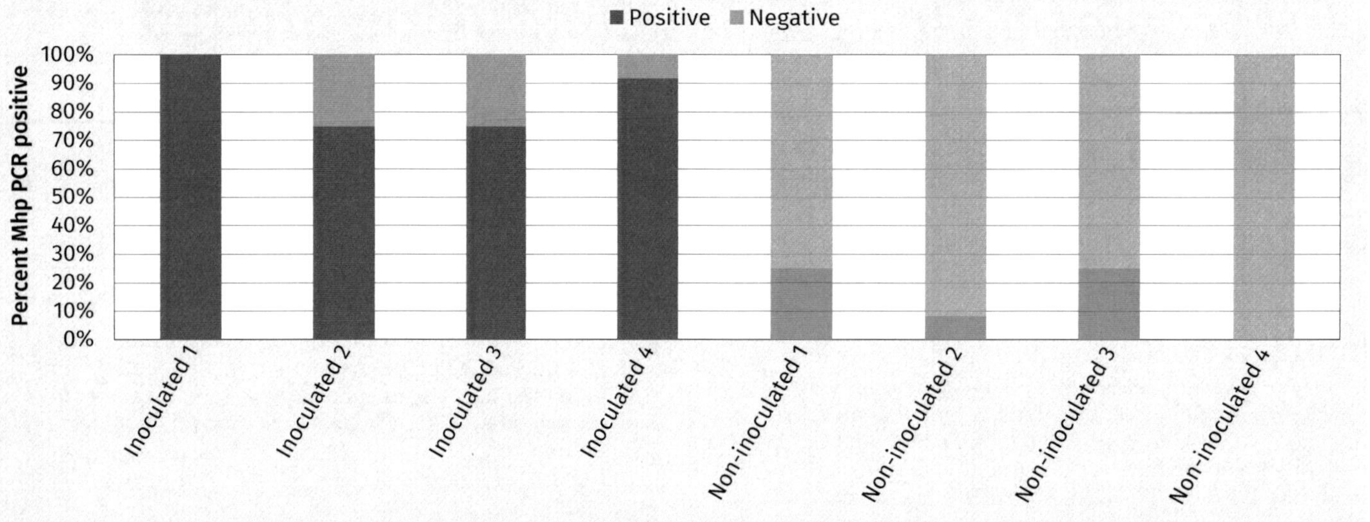

Metabolomic and biochemical profile of piglet serum and processing fluids

Albert Canturri[1]; Fernanda Leal[1]; Qingqing Mao[2]; Carles Vilalta[1]; Chi Chen[2]; Maria Pieters[1]
[1]Department of Veterinary Population Medicine, University of Minnesota, St. Paul, Minnesota;
[2]Department of Food Science and Nutrition, University of Minnesota, St. Paul, Minnesota

Introduction

The use of metabolomics in serum, body secretions and tissues is being increasingly employed to identify interactions between hosts and pathogens, both in human and veterinary research. The discovery of biomarkers through metabolomic analyses of viral, bacterial and parasitic infections can potentially be useful in the diagnosis, control and eradication of infectious agents. However, there is need to explore and determine the metabolomic and biochemical characteristics of normal, uninfected specimens.[1]

The collection of processing fluids, which are the serosanguineous transudate from tail docking and castration of piglets, is a trending sampling strategy in the field and it is commonly used for monitoring and diagnosis of infectious diseases at herd level.[2] Therefore, the objective of the present study was to characterize the metabolomic profile of serum, tail fluid and testicle fluid, both by quantifying their metabolites and by modeling their metabolome differences in terms of free amino acids, free fatty acids and complex lipid species. The biochemistry profile of all specimens was also characterized.

Materials and methods

Serum, testicle fluids and tail fluids were individually collected from piglets in a sow farm, which was negative to *M hyopneumoniae,* PEDV and PPRSV, and separately analyzed by gender. For males, 10 testicle fluid samples (2 males from 5 different litters), 9 serum samples from the same animals and 10 tail fluid samples (pool of 5 tails from other males in each litter) were collected. For females, 9 serum samples and 10 tail fluid samples were similarly collected. These individual samples were analyzed using a liquid chromatography-mass spectroscopy based metabolomics platform, which included fluid preparation, chemical derivatization (both dansyl-chloride-based and 2-hydrazine quinoline-based derivation), data deconvolution processing and multivariate analysis (MDA), followed by targeted and untargeted marker characterization and quantification. An extended chemistry panel of the clinical specimens was performed via automated analyzers at the Veterinary Clinical Pathology Laboratory of the University of Minnesota.

Results and discussion

Testicle fluid and serum shared similar amino acid (AA) profiles, with the same five most relatively abundant AA (i.e. Glycine, Proline, Glutamate, Alanine and Glutamine) although in different order. On a Principal Component Analysis model, the score plot showed that tail fluid were clearly separated from testicle fluid and serum. Based on MDA, testicle fluid and serum samples showed significantly different metabolite profiles, driven by differences in the absolute concentration of free AA and the complex lipids profile. Biochemical analyses mostly revealed quantitative similarities among individual analytes between serum and testicle fluid. However, the quantity of some analytes such as aspartate aminotransferase, alkaline phosphatase or creatine kinase was notably different. For tail fluids, the majority of values were below the reportable range limit of that particular analyte's linearity. No significant differences were detected by gender. The description of the metabolomic profiles of serum and processing fluids might help at setting baseline values when performing studies attempting to identify metabolic changes associated with infections. Furthermore, the data suggests that processing fluids are metabolically rich substrates and they could eventually represent an adequate niche for bacteria to thrive, as their nutritional requirements for maintenance would be fulfilled. These results justify additional research on the metabolic changes associated with common infections at the herd level and on the characterization of the host responses to specific infectious agents.

References

1. Voge NV, Perera R, et al., Metabolomics-Based Discovery of Small Molecule Biomarkers in Serum Associated with Dengue Virus Infections and Disease Outcomes. PLoS Negl Trop Dis. 2016 Feb 25;10(2).

2. Lopez WA, Angulo J, Zimmerman JJ, et al. Porcine reproductive and respiratory syndrome monitoring in breeding herds using processing fluids. *J Swine Health Prod.* 2018;26(3):146-150.

The impact of vaccination strategy on between-farm PRRSV transmission

Jason A. Galvis[1]; Joaquin M. Prada[2]; Cesar Corzo[3]; Gustavo Machado[1]
[1]Department of Population Health and Pathobiology, College of Veterinary Medicine, Raleigh, North Carolina; [2]School of Veterinary Medicine, Faculty of Health and Medical Sciences, University of Surrey, Guildford, United Kingdom; [3]Veterinary Population Medicine Department, University of Minnesota, St Paul, Minnesota

Introduction

Porcine reproductive and respiratory syndrome virus (PRRSV) continues to cause substantial economic losses for the pork industry.[1] Immunization is by far the most utilized PRRSV control strategy, however there is limited information of how different vaccination programs help preventing between-farm PRRSV spread. Therefore, it is imperative to investigate which vaccination strategy is most effective to prevent between-farm transmission. In this study, we developed a between-farm PRRSV spread model to quantify the contribution of three main transmission routes on PRRSV spread, and measure the impact of five PRRSV vaccination strategies on the number of new outbreaks.

Materials and methods

We built a between-farm PRRSV transmission model with observed outbreak data from three production systems within one US state. The model allows PRRSV to spread between farms by three main transmission routes: pig movements, local transmission (e.g. airborne) between proximal farms, and reinfection driven by the resurgence of the virus (re-break). The model accounts for variables affecting transmission, such as PRRSV seasonal effects and the effect of vegetation coverage, which acts as a barrier against local transmission. Using the model, we estimated the most likely route of transmission for each farm type (sow, gilt development unit [GDU], nurseries and finishers). In addition, we simulated five vaccination strategies. Two prophylactic: #1) all piglets vaccinated at weaning; #2) in addition to vaccination at weaning, every pig moved into a GDU or finisher farm received a second vaccine booster. Two reactive: #3) mass vaccination of sows after PRRSV is detected; #4) vaccination of any farm type after PRRSV is detected; and #5) the combination of strategies #2 and #4 ("combined"). In our model exercise, we evaluated scenarios varying the vaccines efficacies (*ve*) from 1% to 90% of whole-herd protection.

Results and discussion

Among the most important routes of transmission by farm type, 59% ad 99% of new sow and GDU farms infections were related to local transmission, respectively, while 80% of infections were related to pig movements in nursery and both routes accounted for almost 50% in finisher farms. Vaccine efficacy (*ve*) 1% was sufficient to reduce 5% of the total number of PRRSV outbreaks (Figure 1). At this efficacy, the combined strategy #5 had the greatest impact, reducing the number of new infections by 14%. Meanwhile, prophylactic strategies #2 and #1 reduced cases by 10% and 6%, respectively, and the reactive strategies #3 and #4 reduced cases by 5%. In terms of between-farm transmission, the combined strategy was the only one to completely contain transmission (*ve* > 80%). However, at similar (*ve* < 15%) prophylactic #2 and reactive #4 successfully controlled more than 25% of the new PRRSV spread events. Finally, prophylactic #1 and reactive #3 strategies had marginal effects on controlling PRRSV spread. This work advances our understanding of PRRSV between-farm transmission and formally estimates the number of prevented PRRSV outbreaks by different vaccination strategies. Our results show that local transmission generated the vast majority of the infections at sow and GDU farms, but nurseries were more likely to be infected by direct incoming movement of infected animals. In finisher farms, both routes were equally important. The best vaccination strategy included the immunization of finisher farms, a PRRSV control strategy that has recently received more attention, but still requires a benefit cost analysis. Vaccination alone was insufficient for completely controlling PRRSV spread. Until a vaccine with an efficacy of at least 80% becomes available, the industry should continue to rely on other strategies, such as reinforcement of on-farm biosecurity. Our model could be a useful tool for veterinarians, allowing them to identify the effect of transmission routes and different vaccination interventions to control PRRSV spread.

Acknowledgement

Morrison Swine Health Monitoring Project. Funding from USDA-NIFA: 2019-07452.

Reference

1. Holtkamp, DJ et al. Assessment of the economic impact of porcine reproductive and respiratory syndrome virus on United States pork producers. *J. Swine Health. Prod.* 2013;72:84.

Figure 1: Estimated reduction of PRRSV infected farm trajectories under five vaccine efficacy scenarios. A) The total number of prevented infected farms and B) the number of prevented infected farms depicted by production type. The y-axis represents the number of infected farms within the 27 weeks and the x-axis the simulated vaccine efficacies. Among the prophylactics strategies farms were vaccinated in the same week of receiving a pig movement and among the reactive strategies vaccination occurred in the same week a farm reported an outbreak. The line plot represents one upper and lower standard deviation and the median value represented as the symbol in the center of the line, values calculated from 50 simulations by each vaccination strategy.

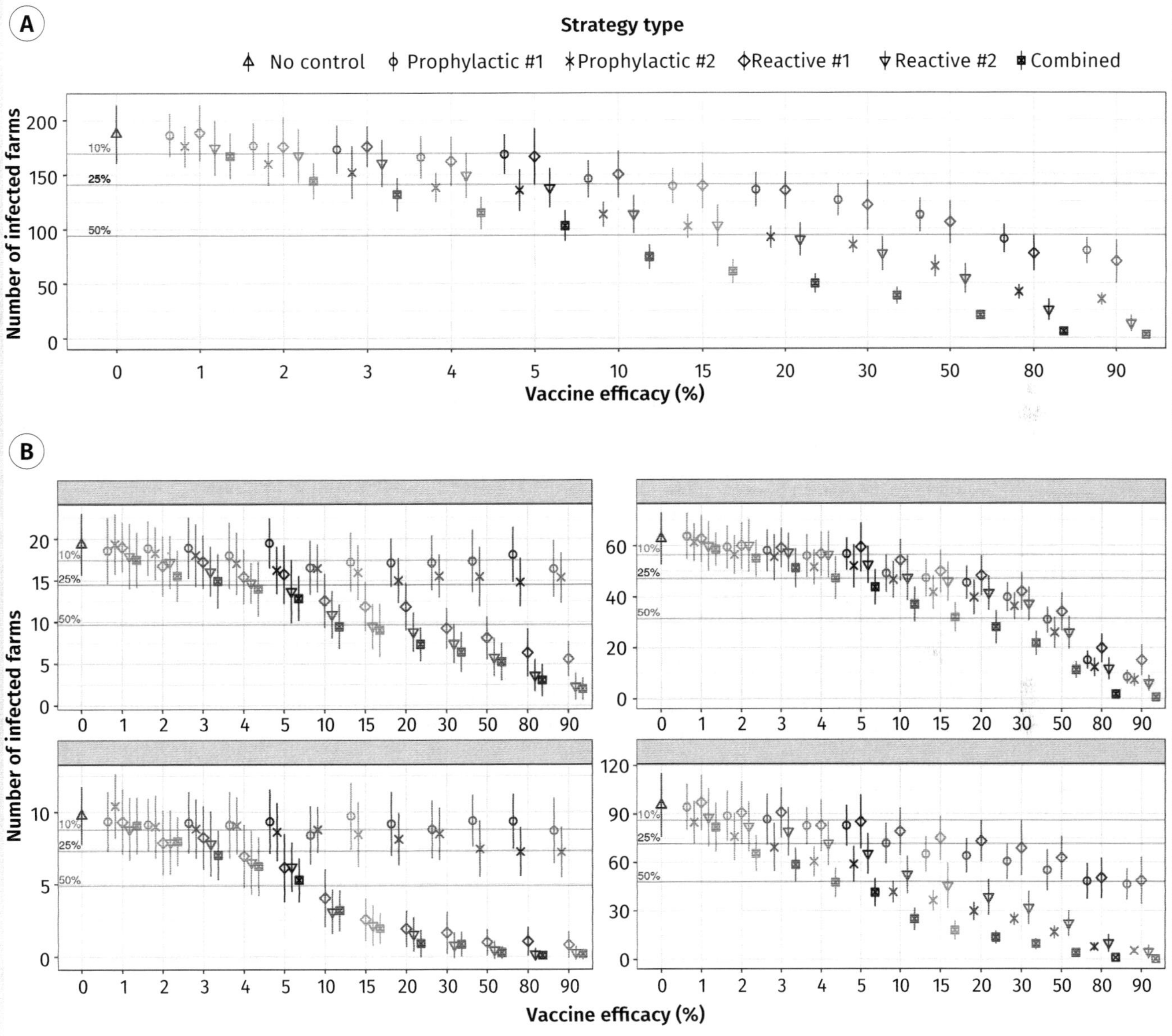

A comprehensive investigation of pig lameness associated to *Mycoplasma hyosynoviae*

M. Surendran Nair[1]; B. Garcia-Morante[1], DVM, MS, PhD; David Pillman[1]; Cathy Carlson[2]; Chi Chen[3]; Maria Pieters[1], DVM, PhD

[1]Veterinary Population Medicine Department, College of Veterinary Medicine, University of Minnesota, St. Paul, Minnesota; [2]Veterinary Clinical Sciences Department, College of Veterinary Medicine, University of Minnesota, St. Paul, Minnesota; [3]Department of Food Science and Nutrition, University of Minnesota, St. Paul, Minnesota

Introduction

Mycoplasma hyosynoviae (*M hyosynoviae*) has become an increasing concern to swine producers as a causative agent of arthritis. This pathogen, however, has been also detected in joints of pigs with no apparent clinical lameness.[1,2] Henceforth, it is still difficult to correlate diagnostic data to the actual disease prevalence in the field. The aim of this work was to investigate the potential association between detection of *M hyosynoviae* in oral fluid (OF) and synovial fluid (SF) samples and the clinical presentation of lameness in growing pigs.

Materials and methods

Five wean-to-finish farms reporting recent history of lameness were enrolled in the study. Oral fluids were collected by pen (9 to 10 samples collected by site). Real-time PCR was performed on each OF to detect *M hyosynoviae*. Lameness scores (0-4) were recorded in pigs from pens where OF were collected.[1] In addition, three finisher pigs, one healthy pig (score 0) and two lame pigs (score ≥ 3) from each farm were euthanized to collect synovial membrane and SF samples from multiple joints. In one farm, an extra healthy pig was sampled, accounting for a total of 16 pigs. The collected synovial membrane and SF samples were evaluated by histopathology and real-time PCR, respectively. The primers and probes were synthetized following the *M hyosynoviae* detection standards from the University of Minnesota, Veterinary Diagnostic Laboratory. Samples were considered positive to *M hyosynoviae* by real-time PCR when Ct ≤ 37. Real-time PCR results as well as histopathology variables were used for tests on differences between lame and healthy pigs by Fisher's exact test.

Results

Mycoplasma hyosynoviae was detected in all OF from the five enrolled farms. However, lameness (score ≥ 2) was observed in 4.42% of the assessed pigs across all five sites. Nine out of the 10 clinically lame pigs were positive to *M hyosynoviae* by real-time PCR in SF, at least in one of the sampled joints. From healthy pigs, three pigs tested positive to *M hyosynoviae* in at least one of the joints sampled. Lameness scores determined by visual evaluation of pigs were not correlated with detection of *M hyosynoviae* in OF and/or SF. Detection of *M hyosynoviae* in SF was also poorly correlated with histological alteration of the synovial membrane, as positive results were obtained in joints with no microscopic arthritic lesions and *vice versa*, both in lame and healthy pigs.

Discussion and conclusions

Despite the understanding of the clinical disease, the true prevalence, incidence, and dynamics of infection of *M hyosynoiviae* remain unclear. Our results suggested that detection of *M hyosynoviae* in SF as well as in OF samples by real-time PCR is not necessarily associated with the presentation of clinical lameness, neither at the individual nor at the population levels.

References

1. Nielsen EO et al. *Mycoplasma hyosynoviae* arthritis in grower-finisher pigs. *J Vet Med A Physiol Pathol Clin Med*. 2001;48:475-86.

2. Nielsen EO et al. Use of a novel serum ELISA method and the tonsil-carrier state for evaluation of *Mycoplasma hyosynoviae* distributions in pig herds with or without clinical arthritis. *Vet Microbiol*. 2005;111:41-50.

Detection and characterization of microbial flora and host metabolites in synovial fluids of pigs with clinical signs of lameness

M. Surendran Nair[1], BVSc, MS, MSCTR, PhD; B. Garcia-Morante[1], DVM, MS, PhD;
C. Carlson[2], DVM, PhD, DACVP; C. Chen[3], BVSc, PhD; M. Pieters[1], DVM, PhD
[1]Veterinary Population Medicine Department, College of Veterinary Medicine, University of Minnesota, St. Paul, Minnesota; [2]Veterinary Clinical Sciences Department, College of Veterinary Medicine, University of Minnesota, St. Paul, Minnesota; [3]Department of Food Science and Nutrition, University of Minnesota, St. Paul, Minnesota

Introduction

The presence of microbiota in joints of humans suffering arthritis has been confirmed, suggesting a potential correlation with degree of disease.[1] Lameness in pigs occurs at different production ages and is associated with an array of risk factors that may differ between production systems. To date, the potential role of inherent microbial flora in pig joints has not been evaluated in understanding lameness development. Thus, this study sought to characterize the synovial microbiota and host metabolic changes in joints of pigs with clinical signs of lameness.

Materials and methods

Five wean-to-finish farms reporting recent history of lameness were enrolled in the study. Three pigs from each farm were selected based on lameness scores:[2] one healthy (score 0) and two lame pigs (score ≥ 3). In one farm, an extra healthy pig was selected, accounting for a total of 16 pigs. Synovial fluid (SF) samples were obtained from eight joints per pig (n = 128). Due to low-biomass composition, 21 SF samples passed the quality control to perform the 16S rRNA sequencing to characterize the microbiota in pig joints. However, the SF from all joints were analyzed for determining host metabolite composition using liquid chromatography-mass spectrometry (LC-MS).

Results

Each SF sample analyzed had a characteristic microbiome profile. However, the dominant phyla were Firmicutes (dark blue), Actinobacteria (red), Bacteroidetes (grey) and Proteobacteria (purple). Abundance of erysipelas and genus from the family of *Pasteurellaceae* were observed in 57% and 43% of the analyzed SF samples, respectively. These taxa could potentially include bacteria associated with lameness such a *Glaesserella* (formerly *Haemophilus*) *parasuis* and *Actinobacillus suis*. Metabolomics analysis revealed that within each farm, there was a significant difference in amino acid composition profiles in SF between the healthy and lame pigs. The altered amino acid biomarkers identified in each farm were found to be associated with the Krebs cycle as well as amino acid metabolism pathways, indicating increased protein catabolism in the affected joints.

Discussion and conclusions

To date, no literature has analyzed intra-articular bacterial nucleic acids in pigs using high-throughput techniques. Overall, this study suggests existence of bacterial nucleic acids in SF samples from both healthy and lame pigs. The potential association between joint microbiota and lameness remains uncertain. Nevertheless, the results from microbial ecology analysis in joints point out the importance of considering co-infections when assessing lameness etiology. Besides, lameness may result on metabolomic changes in SF from pigs of the same farm.

References

1. Zhao Y et al. Detection and characterization of bacterial nucleic acids in culture-negative synovial tissue and fluid samples from rheumatoid arthritis or osteoarthritis patients. *Sci Rep.* 2018;8:14305.

2. Nielsen EO et al. *Mycoplasma hyosynoviae* arthritis in grower-finisher pigs. *J Vet Med Ser A*. 2001;48:475-486.

Effects of early-life antibiotic administration on porcine respiratory microbiota

M. Surendran Nair, BVSc, MS, MSCTR, PhD; B. Garcia-Morante, DVM, MS, PhD; A. Betlach, DVM; Mark Schwartz, DVM; Maria Pieters, DVM, PhD
Veterinary Population Medicine Department, College of Veterinary Medicine, University of Minnesota, St. Paul, Minnesota

Introduction

Current perspectives on the use of antimicrobials in food animals has changed due to the increased knowledge on antibiotic resistances and their effect on the microbiota composition.[1] Still, the impact of antibiotic treatments on the swine respiratory microbiota is poorly understood.[2-4] This study aimed, therefore, to characterize the impact of Tulathromycin administration, at either 4 or 19 days of age, on the diversity of the nasal microbiota over a 56-day period (lactation and post-weaning).

Materials and methods

Eighty-five piglets from litters of 10 different sows of a commercial farm (Midwest US) were randomly allocated to one of the following treatment groups. The control group 1 (n = 29) consisted of piglets that did not receive antibiotics. Groups 2 (n = 27) and 3 (n = 29) received a single intramuscular injection of Tulathromycin, at either 4 or 19 days of age, processing, and weaning, respectively. Tulathromycin was administered at a dosage of 2.5 mg/kg, following manufacturer's recommendations. Nasal swabs were collected from sows post-farrowing and from piglets on days 4 (T1), 12 (T2), 19 (T3), 28 (T4) and 56 (T5) of age. Nasal bacterial composition was assessed by amplifying the V3-V4 regions of 16S rRNA gene and the sequences generated by Illumina MiSeq technology were processed and analyzed by means of R software.

Results

The nasal microbiota composition was dominated by Proteobacteria (42%), Firmicutes (33%), Actinobacteria (12%) and Bacteroidetes (9%), whereas the predominant genera were *Moraxella* (14%), *Actinobacillus* (14%) and *Rothia* (12%). Based on diversity and composition analyses, the microbiome was not significantly different between treatment groups at any of the time points analyzed. Therefore, nasal microbiota was stable regardless of antibiotic treatment. Not with standing, nasal microbiota diversity increased significantly with age, between pre-weaning (T1 to T3) and post-weaning (T4-T5) time points (Figure 1).

Discussion and conclusions

Under the conditions of this study, the antibiotic administration to piglets during the lactation period did not interfere with normal colonization patterns of the nasal microbiota. Overall, the present results point out that the age and the transition to nursery/finisher sites may exert a greater influence over the nasal microbiome of piglets than early-life antibiotic treatments.

References

1. Thomason CA et al. Resident microbiome disruption with antibiotics enhances virulence of a colonizing pathogen. *Sci Rep.* 2017;7:16177.

2. Zeineldin M et al. Microbial shifts in the swine nasal microbiota in response to parenteral antimicrobial administration. *Microb Pathog.* 2018;121:210-217.

3. Correa-Fiz F et al. Antimicrobial removal on piglets promotes health and higher bacterial diversity in the nasal microbiota. *Sci Rep.* 2019;9:6545.

4. Mou KT et al. Shifts in the nasal microbiota of swine in response to different dosing regimens of oxytetracycline administration. *Vet Microbiol.* 2019;237:108386.

Figure 1: Bray-Curtis matrix for beta diversity index. Non-metric multidimensional scaling of Bray-Curtis distances between nasal samples based on microbial abundances. PCoPoints are colored by time point (T1 to T5) to which the samples belonged. The shape of each point indicates the treatment group. G1= negative control; G2= processing treatment; G3= weaning treatment.

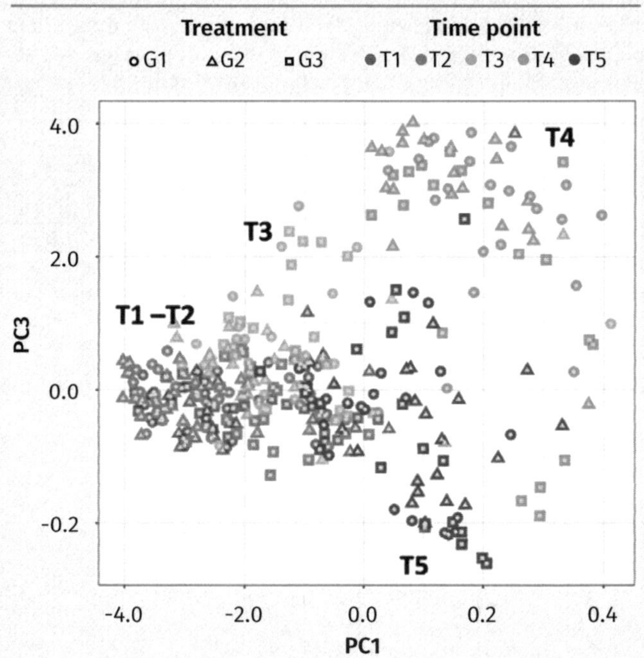

Comparative study on the production parameters, animal health status and drug usage of Hungarian swine farms

László Gombos[1]; László Ózsvari[2]
[1]Agrofeed Ltd. Hungary; [2]University of Veterinary Medicine Budapest, Hungary

Introduction

Antibiotics play a critical role in improving the health of animals and humans. The responsibility of the correct use leads us to reach the one health concept. Antibiotic-free pork production is a good way to create an added-value product. We, swine veterinarians, have the responsibility to manage well the prevention or treatment strategies on farms.

Material and methods

In our study we examined the relationships of the general animal health status and the management of the surveyed Hungarian farms with special regard to the antibiotic usage (AU). The data were collected from 13 large scale farms 2017-2019. By using a questionnaire, we collected the production parameters, the animal health status, (pathogens, the severity of lesions) and AU. The data were checked by farm visits and were analyzed under different farm conditions. (Table 1)

Results

According to the results the most common pathogens were *Mycoplasma hyopneumoniae* (100%), *Lawsonia intracellularis* (93%), swine influenza (85%), *Actinobacillus pleuropneumoniae* (53%) and among the secondary pathogens *Streptococcus suis* (77%) and haemolytic *Escherichia coli* (92%). Only 5 farms had well managed biosecurity system. All farms had a sort of prevention strategies, but only 2 farms controlled their effectiveness. None measured the effectiveness of disinfectants. 31.8% (15.8%-48.1%) of the total veterinary cost was accounted for the AU. Three farms had no antibiotic resistance test results. The most commonly used antibiotics (doxycycline, amoxycillin, enrofloxacin), which were used for the treatment of porcine respiratory disease complex were 60% of the total drug cost. Three farms used injectable antibiotics with new-born piglets for preventive purposes.

Discussion and conclusions

In our study we revealed that detailed information and knowledge about animal health management and the prevailing pathogens on the farm have primary importance in order to manage correctly the problems and reduce the high level of antibiotic use.

Table 1: Presence of the pathogens sorted by the inspected farms

Pathogen / **Farm**	**A**	**B**	**C**	**D**	**E**	**F**	**G**	**H**	**I**	**J**	**K**	**L**	**M**
AU (%) in average of total medication cost	28.2	33.6	48.1	40.5		21.9	28.8		15.8		30		39.4
PRRS			+					+		+			
APP		+	+					+		+	+	+	+
Mycoplasma hyopneumoniae	+	+	+	+	+	+	+	+	+	+	+	+	+
Pasteurella multocida		+		+	+			+	+		+		+
Atrophic rhinitis								+		+	+		
Serpulina hyodisenteriae								+		+		+	
Lawsonia intracellularis	+	+	+	+	+	+	+	+		+	+	+	+
Swine flu	+	+	+	+	+	+	+	+	+		+		+
PCV-2	+	+	+					+	+	+	+	+	+
PPV								+					+
Isospora suis								+					
E coli								+					
Sarcoptes								+					
Mycoplasma hyorhinis				+									

Assessing survival chances of neonatal piglets in field conditions

Jan Jourquin[1], DVM; Frederic Vangroenweghe[1], DVM, MSc, PhD, Dipl. ECPHM; Dries Donkers[2], DVM; Joaquin Morales[3], DVM, PhD; Kodore Camara[4], MSc
[1]Elanco Animal Health, Antwerp, Belgium; [2]Ghent University, Belgium;
[3]PigChamp pro Europa, Spain; [4]Elanco Animal Health, Greenfield, Indiana

Introduction

The causes of decreased birth weight (BW) and their impact on piglet survivability are extensively documented[1] yet using only birth weight as an indicator for survivability may be too simplistic.[2] Piglets born immature at term are unlikely to survive.[3] However, birth weight is a very easy, fast and non-invasive method of assessment. A threshold of 1.13 kg was proposed below which a piglet has lower survival chances until weaning.[4] Head shape and vitality scores were used to define Intra Uterine Growth Restriction (IUGR) in piglets.[3] These scoring systems are subjective, and little is known about the genetic line component.

The objective of this study was to compare different survival criteria in different genetic lines in order to provide a universal advice for survival chance assessment in field conditions.

Materials and methods

Data were collected from three farms located in Belgium (A and B) and Spain (C) with two different genetic lines (Danbred (2 farms) and Topigs (1 farm) represented. All live born piglets (BA) from 80 litters were assessed. The piglets were processed between 2 and 24 hours after birth. BW, rectal temperature (RT) and Crown Rump Length (CRL) were measured. A morphological IUGR score was based on 3 criteria: dolphin head shape, bulging eyes and wrinkles perpendicular to mouth.[2] When at least one of the three criteria were met, the piglet was considered to be IUGR. Body Mass Index (=BW/(CRL)^2 was calculated. From farm C, piglet pre-weaning mortality data were collected. Statistical analysis was performed using mixed or Glimmix procedures where appropriate and Pearson correlation coefficients were calculated between the different parameters.

Results

There was a clear difference between genetic lines for litter size, BW, BMI and IUGR incidence (Table 1). The incidence of IUGR was higher when the average birth weight was lower but the incidence of IUGR in the low weight piglets (< 1.13kg) was remarkably similar between farms. Piglet mortality was high in low weight piglets specifically when they also suffered from IUGR. There was a farm and weight effect for RT, but within the C farm, piglets that died had lower RT compared to those surviving.

The highest correlation with BW was CRL (R = 0.89, 0.87 and 0.70 for farm A, B and C respectively). IUGR was negatively correlated to BW (R = -0.51, -0.68 and -0.43 for farm A, B and C respectively).

Discussion and conclusion

BW remains a universal criterion to assess survival chances in field conditions. Low BW piglets demonstrating IUGR characteristics and piglets with a low RT (< 36°C) between 2 and 24 hours after birth are in a critical situation. They should receive special care or be euthanized for welfare reasons.

References

1. Baxter et al. Theriogenology 69 (2008) 773–783.
2. Hales et al. J. Anim. Sci 91 (2013) 4991-5003.
3. Chevaux et al. Proceedings 21th IPVS Congress 2010 O.175.
4. Feldpausch et al. Transl. Anim. Sci. 2019, 3:633-640.

Table 1: Litter data and IUGR data per farm

	Farm A	Farm B	Farm C
Litters (n)	15	15	50
Genetic line	Danbred	Danbred	Topigs
Parity	2.33	5.07	3.78
BA per litter	18.73[a]	17.20[a]	14.10[b]
Litter weight BA (kg)	21.25[a]	20.17[a]	20.85[a]
Average per litter			
BW (kg)	1.20[a]	1.18[a]	1.51[b]
RT (°C)	38.18[a]	37.71[a]	37.92[a]
CRL (cm)	26.6[a]	25.05[b]	24.51[b]
BMI	16.69[a]	18.37[a]	25.08[b]
IUGR incidence per Farm			
All piglets	41.3[a]	42.6[a]	25.0[b]
BW < 1.13 kg	76.6[a]	76.4[a]	69.9[a]
BW ≥ 1.13 kg	21.4[a]	11.7b[b]	17.4[ab]

[a,b] indicate statistical differences ($P \leq 0.05$) between farms

Farm C individual piglet data

	IUGR survived		**IUGR died**	
Weight threshold (kg)	**< 1.13**	**≥ 1.13**	**< 1.13**	**≥ 1.13**
Piglets (n)	45	93	27	11
BW (kg)	0.93[a]	1.48[1]	0.81[b]	1.52[1]
RT (°C)	37.92[a]	38.10[1]	34.55[b]	36.07[2]
CRL (cm)	20.95[a]	24.41[1]	20.65[a]	24.78[1]
BMI	21.34[a]	24.89[1]	19.28[b]	24.91[1]
	Normal survived		**Normal died**	
Piglets (n)	26	460	5	33
BW (kg)	1.06[c]	1.62[3]	0.91[d]	1.46[4]
RT (°C)	37.65[c]	38.31[3]	32.00[d]	35.31[4]
CRL (cm)	22.10[c]	24.97[3]	23.80[d]	25.53[3]
BMI	21.81[c]	26.11[3]	16.01[d]	22.60[4]

For < 1.13 kg pigs, a and b or c and d indicate statistical differences ($P \leq 0.05$)
For ≥ 1.13 kg pigs, 1 and 2 or 3 and 4 indicate statistical differences ($P \leq 0.05$)

Timing of Draxxin® administration in a PRRSV outbreak may result in positive reproductive outcomes

R. Kiehne, DVM
Swine Vet Center, P.A., St. Peter, Minnesota

Introduction

Porcine respiratory and reproductive syndrome (PRRS) causes reproductive failure in gestating sows. PRRSV replicates and causes pathology at the maternal-fetal interface by apoptosis of chorionic and uterine epithelial layers. There is a significant positive association between the apoptotic events at the maternal-fetal interface and the severity of PRRSV-induced vasculitits in the endometrium. Viral replication, inflammatory lesions in the endometrium and fetal placenta, as well as the in the blood vessels are responsible for reproductive failure. Research in the area of Draxxin's® mode of action against a PRRSV infection indicates that Draxxin® can prevent PRRSV-induced macrophage morphological alterations allowing better macrophage function. Moreover, there is inhibition of damage in inflamed tissues during a PRRSV infections. The early pig conceptus undergoes many changes prior to implantation and initiates inflammation at the maternal endometrial surface in order for proper placentation to occur. If the process of conceptus attachment is not performed perfectly, there are documented cases of complete pregnancy loss, as well as decreased litter sizes by over 25% by term. The objective of this investigation was to evaluate a Draxxin® intervention in gestating sows during a PRRSV outbreak to mitigate clinical signs. The hypothesis was that Draxxin® administration during early gestation may prove to be a disadvantage due to early placental development, but advantageous during later pregnancy to prevent PRRSV-induced reproductive abnormalities that occur at the maternal-fetal interface.

Materials and methods

The investigation was conducted in NE Iowa 2,500 sow farm and prompted when PRRSV was detected on 5/23/19. The first Draxxin® dose (5 mL) was administered on 5/28/19. Live viral inoculation (LVI) was performed on 5/30/19. The second Draxxin® dose (5 mL) was administered on 6/12/19. Therefore, the timeline for Draxxin® administration to the LVI was -2 and 13 days, respectively. Approximately 20% of the sows were not treated with Draxxin® to use as a control. A similar percentage was females were enrolled in each treatment group. When the data set was collected, date of service (i.e. day of insemination) was subtracted from the date of first Draxxin® administration to obtain the day of pregnancy in which Draxxin® was administered to a pregnancy group: Days 0-18 (early establishment of pregnancy); Days 19-29 (early attachment to endometrium); Days 30-49 (increased interdigitation of chorioallantois and endometrium); Day 50-69 (fetal bone ossification begins); Day 70-90 (increased placental growth and vascularization); Day 91-105 (late fetal stage). For the analysis, there was a total of 4 outcome events that were used: 1) Abortions; 2) Culls and Death; 3) Farrowed; and 4) Found open, preg check negative, repeat mating. Number of pigs weaned per sow enrolled in the study was determined. Due to the pilot nature of this investigation, only summary statistics have been performed.

Results and discussion

The largest outcome event was the percentage of sows that farrowed (68.5 vs 64.1% Draxxin® vs. control), followed by sows that aborted (17.4 vs 22.9% Draxxin® vs. control), were not pregnant (6.9 vs. 6.7% Draxxin® vs. control), and finally deads and culls (7.2 vs. 6.3% Draxxin® vs. control). The total number of weaned pigs per enrolled sow was calculated within each pregnancy group: 0-18: 8.46 vs. 8.44 (Draxxin® vs. control); 19-29: 2.76 vs 5.22 (Draxxin® vs. control); 30-49: 5.23 vs 5.90 (Draxxin® vs. control); 50-69: 6.45 vs. 5.52 (Draxxin® vs. control); 70-89: 4.90 vs. 3.29 (Draxxin® vs. control); and 90-116: 2.90 vs. 0.96 (Draxxin® vs. control).

We accept our hypothesis that Draxxin® administration during early gestation would be detrimental to conceptus development, but during later gestation, is advantageous to limit apoptosis at the maternal-fetal interface. If females treated with Draxxin® during early pregnancy (i.e. 0-18) didn't recycle, they had the greatest born live. Draxxin® administered during early pregnancy appears to be increase negative outcome events (i.e. Found open, preg check neg, and repeat matings). However, there appears to be an advantage to administering Draxxin® in a PRRSV+ situation after day 50 of gestation. This advantage is apparent in a reduction in abortions, stillborn and mummified fetuses, as well as total pigs weaned. In conclusion, it is recommended from this pilot investigation that in response to a PRRSV outbreak (or LVI), administration of Draxxin® after day 50 of gestation is advantageous; however, before day 50 of gestation may be a detriment.

PRRS viral diversity in piglets born during an outbreak

M. Kikuti[1], DVM, MPH, PhD; C. Vilalta[1], DVM, PhD; J. Sanhueza[2], DVM, MS, PhD; N. Pamornchainavakul[1], DVM; J. L. Kevill[1], PhD; I. A. D. Paploski[1], DVM, MS, PhD; R. Kienhe[3], DVM; K. VanderWaal[1], BS, PhD; D. Schroeder[1], BS, PhD; C.A. Corzo[1], DVM, MS, PhD
[1]Department of Veterinary Population Medicine, University of Minnesota, Saint Paul, Minnesota;
[2]Departamento de Ciencias Veterinarias y Salud Pública, Universidad Católica de Temuco;
[3]Swine Vet Center, St. Peter, Minnesota

Introduction

Although genetic variability has been estimated to be around 3% at both the within herd and individual level using ORF5 sequences,[1] whole genome variability within a naturally infected herd and within individuals is expected to be higher, more informative, and has yet to be documented. Here, we describe preliminary results of overall, within litter, and within piglet PRRSV genetic diversity on a farm undergoing an outbreak.

Methods

A PRRSV naïve 2500-sow farrow-to-wean farm was selected for this study. The farm broke with PRRSV type 2 after maintaining a naïve status for over three years. As part of the PRRS outbreak management protocol, live-virus inoculation was performed in the entire herd nine days after identifying the newly introduced PRRSV. Two weeks after inoculation, all piglets 3-5 days of age (DOA) were individually ear tagged and bled. Surviving piglets were bled again at 17-19 DOA (pre-weaning). Samples were individually tested for PRRSv by RT-PCR, and all positive samples were submitted for MINION sequencing. Consensus whole genome sequences were generated from reads by mapped to the reference MH651739.[2] ORFs 2-7 were analyzed. Samples that were only partially sequenced in any of the ORFs 2-7 were excluded from further analysis.

Results

A total of 127 piglets from 21 litters were sampled at 3-5 DOA and all animals tested PRRSV RT-PCR positive. All 63 surviving piglets at 17-19 DOA were still PRRSV RT-PCR positive. Complete consensus reads were obtained for 101 samples (54 from 3-5 DOA and 47 from 17-19 DOA). Percent identity amongst samples was overall high for most comparisons (> 99.0%). The lowest overall and within litter percent nucleotide identity was 95.8% found in ORF4, particularly in litters 10 and 12 at 17-19 DOA. The second lowest percent nucleotide identity was the overall 97.7% identity in ORF5a at 3-5 DOA, and overall and within litter 98.5% in ORF5a at 17-19 DOA. Within animal percent identity between both samplings was also assessed, with lower identity also found in ORFs 4 and 5a at 97.2% and 98.5%, respectively. While most sequences differed less than 1% from the overall ORFs 2-7 consensus, only 30.7% (31/101) of all the sequences were identical to the consensus.

Discussion

Although viral diversity within the piglet population is generally small, higher diversity was found in ORFs 4 and 5a. ORF5a overlaps with the beginning of ORF5, where sites under selective pressure were previously described.[3,4] PRRSV mutation rate has been estimated at the order of 10-2/site/year,[5] so the high nucleotide difference between the two samples found might represent turnover of a co-circulating virus that was not detected in the first sampling. These animals were also followed at 40-41, 65-66, and 109-110 DOA and will have viral diversity assessed at these stages as well.

References

1. Goldberg T, Lowe J, Milburn S, Firkins L. Quasispecies variation of porcine reproductive and respiratory syndrome virus during natural infection. *Virology.* 2003;317:197-207.

2. Zhang H, Leng C, Ding Y, et al. Characterization of newly emerged NADC30-like strains of porcine reproductive and respiratory syndrome virus in China. *Arch Virol.* 2019;164(2):401-411. **doi:10.1007/s00705-018-4080-7**

3. Delisle B, Gagnon CA, Lambert M-È, D'Allaire S. Porcine reproductive and respiratory syndrome virus diversity of Eastern Canada swine herds in a large sequence dataset reveals two hypervariable regions under positive selection. *Infect Genet Evol.* 2012;12(5):1111-1119. **doi:10.1016/j.meegid.2012.03.015**

4. Paploski IAD, Corzo C, Rovira A, et al. Temporal Dynamics of Co-circulating Lineages of Porcine Reproductive and Respiratory Syndrome Virus. *Front Microbiol.* 2019;10:2486. **https://www.frontiersin.org/article/10.3389/fmicb.2019.02486.**

5. Hanada K, Suzuki Y, Nakane T, Hirose O, Gojobori T. The origin and evolution of porcine reproductive and respiratory syndrome viruses. *Mol Biol Evol.* 2005;22(4):1024-1031. **doi:10.1093/molbev/msi089**

Detection of influenza A virus in pig farm workers

Gustavo Lopez, DVM; Peter Davies, BVS, PhD; Cesar Corzo, DVM, MS, PhD; My Yang; Marie Culhane DVM, PhD; Montserrat Torremorell, DVM, PhD
Department of Veterinary Population Medicine, College of Veterinary Medicine, University of Minnesota, St. Paul, Minnesota

Introduction

Influenza A virus (IAV) affects many hosts, including pigs and people, which represents a public health concern.[1] Bidirectional transmission of IAV leads to both, disease morbidity and the potential to generate novel IAV strains. US swine herds experience frequent outbreaks of acute influenza that can impact both swine and human health. Despite the recognition that bidirectional transmission of influenza virus occurs between people and pigs, little is known about how frequently this transmission or exposure takes place. Therefore, the objectives of this study were to implement a surveillance system at the farm worker-swine interface, quantify how frequently farm workers tested positive for IAV and characterize the IAV positive samples to assess risk of interspecies transmission.

Material and methods

Seven sow farms located in the US-Midwest were selected for the study. The farms were representative of US commercial farms with an average of 4,000 sows. The farms had a history of IAV infections and 5 of the farms tested IAV positive during the course of the study. We aimed to recruit swine farm workers that worked with pigs at least 2 days a week. After collecting baseline information, we asked each participant to self-collect a nasal swab before entering the farm and at the end of the working day, after completing the daily chores, twice a week for 8 weeks when human seasonal infections were considered prevalent according to the Center of Disease and Prevention (CDC) influenza reports. At each sampling point, participants also collected their body temperature using disposable thermometers and answered a short survey regarding the activities performed during the day, and whether they had influenza-like illness (ILI).

Pigs were also sampled at three time points during the study. Each time 30 nasal swabs were collected from 20 day-old pigs prior to weaning. Farm worker samples were tested individually with an influenza A specific rRT-PCR test that detects both human and swine IAVs.[2] Samples with a cycle threshold < 40 were considered positive and positive samples were selected for virus isolation on Madin-Darby canine kidney cells.[3] Pig samples were tested in pools of 3 using an rRT-PCR that targets the conserved matrix gene of IAV.[4] The study took place during the 2019 and 2020 peak of human seasonal influenza (January-March).

Results

There were 64 workers enrolled from seven participating farms, and a total of 1,813 nasal swabs were obtained from the farm workers. Out of the samples collected, 58 samples (3.2%) tested IAV rRT-PCR positive, 20 (34.4%) of them belonged to samples taken before entering the farm and 38 (65.6%) were collected at the end of the working day (Table 1). From the 64 enrolled workers, 33 of them tested IAV positive at least once during the study. Pigs from five of the seven participating farms were IAV positive at the three sampling points during the course of the study.

Conclusions and discussion

We were able to establish an IAV surveillance system where workers complied with self-sample collection protocols which indicates our ability to perform studies at the pig-human interface. Our results provide evidence that a number of farm workers can test influenza positive when they report to work at swine farms and at the end of the working day. Results obtained at the end of the day are most likely indicative of IAV exposure while working with pigs at the farms. Further characterization of the samples is needed to understand the implications for virus introduction into farms and risk of bi-directional transmission between pigs and people.

Acknowledgement

This study was funded by the National Pork Board.

References

1. Myers KP, Olsen CW, Gray GC. Cases of Swine Influenza in Humans: A Review of the Literature. *Clin Infect Dis.* 2007;44(8):1084-1088. **doi:10.1086/512813**

2. (WHO) WHO. CDC protocol of realtime RTPCR for swine influenza A(H1N1) 28. *Control.* 2009;1(April).

3. Meguro H, Bryant JD, Torrence AE, Wright PF. Canine kidney cell line for isolation of respiratory viruses. *J Clin Microbiol.* 1979;9(2):175-179.

4. Slomka MJ, Densham ALE, Coward VJ, et al. Real time reverse transcription (RRT)-polymerase chain reaction (PCR) methods for detection of pandemic (H1N1) 2009 influenza virus and European swine influenza A virus infections in pigs. *Influenza Other Respi Viruses.* 2010;4(5):277-293. **doi:10.1111/j.1750-2659.2010.00149.x**

Table 1: Number of workers and nasal swabs that tested influenza A positive by RT-PCR by moment of collection

Nasal swabs	No. of workers influenza positive at least once (%)	No. of samples influenza positive (%)
Before work	16/66* (24.2)	20/919^ (2.2)
After work	24/66 (36.3)	38/895 (4.2)
Total	33/66 (50)	58/1,814 (3.2)

* Number of positive people/total number of people

^ Number of positive samples/total number of collected samples

Samples were considered positive when the RT-PCR ct value was < 40

Evaluation of parity, personnel, and cross-fostering in influenza virus infection during the pre-weaning period

Gustavo Lopez[1], DVM; Peter Davies[1], BVSc, PhD; Marie Culhane[1], DVM, PhD; Emily McDowell[2], DVM; Eduardo Fano[3], DVM, MSc, PhD; Christa Goodell[3], DVM, MS, Dipl ACVMP, PhD; Montserrat Torremorell[1], DVM, PhD
[1]Department of Veterinary Population Medicine, College of Veterinary Medicine, University of Minnesota, St. Paul, Minneeosta; [2]Pipestone Veterinary Services, Pipestone, Minnesota;
[3]Boehringer Ingelheim Animal Health USA Inc. Duluth, Georgia

Introduction

Piglets are born naïve to influenza A virus (IAV) but in endemically infected farms they often become infected prior to weaning. Infection of piglets prior to weaning is one of the key factors that contribute to IAV persistence in breeding herds. Therefore understanding how piglets become infected, what factors contribute to piglet infection and what are the most relevant sources of IAV exposure to piglets is key to control IAV. In this study, we evaluated whether changes in management practices during the pre-weaning period affect IAV prevalence. Changes in management practices included farm workers behavior related to the movement of pigs between litters (cross-fostering) and biosecurity measures implemented by the workers while handling piglets. We also evaluated the effect of sow parity in IAV prevalence during the course of lactation.

Material and methods

Three farms known to be IAV positive at weaning were selected for the study. These farms were selected based on their willingness to participate in the study, ability to follow the study protocols, lack of influenza vaccination in pigs prior to weaning and history of piglets testing IAV positive at weaning for 3 months prior to start the study. At enrollment, farrowing rooms were allocated to either "control" or "treatment" groups. Litters within the control rooms were processed (tail-docking and castration), handled and treated according to existing farm protocols. Litters in the treatment rooms were not cross-fostered after processing, and processing or any other handling of the pigs was done by farm personnel changing gloves, and wearing boot covers if personnel was to enter the crates. A total of 28 farrowing rooms distributed across 3 farms were enrolled into the study. Within each farrowing room litters were stratified into young parities (P0, P1 and P2) and older parities (> P3), random selection of litters within each strata was performed. The selected litters were sampled throughout the preweaning period to assess IAV status. A total of 360 litters (120 per farm) were enrolled into the study.

Sample collection protocol is shown in Figure 1. Sows from the selected litters were sampled at farrowing and weaning using a nasal swab. Sampling of the study farrowing rooms environment and litters took place at 1, 8, 13 and 18 days of age of the piglets, approximately. The samples collected were udder skin wipes from study litters and air particle depositions from the study farrowing rooms[1] however, detection of IAV infections can be challenging in endemically infected herds with low prevalence of infection. Methods: In groups of suckling (18-21 days of age. Samples were also collected from materials used in the farrowing rooms and from worker's hands after piglet handling using a cotton gauze impregnated in DMEM transport media.

Samples were tested using rRT-PCR that targets the conserved matrix gene of IAV[2] modify an existing matrix (M. Statistical differences between groups and parities were assessed using chi-square test for categorical outcomes using statistical software.[3]

Results and discussion

Two of the 360 (0.65%) nasal swabs collected from sows at farrowing were IAV positive. At weaning 101/360 (28%) of the study litters were IAV positive. The udder wipes collected from study litters showed a lower IAV prevalence in the treatment group 209/720 (29%) compared to the control group 318/720 (43%). No differences in IAV prevalence were seen between young parity sows 254/676 (37.5%) and older parity sows 273/764 (35.7%). Samples collected from farm worker's hands had 58% (65/111) positivity to IAV. Our results indicate that specific management practices directed at minimizing spread of IAV in pigs prior to weaning may help to decrease prevalence of IAV at weaning.

Acknowledgement

This study was funded by Boehringer Ingelheim Animal Health USA Inc. Duluth, GA 30096.

References

1. Garrido-Mantilla J, Alvarez J, Culhane M, Nirmala J, Cano JP, Torremorell M. Comparison of individual, group and environmental sampling strategies to conduct influenza surveillance in pigs. *BMC Vet Res.* 2019;15:61:1-10. **doi:10.1186/s12917-019-1805-0**

2. Slomka MJ, Densham ALE, Coward VJ, et al. Real time reverse transcription (RRT)-polymerase chain reaction (PCR) methods for detection of pandemic (H1N1) 2009 influenza virus and European swine influenza A virus infections in pigs. *Influenza Other Respi Viruses.* 2010;4(5):277-293. **doi:10.1111/j.1750-2659.2010.00149.x**

3. R Core Team. R: A Language and Environment for Statistical Computing. 2017. **https://www.r-project.org/**.

Figure 1: Sample collection timeline. The boxes represent the age of the litters when samples were collected, and the stars represent the type of sample that was collected.

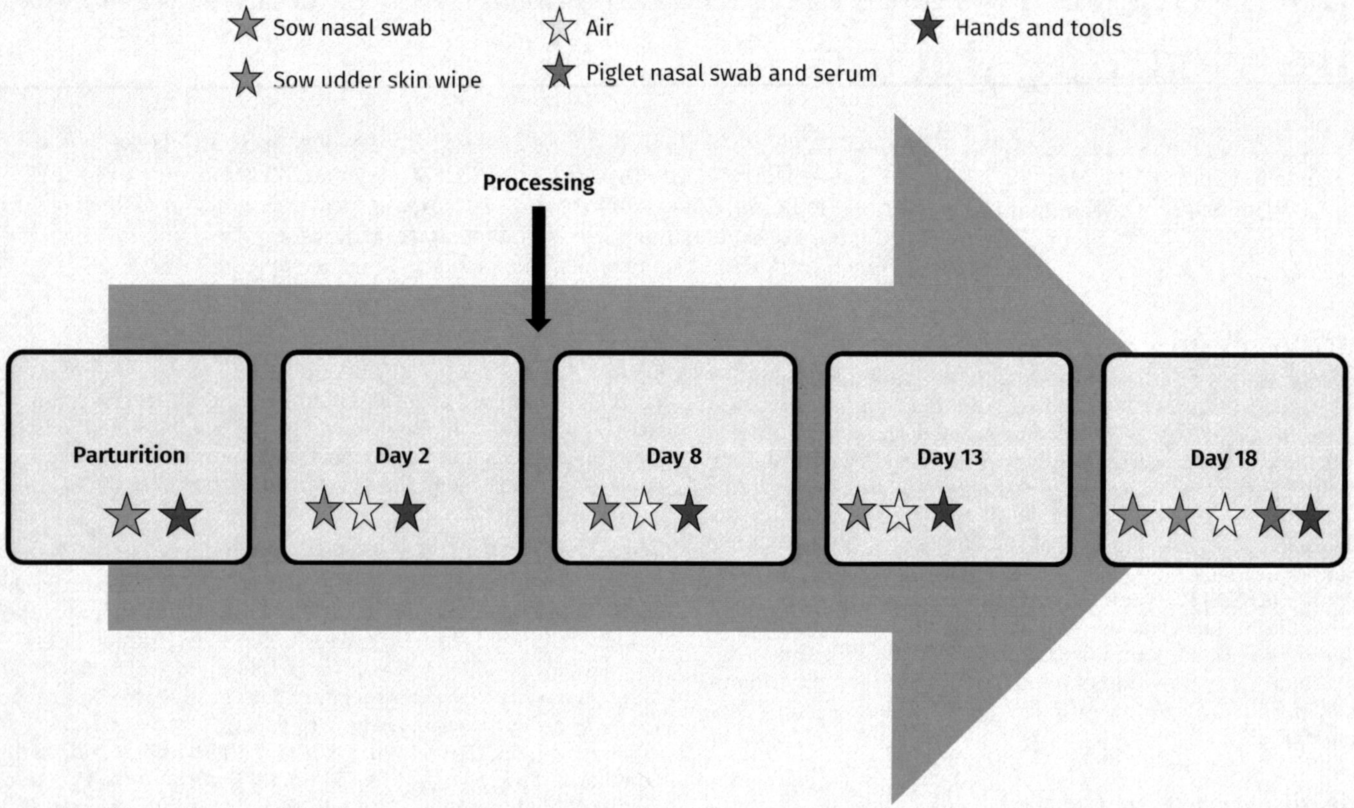

Modeling pig movement data to improve targeted disease surveillance and control plans of infectious diseases

Nicolás C. Cárdenas[1]; Jason A. Galvis[1]; Francisco P. N. Lopes[2]; Joana Voges[2]; Antônio A. R. Medeiros[2]; Gustavo Machado[1]
[1]Department of Population Health and Pathobiology, College of Veterinary Medicine, Raleigh, North Carolina
[2]Secretary of Agriculture, Livestock and Agribusiness of State of Rio Grande do Sul (SEAPI RS), Porto Alegre, Brazil

Introduction

Tracking animal movements over time can fundamentally determine the success of disease control interventions throughout targeting farms that are tightly connected. In swine production, animals are transported between farms based on growth stages, thus it generates time-varying contact networks that may influence the dynamics of disease spread. Still, risk-based surveillance strategies are mostly based on static-time network representations. For that reason, we reconstructed the temporal pig network of one Brazilian state from 2017 to 2018, comprising 351,519 movements and 48 million transported pigs. A spread disease model considering the time-respecting paths for all contact network was designed for targeted high-risk farms that can be hubs on the transmission endemic and upon introduction, exotic diseases such as African swine fever

Materials and methods

The modeling setup was adapted to identify the best network-based metric to be used in the identification of farms to undergo target surveillance. This approach is based on a classic deterministic Susceptible-Infected (S-I) disease spread model. Briefly, this spread cascade model follows the movements' temporality, which is a temporal representation of the cumulative and consecutive contacts among farms, and therefore can be interpreted as the size of an epidemic.[1] The model started from each seed farm in the network at a given time "0", with an infection that was assumed to active for one day. The hypothetical disease spread followed the chronological order of movements over two years. In short, when there was a directed interaction, including the three dimensions of time t, infected farm i and susceptible farms j, disease has successfully been transmitted. These events were counted in favor of each farm i cascade and the triplet t, i, j and added to the temporal links of the stream.

Results and discussion

The results show that targeting the first 1,000 farms ranked by degree would be sufficient to control disease spread. Our finding also suggested that assuming a worst-case scenario in which every movement could transmit disease, pursuing farms by degree would limit the transmission to 29 farms, which is lower than the number of infected farms for random surveillance, with epidemic sizes of 2,593 farms. The distributions of the spread cascades are showed in the figure 1. Our proposed modeling handles the temporal order of movements, calculates the minimum number of farms needed to be targeted and approximates the expected epidemic size in a worst-case scenario where each movement can effectively transmit disease. Finally, the introduction of risk-based approaches may optimize the costs of routine surveillance and can be used in preparation for the introduction and spread of endemic and exotic diseases such as African swine fever.

Acknowledgements

Acknowledgements to the Fundo de Desenvolvimento e Defesa Sanitária Animal (FUNDESA-RS) and by the NC State University, College of Veterinary Medicine, Global Health funds.

Reference

1. Payen, A., L. Tabourier, and M. Latapy, 2019: Spreading dynamics in a cattle trade network: Size, speed, typical profile and consequences on epidemic control strategies. PLoS One.

Figure 1: Final spread cascades sizes by each target-network surveillance strategy. In red, the estimated spread cascade, which had a maximum value of 29 farms while targeting farms at random, resulted in a cascade of 2,593 farms (black).

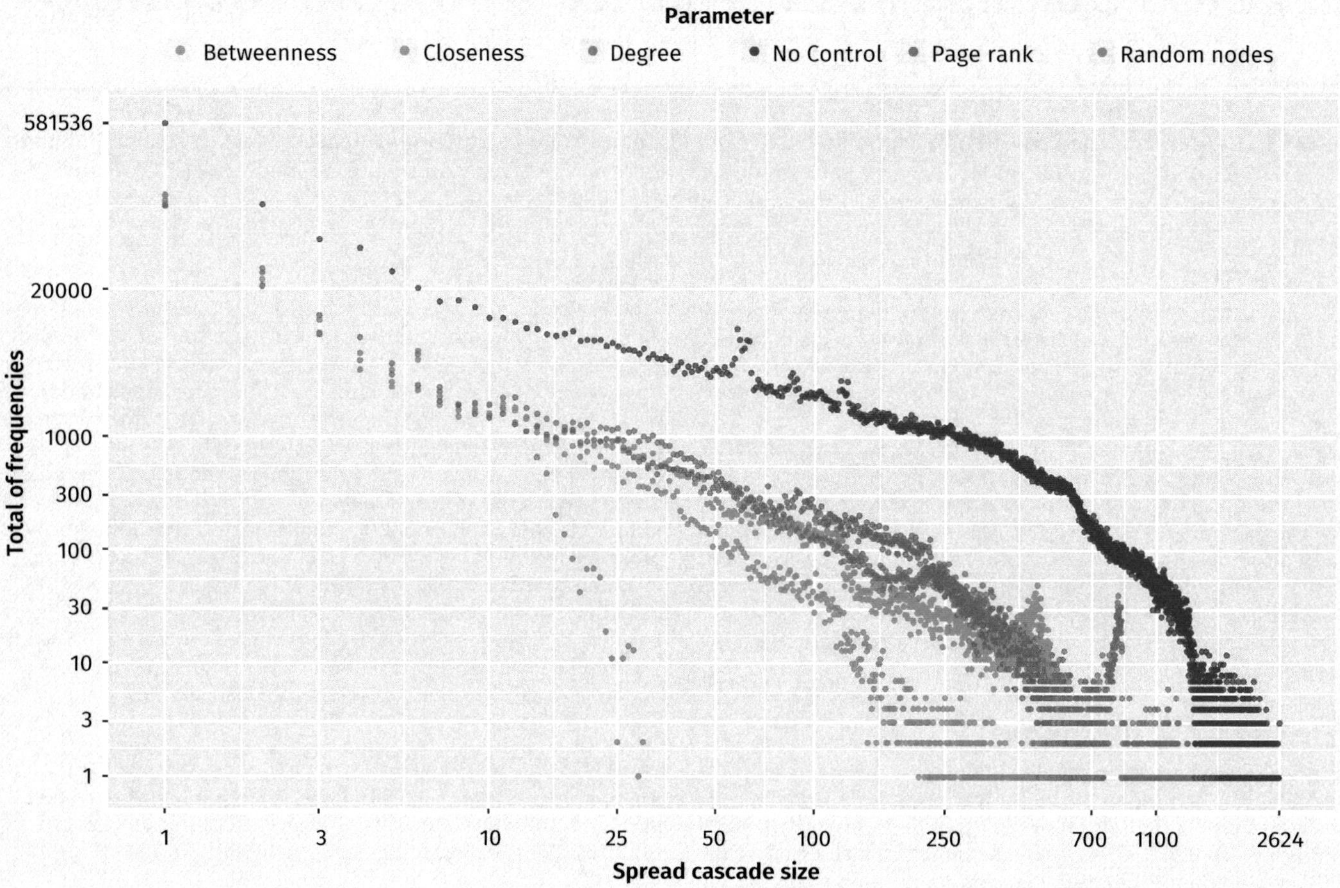

Measuring the effect of disease diagnostic information on the mortality of growing pigs raised under field conditions

Edison S. Magalhães[1]; Pete Thomas[2]; Giovani Trevisan[1]; Derald J. Holtkamp[1]; Christopher J. Rademacher[1]; Jeffrey Zimmerman[1]; Kent J. Schwartz[1]; Eric R. Burrough[1]; Daniel C. L. Linhares[1]
[1]Department of Veterinary Diagnostic and Production Animal Medicine, Iowa State University, Ames, Iowa;
[2]Iowa Select Farms., Iowa Falls, Iowa

Introduction

Sow farm productivity and health status impacts the subsequent performance of the weaned cohorts up until market.[1,2] The increasing implementation of technologies in the swine industry produces a diverse and disperse availability of data streams within production systems (e.g., productivity, health, environment, diagnostic, logistic, and infrastructure). In the absence of connected data streams, producers will likely remain unable to tailor swine health and production management to specific swine populations/pig flows. The integration of disease diagnosis information with these data streams can assist field veterinarians to make data-driven solutions to improve swine herd performance. Thus, the objective of this study was to develop algorithms to aggregate and merge multiple data streams from a Midwestern production system, and to measure the association of disease diagnostic information on wean to finish (W2F) mortality, by analyzing the final data base.

Materials and methods

Disease diagnostic codes (DxCode) are assigned by diagnosticians to each tissue case received at the Iowa State University Veterinary Diagnostic Laboratory (ISU-VDL), based on submission information, laboratory assays, and compatible macroscopic and/or microscopic lesions. SAS scripts were built to match and merge DxCode data to productivity parameters and PRRS status data from breed-to-market, for all Iowa Select Farms (ISF) cohorts of growing pigs marketed from January 2018 to June 2019. The dataset included nursery, single stock W2F, and double stock W2F flows, as well as cohorts originated from single or multiple sow farms. The SAS scripts aggregated and merged productivity, sow farm PRRS status at weaning, and DxCode data into a single consolidated dataset, informing the pathogen(s) diagnosed throughout the growing phase of production for each cohort marketed. Furthermore, statistical analysis was conducted to: **A)** compare the mean W2F mortality for cohorts with DxCode allotted during the growing phase relative to those without any DxCode allotted; **B)** compare the average age of the growing groups when tissue samples were submitted to the VDL and their respective W2F mortality, classifying age as early (first 21 days post-weaning), mid (22-69 days post-weaning), or late age of diagnosis (70+ days post-weaning); **C)** analyze the frequency of tissues submitted to the VDL, within the same cohort throughout the growing phase, classified as either "single" or "multiple" submissions; **D)** analyze the diversity of pathogens identified in each event of tissue submission, classifying as "single etiology" or "multiple etiology"; **E)** compare the mean W2F mortality of cohorts with DxCode assigned for PRRSV across the three aforementioned age groups; **F)** analyze cohorts with DxCode for PRRSV and their retrospective PRRS status at weaning (PRRSV-negative, endemic, or epidemic); **G)** analyze the interaction between cohorts with PRRS DxCode assigned and their retrospective sow farm preweaning mortality (PWM), described here in quartiles averages; **H)** compare the mortality difference among cohorts in 4 different categories of DxCode (PRRSV- and Influenza A virus (IAV)-negative, PRRSV-negative and IAV-positive, PRRSV-positive and IAV-negative, and cohorts with DxCode positive for both PRRS and IAV).

Results

The final consolidated dataset contained records of 1,742 closeouts marketed, where 24.91% (434 closeouts) had at least one tissue submission to the VDL during the growing phase, and the remaining 75% (1,308 closeouts) had no tissue submission. Table 1 summarizes the differences in W2F mortality of the cohorts for each of the aforementioned characteristics measured.

Discussion and conclusions

This study demonstrated the importance of analyzing diagnostic data combined with other data streams. The cohorts with PRRSV DxCode were highly associated with increased W2F mortality. The W2F mortality of weaned groups with negative and epidemic status for PRRSV at weaning was similar when PRRS was diagnosed throughout the growing phase. Furthermore, cohorts originating from sow farms with high PWM (17.3%) had the worst W2F mortality. Co-infections were a common finding and were associated with higher W2F mortality than groups with single etiologies or no etiology. Timing of DxCode assignment was also important, i.e., earlier diagnosis was associated with higher W2F mortality. The present study focused on PRRSV positive DxCode due to high availability of information for this pathogen, concentrating on its interaction with and IAV, and other sow farm risk factors. Findings validated that the breeding herd health status is a good predictor of W2F health.[1-3]

References

1. Alvarez et al. 2015. Prev. Vet. Med. 121: 240–245.
2. De Grau et al. 2005. Can. J. Vet. Res. 69: 241–245.
3. Larriestra et al. 2006. Can. J. Vet. Res. 47: 560-566.

Table 1: Average mortality for each category of W2F groups, based on different comparisons

Characteristic	Category A	Category B	Category C	Category D
(a) DxCode assigned - W2F mort.	Yes – 10.4%[a]	No – 8.1%[b]	-	-
*(b) DxCode age - W2F mort.	Early age – 11.5%[a]	Mid age – 10.1%[b]	Late age – 9.2%[b]	-
*(c) DxCode frequency - W2F mort.	Single – 10.2%[a]	Multiple – 11.4%[a]	-	-
*(d) DxCode diversity - W2F mort.	Single agent – 9.71%[a]	Multiple agent – 11.2%[b]	-	-
*(e) PRRS & age DxCode - W2F mort.	Early PRRS – 14.7%[a]	Mid PRRS – 11.4%[a]	Late PRRS – 10.2%[b]	-
*(f) PRRS DxCode & Status - W2F mort.	Negative – 12.4%[a]	Endemic – 11.4%[a]	Epidemic – 12.9%[a]	-
*(g) PRRS DxCode & PWM - W2F mort	10.6% – 10.9%[a]	13.0% – 11.7%[ab]	14.6% – 11.8%[ab]	17.3% – 13.5%[b]
*(h) PRRS & IAV DxCode - W2F mort	PRRSV-, IAV- 9.45%[a]	PRRSV-, IAV+ 9.14%[a]	PRRSV+, IAV- 11.9%[b]	PRRSV+, IAV+ 12.3%[b]

[ab] Different letters represent statistical difference of W2F mortality.

* Analysis conducted for the groups with DxCode assigned (n = 434).

Feeder pig movements as a potential driver of the spread of PRRS lineage 1 viruses in the US

D. Makau[1], BVetMed, PhD; M. Alkhamis[2], DVM, MPVM, PhD; I. Paploski[1], DVM, MSc, DrPH; C. Corzo[1], DVM, MSc, PhD; K. VanderWaal[1], PhD
[1] Department of Veterinary Population Medicine, College of Veterinary Medicine, University of Minnesota;
[2] Department of Epidemiology and Biostatistics, Faculty of Public Health, Health Sciences Center, Kuwait University, Kuwait

Introduction

Since the initial detection of porcine reproductive and respiratory syndrome virus (PRRSV) in the US, the virus has continued to evolve and spread causing disease and significant economic losses to the swine industry. Despite numerous changes adopted by production systems, PRRS remains a serious challenge to the industry. Around 2014, the emergence of a more virulent PRRSV strain, causing high preweaning mortality and severe clinical disease in sows in the US was reported. The virus was identified as restriction fragment length polymorphism type1-7-4 or lineage 1 (predominantly sub-lineage 1A [L1A]).[1,2] The availability of Bayesian phylodynamic models allows for the investigation of how different factors influence viral spread and evolution. The objective of this study was to determine the factors associated with the viral evolution and spatial diffusion of L1A in a swine dense production area in the US using Bayesian phylodynamic models.

Materials and methods

Data included 993 ORF-5 PRRSV L1A sequences isolated from 410 farms in two production systems in a swine dense pig production area about 220,000 km^2 (~85,000 mi^2) in the US, between 2014 and 2017. The study area was divided into spatial grids over which vegetation cover, environmental temperature, wind and precipitation, farm density and animal movements between grids were summarized. We combined general time reversible, gamma distribution, skygrid and uncorrelated lognormal relaxed clock models in Bayesian evolutionary analysis by sampling trees (BEAST) to estimate the evolutionary rate, the rate and frequency of transmission events between different pairs of grid cells and identify factors that influenced the spatial diffusion.

Results and discussion

We inferred that the initial and only introduction of L1A in the study area was in the first quarter of 2013 and genetic diversity and effective population size of the virus changed seasonally peaking in the summer months. The movement of feeder pigs and grid adjacency enhanced the spread of PRRSV (Figure 1). These findings agreed with other studies but wean pig movements did not significantly influence disease spread in the region.[3,4] Our study underscores the importance of animal movements especially of feeder pigs in the spread of PRRSV. Therefore, these movements should be handled cautiously, and more robust disease monitoring in nursery farms could help in managing the disease. We also highlight the importance of coordinated biosecurity measures to mitigate regional disease spread.

References

1. Wang, X., Marthaler, D., Rovira, A., Rossow, S., & Murtaugh, M. P. (2015). Emergence of a virulent porcine reproductive and respiratory syndrome virus in vaccinated herds in the United States. Virus Research, 210, 34–41. **https://doi.org/10.1016/j.virusres.2015.07.004**

2. Paploski, I. A. D., Corzo, C., Rovira, A., Murtaugh, M. P., Sanhueza, J. M., Vilalta, C., Schroeder, D. C., VanderWaal, K. (2019). Temporal Dynamics of Co-circulating Lineages of Porcine Reproductive and Respiratory Syndrome Virus. Frontiers in Microbiology, 10, 1–23. **https://doi.org/10.3389/fmicb.2019.02486**

3. Jara, M., Rasmussen, D. A., Corzo, C. A., & Machado, G. (2020). Porcine reproductive and respiratory syndrome virus dissemination across pig production systems in the United States. Transboundary and Emerging Diseases, (April), tbed.13728. **https://doi.org/10.1111/tbed.13728**

4. Alkhamis, Mohammad A., Perez, A. M., Murtaugh, M. P., Wang, X., & Morrison, R. B. (2016). Applications of Bayesian Phylodynamic Methods in a Recent U.S. Porcine Reproductive and Respiratory Syndrome Virus Outbreak. Frontiers in Microbiology, 7, 1–10. **https://doi.org/10.3389/fmicb.2016.00067**

Figure 1: Factors affecting the spatial diffusion of PRRSV L1A between spatial grids in a swine dense production area in the US. Factors with a Bayes factor support (BF) > 6, are considered significant and there is strong support for their role in the spatiotemporal diffusion of the virus. Factors with a BF > 3 but < 6 like farm density, may have some influence in the viral diffusion trends observed. Model parameters are classified either as related to the environment, the grid/area or animal movements. (o)-means origin, (d)-means destination

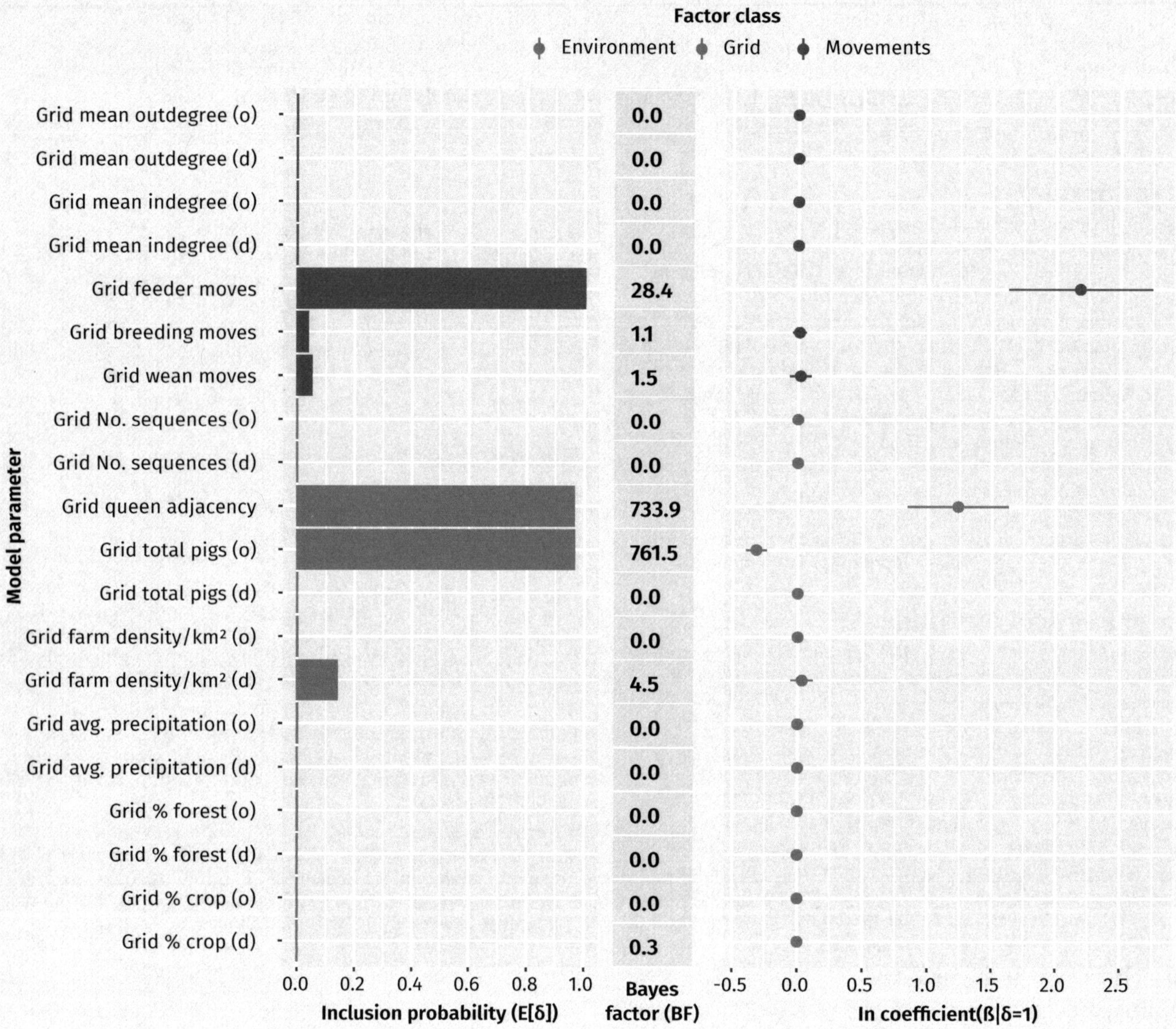

Investigation of *Actinobacillus suis* genomes obtained from clinical cases reveals the emergence of a novel strain

M. Meneguzzi[1]; M. O. Costa[2,3]
[1]University of Minnesota; [2]University of Saskatchewan, Canada; [3]Utrecht University, the Netherlands

Introduction

Actinobacillus suis is a microorganism that colonizes the upper respiratory tract of pigs from suckling to grower and finisher phases often associated to mortality in high health herd in most countries with relevant pork production.[1-4] This pathogen can overcome the host immune response, leading to pneumonia, septicemia, meningitis and arthritis. Previous reports have identified genes that encode for toxins known as RTX (repeats in toxin), such as ApxI and ApXII, in *A suis* isolated from sick pigs. These toxins likely contribute to the virulence of this pathogen.[5,6] Here, our objective was to investigate the molecular epidemiology and virulence profile of *A suis* isolated from septicemic pigs.

Materials and methods

Between 2018 and 2019 four commercial farms in Canada experienced episodes of increased mortality rate due to *A suis* infection in grower pigs. For each farm, samples collected from the lungs (n = 3) or heart (n = 1) were submitted for culture, isolation and phenotypic characterization. Upon obtaining pure *A suis* cultures, whole-genome sequencing was performed (survey group). Five publicly available *A suis* genomes were also included in the analysis (reference group). They were obtained between 1963 and 1991 and were isolated from porcine septicemia cases in Ontario, Canada (n = 4) and Washington, USA (n = 1, Table 1). Phylogenetic analyses performed included the unweighted pair group method with arithmetic mean (UPGMA) and orthologous average nucleotide identity (OrthoANI[7]). An orthoANI value above 95% was used to define bacterial species. The presence of RTX toxins was also investigated by using BLAST-P on each genome against reference proteins from three databases (VFDB, Victors, and PATRIC-curated virulence factors[8]).

Results and discussion

Three (292, 419, and 466) out of the four genomes from the survey group were clustered together and the OrthoANI analysis resulted in an average nucleotide similarity of 100% (Table 1). Interestingly, when comparing these three clonal genomes from the survey group with isolates H89-1173 and H91-0380 from the reference group, an average nucleotide similarity greater than 99.9% was observed. Genome 298 did not cluster with the other three genomes from the survey group. This genome had an average nucleotide similarity value of 99.73% when compared to either of the other three genomes within the survey group. When comparing genome 298 (survey group) to those in the reference group, the highest average nucleotide similarity was 99.79% - isolate H89-1173, a strain isolated in 1989. All genomes analysed carried the virulence genes ApxIA, ApXIIA, ApXIC, and ApXIIC. In conclusion, OrthoANI values above 95% confirmed that isolates from the survey group are indeed *A suis*. Three genomes from the survey group (292, 419, 466) were considered clonal and were very similar to previous Canadian isolates (H89-1173 and H91-0380). Isolate 298 was different from all isolates previously characterized, suggesting the emergence of a novel *A suis* strain associated with clinical disease.

References

1. Wilson R, McOrist S. *Actinobacillus suis* infection in pigs in Australia. *Aust Vet J.* 2000;78(5):317-319. **doi:10.1111/j.1751-0813.2000.tb11778.x**

2. MacInnes JI, Gottschalk M, Lone AG, et al. Prevalence of *Actinobacillus pleuropneumoniae, Actinobacillus suis, Haemophilus parasuis, Pasteurella multocida,* and *Streptococcus suis* in representative Ontario swine herds. *Can J Vet Res.* 2008;72(3):242-248. /pmc/articles/PMC2327245/?report=abstract. Accessed August 5, 2020.

3. Michael J. Yaeger. An outbreak of *Actinobacillus suis* septicemia in grow/finish pigs. *Vet Diagn Invest.* 1996:381-383. **https://journals.sagepub.com/doi/pdf/10.1177/104063879600800318.** Accessed August 12, 2020.

4. Mauch C, Bilkei G. *Actinobacillus suis*, a potential cause of abortion in gilts and low parity sows. *Vet J.* 2004;168(2):186-187. **doi:10.1016/S1090-0233(03)00142-4**

5. Kamp EM, Vermeulen TM, Smits MA, Haagsma J. Production of Apx toxins by field strains of *Actinobacillus pleuropneumoniae* and *Actinobacillus suis. Infect Immun.* 1994;62(9).

6. Burrows LL, Lo RY. Molecular characterization of an RTX toxin determinant from *Actinobacillus suis. Infect Immun.* 1992;60(6).

7. Lee I, Kim YO, Park SC, Chun J. OrthoANI: An improved algorithm and software for calculating average nucleotide identity. *Int J Syst Evol Microbiol.* 2016;66(2):1100-1103. **doi:10.1099/ijsem.0.000760**

8. Wattam AR, Davis JJ, Assaf R, et al. Improvements to PATRIC, the all-bacterial bioinformatics database and analysis resource center. *Nucleic Acids Res.* 2017;45(D1):D535-D542. **doi:10.1093/nar/gkw1017**

Table 1: OrthoANI values from pairwise comparisons between *A suis* genomes. Survey group samples are shown in blue and reference group samples in dark grey.

Sample ID	Group	Country	Year	292	298	419	466	ATCC 33415	ATCC 15557	H89 1173	H91 0380	H91 0406	NCTC 12996
292	Survey	Canada	2018		99.73%	100%	100%	99.81%	99.70%	99.92%	99.92%	99.88%	99.81%
298	Survey	Canada	2019			99.73%	99.73%	99.65%	99.54%	99.79%	99.78%	99.74%	99.66%
419	Survey	Canada	2019				100%	99.81%	99.70%	99.92%	99.92%	99.89%	99.81%
466	Survey	Canada	2018					99.81%	99.70%	99.92%	99.92%	99.89%	99.81%
ATCC 33415	Reference	USA	N/A						99.77%	99.83%	99.85%	99.82%	100%
ATCC 15557	Reference	Canada	1963							99.77%	99.73%	99.73%	99.77%
H89 1173	Reference	Canada	1989								99.98%	99.99%	99.82%
H91 0380	Reference	Canada	1991									99.98%	99.86%
H91 0406	Reference	Canada	1991										99.82%

Persistent infection and protection against Senecavirus A in pigs previously infected with a homologous and a heterologous isolate

G. Preis[1]; A. Buckley[2]; W. Wiese[3], F. Vannucci[3]
[1]Veterinary Population Medicine, University of Minnesota, St. Paul, Minnesota;
[2]Virus and Prion Research Unit, USDA-ARS-National Animal Disease Center, Ames, Iowa;
[3]Veterinary Diagnostic Laboratory, University of Minnesota, St. Paul, Minnesota

Introduction

The emergence of a vesicular disease in pigs in 2014/2015,[1] caused by Senecavirus A (SVA), has been responsible for great concern due to its high similarity with reportable diseases such as foot-and-mouth disease. Despite the recent onset of SVA outbreaks in pig farms from different countries, the virus has been shown to be silently circulating in US herds since at least 1988.[2] Preliminary data demonstrated the presence of virus in the tonsils of naturally infected pigs approximately 90 days after showing clinical disease, which suggests a potential for establishing persistent infections and an asymptomatic carrier state. In addition, there was no evidence whether animals previously exposed to a SVA strain are protected against homologous and heterologous isolates.

Materials and methods

Twenty-eight three weeks old crossbred gilts were allocated in three groups: group HC, inoculated with a historical followed by a contemporary isolate (n = 12); group CC, inoculated with a contemporary isolate in the two different time-points (n = 12); and group N, inoculated with RPMI medium (n = 4). Historical and contemporary isolates were obtained in 1999 and 2017 respectively. One animal from group CC died from an SVA-unrelated cause and was necropsied at 39 days post first inoculation (dpi). Two pigs from group N, three from group CC and four from HC were euthanized and necropsied at 49 dpi. The remaining pigs from groups HC and CC were re-challenged 50 days after the first inoculation. Oral swabs, fecal swabs and sera were collected for RT-qPCR, and sera was also submitted to indirect immunofluorescence (IFA) and serum neutralization (SN) assays. Shedding was determined when either oral or fecal swabs were positive by RT-qPCR. The time-points for collections were 1, 3, 7, 10, 14, 21, 28, 35, 42, 48, 52, 55, 57, 64 and 71 dpi. At day 71, the remaining pigs from all groups were necropsied. Tonsils of the soft palate, submandibular and retropharyngeal lymph nodes from all pigs were submitted for RT-qPCR.

Results and discussion

All animals from groups HC and CC were shedding at day 3 dpi. Shedding started decreasing at days 10 and 14 dpi, with the last detections on days 28 and 35 dpi, for groups HC and CC, respectively. PCR positive serum was detected 3 dpi in both infected groups (12/12 HC and 11/12 CC) and persists until days 28 for group HC and 21 for group CC. Serological IgG response was first detected at 14 and 10dpi in HC and CC groups, respectively.

Two days after the second inoculation, 3/8 oral and 8/8 fecal swabs from both groups were positive by PCR, but there were no positive swabs in the following time-points. Viremia was observed in one pig from each of the HC and CC groups at 7 and 5 days after re-challenging, respectively. All pigs from groups HC and CC had at least one of the three collected tissues positive by RT-qPCR after necropsy, suggesting potential persistent infection, which should be further evaluated.

Viral shedding after the first inoculation persisted up to 28 and 35 days in HC and CC groups, respectively, while viremia was consistently detected for 28 days from HC and 21 days from CC group. However, SVA was only detected two days after the second inoculation (52 dpi), with fewer animals being RT-qPCR positive in oral swabs than in fecal swabs. Since the animals were recently re-challenged, the higher positivity rate in fecal swabs may be due to detection of the inoculum. SN assays in serum collected before the re-challenge demonstrated a higher average titer for the homologous strain used in each group. Additionally, a higher SN titer was seen in both groups when testing against both isolates in serum collected after re-challenge, suggesting a booster effect. These findings suggest that both groups re-challenged with homologous and heterologous strains showed protective immunity, characterized by limited viremia and shedding and increasing SN titers after re-challenge.

References

1. Vannucci, F. A., Linhares, D. C. L., Barcellos, D. E. S. N., Lam, H. C., Collins, J., & Marthaler, D. (2015). Identification and Complete Genome of Seneca Valley Virus in Vesicular Fluid and Sera of Pigs Affected with Idiopathic Vesicular Disease, Brazil. Transboundary and Emerging Diseases, 62(6), 589–593.

2. Knowles NJ, Hales LM, Jones BH, et al. Epidemiology of Seneca Valley Virus: Identification and Characterization of Isolates from Pigs in the United States. *XIVth Meet Eur Study Gr Mol Biol Picornaviruses*. 2006;(January):G2.

Herd prevalence of PRRSV, PEDV, PDCoV and TGEV in pig farm manure pits

J. Montoya Lopez; C. Vilalta; J. Sanhueza; C. A. Corzo
Department of Veterinary Population Medicine, University of Minnesota., St. Paul, Minnesota

Introduction

Viruses such as porcine reproductive and respiratory syndrome (PRRS), porcine epidemic diarrhea (PED), porcine delta coronavirus (PDCoV) and transmissible gastroenteritis (TGE) continue to be present in the United States (US) swine industry generating important losses.[1-3] Factors related to the occurrence of these diseases need to be understood in order to implement intervention strategies and thus stop these viruses from being introduced into swine herds. Once pigs become infected with these viruses, shedding shortly begins and viral particles are shed into the environment through different routes (e.g. manure, urine, exhalation, saliva, blood). Within slatted barns, these viral particles can either exit through exhaust fans or fall into the manure pits, in which they may remain until they leave the barn through manure pumping. It is known for instance that the occurrence of PRRS outbreaks in sow farms has a seasonal pattern, with most cases occurring during the fall and winter,[4] a time in which manure pumping activities are performed by most herds. While PRRSV airborne transmission between barns has been extensively studied in recent years, the role of manure management practices in PRRSV transmission has not received the same level of attention and determinants for PRRS occurrence are yet fully understood. Therefore, the objective of this study was to estimate the barn level prevalence of PRRSV, PEDV, PDCoV and TGEV in manure pits during the fall of 2019.

Materials and methods

This cross-sectional study was conducted in the midwestern US. Since the prevalence of herds positive to these viruses in pit manure is unknown, a default barn level prevalence of 50% with an allowable absolute error of 5% and a 95% level of confidence was used for sample size calculation. Based on this calculation, a total of 385 pig barns were needed to estimate herd level prevalence for these viruses. Pit manure samples were obtained through **A)** Collected directly from manure pits and **B)** From Minnesota environmental agencies who had received manure samples from pig production companies that were submitted for nutrient analysis on a yearly basis from pig producers. Each sample was accompanied by date of collection and county where the farm is located. These samples were submitted to the University of Minnesota Veterinary Diagnostic Laboratory for testing by RT-PCR for PRRSV, PEDV, PDCoV and TGEV. Positive RT-PCR results were further tested by virus isolation. A logistic regression model was performed to determine the relationship between the log odds of being positive for each virus, and sow farm origin status and type of farm (boar stud, sow farm, growing pig farms). Interaction terms were checked using ANOVA type III due to multicategorical variables used in the model.

Results

At the moment of writing, a total of 300 manure samples had been collected and tested. PRRSV was detected in 27 (9%, 95% CI [8.81%, 9.19%]) farms with a median, cycle threshold (Ct) value of 37.56 (range 35.48 to 39.57). PEDV was detected in 48 (16%, 95% CI [15.8%, 16.2%]) farms with a median Ct value of 33.11 (range 24.13 to 37.25). PDCoV was detected in 19 (6.3%, 95% CI [6.14%, 6.46%]) farms with a median Ct value of 34.49 (range 28.65 to 39.75). TGEV was not detected in any sample. No viable virus was found in the analyzed samples, nor samples were sequenced. No statistically significant associations between the outcome and factors were detected in the model.

Discussion

Results from this study confirms the presence of these viruses in manure pits. Unfortunately, the status of the pigs in these barns was unknown, which opens an opportunity for a follow up study. Viability of the virus was not proven; however, sample freeze-thaw cycles and viral dilution (low viral titer) may have played a role into these findings. Therefore, further investigation into virus viability is warranted. PRRSV Ct values tended to be higher compared to PEDV/PDCoV. These differences may be explained by the fact that the latter viruses replicate in the intestinal lining, leading to a large number of viral particles being shed into the environment. This study raises awareness on the presence of viral genetic material and concentration of high impact pathogens in manure pits. More studies need to be conducted to further understand the risk implications of these findings, especially when conducting manure pumping and spreading activities.

Acknowledgements

Signature Program, University of Minnesota. Participating production companies and swine practitioners

References

1. Holtkamp D.J. Assessment of the economic impact of porcine reproductive and respiratory syndrome virus on United States pork producers. J. Swine Health Prod. 2013; 21:72–84.

2. Paarlberg pl. Updated estimated economic welfare impacts of porcine epidemic diarrhea virus (PEDv).2014 April 1:38.

3. Schulz LL, Tonsor GT. Assessment of the economic impacts of porcine epidemic diarrhea virus in the United States. J Anim Sci. 2015 Nov 1;93(11):5111–8.

4. Tousignant SJP, Perez AM, Lowe JF, Yeske PE, Morrison RB. Temporal and spatial dynamics of porcine reproductive and respiratory syndrome virus infection in the United States. Am J Vet Res. 2015 Jan;76(1):70–6.

PRRSV farm prevalence in near-to-market-pigs in the United States

J. Montoya Lopez; M. Kikuti; C. Vilalta; J. Sanhueza; C. A. Corzo
Department of Veterinary Population Medicine, University of Minnesota, St. Paul, Minnesota

Introduction

PRRSV research on the growing pig population is scarce despite the fact that this population accounts for the majority of pig farms in the United States (US) swine industry. Basic epidemiologic features related to measurement of disease (e.g. counts, prevalence, rates and ratios) and the determinants of health and disease need to be unveiled and characterized in order to continue to make progress towards whole system and regional PRRSV prevention, control and elimination. Here, we aim to understand the farm level PRRSV prevalence of near-to-market pigs to begin unveiling the potential role that growing pig sites may play on the epidemiology of this disease.

Materials and methods

This cross-sectional study was conducted in main pig production regions in the US Since the PRRSV farm prevalence within this population was unknown, the study was designed to estimate a 50% farm prevalence with an allowable error of 10% which yielded a total of 95 farms to be sampled. Within each finishing pig farm, a total of 30 blood samples per farm were randomly collected enabling us to be 95% confident of detecting at least 1 positive pig when the within herd prevalence was at least 10%. Farm participation was voluntary, and samples were collected, regardless of clinical signs from near-to-market finishing pigs (≥ 20 weeks old) between October of 2018 and November of 2019. Serum samples were tested for North American (NA) and European (EU) PRRSV by RT-PCR in pools of 5 at the University of Minnesota Veterinary Diagnostic Laboratory. Positive pools with a Ct of 33 or below were opened for individual testing and ORF5 sequencing.[1] Additional data was collected for every farm through a farm characteristics survey which included location, vaccine usage, biosecurity measures, and mortality management practices in an attempt to explore factors associated to the outcome.

Results

A total of 85 farms from 19 companies or veterinary clinics located in 12 states participated in this study. Out of 510 pools, 31 pools (6.08%; 95% CI [4.17%, 8.52%]) tested positive for the NA PRRSV with an average cycle threshold (Ct) value of 35.87 (SD ± 2.33) and a median 36.83 (IQR 3.35), respectively. For the EU PRRSv, 5 pools (0.98%; 95%CI [0.03%, 2.27%]) tested positive with an average Ct value of 34.26 (SD ± 0.64) and a median of 34.21 (IQR 1.19). Six (5 for NA virus and 1 for EU virus) sequences were obtained from positive samples.

From the 19 participating companies, a total of 9 had at least one positive farm. Out of the 85 participating farms, 19 (22.35% 95%CI [14.03,32.69]) and 2 (2.35% 95%CI [0.29, 8.24]) farms had at least one positive pool for NA and EU PRRSV, respectively. No farm tested positive for both types of PRRSV. Within the positive farms, the average number of positive pools was 1.63 (SD ± 0.466). At the moment of writing, the risk factor and sequence analysis had not been completed.

Discussion

Results from this study raise awareness from a risk perspective due to the fact that approximately a quarter of the finishing sites with near-to-market-pigs in the US had at least one viremic pig. The presence of viremic pigs amongst randomly selected animals in farms that were not targeted specifically towards PRRSV detection highlights the potential importance of this population in disease transmission. More studies are needed to further understand the role that PRRSV positive growing pig farms play in the epidemiology of this disease.

Reference

1. Zhang J, Zheng Y, Xia X-Q, Chen Q, Bade SA, Yoon K-J, et al. High-throughput whole genome sequencing of Porcine reproductive and respiratory syndrome virus from cell culture materials and clinical specimens using next-generation sequencing technology. J VET Diagn Invest. 2017 Jan;29(1):41–50.

Characterization of *Glaesserella parasuis* strains circulating in US swine herds using whole genome sequencing

R. Mugabi[1]; G. Li[1]; N. Macedo[1]; B. Arruda[1]; O. Sahin[1]; R. Main[1]; M. Gottschalk[3]; D. Tucker[4]; M. J. Clavijo[1,2]
[1]VDPAM, Iowa State University, Ames, Iowa; [2]PIC®, Hendersonville, Tennessee; [3]University of Montreal, Saint-Hyacianthe, Canada; [4]University of Cambridge, Cambridge, United Kingdom

Introduction

Glaesserella parasuis (GPS) is one of the most important bacterial pathogens for the swine industry. Control of this pathogen requires timely detection, strategic use of antimicrobials, and effective vaccines. Although serotyping is routinely done, it has pitfalls. For example, some strains are untypable and its discriminatory power is low. Whole genome sequencing (WGS) offers more genetic detail and discrimination between isolates. From WGS data, phylogeny based on the core genome, serotype, (MLST) profile, putative virulence genes, and AMR profiles are determined for each isolate. This combined with relevant metadata (i.e. sites, age of the pig, organ of isolation, date), can help determine the clinical relevance of an isolate, thus, enhancing control and prevention programs. The objective of this study was to perform genetic characterization of *G parasuis* isolates obtained from diseased pigs.

Materials and methods

A total of 97 GPS isolates submitted to the VDL between 2015 and 2020 majority from USA swine farms were whole genome sequenced using Illumina Miseq platform and analyzed using an in-house bioinformatic pipeline. Isolates originated mainly from lung (n = 41), heart (n = 23), brain (n = 6), and fibrin (n = 6).

Results and discussion

Phylogenetic analysis revealed seven different clusters with cluster 7 lacking most of the virulence-associated trimeric autotransporters (*vtaA*) genes. Clusters 2 and 6, possessed a relatively large number of *vtaA* genes. Majority of cluster 1 isolates were serotype 7, whereas cluster 7 had multiple serotypes (2, 3, 6, 8,9, 15). Some serotypes were distributed in multiple clusters supporting the poor resolution of serotyping. The most prevalent serotypes were; 7 (n = 27), 4 (n = 11), 13 (n = 14), 4 (n = 11), 2 (n = 9), 5/12 (n = 8), 1 (n = 4), 3 (n = 1), 8 (n = 1), 15 (n = 1) and a few were untypeable by WGS. Previous studies have reported serotype 4, 5, 7, 13 as being frequently isolated from diseased pigs in North America. The overabundance of serotype 7, a serotype previously thought as non-virulent, could be due to submission bias, but could also reflect a strain associated with disease. MLST revealed 45 different STs including ST454 (n = 18), ST6 (n = 8), and ST415 (n = 6) and the rest with 1-3 isolates each. It is believed that strains that possess a high number of these genes especially group 1 and 2, are more pathogenic than those with fewer *vtaA* genes. Genomic analysis revealed that ¾ of the isolates possessed at least 4 of the 13 *vtaA* genes. Some strains from serotypes 3, 7, 5/12 and 13, carried most of the *vtaA* genes. Very few isolates carried AMR genes, which coincides with phenotypic susceptibility observed at the ISU VDL. Eight AMR genes were identified including *Aph3"la* (n = 1), ^{bla}ROB-1 (n = 1), *strA* (n=1), *sul2* (n = 1), *blaZ* (n = 15), *ermA* (n = 4), *spc* (n = 8) and *tetM* (n = 8).

Conclusion

Knowledge generated with WGS important for autogenous vaccine development especially for pathogens where commensal and pathogenic strains can co-exist like GPS. The repertoire of AMR genes can also guide veterinarians in therapeutic decision making which may improve antimicrobial stewardship.

Comparison of highly pathogenic porcine reproductive and respiratory syndrome virus 2 (HP-PRRSV-2) transmission between conventional needles and a needle-free vaccinator in experimental pigs

A. Madapong[1], PhD; K. Saeng-chuto[1], PhD; A. Tantituvanont[2], PhD; D. Nilubol[1], DVM, PhD
[1]Department of Veterinary Microbiology, Faculty of Veterinary Science, Chulalongkorn University, Bangkok, Thailand
[2]Department of Pharmaceutic and Industrial Pharmacies, Faculty of Pharmaceutical Sciences, Chulalongkorn University, Bangkok, Thailand

Introduction

In modern swine farming, consecutive vaccination of swine without needle changes is common. It has been shown that some viruses can be transmitted in other species iatrogenically through this method.[1,2] The objective of the trial was to ascertain if iatrogenic transmission of HP-PRRSV-2 could be replicated in a consecutive vaccination model with both needles and an intradermal needle free vaccinator.

Materials and methods

Forty-two, 3-week-old, PRRSV-free pigs were allocated into 5 groups with 3 pigs each; IM/High, ID/High, IM/Low, ID/Low and NoChal. At 0 days post-challenge (DPC), the IM/High and ID/High groups were intranasally inoculated with 4 ml of HP-PRRSV-2 (10^6 $TCID_{50}$/mL, 2 mL/nostril). The IM/Low and ID/Low groups were intranasally inoculated with 4 mL of HP-PRRSV-2 (10^3 $TCID_{50}$/mL, 2 mL/nostril). At 7 DPC (0 Days post-injection; DPI), the IM/High and IM/Low groups were sham vaccinated intramuscularly injected with 1 mL of adjuvant (Diluvac Forte®) using conventional needles (gauge 21, 1"). The ID/High and ID/Low groups were sham vaccinated intradermally with 0.2 mL of adjuvant (Diluvac Forte®) using an intradermal needle-free device (IDAL® 3G). Subsequently, the same needle or needle-free device were used to inject the same volume of adjuvant to sentinel pigs with the same route of injection for each group (1:2 ratio). The NoChal group served as a control. Blood samples were collected at 0, 7, 14, 21 and 28 DPC. PRRSV viremia and seroconversion were evaluated (Figure 1).

Results and discussion

The results showed that the IM/High and ID/High groups had the highest viremia at 7 DPC. Meanwhile, the IM/Low and ID/Low groups had the highest viremia at 14 DPC. Seroconversion was first detected at 14 DPC in the IM/High and ID/High groups. However, the IM/Low and ID/Low groups were seroconverted at 28 DPC. In the sentinel pigs of IM/High and IM/Low groups seroconversion was detected at14 DPI. In contrast, no seroconversion was detected in the sentinel pigs of both ID/High and ID/Low groups throughout of experiment. Our findings confirm the potential of PRRSV transmission through conventional needle during vaccination. On the other hand, the use of an intradermal needle-free vaccinator appeared to inhibit HP-PRRSV-2 transmission and could be used for alternative route of vaccination to enhance internal biosecurity on farm. We note that there are many other possible routes of iatrogenic virus transmission such as contaminated hands and farm boots, and thus this measure should be used in conjunction with other biosecurity measures.

References

1. Darpel KE, Barber J, Hope A, Wilson AJ, Gubbins S, Henstock M, Frost L, Batten C, Veronesi E, Moffat K, Carpenter S, Oura C, Mellor PS, Mertens PP (2016) Using shared needles for subcutaneous inoculation can transmit bluetongue virus mechanically between ruminant hosts. Sci Rep 6:20627.

2. Qi X, Qian YH, Bao CJ, Guo XL, Cui LB, Tang FY, Ji H, Huang Y, Cai PQ, Lu B, Xu K, Shi C, Zhu FC, Zhou MH, Wang H (2013) Probable person to person transmission of novel avian influenza A (H7N9) virus in Eastern China, 2013: epidemiological investigation. BMJ 347:f4752.

Figure 1: Schematic illustration of the present study.

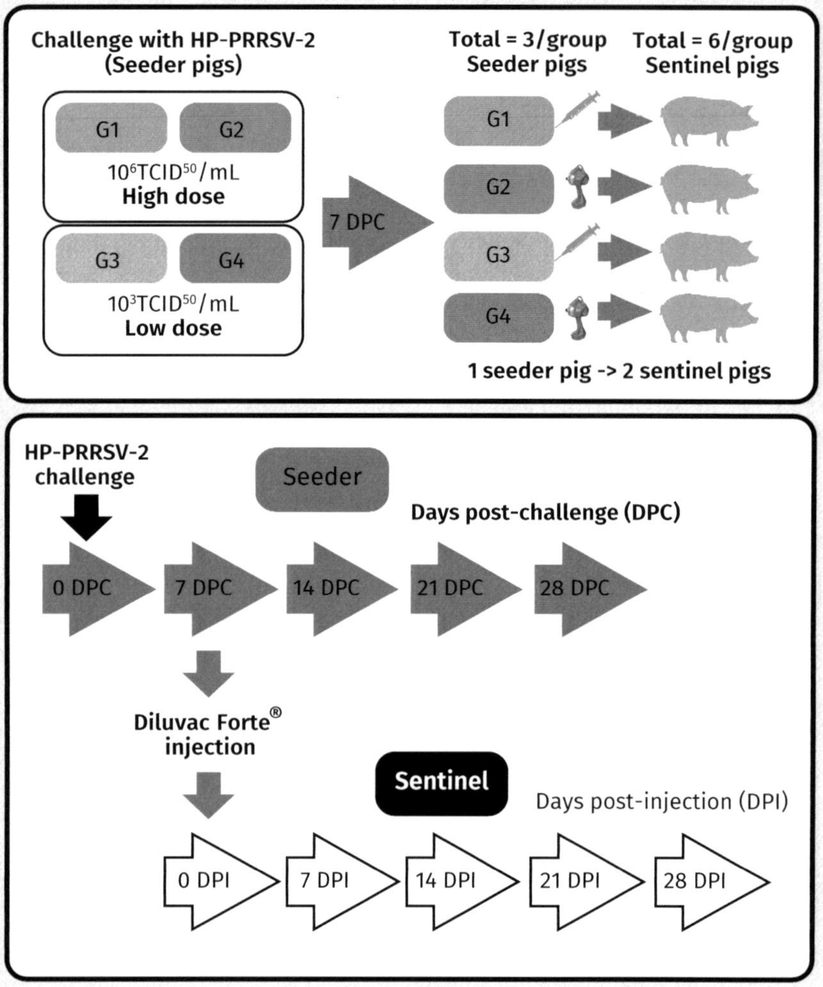

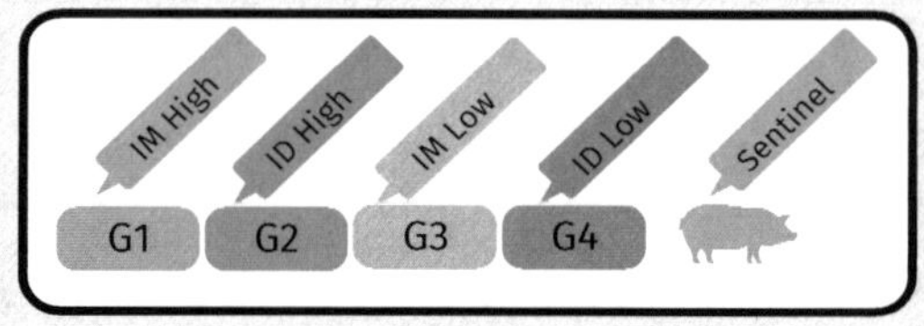

- Seeder pigs were challenged at 0 days post-challenge (DPC) at 3 week of age.
- At 7 DPC, seeder pigs were intramuscularly (IM) or intradermally (ID) injected with Diluvac Forte® using conventional syringe with needles and IDAL® needle-free vaccinator.
- The same needles/IDAL® needle-free vaccinator were used to inject sentinel pigs.
- Ratio of seeder and sentinel pigs was 1:2, 1 seeder pig to 2 sentinel pigs.
- Blood samples were collected, sera were separated and assayed for seroconversion using ELISA kit (IDEXX X3 Ab, IDEXX Laboratories, USA)

Immune response and protective efficacy of concurrent vaccination with porcine reproductive and respiratory syndrome virus 2 (PRRSV-2) modified live vaccine and porcine circovirus 2 (PCV2) in a co-challenge of HP-PRRSV-2 and PCV2

A. Madapong[1], PhD; K. Saeng-chuto[1], PhD; A. Tantituvanont[2], PhD; D. Nilubol[1], DVM, PhD
[1]Department of Veterinary Microbiology, Faculty of Veterinary Science, Chulalongkorn University, Bangkok, Thailand
[2]Department of Pharmaceutic and Industrial Pharmacies, Faculty of Pharmaceutical Sciences, Chulalongkorn University, Bangkok, Thailand

Introduction

The objectives of this study were to investigate the immune response and protective efficacy of concurrent vaccination with PRRSV-2 MLV and PCV2 vaccines, applied both intradermally and intramuscularly against co-challenge with HP-PRRSV-2 and PCV2. Previous work has ascertained that concurrent vaccination is efficacious[1] but did not assess if the additional variable of different routes of vaccination affected immune responses.

Materials and methods

Sixty-three, 3-week-old, PRRSV-free pigs were allocated into 7 groups with 9 pigs each: IM0/0, ID0/0, IM0/7, ID0/7, TF0/0, NV/CH and NV/NC. The IM0/0, IM0/7, ID0/0 and ID0/7 were intramuscular (IM) and intradermal (ID) vaccinated with PRRSV-2 MLV (Prime Pac® PRRS, MSD Animal Health The Netherlands) and PCV2-based vaccine (Porcilis® PCV M. hyo, MSD Animal Health, The Netherlands) at 0 and 7 days post-vaccination (DPV), respectively. The TF0/0 group was IM vaccinated with 3FLEX® (Boehringer Ingelheim, Germany) at 0 DPV. The NV/CN and NV/NC groups were left non-vaccinated. At 28 DPV, all groups were intranasally inoculated with HP-PRRSV-2 (106 TCID50/mL) and PCV2 (105 TCID^{50}/mL). The NV/NC group served as a control. Sera were separated and assayed for PRRSV- and PCV2-specific antibodies. Peripheral blood mononuclear cells (PBMC) were isolated and used for in vitro stimulation. IL-10 production and IFN-γ-secreting cells (SC) were evaluated. PRRSV and PCV2 viremia were quantified using RT-qPCR. Pigs were necropsied at 7 days post-challenge (DPC) and lung lesion was evaluated. PRRSV- and PCV2-specific antigens in tissues were analyzed using IHC.

Results and discussion

Regardless of immunization route, concurrent vaccination had no effect on PRRSV-specific antibody response but failed to increase PCV2-specific antibody response following vaccination. Vaccination at day 0/0 induced more IL-10 production compared to that of day 0/7, and IM groups had significantly higher IL-10 levels than that of the ID groups. The TF0/0 group had highest IL-10 level than that of the other vaccinated groups at 7 and 14 DPV. Additionally, the ID groups had relatively higher IFN-γ-SC than that of the IM groups. Following challenge, PRRSV and PCV2 viremia, lung lesions, and PRRSV- and PCV2 antigens in tissues were decreased in all vaccinated groups regardless of vaccination routes compared to the non-vaccinated controls. The IM0/7 group was the lowest compared to the other vaccinated groups (Figure 1). The study demonstrates that concurrent vaccination had no interference with the development of specific humoral and cell-meditated immunity and is associated with partial protection against co-infection with HP-PRRSV and PCV2.

Reference

1. Martelli P, Ardigo P, Ferrari L, Morganti M, De Angelis E, Bonilauri P, Luppi A, Guazzetti S, Caleffi A, Borghetti P (2013) Concurrent vaccinations against PCV2 and PRSSV: study on the specific immunity and clinical protection in naturally infected pigs. Vet Microbiol 162:558-571.

Figure 1: The mean genomic copies of PCV2 DNA in blood. Different letters indicate significant difference ($P < 0.05$) between groups at each days.

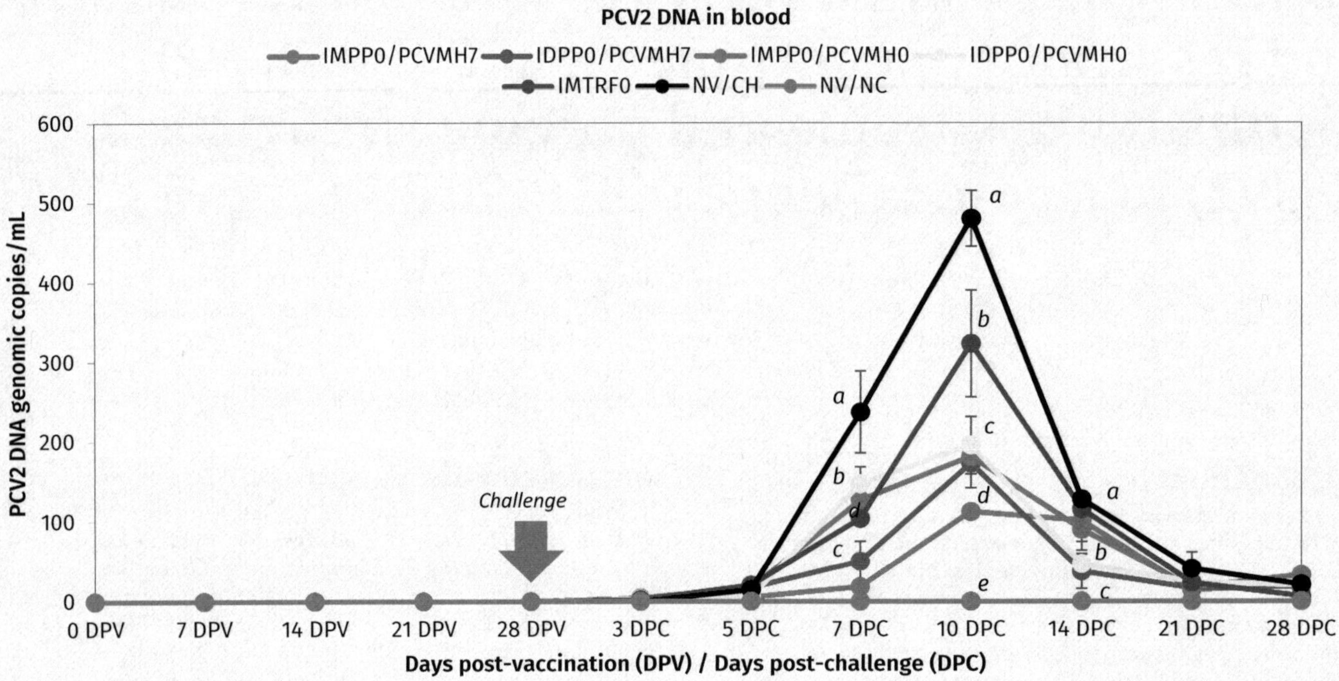

Estimating farm-level reproduction numbers for PRRSV using sequence-based transmission tree analysis

N. Pamornchainavakul, DVM; I. Paploski, DVM, MSc, PhD; D. Makau, BVetMed, PhD;
C. Corzo, DVM, MS, PhD; K. VanderWaal, PhD
College of Veterinary Medicine, University of Minnesota

Introduction

The reproduction number (R) is a fundamental value characterizing a pathogen's ability to spread within populations and is defined as the number of individuals that are infected by a single case. Many disease monitoring programs, however, archive data at the farm-level rather than individual-level, and management interventions are often implemented at the farm level. This creates the need to evaluate a pathogen's transmissibility at the farm-level to better trace disease spread and evaluate the success of control measures. Using transmission tree analysis, we quantify the farm-level R of porcine reproductive and respiratory syndrome (PRRS), which is more compatible with the scale at which disease dynamics are analyzed.

Materials and methods

Utilizing 943 open reading frame 5 (ORF5) sequences of PRRSV from 651 farms in a single geographic region, we identified three clades of closely related sequences (mean 98.4% nucleotide identity based on ORF5) that were collected from 151 farms between 2014 and 2017. We constructed time-scaled phylogenies for these three clades, all of which belonged to sub-lineage L1A (consisting primarily of RFLP-type 1-7-4).[1] Using the resulting trees, average generation time, and Bayesian inference, transmission networks were built to infer pig-to-pig infection chains, including inference on the number of unsampled individuals within the chain.[2] Farm-level transmission events were inferred for any pair of sequences for which the infection chain transitioned between farms. We then summarized infection chain length according to movement pathlength (how many steps apart farms were in the animal movement network). The median length of infection chains for farms directly connected by movements was then used as a threshold for defining direct farm-to-farm transmission. The number of inferred direct transmission events originating from each source farm was used to calculate farm-level R.

Results

Infection chain length was significantly correlated with movement network pathlength (rho=0.53, $P < 0.05$). For farms directly connected by movements, the median infection chain lengths for direct farm-to-farm transmission were 28, 35, and 46 pig-to-pig transmission events for each clade, respectively. Overall farm-level R had a median of 1 and a maximum of 5 (IQR = 1-2). Most of the inferred transmission events (155 out of 174) could not be attributed to direct animal movement, while 3% and 15% of events might occurred via semen and local area spreading, respectively (Figure 1).

Discussion

For these PRRSV variants, a farm is expected to infect a median of one other farm and animal movement was responsible for only 10% of total transmission events. This approach provides supportive information for epidemiologic assessments and a clearer depiction of transmission risk than solely interpreting evolutionary history or animal movement data.

References

1. Paploski IAD et al. Temporal Dynamics of Co-circulating Lineages of Porcine Reproductive and Respiratory Syndrome Virus. *Front Microbiol.* 2019;10:2486.

2. Xavier Didelot et al. Genomic Infectious Disease Epidemiology in Partially Sampled and Ongoing Outbreaks. *Molecular Biology and Evolution.* 2017;34(4):997–1007.

Figure 1: Proportion of movement pathlength types for direct farm-to-farm transmission.

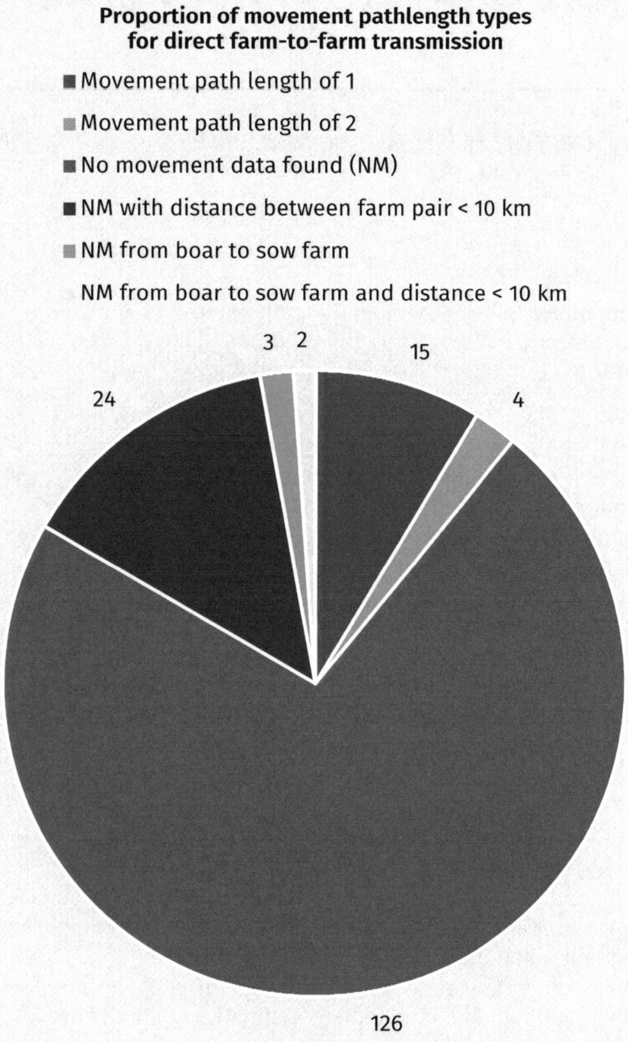

Estimating viral population size of PRRSV sublineages based on genetic diversity

I. Paploski[1], DVM, MSc, DrPH; D. Makau[1], DVM, MSc, PhD; N. Pamornchainavakul[1], DVM; D. Schroeder[1,2], BSc, PhD; A. Rovira[1], DVM, MSc, PhD; K. VanderWaal[1], BS, PhD
[1]Department of Veterinary Population Medicine, University of Minnesota, St. Paul, Minnesota;
[2]School of Biological Sciences, University of Reading, Reading, United Kingdom

Introduction

Porcine reproductive and respiratory syndrome virus (PRRSV) is comprised of genetic variants that evolve in parallel and complicate efforts to control the disease.[1] We hypothesize that, at the population level, partial cross-immunity against differing strains might lead to sequential turnover in the dominant phylogenetic lineage through time. In the US, most field viruses belong to lineage 1, which is highly diverse, suggesting that sub-lineages exist.[2] Our objective is to describe sub-lineages within lineage 1 and quantify their occurrence through time as a first-step towards understanding multi-strain dynamics.

Methods

We performed a discriminant principal component analysis on 7,252 lineage 1 ORF5 sequences. Clusters were identified through a k-means analysis of the principle components that accounted for 80% of the genetic data variance. The optimal number of clusters was considered the smallest BIC that preserved the known sub-lineage assignment of 75 reference sequences. We estimated the viral population size through time for all sub-lineages using the Skygrid Coalescent Model.

Results

A total of eight (L1A to L1H) sub-lineages were identified, including three previously undescribed. In addition, the emergence of new sub-lineages occurred on average every 4 years. In 2018, approximately 40% of all sequences belonged to sub-lineage L1H (mostly RFLP type 1-8-4) and 33% to L1A (mostly types 1-7-4 and 1-4-4). However, 13 years ago, the dominant sub-lineages were L1B (mostly type 1-18-2) and L1F (mostly type 1-8-4). Estimated population size of sub-lineages through time exhibited a wave-like pattern; each sub-lineage emerged, diversified, and subsequently declined, creating a pattern of sequential dominance of different lineages through time (Figure 1). Estimating population size based on genetic diversity and a coalescence model allowed us to overcome challenges related to uneven sequencing effort through time, and provides a more accurate picture of the abundance of different sub-lineages than relying in sequence occurrence (count) data alone.

Discussion

By classifying PRRSV lineage 1 sequences into sub-lineages, we were able to document the periodic emergence of new sub-lineages through time, which is consistent with our hypothesis of lineage turnover. This work lays the foundation for research on when and why different sub-lineages evolve. One potential driver for this involves immunological cross-reactivity in the immune response elicited by exposure to different sub-lineages. More research on this topic is needed, and will help shed light on how PRRSV evolves, adapts, and persists in US pig populations.

References

1. Shi M et al. Phylogeny-based evolutionary, demographical, and geographical dissection of north american type 2 porcine reproductive and respiratory syndrome viruses. *J Virol.* 2010; 84 8700-8711.

2. Paploski IAD et al. Temporal Dynamics of Co-circulating Lineages of Porcine Reproductive and Respiratory Syndrome Virus. *Front Microbiol.* 2019; 10:2486.

Figure 1: Estimated population size through time for different sublineages within PRRSV lineage 1.

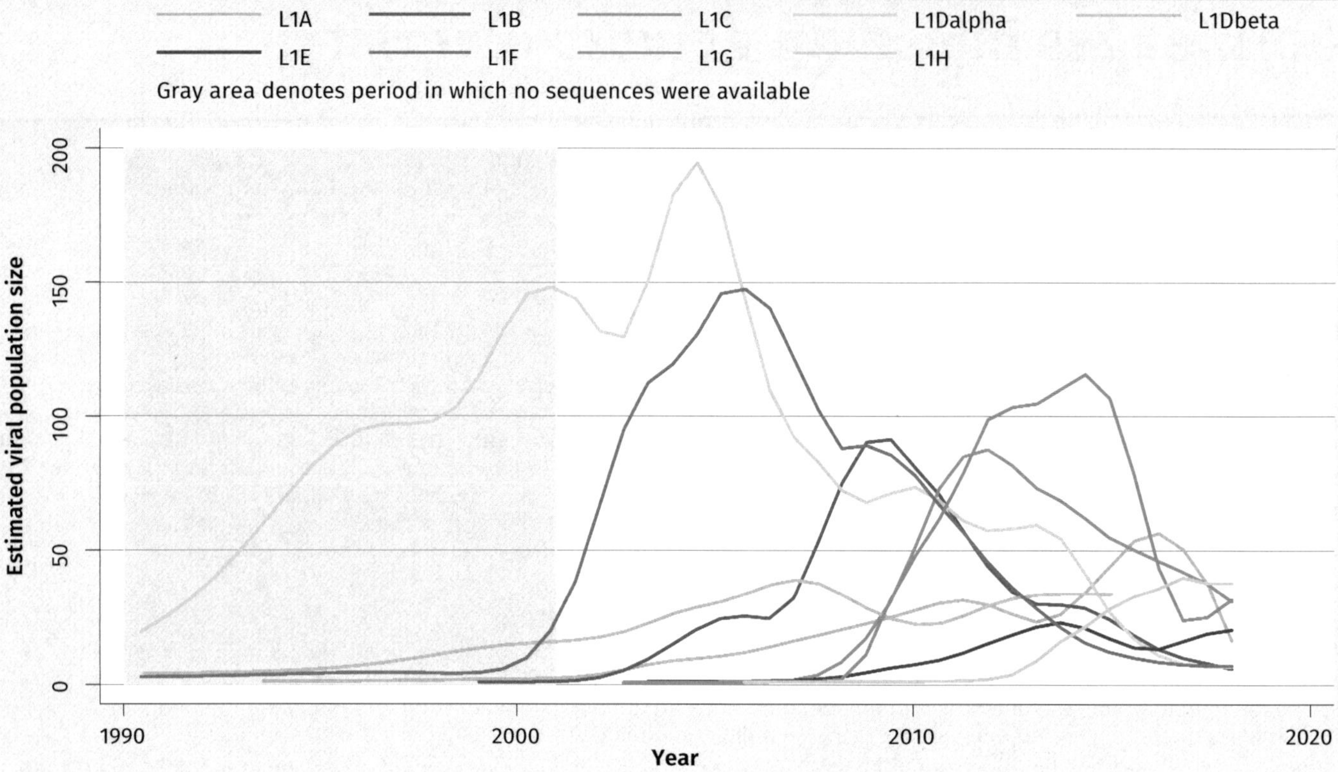

Prevalence of postweaning diarrhea caused by *Escherichia coli* in Spain

L. Pérez, DVM, MSc; M. Claver, DVM, BSc; P. J. Sánchez, DVM, MSc; J. A. González, DVM, BSc
Elanco Animal Health, Alcobendas, Spain

Introduction

Post-weaning diarrhoea (PWD) caused by *Escherichia coli (E coli)* remains a major cause of economic losses for the pig industry due to mortality, morbidity, decreased growth rate and cost of medication.[1] PWD usually occurs a few days after weaning causing dehydration, loss of body conditions and mortality.[2] However, in case of Spain, there has been a shift in the dynamic of infection, and clinical signs are mainly observed between the second and the third weeks after weaning.

Material and methods

A total of 193 pig farms with acute cases of PWD were sampled between January 2018 (n = 53) and December 2019 (n = 140). These farms were located in Spain. Animals were selected based on PWD clinical signs (diarrhoea, dehydration and increased mortality). A total of 579 four to eight-week old pigs within the first 24 hours of the acute phase of the disease were sampled and three rectal swabs were collected from three different pigs on each farm. Swabs with fecal matter were inserted into Amies transport medium in airtight screw cap plastic vials and were submitted to the laboratory (Exopol; Zaragoza, Spain) for *E coli* diagnosis. A multiplex PCR was carried out directly from 3 pooling of fecal samples contained in rectal swabs. PCR was performed to detect adhesion factor and toxin genes, including F4 (K88) and F18 fimbriae, adhesin involved in diffuse adherence (AIDA-1), heat-stable and heat-labile enterotoxins (STa, STb, LT), shigatoxin (STx2e), enteroaggregative heat-stable toxin (EAST-1) and, *E coli* attaching and effacing (eae).

Results

Only 1.6% (3 out of 193) of farms were negative for all the *E coli's* virulence factor genes. The total prevalence of genes for adhesion factors and toxins of *E coli*, Rotavirus type A and PEDV among 193 farms with PWD cases in 2018 and 2019 are shown in Table 1. AIDA-1 gene was the most commonly found (70.1%) followed by F18 (61.8%) and F4 (48.8%). The most prevalent toxin was EAST-1 (97.4%) followed by STb (86.1%) and STa (78.0%). Rotavirus type A was found in 64.6% of the PWD cases and PEDV in 11.4%. Enteropathogenic *E coli* (EPEC) was present in 75.8% of isolates.

Discussion and conclusion

The present study searched for F4, F18 and AIDA-1 adhesion factor genes. Results showed that the most prevalent adhesin gene was AIDA-1 (70.1%) followed by F18 (61.8%) and F4 (48.8%). In relation to F4 and F18 fimbriae, another Spanish study carried out in 2015 found that F18 gene was the most prevalent in comparison to F4, F5, F6 and F41.[3] The most prevalent enterotoxin gene has been EAST-1 (97.4%). Other studies are in accordance with the present study and showed EAST-1 as the most prevalent.[4] The present study shows the high prevalence of ETEC pathotype, which is able to cause PWD. There is a high degree of variability in the *E coli* virulence factors within country as well as differences between countries. It is necessary to conduct a rigorous diagnosis in order to characterize the *E coli's* virulence factors and find the most appropriate solution in each PWD case, since it is described that some virotypes are more severe and show a higher resistance to certain antibiotics.[5]

References

1. Fairbrother JM et al. *Escherichia coli* in postweaning diarrhea in pigs: an update on bacterial types, pathogenesis, and prevention strategies. *Animal Health Research Reviews*. 2005; 17-39.

2. Fairbrother JM et al. Colibacillosis. *Diseases of Swine*. 2012; 723–47.

3. Sánchez PJ et al. Prevalence of virulence factors in *Escherichia coli* isolated from pigs with post-weaning diarrhoea in Spain. IPVS 2016; 467.

4. Li S et al. Prevalence and characterization of virulence genes in *Escherichia coli* isolated from piglets suffering post-weaning diarrhoea in Shandong Province, China. 2020. *Vet Med Sci*. 69-75.

5. Congreso de Producción Porcina en Resistencia. 2016. Available at: **www.engormix.com/pig-industry/articles/recent-trends-virulence-antimicrobial-t39746.htm**. Accessed 14 October 2016.

Table 1: Total prevalence (%) of genes for adhesion factors and toxins of *E coli*, rotavirus type A and porcine epidemic diarrhea virus (PEDV) among 193 farms with PWD cases between 2018 and 2019.

	Virulence factor								
Pathogen	**Adhesion factor**			**Toxins**					
	F4	F18	AIDA	Sta	Stb	LT	Stx2e	EAST-1	eae
Percentage of *Escherichia coli's* virulence factors	48.8	61.8	70.1	78.0	86.1	66.8	22.4	97.4	75.8
Percentage of rotavirus A					64.6				
Percentage of PEDV					11.4				

Comparison of methods for *Mycoplasma hyopneumoniae* DNA detection in oral fluids

A. P. Poeta Silva[1], DVM, MS; G. Storino[2], DVM; F. Matias Ferreyra[1], DVM, MS; J. Miller[1]; K. Harmon[1], PhD; P. Gauger[1], DVM PhD; W. Witbeck[3]; K. Doolittle[3]; S. Zimmerman[3], DVM, MS MBA; R. Derscheid[1], DVM, PhD; M. Clavijo[1], DVM, PhD; B. Arruda[1], DVM, PhD; J. Zimmerman[1], DVM, PhD
[1]Veterinary Diagnostic and Population Animal Medicine, Iowa State University, Ames, Iowa;
[2]Escola de Agricultura e Ciências Veterinárias, Universidade Estadual de São Paulo, SP, Brasil;
[3]IDEXX Laboratories Inc., Westbrook, Maine

Introduction

A challenging aspect of *Mycoplasma hyopneumoniae* (*MHP*) prevention is detection of infected animals. Theoretically, detection of *MHP* DNA by PCR should offer the highest likelihood of detection. *MHP* PCR testing of tracheal samples has been described as the preferred sample for this purpose, but sample collection is labor-intensive, requires trained personnel, and is stressful for the animals. Conversely, pen-based oral fluid samples are easily collected and do not require restraint of pigs. Therefore, the objective of this study was to compare the detection of *MHP* DNA in pen-based oral fluids from animals of known *MHP* infection status.

Materials and methods

Pen-based oral fluid samples of known status were collected from 5 groups of 6-week-old pigs from day post inoculation (DPI) -4 through DPI 59. The negative control group consisted of 3 pigs; the 4 treatment groups contained 9 pigs each. Treatment groups differed in the number of pig inoculated with *MHP* (lung homogenate, 1×10^5 CFU/mL *MHP* 232) at time zero: 1/9, 3/9, 6/9, and 9/9 pigs. Tracheal swabs (TS) were collected to monitor individual pig *MHP* infection at DPIs -3, 3, 7, 10, 14, 21, 24, 28, 35, 38, 45, 52 and 59. Oral fluid samples (n = 322) were collected daily for 59 days post inoculation (DPI). Extraction methods included (1) MagMAX™-96 Pathogen RNA/DNA kit, Applied Biosystems™, Carlsbad, CA; and (2) IDEXX RealPCR* DNA/RNA Magnetic Bead Kit, IDEXX Laboratories Inc., Westbrook, ME. PCRs were (1a) TaqMan® Fast Virus 1-Step Master Mix (Life Technologies, Carlsbad, CA) and primer/probe described for Mhp183; (1b) TaqMan® Fast Virus 1-Step Master Mix with the addition of AmpliTaq® 360DNA Polymerase (5U/uL) (ThermoFisher Scientific), and primer/probe described for Mhp183; and (2) RealPCR* Master Mix and RealPCR* M hyo DNA Mix, IDEXX Laboratories Inc. PCRs were run to 40 cycles using the Applied Biosystems® 7500 Real-Time PCR (ThermoFisher Scientific).

Results and discussion

Using Extraction 1 and PCR 1, 18 of 20 inoculated pigs were PCR positive (tracheal swabs) for *MHP* DNA on DPI 3. All pigs were PCR positive for *MHP* DNA in tracheal swabs on or before DPI 59. For oral fluids, a comparison of positivity rates among protocols by mixed logistic regression showed significant differences among procedures, with the highest positivity rate achieved using Extraction 1 and PCR 2 (173 positives among 322 samples), followed by Extraction 1 and PCR 1b (148 of 322), Extraction 2 and PCR 2 (134 of 322), and Extraction 1 and PCR 1a (109 of 322). Significant differences among procedures in the detection of *MHP* DNA suggests that further improvements in laboratorial methods may be possible.

Determining infectivity of pork from PEDV positive pigs of different ages

Loni Schumacher, DVM, MS; Rawal Gaurav, DVM, MS; Ethan Aljets; Deaglan Riordan; Jianfeng Zhang, PhD; Jinhui Zhu, MS, PhD; Wannarat Yim-Im, MS; Drew Magstadt, DVM, MS; Phillip Gauger, DVM, MS, PhD; Jianqiang Zhang, MD, MS, PhD
Department of Veterinary Diagnostic and Production Animal Medicine, College of Veterinary Medicine, Iowa State University, Ames, Iowa

Introduction

Porcine epidemic diarrhea virus (PEDV) is an alphacoronavirus and is the causative agent of porcine epidemic diarrhea (PED), an enteric disease affecting global swine industries.[1] Since its emergence in the US, some importers of American pork voiced concerns about the presence of PEDV in meat. However, there have been only limited reports and data regarding PEDV in muscles of neonatal pigs and no data is available in weaned, grower, and finisher pigs. Even in neonatal piglets, there are only viral RNA data on muscle samples detected by PEDV polymerase chain reaction (PCR). In a previous study, we evaluated the detection of PEDV in muscle of PEDV-inoculated 3-week-old ("weaned"), 8-week-old ("grower"), and 23-week-old ("finisher") pigs as tested by PCR. However, PCR results cannot distinguish non-infectious virus from infectious virus and virus isolation from cell culture is not sensitive enough to assess PEDV infectivity. Therefore, we aimed to employ the sensitive young pig bioassay in the current study to determine if pork from PEDV infected pigs contains infectious PEDV for the first time.

Materials and methods

Ten-day-old PEDV naïve pigs (N = 42) were divided into 14 groups (3 pigs/group) and orogastrically inoculated with the following inocula. Group 1: virus-negative culture media; Group 2: non-S INDEL PEDV culture isolate USA/IN19338/2013 (CT = 11.4, 105 TCID50/pig); Groups 3-5: three muscle homogenates (CT 26.3, 30.8 and 29.9) from three previous PEDV-inoculated weaned pigs; Groups 6-8: three muscle homogenates (CT > 40 for each) from three previous PEDV-inoculated finisher pigs; Groups 9-11: three ileum homogenates (CT 22.1, 21.7 and 18.4) from the same three previously inoculated weaned pigs as described in Groups 3-5; and Groups 12-14: three ileum homogenates (CT 27.0, 24.3 and 24.2) from the same three previously inoculated finisher pigs as described in Groups 6-8. Fecal swabs were collected daily until end of study (7 dpi). At 7 dpi, all piglets were euthanized and necropsied. Fresh and formalin-fixed small intestines (jejunum and ileum) were collected. Fecal swabs and small intestines were tested by a PEDV nucleocaspid gene-based quantitative real-time RT-PCR as described previously.[2]

Results and discussion

All pigs in Group 1 were bioassay negative and all pigs in Group 2 were bioassay positive. All pigs inoculated with muscle homogenate either from weaned pigs (Groups 3-5) or from finisher pigs (Groups 6-8) were bioassay negative and lacked detectable PEDV in feces and small intestines. All pigs inoculated with ileum homogenate from weaned pigs (Groups 9-11) were PCR positive by 2 dpi and had fecal shedding to the end of study. Interestingly, among the three groups of pigs inoculated with ileum homogenates from finisher pigs (Groups 12-14), only one group (Group 13) was bioassay positive. Based on these results, pork from PEDV infected pigs does not contain infectious virus as determined by the young pig bioassay model. Information gained from this study may help address safety concerns in the market about PEDV infectivity of meat.

References

1. Jung K, et al. Porcine epidemic diarrhea virus infection: Etiology, epidemiology, pathogenesis, and immunoprophylaxis. *Vet J.* 2015;204(2):134-43.

2. Thomas J, et al. Effect of porcine epidemic diarrhea virus infectious doses on infection outcomes in naïve conventional neonatal and weaned pigs. *PLoS ONE.* 2015;10(10): e0139266.

On the ability to predict the farm-level risk of PRRSV outbreaks by ensemble machine learning modeling and biosecurity practices

Abagael Sykes[1]; Derald J. Holtkamp[2]; Broc Mauch[2]; Daniel Linhares[2]; Gustavo S. Silva[2]; Gustavo Machado[1]
[1]Department of Population Health and Pathobiology, North Carolina State University; [2]Veterinary Diagnostic and Production Animal Medicine Department, College of Veterinary Medicine, Iowa State University

Introduction

Biosecurity and immunization strategies have proven successful in reducing porcine reproductive and respiratory syndrome virus (PRRSV) spreading events.[1] However, compliance with recommended biosecurity proposals is precluded by their high cost and, more importantly, the lack of quantitative methods to measure of the impact of biosecurity in reducing the risk of PRRSV occurrence. This work aimed at quantifying PRRSV introduction risk by calculating the contribution of individual biosecurity measures on individual farms PRRSV status.

Materials and methods

A total of 158 US breeding farms with and without reported cases of PRRSV for the past 5 year, responded to a short biosecurity survey with 44 questions related with biosecurity practices and farm demography (referred herein as features).[2] A set of machine learning algorithms were used individually and in combination to: calculate the probability of PRRSV case at each farm, and to quantify how each biosecurity feature played a role in shaping PRRSV risk at farm level (calculated by Shapley values, [phi]).[3] Shapley values calculate the importance of a feature by assessing the model prediction with and without the feature. For example, in a PRRSV positive farm, Shapley values greater than 0 identify features which contribute to the risk of continued infection or exposure within the farm, while negative values identify those features which contribute to lowering the exposure risk.

Results

Overall, the performance of individual machine learning algorithms varied from 65.42% under gradient boosting machine [4] to 80.23% for random forest;[5] while the ensemble of all the algorithms performed better with a score of 85.54%, 5.31% better than the best individual model. Figure 1 present an example of a PRRSV positive farm to show the contribution of each feature on PRRSV status. Further, calculating Shapley values unpacks the importance of each biosecurity practice in terms of PRRSV exposure risk in an intuitive way that can guide improvements in specific biosecurity practices. For example, in Figure 1 farm characteristics such as the large capacity of breeding females in a 3-mile radius of the site, had a large contribution to its positive status. In contrast, the distance of the site from the public road (880m) lowers the exposure risk of this farm. Additionally, by collating the predicted probability of PRRSV occurrence across the farms, sites could be benchmarked against each other to identify those at an increased risk of exposure.

Conclusions and discussion

In summary, we calculated the risk for PRRSV introduction of individual farms and quantified how each biosecurity feature contributed towards a farm reporting to be positive or negative for PRRSV. The tool developed in the present study is suitable for identifying where efforts should be focused in biosecurity actions forming part of disease control programs.

References

1. Rathkjen, P.H. et al. 2017. Acta Vet Scandinavica 59:4.
2. Silva, G.S. A et al. 2019. Prev Vet Med 171:104749.
3. Shapley, L. S. 1953. PNAS, 39(10), 1095.
4. Friedman, J.H. 2001. The Annals of Statistics.
5. Breiman, L. 2001. Mach. Learn. 45, 5-32.

Figure 1: Feature value contributions of each biosecurity and farm feature for a PRRSV positive farm example. These were used to calibrate the models based on Shapley values (phi).

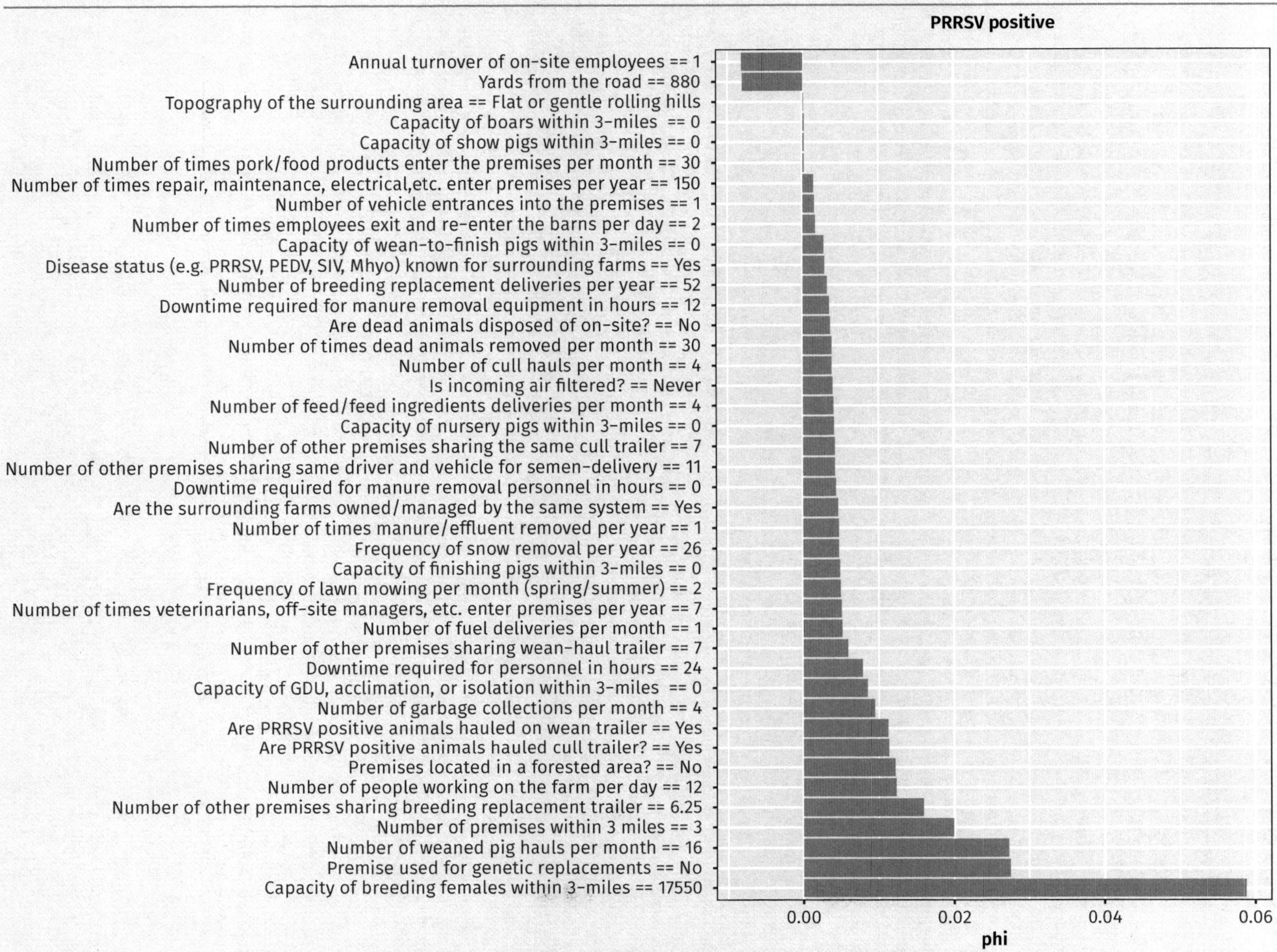

Use of aggregated diagnostic data to monitor pathogen detection and alert swine stakeholders about relevant changes

G. Trevisan[1], DVM, MBA; E. Magalhães, DVM; R. Main[1], BS, DVM, PhD; J. Torrion[2], DVM, MS, PhD; J. Christopher-Hennings[3], DVM, MS; J. Henningson[4], DVM, PhD; D. C. L. Linhares [1], DVM, MBA, PhD
[1]Iowa State University, Ames, Iowa; [2]University of Minnesota, Saint Paul, Minnesota;
[3]South Dakota State University, Brookings, South Dakota; [4]Kansas State University, Manhattan, Kansas

Veterinary diagnostic testing results guide clinical interventions for the continuous improvement of animal health and welfare. Animal health, welfare, and productivity of animals' populations might be severely affected by infectious diseases. Swine veterinarians work hard to maintain and improve animal health. Veterinarians in the USA routinely submit thousands of samples to veterinary diagnostic laboratories for disease diagnosis, pathogen monitoring, and surveillance. Individual diagnostic laboratory reports are immediately useful to manage, intervene, and improve animal or herd health. However, aggregated historical laboratory data integration and visualization are increasingly valued by clinicians and decision-makers to identify changes in various animal populations' health status over time and geographical space. Our team has collaborated to develop an inter-VDL data communication and collation project named Swine Disease Reporting System (SDRS).[1,2] Aggregated submission information and PCR test results submitted for testing at participant VDLs for five porcine endemic pathogens (PRRSV, PEDV, PDCoV, *M hyopneumoniae*-MHP, and TGEV), in the United States, were included into the SDRS database. All together, participant VDLs account for more than 95% of porcine samples tested at the National Animal Health Laboratory Network (NAHLN). Thus, this work aims to summarize the monitoring capabilities currently used to monitor the SDRS data for endemic pathogen detection and inform stakeholders on changes in the pattern of pathogen detection.

Macroepidemiological aspects of agent detection were described.[1] A seasonal detection of PRRSV, PEDV, PDCoV, measured by the percentage of positive cases (positive/total × 100), with the highest detection during colder months, was revealed. Additionally, an increase in the viral load on tested samples, measured by average Ct values of positive cases, was inversely proportional to the percentage of positive results. A seasonal trend of detection for MHP with higher detection during fall months (September-November) was encountered; however, no significant seasonal change in Ct values was identified. Overall, seasonal detection could be modeled and used to forecast the subsequent year's weekly expected detection for the monitored agents, keeping veterinarians posted for changes in agent detection patterns. TGEV was rarely detected after 2015, despite thousands of PCR tests being done. Veterinarians might consider the seasonal effects of detection and viral load to reinforce biosecurity and biocontainment measures to reduce potential disease spread across farms.

Monitoring overall detection is informative at a macroepidemiological level. An approach using historical data to define state-specific baselines and forecast the subsequent year's monthly agent-specific expected detection was implemented. Changes from baseline are reported as a) no change when results were within 2 standard deviations (STD) from baseline; b) changes between 2-3 STD from baseline; c) changes of at least 3 STD above or below the baseline. Regional differences for changes of at least 3 STD were encountered in the year 2020. As of November 2020, PRRSV detection was above 3 STD during June in South Dakota (SD) and Minnesota (MN), during July in MN, Missouri (MO), and Indiana (IN), during August in MO, during September in Nebraska (NE), MO, and IN. PEDV detection was above 3 STD during February and March in NE. PDCoV detection was above 3 STD during January and February in Oklahoma, during April in North Carolina (NC), during May in NC and Ohio, and during June in NC.

An online structured data visualization tool is freely available to the industry (i.e., no cost, no password) at **www.fieldepi.org/SDRS**. Monthly, the relevant findings are incorporated into a written, audio, and video reports freely distributed to stakeholders. The SDRS represents a valuable asset for foreign animal disease preparedness. This system has the capability and know-how to receive and process data from different VDLs. The SDRS platform can be adapted to capture, process, and inform additional agents' detection if the needs arise. Historical data on pathogen detection generates insightful information for stakeholders. The real value lies in the ability to collate different data sources and use it to actively and on a near real-time basis monitor pathogen detection and activity across regions, thus informing stakeholders of relevant changes in the patterns of pathogen detection.

References

1. Trevisan G et al. Macroepidemiological aspects of porcine reproductive and respiratory syndrome virus detection by major United States veterinary diagnostic laboratories over time, age group, and specimen. PLoS One. 2019;14(10):e0223544.

2. Trevisan G et a. Prediction of seasonal patterns of porcine reproductive and respiratory syndrome virus RNA detection in the US swine industry. J Vet Diagn Invest. 2020.

How to coach the swine farmers to omit antibiotic treatment in piglets vaccinated against post-weaning diarrhea: the example of the fecal scorecard

F. Vangroenweghe, DVM, MSc, PhD, PhD, Dipl. ECPHM
Elanco Animal Health, BU Food Animals, Plantijn en Moretuslei 1, 2018 Antwerpen, Belgium

Introduction

Post-weaning *Escherichia coli* diarrhea (PWD) remains a major cause of economic losses for the pig industry. PWD, caused by enterotoxigenic *E coli* (ETEC), provokes mild to severe watery diarrhea (5-10 days post-weaning). Therapy to combat PWD consists of antibiotic treatment in combination with high doses of ZnO (3,000 ppm). An oral live bivalent *E coli* F4/F18 vaccine has become available, which reduces the impact of PWD provoked by F4-ETEC and F18-ETEC through reduction of clinical signs of PWD and decreased excretion of pathogenic *E coli* post-weaning. However, following vaccination, a temporary short-term change in fecal score may still occur, which does not evolve further to clinical diarrhea. The objective of the current study was to evaluate the behavior of swine farmers concerning early treatment of post-weaned piglets vaccinated pre-weaning with a post-weaning diarrhea vaccine to reduce the incidence and severity of post-weaning diarrhea using a fecal scorecard to assess fecal consistency in the critical period post-weaning.

Materials and methods

A fecal scorecard was developed based on the fecal score described by Nadeau et al. (2017). The essentials of the scorecard were the scores 2 – 3 – 4, described as mild – moderate – severe fecal scores. Following application of the oral live bivalent *E coli* F4/F18 vaccine, no fecal scores belonging to score 3 and 4 have been observed.[1] Therefore, it is of critical importance that swine farmers who observe a fecal score of 2 do not immediately take action and treat the piglets, since experience has learnt that this score is temporary and short-term, evolving within 24-36 h towards fecal score of 1 or 0. Using the fecal scorecard, treatment behavior of swine farmers (n = 20) was monitored during the critical 7 days post-weaning when clinical signs of post-weaning diarrhea occur. The number of farms postponing or omitting antimicrobial treatment was counted on the total number of swine farms (n = 20) participating in the study.

Results

Analysis of 20 farms showed that managing expectations and clear communication on vaccine experiences and treatment needs results in 85% of the farmers (n = 17) postponing and eventually omitting premature treatment of their vaccinated piglets.

Discussion and conclusions

Managing expectations and clear communication with swine farmers on the performances of an advised product use are key essential to obtain a significant reduction in antibiotic use. In order to allow the swine farmer, his veterinarian and the technical consultant to discuss on fecal scores obtained during the critical period following weaning, the use of a standardized fecal score is essential. The practical use of the developed fecal scorecard thus prevents swine farmers from early, unnecessary antibiotic treatment through coaching discussion with his advisors. In the scope of further antibiotic reduction, it is key to improve communication concerning the specific criteria to decide when to start an antibiotic treatment in a particular batch of pigs. Only through this approach, sustainable reduction in antibiotic treatment will be obtained.

Reference

1. Nadeau E, Fairbrother JM, Zentek J, Bélanger L, Tremblay D, Tremblay C-L, Röhe I, Vahjen W, Brunelle M, Hellmann K, Cveji D, Brunner B, Schneider C, Bauer K, Wolf R, Hidalgo A. 2017. Efficacy of a single oral dose of a live bivalent *E coli* vaccine against post-weaning diarrhea due to F4 and F18-positive enterotoxigenic *E coli*. *The Veterinary Journal* 226, 32-39.

Figure 1: Fecal scorecard with score explanation and visual features to obtain a clear communication what to expect from a preventive post-weaning diarrhea vaccination protocol in order to prevent unnecessary treatment.

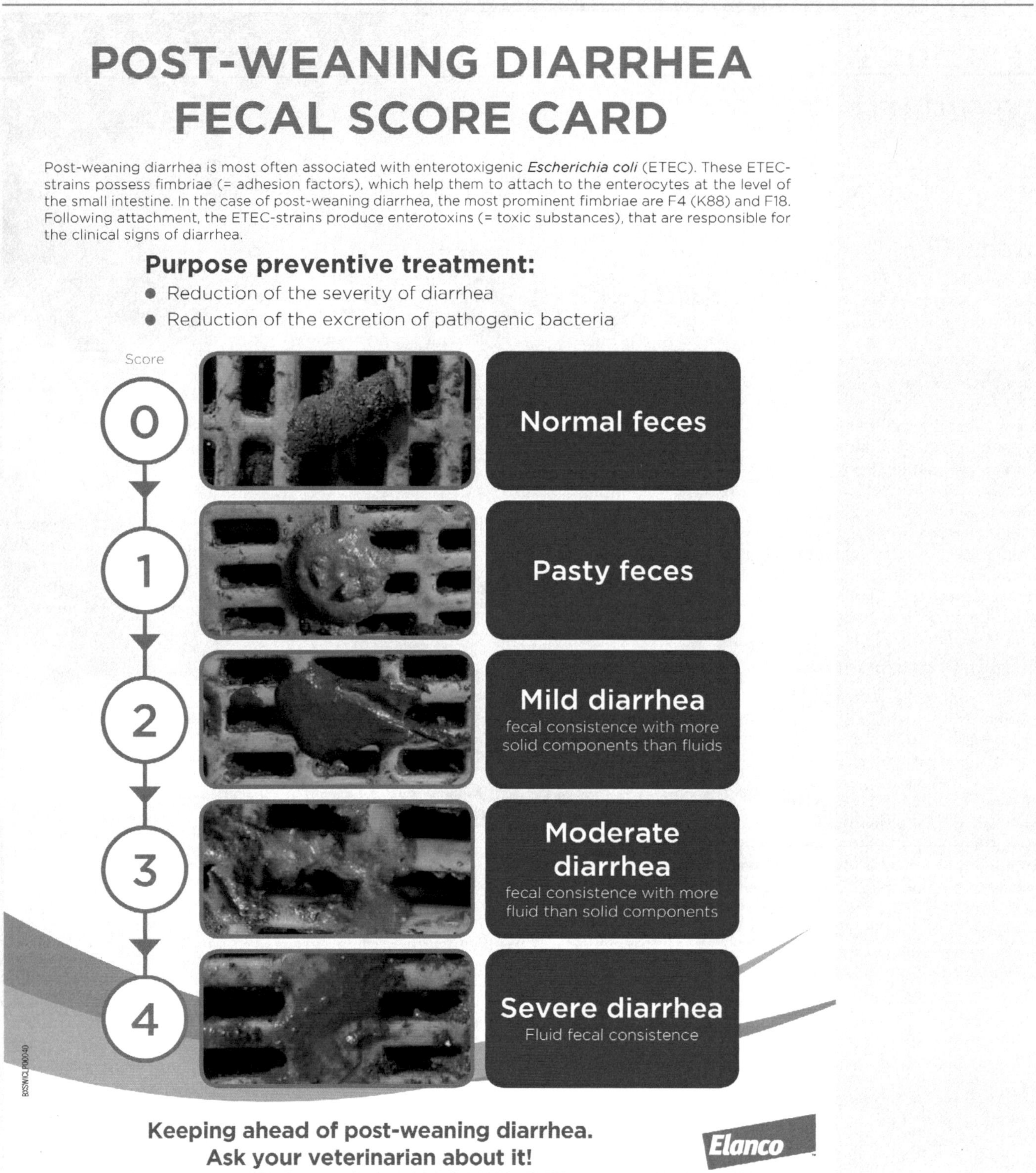

Using neonatal rectal temperature for evaluation of farrowing room management: a practical approach

F. Vangroenweghe[1], DVM, MSc, PhD, PhD, Dipl. ECPHM; J. Jourquin[2], DVM
[1] Elanco Animal Health, BU Food Animals, Plantin en Moretuslei 1a, 2018 Antwerpen, Belgium
[2] Elanco Animal Health, R&D, Plantin en Moretuslei 1a, 2018 Antwerpen, Belgium

Introduction

Good management around farrowing is crucial for a swift start of the newborn piglets. In practice, most sow farms operate at high standard levels of management, with special focus on the farrowing department and on the perinatal period. Independent of the reason, hypothermia of piglets in the first 24 hours after farrowing has a negative impact on their survival chances and development. The objective of the current study was to compare rectal temperature (RT) of neonatal piglets between farms that use similar genetics and that are under similar climatological conditions thus evaluating the impact of the farrowing room conditions on the newborn piglets.

Materials and methods

Two farrow-to-finish herds in Flanders (Belgium) were selected based on genetic mother lines and availability of newborn piglets on the same day. All live born piglets from 15 randomly chosen sows per farm were included in the study. Piglets' RT (expressed in °C) and body weight (BW; expressed in g) were measured between 2 and 24h post-farrowing and litter data (total born (TB), life born (LB) and sow parity) were collected. The number and proportion of piglets below 1.130kg, considered piglets at risk[1] was calculated. Farm and sampling details are provided in Table 1. Statistical analysis on RT and BW was carried out with JMP 14.0 using a *t*-test to compare between farms, litters and parities.

Results

A total of 532 life born piglets, 269 and 263 from Farm A and Farm B, respectively – from a total of 30 litters, 15 litters from each respective farm – were included in the study (Table 1). Litter size, LB per litter and average BW did not significantly differ between farms, although LB and BW were significantly different ($P < 0.05$) between sows within both farms. The variation in piglet BW was broader in Farm B (288-2295g) compared to Farm A (441-1945g). There was a significant difference ($P < 0.05$) in proportion of piglets below 1.130 kg between farm A (27.9%) and Farm B (33.5%). Also RT was significantly lower for the piglets of Farm B (37.72°C) compared to Farm A (38.17°C), although assessment were performed on the same day under identical climatological conditions. The relation between BW and RT could be fitted in the following formula: $[RT = 22.651253 + 2.1739104 * Log_{10}(Weight)]$ ($P < 0.0001$). Remarkedly, in Farm B, piglets from sows in a specific location within the farrowing room systematically had lower RT compared to the average farm RT.

Discussion and conclusions

The current study demonstrated that random assessment of RT from several litters of newborn piglets within 24 h post-farrowing can be informative for issues regarding farrowing room management. In Farm B, assessment of newborn piglet RT revealed a systematic problem in the pens near the inspection corridor. Further investigation pointed to cold air entering through a poorly closed doorway, resulting in undercooled piglets in the pens in the immediate proximity of the cold air flow. Therefore, corrective measures were suggested in order to optimize piglet RT in the pens affected by this local climatic condition. Piglet BW can be considered an indicator of overall piglet quality and survival. Based on the significant correlation between BW and RT, low BW piglets are more vulnerable to early perinatal farrowing room management issues. In conclusion, assessment of new piglets RT can help to evaluate optimal farrowing room management and eventually reveal undetected mistakes that might impact piglet quality and survival rate.

Reference

1. Jourquin J., Morales J. 2018. Impact of runting on colostral intake, survival chance and development. *Proceedings 10th ESPHM*, Barcelona, Spain. REP-OP-1, p. 107.

Table 1: Farm, sampling details and evaluation parameters – rectal temperature (expressed in °C; average ± SEM) and body weight (expressed in g; average ± SEM) – in newborn piglets from Farm A and B

Parameter	Farm A	Farm B	*P*-value
Number of sows	600	650	N/A
Batch Management System (BMS)*	4-week BMS	4-week BMS	N/A
Sow genetic line	DanBred	DanBred	N/A
Boar genetic line	Duroc	Belgian Piétrain	N/A
Region	Flanders	Flanders	N/A
Date of data collection	January 23	January 23	N/A
Rectal temperature (°C)	38.17 ± 0.07	37.72 ± 0.07	< 0.0001
Live born	17.93 ± 0.80	17.53 ± 0.83	> 0.05
Average body weight (g)	1,200 ± 19.1	1,162 ± 19.3	> 0.05
Minimum-maximum weight (g)	441 - 1945	288 - 2295	N/A
Piglets below 1,130 g (n, %)	75 (27.9%)	88 (33.5%)	< 0.05

* Interval between farrowing batches; N/A not available

Characterization and comparison of PRRSV ORF5 sequences between clinical samples and isolates obtained in MARC-145 or/and ZMAC cell lines

Wannarat Yim-im, MS; Haiyan Huang, BS; Phillip Gauger, DVM, MS, PhD; Karen Harmon, PhD; Jianqiang Zhang, MD, MS, PhD
Veterinary Diagnostic and Production Animal Medicine, Iowa State University, Ames, Iowa

Introduction

Porcine reproductive and respiratory syndrome (PRRS) is an economically important swine disease. The causative agent, PRRS virus (PRRSV), includes PRRSV-1 and PRRSV-2. Commercial vaccines do not always provide effective protection against genetically diverse PRRSV strains. Autogenous vaccines against circulating strains are still commonly used. For autogenous vaccine production, the first step is to conduct PRRSV virus isolation (VI) in MARC-145 and/or ZMAC cell lines to obtain a PRRSV isolate.[1] Is the virus isolate obtained in cell culture always same as the virus present in clinical samples? This question has not been thoroughly investigated thus far. The objective of this study was to compare the RFLP patterns, lineage information, and percent nucleotide (nt) identity of ORF5 sequences of PRRSV in clinical samples and cell culture isolates.

Materials and methods

From 2010 to 2020, 1,002 PRRSV-2, 16 PRRSV-1, and 3 both PRRSV-1 and PRRSV-2 PCR-positive clinical samples, whose ORF5 sequences are available and from which PRRSV isolates were obtained in MARC-145 and/or ZMAC cells, were included in this study. These included 564 serum, 406 lung, and 51 other specimen types of clinical samples, and their corresponding cell culture isolates. ORF5 sequences were determined on all of these cell culture isolates in this study. Subsequently, ORF5 sequences of PRRSV in clinical samples and cell culture isolates were compared in terms of RFLP patterns, genetic lineage information, and percent nucleotide identity.

Results and discussion

The results are summarized and discussed in several categories.

Category 1. Sixteen PRRSV-1 PCR-positive clinical samples had 99.8-100% ORF5 nt identity compared to isolates obtained in cell cultures, regardless of isolation in MARC-145 or ZMAC cells. 956 PRRSV-2 PCR-positive clinical samples (95.4%) had same RFLP patterns, same genetic lineages, and 98-100% ORF5 nt identity compared to isolates obtained in cell cultures (MARC-145, ZMAC, or both). In this category, viruses in clinical samples and cell culture isolates are considered to match each other.

Category 2. Twenty-four PRRSV-2 PCR-positive clinical samples (2.4%) had 98.6-99.8% ORF5 nt identity and same genetic lineages but different RFLP patterns compared to virus isolates. In these cases, mismatched nucleotides occurred at cutting sites of HincII or SacII of the virus isolates, resulting in RFLP pattern changes. In this category, viruses in clinical samples and cell culture isolates are considered similar to each other.

Category 3. Two PRRSV-2 PCR-positive samples (0.2%) had same RFLP patterns and genetic lineages but diverse ORF5 nt identity (90.8 and 96.5%) compared to virus isolates obtained in MARC-145 (VI not attempted in ZMAC). Mismatched nucleotides between clinical samples and PRRSV isolates occurred outside the enzymatic cutting sites, having no effect on RFLP patterns.

Category 4. Sixteen PRRSV-2 PCR-positive samples (1.6%) had 83.7-96.4% ORF5 nt identity, different RFLP patterns, and different genetic lineages compared to cell culture isolates. Additional work is needed to further characterize samples in this category.

Category 5. Four PRRSV-2 PCR-positive clinical samples (0.4%) were found to contain two different PRRSV-2 strains as revealed by ORF5 sequencing. The first clinical sample included RFLP 1-5-3 and 2-5-2 viruses (83.7% nt identity); VI was not conducted in MARC-145 and VI in ZMAC resulted in isolation of only 2-5-2 virus. Each of the second and third clinical samples included RFLP 1-2-4 and 2-5-2 viruses (84.5% nt identity); VI in MARC-145 resulted in isolation of 2-5-2 virus and VI in ZMAC led to isolation of 1-2-4 virus. The fourth clinical sample included RFLP 1-7-4 and 1-4-4 viruses (87.3% nt identity); VI in MARC-145 only resulted in isolation of 1-4-4 virus whereas the virus isolates obtained in ZMAC included both 1-7-4 and 1-4-4 viruses. In this category, the clinical samples were confirmed to contain two PRRSV-2 strains but VI in MARC-145 and ZMAC could have different outcomes.

Category 6. For three PRRSV-1 and PRRSV-2 PCR-positive samples, VI in ZMAC led to isolation of both PRRSV-1 and PRRSV-2 that respectively had 99.7-100% nt identity with the viruses in clinical samples; however, only either PRRSV-1 or PRRSV-2 but not both was isolated in MARC-145 and the isolates had 99.8-100% nt identity to the viruses in clinical samples.

In conclusion, in most occasions (97.8%), PRRSV isolates obtained in cell cultures (either MARC-145 or ZMAC or both) had very similar ORF5 sequences to the viruses in clinical samples. However, occasionally, PRRSV isolates obtained in cell culture had different sequences and RFLP patterns to the viruses in clinical samples. In addition, PRRSV isolates obtained in MARC-145 and ZMAC could occasionally have different sequences. It would be prudent to sequence the PRRSV cell culture isolates before using them for autogenous vaccine production or other characterizations.

Reference

1. Yim-im et al. Comparison of ZMAC and MARC-145 cell lines for improving porcine reproductive and respiratory syndrome virus isolation from clinical samples. J Clin Microbiol. (In Press).

Effects of Bioplex® organic trace minerals in gestating and lactating sow diets: a review

S. A. Elliott, MS
Alltech, Inc., Nicholasville, Kentucky

Introduction

Trace minerals can have a direct impact on sow health and reproductive efficiency. Litter size, piglet health and birth weight can all be affected by nutritional and trace mineral status during gestation. Mahan found that during the last 14 days of gestation, approximately 50% of the total minerals (macro and micro) are retained in the body of developing pigs, and Mahan also stated that the critical stage for mineral nutrition in sows appears to be during late gestation and lactation, when the nutritional demands are extremely high.[3] This clearly indicates the importance of optimal trace mineral status in order to avoid possible depletion that can lead to increased oxidative stress, resulting in reduced milk production, impaired reproductive performance and even decreased longevity in sows.[6] Supplementation with organic trace minerals has been shown to improve sow reproductive performance[4] and increase both the trace mineral concentration in sows' milk and the quantity of trace minerals transferred across the placenta to the fetus.[5] Due to the increased bioavailability of Bioplex® trace minerals, they can be included at lower levels to help offset the negative excessive mineral excretion and inefficiency associated with the overuse of inorganic trace mineral (ITM) sources.

Results and discussion

This review evaluates several studies looking at the impact on reproductive efficiency and subsequent litter performance when comparing typical ITM industry diets with equal or lower inclusion rates of Bioplex minerals (Table 1).

Reproductive efficiency and born alive

When looking at reproductive efficiency over six parities, Peters and Mahan reported a greater ($P < 0.05$) number of pigs born (total and live) when sows were supplemented with Bioplex (at NRC levels) rather than with ITM at higher industry levels.[4] Berthechini et al reported no statistical difference in live pigs born when comparing low Bioplex treatment to high inorganic treatment groups (the low group representing 50% of the level of the high group) but noted significantly ($P < 0.05$) higher birth and weaning weights for piglets fed Bioplex diets.[1] The high Bioplex treatment group resulted in significantly greater number of piglets born alive ($P < 0.05$). In the Ma et al, study, sows received either industry levels of ITM or 80% of those levels in Bioplex form.[2] When compared to ITM, sows that received Bioplex had numerically higher numbers of live pigs born per litter. Peters and Mahan compared ITM and Bioplex at either NRC or higher industry levels (IND) over four parities.[3] The number of pigs born (total and live) was greater ($P < 0.05$) for sows fed NRC levels compared to those fed IND levels. There were no statistical differences between the mineral sources, but there was a numerical difference in piglets born alive to Bioplex sows at IND levels compared to ITM at IND levels, and Bioplex sows fed NRC levels resulted in the highest numerical number of piglets born alive.

Weaning weight

In regard to weaning weight, better results were found for Bioplex ($P < 0.05$) regardless of supplementation level. Similar results were reported by Ma et al.[2]; compared to the high inorganic treatment, the low Bioplex treatment significantly increased ($P < 0.05$) the number of piglets with birthweights > 1 kg by 6.5%, and litter weaning weight was increased 15.8% ($P < 0.05$).[2] Peters and Mahan found that due to an increase in litter size for Bioplex sows, their litter birth weights were heavier ($P < 0.05$).[4]

References

1. Bertechini AG, Fassani EJ, de Brito JÁG, Barrios PR. Effects of dietary mineral Bioplex in pregnant and lactating sow diets on piglet performance and physiological characteristics. *Rev Bras Zootec* 2012;41:624-9. **https://doi.org/10.1590/S1516-35982012000300022**

2. Ma L, He J, Lu X, et al. Effects of low-dose organic trace minerals on performance, mineral status, and fecal mineral excretion of sows. *Asian-Australalas J Anim Sci.* 2019;33(1):132-138. **https://doi.org/10.5713/ajas.18.0861**

3. Mahan, D. (2006) The changing mineral status of high producing sows: What are their needs and when are the critical periods. Midwest Swine Nutrition Conference, Indianapolis, IN.

4. Peters JC, Mahan DC. Effects of dietary organic and inorganic trace mineral levels on sow reproductive performances and daily mineral intakes over six parities. *J Anim Sci* 2008;86: 2247-60. **https://doi.org/10.2527/jas.2007-0431**

5. Peters JC, Mahan DC, Wiseman TG, Fastinger ND. Effect of dietary organic and inorganic micromineral source and level on sow body, liver, colostrum, mature milk, and progeny mineral compositions over six parities. *J Anim Sci* 2010;88: 626-37. **https://doi.org/10.2527/jas.2009-1782.**

6. Zhao Y, Flowers WL, Saraiva A, Yeum K-J, Kim SW. Effect of heat stress on oxidative stress status and reproductive performance of sows. *J. Anim Sci* 2011, 89(2):108.

® Registered trademark of Alltech, Inc.

Table 1: Summary of trial findings when comparing Bioplex organic minerals to inorganic mineral sources on live pigs born/litter, weight at birth and weight at weaning

	Live pigs born per litter			Piglet weight at birth (g)			Weight at weaning (g)		
	Inorganic	Bioplex	Difference	Inorganic	Bioplex	Difference	Inorganic	Bioplex	Difference
Bertechini et al (2012)									
Low	10.8^{b}	11.9^{ab}	1.1	1496^{b}	1636^{a}	140	6176^{c}	7639^{a}	1463
High	11.2^{ab}	12.18^{a}	0.98	1516^{b}	1677^{a}	161	6915^{b}	7832^{a}	917
Ma et al (2019)	11.05	11.28	0.23	1430^{a}	1490^{b}	60	6480	6720	240
Peters and Mahan (2008)	10.44	11.08	0.64	1720	1660	-60	6330	6330	-
Peters and Mahan (2005)									
NRC	11.6	11.8	0.2	1600	1640	40	6160	6400	240
Industry	10	11.1	1.1	1740	1630	-110	6460	6280	-180
Average	10.85	11.56	0.71	1584	1622	38.50	6420	6867	536

Numbers in rows followed by different letters represent differences ($P < 0.05$) by Tukey test

Dietary sodium diformate in sows during late gestation and lactation: performance improvements in suckling piglets under European conditions

C. Lückstädt[1]; J. Paulsrud[2]
[1]ADDCON GmbH, Bitterfeld-Wolfen, Germany; [2]Apiam Solutions LLC, Sioux City, Iowa

Introduction

The application of organic acids and their salts to diets for pigs has been studied extensively for more than 50 years.[1] Despite its well-documented effects, data on its impact in sows during late gestation and lactation, and their subsequent effects on suckling piglets under commercial conditions is scarce.

Materials and methods

This study tested the effects of sodium diformate (Formi NDF, ADDCON) fed to sows on their suckling piglets. 20 sows (Landrace × Big Yorkshire) on a commercial farm in Eastern Europe were equally divided into two groups, containing 10 sows each. They were fed from one week before farrowing till the end of weaning (day 27) on a typical lactation diet (corn-wheat-soy based), containing either 0.8% of NDF, or a negative control which did not contain the additive. The lactation diet was fed *ad libitum* and daily feed intake measured. The following parameters were monitored: litter weight of new-born piglets, the nursing mortality, the number of weaned piglets, the individual body weight of weaned piglets and the litter weight during weaning. Data were analysed using the t-test and a significance level of 0.05 was used in all tests.

Results and discussion

Feed during the trial was well accepted by both groups. Sows fed the NDF-diet had a noticeably higher feed intake compared to sows from the negative control group (+460 g/day, Table 1). During farrowing the litter weight of piglets at birth differed significantly ($P < 0.01$) by more than 5.5 kg, for the NDF-fed sows. Furthermore, uniformity of piglet and litter weight was improved. During weaning at day 27, the number of weaned piglets differed significantly (10.5 and 11.5 for Control- and NDF-group; $P = 0.03$), due to a higher mortality during nursing in the control-group ($P < 0.1$). The weaning weight of the piglets was only numerically influenced (60 g heavier piglets in the NDF-group). However, due to the increased number of weaned piglets, the litter weight during weaning of NDF-fed sows was also higher ($P < 0.1$) by more than 8.6 kg. Again, uniformity of piglet and litter weight was improved, leading to a better manageable pig production.

As previously reported for potassium diformate[2,3], it can be also concluded from this study, that the inclusion of sodium diformate to the lactation diet of sows acts as a performance enhancer for suckling piglets under commercial conditions in Europe.

References

1. Lückstädt C and Mellor S 2011. Recent Advances in Animal Nutrition – Australia. 18,123-130.

2. Lückstädt C 2011. Advances in Animal Biosciences. 2(1),145.

3. Øverland M, Bikker, P and Fledderus J. 2009. Livestock Science 122, 241-247.

Table 1: Performance parameter of piglets from sows fed with or without sodium diformate (Formi NDF)

Parameter	Control (n = 10)	0.8% NDF (n = 10)	P-value
Daily feed intake sow [kg]	3.69	4.16	[+0.46 kg]
Litter weight at birth [kg]	15.90 ± 5.21	21.44 ± 3.69	0.009
Nursing mortality piglets [n]	1.2 ± 1.33	0.5 ± 0.81	0.096
Piglets weaned, per sow [n]	10.5 ± 1.2	11.5 ± 0.8	0.027
Weight of weaned piglets [kg]	8.31 ± 0.94	8.37 ± 0.66	0.45
Litter weight at weaning [kg]	87.70 ± 16.08	96.35 ± 11.29	0.102

Sodium salicylate vs. 12.5% liquid aspirin pharmacokinetics study

N. Winkelman[1], DVM; A. Mueller[1], DVM; R. Winscher[1], CVT; G. Weaver[2], DVM
[1]Swine Services Unlimited Inc., Rice, MNinnesota; [2]Aurora Pharmaceutical, Inc., Northfield, Minnesota

Introduction

Aspirin products are widely used in the rearing of swine mostly for fever and inflammation reduction in the case of infectious disease, but also for less-descript reasons such as maintenance of water and feed intake and maintenance of general health throughout the growing period. In the United States, data are lacking that supports the pharmacokinetic differences between sodium salicylate and liquid aspirin when administered orally via the drinking water to swine. The goal of this study is to determine plasma levels achieved within the first twenty-four hours of Oral-Pro® Sodium Salicylate 48.6% (OPSS 48%) and Liquid Aspirin 12.5% (LA 12.5%) when administered a single oral treatment at a dose equivalent to label instructions orally.

Materials and methods

The study was divided into two phases. All pigs were weighed at the beginning of each phase. Oral-Pro® Sodium Salicylate 48.6% (Aurora Pharmaceutical, Inc.) was administered by single oral dose by stomach tube at the rate of 3.23 mL per 100# body weight. Liquid Aspirin 12.5% was administered orally by single oral dose by stomach tube at the rate of .97 mL per 100# body weight.

Phase I: Pigs 1-5 were assigned to Treatment 1 and treated with OPSS 48.6% by a single oral dose by stomach tube for 1 day. Blood was collected at the following time points: 0 hr., 0.5 hr., 1 hr., 2 hr., 3 hr., 6 hr., 12 hr., 24 hr. Pigs 6-10 were assigned to Treatment 2 and treated with LA 12.5% ASA by a single oral dose by stomach tube for 1 day. Blood collections occurred at the same time points as with Treatment 1. After the completion of the 24 hr. blood collections, blood collection and treatment were ceased, and the animals went through a 4-day washout period.

Phase II: Crossover design starts after the completion of the 4-day washout period. Pigs were weighed on the day of beginning of Phase II. Pigs 1-5 were assigned to treatment with LA 12.5% by single oral dose at the same dosage rate as in Phase I and the same sample time points were collected via jugular venipuncture. Pigs 6-10 were assigned to treatment with OPSS 48.6% by single oral dose at the same dosage rate as in Phase I and the same sample time points were collected via jugular venipuncture.

Results and discussion

Both phases of the study produced similar results. Given a single oral dose, OPSS 48% treated animals reached blood levels of salicylate considered therapeutic for humans (20-50 ug/mL) within 15 minutes of treatment and maintained above therapeutic levels for about 5 hours after initiation of treatment. Conversely, LA 12.5% treated animals didn't achieve therapeutic blood levels at any of the time points measured. The average blood level of salicylate achieved over the 24-hour testing period in the OPSS 48.6% treated animals was 30.9 ug/mL in Phase 1 and 31.4 ug/mL in Phase II – both above therapeutic levels. Conversely, the average blood level of salicylate achieved over the 24-hour period in the LA 12.5% treated animals was 1.4 ug/mL for both phases of the study – both well below therapeutic levels.

OPSS 48.6% treated animals achieved a C-max of 62.7 ug/mL in Phase 1 and 68.9 ug/mL in Phase 2 – both at 2 hours post-initiation of treatment. LA 12.5% treated animals achieved a C-max of 3.9 ug/ml in Phase 1 and 3.6 ug/mL in Phase 2 – both at 1-hour post-initiation of treatment. The C-max achieved in the LA 12.5% treated animals was well below therapeutic levels (Figure 1).

When given a single oral treatment at a dose equivalent to label instructions, Oral Pro Sodium Salicylate 48.6% treated animals achieved therapeutic blood levels within 1/2 hours and maintained them for approximately 5 hours, whereas, Liquid Aspirin 12.5% treated animals were never close to achieving therapeutic levels.

Figure 1: Plasma level bioequivalence study.

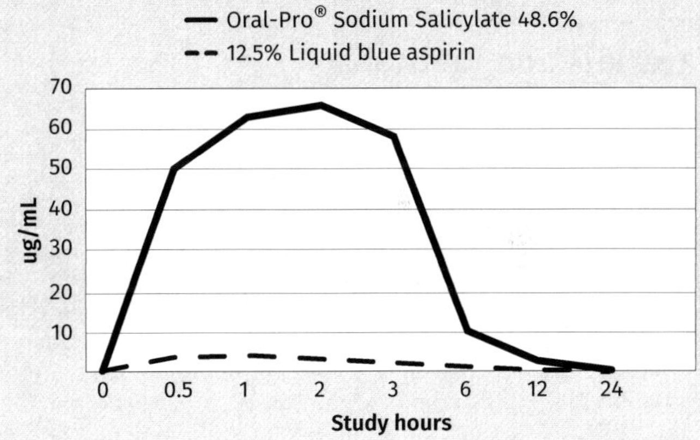

Comparison of efficacy between two PCV2 vaccines under PCV2d experimental exposure

Eduardo Fano[1], DVM MC PhD; Jana Morgan[1], DVM; Tyler Bates[1], BS; Lauren Glowzenski[2], DVM
[1]Boehringer Ingelheim Animal Health USA Inc., Duluth, Georgia; [2]TriOak Foods, Columbus Junction, Iowa

Introduction

Porcine circovirus type 2 (PCV2) is the primary causative agent behind porcine circovirus–associated disease (PCVAD), one of the most significant diseases affecting swine producers around the world. Since the introduction of PCV2 vaccines into the US market in 2006, the benefit of vaccination has been clearly demonstrated. Practitioners and producers have special interest in best vaccination protocols conferring protection against prevalent PCV2 strains. By their nature, DNA viruses like porcine circovirus Type 2 (PCV2) mutate more slowly than RNA viruses. Still, relative to other DNA viruses, the mutation rate of PCV2 is one of the highest, giving rise to the shifting strains the swine industry has been grappling with in recent years. PCV2d is the most prevalent variant of the virus in the US Swine industry. Several studies evaluating the efficacy of Ingelvac CircoFLEX® in clinical and field situations have shown it has been efficacious under PCV2d challenge as a one dose vaccine applied around weaning. The objective of this study was to assess the comparative efficacy of Fostera® PCV MH Gold (1 dose) to FLEXcombo® under PCV2d experimental challenge.

Materials and methods

The experimental animals were housed in a 1200 head wean-to-finish commercial swine facility. Pigs were housed in a barn appropriate for breed and age and were kept under similar conditions of climate, ventilation, temperature, humidity and light. A total of 1568 pigs were blocked by weight and randomly assigned to one of three treatments, T1, T2 and T3-NVC (Non-vaccinate control). The pigs were approximately 21 days of age at weaning and the barn was filled over the course of 1 week. Pigs were individually weighed and ear tagged (trial day 0). On trial day 1, pigs were allocated to treatment groups and administered their treatments (Table 1). On trial day 1, T1 and T2 groups received a single dose of vaccine. On trial day 42, all pigs were inoculated with 2 mL IM of 10^5 TCID50/mL virulent PCV2d challenge material. All pigs were weighed at first marketing cut of the growing period at 25 weeks of age (trial day 154, off test). Performance data included day 0 and 153 weights as well as day 0-153 average daily gain. Growth performance data was analyzed using Two-Sample t-Test. Mortality and treatment data was analyzed using Pearson's chi square. Non-vaccinates were removed from overall statistical analysis, they were considered just as control pigs used for challenge confirmation.

Results and discussion

The effect of vaccination protocol in growth performance is shown in Table 1, the effect on mortality and treatments rates are shown in Table 1 as well. There was significant difference in growth performance between the two vaccines (T3-NVC group removed from statistical analysis). The PCV2a based vaccine (T1 - FLEXcombo®) showed a better ADG than the T2 vaccine (PCV2a and PCV2b based vaccine). There were no significant differences in mortality, whereas the treatment rate was significantly lower in T1 pigs. PCV2 viremia was detected in T3-NVC on trial days 56 and 113 and molecular characterization indicated PCV2d infection. A successful PCV2d challenge was confirmed by this information and the low ADG and high treatments rate in the NVC group. Pigs vaccinated with both vaccines in this study showed no clinical signs of PCVAD. Under the conditions of this field study (experimental infection and large number of animals), there was no evidence of a better cross protection response from the PCV2a and PCV2b based vaccine when compared to FLEXcombo®, a PCV2a based vaccine.

Table 1: Treatment description and growth performance, mortality and treatment rate by vaccine treatment group

Treatment description	T1	T2	T3-NVC
Vaccine	Ingelvac CircoFLEX® (FLEXcombo®)	Fostera® PCV MH Gold	Non-vaccinated challenge control group
PCV2 variants included in the vaccine	PCV2[a]	PCV2a PCV2b	NA
Number of pigs (n)	703	706	102
Protocol description	1-full dose at weaning	1-full dose at weaning	NVC pigs used for challenge confirmation
Parameter	**T1**	**T2**	**T3-NVC****
Mean ADG (lb/day)	1.74 a*	1.72[b]	1.46
ADG % CV	9.93%	11.10%	17.61%
Mean wean weight (lb)	13.57	13.53	10.09
Final mean weight (lb)	247.89[a]	244.77[b]	207.89
Total mortality (%)	3.74%	4.14%	4.63%
Treatments (%)	18.9%[a]	24.0%[b]	49.1%

* Means with different superscripts indicate difference at $P \leq 0.05$

** Non-vaccinates removed from overall statistical analysis

Development of *Streptococcus suis* challenge models and their use in identifying adjuvants that can increase the protective efficacy of autogenous vaccines

E. Bumgardner, MS; D. McKeown, BS; D. Stine, DVM PhD
Cambridge Technologies, Veterinary Diagnostics and Research Department, Worthington, Minnesota

Introduction

Streptococcus suis is an emerging problem in growing pigs, often causing lameness and CNS symptoms that can lead to mortality in affected herds. The importance of this pathogen is increasing in recent years as the isolation of antibiotic resistant strains has become a cause for concern.[1] Of the serotypes that are know to infect swine, serotype 2 and ½ strains are amongst the serotypes identified as most virulent.[2] Both of these serotypes are widespread and solutions are urgently needed that can help prevent infection. In the work presented here we undertook a study designed to compare the effects of several adjuvants on the protective efficacy of autogenous bacterins.

Materials and methods

Sixty freshly weaned animals were obtained from a commercial source and housed in our research barn. Pigs were immunized twice, at three and five weeks of age. Treatments were delivered intramuscularly using a 2cc dose of the vaccines described in Figure 1. Two weeks after the final immunization, pigs were challenged IV/IP by delivering a 1cc dose of serotype ½ *S suis* culture at a titer of 3.7*10^9 cfu/mL via each route. Pigs were monitored and scored for clinical signs for a period of four days post-challenge using a scale numbered from 0 to 4; 0 (healthy), 1 (mild lameness), 2 (severe lameness), 3 (CNS symptoms), 4 (dead). If pigs exhibited CNS symptoms they were humanely euthanized and scored a 3 for the remainder of the observation period. When the study was complete, the sum of clinical scores obtained from each animal was determined and the average was summarized (Figure 1). Scores were analyzed using a one-factor ANOVA at a significance level, $\alpha = 0.05$ to determine whether any group scored significantly higher or lower than the non-vaccinated controls.

Results and discussion

The results of the study (Figure 1) demonstrate that Adjuvant A exhibited the best level of protection when compared to the others we evaluated with Antigen 2. However, none of the vaccines that utilized Antigen 2 were able to provide protection that differed significantly from the non-vaccinated controls. Interestingly, the group that received antigen 1 in combination with adjuvant A was significantly protected from clinical disease, exhibiting only mild lameness in three pigs, for a single day each, over the course of the observation period. Further work will be done to evaluate the protective capacity of this vaccine against other serotypes.

References

1. Holmer I et al. Antibiotic resistance in porcine pathogenic bacteria and relation to antibiotic usage. *BMC Vet Res.* 2019;15:449.

2. Estrada A. et al. Serotype and genotype (Multilocus Sequence Type) of *Streptococcus suis* isolates from the United States serve as predictors of pathotype. *J. Clin. Microbiol.* 2019;57:e00377-19.

Figure 1: The sum of clinical scores from each animal was used to determine the average for each group over the course of challenge. The maximum possible score was 16 and the minimum possible score, 0. (*) denotes statistical significance versus the non-vaccinated control group.

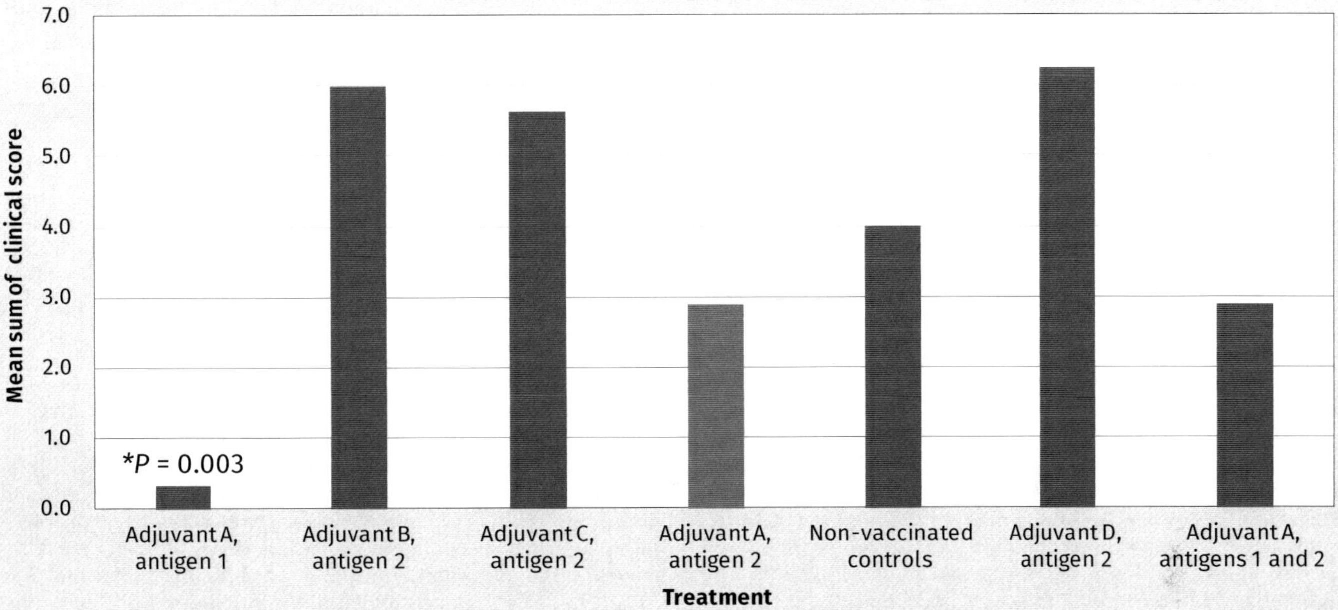

Labor-free fly control ClariFly 267 (diflubenzuron) achieves 99% reduction in housefly (*Musca domestica*) populations in both deep pit and shallow pit swine facilities

Kirk Dailey, BS; Gene Spellman, BS
Central Life Sciences, Schaumburg Illinois

Introduction

Despite extensive investments in biosecurity measures and protocols, farms still have unexplainable cases of viral diseases. Fly control is now recommended as an important plank of swine biosecurity protocols and disease risk reduction.[1-3] Flies can carry and transmit important swine disease viruses such as porcine reproductive and respiratory virus (PRRSV),[1] porcine epidemic diarrhea virus (PEDV),[2] Senecavirus A (SVA)[3] and now African swine fever virus (ASFV).[4] Flies can carry infectious bacteria,[5] and antibiotic-resistant bacteria.[6] Flies can travel up to four miles.[7] Fly control in swine till now has been labor-intensive, sporadically effective, non-coordinated, using multiple products with harsh chemicals. ClariFly 267 (diflubenzuron) insect growth regulator larvicide was developed and approved for swine as an in-feed fly control product. Earlier research showed ClariFly 267 in-feed larvicide achieved a 99% reduction in housefly populations in swine finishing facilities with deep pit manure handling facilities.[8] Questions remained if ClariFly 267 would perform as well in facilities with shallow pit manure facilities. The purpose of this study is to show similar fly control performance in shallow pit manure facilities with ClariFly 267.

Materials and methods

A 10,000 sow farrow-to-finish operation in the Midwest was selected. ClariFly 267 (diflubenzuron) in-feed larvicide was incorporated within the typical ration at the label level of 0.20 mg ClariFly 267 per kg of body weight per day for farrowing, and 0.10 mg ClariFly 267 per kg body weight in gestation, for the eight week study. Feed samples were collected at the start and mid-point of the study. The feed samples were assayed for proper levels of ClariFly 267. To count flies and monitor the fly population reduction, Starbar EZ Trap sticky fly traps were hung in each study room. Traps were placed approximately seven feet above the floor. Fans and air vents were avoided for trap placement to avoid air movement interfering with fly activity. Traps were hung for 1 week and replaced with new, unused traps for the next week of the study. Weekly trap collection, fly counting and new trap placement was repeated throughout the study.

Results

After five weeks on ClariFly 267 in-feed larvicide, the fly populations, as represented by the flies caught and counted on the traps collected weekly during the study, were reduced more than 99% by the end of week five versus the end of week one, showing that ClariFly 267 was very effective in reducing fly populations in these shallow manure pit facilities. Orderly placing weekly traps next to each other offers a visually striking picture of the reduction in fly population achieved with ClariFly 267 (Figure 1).

Discussion

ClariFly 267 in-feed larvicide proved extremely effective in reducing housefly populations in this shallow pit manure study, achieving a 99% reduction in fly counts in each room. ClariFly 267 in-feed larvicide offers fly control options in a labor-free fashion, since ClariFly 267 is mixed in the swine ration. The pigs do the work, and are effective in delivering ClariFly 267 where it is needed-in the manure. ClariFly 267 is a chitin-synthesis inhibitor. Chitin is a key component of the fly's exoskeleton. Inhibiting chitin synthesis blocks further development of the fly in the larval stage. ClariFly 267 does not kill adult flies. Adult flies are prevented from developing from earlier larval stages. Existing adult flies eventually die off and are not replaced when ClariFly 267 is fed.

It is important to show patience when feeding ClariFly 267, as adult flies are still visible during the initial feeding stages. ClariFly 267 offers a convenient and effective option to achieve fly control in swine biosecurity protocols. Fly control also provided an improved working environment for pigs and employees, as reported by the study cooperator. More research is needed in the area of improved working environment and animal welfare to quantify the benefits resulting from improved fly control.

References

1. Otake S, Dee SA et al. Transmission of porcine reproductive and respiratory syndrome virus by houseflies, (Musca domestica Linnaeus). *Vet Rec* 2003. 152: 73-76.

2. Allison G, et al. PEDV positive bioassay reveals houseflies (Musca domestica) can transmit infectious PEDV to pigs. *Proceedings 50th AASV 2019.*

3. Allison G, et al. PEDV, PRRSV, and Seneca Valley virus PCR positive flies (Musca domestica). What now? *Proceedings 49th AASV 2018.* 394-396.

4. Olesen, A, et al. Infection of pigs with African swine fever virus via ingestion of stable flies (Stomoxys calcitrans) *Transboundary Emerging Diseases 2018,* 65: 1152-1157.

5. Khamesipour F, et al. A systematic review of human pathogens carried by the housefly (Musca domestica L.). *BMC Public Health 2018,* 18:1049.

6. Zurek L, et al. Insects Represent a Link between Food Animal Farms and the Urban Environment for Antibiotic Resistance Traits. *Appl Environ Microbiol. 2014 Jun*; 80(12): 3562–3567.

Figure 1: Results from shallow pit manure building.

7. Nazni WA, Luke H et al. Determination of the flight range and dispersal of the house fly, Musca domestica (L.) using mark release recapture technique. *Trop Biomed 2005 Jun*;22(1):53-61.

8. Spellman, G, et al. New labor-free fly control ClariFly 267 (diflubenzuron) achieves 99% reduction in Housefly (*Musca domestica)* populations in swine facilities. *Proceedings 51st AASV 2020*. 175-176.

Evaluation of a combination of benzoic acid and nature-identical flavorings to reduce the ASF viral contamination of complete feed

H-X. Zhai[1], PhD; M. C. Walsh[2], PhD; S. Hough[3], DVM; J. R. Bergstrom[3], PhD
[1]DSM (China) Animal Nutrition Research Center, Bazhou, China; [2]DSM Nutritional Products, Kaiseraugst, Switzerland; [3]DSM Nutritional Products, Parsippany, New Jersey

Introduction

Feed has been identified as a valid vector for transmitting African swine fever virus (ASFV) in lab settings. The combination of benzoic acid and nature-identical flavourings (VVC) was recently confirmed to mitigate the risk of viral transmission via feed, but its effects against ASFV had not been explored in that study.[1] The aim of the present *in-vitro* study was to evaluate the potential of VVC to reduce the risk of ASFV transmission via feed.

Materials and methods

Four samples of corn and soybean meal pig feed were mixed individually in a miniature blade mixer either without addition of feed additives for the negative control (NC) and positive control (PC) treatments or with 0.5% benzoic acid (BA) for BA treatment or with the combination of 0.5% BA and 0.02% nature-identical flavourings for VVC treatment. These samples were gamma-irritated at 25 kGy before further operations in a biosafety level (BSL)-3 laboratory at South China Agricultural University (Guangzhou, China). Four 10-gram sub-samples were weighed into centrifuge tubes. Each of the sample tubes of the PC, BA and VVC treatments was inoculated with 1 mL $10^{5.3}$ 50% hemadsorption doses (HAD_{50})/mL ASFV (ASFV/China/GZ201801). The sample tubes were continuously rotated at 1,200 rpm/min to prevent clotting. On day 1, 3, 6 or 9 post inoculation (PI), each sample tube per treatment was taken for hemadsorption test (rosette formation) and qPCR quantification. Ten mL PBS were added to each sample tube for collection of supernatant after centrifugation. The 100-fold diluted supernatant was used to infect porcine alveolar macrophages in plate wells for 2 h before the addition of RPMI 1640 medium. After 24-h incubation, qPCR test was performed and red blood cells were added for the hemadsorption test.

Results and discussion

The NC samples containing no feed additives tested negative for hemadsorption and were confirmed to be free of ASFV DNA by qPCR on all sampling days while the PC samples were all positive and had an initial Ct value of 30.4 on day 1 PI and then remained relative stable from day 3 to 9 PI (34.3-34.9; Table 1). The inclusion of BA resulted in a decrease from 4 positive hemadsorption tests on day 1 PI to 3 on day 3 and then complete negativity on day 6 and 9 PI. Complete negative tests were achieved already on day 1 PI with VVC. These results suggest that the addition of the natural identical flavourings could accelerate the viricidal effect of BA by 3 days. The qPCR tests showed the samples positive for ASFV could change into negative on day 9 PI with the inclusion of BA or VVC. The significantly higher Ct value (38.2) on day 3 PI for VVC relative to BA (34.8) also indicates an accelerating effect of the flavourings in degrading the DNA of ASFV. Enhanced degradation of porcine epidemic diarrhoea virus RNA in feed was also reported for this mixture in comparison to BA alone.[2]

References

1. Dee et al. An evaluation of additives for mitigating the risk of virus-contaminated feed using an ice-block challenge model. Transbound Emerg Dis. 2020:1.

2. Gebhardt et al. Determining the impact of commercial feed additives as potential porcine epidemic diarrhea virus mitigation strategies as determined by polymerase chain reaction analysis and bioassay. Transl Anim Sci. 2019;93.

Table 1: Effect of benzoic acid and the combination of benzoic acid and nature-identical flavorings on hemadsorption and ASFV detection by qRT-PCR

	Hemadsorption test‡				qRT-PCR test (Ct value)¶			
Treatment§	1 dpi	3 dpi	6 dpi	9 dpi	1 dpi*	3 dpi	6 dpi	9 dpi
Negative control	- - - -	- - - -	- - - -	- - - -	0	0	0	0
Positive control	+ + + +	+ + + +	+ + + +	+ + + +	30.4^{e}	34.3^{d}	34.9^{d}	34.5^{d}
Benzoic acid	+ + + +	+ + + -	- - - -	- - - -	35.6^{cd}	34.8^{d}	39.4^{ab}	40.5^{a}
Benzoic acid plus nature-identical flavorings	- - - -	- - - -	- - - -	- - - -	35.2^{cd}	38.2^{b}	37.3^{bc}	40.6^{a}

‡ (+) indicates rosette formation and (-) indicates no rosette formation. There were four replicates per treatment each represented by their own +/- symbol.

* Main effect of day on Ct value, n = 9, SEM = 0.26, $P < 0.01$;

§ Main effect of treatment on Ct value, n = 12, SEM = 0.22, $P < 0.01$;

¶ Treatment × day interaction effect on Ct value, n = 3, SEM = 0.44, $P < 0.01$;

a-e Means within interaction lacking a common superscript differ ($P < 0.05$).

Validation of mRNA quantification of tight junction proteins and cytokines in fecal samples as intestinal integrity biomarkers in swine: a preliminary experience

G. Ramis[1], DVM PhD Dip ECPHM; C. Cabrera-Gómez[1], DVM; C. Sanmartin[2], DVM; G. Cano[2], DVM; I. Riu[2], DVM; J. Sanmartin[2], DVM; P. J. Sánchez-Uribe[3], DVM
[1]Dpto, Producción Animal, Facultad de Veterinaria, Universidad de Murcia, Spain;
[2]Optimal Pork Production Group; [3]Elanco Animal Health

Introduction

The swine industry is currently involved in a hard mission to reduce the usage of antimicrobials. Different strategies are being developed to substitute these molecules and to preserve the health of animals. Intestinal health is one of the major concerns, and veterinarians need indicators of the intestinal integrity in different phases of production. The integrity of tight junctions among the epithelial cells is necessary for maintaining the health of piglets and to avoid the undesirable phenomenon of bacterial translocation and abnormal paracellular permeability.[1,2] Commonly, intestinal integrity has been assessed by measuring the presence of biomarkers in sera. In this work, we have validated the use of gene expression to assess intestinal integrity, and evaluated the relative quantities in pigs from four different farms.

Material and methods

Paired samples (feces and intestinal tissue) from 4 animals were obtained. Tissue and feces were preserved in RNAlater (Life technologies), and unpreserved feces were also obtained. Total RNA was isolated by commercial kit, and reverse transcription was performed to produce cDNA only from mRNA. Primers for zonulin 1 (ZON), Occludin (OCL), calprotectin (CALP), as intestinal integrity markers, were designed, and previously published primers for IFN-γ and TGF-β were used as immune stimulation indicators. Two different commercial kits for RNA isolation were tested. The integrity of mRNA in feces, compared to tissue and the performance of RNA isolation kits was done using β-actin and IFN- γ genes. Results were normalized with β-actin, and the relative quantification was made correcting to primers efficiency as previously published.[3] Fecal and sera samples from four farms ($n = 30$ each) were obtained, and the farms were selected on the basis of an objective classification of facilities, management and health indicators. Farms were classified as apparently low health level (LH; A and B) and apparently high health level (HH; C and D). Data obtained from four farms were compared to ELISAs for zonulin-1, calprotectin and endotoxin. The gene expression for cytokines was only done for A and C farms as examples. The data were analyzed by means of median comparison tests, and the frequency was analyzed by Chi-squared test with analysis of adjusted residues.

Results and discussion

There was a high correlation ($R = 0.856$, $P < 0.001$) among the quantification of β-actin in feces and intestinal tissue (regardless the intestinal portion; jejunum, ileum or colon). Also, there was a high correlation among preserved and not preserved feces for both genes ($R = 0.928$, $P < 0.001$; $R = 0.868$, $P > 0.001$, respectively), and there was also a high correlation among the results obtained with each one of the two commercial kits for RNA isolation ($R = 0.813$, $P = 0.002$).

The results from the four farms and the five biomarkers are shown in Figure 1. We found significant differences for CALP comparing farms with the A farm having the lowest expression and the C farm having the highest. OCL and ZON showed differences among A and D compared to B and C with the lowest expression for farm D.

There was correlation between CALP gene expression and endotoxin by ELISA ($R = -0.311$, $P = 0.005$), and between ZON gene expression and by ELISA ($R = -0.287$, $P = 0.01$), which confirms that the loss of intestinal integrity results in increased quantity of protein in blood and decreased quantity of mRNA for the proteins. The percentage of positive samples was also different depending on the farms. Generally, farm D had a lower frequency of positive samples than expected ($AR = -3.4$ for CALP, $AR = -5.2$ for OCL and $AR = -5$ for ZON, $P < 0.001$). The frequency of positives for cytokines was significant different comparing farms A and C (Figure 1).

In this case, farms A and C fulfilled the expectation since they were qualified as LH and HH and all indicators confirmed the on-farm qualification. However, results from farm B and farm D were not as expected since farm B (qualified *a priori* as LH) was similar to farm C in some parameters, and farm D (qualified *a priori* as HH) had results similar to farm A. The geographic area where farm D was located experienced elevated environmental temperatures during the three days prior to sampling, and it is hypothesized that this could have had an impact on the measures of intestinal integrity. The values of endotoxin, only significantly higher in farm A, confirm that the effect shown by farm D could be transient and not chronic.

References

1. Zhang et al. Resveratrol alleviates heat stress-induced impairment of intestinal morphology, microflora, and barrier integrity in broilers. Poultry Science. 2017. 96(12):4325–4332.

2. Ferrando, M. L., & Schultsz, C. A hypothetical model of host-pathogen interaction of *Streptococcus suis* in the gastro-intestinal tract. Gut Microbes. 2016. 7(2).

3. Pfaffl MW. A new mathematical model for relative quantification in real-time RT-PCR. Nucleic Acids Res. 2001. 29(9):e45.

Figure 1: A) Comparison of relative quantitation for zonulin, occludin and calprotectin, B) Comparison of relative quantitation for TGF-β and IFN-γ, C) Percentage of samples amplifying for zonulin, occludin and calprotectin, D) Percentage of samples amplifying for TGF-β and IFN-γ.

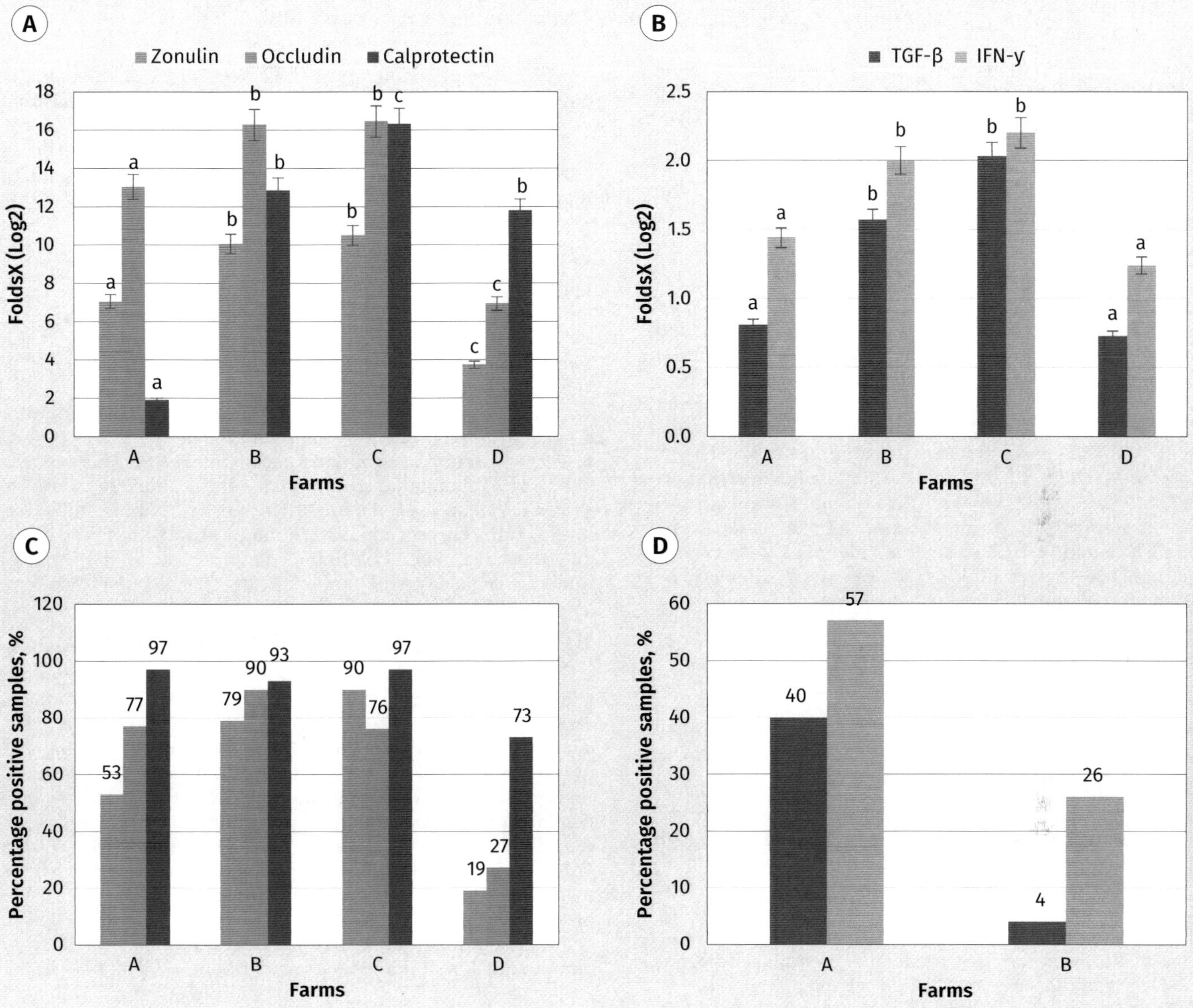

Software integrations provide easy access to data

J. Tyler Holck, DVM, MS, MBA
Veterinary Technical and Business Consultant, Feed His People LLC

Introduction

Easy access to diagnostic data, as well as feed and medication information, is important to help veterinarians make sound herd health decisions. This paper will describe the innovative collaboration of cloud-based software solutions, LabLINK HIMS™ from GVL and Feed Allocation System (FAS) from PrairiE Systems.

Streamline access to information

LabLINK HIMS aggregates over 50,000 diagnostic cases per year and FAS handles over 20,000 groups of pigs placed per year (Figure 1). Having easy access to diagnostic, medication, feed and performance information in one place allows veterinarians to easily review and make herd health decisions. The integration between LabLINK HIMS and FAS imports the diagnostic data from LabLINK HIMS to make it easily accessible for the herd health team to view in FAS. After the diagnostic data has been imported into FAS, it can be mapped to a farm or group, allowing the veterinarian to view only the diagnostic cases relevant to that farm or group. The LabLINK HIMS data is provided via a high-level view of the case and relevant information. The raw data can also be exported directly from FAS for further analysis.

Software to support and document the VCPR[1]

Maintaining documentation to support a valid veterinarian-client-patient relationship (VCPR) in one place can be a challenge. A valid VCPR must meet the following conditions:

1. The licensed veterinarian has assumed the responsibility for making medical judgements regarding the health of the patient(s) and the need for medical therapy and has instructed the client on a course of therapy appropriate to the circumstance.
2. There is sufficient knowledge of the patient(s) by the veterinarian to initiate at least a general or preliminary diagnosis of the medical condition(s) of the patient(s).
3. The client has agreed to follow the licensed veterinarian's recommendations.
4. The licensed veterinarian is available for follow up evaluation or has arranged for emergency care coverage.
5. The veterinarian provides oversight of treatment.
6. A timely physical examination of the patient(s) has taken place or the veterinarian is personally acquainted with the keeping and care of the patient(s) by virtue of medically appropriate and timely visits to the operation where the patient(s) is (are) kept, or both.
7. Patient records are maintained.

Through the integration of LabLINK HIMS and FAS, the veterinarian and production teams are connected, with easy access to view a farm or group's veterinary feed directives (VFDs), record illness information and view diagnostics all in one place. This assists with reviewing and maintaining patient records as required by VCPR.

GVL – LabLink HIMS

LabLINK HIMS is a real-time diagnostic solution that enhances herd health management for disease monitoring, analysis and pathogen response. Access to real-time diagnostic data and subsequent analysis tools is critical in modern pork production. When epidemics like PRRS strike, LabLINK HIMS can make a significant difference by providing real-time results which allow for quicker decisions and immediate action. Data sets can be customized for a specific organization's needs. Information such as years of historical data, flows and premise identification are tailored to meet each client's specific production needs. This customization leverages the current and historic diagnostic database and enables effective animal health information monitoring. For more information on GVL and LabLink HIMS, visit **globalvetlink.com**.

PrairiE Systems – Feed Allocation System

Feed Allocation Systems simplifies and clarifies the feed ordering process while reducing the number and duration of feed outages. Clients will spend less time ordering and more time analyzing your nutritional program. With FAS, clients can track VFDs, withdrawal periods and process verification, all quickly and easily through the web-based application. In addition, producers can document oral and injectable treatment information and daily mortalities at the group level. FAS is fully integrated with GVL via FASLink®. FAS Mobile allows clients to quickly view critical information from any device and enables the farm staff to place feed orders, track and enter inventory, and communicate directly with company administration. For more information on PrairiE Systems and FAS, visit **prairiesystems.com**.

Reference

1. Principles of veterinary medical ethics of the AVMA. August 2019. Available at: **https://www.avma.org/resources-tools/avma-policies/principles-veterinary-medical-ethics-avma#III**. Accessed 2 November 2020.

Figure 1: Growth of annual digital diagnostic and group placement data.

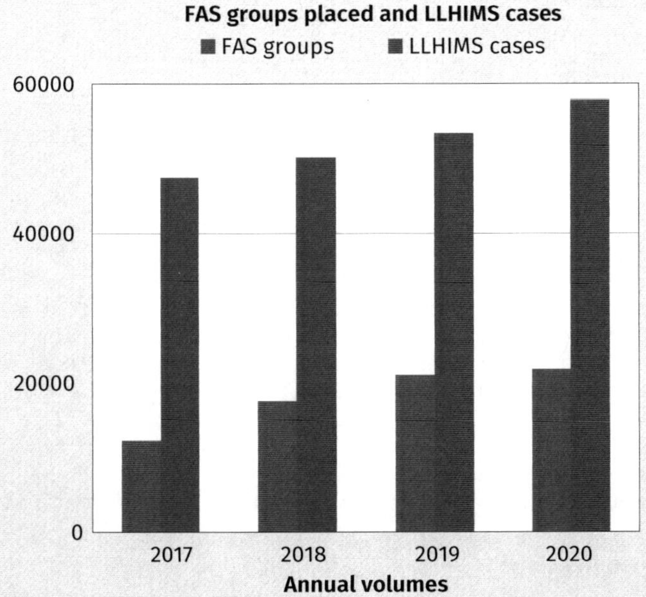

Understanding IAV-S diversity over time using the SEQUIVITY® Dashboard

Channing Sebo, DVM; Jeremy Maurer, VMD
Merck Animal Health, De Soto, Kansas

Introduction

Managing the disease impact and economic losses associated with Influenza A virus in swine (IAV-S) continues to be a challenge for the global swine industry. Prevention of IAV-S can be difficult given ease of transmission, the reverse zoonotic potential and high genetic mutation rate of the virus. Vaccination against IAV-S is especially challenging due to the extensive diversity of hemagglutinin (HA) proteins present in the field due to antigenic shift and drift.[1] Success with vaccination is heavily dependent on sequence analysis that enables strain matching of the HA protein to the field strains that are causing disease.[2,3] Merck Animal Health (MAH) uses the SEQUIVITY® Dashboard to assist customers with sequence management and prescription vaccine design.[4]

Swine veterinarians most commonly use a "reactive approach" to IAV-S vaccination, implementing custom vaccines based on the strain(s) circulating within the farm or production system. Vaccination is then typically applied only after the infection is endemic. This is a costly strategy with many weeks of production hindered waiting for the vaccine to be produced and administered. Using the SEQUIVITY® Dashboard, veterinarians can move towards a more "proactive approach" to vaccination and prevention. SEQUIVITY® vaccines can be tailored to account for herd specific as well as at-risk strains by utilizing the Dashboard's geospatial features. This preventative approach allows MAH customers to make more informed decisions on strain selection in vaccine design.

Methods and results

MAH actively monitors IAV-S diversity and trends overtime in the US swine population to assist in vaccine planning. Recently, an analysis was performed to evaluate the H1 and H3 sequence trends in the US over the past 10 years through October 2020. The data utilized all the IAV-S sequences present in the SEQUIVITY® Dashboard - which is comprised using both customer data and the NIAID Influenza Research Database (IRD).[5] The analysis is presented in Figure 1 as percentages of the total number of sequences in the database that year. The total number of sequences analyzed each year ranged from 438-2140.

Over this ten-year period, dynamic changes in the strains circulating in the US have been observed. In 2010, there was no evidence of H3 human-like or H1 alpha circulation. H1 beta strains have had low levels of circulation throughout the 10-year period, never representing more than 6.8% of the sequences. H1 delta strains have shown dynamic changes as well. In 2010 Delta 1b was the most dominant of the Deltas, persisting through 2013 with a sharp decline in 2014-the same year H3 human-like strains prevalence dramatically increased; Delta 1a strains became highly prevalent in 2015, with dominance persisting through 2017 and declining in recent years; In 2018 H1 Delta 2 strains increased in prevalence and peaked in 2019. H1 gamma circulation remains one of the most consistent and prevalent strains in circulation and was in the top three strains from 2011 to present. H1 pandemic strains have been increasingly more dominant from 2019 to present, with 2020 prevalence of 16.57%. In 2020 the top three strains in circulation are H3 Cluster I-IV (primarily Cluster IVa strains), followed by H1 gamma, and finally H1 pandemic strains.

Conclusion

In conclusion, the complex pathogen dynamics that exist within swine populations require producers and veterinarians to have access to real-time analytics to better combat and prevent disease within their herds. Monitoring IAV-S strain dynamics is an important factor in understanding the risk and prevalence of different strains across populations and ultimately to implore better proactive vaccine implementations. The SEQUIVITY® Dashboard is currently used extensively by US and Canadian swine veterinarians to monitor IAV-S and other pathogens to make proactive prevention strategies and mitigate financial and disease impacts.

References

1. Vincent AL et al. Swine Influenza Viruses: A North American Perspective. *Adv V Res.* 2008: 127-154.

2. Vander Veen RL et al. Alphavirus replicon vaccines. *Ani Health Res Rev.* 2012: 1-9.

3. Tamura S et al. Mechanisms of Broad Cross-Protection Provided by Influenza Virus Infection and Their Application to Vaccines. *Jpn J Infect Dis.* 2005: 195-207.

4. Sebo et al. Influenza A vaccination using the SEQUIVITY™ Technology. AASV. 2019.

5. Data was obtained from the NIAID Influenza Research Database (IRD) [Zhang Y, et al. (2017)] through the web site at **http://www.fludb.org.**

Figure 1: IAV-S diversity by strain by year (2010 - 2020).

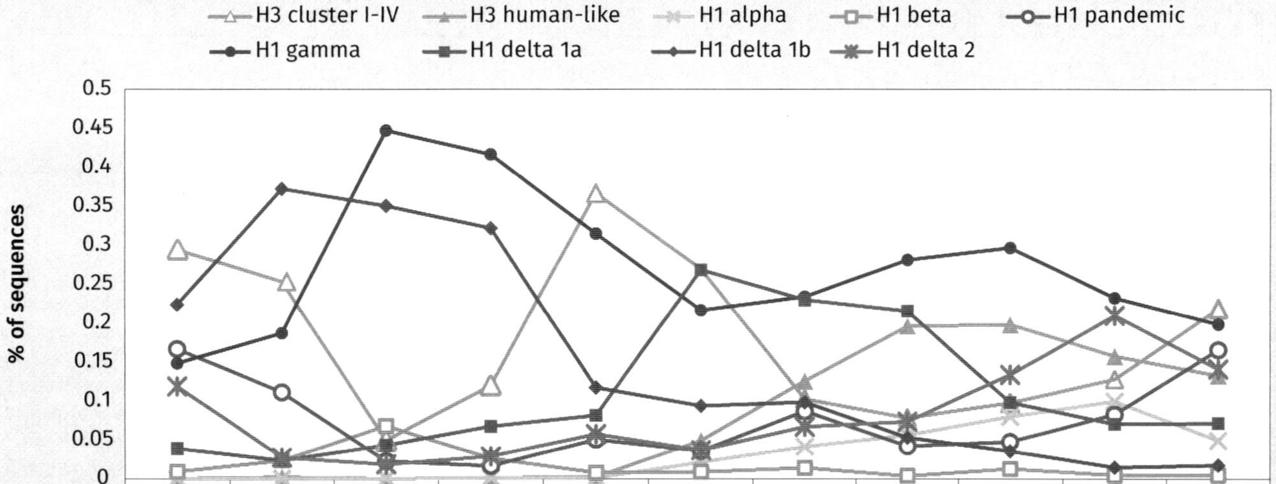

	2010	2011	2012	2013	2014	2015	2016	2017	2018	2019	2020 (through October 2020)
H3 cluster I-IV	29.45%	25.34%	4.95%	11.97%	36.67%	26.96%	10.26%	7.83%	9.77%	12.79%	21.86%
H3 human-like	0.00%	0.19%	0.00%	0.14%	0.23%	4.93%	12.57%	19.61%	19.81%	15.76%	13.28%
H1 alpha	0.00%	0.00%	0.00%	0.14%	0.11%	2.13%	4.14%	5.63%	7.99%	9.97%	4.94%
H1 beta	0.91%	2.34%	6.80%	2.61%	0.80%	0.87%	1.41%	0.37%	1.26%	0.47%	0.47%
H1 pandemic	16.67%	11.11%	2.27%	1.65%	5.06%	3.57%	8.68%	4.21%	4.77%	8.28%	16.57%
H1 gamma	14.84%	18.71%	44.74%	41.68%	31.49%	21.64%	23.41%	28.13%	29.67%	23.24%	19.86%
H1 delta 1a	3.88%	2.34%	4.33%	6.74%	8.16%	26.86%	22.99%	21.56%	9.81%	7.06%	7.17%
H1 delta 1b	22.37%	37.23%	35.05%	32.19%	11.72%	9.37%	9.84%	5.31%	3.55%	1.46%	1.76%
H1 delta 2	11.87%	2.73%	1.86%	2.89%	5.75%	3.67%	6.70%	7.36%	13.36%	20.98%	14.10%
Total sequences	**438**	**513**	**485**	**727**	**870**	**1035**	**1209**	**1902**	**2140**	**2126**	**851**

Monitoring operational efficiency of boar semen production systems with the new Prism10 software package

M. Barros[1], DVM; T. Langer[2], BS; S. Trulen[3], BS; J. Quackenbush[3], BS; R. Grossfeld[2], PhD, Dr.; Dr. C. Simmet[2], DVM
[1]Minitub do Brasil Ltda, Porto Alegre, Brazil; [2]Minitube GmbH, Tiefenbach, Germany; [3]Minitube USA, Verona, Wisconsin

Introduction

In swine industry, the use of software to assist production management has been shown to be increasingly important. The main advantages of using software are: expanding and speeding up the extraction of management information, automating processes and integrating equipment used in the production routine, improving decision making and response time to problems, reducing rework and consequently costs, controlling a large volume of information and guarantee its traceability. In the case of boar semen production systems, there is no minimum size or reference number of animals from which the implementation of management software would be necessary. The recommendation is that all studs, regardless of their size, work with software tools for business management. The main focus has been for a long time on maximizing productivity per animal in herds, but other relevant information has been demanded by the users. In addition to production, the genetic value of animals has become one of the main points to be explored and managed within the boar studs, as well as data related to the quality control of processed semen and post-production information.

Prism10 software

Minitube's Prism10 software combines not only Automation, Traceability and Integration inside the boar studs, but also provides high flexibility to process any required product, a user-friendly interface and the possibility to extract all needed data through dynamic reports and dashboards (Figure 1).

The features and versatility of Prism10 web-based application allow the users to process ejaculates in several ways, such as single-sired, pools and split parts. The seminal doses, main product of the boar studs, need to be produced using different packaging methods, with various volumes and sperm concentrations for traditional and post cervical inseminations. Furthermore, each user can work with customized preferences in the software, which means that it will suit the requirements of users worldwide. All development and technical support are provided by Minitube's in-house team, with technicians located at different countries who ensure continuous improvements and security for the boar semen industry.

Conclusion

The field of porcine reproduction technology needs a modern lab management software that meets each individual demand in the boar studs. In terms of technology, the continuous increase in the use of Smartphones and Tablets devices has required the development of mobile applications, facilitating access to information through dynamic reports and graphs, from anywhere and anytime, over the internet. Besides the benefits in the production workflow, Prism10 will provide better relationship between managers and operational areas, significantly improving companies' results, aiming at strategic advantages and increased competitiveness.

Figure 1: Examples of dashboards in Prism10 software.

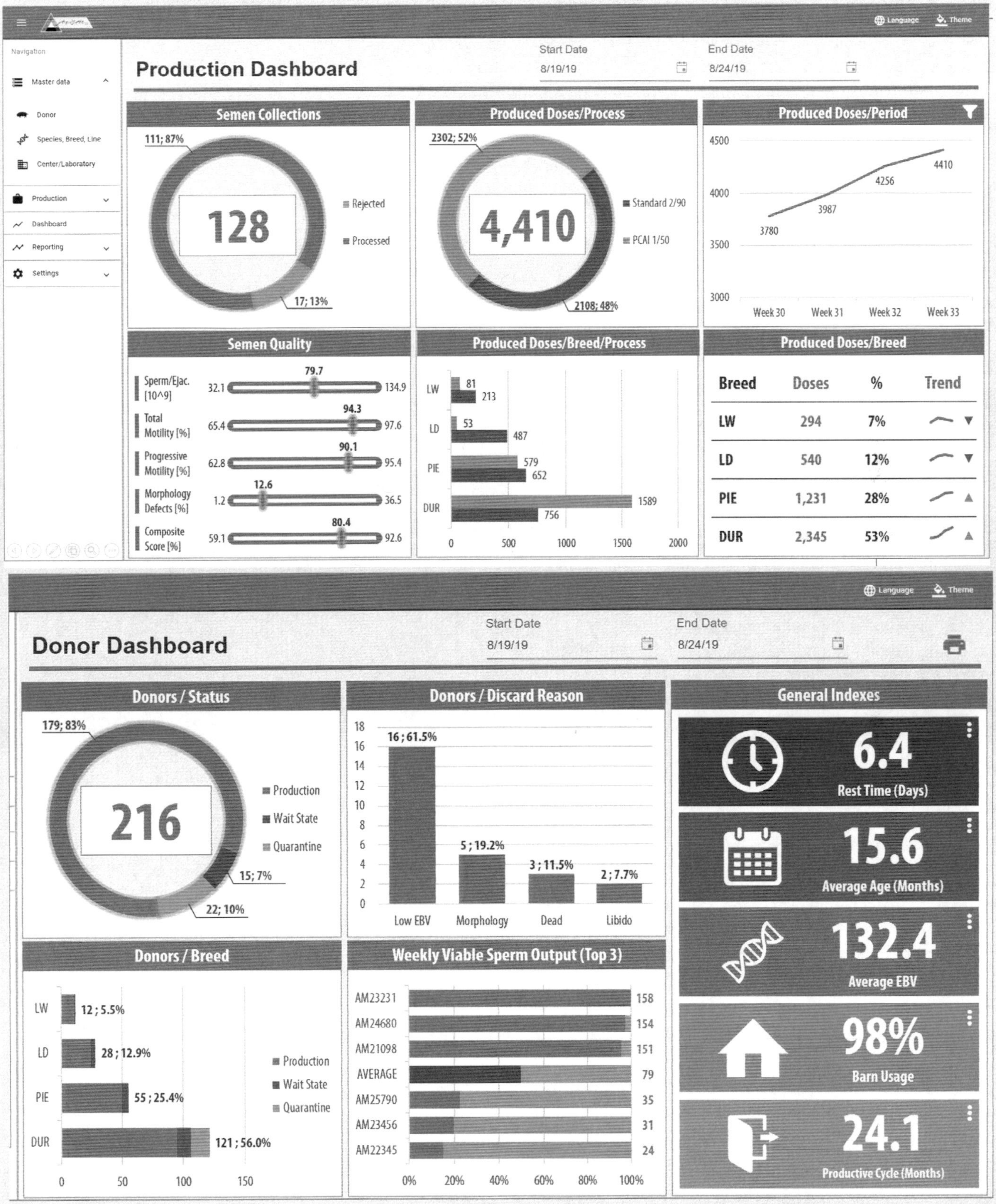

Evaluation of the effect of a *Salmonella* vaccine on production performance in a herd with salmonellosis

Fernando Leite[1], DVM, MS, PhD; Steve Tousignant[1], DVM, PhD; Greg Cline[1], DVM; Tyler Bates[1], BS; Amy Jacobs[1], BS; Adam Mueller[2], DVM
[1]Boeringer Ingelheim Animal Health USA Inc., Duluth, Georgia; [2]Swine Services Unlimited Inc., Rice, Minnesota

Introduction

Salmonella continues to be a health challenge in many swine herds in the US. This study investigated the effect of a *Salmonella* vaccine in a herd with a clinical history of salmonellosis. This was a finishing herd located in Minnesota that received pigs at approximately 9 weeks of age which had experienced diarrhea and diagnosed salmonellosis in previous turns.

Materials and methods

Pigs from one nursery site were transported to the finisher barn. Two rooms were filled with 1,060 and 1,055 pigs each. On arrival, truck weights were recorded to have a starting weight for each room. After two weeks one room received Enterisol® Salmonella T/C vaccine in the drinking water. Sixty pigs from each room (treatment) were tagged and had blood and fecal samples collected at 3 different time points until market at approximately 4-5 week intervals. These samples were submitted for *Salmonella* ELISA and enrichment culture, to investigate disease dynamics over time. Close out data, mortality, average daily weight gain and culls were assessed in both rooms.

Results and discussion

Prior to vaccination, both groups had 38% (VAC) and 30% (Non-VAC) pigs positive for *Salmonella* serum antibodies, this difference was not significant. The proportion of pigs with antibodies increased to 72% and 73% in the two subsequent collections in the non-vaccinated group, indicating field exposure to *Salmonella*. The increase in serum antibody response in subsequent collections was greater in the vaccinated group, with 98% and 100% of pigs having serum antibodies. The difference in the number of pigs with serum antibodies between groups was significant ($P < 0.05$). Prior to vaccination, both groups had 6.7% of pigs positive for *Salmonella* in feces. At the two subsequent collections, the non-vaccinated group had 5% and 22.4% of pigs positive for *Salmonella* in feces. In the vaccinated group, the two subsequent collections had 0% and 19.7% of pigs shedding *Salmonella*. From a total of 21 *Salmonella* positive culture samples in the non-vaccinated group, 19 were of serogroup B and 2 were of serogroup E. All of the 16 positive culture samples from the vaccinated group were of serogroup B. These results suggest that this herd faced a challenge with *Salmonella* which was primarily of serogroup B strains. The fact that nearly all of the pigs vaccinated with Enterisol® Salmonella T/C seroconverted versus only 73% of non-vaccinates ($P < 0.05$) indicates that a different and more homogenous immune response was induced in the vaccinated group than in the group that did not receive the vaccine. No significant differences were noted in the number of pigs shedding *Salmonella* during the trial, although these only represented qualitative and not quantitative results.

When evaluating production performance, days on feed ranged from 103 days to 129 days in vaccinates and 109 to 129 days on feed in non-vaccinates. The average daily weight gain was 2.323 lbs for vaccinates and 2.266 lbs for non-vaccinates for the period on feed (Figure 1). The adjusted final weight difference at 103 days on feed was 6.18 pounds between both groups, with vaccinates being the heaviest. No significant differences in cull rates or mortality were noted between both groups.

Conclusion

This study evaluated the effect of vaccination for *Salmonella* in a herd with a history of salmonellosis. A limitation of this study was that the vaccine was only implemented in one room and weights were not measured individually. Nonetheless results indicate a positive trend of vaccination on growth performance which should be confirmed in further trials.

Figure 1: Average daily weight gain in animals that received Enterisol® *Salmonella* T/C (vaccinated) and non-vaccinated animals.

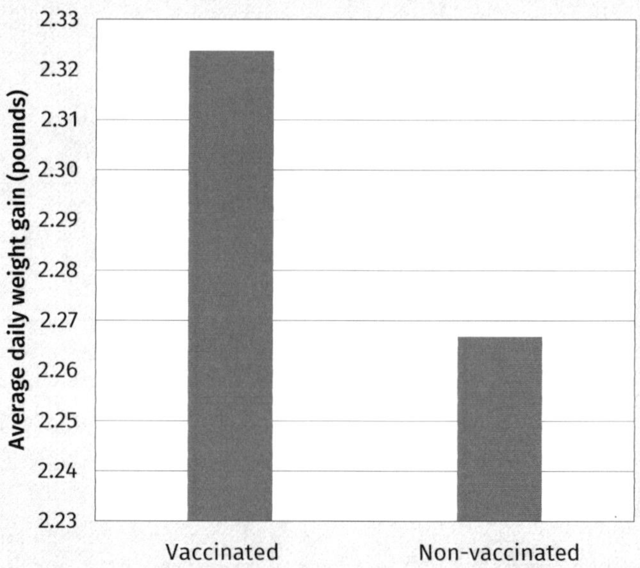

Continual improvement - the case of the PRRS oral fluid ELISA

A. Henao-Diaz[1,3], DVM, PhD; M. Zhang[2], MS; L. Giménez-Lirola[1], PhD; P. Gauger[1], DVM, PhD; D. Baum[1], DVM, PhD; M. J. Clavijo[1,3], DVM, PhD; M. Rotolo[3], DVM, PhD; E. Ramirez[4], DVM, PhD; R. Main[1], DVM, PhD; J. Zimmerman[1], DVM, PhD
[1]Veterinary Diagnostic and Production Animal Medicine, Iowa State University, Ames, Iowa; [2]Statistics, Iowa State University, Ames, Iowa; [3]PIC North America, Hendersonville, Tennessee; [4]Grupo Porcícola Mexicano Kekén, Yucatán. México

Introduction

PRRSV surveillance requires an easily collected specimen and accurate diagnostic tests. Oral fluid (OF) is the most common surveillance specimen in the US[1] and both RT-PCR and OF ELISA have been adapted to the matrix. Detection of PRRSV RNA provides the advantage of early detection; PRRSV antibody provides the advantages of lower cost and longer duration of detection. In routine surveillance, unexpected positives cause disruption in the workflow on the farm and undermine client confidence in the laboratory. Hence, diagnostic specificity (DX-SP), not diagnostic sensitivity (DX-SE), is the major consideration.

Under experimental conditions, the DX-SE and DX-SP of the PRRS OF ELISA were estimated at ~100%[2] using the manufacturer cut-off S/P 0.4. Herein, we evaluate the DX-SP of the PRRS OF ELISA using experimental and field samples and provide and alternative cut-off for surveillance.

Materials and methods

Two sets of OFs samples were used: **Set 1:** 596 OFs of known negative and positive status, generated under experimental conditions. **Set 2:** 1574 OFs from 94 presumed PRRSV-naïve sites submitted for routine testing at the Iowa State University Veterinary Diagnostic Laboratory.

All samples were tested on the PRRS oral fluid Ab ELISA (IDEXX Laboratories Inc.) and the data analyzed using non-parametrical statistical procedures:

- DX-SE and DX-SP analyzed by receiver operating characteristic curve (ROC) analysis (Set 1);

- ELISA S/P by pig age analyzed using linear regression and Tukey's box plot (Set 2);

- Effect of alternative cut-offs on positivity rates analyzed using Cochran's Q (Sets 1, 2)

Results and discussion

Set 1: at a cut-off of S/P ≥ 0.4 DX-SE and DX-SP were both > 99.4%. At a cut-off of S/P ≥ 1.0 DX-SE and DX-SP were both > 96.5%. Higher S/P cut-offs did not improve DX-SP (99.4%) because of the limited number of negative samples (n = 167), but likewise had minor impact on DX-SE. Cut-offs of S/P ≥ 0.4 and ≥ 1.0 classified 50% of pigs (6/12) as positives at 10 and 11 day post vaccination, respectively.

Set 2: Figure 1 shows the 99th percentile S/P response by pig age or category. "Extreme S/P outliers" (n = 46; Tukey's box plot) were associated with specific categories, i.e., 9, 24, 25 weeks of age, and gestation.

Routine PRRSV surveillance demands near-perfect diagnostic specificity, even at the cost of diagnostic sensitivity. Using a cut-off of S/P ≥ 1.0 significantly reduced false positives and minimally affected DX-SE. Thus, this cut-off is recommended to interpret PRRS OF Ab ELISA results from presumed negative sites to reduce the false alarms.

Notably, extreme S/P responses were associated with PRRSV-naïve pigs of specific ages and/or in gestation. Work in progress is expected to identify the cause(s) of this non-random pattern in antibody detection.

For more information consult: Henao-Diaz A, et al. Adapting a porcine reproductive and respiratory syndrome virus (PRRSV) oral fluid antibody ELISA to routine surveillance. *Prev. Vet. Med. (in press).*

Acknowledgments

Authors declare no conflicts of interest (COI). JZ has consulted with IDEXX Laboratories, Inc. independent of this research; consulting has been reviewed and approved by ISU in accordance with its COI policies. PRRSV ELISA kits were provided by IDEXX Laboratories, Inc.

References

1. Trevisan G, et al. PLoS One. 2019;14:1–16.
2. Henao-Diaz A, et al. Res Vet Sci. 2019;125:113–8.

Figure 1: PRRS OF ELISA S/P results (99th percentile). Oral fluid samples (n = 1574) from 94 PRRSV known negative sites (Set 2) by pig age and categories.

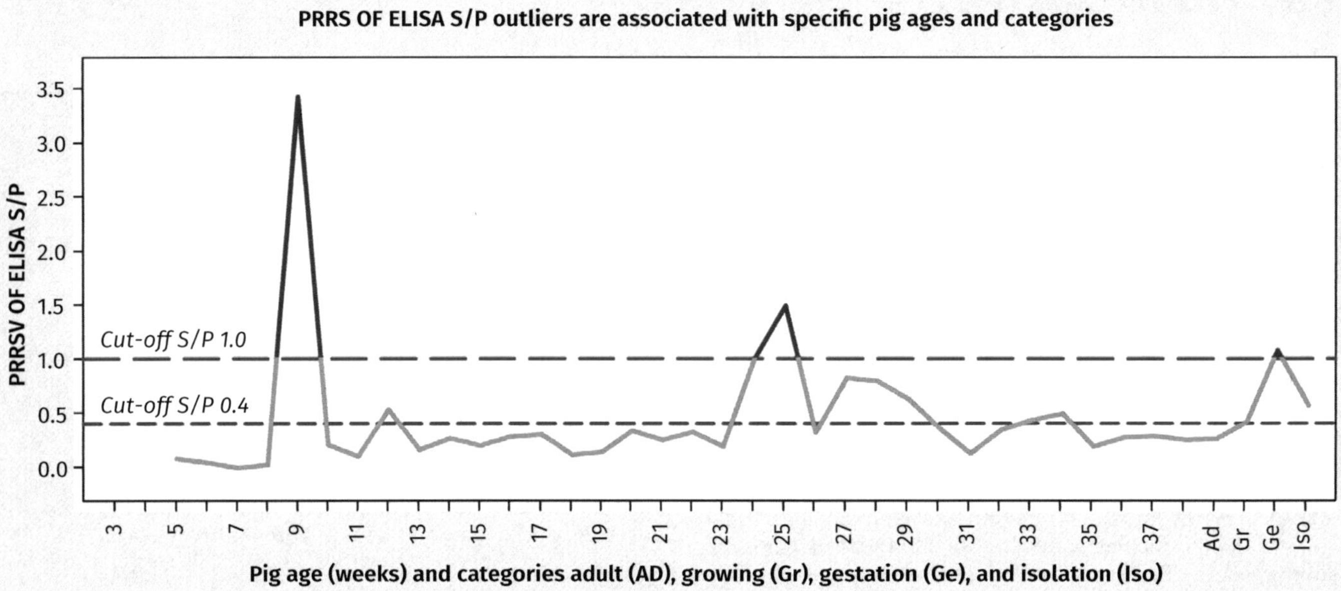

The use of medium chain fatty acids to reduce porcine reproductive and respiratory syndrome virus replication in MARC-145 cells

S. A. Crowder[1], PhD candidate; R. M. Pogranichniy[2], DVM; B. De Rodas[3], PhD; T. P. Karnezos[4], PhD; J. S. Radcliffe[1], PhD
[1]Dept. of Animal Science, Purdue University, West Lafayette, Indiana; [2]Department of Diagnostic Medicine and Pathobiology, Veterinary Diagnostic Laboratory, Kansas State University College of Veterinary Medicine, Kansas; [3]Purina Animal Nutrition LLC., Arden Hills, Minnesota; [4]Land O'Lakes Inc., Arden Hills, Minnesota

Introduction

The swine industry has been affected by Porcine Reproductive Respiratory Syndrome Virus (PRRSV) for over 30 years.[1] Genetic diversity in PRRSV has reduced vaccine efficacy and cross protection between strains.[2] Recent experiments have examined the antiviral properties of medium chain fatty acids (MCFAs) with recent research focusing on feed biosecurity and viral diseases. PRRSV is of the highest economic importance to the Pork Industry today with an estimated $660 million in annual production losses.[3] Activated or free MCFAs have been used in different combinations and concentrations to optimize pig performance. Therefore, the objective of this experiment was to evaluate the efficacy of different combinations of MCFAs to reduce viral replication of the PRRSV.

Materials and methods

MARC-145 cells, are highly permissive to PRRSV replication, were maintained in Dulbecco modified Eagle medium (DMEM) containing 10% fetal bovine serum and 50µg of gentamicin. Type II North American PRRSV strain P129 and Type I European PRRSV Strain Lelystad were propagated in MARC-145 cells to prepare the virus stocks for the present study. MCFAs were used in varying concentrations ranging from 50µg/mL to 500µg/mL and compared to control cells and TCID50/mL was calculated for all treatment concentrations and non-treated controls. PPRSV replication in the cells was detected using the FITC labeled IgG anti-PRRSV monoclonal antibody (SDOW- 17F, RTI Labs, Brookings, South Dakota).

Results

Results from this experiment demonstrate that combinations of MCFAs were able to reduce replication of both strains (type I Lelystad and type II P129) of PRRSV in the MARC-145 cells. MARC-145 cells incubated with combinations of MCFAs prior to and after inoculation with type I Lelystad strain or type II P129 strain of PRRSV reduced ($P < 0.05$) PRRSV replication at concentrations above 200µg/mL. Incubation of MARC-145 cells with caprylic and capric acids prior to and after inoculation with type I Lelystad strain or type II P129 strain of PRRSV resulted in a 1.9 to 4.1 Log reduction ($P < 0.05$) in PRRSV replication. MARC-145 cells incubated with caproic, caprylic, capric, and lauric acids prior to and after inoculation with type I Lelystad strain or type II P129 strain of the PRRSV resulted in a 1.8 to 4.3 Log reduction ($P < 0.05$) in PRRSV replication. Additional studies are needed to confirm these results in an in-vivo model. Further research is also needed to determine the mode of action of individual and combinations of MCFAs on PRRSV replication.

References

1. Brar MS, Shi M, Hui RK, Leung FC. Genomic evolution of porcine reproductive and respiratory syndrome virus (PRRSV) isolates revealed by deep sequencing. *PLoS One*. 2014;9(4):e88807. Published 2014 Apr 3. **doi:10.1371/journal.pone.0088807**

2. Taeyeon Kim et al. Comparison of Two Commercial Type 1 Porcine Reproductive and Respiratory Syndrome Virus (PRRSV) Modified Live Vaccines against Heterologous Type 1 and Type 2 PRRSV Challenge in Growing Pigs. 2015 Jun;22(6):631-40. *Epub* 2015 Apr 8 **doi: 10.1128/CVI.00001-15**

3. Holtkamp DJ, et al. Assessment of the economic impact of porcine reproductive and respiratory syndrome virus on United States pork producers, *Journal of Swine Health and Production*, 2013;21(2):72-84.

LeeO™ individual animal data system to track and analyze birth to harvest information

Annie Lerner [1], PhD; Alyssa Betlach[2], DVM; Mark Schwartz[1], BS, Jeff Ley[3], BA; Tyler Holck[4], DVM, MS, MBA
[1]Schwartz Farms, Sleepy Eye, Minnesota; [2] Swine Vet Center, St Peter, Minnesota;
[3]Prairie Systems, Spencer, Iowa; [4]Feed His People, LLC, Gilbert, Iowa

Introduction

LeeO™ is an individual animal tracking system designed for paperless data collection in "real time", supporting precision pork in the commercial pork supply chain.[1] Utilizing radio frequency identification (RFID), LeeO™ scans tags and records individual animal weights quickly and effortlessly with a proprietary LeeO™ O Reader and bluetooth technology connecting to automated individual scales. This paper describes the use of the LeeO™ system to track and analyze individual animal birth to harvest information in a commercial swine production system.

Materials and methods

Litters born from a total of 278 dams (PIC Camborough; Parity 1-3) from a high-health sow farm in the upper Midwest were used in this assessment. Pregnant dams were moved to the farrowing room at day 112 of gestation and QR codes on the farrowing cards were read with the LeeO™ O Reader to register the sow, her parity and history. Piglets (n = 3,325) were tagged with a LeeO™ UHF tag at birth and registered into LeeO™, which automatically recorded piglet birth date, sire, and dam information. Gender and birth weights were also collected at birth using the LeeO™ piglet scale. Stillbirths and mummies were also entered into LeeO™ along with pre-wean mortalities.

At weaning, a subset of the piglets (n = 2,038) was moved to a wean-to-finish barn and individually weighed upon arrival using the LeeO™ piglet scale. Treatments and mortalities were also collected in LeeO™ during the growing phase. Pig weights were recorded using a LeeO™ finishing scale prior to slaughter. Pigs were then shipped to a commercial packing plant for harvest where the LeeO™ tags were linked to harvest ids using the LeeO™ mobile app. The primary function of this commercial research trial was to evaluate multiple genetic lines and nutritional strategies, however, use of the LeeO™ system allowed for data capture beyond the solely experimental treatments. Growth performance data were analyzed using a linear regression model in R studio® version 3.6.1.[2]

Results and discussion (Table 1)

While data from the internal research study will remain proprietary, the information gathered was beneficial in distinguishing differences in genetics and nutritional program from birth to harvest. Results for this abstract will focus on intriguing data that was gathered and assessed through the collection of individual piglet information. A total of 3,325 piglets were born alive and registered into this assessment using LeeO™, including 1,766 barrows and 1,559 gilts. The average birth weight for all piglets was 3.1 lb with a birth weight range from 0.9 lb to 6.5 lb with 7.3% weighing < 2 lbs. Pre-weaning mortality for piglets weighing < 2 lbs was 21.0%, compared to only 4.1% for pigs weighing ≥ 2 lbs at birth.

Birth weights were used to group piglets (n = 2,038) into 5 categories in the wean-to-finish assessment as described in Table 1. Differences in pre-wean ADG, wean weight, wean-to-finish ADG, market live weights, and hot carcass weights were observed based upon birth weight category of the pigs placed in the study. While birth weight category did not impact WTF mortality or percent full value pigs in this assessment starting at weaning, based upon differences in pre-wean mortality, birth weight would be expected to impact birth-to-market mortality and full value determination from birth. Individual data collection was possible due to high LeeO™ tag retention which was 99.4% from birth to final finish weight and greater than 99% through the harvest process.

Conclusions

The LeeO™ individual animal data system was very effective in gathering data from birth to harvest and allow for analysis of the impacts of birth weight on wean-to-finish performance. Additional applications for LeeO™ individual animal data include more precise feed budgeting in early nursery and marketing at harvest, and better decision making for sow and progeny productivity in commercial pork production.[1]

References

1. ter Linde P et al. The value of LeeO, an individual animal identification system adding value to the pork supply chain the Netherlands. AASV, 2020; 337.

2. R Core Team, 2019. R: A language and environment for statistical computing [Computer software]. R Foundation for Statistical Computing, Vienna, Austria. **http://www.r-project.org/.**

Table 1: Results – birth weight impact on wean-to-finish performance

Parameter	BW < 2.0 lbs	BW 2.0-2.5 lbs	BW 2.5-3.0 lbs	BW 3.0-3.5 lbs	BW > 3.5 lbs	*P*-value (SE)
Number of piglets born	128	311	511	489	599	
Average parity	1.4	1.4	1.5	1.5	1.6	
Birth weight (lbs)	1.75[a]	2.26[b]	2.77[c]	3.24[d]	4.01[e]	< 0.01 (0.3)
Pre-wean ADG (lbs)	0.38[a]	0.41[a]	0.44[b]	0.46[b]	0.49[c]	< 0.01 (0.1)
Wean weight (lbs)	9.75[a]	10.60[b]	11.71[c]	12.58[d]	14.24[e]	< 0.01 (0.3)
WTF ADG (lbs)	1.61[a]	1.67[b]	1.68[c]	1.72[d]	1.72[d]	< 0.01 (0.2)
Market live weight (lbs)	268.14[a]	276.54[b]	277.38[b]	281.16[bc]	279.88[bc]	< 0.01 (2.6)
Hot carcass weight (lbs)	204.45[a]	208.47[ab]	208.97[ab]	211.11[b]	210.82[b]	< 0.01 (2.2)
WTF mortality (%)	8.6	8.0	7.6	8.8	5.3	
Full value pigs (%)	86.7	89.7	90.4	88.9	91.7	

[a-e] Rows with different letters are significantly different ($P < 0.05$)

Effects of genetic capacity for protein deposition on N-efficiency in growing pigs

L. M. G. Verschuren[1,2], MSc; G. P. Binnendijk[2], BSc; R. Bergsma[1], PhD; J. R. Dunkelberger[3], PhD;
C. M. C. van der Peet-Schwering[2], PhD; A. J. M. Jansman[2], PhD
[1]Topigs Norsvin Research Center B.V., Beuningen, The Netherlands;
[2]Wageningen Livestock Research, Wageningen, The Netherlands; [3]Topigs Norsvin USA, Burnsville, Minnesota

Introduction

In recent decades, a rising global population and related increase in demand for animal protein products prompted a major expansion of the animal production sector. This also forced the sector to produce more livestock products with fewer resources and, thus, to focus on increased production efficiency. Resource efficiency in pigs is largely related to feed and nutrient efficiency. Feed efficiency is determined by both energy and protein conversion efficiency. Nitrogen (N) is provided to pigs mainly as dietary protein and (free) amino acids (AA). N-efficiency in pigs can be improved by adjusting dietary protein and AA supply to the amino acid requirements of an individual pig.[1,2] The requirements depend in part on genetic characteristics of the individual. Results of a previous selection experiment showed that divergent selection for feed efficiency simultaneously resulted in differences in efficiency in dietary energy utilization.[3] The effect on N-efficiency, however, was not clearly proven.[3] The aim of the present study was to evaluate the impact of genetic capacity for protein deposition (estimated breeding value, EBV; low vs. high) on N-efficiency in growing pigs in an N-balance study.

Materials and methods

Two groups of 20 male pigs [TN Tempo sire] x [York x Dutch Landrace dam]) were selected with a low (1.04 kg) or high body weight (BW) (1.80 kg), and within birthweight group with a low or high EBV (-2 and +11 g/d, respectively, relative to an average crossbred pig with a protein deposition of 146 g/d). EBV for protein deposition was derived from genotyping the individual pigs with a 50K single nucleotide polymorphism (SNP) chip and genomic prediction of protein deposition capacity. Pigs were subjected to N-balance measurements using quantitative urine and feces collection in two sequential periods of 5 d. Pigs (14 weeks of age at the start of the adaptation period) were fed a protein sufficient (CP 163 g/kg; 100%) diet or a dietary treatment providing only 70% of the amount of protein relative to the reference treatment in a change-over design. Diets contained casein, wheat gluten meal, and potato protein as protein sources and were fed at 2.8 times the maintenance requirement for energy, equally divided over two meals per day. Low and high BW pigs weighed 53 kg and 63 kg, respectively, at the start of the adaptation period prior to the first balance period. Pigs were housed in metabolic cages from 11 d prior to the start of the first balance period. Results for N-metabolism were analyzed statistically using ANOVA with sequence of dietary treatment, birth weight, breeding value for protein deposition, dietary protein supply, and their interactions as factors.

Results and discussion

N-balance parameters for the experimental treatments are shown in Table 1. A high breeding value for protein deposition reduced urinary N-excretion and increased N-retention and N-efficiency (all $P < 0.05$). Restriction of the dietary protein supply reduced urinary N-excretion to a higher extent in pigs with a low EBV compared to pigs with a high EBV for protein deposition. Results also show that the difference in N-efficiency (% of N-intake) between pigs fed the two dietary treatments was higher in pigs with a low EBV (5%) than pigs with a high EBV (1.1%) for protein deposition.

N-retention and N-efficiency were similar in pigs with a high and low EBV when fed the protein restricted diet but higher in pigs with a high EBV than pigs with a low EBV when fed a protein adequate diet. This suggests that pigs with a high EBV for protein deposition may have higher requirements for dietary protein and amino acids compared to pigs with a low EBV. The difference in EBV between the high and low EBV pigs was 13 g/d. On the protein adequate treatment, N-retention in the high and low EBV pigs was 28.3 and 26.1 g/d respectively, equivalent to a protein deposition of 177 and 163 g/d, respectively. Thus, the expected difference in protein deposition between pigs with a high and low EBV for protein deposition was realized when they were fed the protein adequate treatment. In conclusion, a higher EBV for protein deposition was associated with superior N-efficiency, providing evidence that genetics plays a role in N-efficiency.

Acknowledgement

This study is part of the Feed-a-Gene Project, funded from the European Union's H2020 Program under grant agreement no. 633531.

References

1. Ferket PR et al. Nutritional strategies to reduce environmental emissions from nonruminants. *Journal of Animal Science*. 2002;80.

2. Pomar C et al. The impact of daily multiphase feeding on animal performance, body composition, nitrogen and phosphorous excretions, and feed costs in growing-finishing pigs. *Animal*. 2014;8.

3. Gilbert H et al. Review: Divergent selection for residual feed intake in the growing pig. *Animal*. 2017;11:9.

Table 1: Effects of estimated breeding value (EBV) for protein deposition and dietary protein supply (adequate vs. restricted) on N-balance parameters (g/d) and N-efficiency (% of N-intake) in growing pigs

	Adequate			Restricted		*P*-values[1]		
	High EBV	**Low EBV**		**High EBV**	**Low EBV**	**EBV**	**Diet**	**EBV*Diet**
N-intake	51.55			37.69		NS	***	NS
N-feces			3.84			NS	NS	NS
N-urine	19.08	22.23		12.45	12.85	**	***	**
N-retention	28.26	26.07		21.15	21.03	NS	***	*
N-efficiency	55.2	50.2		56.3	55.2	**	**	**

[1] *P*-values:
* $P < 0.1$
** $P < 0.05$
*** $P < 0.001$

Surveillance of microbial virulence genes in over 5000 pigs reveals associations between gene abundance, health status and production stage

E. Galbraith[1], MS; K. Leistikow[1], BS; S. Son[1], BS, MBA; J. Lee[2], MS; M. King[1], PhD
[1]Microbial Discovery Group, Franklin, Wisconsin; [2]United Animal Health, Sheridan, Indiana

Introduction

Understanding relationships between enteric pathogen abundance, health status and production stage is crucial in development of strategies to prevent disease, reduce morbidity and mortality, and improve the efficiency of pig production. Here, a rectal swab surveillance expands upon an earlier pilot study[1] utilizing a quantitative PCR approach to characterize the enteric pathogen gene profiles of pigs from over 200 flows between 2016 and 2019. This extended analysis aimed to confirm associations between health status or production stage and genes from *E coli, C perfringens, C difficile* and *Salmonella,* and to identify new patterns of association with additional gene targets.

Materials and methods

Rectal swabs were obtained from 5020 pigs classified as healthy or sick across production stages, including sows, newborn and late lactation piglets, nursery, and grow finish animals. Genomic DNA and total RNA were extracted from swab samples and tested with a panel of qPCR assays for genus- or species-specific marker and virulence genes unique to bacterial, viral, and fungal pathogens of interest. Gene data were log-transformed, then one-way ANOVA followed by Tukey's pairwise comparisons were performed to assess differences in gene quantity between healthy pigs and those exhibiting disease symptoms, and between pigs at different stages of production.

Results and discussion

In this meta-analysis, sick pigs harbored higher quantities of multiple opportunistic bacterial- and fungal-associated genes. Markers found in enterotoxigenic (*K88, K99, F18, LT, STa*), enteropathogenic (*eaeA*) and shigatoxigenic (*stx2*) *E coli, Salmonella* (*invA*) and *Clostridium* species (*cpa, cpb2, tcdA,* and *tcdB*) were elevated in sick pigs ($P < 0.05$). *L intracellularis,* detected frequently in this surveillance, was found in higher quantities in sick pigs ($P < 0.05$), supporting previous evidence of gut microbiome disruption by the ileitis-associated pathogen.[2] Elevation of *Aspergillus* gene quantities in sick pigs ($P < 0.05$) alludes to a possible role for fungi in gastrointestinal disease. A stage-associated progression was observed in several microbial populations. *C perfringens* and *C difficile* toxin genes were most abundant during the first week of life ($P < 0.05$) and decreased together with age. In contrast, a *Campylobacter coli* marker increased in the neonatal piglet, reaching greatest quantities just before weaning. *Salmonella* genes were detected in highest quantities during the first few weeks of the nursery phase, and several enterotoxigenic *E coli* genes *LT, STa, K99* and *F18,* correlated uniquely to nursery animals several weeks post-wean ($P < 0.05$). *L intracellularis* and rotavirus genes lacked clear patterns of association with stage, and a rise in *Aspergillus* gene quantity may reflect increasing feed exposure. Surveillance gene data provide producers with a deeper understanding of the complex microbial interactions connected to pig development and health.

References

1. Leistikow K et al. Meta-analysis of enteric pathogen genes identified in 122 swine flows across the United States. Proceedings AASV Annual Meeting. 2019;196.

2. Borewicz, KA et al. Changes in the porcine intestinal microbiome in response to infection with *Salmonella enterica* and *Lawsonia intracellularis. PLoS ONE.* 2015; 10(10): e0139106.

Evaluation of the effect of a Sequivity® porcine circovirus 3 vaccine on sow farrowing performance and piglet survivability

T. Specht[1], DVM; A. Johnson[2], DVM; E. Guzman[2], MS; K. Vonnahme[2], PhD;
D. Amodie[2], MS; L. Galina-Pantoja[2], DVM, PhD
[1]Heimerl Farms, Johnstown, Ohio; [2]Zoetis, Parsippany, New Jersey

Introduction

Porcine circovirus 3 (PCV3) has been identified as a cause of reproductive failure, encephalitis and myocarditis in perinatal piglets, porcine dermatitis and nephropathy syndrome, and periarteritis in swine in the United States.[1] While most infections are subclinical, PCV3 appears to be relevant in occasional cases of mummies, abortions, myocarditis and vasculitis.[2] The Sequivity® PCV3 vaccine has been used by several swine producers who are concerned that PCV3 is causing problems in their systems; however, to our knowledge, no head to head study has shown a benefit of vaccinated animals over a non-vaccinated control group.

The objective of this study was to evaluate the effectiveness of a Sequivity® PCV3 vaccine to improve sow reproductive performance and piglet survivability in a farm with clinical issues suspected to be attributed to PCV3.

Materials and methods

This study was conducted in a 2400-head, mixed parity sow farm that had been previously confirmed positive for PCV3 via real-time polymerase chain reaction (RT-PCR) run on processing fluids. Within a parity (i.e. first parity, second through fourth parity, or fifth through eighth parity), all breeding females on site were randomly assigned into one of two treatment groups, T1: phosphate buffered saline (PBS) or T2: Sequivity® PCV3. Treatments were administered to all breeding females on site according to their treatment group on a single day, which was determined as day 0 of the study. For gilt entries after the study began (which occurred on study days 23, 93, and 183), all incoming gilts were randomly assigned to either T1 or T2, and treatments were administered at arrival to the sow farm. Within the farrowing house, after the study began, cross-fostering was only allowed within sows of the same treatment group and when possible, within the same parity group.

Performance data were collected for 34 weeks via MetaFarms; the following key performance indicators (KPIs) were evaluated: average mummified, average stillborn, pre-wean mortality %, % females bred < 7 days, aborts, farrowing rate, average liveborn, pigs weaned/sow, and female mortality.

Statistical analyses

Key production indicators (KPIs) were monitored over time through weekly reports, and statistical process control (SPC) charting along with descriptive statistics were performed for each. All the analyses were conducted using Minitab™ 18.

Results and discussion

Results from the descriptive statistics analysis are displayed in Table 1. Under the conditions in this study, KPIs showed that performance observed was similar between non-vaccinated sows and sows vaccinated with the Sequivity® PCV3 vaccine. These results are consistent with the SPC charting results, in which KPIs were found to be similar between the two treatment groups. While more work should be done to understand the role PCV3 plays in causing disease—both in sows and growing pigs—for the sow farm investigated in this project, no benefit was found in the use of a Sequivity® PCV3 vaccine.

References

1. Arruda, B et al. PCV3, a newly discovered swine pathogen. 50th Annual Meeting of the American Association of Swine Veterinarians, 2019.

2. Rovira, A et al. Frequency of PCV3 detection and association with clinical signs and lesions. Transboundary and Emerging Diseases Committee, 50th Annual Meeting of the American Association of Swine Veterinarians, 2019.

Table 1: Sow reproductive performance and piglet survival KPIs

Category	KPI	Mean ± SD[1]		Median (range)	
		T01: PBS	T02: Sequivity® PCV3	T01: PBS	T02: Sequivity® PCV3
Sow reproductive performance	Mummies (mean), n	0.3 ± 0.1	0.3 ± 0.1	0.3 (0.1 - 0.6)	0.3 (0.1 - 0.6)
	Stillborn (mean), n	0.9 ± 0.2	0.9 ± 0.2	0.9 (0.5 - 1.6)	0.9 (0.6 - 1.5)
	Females bred <7d, %	90.4 ± 12.4	90.3 ± 11.6	96.3 (53.5 - 100)	95.8 (65.1 - 100)
	Abortions, n	1.8 ± 1.5	1.5 ± 0.0	1.5 (0.0 - 6)	2.0 (0.0 - 3)
	Farrowing rate, %	92.8 ± 6.2	93.3 ± 5.5	95.4 (76.4 - 100)	93.4 (80.4 - 100)
	Liveborn (mean), n	14.0 ± 0.5	14.1 ± 0.6	14.1 (13.1 - 14.8)	14.2 (13.0 - 15.5)
	Sow mortality, %	1.1 ± 1.4	1.0 ± 1.2	0.0 (0.0 - 5.0)	0.0 (0.0 - 4.8)
Piglet survivability	Prewean mortality, %	10.1 ± 2.5	9.9 ± 2.2	9.5 (5.4 - 15.8)	9.9 (4.7 - 15.9)
	Pigs weaned/sow, n	12.7 ± 0.5	12.8 ± 0.4	12.8 (11.7 - 13.4)	12.9 (11.7 - 13.4)

[1] SD = standard deviation

Farmers under fire

Kay Johnson Smith, President and CEO
Animal Agriculture Alliance

Activist campaigns targeting modern farming and ranching continue to increase in both number and intensity. Legislative and ballot initiatives, negative public relations campaigns utilizing "undercover" videos, paid media stories blaming the COVID-19 pandemic on animal agriculture, mass on-farm protests, theft of animals, targeting company executives at their homes and public shaming of companies through activist scorecards such as the World Animal Protection report "Quit Stalling" or the "Business Benchmark for Farm Animal Welfare"[1] are just a few of the tactics used to try to create distrust among consumers and to intimidate farms, retailers, restaurants, food service companies, investors in agricultural and food businesses, policy makers and even veterinarians.

These pressure campaigns have driven many food companies to make supplier policies about farm animal housing systems or the use of antibiotics, hormones or GM feed ingredients in order to distinguish themselves in the marketplace and/or to protect their brand. Likewise, many policy makers – especially in states that allow ballot initiatives - are under pressure to pass legislation restricting production practices and even the sale of animal protein products unless they meet the same requirements in the legislation. Even churches are not immune to activist campaigns as activists cherry-picked biblical scriptures or take them out of context, creating guilt for eating meat or buying products from large companies. Campaigns such as the Humane Society of the US's Every Living Thing[2] campaign works to drive churches to develop policies to guide their congregations on food and animal care matters, of course encouraging them to reduce eating meat or forgo it altogether.

In January 2019, the EAT-Lancet Commission on Food, Plant, Health encouraged people to drastically cut their intake of animal protein to improve their health and to save the planet by reducing greenhouse gas emissions.[3] This led to numerous subsequent initiatives and animal rights activists using this as another confirmation to support their anti-meat agendas. In 2020, the well-known publication *The Guardian* received hundreds of thousands of dollars[4] to promote false and misleading stories about animal agriculture, including linking Covid-19 and other pandemics to so-called "factory farming" while others such as *The Hill* have animal rights activists as contributing "journalists".[5]

Even worse, newer activist groups have been utilizing more extreme intimidating and illegal tactics than the US has seen in the past 15 years since the passage of the Animal Enterprise Terrorism Act in 2006.[6] Much like the violent protestors causing unrest in several large cities in recent months, these animal rights activists feel justified in disrupting business on farms and processing plants through mass protests, trespass, theft of animals, blocking gates by chaining themselves to them, targeting employees to install malware on computer systems and even wiretapping – all with total disregard to biosecurity and the health and safety of the animals. Some have gone as far as targeting farm owners and others at their personal homes.[7]

The activists use these campaigns to drive their anti-meat/anti-animal agriculture agenda, all the while fundraising on every conflict - to the tune of hundreds of millions of dollars annually. All tactics noted above are accompanied with a Donate Now pop up or button on their websites, and in particular the use of videos has proven effective. As the founder of extremist group Mercy for Animals said, "The lifeblood of Mercy For Animals is our undercover investigations of animal agriculture."[8] Animal rights activism in the US is big business., with just the top 30 or so organization alone (out of hundreds) bringing in over $500 million annually. Of that, very little supports direct animal care while most goes toward campaigns, staff salaries/benefits and fundraising.[9]

The Animal Agriculture Alliance is a unique nonprofit organization based in Arlington, VA (USA) that has monitored and reported on animal rights activists and their threats to animal agriculture since 1987. Insights about these groups and their campaign tactics will be shared, as well as recommendations for steps to take to help protect farms, processing facilities and the companies and individuals who work with these businesses.

What can you do?

- Ensure all employees hired are legitimately there to work for you; check references; ensure seasoned staff work with new hires closely and monitor for "red flags".
- Thoroughly train employees on your animal care expectations and proper handling; have all employees sign agreements to follow your guidelines and report any concerns of animal mistreatment.
- Implement a culture of care to ensure proper animal handling by professionalizing related jobs.
- Know your customers. Engage all along the supply chain.
- Ensure farms cultivate their relationships with their veterinarian and local law enforcement.
- Help consumer-facing brands understand the issues, as well as the risks of negotiating with animal rights organizations.
- Establish yourself as a resource for information about farm animal care with your legislators (at all levels), public officials, church leaders and community leaders – before issues arise.
- Become media trained and be willing to talk to media to help explain complex issues.

Take away the mystery

- Have an online presence: Website, Facebook, Twitter, Instagram, Pinterest – ensure customers and neighbors can find you online – otherwise, they will see only the activists' portrayal of your farm/business.
- Share existing resources from credible groups on hot topic issues on social media or with your network. The Animal Agriculture Alliance website offers numerous free resources on farm animal welfare, advancements in animal agriculture sustainability, antibiotic use and animal nutrition, as well as infographics on complex issues that can be easily shared.[10]
- Host an open house or provide virtual tours.
- Participate in community venues (Lions, Rotary, Garden, Chamber of Commerce, etc.) to talk about your business and answer questions about food and farming.

Important Animal Agriculture Alliance resources

- **Farm animal welfare programs by species** – convenient access to all major US farm animal species' animal care guidelines available on the Alliance website, **www.animalagalliance.org**. See Figure 1.
- ***Sustainability Impact Report*** - to share with policy makers, community leaders to demonstrate agriculture's commitment to animal care, food safety, the environment and continuous improvement. Available on the Alliance website, **www.animalagalliance.org**. See Figure 2.
- **Meat Matters materials and campaign** – designed to counter various anti-meat campaigns. Includes poster, white papers on nutrition and the environment and a Proud Omnivore Pledge. Available on the Alliance website, **www.animalagalliance.org**. See Figure 3.
- **Talking points and infographics on major "hot topic" issues** – designed to help talk about complex issues and to share on social media. Available on the Alliance website, **www.animalagalliance.org**. See Figure 4.
- **Faith outreach toolkit and *What Would Jesus Really Eat?* book** – resources designed to help explain biblical support for meat eating from an ethical and religious perspective. Toolkit available to Alliance members and book available for purchase by anyone on Alliance website, **www.animalagalliance.org**. See Figure 5.
- **College Aggies Online program** – college student "ag-vocacy" training program. More information can be found on Alliance website, **www.animalagalliance.org**. See Figure 6.
- **Animal Ag Allies** – farmer, rancher, veterinarian "ag-vocacy training program. More information can be found on Alliance website, **www.animalagalliance.org**. See Figure 7.
- *Alliance Link* free monthly e-newsletter. Sign up on Alliance website, **www.animalagalliance.org**.
- **Alliance social media outreach** – all stakeholders are invited to follow the Alliance. More information can be found on the Alliance website, **www.animalagalliance.org**.

References

1. Business Benchmark on Farm Animal Welfare 2019 report. 2019. Available at **https://www.bbfaw.com/** Last accessed November 12, 2020.

2. Every Living Thing website. 2015. Available online at **https://www.everylivingthing.com/** Last accessed November 12, 2020.

3. Eat Lancet Commission on Food, Plant, Health. 2019. Available online at: **https://eatforum.org/eat-lancet-commission/** Last accessed November 25, 2020.

4. Open Philanthropy Project grants to *The Guardian*. 2017-2020. Available online at: **https://www.openphilanthropy.org/giving/grants?field_focus_area_target_id_selective=531** Last accessed November 25, 2020.

5. *The Hill* Opinion Contributor Gene Baur, also President and co-founder of animal rights activist group Farm Sanctuary. 2020. **https://thehill.com/opinion/energy-environment/486398-coronavirus-and-the-karmic-interconnectedness-of-humans-animals Last accessed November 25**, 2020.

6. US Government Printing Office November 27, 2006. Available online at: **https://www.congress.gov/109/plaws/publ374/PLAW-109publ374.htm** Last accessed November 12, 2020.

7. The Smithfield Times, August 18, 2020. Available online at: **https://www.smithfieldtimes.com/2020/08/18/vegan-protestors-target-town-ceo/** Last Accessed November 27, 2020.

8. National Conference to End Factory Farming. 2011. Available online at: **https://www.progressivedairy.com/topics/management/undercover-video-scandal-why-it-could-happen-to-you** Last accessed November 12, 2020.

9. Help Pet Shelters. 2020. Available online at: **https://www.helppetshelters.com/charity-profiles/** Last accessed November 12, 3030.

10. Animal Agriculture Alliance. 2020. Available at: **https://animalagalliance.org/** Last accessed November 27, 2020.

Figure 1: Animal welfare program by species.

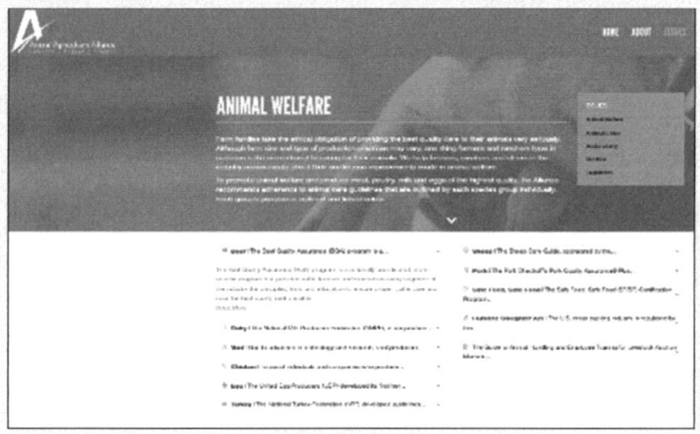

Figure 2: Sustainability Impact report.

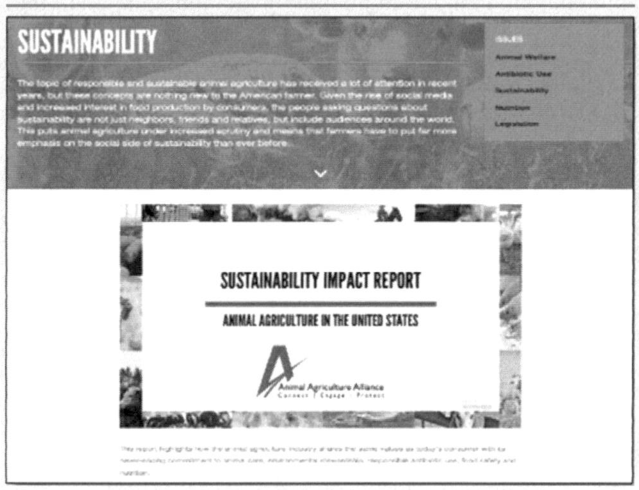

Figure 3: Meat matters materials.

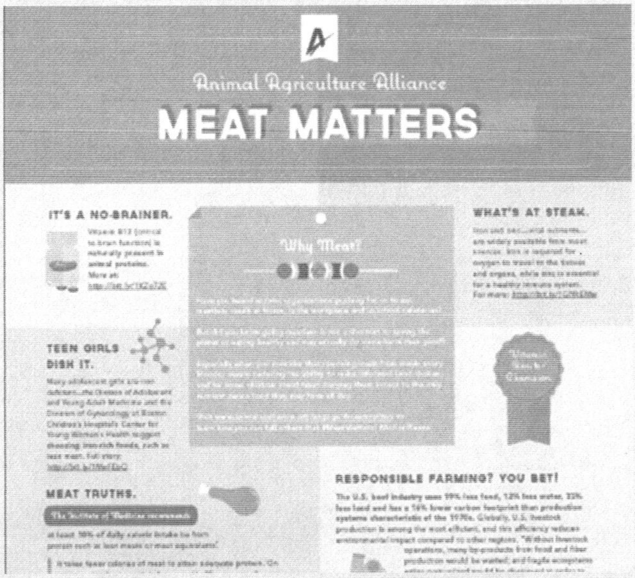

Figure 4: Talking points and infographics on major "hot topic" issues.

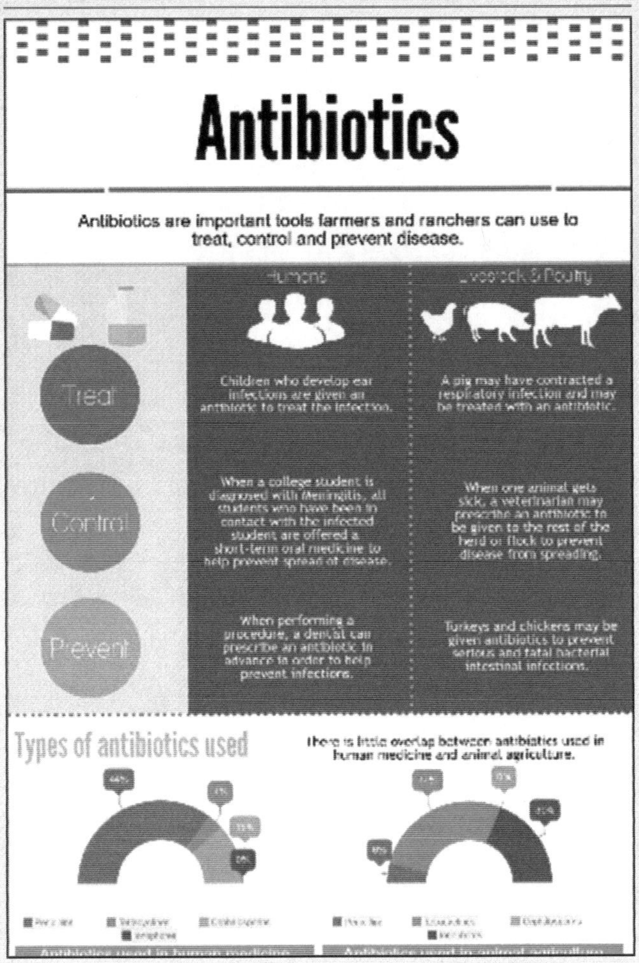

Figure 5: Faith outreach toolkit.

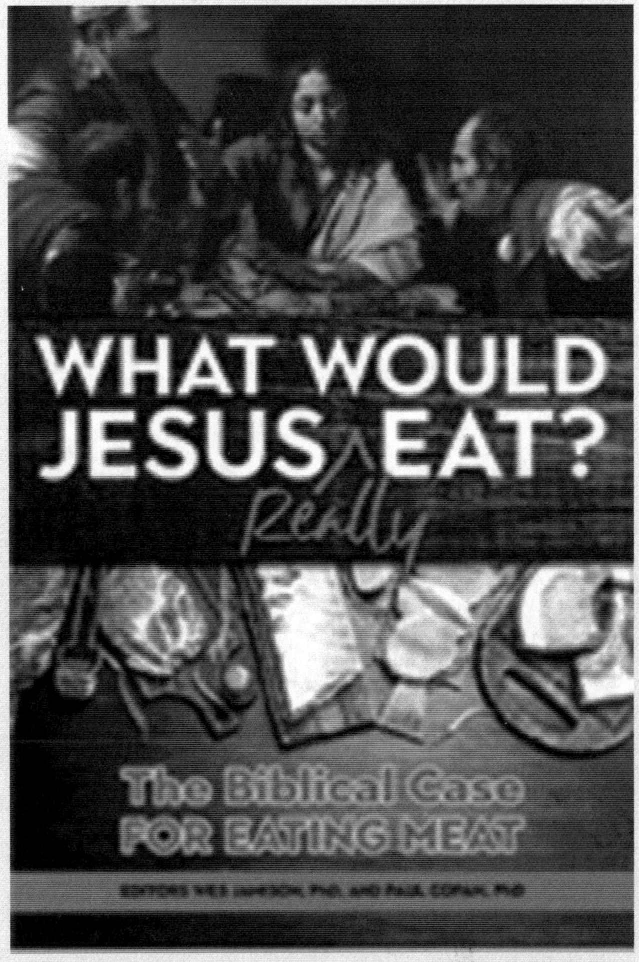

Figure 6: College Aggies Online, a college student ag-vocacy training program available at **animalagalliance.org.**

Figure 7: Animal Ag Allies, a farmer, rancher, and veterinarian ag-vocacy training program available at **animalagalliance.org.**

Sustainable intensification: how to produce more animal protein without depleting natural resources

Frank Mitloehner, PhD
Department of Animal Science, University of California, Davis, California

The global livestock production is projected to double the current rate by 2050[6] and the majority of this growth will be occurring in the developing world (Wood et al., 2006). The assessment of the holistic impacts of food animals in the context of global and regional environmental policy and food security becomes imperative. Much of the growth in the global livestock sector will occur in areas that are currently forested (i.e., parts of South America, South East Asia, Africa). It has been well established that significant reductions of carbon sequestering forests will have large effects on global climate change.

Livestock production in most countries of the developed world (e.g., United States and Europe) has a relatively small GHG contribution within the overall carbon portfolios, dwarfed by large transportation, energy, and other industry sectors. In contrast, livestock production in the developing world can be a dominant contributor to a country's GHG portfolio, due to the developing world's significantly smaller transportation and energy sectors. In the United States, transportation accounts for at least 27% of total anthropogenic GHG emissions compared to roughly 9% for all of agriculture, which includes 3.9% associated with livestock production. However, in countries like Paraguay, the trend is likely reversed because of Paraguay's much smaller transportation and energy sectors, and a relatively large livestock sector, which might contribute to more than 50% of that county's carbon footprint.

The fact that land-use changes associated with livestock (i.e., forested land converted to pasture or cropland used for feed production) are a significant source of anthropogenic GHGs in Latin America and other parts of the developing world is apparent. However, it is likely that any kind of land-use change from the original forestland will lead to great increases in global warming. FAO (2006) attributes almost half of the climate-change impact associated with livestock to the change of land-use patterns. Latin America has the greatest pool of 'unused but suitable' land that is currently covered by forests but could be turned into agricultural crop or livestock production.[1] In 2000, Latin America had 203 million hectares arable land in use and 863 million hectares of unused land suitable for cropland (19% in use). [1] Over the same time span, developed countries had 387 million hectares arable land in use and 487 million hectares of unused land suitable for cropland and livestock (44% in use). [1] Transformation of land from forest to agriculture has occurred in the developed countries centuries ago to make way for industrialization and general societal wealth. Not surprisingly, numerous developing countries are currently attempting to develop their economies by turning economically marginal land into production.

The United States and most other developed countries have not experienced significant land-use change practices around livestock production within the last few decades. Instead, over the last 25 years forestland has increased by approximately 25% in the United States and livestock production has been intensified (concentrated geographically), thus reducing its geographical footprint. Modern livestock production has experienced a marked improvement of efficiencies, leading to significantly decreased numbers of animals to produce a given amount product that satisfies the nutritional demands by society.[2] Intensification of livestock production provides large opportunities for climate change mitigation and can reduce greenhouse gas emissions from deforestation, thus becoming a long-term solution to a more sustainable livestock production.

Comparing numerous international reports[3-5] shows large agreement with respect to emission predictions from most livestock related categories. There is general consensus that as a direct GHG category, enteric fermentation in ruminants and manure management are the most important categories within livestock production. Categories like on-farm fuel use or feed production are dwarfed by emissions coming from the animals and their manure.

Overall, growing demands for animal protein could strongly increase GHG emissions from agriculture. However, knowledge exists to improve efficiencies in livestock production, which dramatically reduces GHG per unit of production. What is called for is sustainable intensification in animal agriculture, coupled with technology transfers from developed to developing countries, to supply a growing demand for animal protein using sustainable and modern production practices.

References

1. Bruinsma, J. (2003). World Agriculture: Towards 2015/2030, an FAO Perspective. Earthscan, London, UK.

2. Capper, J. L., Cady, R. A., and Bauman, D. E. (2009). The Environmental Impact of Dairy Production: 1944 compared with 2007. *Journal of Animal Science*. 1910 10.2527/jas.2009-1781.

3. CEC. (2005). Inventory of California Greenhouse Gas Emissions and Sinks: 1990 to 2002 Update. California Air Resources Board. **http://www.climatechange.ca.gov/inventory/index.html**.

4. EPA. Holtkamp, J., Hayano, D., Irvine, A., John, G., Munds-Dry, O., Newland, T., Snodgrass, S., and Williams, M. (2006). Inventory of U.S. Greenhouse Gases and Sinks: 1996–2006. Environmental Protection Agency. **http://www.epa.gov/climatechange/emissions/downloads/08_Annex_1-7.pdf.**

5. FAO. Steinfeld, H., Gerber, P., Wassenaar, T., Castel, V., Rosales, M., and de Haan, C. (2006). Livestock's Long Shadow. Food and Agriculture Organization of the United Nations. **http://www.fao.org/docrep/010/a0701e/a0701e00.HTM**.

6. Smith, P., Martino, D., Cai, Z. C., Gwary, D., Janzen, H., Kumar, P., McCarl, B., Ogle, S., O'Mara, F., Rice, C., Scholes, B., Sirotenko, O. (2007). Policy and Technological Constraints to Implementation of Greenhouse Gas Mitigation Options in Agriculture. *Agriculture Ecosystem and Environment*. 118, 6–28.

Your right and responsibility to be well

Elizabeth B. Strand, Ph.D., LCSW Endowed All Creatures Great and Small Clinical Associate Professor, Founding Director, Veterinary Social Work
University of Tennessee College of Veterinary Medicine and Social Work Knoxville, Tennessee

Introduction

A right is a freedom and a responsibility is a duty. As esteemed veterinary professionals you have the ultimate freedom to be well and you also have the duty to ensure it for the common good of the profession. This is not easy because the practice of veterinary medicine is associated with high levels of burnout, lower wellbeing, and psychological distress, especially for younger veterinarians.[1,2] This comes from interacting with owners of animals, animal disease outbreaks, long work hours, student debt, conflict with colleagues, animal death, personality traits, moral stress, and negative public perception.[3-5] In fact veterinarians report higher levels of burn out than physicians regardless of working fewer hours than physicians, and also higher rates of suicidal thoughts and behaviors.[2] In the face of this it is helpful to consider wellbeing as a competency that takes skills and work to maintain. This brief session will offer three categories to attend to in developing this competency:
1) Thoughts and valence 2) Breath, 3) Connection.

Thoughts and valence

Becoming aware of your habits of thought (e.g. "thinking errors") that maintain stress and identifying valence (e.g. "emotion") are associated with increased resilience and mental health.[6,7] This is a skill that takes self-awareness- knowing what is going on inside yourself. Taking time every day to notice the quality of your thoughts and emotions is a necessary skill to master in practicing your rights and responsibilities to be well.

Common thinking errors

Learning some key thinking errors and recognizing them when they occur is a first step. Here is a brief overview of some common thinking errors.[8] *All-or-nothing* thinking is seeing things as being black or white, a success or a failure, good or bad. *Overgeneralizing* is taking one particular event and generalizing it to the rest of your life. *Filtering-out-the-positive* is zooming our attention in on only the bad aspects of a situation and failing to give at least equal attention to the positive aspects. *Mind-reading* is assuming you know what's going on in someone else's mind. *Catastrophizing* is thinking things are much worse than they actually are. *Emotional reasoning* is assuming something is true because you feel it is true. *Labeling* is putting people and experiences into categories that are often based on isolated and often negative incidents. Labeling robs us of noticing improvements in ourselves and others. *Fortune-telling* is assuming you know what will happen in the future. *Personalization* is making a situation about you when it does not actually involve you. *Unreal ideal* is making unfair comparisons about ourselves and other people that can ruin our motivation.

Valence

Identifying emotions that you have throughout the day is an essential step to managing emotions so they support your resilience and mental health instead of erode them. Most of us are not taught to recognize and communicate our feelings so this can be a challenge at first. Emotions are important because they point to our needs. If we feel guilty we may need to make amends. If we feel sad we may need to work to forgive ourself or someone else. If we feel angry we may need to set a boundary somewhere in our life. If we feel happy we may need to recognize what is making us happy and work to have more of the activity in our life. If we feel grateful we may need to express that gratitude to someone so that we strengthen our connections with others and practice savoring the gratitude which helps with resilience.[9] Learning words to describe emotions is part of attaining granularity in emotional literacy. The better and more precisely we can identify feelings the better we are at managing them. An emotions wheel[10] is a good place to start the process of emotional literacy. Emotional literacy is the ability to state simply " I feel ...." and then actually experience the feeling in your body.[7]

Breath

All effective stress management strategies include breathing as an important part of the technique. The breath is the one part of the autonomic nervous system that we both have control over and that also continues to function without our conscious control. It also has the most priority in our biological system. If you were to simultaneously stub your toe and choke on a pork tenderloin your brain would prioritize your choking over the pain in your toe because the breath is so necessary for survival. Therefore voluntarily controlling your breath is the most efficient and powerful way to manage stress, because the brain listens to what the breath says about the state of the body.[11] There are many breathing techniques that have been shown to affect the autonomic nervous systems stress response.[12] One breathing technique that has been shown to balance the parasympathetic and the sympathetic branches of the autonomic nervous system is coherent breathing. Coherent breathing is five breaths per minute equal breath in and out through the nose. This breathing technique has been effective in the treatment of anxiety depression and post-traumatic stress disorder in many populations.[13-15] Coherent breathing creates a state of physiological coherence that can be observed in heart rate variability. Coherent heart rate variability is correlated with the higher order emotions such as love, awe, peace, and appreciation.[16,17]As little as seven minutes of coherent breathing can balance the stress response very efficiently.[11] Thus learning the skill of coherent breathing and practicing it every day is an important ingredient in your right and responsibilities to be well.

Right connection

Sometimes it's hard to talk to family and friends about the work you do as a veterinarian. As we experience burnout we may notice not wanting to tell people what we do for a living because we don't want to hear their comments about our work. Perhaps we don't want them to ask us for unsolicited advice or perhaps we don't want to hear them say, "I could never do that I love animals too much." This may be especially difficult if we have had to work with distressed clients or euthanized sick or healthy animals in a particular day. Even so social connection is one of the most protective things that one can have for mental health and resilience.

Here is a series of questions that can be used to engage in a structured veterinary well-being debrief. This debrief technique can be used with a team, a colleague, mentor, or friend, or even individually to help ourselves connect with our honest human experience of veterinary work. The questions are: 1) What is keeping me up at night or what can I not stop thinking about from today or this week, 2) What did I do well in that situation? 3) What do I wish I had done differently in that situation? 4) What did I learn in that situation? 5) What do I feel grateful for in this situation or just generally right now? 6) Was there anything humorous that occurred during the management of the situation?

If a person gets to questions five or six and they are not able to identify anything they feel grateful for or anything that made them laugh, it may be an indicator that they need to ramp up self-care and perhaps consult with someone to help them work through any emotions or thoughts they are having about their experiences.

Conclusion

Recognizing thoughts and feelings, having a daily practice of coherent breathing, and engaging in the veterinary wellbeing debrief regularly are important ingredients for enjoying your right and attending to your responsibility for well-being. This effort will have positive impacts for you personally but also for the greater good of the profession. The AASV is committed to your well-being and has developed an excellent while being resource page,[18] video on well-being, and his even established a support approach called HEARD Vet. HEARD Vet started as a small group of mentors making themselves available in a support group for AASV members. HEARD Vet is a growing offering and your input and participation is what will make the difference for you and your colleagues well-being overtime. So if you're interested in contributing to HEARD Vet please reach out to Sue Schulteis (**aasv@aasv.org**,) Abbey Cannon (**canon@aasv.org**), and/or Elizabeth Strand (**estrand@utk.edu**).

References

1. Volk, J. O., Schimmack, U., Strand, E. B., Lord, L. K. & Siren, C. W. Executive summary of the Merck Animal Health Veterinary Wellbeing Study. J. Am. Vet. Med. Assoc. 252, 1231–1238 (2018).

2. Volk, J. O., Schimmack, U., Strand, E. B., Vasconcelos, J. & Siren, C. W. Executive summary of the Merck Animal Health Veterinarian Wellbeing Study II. J. Am. Vet. Med. Assoc. 256, 1237–1244 (2020).

3. Robbins, J. A., Franks, B., Weary, D. M. & von Keyserlingk, M. A. G. Awareness of ag-gag laws erodes trust in farmers and increases support for animal welfare regulations. Food Policy 61, 121–125 (2016).

4. Fernandes, J. et al. Addressing Animal Welfare through Collaborative Stakeholder Networks. Agriculture-Basel 9, (2019).

5. Moses, L., Malowney, M. J. & Wesley Boyd, J. Ethical conflict and moral distress in veterinary practice: A survey of North American veterinarians. J. Vet. Intern. Med. 32, 2115–2122 (2018).

6. Meichenbaum, D. Stress inoculation training: A resilience-engendering intervention. in Treating PTSD in military personnel: A clinical handbook (2nd ed.) (ed. Moore, Bret A. [Ed]; Penk, Walter E. [Ed]) vol. 2 136–150, Chapter xvii, 462 Pages (The Guilford Press (New York, NY, US), 2019).

7. Barrett, L. F. How Emotions Are Made: The Secret Life of the Brain. (Mariner Books, 2017).

8. Morin, A. 10 Thinking Errors That Will Crush Your Mental Strength. Psychology Today (2015).

9. Moskowitz, J. T. et al. Measuring Positive Emotion Outcomes in Positive Psychology Interventions: A Literature Review. Emot. Rev. **doi:10.1177/1754073920950811**.

10. The Gottman Institute. Printable Feeling Wheel. **https://www.gottman.com/blog/printable-feeling-wheel/** (2020).

11. Brown, R. P. & Gerbarg, P. L. The healing power of the breath: Simple techniques to reduce stress and anxiety, enhance concentration, and balance your emotions. vol. 168 168 Pages (Shambhala Publications (Boston, MA, US), 2012).

12. Zaccaro, A. et al. How Breath-Control Can Change Your Life: A Systematic Review on Psycho-Physiological Correlates of Slow Breathing. Front. Hum. Neurosci. 12, 353 (2018).

13. Gerbarg, P. L. et al. The Effect of Breathing, Movement, and Meditation on Psychological and Physical Symptoms and Inflammatory Biomarkers in Inflammatory Bowel Disease: A Randomized Controlled Trial. Inflamm. Bowel Dis. 21, 2886–2896 (2015).

14. Gerbarg, P. L., Brown, R. P. & protected], [email. Integrative treatments for masked anxiety and PTSD in highly sensitive patients. in Evidence-based herbal and nutritional treatments for anxiety in psychiatric disorders (ed. Camfield, David [Ed]; McIntyre, Erica [Ed]; Sarris, Jerome [Ed]) vol. 208 155–167, Chapter xiv, 208 Pages (Springer International Publishing (Cham, Switzerland), 2017).

15. Scott, T. M. et al. Psychological Function, Iyengar Yoga, and Coherent Breathing: A Randomized Controlled Dosing Study. J. Psychiatr. Pract. 25, 437–450 (2019).

16. McCraty, R., Atkinson, M., Tiller, W. A., Rein, G. & Watkins, A. D. The effects of emotions on short-term power spectrum analysis of heart rate variability. Am. J. Cardiol. 76, 1089–1093 (1995).

17. McCraty, R. New Frontiers in Heart Rate Variability and Social Coherence Research: Techniques, Technologies, and Implications for Improving Group Dynamics and Outcomes. Front Public Health 5, 267 (2017).

18. AASV Veterinary Wellbeing **https://www.aasv.org/Resources/Wellbeing**, Accessed Nov. 12, 2020.

Preparing for a low-zinc world

Mike Tokach; Jordan Gebhardt; Jason Woodworth; Joel DeRouchey; Bob Goodband
Applied Swine Nutrition Team, Kansas State University

Why do we use high levels of ZnO?

Since the first publication demonstrating the value of ZnO in reducing post-weaning diarrhea[1], zinc oxide use has become a core component of most nursery diet programs. When included in the diet at 1,500 to 3,000 ppm for the first 2 to 3 weeks after weaning, Zn from ZnO improves feed intake and growth performance.[2-5]

The modes of action for Zn are numerous (Table 1). Which mode of action is ultimately responsible for reduction in diarrhea and improvement in feed intake and growth rate is unknown, but is likely a combination of the multiple actions that zinc has in the pig. To improve understanding of this complex area, the modes of action can be grouped into four categories: high zinc requirement; gut protection; antibacterial; and direct impacts on feed intake.

Evidence suggests that the young, weaned pig may simply have a higher zinc requirement than previously thought. Zinc is a key component of many enzymes and metalloenzymes necessary nutrient metabolism and immunity. To achieve zinc homeostasis, high zinc needs to be fed for the pig to maintain plasma zinc at levels achieved in unweaned pigs.[6]

The gastrointestinal tract of the pig is still relatively immature at weaning in commercial production. Pharmacological levels of dietary Zn appear to aid in the development process. Feeding of high Zn concentrations reduces the releases of pro-inflammatory histamine by inhibiting proliferation and activation of mast cells.[7] Moeser et al[8] has demonstrated that mast cell activation is one of the main drivers of increased intestinal permeability and diarrhea in pigs weaned at young ages. Thus, Zn would be expected to provide more value in pigs weaned at younger pigs than older pigs due to the inhibition of mast cells. Zn also aids in maintaining normal morphology of gastrointestinal tract after weaning.[9,10] High dietary Zn also increases expression of IGF1 and its receptors, which regulate growth and differentiation of intestinal cells.[11]

Most think of pharmacological levels of zinc purely as an antibacterial. High Zn levels does prevent attachment of pathogenic bacteria in intestine.[12] Zinc also increases anti-inflammatory cytokines and tight junctions and reduces pro-inflammatory cytokines of weaned pigs.[13] The improvement in tight junction status may be responsible for the reduced intestinal permeability that has been found to prevent translocation of pathogenic bacteria, such as *E coli* and Enterococcus spp. in mesenteric lymph nodes of the small intestine.[14] Zinc does not actually kill or reduce the number of *E coli* in vivo or in vitro.[15,16] Thus, the decrease in translocation of the E coli is responsible for the reduction in mortality instead of a reduction in *E coli* itself.

Zinc also has been shown to have direct effects on feed intake by regulating secretion of brain-gut peptides that stimulate feed intake.[17]

Finally, another impact of feeding pharmacological levels of Zn is a change in the gut microbiome. Although microbiome studies often indicate that increases in microbiota diversity is associated with increased growth, dietary ZnO decreases microbiota diversity and increases the abundance of Firmicutes, such as the family Ruminococcaceae, in the colon.[18] Other recent research also found shifts in microbiota of weaned pigs to greater Firmicutes with feeding of ZnO[19]; however, it is unknown whether this is a beneficial or negative effect of ZnO.

Where do we use high levels of ZnO today?

Pharmacological levels of Zn are most commonly fed from weaning to 25 lb (12 kg) wherever it remains legal to feed high levels of Zn. In rare cases, use is extended to 30 lb (14 kg) in situation where *E coli* becomes an issue when Zn is removed from the diet. This need becomes especially important when the production system does not want to use antibiotics during this stage. Although dosages can vary, Zn from ZnO is often used at 3,000 ppm until 15 lb (7 kg) and 2,000 ppm from 15 to 25 lb (7 to 12 kg). In the rare cases that it is included in the diet from 25 to 30 lb, it is fed at 1,000 ppm.

High concentrations of Zn should not be fed for longer periods because older pigs do not garner the benefit in growth performance or immune status that is found in younger pigs. For comparison sake, the normal Zn inclusion rate for late nursery pigs is approximately 110 ppm. Thus, high Zn levels are only included in the first 15 to 20 lb of feed that a pig receives and not used in the remaining 600 to 700 lb of feed. Thus, pharmacological levels of Zn are fed in 2.5 to 3.5% of the total feed consumed by the pig.

Why are others concerned about ZnO use in nursery pig diets?

Three main concerns have been raised about feeding of pharmacological levels of ZnO: 1) antimicrobial resistance; 2) heavy metal pollution of the environment and 3) toxicity. Use of high levels of ZnO will be banned in the European Union in 2022. Their major concerns were environmental accumulation of zinc and its associations with development of antimicrobial resistance.[20]

Because genetic coupling between genes of heavy metal resistance and antibiotic resistance is sometimes associated, antimicrobial resistance has been raised as a concern. Thus, selection of bacteria for Zn resistance can lead to bacterial resistance to some antibiotics. Ciesinski et al[21] found that the percentage of *E coli* that were resistant to multiple antibiotics (multi-resistant *E coli)* was increased by feeding ZnO for 28 days after weaning. This increase did not occur when ZnO was only fed for 14 days. The greatest concern with feeding of high zinc has been the potential for selection for methicillin-resistant *Staphylococcus aureus* (MRSA) due to co-location of the methicillin-resistance gene and zinc-resistant gene in staphylococcal chromosome. Feeding of high levels of ZnO has been found to increase the prevalence and persistence of MRSA in nursery pigs.[22] Thus, minimizing the feeding duration of ZnO is important.

Environmental heavy metal concerns are often raised when discussing Zn because a majority of the pharmacological levels of zinc being fed is excreted by the pig. If high levels of Zn are fed for too long of a period or past the recommended 25 lb BW, this can become a concern. For most pig production sites around the world, this is a minor issue when Zn is used appropriately.

Belkova et al[23] commented that Denmark is often used as an example where in acid and sandy soils,[24] zinc levels had exceeded the predicted no-effect concentrations (PNECs) and was nearing the point of becoming a toxic hazard. Belkova expands that Denmark produces 32 million weaner pigs (11 times the average of other EU countries) on a very small land mass. Thus, heavy metal contamination from Zn may be an issue in their situation, but not in most countries around the world. As stated earlier, high Zn is only used in 2.5 to 3.5% of the feed that a growing pig consumes. Finally, much of the land used for crops or grazing is declining in zinc concentrations rather than increasing.

Feeding of high levels for longer than 3 to 4 weeks has been implicated with toxicity, but the only known publication on toxicity with ZnO is a situation where pigs mistakenly had 8,000 ppm ZnO mixed into their feed for 8 months.[25] The high Zn intake and low dietary Cu led to a Cu deficiency. No other long term negative effects have been detected with ZnO.[26] Feeding of other Zn sources, such as Zn carbonate can cause toxicity at high levels.[27] High intake of Zn can lower utilization of other minerals, such as copper, iron, or selenium because of overproduction of transport proteins that bind the metals. High Zn has been shown to reduces digestibility of Ca, P, Na, K, Cu.[28,29] These researchers indicate that lowering ZnO to 2,000 ppm or less in diets with high phytase may be appropriate. There is an interaction with phytase where Zn can bind to phytate molecule making it less available for the phytase. This may also result in an overestimation of available P when high levels of Zn are included in the diet when relying on high P release from phytase. Finally, there are many sources of Zn and some have impurities that must be controlled to prevent problems.

What needs to be done if we cannot use ZnO?

If the ability to use pharmacological concentrations of Zn from ZnO is lost, there are several things that a production system may need to consider to control postweaning diarrhea. The first step will be to learn from others around the world that have already begun this process. Many of the changes outlined below are from experiences learned from those that have removed or are in the process of removing pharmacological Zn from their diets.

Several management changes that should be considered are outlined in Table 2. Increasing weaning age to have a more mature digestive system must be considered. The reduction in mortality[30,31] that comes with increasing weaning age makes this an important consideration. Adam Moeser's work demonstrating that the mode of action is largely gut mediated suggests that increasing weaning age should greatly reduce the need for ZnO use. Sanitation to minimize bacteria remaining in the environment is always important; however, the importance becomes enhanced when ZnO is not available to help control the pig's response to the bacteria. Because stressors are additive in their negative impacts, reduction in other management stressors also provides benefit. Steps such as provision of creep feed or socialization of pigs prior to weaning can lessen the stress that the pig experiences at weaning. Proper temperature, ventilation, and immunization are critical steps in lessening other stressors.

Because some pigs are resistant to *E coli* or nursery feed intake patterns that make them less susceptible to problems with *E coli*, genetic changes may be necessary in some production systems if they cannot control postweaning diarrhea by other means.

Numerous dietary changes have been shown to be helpful to reduce postweaning diarrhea. A review by Pluske et al[32] provides an excellent summary of dietary changes that should be considered for optimal gut health. Many of their recommendations are included in Table 2. The dietary changes revolve around minimizing the amount of undigested material available for bacterial growth, increasing passage rate and lowering gastric pH to limit bacterial growth, improving immune function, and providing other additives that may have antibacterial effects. For example, pharmacological levels of copper and zinc are usually not additive in their response; however, if high levels of zinc are not included in the diet and pharmacological levels of copper remain legal, copper can provide benefit at growth promoting levels. Similarly, if ZnO cannot be fed at high levels, other zinc sources may provide more benefit than ZnO when the inclusion rate is limited to 500 ppm or less. Part of antibacterial effect of ZnO is thought to be due to its ability to generate hydrogen peroxide from its surface.[33] Similar to the benefit of reducing feed particle size increasing surface area and opportunities for enzymes to improve digestibility, increases in Zn surface area through lower particle size increases the efficiency of hydrogen peroxide production and antibacterial activity.[34] Some researchers[35,36] found that lower inclusion rate of porous and nanoparticles of ZnO could achieve similar performance to higher inclusion rate of normal ZnO; however, more research is required to validate these responses.

Finally, water quality and pH also should be evaluated. If pH is over 7.0, reducing pH with water acidification is often necessary to control postweaning diarrhea. Field reports from Denmark suggest lowering water pH to 5.0 may be beneficial when hemolytic *E coli* is a problem. If the water is high in sulfates or total mineral content, it can result in loose watery feces and the visual appearance of diarrhea. High sulfate water does not appear to have a major impact on pig growth; however, it may increase pig treatments and frustration in the barn due to inability to control the loose stools that occur. In these situations, obtaining an alternative water source may be the preferred solution.

What should we do?

Be proactive in educating pork producers, employees, government officials, and others on why we use Zn, the short period that high levels of Zn are fed, and the high Zn requirement of young pigs. We need to extol those with the power to legislate to not let bad policy sacrifice the welfare of the pig. Do our part in limiting the use of pharmacological level of Zn by not using high levels of Zn outside of the immediately after weaning area. This will help reduce total Zn use. Explore or support the exploration of alternatives to use of ZnO and learn from others on approaches that they are taking to reduce Zn usage.

References

1. Poulsen HHD. Zinc oxide for weaned pigs. In: Proceedings of the 40th annual meeting of the European Association for Animal Production. *EAAP Publications*, Dublin. 1989; 265-266.

2. Smith JW II et al. Effects of the interrelationship between zinc oxide and copper sulfate on growth performance of early-weaned pigs. *J Anim Sci*. 1997;75:1861-1866. **doi: 10.2527/1997.7571861x**

3. Smith JW II et al. The effects of increasing zinc oxide on growth performance of weanling pigs. *Prof Anim Sci*. 1999;14:197-200.

4. Hill GM et al. Effect of pharmacological concentrations of zinc oxide with or without the inclusion of an antibacterial agent on nursery pig performance. *J Anim Sci*. 2001;79:934-41. **doi: 10.2527/2001.794934x**

Table 1: Modes of action for ZnO

Young, weaned pig may have high zinc requirement
• Zinc is a key component of many enzymes and metalloenzymes necessary nutrient metabolism and immunity.
• Zinc homeostasis (need high Zn inclusion to maintain plasma zinc at levels achieved in unweaned pigs (Davin et al., 2013).
Protects immature gut
• Reduces the releases of pro-inflammatory histamine by inhibiting proliferation and activation of mast cells (Kim et al., 2012).
• Helps maintain normal morphology of gastrointestinal tract after weaning (Carlson et al., 1998; Lie at al., 2001).
• Increases expression of IGF1 and its receptors, which regulate growth and differentiation of intestinal cells (Li et al., 2006).
Antibacterial
• Prevents attachment of pathogenic bacteria in intestine (Starke et al., 2012).
• Zinc increases anti-inflammatory cytokines and tight junctions and reduces pro-inflammatory cytokines of weaned pigs (Peng et al, 2020).
• Reduce intestinal permeability to prevent translocation of pathogenic bacteria, such as *E coli* and *Enterococcus* spp. in mesenteric lymph nodes of the small intestine (Huang et al., 1999).
○ Zinc does not actually kill or reduce the number of *E coli* in vivo or in vitro. Some thought that the effect of zinc is systemic via blood than direct gastrointestinal effects (Jensen-Waern et al., 1998; Roselli et al., 2003).
Increases feed intake
• Regulates secretion of brain-gut peptides that stimulate feed intake – Li et al., 2010

5. Shelton NW et al. Effects of copper sulfate, tri-basic copper chloride, and zinc oxide on weanling pig performance. *J Anim Sci.* 2010;89:2440-2451.

6. Davin R et al. Effect of weaning and in-feed high doses of zinc oxide on zinc levels in different body compartments of piglets. *J Anim Phys Anim Nutr* 2013; 97 6–12. **doi: 10.1111/jpn.12046**

7. Kim JC et al. Nutrition and pathology of weaner pigs: Nutritional strategies to support barrier function in the gastrointestinal tract. *Anim Feed Sci Technol.* 2012;173:3-16

8. Moeser AJ et al. Weaning stress and gastrointestinal barrier development: Implications for lifelong gut health in pigs. *Anim Nutr.* 2017;3:313-321. **https://doi.org/10.1016/j.aninu.2017.06.003**

9. Li BT et al. Small intestinal morphology and bacterial populations in ileal digesta and feces of newly weaned pigs receiving a high dietary level of zinc oxide. *Can J Anim Sci.* 2001;81:511–6.

10. Carlson MS et al. Effect of pharmacological zinc on intestinal metallothionein concentration and morphology in the nursery pig [abstract]. *J Anim Sci.* 1998;76:Suppl 1:57.

11. Li X et al. Dietary supplementation with zinc oxide increases IGF-I and IGF-I receptor gene expression in the small intestine of weanling piglets. J Nutr. 2006;136:1786–1791, **doi.org/10.1093/jn/136.7.1786**

12. Starke I et al. Dietary zinc oxide leads to short and long term modification in the intestinal microbiota of piglets. XII International symposium on digestive physiology in pigs. May-June 2012, USA.

13. Peng P et al. The effects of dietary porous zinc oxide supplementation on growth performance, inflammatory cytokines and tight junction's gene expression in early-weaned piglets. *J. Nutr Sci Vitaminol.* 2020;66:311-318.

14. Huang SX et al. Effect of dietary zinc supplementation on Escherichia coli septicemia in weaned pigs. *Swine Health Prod.* 1999;7:109-111.

15. Jensen-Waern M et al. Dietary zinc oxide in weaned pigs – effects on performance, tissue concentrations, morphology, neutrophil functions and faecal mikroflóra. *Res Vet Sci.* 1998;64:225-31.

16. Roselli M et al. Zinc oxide protects cultured enterocytes from the damage induced by *Escherichia coli*. *J Nutr.* 2003;133:4077-82.

17. Li X et al. Mechanisms involved in the growth promotion of weaned piglets by high-level zinc oxide. *J Anim Sci Biotech.* 2010;1:59-67.

18. Rattigan R et al. Effects of reducing dietary crude protein concentration and supplementation with laminarin or zinc oxide on the faecal scores and colonic microbiota in newly weaned pigs. *J Anim Physiol Anim Nutr.* 2020;00:1–13. **doi: 10.1111/jpn.13428**

19. Wei X et al. ZnO modulates swine gut microbiota and improves growth performance of nursery pigs when combined with peptide cocktail. *Microorganisms.* 2020;8:146; **doi:10.3390/microorganisms8020146**

20. European Commission implementing Decision of 26.6.2017, C (2017) 4529 Final. 2017. **https://ec.europa.eu/health/documents/community-register/2017/20170626136754/dec_136754_en.pdf**

Table 2: Changes to consider for a low Zn world

Management changes

- Increase weaning age
- Ensure adequate sanitation
- Reduce management stressors

Genetic changes

- Use of genetics with *E coli* resistance or more robust GI tract

Dietary changes

- Reduce dietary crude protein as much as possible with use of feed grade amino acids maintaining a maximum digestible lysine to crude protein ratio of 6.35%.
    - Dietary lysine will likely need to be below the requirement of most modern genotypes.
- Increase amino acid ratios relative to lysine when diets are below the lysine requirement. For example: 60% Met&Cys:Lys; 21% Trp:Lys; 65% Thr:Lys; 72% Val:Lys; 55% Ile:Lys.
- Use highly digestible protein sources immediately after weaning.
- Include high levels of dietary phytase to improve protein digestibility and lower calcium content
- Minimize soluble nonstarch polysaccharide levels
- Include a source of insoluble nonstarch polysaccharides (ex. lignocellulose, oat hulls, coarse wheat bran)
- Consider grinding grains coarser
- Consider inclusion of xylanase
- Decrease buffering capacity of the diet by lowing calcium levels
- Review vitamin, antioxidants, and fatty acid levels required for optimal immune function
- Review inclusion of other feed additives if lower levels of ZnO must be fed
    - Add copper to near the legal limit
    - Consider another zinc source that may have improved efficacy relative to ZnO when added at lower levels (ex. Zn nanoparticles)
    - Inclusion of acidifiers, short or medium chain fatty acids, or other antibacterial additives

Water quality and pH

- Lower water pH if over 7.0
- High mineral water will increase visual loose stools

21. Ciesinski L et al. High dietary zinc feeding promotes persistence of multi-resistant E.coli in the swine gut. *PLoS ONE.* 2018;13(1):e0191660. **https://doi.org/10.1371/journal.pone.0191660**

22. Slifierz MJ et al. Zinc oxide therapy increases prevalence and persistence of methicillin-resistant staphylococcus aureus in pigs: A randomized controlled trial. *Zoonoses Public Health,* 2015;62:301–308. **doi: 10.1111/zph.12150**

23. Belkova, J., E. Vaclavkova, and P. Horky. 2018. The effect of zinc in the diet for weaned piglets and the possibilities of substitution – Review. *Research in Pig Breeding* 12(2):1-5.

24. Burch D. Zinc oxide – for the prevention of post-weaning enteritis/diarrhoea – an EU update. [Online] available at **https://www.pig333.com/articles/eu-zinc-oxide-forthe-prevention-of-post-weaning-enteritisdiarrhoea_12251/**

25. Komatsu T et al. Chronic pancreatitis in farmed pigs fed excessive zinc oxide. *J Vet Diag Inv.* 2020;1-6. **DOI: 10.1177/1040638720944368**

26. Hill GM and ER Miller. Effect of dietary zinc levels on the growth and development of the gilt. *J Anim Sci.* 1983;57:106-113.

27. NRC. Nutrient requirements of swine. 11th ed. Washington: *Natl. Acad. Press*; 2012.

28. Walk CL et al. Evaluation of the effects of pharmacological zinc oxide and phosphorus source on weaned piglet growth performance, plasma minerals and mineral digestibility. *Anim.* 2015:9:1145–1152. **doi:10.1017/S175173111500035X**

29. Blavi L et al. Effects of zinc oxide and microbial phytase on digestibility of calcium and phosphorus in maize-based diets fed to growing pigs. *J. Anim. Sci.* 2017;95:847–854. **doi:10.2527/jas2016.1149**

30. Main RG et al. Increasing weaning age improves pig performance in a multisite production system. *J Anim Sci.* 2004;82:1499–1507. **http://doi.org/10.2527/2004.8251499x**

31. Rosero DS et al. Influence of wean age and disease challenge on progeny life-time performance. 2016 Leman Swine Conference. **https://drive.google.com/file/d/0B8SHqmX_vRmVWnFLeG45WDV2RFE/view**

32. Pluske JR et al. Gastrointestinal tract (gut) health in the young pig. *Anim Nutr.* 2018;4:187-196. **https://doi.org/10.1016/j.aninu.2017.12.004**

33. Sawai J et al. Hydrogen peroxide as an antibacterial factor in zinc oxide powder slurry. *J Ferment Bioengineering.* 1998;86:521–2. **http://dx.doi.org/10.1016/S0922-338X(98)80165-7.**

34. Yamamoto, O. Influence of particle size on the antibacterial activity of zinc oxide. *Internat J Inorganic Materials.* 2001;3:643–6.

35. Wang C et al. Zinc oxide nanoparticles as a substitute for zinc oxide or colistin sulfate: Effects on growth, serum enzymes, zinc deposition, intestinal morphology and epithelial barrier in weaned piglets. *PLoS One.* 2017;12:e0181136. **doi:10.1371/journal.pone.0181136**

36. Long L et al. Comparison of porous and nano zinc oxide for replacing high dose dietary regular zinc oxide in weaning piglets. *PLoS ONE* 2017;12:e0182550. **https://doi.org/10.1371/journal.pone.0182550**

PRRS CLAMP: molecular diagnostic tools to distinguish wild-type and vaccine strains of PRRSV

Phillip Gauger, DVM, PhD; Karen Harmon, PhD
Iowa State University Veterinary Diagnostic Laboratory,
Department of Veterinary Diagnostic and Production Animal Medicine, Ames, Iowa

Introduction

Porcine reproductive and respiratory syndrome virus (PRRSV) continues to challenge the swine industry with production losses due to respiratory disease, abortions, excess morbidity and elevated mortality.[1] In addition, increased production costs can be attributed to vaccine expense, treatment costs and diagnostic fees.[2,3] The virus has become more difficult to control in spite of aggressive biosecurity methods, multiple commercial vaccines, live virus inoculation methods and the implementation of vaccination protocols.[4,5] Modern swine production practices, virus transmission through pig movement and rapid evolution and genetic diversity of the virus have also contributed to the difficulties associated with controlling PRRSV in production systems.[6,7] Collectively, PRRSV challenges have resulted in an increased use of molecular diagnostic tests and sequencing to help control transmission, clinical disease and monitor virus epidemiology.

One particular challenge facing the swine industry has evolved through the use of modified live virus PRRSV (MLV-PRRSV) vaccines and the inability of diagnostic tests to distinguish between detection of wild-type PRRSV (WT-PRRSV) and/or vaccine-like strains of the virus. PRRSV monitoring through reverse transcription real-time PCR (PCR) has become a routine process to help track virus epidemiology and reduce transmission through movement of infected replacement stock or commercial swine with PRRSV infection. All strains of MLV-PRRSV have genetically unique strains of the virus although none were created as differentiating infected from vaccinated animals (DIVA) products. Vaccine-like MLV-PRRSV can replicate in vaccinated animals for long periods of time, are capable of transmission to unvaccinated swine, genetically evolve through point mutations, insertions and/or deletions in the genome, and may experience recombination with WT-PRRSV.[4,8-11] The overall epidemiology or genetic ecology of PRRSV has also become more complicated due to these factors.

In spite of highly sensitive and specific molecular diagnostic assays, multiple tests or sequencing assays may be necessary to distinguish between the presence of MLV-PRRSV or WT-PRRSV when mixed infections are suspected or to sufficiently characterize the strain and/or epidemiology of the virus. New molecular testing methods have been developed to not only detect PRRSV in a specimen, but also characterize the presence of different PRRSV strains in a production system and attempt to distinguish between detection of an MLV-PRRSV, WT-PRRSV or multiple strains of the virus if they are present in a diagnostic specimen or population sample types that are likely to contain mixed PRRSV infections or vaccine-like strains of the virus.

The objectives of this proceedings are to describe current molecular diagnostic assays available to detect PRRSV in swine diagnostic specimens. This includes the molecular assays that are available, suggestions when they may be used, interpretation of the results, and how they may be utilized to determine if an MLV-PRRSV or WT-PRRSV is present.

PRRSV screening PCR assay

PRRSV screening PCR assay design

The first step in a diagnostic process if PRRSV is suspected in the history or included in the differential diagnosis, is to detect PRRSV if it is present in a specimen regardless if the goal is monitoring, surveillance or diagnosing clinical disease. This requires a PCR assay that is sensitive enough to detect all known strains of PRRSV circulating in swine but also specific enough to exclude detection of related virus that are not a species of PRRSV. This assay is commonly referenced as a PRRSV screening PCR because it has the ability to "screen" sample(s) for the presence of any PRRSV regardless of the strain and in spite of genetic divergence that is so commonly observed in the type 2 as well as type 1 species. Developing a screening PCR that can detect all strains of PRRSV is difficult due to the broad genetic diversity of the virus circulating in swine. However, commercial PCR assays are available that provide sensitive, broad detection of PRRSV without sacrificing specificity of the assay.

Commercial PRRSV PCR screening assays are available from ThermoFisher Scientific, Tetracore, Qiagen and IDEXX. Although each assay incorporates the use of different PCR reagents, they are designed to target conserved regions of the PRRSV genome that are less likely to demonstrate mutations that may cause loss of detection through primer/probe mismatches to the target. The ORF6 and ORF7 genes are typical targets for a PRRSV screening PCR assay and it is common that these tests contain multiple combinations of forward and reverse primers and probes to achieve sufficient detection. Real-time PCR is reported with a cycle threshold (Ct) value that indicates the cycle time, usually between 3 and 37 out of 40 cycles, when detection of the target occurred based on threshold and baseline settings in the computer software. It is important to remember the Ct value is inversely proportional to the quantity of virus in the sample although it provides only a semi-quantitative impression. However, the Ct value is useful to evaluate the clinical significance of a positive result or to assess the likelihood of success of subsequent testing such as sequencing, and it may help evaluate if an outcome represents a potential false positive result.

PRRSV screening PCR result interpretation

A PRRSV screening PCR positive result may represent a true positive or could be a false positive result for several reasons, including cross-contamination or laboratory error. Although less likely, it could represent detection of a target that is not PRRSV suggesting issues with specificity. When a false positive result is suspected, the first priority is to retest the sample that was submitted to the laboratory to confirm the original result prior to any major decisions at the farm or production site. If a retest is negative, suggesting the original result is a false positive, consider laboratory error or contamination and also evaluate if errors may have occurred at the site of sample collection. However, low concentrations of virus in a sample (Ct values ≥ 35) do not consistently repeat the original Ct value and often retest negative (repeatability is low). All of these factors should be considered when evaluating the original and any retest results. Additional samples collected from the same population of animals may be necessary to confirm the PRRSV status if retesting results are inconclusive.

PRRSV PCR false negative results are rarely observed although they may occur. Causes of a false negative results may include laboratory error (reporting the incorrect results), factors influenced at the farm level (animal selection, sample selection, stage of the disease process), factors associated with the assay (incorrect test, reagents, limit of detection), and rarely considered are factors associated with the PCR target (primer/probe mismatches and mutations in regions of the target genome that affect detection). If false negative results are suspected, request a retest prior to submitting additional samples.

Most commercial PRRSV PCR assays detect and differentiate types 1 and 2 but are not designed for any further characterization of the virus nucleic acid in the sample. These screening PCRs are not capable of differentiating different genetic strains of the virus or suggesting if a specific strain, RFLP or lineage is present. Additional molecular testing is necessary to answer the subsequent diagnostic questions. Remember that Ct values are semi-quantitative and provide a perception of virus quantity in a sample that should be evaluated prior to additional testing to assess if success is likely or perhaps not worth the added expense.

PRRSV screening PCR post-testing considerations

If a PRRSV screening PCR assay is positive, additional testing may include 1) Sanger or ORF5 gene sequencing to evaluate epidemiology of the virus; 2) Vaccine-like specific PCR if an MLV-PRRSV is suspected and confirmation is important to the diagnostic question; and 3) CLAMP sequencing if both MLV-PRRSV and WT-PRRSV co-infection is suspected, consistent with the history of MLV vaccine use, and detection of either virus is important to the diagnostic question.

PRRSV ORF5 gene sequencing or Sanger gene sequencing

PRRSV ORF5 gene sequencing assay design

If the PRRSV screening PCR is positive, additional diagnostic questions may include monitoring the epidemiology of the virus and ensuring the PRRSV detected in the most recent PCR test is genetically similar to the PRRSV circulating on the farm or perhaps a new strain of the virus. PRRSV ORF5 gene sequencing, or traditional Sanger sequencing, is an appropriate method following a positive screening PCR to help characterize the virus and has been the standard method to evaluate PRRSV epidemiology for several years. PRRSV ORF5 sequencing is a molecular diagnostic assay based on amplification of the gene followed by the selective incorporation of chain-terminating dideoxynucleotides to determine the sequence of nucleotides within the gene. PRRSV ORF5 Sanger sequencing requires two steps. The first process is gene amplification using a conventional (not real-time) PCR process with forward and reverse primers that target complimentary nucleotides flanking the ORF5 to ensure the entire gene is included in the sequence. Amplification is followed by the chain-terminating dideoxynucleotide step to determine the positions of the A, T, C and G bases of the ORF5. This process is conducted overnight at the ISU DNA Facility on main campus and raw sequencing data is available the following day for staff to complete analysis and release the corresponding data in the ISU VDL reports.

PRRSV ORF5 gene sequencing result interpretation

Sequencing success can be influenced by the concentration of the virus in the sample although sequencing the PRRSV ORF5 gene is typically more successful at lower concentrations of the target compared to viruses that require sequencing larger genes (> 2,000 nucleotides). Type 1 PRRSV, or the European strains, are 606 nucleotides in length. Based on the PRRSV screening PCR Ct value, success may be determined by the following: < 25 CT Likely; 25-28 CT Maybe; > 28 CT Doubtful. Type 2 PRRSV, or the North American strains, are 603 nucleotides in length. Based on the PRRSV screening PCR Ct value, success may be determined by the following: < 30 CT Likely; 30-34 CT Maybe; > 34 CT Doubtful. Gene/Sanger sequencing does not differentiate if a virus is MLV-PRRSV or WT-PRRSV nor does it differentiate between different strains of the virus and Sanger sequencing is rarely capable of detecting more than one sequence/virus in a sample. The gene amplification process is more likely to preference sequencing the strain of PRRSV that is present in the highest concentration. Rarely does gene sequencing data indicate the presence of a mixed PRRSV infection although poor quality raw sequencing data may suggest a co-infection is possible. Gene/Sanger sequencing is not capable of indicating if a PRRSV is a recombinant between two different strains of the virus.

In addition to virus concentration affecting sequencing success, the presence of multiple viruses in a sample, inhibitors of amplification or sample type may influence outcomes. Occasionally, there is no apparent reason why a PRRSV ORF5 cannot be sequenced regardless of concentration and these sequencing attempts are reported as negative. Ambiguous base is a term that indicates more than one nucleotide was detected at a position in the gene. Ambiguous bases are designated by a letter that is specific to the combination of nucleotides that are present. When more than 5 ambiguous bases are detected in an ORF5, the result should be interpreted with caution as any analysis or comparisons of sequence homology and/or phylogenetic analysis may be inaccurate. The ISU VDL will create a new extract from the original sample prior to sequencing. This process provides a retesting opportunity that may suggest the original PRRSV screen PCR was a false positive if the gene could not be amplified indicating a retest of the original sample using the PRRSV screening assay is warranted. In addition, low concentrations of virus and mutations in the region of the forward and/or reverse primers that are used to amplify the ORF5 gene may negatively impact gene amplification, or cause poor sequence quality. Similar to real-time PCR, gene amplification requires similar levels of quality control to ensure the assay is working properly and producing acceptable results.

Diagnostic reports with ORF5 sequence data will also include a comparison of nucleotide identity with the MLV-PRRSV vaccine strains. Although arbitrary, nucleotide identity ≥ 99% is reported a vaccine-like sequence and < 99% is reported as a WT-PRRSV.

However, caution is encouraged when interpreting the origin of an ORF5 sequence that is approximately 96%-98.9% similar to an MLV-PRRSV. This range of nucleotide identity may represent a drifted MLV-PRRSV that has mutated in the ORF5. Whole genome sequencing would be an appropriate method to determine if a virus is in fact a WT-PRRSV or MLV-PRRSV.

PRRSV ORF5 gene sequencing post-testing considerations

Sanger sequencing results likely represent the PRRSV in the highest concentration in a sample although primer homology may also contribute to strain preferences when infections are mixed, which most often occurs between MLV-PRRSV and WT-PRRSV. If an MLV-PRRSV is generated from Sanger sequencing and clinical impressions suggest a co-infection with a WT-PRRSV is possible, consider PRRSV CLAMP sequencing. If a PRRSV screening PCR positive result and clinical history or presentation suggest the virus likely represents a vaccine strain, the vaccine-like specific PCR is suggested based on the vaccine that is used in the production system.

PRRSV vaccine-like reverse transcription real-time PCR

MLV-PRRSV vaccine-like PCR assay design

The vaccine-like specific PCR, MLV-PRRSV, is suggested for use in a PRRSV monitoring protocol in combination with other diagnostic tests (PRRSV screen PCR and/or sequencing) or it may be used when the history, clinical impressions, and vaccine use is well-known to justify the use of this assay. Individual MLV-PRRSV PCRs are available for all 6 of the commercial PRRSV vaccines that includes BIVI Ingelvac PRRSV MLV, BIVI Ingelvac PRRSV ATP, Zoetis Fostera PRRSV, Merck PrimePac PRRSV, Elanco Prevacent PRRSV and a new vaccine from Pharmgate known as PRRSGard. The vaccine-like PCR is a reverse transcription real-time PCR that contains a single set of primers and probe (single-plex assay) that target a complimentary region of the ORF5 gene or a region beyond the ORF5 gene, depending on the assay. Due to this constraint, the vaccine-like PCR is less sensitive compared to the PRRSV screening PCR assay and Ct values ≥ 32 are likely to be negative regardless if the target is present.

MLV-PRRSV vaccine-like PCR result interpretation

The most important aspect of the MLV-PRRSV PCR is that it is NOT A DIFFERENTIAL ASSAY; therefore, the test will only detect the presence of a vaccine-like strain in a sample, but does not exclude the presence of another strain of PRRSV, including WT-PRRSV. A PCR assay can be developed to be specific to a single strain of PRRSV but it is not possible to create a PCR assay that excludes a single strain of the virus. Ultimately, these PCR assays are only capable of detecting if a vaccine-like strain is present in a sample.

The MLV-PRRSV PCR Ct result should be interpreted in conjunction with the Ct value from the PRRSV screening PCR assay. A difference of 6 Ct, or more, may suggest two viruses are present in the sample and warrants additional testing or CLAMP sequencing. An advantage of the vaccine-like PCR is they can detect a strain of MLV-PRRSV within a broader range of nucleotide identity, approximately 97% - 100%, and provides an opportunity to confirm a drifted vaccine-like ORF5 sequence reported as a WT-PRRSV.

MLV-PRRSV vaccine-like PCR post-testing considerations

Collectively, the MLV-PRRSV PCR is another diagnostic tool that can be strategically used to answer specific PRRSV diagnostic questions. If WT-PRRSV and MLV-PRRSV are suspected in a sample, the CLAMP sequencing assay is the best approach to potentially confirm. Some PRRSV monitoring protocols routinely use the vaccine-like PCR in association with the CLAMP sequencing assay to reduce the risk of moving pigs infected with a WT-PRRSV when vaccinated pigs are PCR positive for the virus. In addition, if there are inconsistent results between a negative vaccine-like PCR and corresponding MLV-PRRSV ORF5 sequence, the presence of a potential recombinant PRRSV strain should be considered, which requires whole genome sequencing to confirm.

PRRSV CLAMP sequencing assays

PRRSV CLAMP sequencing assay design

The PRRSV CLAMP sequencing assay utilizes a process similar to Sanger sequencing; however, during the ORF5 amplification PCR process, a CLAMP oligonucleotide that is homologous to a region specific to a vaccine strain of PRRSV is included in the reaction that will prevent amplification, and detection, of the MLV-PRRSV ORF5 region. This method of targeted inhibition of MLV-PRRSV ORF5 amplification to enhance detection of sequences of lower concentration or suboptimal primer complementarity was adapted from method used to sequence specific types of human cancer and inform appropriate treatments.

The CLAMP sequencing assay is currently available for three of the commercial MLV-PRRSV vaccines. These are named accordingly BIVI Ingelvac MLV CLAMP, Zoetis Fostera MLV CLAMP, and Elanco Prevacent MLV CLAMP. The CLAMP sequencing assay is less sensitive than the PRRSV screening PCR and Ct values may suggest the success based on the following: < 28 CT likely; 28-20 CT maybe; > 30 CT doubtful. Therefore, CLAMP sequencing may be negative although PRRSV screening PCR is positive based on concentration of the target in the sample.

PRRSV CLAMP sequencing result interpretation

Interpretation of the CLAMP sequencing result is incomplete without Sanger (No-CLAMP) sequencing, thus both assays are conducted simultaneously at the ISU VDL. Therefore, detection of a WT-PRRSV (CLAMP result) and a vaccine-like MLV-PRRSV (No-CLAMP result) is the ultimate indication of a mixed infection. However, similar to any diagnostic assay, results have their nuances that require careful interpretation. A CLAMP sequence may be vaccine-like if it represents a different vaccine strain than what the test was designed to target. An example may include a CLAMP result that is Fostera vaccine strain when using the Ingelvac CLAMP sequencing assay. A WT-PRRSV may be detected in both the CLAMP and No-CLAMP assays if an MLV-PRRSV is not present in the sample.

In addition, CLAMP sequencing may fail to prevent amplification of an MLV-PRRSV sequence when concentration of the target is too high for complete "clamping" or when potential mis-matches between the target and the CLAMP oligonucleotide are present. Finally, the CLAMP sequencing process may prevent amplification of WT-PRRSV when the ORF5 sequence homology is in the range of 96% - 98.9%. Thus, CLAMP results are negative, which may be considered another benefit of this assay when the goal is to distinguish if a true WT-PRRSV or drifted MLV-PRRSV are present. Interpret all CLAMP results with knowledge of the PRRSV screening PCR result, the suspected concentrations of the virus based on Ct value, and the complete vaccine and clinical history of the farm. The possible outcomes of the CLAMP and No-CLAMP sequencing are represented in Table 1 although some combinations of CLAMP/No-CLAMP outcomes may require further analysis.

Finally, another scenario that can yield unexpected results is PRRSV genetic recombination. Similar to discrepant results between ORF5 Sanger sequencing and vaccine-like PCR, if CLAMP sequencing outcomes and vaccine-like PCR results are inconsistent or conflicting, consider the possibility a recombinant PRRSV may be present in the sample. To verify that genetic recombination has occurred, whole genome sequencing can be useful.

PRRSV CLAMP sequencing post-testing considerations

Collectively, the CLAMP/No-CLAMP sequencing assay is the best method available to detect a mixed PRRSV infection between an MLV-PRRSV and WT-PRRSV. The assay is rapid and less expensive compared to next generation sequencing. Unfortunately, there is no method to detect and sequence the ORF5 of two WT-PRRSV unless multiple primers can amplify separate gene sequences, which is unlikely and rarely occurs. When a recombinant virus is suspected, or when an ORF5 sequence is 96% - 99% homologous to a vaccine strain and needs confirmation, it is suggested to use next generation or whole genome sequencing methods for this purpose.

Next generation sequencing or whole genome sequencing

Next generation sequencing (NGS) offers a range of processes that can provide the whole genome sequence (WGS) of a specific virus, detect the presence of novel viruses, or conduct deep sequencing of the microbiome. For routine diagnostic purposes, PRRSV whole genome sequencing may be an appropriate option to answer a diagnostic question or monitor the epidemiology of a PRRSV circulating in a production system. Specific discussions of NGS is beyond the scope of this proceedings and is offered in a separate proceedings paper.

Conclusion

Multiple molecular diagnostic assays are available for detecting, sequencing and characterizing PRRSV. In addition, the CLAMP sequencing assay is an appropriate method to determine if MLV-PRRSV and WT-PRRSV mixed infections are present. Although no molecular diagnostic test is perfect, these tests and combinations of different assays improve the probability of detecting and characterizing a strain or multiple strains of PRRSV present in a group of pigs and helps reduce the risks associated with moving pigs that are unknowingly infected with WT-PRRSV, especially when MLV-PRRSV is present in the same population.

References

1. Zimmerman JJ, Dee SA, Holtkamp DJ, et al. Porcine Reproductive and Respiratory Syndrome Viruses (Porcine Arteriviruses) In: Zimmerman JJ, Karriker LA, Ramirez A, et al., eds. *Diseases of Swine*. 11 ed. Hoboken, NJ: Wiley Blackwell, 2019;685-708.

2. Holtkamp DJ, Kliebenstein JB, Neumann EJ, et al. Assessment of the economic impact of porcine reproductive and respiratory syndrome virus on United States pork producers. *Journal of Swine Health and Production* 2013;21:72-84.

3. Nathues H, Alarcon P, Rushton J, et al. Cost of porcine reproductive and respiratory syndrome virus at individual farm level - An economic disease model. *Prev Vet Med* 2017;142:16-29.

4. Renukaradhya GJ, Meng XJ, Calvert JG, et al. Live porcine reproductive and respiratory syndrome virus vaccines: Current status and future direction. *Vaccine* 2015;33:4069-4080.

5. Meeusen EN, Walker J, Peters A, et al. Current status of veterinary vaccines. *Clin Microbiol Rev* 2007;20:489-510, table of contents.

6. Kimman TG, Cornelissen LA, Moormann RJ, et al. Challenges for porcine reproductive and respiratory syndrome virus (PRRSV) vaccinology. *Vaccine* 2009;27:3704-3718.

7. Paploski IAD, Corzo C, Rovira A, et al. Temporal Dynamics of Co-circulating Lineages of Porcine Reproductive and Respiratory Syndrome Virus. *Frontiers in Microbiology* 2019;10.

8. van Geelen AGM, Anderson TK, Lager KM, et al. Porcine reproductive and respiratory disease virus: Evolution and recombination yields distinct ORF5 RFLP 1-7-4 viruses with individual pathogenicity. *Virology* 2018;513:168-179.

9. Wang A, Chen Q, Wang L, et al. Recombination between Vaccine and Field Strains of Porcine Reproductive and Respiratory Syndrome Virus. *Emerg Infect Dis* 2019;25:2335-2337.

10. Yu F, Yan Y, Shi M, et al. Phylogenetics, Genomic Recombination, and NSP2 Polymorphic Patterns of Porcine Reproductive and Respiratory Syndrome Virus in China and the United States in 2014-2018. *J Virol* 2020;94.

11. Botner A, Strandbygaard B, Sorensen KJ, et al. Appearance of acute PRRS-like symptoms in sow herds after vaccination with a modified live PRRS vaccine. *Vet Rec* 1997;141:497-499.

Table 1: PRRSV CLAMP and No-CLAMP sequencing result combinations that are possible with the BIVI Ingelvac MLV CLAMP, Zoetis Fostera MLV CLAMP, or Elanco Prevacent MLV CLAMP

		CLAMP sequencing	
		Positive	**Negative**
No-CLAMP (Sanger Sequencing)	Positive	Wild type PRRSV	Vaccine-like PRRSV
	Negative	Should not happen	Negative

What can we learn from a PRRS whole genome sequence?

Albert Rovira
Veterinary Diagnostic Laboratory, University of Minnesota, Saint Paul, Minnesota

Porcine reproductive and respiratory syndrome virus (PRRSV) is a single stranded RNA virus with an approximate genome size of 15,000 nucleotides. This makes it a medium-size virus, ten times bigger than small viruses like porcine circovirus type 2 (PCV2) and ten times smaller than large viruses like African swine fever virus. Genetic variability within PRRSV is remarkable and traditionally different strains have been identified and differentiated by sequencing the Orf 5 gene of the virus. This sequencing method is currently available at diagnostic laboratories with high precision and reproducibility, and quick turnaround time. Other advantages of Orf 5 sequencing are its low cost and the extensive databases available for sequence comparisons. However, the Orf 5 gene is only 600 nucleotides long, thus representing only 4% of the genome. In some cases, it may be useful to sequence the whole genome of the virus. The obvious advantage of sequencing the whole genome is the ability to analyze virus changes on parts of the genome other than the Orf 5. However, whole genome sequencing (WGS) is expensive, less sensitive, less reliable and slower than Orf 5 sequencing. In addition, interpretation of the results is more complicated and there are much fewer sequences available to make comparisons. However, all these limitations are being addressed by new advances in next generation sequencing technology and bioinformatics pipelines, and WGS is quickly becoming more reliable, fast and easy to interpret, and less costly.

In general, WGS can be a good tool in cases where the result of Orf 5 is unclear or where the result of Orf 5 sequencing does not match the clinical presentation. Here are some examples:

Example 1: WGS was used in a case where multiple abortions were experienced in a sow farm that was positive and stable with vaccination for PRRSV. In that case, Orf 5 sequencing suggested a vaccine-like virus, 98% similar to a vaccine that had not been used in that farm for the past two years. Therefore, the Orf 5 sequencing results (vaccine) did not explain the clinical picture of abortions. WGS revealed that the strain that was apparently causing the abortions was not a vaccine strain. Instead, it was a new strain resulting from several recombination events between two vaccine strains, one that had been used in the farm in the past and one that was being used at the time of the outbreak. It is possible that the combination of different parts of the genome from two attenuated strains resulted in a virulent recombinant strain.

Example 2: This is a similar case presentation with a different outcome. In this case, Orf 5 indicated a vaccine-like strain, 99.0% similar to a commercial vaccine. WGS confirmed the results of the Orf 5 sequencing, showing 99.3% similarity to the vaccine strain at the genome level. Therefore, in this case a vaccine-like strain appears to be associated with abortions. If this is true, there may be other factors that facilitated the unusual behavior of this attenuated strain (timing of vaccination, dosage, co-infections). Alternatively, the small changes observed in the genome (0.7%, or 105 mutations) may have had an effect in the virulence of this strain.

Example 3: A sow farm that went through an outbreak with a 1-7-4 PRRSV strain, did mass serum inoculation and herd closure to eradicate it. After eight months of negative processing fluids testing, another PRRSV outbreak was experienced, characterized by three weeks of abortions, double sow mortality and increased pre-wean mortality. Orf 5 sequencing of PRRSV from serum of a sow that aborted showed a 1-7-4 strain, 98.5% similar to that from the previous outbreak. This result suggested a very similar strain, which did not explain why the sows that had been recently inoculated were experiencing clinical disease again. Therefore, WGS was attempted from the recent outbreak and also from stored samples from the previous outbreak. A whole genome was obtained from the recent samples but only a partial genome was obtained from the previous outbreak. This partial genome was composed of nine fragments of different sizes. While this outcome was not ideal, it was enough to show that the two strains were different and differences were concentrated in the first part of the genome: similarity for the three fragments from the first part of the genome ranged from 82% to 90%, while similarity in the remaining six fragments ranged from 96% to 100%. Therefore, the strain obtained in the second outbreak may be a recombinant between the strain obtained in the first outbreak and an unknown strain. Alternatively, it could be a lateral introduction. The differences found in the first part of the genome may explain the lack of cross protection observed in this case.

Summary

These examples highlight the benefits of using WGS in cases where Orf 5 sequencing does not fully answer the diagnostic questions we had. In addition to these cases, it would be a good practice to run WGS on any major outbreaks observed in a sow herd. This will help build a collection of previous PRRSV sequences that will be readily available for comparisons in the future. Stored samples from previous outbreaks are not available in most of the cases and, when available, they do not have the same quality as fresh samples. Therefore, routinely sequencing the genome of major outbreaks as they occur is recommended to build and maintain a strong database for future outbreak investigations in that particular farm and in other potential contact farms.

Friend or foe: what next generation sequencing can tell you about the endemic agents in your herd

Maria J. Clavijo[1,2], DVM, PhD; Robert Mugabi[1], DVM, MS, PhD; Ganwu Li[1], MS, PhD
[1]Veterinary Diagnostic and Production Animal Medicine, Iowa State University, Ames, Iowa;
[2]PIC®, Hendersonville, Tennessee

Introduction

Streptococcus suis (S suis) and *Glaesserella parasuis (GPS)* are considered frequent causes of septicemic disease in pig populations, creating a significant impact on the health and well-being of the pig and on the systems productivity. In fact, recent data from the Iowa State Veterinary Diagnostic Laboratory show that the diagnosis and detection for all three agents has increased in the last couple of years.[1] These agents are considered normal colonizers of the upper respiratory tract of pigs and both pathogenic and commensal strains can co-exist within and animal, complicating diagnosis. Control of pathogenic strains hinges on proper diagnosis, minimizing triggering or predisposing factors, effective vaccines based on the farm's strains, and strategic use medication. Next-generation sequencing provides the best resolution for strain characterization unlike conventional typing methods, and when combined with other relevant information (i.e. clinical history, histopathology and sourcing) can serve as a robust surveillance system to monitor these bacterial agents in flows. The objective of this presentation is to share some of the lessons learned on the use of next generation sequencing to improve the control of endemic agents in the field.

Lessons learned - *S suis*

A total of 531 *S suis* isolates submitted to the ISU-VDL were whole-genome sequenced and analyzed. These were submissions from clients who submitted isolates (2014-2020) from disease associated cases (mainly neurological or lameness cases). A total of 25 different serotypes were identified, 1 (n = 78), 2 (n = 45), 9 (n = 32), 5 (n = 27), 23 (n = 46), 1/2 (n = 24), 8 (n = 19), 7 (n = 18), 14 (n = 16), 4 (n = 14), 3 (n = 11). The rest of the serotypes 10, 11, 12, 13, 15, 16, 18, 19, 21, 24, 28, 29, 30 and 31 were represented by 1 to 8 isolates (Figure 1). A total of 183 isolates were untypeable representing 34.5% of the isolates (Figure 1). Multi-locus sequence typing (MLST) identified 36 sequence types (STs) with the predominant being ST1 (n = 140), ST28 (n = 70), ST108 (n = 30), ST977 (n = 22), ST25 (n = 12), ST87 (n = 14), ST94 (n = 13), and ST961 (n = 12) (Figure 1). Although certain STs were more associated with specific serotypes, for example serotype 1 (n = 78) and serotype 14 isolates were almost exclusively ST1, ST diversity was noticed in 7/25 serotypes with the number of STs ranging from 2-6 per serotype. These serotypes included serotype 2 and 1/2, 3, 4, 5, 7, 9 and 23. But within these diverse serotypes, a given ST usually predominated (26/45 serotype 2 isolates were ST28). It is important to highlight that ***serotyping can tell an incomplete picture on the clinical relevance of an isolate.*** For example, in our database, we identified three distinct genotypes of serotype 2, that given other genetic characteristics (i.e. ST, distribution of VFs and AMR genes) can either be characterized as high, low or medium pathogenicity. Thus, further characterizing these strains is a critical step for the identification and management of clinically important isolates within a flow. Finally, a total of 147 (27.7%) of the isolates could not be typed by MLST, and likely represent novel STs.

Within our database roughly 10% of isolates are untypeable by both serotyping (conventional and WGS) and MLST. The majority of these isolates either originated from polymicrobial cases (i.e. viral and bacterial agents) or the isolation of *S suis* did not align with lesions present in the tissue, suggesting possible contamination. This indicates that these strains potentially represent opportunistic or commensal strains. Thus, it is imperative that selected pigs for submission are exhibiting acute clinical disease and have not been treated with antibiotics. From these, we recommend focusing on affected systemic tissue (i.e. brain, csf, joint swabs/tissue or serosal surfaces). It is also important to avoid using captive bolt or blunt force trauma, since those methods can increase the likelihood of cross-contamination.

Strains that are untypeable by both methods tend to additionally have two unique genetic features: ***1) Overabundance of antimicrobial resistance genes (AMR)*** and ***2) a lack of putative virulence factors,*** specifically they lack the traditional *mrp*, *epf*, and *sly* virulence-associated genes. Still, the clinical relevance of these isolates is yet to be determined. In contrast, majority of strains with a defined serotype and ST usually carried at least 1 of the traditional virulence markers (i.e. *mrp*), harbored a significant amount of putative virulence factors and had significantly less AMR genes. The low levels of antimicrobial resistance noted at the genomic level are in alignment with the low levels of phenotypic resistance observed at the ISU VDL. Currently, distribution of AMR genes aids in determining the clinical relevance of strains.

While the data show a wide diversity of *S suis* strains present in NA swine herds, usually one or two strains predominate within a farm or flow in disease associated cases, and ***these same strains tend to persist over time.*** Finally, in our experience, it is important that ***repeated tissue submissions be carried*** out to increase the likelihood of capturing the clinically relevant strains and have a good understanding of the diversity and frequency of disease associated strains found in a farm.

Lessons learned - *G parasuis*

Over 100 *GPS* isolates submitted to the VDL between 2015 and 2020 have been whole genome sequenced using Illumina Miseq platform and analyzed using an in-house bioinformatic pipeline. Isolates originated mainly from lung (n = 41), heart (n = 23), brain (n = 6), and fibrin (n = 6). The most common serotypes identified were 7, 13, 4, 2, 5/12 and 1, with a clear overrepresentation of serotype 7 strains, previously thought to be avirulent. Similar to *S suis*, multiple serotypes (i.e. 7, 4 and 2) were distributed in multiple clusters, suggesting that further genetic characterization

Figure 1: Distribution of MLST sequence type (ST) by *S suis* serotype.

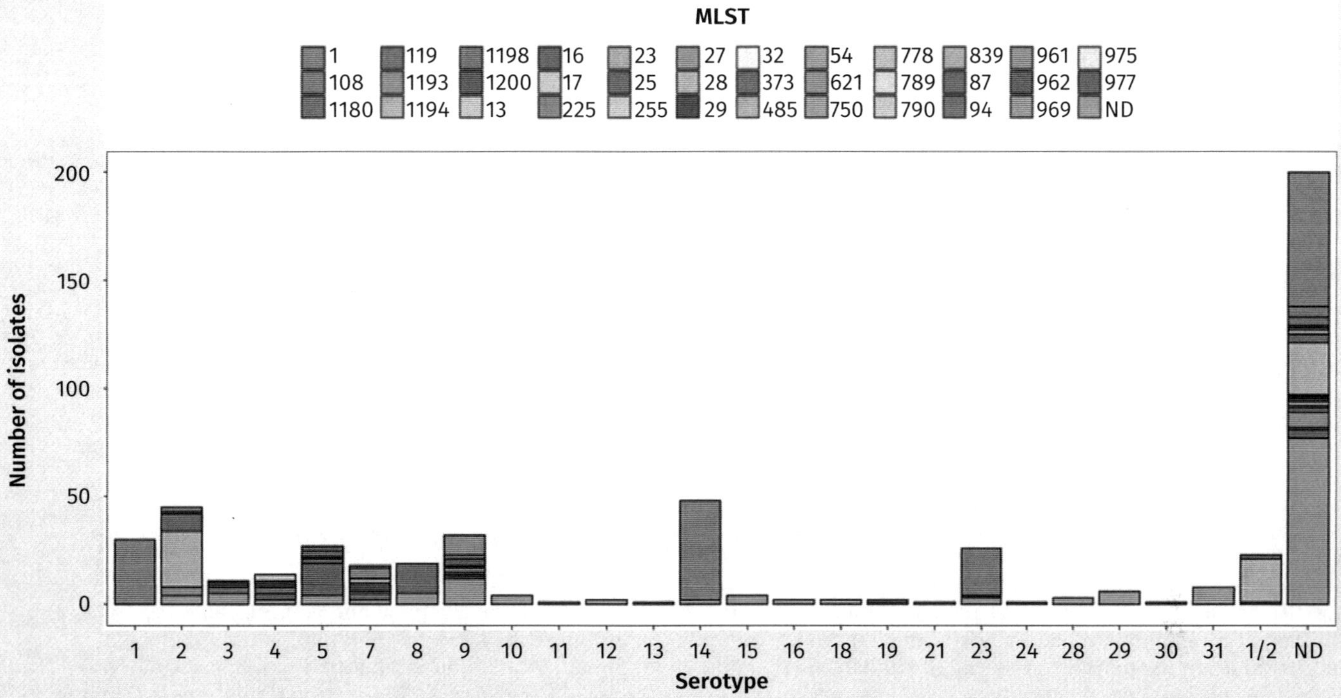

is needed for proper identification and determining their relevance. MLST revealed 45 different STs including ST454 (n = 18), ST6 (n = 8), and ST415 (n = 6) and the rest with 1-3 isolates each. Interestingly, most strains in the database were assigned novel STs by pubmlst, highlighting potential genetic differences between strains from Europe, Asia and North America. In terms of virulence-associated markers, the majority of isolates from serotype 7, 3, 4, 13, 5/12, carried at least 5/9 of group 1 vtaA genes (trimeric autotransporters) and most carried group 3 vtaA genes. While serotype 6 is typically regarded as non-pathogenic, this serotype was detected in multiple pigs from one flow experiencing *GPS*-associated disease and carried most of vtaA group 1 genes, suggesting that not all serotype 6 strains are non-pathogenic. In contrast, only 3 isolates in the database (serotypes 8 and 15) did no carry any of the vtaA genes and clustered with reference strains from serotype 6, 10, 3, 9 and 8, previously described as non-pathogenic. In terms of AMR genes, most isolates were positive for *bcr, aph(3")-Ib, bacA, ksgA*, and *sul2* genes which are markers for bicylomycin, streptomycin, bacitracin, kasugamycin, and sulfonamides, respectively.[2]

Conclusions

Cutting-edge molecular technologies, such as next generation sequencing, offer novel insights into the molecular epidemiology of systemic bacterial agents. Swine practioners can utilize this information to identify the most frequent strains found in disease-associated cases, improve vaccine candidate selection or antimicrobial to use on farm, and optimize mixing of pig populations. This is a continuous process that depends on the relevance of the isolates and systemic collection and analysis.

References

1. Silva et al., 2021. Detection and diagnostic trends of five swine endemic bacterial pathogens (2010-2019). *AASV Proc*. 2021.

2. Mugabi et al., 2021. Characterization of *Glaesserella parasuis* strains circulating in US swine herds using whole genome sequencing. *AASV Proc*. 2021.

Time back to market comparison of different approaches of addressing undercover videos released by activists

Sarah Probst Miller[1,2], **DVM**
[1]AgCreate Solutions, Inc; [2]Pork Avenue Training Portal

Introduction

Over the past decade, AgCreate Solutions, Inc has worked with companies in situations where the need for improvement has been uncovered. Sometimes this need to change is stimulated from the inside during self-assessment or goal-setting exercises; and in other cases, particularly in the case of an undercover video, public accusation, and request/demand to change is unsolicited. In the cases where an employee captures an undercover video, the company experiences feelings of betrayal alongside the damaging visual accusations. They experience intense scrutiny and judgement from the local and national communities. At the same time, it is activist strategy to simultaneously attack the Brand (Packer/Retailer/Food service). To protect their brand, it has become common for the packer/retailer/food service to suspend purchases from farms when a video shows behavior out of line with their beliefs or protocols.

These experiences and feelings alongside the loss of a market put a farm leadership team into crisis mode. How a team reacts to this crisis can either further harm the farm's reputation or create the first step to rebuilding reputation and re-earning the trust of the end user, and very importantly, their packer or any packer. In the past this issue was isolated to the packer who was purchasing animals from the farm in question, today this can quickly spread to all packers who might purchase the animals. Activist strategy is now to inform other packers of the accused farm's actions to eliminate alternate routes to market further complicating any plans to regain market access.

How farms handle their response appears to have an impact on the packer's willingness to reengage with the accused farm and ultimately get them back to normal operations. Dallas Hockman, VP of Industry Relations for the National Pork Producers Council (NPPC) states, "A crisis that is not managed well can wipe out decades of hard work and company value in a matter of hours. A well-managed crisis plan confirms that your company has the processes and procedures in place to address almost any issue that may develop." The real question that has to be asked is "What were you doing prior to the event occurring, that demonstrates that you had thought about it, had a plan in place and a commitment to address the issue if something happens?." Hockman is known for saying, "I know what you are against, what are you for?"

As we have been hired to help companies in these situations, we have observed that certain types of responses appear to further slow or in some cases terminate the farmer's ability to regain market access and other responses seem to work towards creation of agreed upon means to achieve market access. This paper reports our observations and related reflections regarding these experiences. We are including reflections specifically on those situations where based on the video content released, the producers immediately lost market access to their packer.

Comparison

In this comparison exercise, I am going to highlight a few key decision differences between approaches to specific situations that can occur during a crisis like this. In this comparison, companies immediately lost the right to sell pork. Part of their goal was to earn the right to sell pork back while regaining industry and community trust. I will speak generally to protect the companies involved in the variety of experiences.

Our role and perspective influence

Our role in these crises range from:

- Immediate involvement and planning with the owner and senior staff of farm
- Development of media statements alongside NPPC and National Pork Board (NPB), state veterinarians, and packers if included
- On-farm assessment via 3rd party investigation of accusations
- Evaluation of competency baselines related to accusations
- Creation of training plans
- Evaluation of competencies post training over time
- Involvement in proofing additional media statements as needed

Our perspective is based on our experience gained during consultation with farms impacted by such videos, not a full investigation of all undercover video situations. In this paper, we comment on the aspects we have found to be most important in our experience.

The communication plan

Emphasizing the importance of transparency

In the NPPC document entitled *Crisis Planning for your Farm*, one of the first recommendations is to identify a spokesperson and first and foremost be honest and open. The document advises, "Nothing generates more negative media coverage than a lack of honesty. Be transparent. Being as open and transparent as possible can help stop rumors and defuse a potential media frenzy. This transparency must be projected through all communication channels: news interviews, social media, internal announcements, etc."

When we compare different responses by farms, and the desire to regain market access, we would add packers and employees to the list of entities with whom to be transparent. In the variety of responses our team was involved in, farms that involved packers in all discussions on media statements alongside industry experts from the NPPC and NPB and state veterinarians, were able to regain market access faster than those who did not. This inclusion fostered the spirit of transparency and "being in this together."

In our experience, farms that chose to not invite or include their packer in these initial media responses and response planning did not regain market access with the same packer they were selling to when the video was released.

Hockman further comments on this in the NPPC document entitled *Crisis Planning for Your Farm*, "In today's market there is no question, being transparent, quick to address, and commitment to improve is the best option [regarding packer relations]. With Brand Management extremely critical, packers put farms on hold till the issue is addressed and plan of action developed. At NPPC, our goal is to work with farmers and their advisors to shorten the reaction time as much as possible."

As a first response immediately post release, there are a few things critical to acknowledge in our experience. First, it is important to state that your company has zero tolerance for animal abuse and to practice this. Second, it is critical to verbalize and ensure that your team is first and foremost ascertaining that all pigs and people are in a safe spot. Lastly, it is important to state that you are assessing if actions occurred outside of company policy (or industry standards—if the farm does not have clear company policies). If abuse is identified, you must take disciplinary action. It is also important to be cognizant of what can feel like dishonesty to the listener.

Examples that can augment distrust include any immediate denial pre-investigation or subjective statement emphasizing that what was shot could not have really happened. "This video wasn't even shot on our farm." Perhaps this is true; but a statement like this is interpreted as both denial and a lack of willingness to admit that your team may have done something wrong or that something could have happened outside of protocol. Other strategies that feel more like "smoke and mirrors" would be subjective testimonials regarding a farm or farmer's integrity and positive community involvement. Any attempt to claim "goodness" or "perfection" subjectively in a media statement or letter, can further augment distrust and potentially alienate all parties receiving further communications as they come to expect diluted or shielded defenses from your team verses transparency. Avoid this. Instead commit to seriously assessing the situation with true concern for the wellbeing of people and pigs alongside making and implementing a plan based on opportunities identified.

Another important immediate action that may or may not be a part of your media statement, but could empower you to answer follow up questions well and show the packer that you are taking this seriously and taking action as needed is to suspend any people in the video being investigated for egregious or willful abuse as well as immediately evaluating out of compliance actions that may not fall under the category of abuse. Egregious abuse is intentional action done knowing that the action is wrong. This always should result in immediate dismissal. Examples include hitting, kicking, throwing, or purposefully dropping an animal. It is possible that actions will be featured on a video where the actions are inappropriate, out of compliance, but not egregious. In other words, they are done with no knowledge or purposeful intent to cause harm to the animal. An example might be setting a pig down without two feet on the ground and the pig bumping its nose along the way, animals left with no water overnight, or a broken gate left in disrepair that a sow hurts herself on. These situations should be evaluated with a self-assessment of your team's training alongside evaluation of why the actions occurred. While egregious actions should result in immediate dismissal, actions that do not intend harm, but inadvertently cause harm, minimally warrants interventional training and confirmed rerouting of behavior. Employees placed under suspension should then undergo more serious investigation to determine if the suspension transitions to dismissal or targeted retraining. In our experience, farms which immediately suspended potential company policy violators, dismiss abusers, communicate these actions to packers, and instigated immediate investigation of these situations with decisive and appropriate disciplinary action were able to regain access to market. Whereas on the other extreme, farms that did not suspend employees who allowed items viewed as egregious and instead utilized retraining or a warning as enough did not regain access to their previous market, but rather had to seek outside markets in the near and long term. Return to any market was subsequently delayed or never occurred with their previous packer.

How you transparently communicate to employees matters too. They just had a colleague video them or their fellow colleagues unknowingly on farm. They also feel betrayed, angry, and in some cases "dirty" or "guilty by association." The community may treat them as such. Emphasizing your desire to assess and be fair yet in line with your zero tolerance of animal abuse is critical. It is tempting to demonize the activist or person who captured the video. "These people broke into our facility and will be prosecuted," may be a true statement, but what the employee hears is "as long as no one sees what we do, it doesn't matter."

Our experience would show that sharing the media statement going out to the world with employees helps give them words to share at home or in the community. "We do not tolerate animal abuse. Actions occurred on our farm outside of company policy; and we are committed to assessing and implementing change." This helps them feel like they are working at a farm committed to doing the right thing. Your employees will be asked by their neighbors or grandma what is going on. Giving them the same media statement you are using can help them know what to say. When you say, that the farm was "attacked by an undercover video" and instead "plead victim," there are two additional risks: 1) Employees may not take any actions they were a part of that were outside farm protocol seriously thus undermining your ability to implement and verify real change. 2) They repeat the victim statement to others which will sound like denial and a lack of commitment to transparency and best animal care. This can further undermine your standing in the community. Depending on who they talk to, what they say and how they say it could potentially further undermine your ability to regain market access.

Tips for internal or external immediate veterinary/welfare specialist assessment

Making sure all animals and people are in a safe spot also needs to be an immediate action. Involvement of your veterinary team and/or animal welfare specialists is an important part of this. Being able to state that a veterinarian or animal welfare specialist is on site ensuring animals safety can be a strong statement. Some have utilized a 3rd party veterinarian or 3rd party welfare specialist immediately or later in the process. In these circumstances, the company has been able to state that a 3rd party veterinarian/welfare specialist is on site assessing care and ascertaining that all pigs are in a safe spot which can be important if the internal veterinarian/welfare specialist is potentially viewed as too close to the situation.

If you are the 3rd party veterinarian/specialist or internal veterinarian/specialist evaluating the farm post release of video depicting abuse, in our experience, it has been good to first make a list of all human competencies the farm has been accused of in the video. Even if the accusations are within industry standards, we would recommend including them on the list. For example, if in the first scene of the video, they show a bucket of stillborn pigs shot on a farm (maybe not even yours), the farm is being accused of inappropriately disposing dead or causing dead pigs via inappropriate care, then the competencies to assess may include:

- Employees collect dead daily. Employees dispose of dead daily according to industry standards. Employees offer daily observation and care to piglets. Employees attend to farrowing sows to endure health and optimize survival of pigs. Identification and treatment of sick piglets occurs every day and there is record of this.

If the next scene shows pigs being weaned and piglets tossed from one employee to the next as a part of weaning followed by a scene of your employees waving arms, making constant loud noise, weaned pigs piling in a wet hallway, and running of weaned pigs over a dead pig, competencies to assess would then include:

- Employees do not toss or throw pigs during weaning. No pigs are airborne during observed movement. Pigs have two legs on ground prior to be released from being held. Low pressure handling skills employed. Sorting boards are used to move pigs. Pigs are taken in appropriate group sizes. Hallway cleared of distractions and dead pigs prior to movement of pigs. Wet spots are eliminated, and drying agent put down on slick spots prior to movement. Noise is intermittently used. Pressure is appropriate. Etc.

If the next scene shows holes in flooring and pigs walking over these holes or falling in these holes, the competency could be, "Farm completes timely repair of flooring."

There will likely be items visually included that are within industry standards, but considered controversial or minimally, difficult to watch on a video like use of blunt force trauma. In some cases, farms have assessed different euthanasia methods or been asked to implement different euthanasia methods by their packers because of these videos and subsequent customer pressure. Farms that regained market access were willing to adapt to these concerns and requests and prove implementation, giving the packer the ability to say to their customers, "Farm A has eliminated blunt force trauma as a means of euthanasia on farm."

Veterinary/specialist assessment on farm should include doing these tasks with employees on site. Asking staff to pretend you are a new employee and teach you how to wean pigs can give you additional insight into what competencies are being handed down to the newest employees on the site. Veterinary/specialist assessors can then mark these accused competencies as yes/no present to give a baseline assessment of skills. For example, post release of one of the videos we were asked to assess, the farm was weaning the next day. As an assessment team, we made sure, we were there to wean with the farm. As a part of the assessment, we requested, "Please just treat me like I am a new employee and teach me how to wean pigs with you." We had the list of competencies we were concerned with, but we were also intent to understand what people knew and how they applied these skills.

Proactively using measurements in the Common Swine Industry Audit (CSIA) alongside the video competency assessment can also help prepare the farm for the likelihood of the packer and other organizations requesting that a CSIA audit be completed as a part of regaining market access.

Whether you are an internal or external evaluator, when you create your medical/welfare report, we recommend sticking to the numbers and facts. In your summary, especially if you know the farmer, you may be tempted to defend his or her reputation in your language. For example, "Farmer Smith is an excellent member of the community; and in my experience has always given great care to his pigs." It is almost certain that this will be viewed as subjective, biased, and potentially dishonest especially in the face of an undercover video where items of great concern are revealed. Statements like this could subsequently devalue the belief in your objectivity for the on-farm assessment. Instead, work to state facts objectively and assess numerically when possible. To create a quantitative assessment of the competencies a farm is being accused of, we recommend creating a chart of competencies that reflect accusations from the video, mark these yes/no (1/0) and create a percentage of accused competencies that are confirmed to be present verses, confirmed not to be present.

For example, for the competency sort boards are used to move pigs. As you wean pigs, you watch to see if sort boards are used consistently and well. If yes, this competency receives a YES or a score of 1. But on the next competency, if the farm takes a whole room of pigs verses smaller groups as they move them down the hallways, the competency of "Pigs are taken in appropriate group sizes." Receives a NO or a score of 0. If you have assessed 30 competency accusations from the video, and 20 competencies are found to be present and 10 found to be opportunities, you then can say in your report that this farm had a 66% positive expression of competencies related to those of concern. And 33% of competencies of concern were found to have opportunity and need additional training, follow up, and verification. Who do you report it to? If your goal is regaining market access, you must involve your packer.

A statement to the media or others may go like this. "Today a 3rd party veterinarian visited Farm A to assure that all pigs and people were in a safe spot. They also assessed competencies of concern. They did observe some competencies outside of our farm's protocol. Appropriate disciplinary action has been taken. In addition, we are implementing a retraining plan for missing and opportunity competencies in remaining staff. This training will occur immediately and be reinforced with mentorship over the next couple weeks. Training will involve both internal and external verification of proper implementation to further ensure accountability."

Based on amount of opportunities that are found, creating a targeted training plan to implement to rectify the opportunities is critical. A good veterinary medical record could include a training timeline alongside the date when your reassessment will occur. In the case of an urgent missing competency, perhaps the redirection and training are immediate, and reassessment scheduled the following week. Creating a short- and long-term training plan based on opportunity competencies alongside an intervention plan based on proactive measurements in a CSIA can start a farm outside of company protocol on a path to reform. Scheduling and reporting the next anticipated time of competency remeasurement is also critical in helping the farm regain market access. Do share this plan and associated timeline with your packer and ask for input.

In our experience, farms utilizing this strategy of immediate competency assessment via objective and quantitative veterinary/specialist assessment that were able to involve their packer in assessing the medical report and create a plan with their packer to implement training and change assessment, were able to work with their packer to outline a time to return to market under their

Table 1: Example action list and baseline assessment scores versus post training assessment scores: opportunities for immediate actions and employee training were identified from the on-farm assessment by an external veterinarian

External verification-all farms observe the following actions	Baseline assessment score	Post training assessment score
Training and handling boars		
Handler properly positioned sort-board between self and boar	1	1
Two people present if boar in pen with people	1	1
Farm has plan for boar culling	1	1
Farm safely loads boar to boar movement equipment	1	1
Moving pregnant sows and gilts		
Hallway is prepped and distractions removed	0	1
No dead pigs in hallway	1	1
Sows are moved calmly and slowly	1	1
Sows are moved in groups of 3 to 5	0.5	1
Pressure is appropriate	0	1
Noise is intermittent, calm, and not constant	1	1
Employees use appropriate language when moving animals	0	1
Weaning sows		
Hallway is prepped and distractions removed	1	1
No dead pigs in hallway	1	1
Sows are moved calmly and slowly	0.5	1
Sows are moved in groups of 3 to 5	1	1
Pressure is appropriate	1	1
Noise is intermittent calm and not constant	1	1
Employees use appropriate language when moving animals	1	1
Weaning pigs		
Hallway prepped and not slippery, distractions removed	1	1
No dead pigs in hallway	1	1
Pigs are moved calmly and slowly	0	1
Pigs are moved in properly sized groups per farm protocol	0	1
Pressure is appropriate	0	1
Noise is intermittent, calm, and not constant	0	1
Employees use appropriate language when moving animals	0	1
Euthanasia		
No pigs with BCS of 1 observed in random sample of farrowing	1	1
No pigs with BCS of 1 observed in random sample of gestation	1	1
Farm has approved method of euthanasia in place	1	1
There is a log for cleaning of captive bolt gun	0	1
Captive bolt gun is cleaned, disinfected, and stored properly when not in use	1	1
Designated euthanasia employee can demonstrate method to confirm death	0.5	1
Total score	20.5	31
Total possible score	31	31
Competency percentage	66%	100%
Percentage point increase from baseline		34%
Percentage increase from baseline		52%

packer's brand. Farms who used internal assessments to claim all was well without factual competency assessments did not regain market access. In these situations, when outside assessors were involved later verses immediately, this step was acknowledged as good; but trust appeared to be damaged in these circumstances, and it was not enough to regain the right to sell pork under their previous market brand.

Proving change implementation

Holding farms accountable to the plan must be driven top down in the organization. It is likely your packer will request that representatives from their company audit your facilities alongside other 3rd party auditors. These representatives must be welcomed. Give these representatives access to assessment and training you have done. Update them on any concluded or ongoing investigations. Ask for their input. Respond and work with them on suggestions. Explain any differences of approach. If possible, implement your internal or 3rd party veterinary assessment and training and change plan prior to these audits. Then their company assessments can be a part of a growing amount of data proving accountability and regaining trust. Engaging a 3rd party to further assess lasting change on an agreed upon schedule also can help rebuild trust. Planning and reporting on these assessments together can be a part of presenting a long-term plan to work with the packer to further ensure that an event like this will be proactively prevented and that you are committed to holding the team accountable. Saying things like, "This happened under our watch.," is an honest and transparent part of admitting something went terribly wrong. Do add, "We are committed to doing everything in our power to ensure that it doesn't happen again." This dual statement appears to be an important part of further repairing the packer/farmer relationship or gaining access to market from a different packer aware of your situation.

Additional discussion

Denying market access in the face of an undercover video has become a part of brand protection for packers. Understandably holding market hogs has welfare implications as well. The American Association of Swine Veterinarians (AASV) recently adopted a position statement regarding crisis situations that might result in stop movement. It reads:

- In stop movement situations, it is important to do what is right for the pigs. Stop movements can cause animal welfare issues.
- We acknowledge that in certain situations, stop movements are in the best interest to protect the health and welfare of the larger swine population and food supply. Stop movement situations with indefinite time frames will cause welfare issues.
- In considering these stop movement decisions, we encourage plans that keep the pigs' welfare at the forefront of decision making.

While these cases referenced did not have this position statement in place at the time of their situations to reference, we believe this statement could and should be a part of conversations with packers to do what is right for the pig. We encourage that the welfare of the hogs in the stop movement order be a part of transparent discussions with the packer.

References

1. AASV Position Statement: Pig Welfare During Stop Movement Situations. **https://www.aasv.org/aasv/position-stopmovement.html.**

2. Hockman, Dallas, Crisis Planning for Your Farm, National Pork Producers Council.

3. Grandin, Temple, How to Define the term Egregious Animal Abuse and Minimize Excitement or Discomfort that are used by the FSIS/USDA for Humane Slaughter Regulatory Enforcement, **https://www.grandin.com/USDA.FSIS.humane.slaughter.act.enforcement.html.**

4. Pork Avenue Training Portal. 2018. Available at **www.porkavenuetraining.com.**

5. Probst Miller S, Ketchem R. The impact of trained and untrained personnel on product implementation and subsequent pig health. 48th American Association of Swine Veterinarians' Annual Meeting. 2017;165-166.

6. Probst Miller, S., Krantz, S. 2019. Inspiring a verified culture of care post-identification of abuse on an undercover video using a baseline assessment of expressed competencies compared to post training measurement of expressed competencies.

7. Probst Miller, S. 2020. Ensuring that a company's vision and mission and core values are living on farm via pre-training, targeted training, and post-training animal care competency measurements.

8. Probst Miller, S. 2016. Impact of Two Training Implementation Plans on Post-Training Competency Expression of Personnel on Farm. AASV Proceedings, 2016.

California Proposition 12: What are the specifics and how can producers manage?

T. D. Parsons, VMD, PhD, DACAW
Swine Teaching and Research Center, School of Veterinary Medicine,
University of Pennsylvania, Kennett Square, Pennsylvania

Background

California residents, with an overwhelming majority, voted in Proposition 2 or the original Prevention of Farm Animal Cruelty Act into law in 2007. This bill called for the elimination of confinement housing in laying hens, veal calves, and pregnant sows. It specified that the animal must be able to lie down, stand up, and fully extend his or her limbs; as well as being able to turn around freely. Several animal industry groups as well as the American Veterinary Medical Association vigorously opposed the ballot initiative at the time. Shortly after the implementation of Prop 2, it became clear that the new legislature would have limitations.

1) The first challenge was the possibility of animal products such as eggs, veal and pork sold in California could be sourced from a state without similar restrictions. This obviously put California raised products at a potential financial disadvantage.

2) The second area of confusion in the original bill was the distinction that it covered pregnant pigs (gilts or sows). There are many different ways to detect pregnancy with differing degrees of sensitivity as to be able to tell when a sow was pregnant. So the question remained exactly when during the production cycle would bred females need to comply with the California regulations.

3) Finally, the third challenge was that no space allowance was specified in the original bill, only that they might turnaround unimpeded by their crate, cage or pen. Therefore, animals in theory might be able be moved from an individual gestation stall to a pen setting that allowed for the movement stipulations but might provide less available space to the animals than in stalls.

Take two

A new referendum was put forth in 2018 called Proposition 12 to amend the original Prevention of Farm Animal Cruelty Act and address some of the confusion/limitations of the original Proposition 2.

1) The new bill extended its coverage to not only animals raised in California but also to those animals being reared elsewhere whose meat, eggs or milk would be sold in California.

2) It also specified that the new housing requirements would apply to all breeding females - any female pig of the porcine species kept for the purpose of commercial breeding who is six months or older or pregnant.

3) Furthermore, the new bill required these female pigs to be reared at space allowance of 24 square feet per sow. Again, despite opposition from agricultural groups, this referendum easily passed and is slated to take effect January 1st, 2022. Promulgation of Proposition 12 yielded California Regulations for Health and Safety Code Division 20: Miscellaneous Health and Provisions, Chapter 13.8 Farm Animal Cruelty [25990-29994] and can be accessed at **https://www.cdfa.ca.gov/ahfss/pdfs/Prop12_HSC.pdf.**

Implications for producers

1) The impact of the Proposition 12 applies to not only pigs in California but also to any sow whose offspring yields pork sold in California. California residents represent the largest consumer of agricultural goods in the US. If California were a country, rather than a state, California would the 5th largest economy in the world, with a GDP only less than the US, China, Japan, and Great Britain, but larger that France or Germany. Most food purveyors are not willing to forgo their business in Californian and thus there is pressure on swine producers to find ways to meet the Proposition 12 requirements. The California Department of Food and Agriculture has been charged with ensuring compliance to Prop 12. They lack the resources to do so on their own and will need to rely on third party audits to verify that farms systems sourcing pork to California are compliant. Details regarding the audit process such as which organizations will be available to audit and how often a farm needs to be audited remain to be finalized.

As of the writing of this paper, legal action challenging Proposition 12 is unresolved. The 9th Circuit Court of Appeals, a federal court covering the western US, recently decided against the North American Meat Institute's challenge to the constitutionality of the Proposition 12. The lawsuit argued that Prop 12 violates the Commerce Clause of the US Constitution that prevents trade barriers between and among the states. The court concluded that Prop 12 "does not have a discriminatory effect because it treats in-state meat producers the same as out-of-state meat producers." This case was filed originally in California's central district court. A similar suited filed jointly by the National Pork Producers Council and the American Farm Bureau is still pending in different court, California's southern district. The final verdict on the Proposition 12 remains unknown until all appeals are exhausted. However, if Proposition 12 stands, then producers wanting access to the California market need to be working on changes given that it could take effect in a little more than a year from now.

2) The provision for the housing requirements of the bill to cover all breeding females is perhaps the most challenging. Prop 12 does provide exemptions for sows from 5 days prior to farrowing until they are weaned. However, in most production systems, weaned sows are housed in individual breeding stalls to protect both the sows and the barn staff from the unwanted riding by sows in estrus. The new law prohibits stalls and requires farmers to group house sows after weaning or potentially house them in individual pens that meet the stringent requirements of Prop 12. Scientific findings do not support this provision of the bill, as in general, there is simply a dearth of information on the welfare and productivity of animals in this post-weaning period. Previous work done by our group looked at group housing of post-weaned sows and documented a significant increase in skin lesions in the group housed sows compared to crated sows.[1] These lesions were associated with agonistic interactions between the group-housed

animals, but did not seem to affect fertility. Work done in Australia also found similar increases in skin lesions as well as increased stress levels and greater variability in reproductive behavior between group housed versus stall housed weaned sows.[2] The increase in skin lesions experienced by the group weaned sows in the first seven days post-weaning represents an extended of period time for which the sows are painful and likely experiencing compromised welfare that might also impact productivity. However, it can be argued that these compromises to physical welfare must be balanced against whatever psychological welfare benefits arise in the pen from the sow not being confined in a stall during this time. Unfortunately, little objective information is available to address this question.

3) The increased space allowance for sows to 24 square feet per head represents a significant change for most producers which currently house sows at anywhere from 16 to 20 square feet depending on upon the housing system. The immediate implication is a loss of sow inventory of anywhere from 17 to 33% with obvious financial impact. These challenges are exacerbated furthermore by the subsequent underutilization of farrowing facilities if the gestating sow population is reduced. Finally, immediately post-weaned sows may require more than 24 square feet per sow to house, depending upon what solution producers elect, and this promises to drive down sow inventory another couple of percent.

Ways forward

Unless the courts invalidate Proposition 12, producers whose pigs will supply pork to California will need to start to adapt their farms to the new requirements. Here are some things to consider:

1) As compliance mechanisms become better defined, farmers will need to familiarize themselves with both the requirements of Prop 12 as well as those organizations providing certification.

2) Modifications to existing farms will be required to develop husbandry systems that meet the Proposition 12 requirements. This includes the safe and effective estrous detection and insemination of post-weaned sows. Two initial options seem possible.

A) The first would be to wean sows into small pens. While individual sow pens might likely optimize reproductive tasks, it will require much space. Proposition 12 requires that the enclosure must not prevent a breeding pig from lying down, standing up, fully extending its limbs, or turning around freely. Thus, a minimal pens size would likely need to be 6 × 8 feet and perhaps greater for larger sows. At 48 square feet per sow, this is twice the space allowance required by Prop 12. Individual farms will need to decide if they can sacrifice the space to allow for individually housed weaned sows. An alternative might be to employee a larger, but still small pen that houses multiple sows and provides better space utilization. For instance, a 10 x 10 or 8 x 12 foot pen with 4 sows promises enough room for convenient and safe breeding, to meet the Prop 12 movement requirements, but without greatly exceeding the Prop 12 space allowances (eg staying at 24-25 square feet per sow). Existing barn constraints ultimately will dictate what size pen is most practical.

B) The other approach would be to look at larger pens that would include free access stalls for feeding and breeding. The bill allows exceptions for temporary periods for animal husbandry purposes for no more than six hours in any 24-hour period, and no more than 24 hours total in any 30-day period. These exceptions would be sufficient to allow for breeding and daily feeding of animals. The free access stall could provide 14 square feet of space and thus another 10 square feet per head of pen space would need to be allocated to be compliant. For instance, a 24 × 20 pen with two rows of free access stalls separated by a 10 foot loafing area would accommodate 20 sows as per the Proposition 12 requirements. Larger groups of animals could be achieved in this approach simply by designing a longer pen (eg 20 × 40 feet for 40 sows). Again exact pen layout will be determined by the details of the existing facility that is being renovated.

3) Once a Prop 12 compliant breeding area is established, changes to gestation areas also will be required on most farms to ensure the 24 square foot of space allowance per sow. Here are several possible scenarios.

A) Existing pre-implantation pen gestation farm – Sows will gestate in existing pen gestation facilities albeit at a reduced stocking density.

B) Existing post-implantation pen gestation farm – Crates used for post-implantation housing (4-5 weeks of inventory) need to be replaced with pen gestation, but existing pen gestation housing can be used albeit at reduced stocking density.

C) Conventional crated facility – Crates will need to be replaced and a decision made regarding what type of pen gestation will be implemented using 24 square feet per sow.

D) It is also possible that sows are weaned into pens where they are bred and remain in the same pens for the duration of a gestation. There is some simplicity in this approach but it comes at the cost of even further reduced space utilization given some expected fallout of sows from breeding groups.

Each of these options will result in underutilization of farrowing facilities as sow inventory in gestation will be reduced compared to the pre-Prop 12 flow of the barn. In some cases, it may be desirable to use this additional space in farrowing to increase wean age and potentially wean an older, larger piglet. Another consideration might be to push back gilt breeding activities (or even some gilt gestation) to the gilt developer in order to help hold gestation inventory and maintain better farrowing house utilization. Additional gestation capacity could be gained through construction as another way to maintain farrowing house utilization. Finally, construction of new Prop 12 purpose-built facility might be the solution for some farms if they have non-California markets for animals sourced from their existing barns. It is possible that some older, less competitive sow facilities may go-dark in this transition as pork producers strive to meet the Prop 12 requirements. As seen with other housing changes, in some cases new purpose built sow farms designed to meet new requirements may be more cost-competitive than efforts to retrofit existing barns.

As necessity is the mother of invention, new creative solutions in addition to those described here for Prop 12 compliance likely will emerge. Regardless which scenario applies, producer costs per wean pig will likely increase at least by 15 to 30% based solely on the opportunity cost of reduced sow inventory and perhaps more depending upon the cost of renovations and or any production losses associated with the new practices. The task for veterinarians will be to help producers making these changes identify the optimal solution for their specific scenario to meet California's Proposition 12.

References

1. Pierdon, M.K. and T.D. Parsons. Effect of familiarity and mixing method on gestating sow welfare and productivity in large dynamic groups. Journal of Animal Science. 96:5024-5034 (2018).

2. Rault,J.-L., R. S. Morrison, C. F. Hansen, L. U. Hansen and P. H. Hemsworth. Effects of group housing after weaning on sow welfare and sexual behavior. Journal of Animal Science. 92:5683-5692 (2014).

Pain, longevity and lameness

J. Deen, DVM, MSc, PhD, DACAW
Swine Group, College of Veterinary Medicine, University of Minnesota, St Paul, Minnesota

Introduction

The AVMA states that "Animal welfare means how an animal is coping with the conditions in which it lives".[1] Measures of that ability to cope cover a wide area of signalment, but can be broadly broken down into observations of behavior, biologic performance and physiology/pathology. Much of what we expect in animal care is regular and expert application of these aspects in identifying those animals that have difficulty coping.

In our pig barns we have many individuals who are particularly adept at identifying pigs who need intervention to improve their welfare. They rely on a combination of absolute measurements along with relative comparisons of behaviors and condition. Whether it be the way that pigs breathe, walk or interact, that information is combined with evaluations of the condition of the pigs. In addition, there are those pigs that fail to cope to the point that they die, and these often offer the opportunity to do further detailed examinations of the pathologies at work in that population.

My argument is that these are welfare experts, as they have spent a great deal of time studying pigs and the interventions that make them better. In comparison to the scientific literature, they really just lack two major adjuncts. The first is a record of their observations and the second is a reliance on an intermediate translation of behaviors and conditions to descriptors of affective states. This paper is a discussion on methods to claim and exhibit the abilities that farms have to monitor and improve the welfare of pigs on their farms, using the examples of pain, longevity and lameness.

Models of welfare management in the barn

When we examine deviations in the behavior and performance of pigs, we can divide the major coping challenges across four major sources:

- Ambient temperatures, with coping measures including piling, panting and reduced feed intake.
- Flooring, with lameness and, more importantly, reluctance to walk being important indicators
- Other pigs, with fighting and inability to access sleeping areas, feed and water being signs
- Disease, with inappetence and reluctance to move often being significant signs

Where possible, these challenges are managed to allow pigs to cope. This can be through facility design, pathogen control and providing supplemental resources such as heat and increased fats in diets. We can also address some of these needs by ensuring that the requirements are consistent across a population of pigs. For instance, the size of pigs greatly affects the ability to cope with changes in ambient temperatures.

Where we address the needs of the whole population, we often have detailed records. For instance, ambient temperatures are recorded in detail, as are ventilation responses. Where we fall down is in individual interventions, first of all when and if those individual animal interventions occur, and, more importantly, the successes of those interventions. Treatments are often not recorded at the individual pig level, and if recorded, responses to treatment are even less likely to be entered into a record-keeping system.

Moreover, many interventions in welfare literature focus on outcomes that presume a knowledge of the affective state, states such as pain, hunger and boredom. In most cases, attribution of welfare concerns to specific affective states is not an aim of caregivers. This can be for good reasons. First, the assessment of affective states can be subjective, and the states vary in their biologic utility. Secondly, it is subject to anthropomorphism as the only descriptors we have are within our own affective states. Thirdly, the existence of affective states that do not drive the animal to a better level of biologic functioning are suspect.[2] Though these issues are real, the reluctance to discuss and interpret affective states has compromised the ability of producers to enter a wider dialogue.

Pain, longevity and lameness

The most evident affective state with biologic outcomes is the state of being in pain. It can be argued that it is the most likely state that leads to direct and distinct behaviors. Moreover, it may be the only state in pigs that can be directly controlled with pharmacologic interventions, insomuch as pain has been defined as a state that occurs without analgesia. Using this definition, we find that the great majority of lameness can be addressed through the use of such analgesics. Moreover, we find that there is a number of different indications that pain is addressed through coping mechanisms that are often only partly successful. The largest indication is that sow longevity is significantly reduced in lame sows.[3] This can be through direct effects, or indirectly through other effects including reproductive failure. When combined with other insults, the effects are magnified, such as a lame lactating sow during the summer.

This is an area where producers can lay out their approach to addressing this concern. Records of lameness, evaluations of interventions and monitoring of heard health and productivity can all be indications of expert response. In addition, the use of analgesia, when appropriate, can also be evaluated in more detail and recorded as one of many successful interventions.

It should not be surprising that though the painful condition of lameness, with its well recorded biologic outcomes and distinct behaviors is a concern to producers, much of the public focus has been on castration. It should be evident that the effects of castration are much less than lameness. The extent of changes in behavior, the compromises in biologic performance and the real effects on longevity point to the need to have a much stronger focus on lameness. One of the challenges is to create the correct focus within and on our industry for those aspects that are of actual concern.

References

1. AMVA, **https://www.avma.org/resources/animal-health-welfare/animal-welfare-what-it**

2. Curtis SE. Performance indicates animal state of being: A Cinderella axiom? PAS. 2007 Dec 1;23(6):573-83.

3. Anil SS, Anil L, Deen J. Effect of lameness on sow longevity. JAVMA. 2009 Sep 15;235(6):734-8.

Emotion and cognition in pigs: what does the science tell us?

Kristina Horback
Center for Animal Welfare, Department of Animal Science, University of California, Davis, California

Introduction

For many years, the predominant method of assessing animal welfare focused on overt physical ailments, such as illness, lameness, and body condition, while ignoring the emotional state of the animal. This is due in large part to the difficulty in validating assumed emotional experience of animals. The scarcity of research on emotions in animals may also be due to the reluctance of certain scientific communities from accepting a seemingly controversial topic: that animals experience subjective emotions. This reluctance, however, is in paradox to the decades of considerable experimental work using animals as biomedical models to investigate the mechanism of action for pain relief, drug-use (i.e., addiction), and chronic stress (i.e., PTSD).

Emotional responses evolved to guide both human and non-human animals in achieving survival goals and increase fitness (i.e., seek resources and avoid harm).[1] Through repeated paired associations, organisms learn which elements in their environment may benefit or threaten their fitness. The adaptive significance of emotions is the result of three major systems: physiological arousal (e.g., run from predator), behavioral tendencies and expressions (e.g., social benefits), and cognitive changes (e.g., motivation and planning for future).

Defining and identifying emotions in swine

There are multiple types of emotions an organism may experience based on a difference in valence (positive to negative) and arousal (high to low).[2] For example, some emotions may be high arousal but differ in their valence (excited versus rage), whereas others may be of similar valence but differ in their intensity (content versus excited) (Figure 1). While an emotional response is caused by a particular stimulus, and will fade once the stimulus is removed, affective states are longer lasting, diffuse moods which are the result of accumulated experience (e.g., current environmental stimuli and previous experience).[3]

There is currently no universal method to gauge the emotional state in non-human animals objectively. The evaluation of emotions in swine can be deduced from physiological indicators (e.g. heart rate variability, hypothalamic-pituitary-adrenal, and, sympathetic-adrenal-medullary activity) and behavioral indicators (e.g. approach/avoidance behavior, vocalization quality/quantity, body postures). However, interpretation of these measurements is confounded by the fact that these metrics could reflect an emotion of both positive and negative valence. For example, we can measure an increase in heart rate and a spike in cortisol levels, as well as an increase in locomotor activity (e.g., pacing behavior), in a pig that is experiencing both a high arousal negative state (e.g., anxiety) and a high arousal positive state (e.g., excitement). This is why researchers must take a comprehensive approach including physiological, behavioral and cognitive measurements when evaluating the emotional experience of non-human animals (Table 1).

Cognitive indicators of emotional state in swine

Classical and operant conditioning

Interpretation of biological and behavioral measurements can be difficult because they could reflect an emotion of both positive and negative valence. For example, an increase in heart rate or surge in cortisol or adrenaline all indicate high arousal, but can be associated with escape from predation (negative valence) or with the anticipation of a reward (positive valence). Given the limitations of physiological and behavioral assessment of animal emotions, novel cognitive methods are carried out to evaluate how the emotional state of a pig may alter its cognitive processing (e.g., decision making, attention, appraisal and memory). Animal welfare scientists may measure the conditioned response of an animal to infer a subjective pain or pleasure experience. For example, when pigs are given a highly-valued reward (e.g., food, analgesic drugs) only when they are placed in a specific environment, a stimulus-response association can be made called a "conditioned place preference". Conversely, pigs may be administered a punishment (e.g., electric shock) only when placed in a different specific environment in order to establish a "conditioned place aversion".

Cognitive bias testing

Cognitive bias testing allows researchers to infer how an individual's emotional state influences information processing, such as the amount of attention given to an unknown or threatening stimulus, or, the evaluation of ambiguous stimuli (is that unknown object a threat or not a threat?).[4] Animals can be classified

Figure 1: The circumplex model of affect, with emotions placed on axis based on level of arousal and valence.

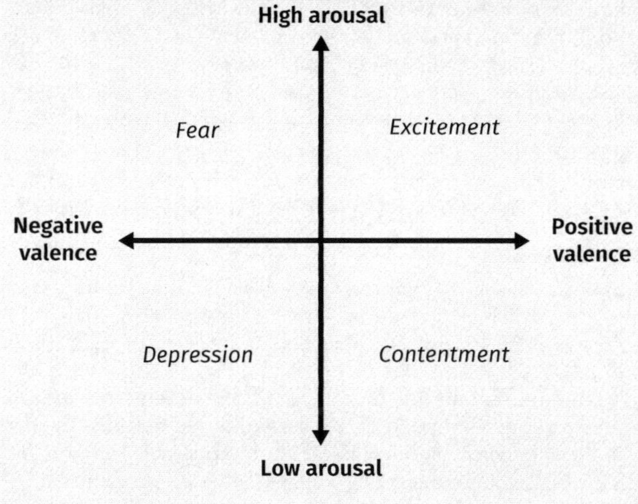

Table 1: The characteristics of the three types of measurements used to evaluate emotions in animals and examples of each during a high arousal-negative valence state (e.g. animal perceives a threat; fear) and a high arousal-positive valence state (e.g. animal anticipates reward; excited)

Measurement	Characteristics	High arousal, negative valence (eg: fearful)	High arousal, positive valence (eg: excited)
Physiological	A visceral state of arousal triggered by the autonomic nervous system	↑ heart rate ↑ cortisol, adrenaline	↑ heart rate ↑ cortisol, adrenaline
Behavioral	An external expression including facial expressions, body posture, activity, and vocalizations	↓ Activity (freeze) ↑ Activity (flight) Alert ear posture ↑ vocalization	↑ Activity Alert ear posture ↑ vocalization
Cognitive	Alteration to processing of environmental stimuli, related to appraisal, attention and memory	Perceive ambiguous stimuli as threatening; do not approach	Perceive ambiguous stimuli as non-threatening; quickly approach

as "optimistic" or having a positive affective state if the animal displays behaviors that suggest an increased expectation of reward in the face of ambiguous stimuli. Alternatively, animals can be classified "pessimistic" or having a negative affective state if the animal displays behaviors that suggest an increased expectation of punishment in the face of ambiguous stimuli. Research on human and non-human animals indicates that negative mental states, such as anxiety or depression, can induce pessimistic judgments of ambiguous stimuli. For example, barren, crowded or unpredictable housing conditions can elicit negative cognitive biases, and therefore assumed negative affective states (e.g., anxiety, stress), in piglets.[4-6]

Emotional contagion

In the animal emotion literature, researchers often refer to this "spread" of an emotional experience from one animal to other observing animals as an "emotional contagion". The perception of an emotional state in a conspecific, through sight, sound, or smell, can lead to physiological and behavioral changes in the observer which mirrors that of the acting animal. The strength of an emotional contagion appears to be influenced by previous experience of the observer and the relationship between actor and observer. For example, it has been reported that when piglets observe another piglet in a stress-induced context (e.g., physical restraint) they will show an increase in behavioral indicators of fear. However, when the observer pigs had previous experience with restraint themselves, they reacted more strongly to observing a conspecific in restraint than the pigs which have never experienced restraint.[7]

Applied emotional and cognition research in swine

Commercial swine are exposed to a range of routine management procedures which can trigger negative affective states, such as fear, distress, frustration, or boredom. Acute bouts of stress may be beneficial for the animal as they can allow the animal to exercise adaption behavior (i.e., allostasis[8]). Chronic negative states like "distress", however, can be very detrimental for the animal's physiological and psychological health. Distress occurs when an animal cannot escape from or adapt to internal or external stressors. Animals which experience chronic negative emotional states often decrease locomotor activity, feeding behaviors, and display physiological changes, such as altered circadian rhythm, body temperature or body weight.[9] In order to improve an animal's welfare, we must not only reduce the severity and quantity of adverse situations which can result in chronic negative affective state, we must also create an environment which allows animals to meet the demands, and thus, experience positive affective states. The absence of negative emotions does not automatically ensure positive emotions, and it is relevant to find and use indicators of positive emotions in animals. The challenge lies in identifying and validating physiological, behavioral, and cognitive indicators of both negative and positive emotional states in animals. Advances in modern science, from psychology to neuroscience, will enhance our ability to recognize emotions in swine; which in turn may improve the quality of life for these animals.

References

1. Oatley K & Jenkins J. Understanding emotions. Oxford: Blackwell. 1998.

2. Russell J. A circumplex model of affect. *J Pers Soc Psychol.* 1980; 39.

3. Russell, J. Core affect and the psychological construction of emotion. *Psychol Rev.* 2003; 110.

4. Mendl et al. Cognitive bias as an indicator of animal emotion and welfare: Emerging evidence and underlying mechanisms. *Appl Anim Behav Sci.* 2009; 118.

5. Douglas et al. Environmental enrichment induces optimistic cognitive biases pigs. *Appl Anim Behav Sci.* 2012; 139.

6. Scollo et al. Does stocking density modify affective state in pigs as assessed by cognitive bias, behavioural and physiological parameters? *Appl Anim Behav Sci.* 2014; 153.

7. Goumon S & Špinka M. Emotional contagion of distress in young pigs is potentiated by previous exposure to the same stressor. *Anim cogn.* 2016; 19.

8. Korte et al. A new animal welfare concept based on allostasis. *Physiol behav,* 2007; 92.

9. Boissy et al. Genetics of fear in ruminant livestock. *Livest Prod Sci.* 2005; 93.

Practical solutions for enrichment

Meghann K. Pierdon; VMD, DACAW
University of Pennsylvania School of Veterinary Medicine, Kennett Square, Pennsylvania

To discuss enrichment, we must first define animal welfare. Different definitions are used, such as how well an animal is coping or whether an animal has what it wants, but as a scientific concept it has several facets. First, it is fundamentally about what the animal is experiencing. It is a characteristic of an animal and not something that we, as caretakers, give to that animal. Second, it has multiple components. Animal welfare can be described as three complementary components that overlap. First, the function or physiology of the animal which incorporates health and productivity. Second, natural behavior which is the species-specific character of an animal. Finally, the feelings, or emotional state of the animal. The welfare of the animal is encompassed by the overlap of these realms. Environmental enrichment is when amendments are made to an environment that are important to the animal and allow the animal to more fully express their species-specific character.[1] To understand how best to provide enrichment then we need to understand first what an animal does, and next, what the animal perceives.

One method to understand the species-specific character of the pig, which is important for welfare and enrichment, is to compare the behaviors of pigs in a semi-natural environment to captive pigs. In the most well-known example, Stolba and Wood-Gush took pigs from modern farms and put them in an extensive setting and created an ethogram, or catalogue of behaviors, by observing the pigs and categorizing their behaviors. The amount of time spent in foraging like activities was over 60% of their day.[2] Looking at the ethogram for pigs in different housing systems that are more barren, we see they spend 60% of their day lying inactive.[3] This points to a difference in the behavioral diversity and activity of pigs in a barren environment compared to a natural environment. More information is needed to understand what motivations exist or which aspects of the barren environment, if any, the pigs find frustrating. We can see what is missing from their behavioral repertoire, but we don't know its value to the pig. To understand what matters to pigs, preference and motivation testing as well as cognitive bias testing can give a pig's eye view of enrichment.

Space is an important aspect of the environment and we need to understand the importance of space to the pig. Space must be limited severely to impact growth. Gonyou et al., developed an equation for the relationship between space and growth. When this relationship is applied, a 90 kg pig would need less than .65 square meters in order to limit growth.[4] To better understand how space impacts pigs, physiological measures give more insights. Scollo et al., compared a low square footage and a high square footage group and found that the pigs in the low square footage group had higher fecal glucocorticoids. Injury scores also increased with decreased square footage though cognitive bias was not different based on square footage.[5]

Straw is a well-studied source of enrichment for pigs. In a comprehensive review by Tuyttens, pigs prefer straw over concrete floors and other point type enrichment. Access to straw increases activity level and behavioral diversity in pigs including increased exploratory and manipulative behaviors. Some studies also document decreased undesirable behaviors.[6] Other research focuses on the appropriate provision of straw. When provided as bedding, pigs interact with straw more than if it is provided in small amounts or mixed in feed.[6] Pigs also interact with straw longer if it is long instead of chopped and some research shows the chopped straw can lead to increased aggression.[7] In some studies, pigs prefer other substrates to straw such as peat, mushroom compost, and saw dust.[8] Straw, however, serves multiple purposes for pigs such as gut fill, thermoregulation, and foraging like behavior. Straw meets the definition of enrichment but is incompatible with many modern facilities manure removal systems so point enrichment may be a valid alternative.

Chains are a commonly provided source of point enrichment. Pigs may interact with chains if they are provided correctly. An experiment by Fedes and Fraser showed that objects with loose ends are interacted with more than those that are in a loop.[9] The addition of non-destructible objects such as balls and pipes to the ends of chains, anecdotally reduces their use[10] but lowering the chain so that pigs may interact with it in a motion that mimics rooting, adding branches to the chain so that more than one pig can interact with the chain simultaneously, and affixing multiple chains to the floor can increase how often pigs interact with the chain.[11] Though studies comparing chains, which are indestructible, to objects which are destructible, show pigs prefer other objects in most studies. This includes rope,[12, 13] cloth strips[14, 15] and rubber hose.[15] Therefore, many welfare experts conclude that chains are not a valid form of enrichment for pigs.[10]

There are several recent studies examining point enrichment in sows and boars. Horback et al., provided three types of point enrichment to sows on a 5000 head commercial sow farm where the sows were housed in large group pens. They were provided a rubber chew toy (Bite-Rite, Ikadan System, Denmark), an altered toy where the rubber was replaced with cotton rope, and a wood block in a PVC tube. The sows spent more time interacting with the rope toy and maintained the level of interaction over the 2 weeks of monitoring. There was the least interaction with the wood. Over time the interest in the rubber and the wood waned more than the interest in the rope.[16] Sirovica et al., did a similar study in individually housed boars providing a rubber chew toy or cotton rope. Boars preferred the cotton rope and spent less time performing stereotypical pen manipulation when provided with the rope.[17] These studies are consistent with other research that shows that pigs prefer materials that are manipulable and destructible.

Social housing is often considered enrichment for social animals. Kirkden et al., designed a motivation test in sows to examine how hard a sow would work to get to a pen with 2 other sows. Only half of the sows worked as hard as they would work to get the last 1/16 of their daily feed.[18] In a study by Matthews (in reference list), growing pigs were trained to a panel pressing task. The pigs generated a demand curve that was elastic for opening the door to go in with other pigs.[19] Interpretation of these studies must be done considering the challenge of studying social connections between animals. First, there are factors the researcher may not be aware of such as which animals are preferred partners and which animals a pig might want to avoid. Second, other factors need to be considered such as how long a pig has been deprived of social housing and the effects of novelty and familiarity.

Complexity in the environment can serve as enrichment. An indication that pigs value complexity is contra-freeloading, where an animal works for food despite the ready availability of the same food. Scientists hypothesize that this is adaptive in the wild in that the information gained is useful for future foraging. De Jonge et al., demonstrated contra-free loading in pigs.[20] Potential sources of complexity include different housing designs. Fraser et al., designed a multi-tiered pen for weaned pigs and showed that they spent time in all areas of the pen.[21] Such complex housing systems allow pigs to choose to socialize or isolate as well as choose different microclimates for improved thermoregulation. Another example of complexity is allowing pigs to make choices. Bench and Gonyou allowed early weaned piglets to choose the temperature of their environment and found that they decreased the temperature at night and as they aged.[22] Other sources of cognitive enrichment can be different training modalities. In farms that use electronic sow feedings stations, pigs need to be trained to use the feeders. In a study by Zebunke et al., the sows were trained to a unique auditory signal for feeding. The trained animals were less fearful and more exploratory indicating that such a training regime may provide enrichment and improved welfare.[23]

Finally, it is important to examine the impact of enrichment on welfare. A thorough review is provided by van der Weerd and Day where they find that there is little impact on the physiological aspects of welfare including no impact on productivity, meat quality, or cortisol levels in the majority of studies included in their analysis.[24] There are a few studies however that show changes in the brain when enrichment is provided.[14,25] Behavior does seem to be influenced by enrichment were we see a decrease in fear behaviors in behavioral tests and decreased aggression toward pen mates including decreased belly nosing and tail biting in some studies.[24] Most studies show no effect of point enrichment on ease of handling though in some studies handling ease decreases in enriched pigs, likely because of increased interactions with humans, therefore handling may be influenced by the type of enrichment provided.[24]

One area where the swine industry should look to the future is the provision of positive welfare and the influence of enrichment on the third realm of welfare, positive emotional states in pigs. In a study of cognitive bias in pigs by Douglas et al., gilts were housed in either an enriched or a barren environment. They trained the gilts to a positive and negative cue, then measured the proportion of approaches and latency to approach an ambiguous cue. Gilts in the barren pen were more likely to treat the ambiguous cue as a negative. Gilts then became more optimistic when they were moved to the enriched pen. Interestingly, those gilts that were originally in the enriched pen and then moved to the barren pen seemed to have a stronger reaction than when they started in the barren pen and moved to the enriched pen. This establishes that enriched environments are perceived by pigs as positive as reflected by a positive cognitive bias in enriched environments.[26] It is important to consider such impacts on the pigs when implementing and developing enrichment for pigs in commercial environments.

References

1. Newberry, R. C. Environmental enrichment: increasing the biological relevance of captive environments. *Appl Anim Behav Sci* 1995; (44): 229–243.

2. Stolba, A., and Wood-Gush D. G. M. The behaviour of pigs in a semi-natural environment. *Anim Sci J* 1989; 48(2): 419-425.

3. Guy J. H., Rowlinson P., Chadwick J. P. Behaviour of two genotypes of growing–finishing pig in three different housing systems. *Appl Anim Behav Sci* 2002; 75(3):193-206.

4. Gonyou, H. W., Brumm, M. C., Bush, E., Deen, J. Application of broken-line analysis to assess floor space requirements of nursery and grower-finisher pigs expressed on an allometric basis. *J Anim Sci* 2006; 84(1): 229-235.

5. Scollo, A., Gottardo, F., Contiero, B., et al. Does stocking density modify affective state in pigs as assessed by cognitive bias, behavioural and physiological parameters? *Appl Anim Behav Sci* 2014; 153: 26-35.

6. Tuyttens, F. A. M. The importance of straw for pig and cattle welfare: a review. *Appl Anim Behav Sci* 2005; 92(3): 261-282.

7. Lahrmann, H., Oxholm, L., Steinmetz, H. The effect of long or chopped straw on pig behaviour. *Animal,* 2015; 9(5): 862-870.

8. Beattie, V. E., Walker, N., Sneddon, I. A. Preference testing of substrates by growing pigs. *Anim Welfare* 1998; 7: 27–34.

9. Feddes J, Fraser D. Non-nutritive chewing by pigs: Implications for tail-biting and behavioral enrichment. *Transactions of the ASAE.* 1994; 37(3):947–50.

10. Bracke, M. B., & Koene, P. Expert opinion on metal chains and other indestructible objects as proper enrichment for intensively-farmed pigs. *PloS one* 2019; 14(2): e0212610.

11. Courboulay, V. Utiliation de différentes modalités de mise à disposition de chaînes pour des porcs en engraissement: impact sur le comportement et les lésions des animaux. Journées *Recherche Porcine* 2011; 43: 83-184.

12. Horrell, I., Ness, P. A., Enrichment satisfying specific behavioural needs in early-weaned pigs. *Abstract 28th International Congress of the ISAE, Appl. Anim. Behav. Sci* 1995; 44: 264.

13. Guy, J. H., Meads, Z. A., Shiel, R. S., & Edwards, S. A. The effect of combining different environmental enrichment materials on enrichment use by growing pigs. *Appl Anim Behav Sci* 2013; 144(3-4): 102-107.

14. Grandin, Temple. Effect of rearing environment and environmental enrichment on behavior and neural development in young pigs. *Diss. University of Illinois at Urbana-Champaign,* 1989.

15. Hill J. D., McGlone, J. J., Fullwood, S. D., Miller, M. F. Environmental enrichment influences on pig behavior, performance and meat quality. *Appl Anim Behav Sci* 1998; 57: 51-68.

16. Horback, K. M., Pierdon M. K., and Parsons T. D. Behavioral preference for different enrichment objects in a commercial sow herd. *Appl Anim Behav Sci* 2016; 184: 7-15.

17. Sirovica, L. V., Creamer, M., & Horback, K. M. Preference for and behavioural response to environmental enrichment in a small population of sexually mature, commercial boars. *Anim Welfare* 2019; 28(3): 271-278.

18. Kirkden, R. D., and Pajor E. A. Motivation for group housing in gestating sows. *Anim Welfare* 2006; 15(2): 119-130.

19. Matthews, L. R., and Ladewig J. Environmental requirements of pigs measured by behavioural demand functions. *Anim behav* 1994; 47(3): 713-719.

20. De Jonge, F. H. Tilly, S. L., Baars, A. M. On the rewarding nature of appetitive feeding behaviour in pigs (Sus scrofa): do domesticated pigs contrafreeload? *Appl Anim Behav Sci* 2008; 114(3): 359-372.

21. Fraser D., Phillips P. A., Thompson B. K. A test of a free-access two-level pen for fattening pigs. *Anim Sci J.* 1986; 42(2): 269-274.

22. Bench C. J., Gonyou H. W. Temperature preference in piglets weaned at 12–14 days of age. *Can J Anim Sci.* 2007; 87(3): 299-302.

23. Zebunke, M., Birger P., and Langbein J. Effects of cognitive enrichment on behavioural and physiological reactions of pigs. *Physiol behav* 2013; 118: 70-79.

24. van de Weerd, H. A., & Day, J. E. A review of environmental enrichment for pigs housed in intensive housing systems. *Appl Anim Behav Sci* 2009; 116(1): 1-20.

25. M. K. Jarvinen, J. Morrow-Tesch, J. J. McGlone, T. L. Powley Effects of diverse developmental environments on neuronal morphology in domestic pigs (Sus scrofa) *Dev. Brain Res* 1998; 107: 21-31

26. Douglas, C., Bateson, M., Walsh, C., Bédué, A. Environmental enrichment induces optimistic cognitive biases in pigs. *Appl Anim Behav Sci* 2012; 139(1): 65-73.

Validating the effectiveness of alternative euthanasia techniques using penetrating captive bolt guns in mature swine

Scott A. Kramer[1]; Brooklyn K. Wagner[2]; Ivelisse Robles[2]; Steve Moeller[3]; Andrew Bowman[4]; Justin Kieffer[3]; Andréia Gonçalves Arruda[4]; Michael Cressman[3]; Monique Pairis-Garcia[2]
[1]United States Department of Agriculture-Animal and Plant Health Inspection Service, Streetsboro, Ohio;
[2]North Carolina State University, College of Veterinary Medicine, Raleigh, North Carolina;
[3]The Ohio State University, Department of Animal Science, Columbus, Ohio;
[4]The Ohio State University, College of Veterinary Medicine, Columbus, Ohio

Disclaimer: The views expressed in this article do not necessarily represent the views of the United States Department of Agriculture-Animal and Plant Health Inspection Service.

Introduction

Euthanasia is the process by which animals are rendered insensible and effectively achieve death through brain function loss and cardiac arrest.[1] Conducting humane euthanasia is a moral obligation for all individuals involved in the swine industry and requires that those working with pigs are appropriately trained to euthanize using techniques that are safe, effective and reliable.[1] Several euthanasia techniques have been identified and approved for use in swine. On commercial swine farms in the United States, the most commonly used euthanasia techniques to date on mature pigs include captive bolt gun, carbon dioxide, electrocution and gunshot.[2]

In mature swine, defined as sexually mature gilts, sows, and boars euthanasia is often achieved through physical disruption of brain activity through the application of a penetrating captive bolt gun.[3] The bolt gun is placed firmly in the frontal position against the forehead of the pig at midline and insensibility and death is achieved by concussive damage and trauma to the brainstem and cerebral hemisphere. Although this method is typically effective, efficacy can be directly impacted by the ability of the bolt to penetrate the skull. As pigs age, the thickness of the skull increases, the frontal sinuses expand, and some breeds may develop a distinct bony ridge on the midline of the forehead.[3,4]

Investigating penetrating captive bolt gun euthanasia at alternative locations to the standard frontal position in live, mature swine is needed Therefore, the objective of the present study was to evaluate the efficacy of euthanasia at two alternative cranial locations (temporal [TEM] and behind-the-ear [BTE]) utilizing two, heavy duty, extended bolt penetrating captive bolt gun equipment styles (cylinder [C] and pistol [P] style; Figure 1).

Materials and methods

The present study was carried out in two stages in accordance with the Guide for the Care and Use of Agricultural Animals in Research and Teaching. The protocol was approved by The Ohio State University Institutional Animal Care and Use Committee (#2020A00000031). Considering there was no data available at the time this study was complete on the effectiveness of the treatments examined herein for swine, sample size for this study was determined using a combination of ethical and financial considerations.

Stage one: Verification of techniques' effectiveness in anesthetized mature swine

Study animals and housing

Thirty-six commercial, mixed breed, mature sows (n = 18; weight: 239.78 ± 14.74kg) and boars (n = 18; weight: 344.17 ± 22.94kg) were purchased from local commercial vendors. Pigs were transported and temporarily held (< 24 hours) at The Ohio State University Meat Science Laboratory where they were individually weighed and identified using permanent marker. The objective of stage one was to verify each technique's effectiveness in unconscious, anesthetized animals.

Euthanasia equipment

Euthanasia was performed using two styles of penetrating captive bolt guns. The cylinder style (C) and pistol style (P) are both Super Heavy-Duty Jarvis Power Actuated Stunners (BUNZL Processor Division/Koch Supplies, Riverside, MO). For the present trial, a 6.0 and 4.0 grain cartridge load was used for the C and P style equipment, respectively.

Cranial location

Each captive bolt style was evaluated at three cranial locations. The TEM location was defined as three cm (1.18 inches) posterior to the lateral canthus of the eye within the plane between the lateral canthus and the base of the ear. The BTE location was defined as directly caudal to the pinna of the ear on the same plane of the eyes with the plane targeted toward the middle of the opposite eye (Anderson et al., 2019).[4] The FRO location was defined at the intersection of imaginary diagonal lines extending from the top of the attachment of each ear to the medial canthus of the opposite eye at an angle resulting in a bolt path directed toward the brain stem (Fig 1). The FRO location was used as the experimental control, given that FRO is an approved location for penetrating captive bolt euthanasia in swine (AVMA, 2020).

Both C and P styles were evaluated at each cranial location, resulting in six treatment combinations. Within stage one, three sows and three boars were randomly assigned and evaluated for each treatment (n = 36 total).

Figure 1: The images represent the recommended placement of a penetrating captive bolt gun for A: Frontal B: temporal and C: behind-the-ear approaches for euthanasia procedures in mature swine. Red dots indicate placement of the penetrating captive bolt gun muzzle. Blue dashed lines represent reference landmarks.

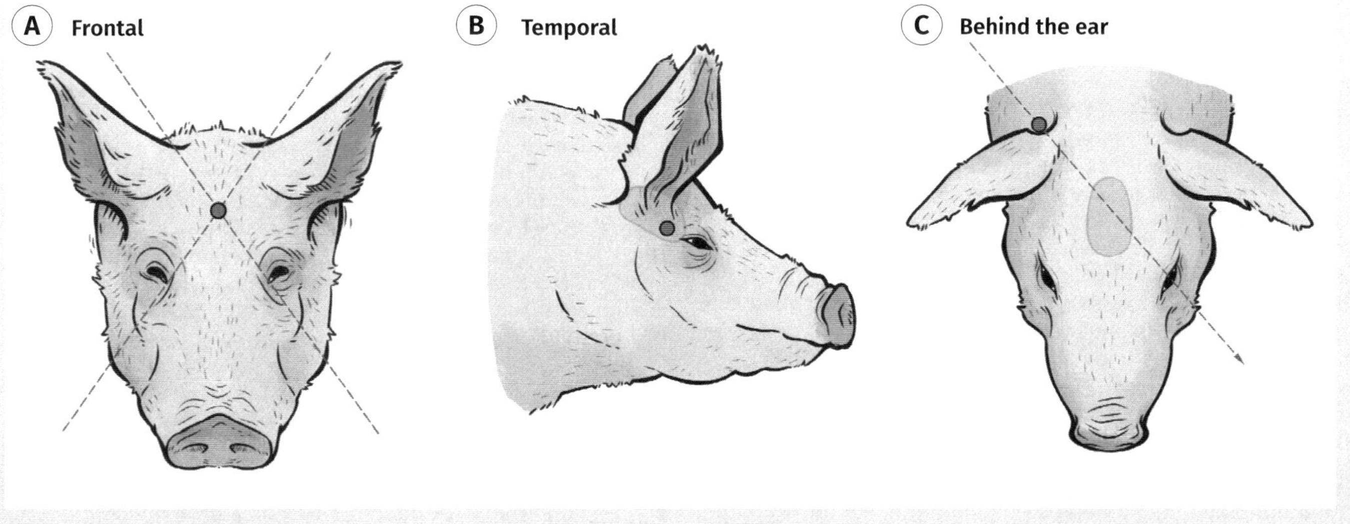

Evaluation of treatment efficacy

Upon application of euthanasia treatment, the following measures were used to assess treatment efficacy:

Confirmation of death

Confirmation of death was defined as any pig that achieved cardiac and respiratory arrest within ten min of treatment application. Cardiac arrest was confirmed by the lack of a detectable rhythmic heart rate using a pulse oximeter (SurgiVet® Advisor® Vital Signs Monitor Code V9204, Smiths Medical ASD, Inc., St. Paul, MN 55112) and respiratory arrest was confirmed by visual assessment of respiratory cessation of the chest cavity. Any pig that did not achieve a confirmed death ten min after treatment administration was euthanized using a commercial grade mobile electrocution stunner (Add stunner information here) applied across the head behind the eyes for 10 s and from head to heart for 10 s.

Skull thickness

Skull thickness was collected by measuring the distance from the edge of the cranial cavity (skull to skin interface of the parietal bone) to the inner bone lamina encompassing the cerebrum.

Hemorrhage and traumatic brain injury (TBI) scoring

Hemorrhage scores were assigned to four regions of the brain as a 0 (no hemorrhage) or 1 (hemorrhage present) score for each region. Traumatic brain injury (TBI) scores were assigned to each of six anatomical landmarks in the brain using a 3-point scale (0 = normal; 1 = some abnormalities; 2 = grossly abnormal, unrecognizable). Scoring systems were adapted.[2]

Upon completion of stage one, each equipment style by cranial location treatment was verified for efficacy within a sex. Efficacy was defined as all three animals within a subclass meeting the confirmation of death requirements within 10 min post-euthanasia application. Treatments that did not achieve this verification were excluded from stage two of the trial. This evaluation was done independently for each sex; therefore, one treatment could had been considered efficacious for one sex, but not for the other.

Stage two: Validation of alternative euthanasia techniques in conscious mature swine

Study animals and housing

Forty-two commercial, mixed breed, mature sows (n = 28; weight: 247.15 ± 16.96kg) and boars (n = 14; weight: 307.94 ± 46.32kg) were purchased from commercial vendors. Pigs were transported and temporarily held (< 24 hours) at The Ohio State University Meat Science Laboratory and managed as described in stage one. The objective of stage two was to evaluate the efficacy of each verified technique in conscious swine.

Euthanasia procedure

The same trained investigator from stage one completed all euthanasia events for stage two. Conscious swine were randomly assigned a treatment, weighed, and restrained using a common pig snare prior to implementation of the euthanasia treatment. A total of four treatments (C-FRO, C-TEM, C-BTE, and P-BTE) were evaluated for sows (7 sows/treatment) and two treatments (C-FRO and C-BTE) were evaluated for boars (7 boars/treatment) for a total of 56 animals.

Confirmation of death and post-mortem pathology (skull thickness, hemorrhage, and TBI scores) were conducted in the same manner as described for stage one.

Statistical analysis

Statistical analysis was conducted using STATA 14.2 (College Station, Texas). The proportion of attempts resulting in confirmation of death was calculated for stage one and two.

Data from all animals (78 animals; 32 boars and 46 sows) were combined in order to investigate the association between treatment and hemorrhage and TBI scores separately. As such, two separate multivariable linear regression models were built using the total sum of hemorrhage scores (across the four regions examined) and the total sum for TBI scores (across the six regions examined) as outcomes (response variables). The main predictors considered for both outcomes were the treatments, specified by the equipment style (C or P), and cranial location (TEM, BTE, FRO). Sex (boar or sow), weight (kg), and whether the animal was anesthetized or not were offered to the model as covariates.

Model-building steps included first checking for linearity between the continuous variable (weight) and the outcome, checking for collinearity between predictors, and construction of univariable models. The cut-off used to declare two variables as highly collinear was 0.80; and the cut-off value for variables from univariable models to proceed to the full model was $P < 0.20$. Sex and weight were perfectly collinear and, therefore, only one of those variables (sex) was carried forward during model construction. Finally, the two final multivariable models were built using a backwards stepwise approach and using a parsimonious approach. Variables were deemed as confounders if they changed any coefficient in the model by 20% or more and, in this case, were retained in the final models. Finally, the interaction between cranial location and equipment type was added to the final model. Final statistical significance was declared at $P \leq 0.05$; and a trend was declared at $0.05 < P \leq 0.10$.

Results and discussion

For stage one, four treatments were successful in sows and two treatments were successful in boars (defined as any pig that achieved cardiac and respiratory arrest within ten min after treatment application). When evaluating treatment efficacy by sex, the C style equipment was 100% effective in achieving death when applied to all cranial locations (FRO, BTE and TEM) for sows; however, the P style equipment was only 100% effective when applied at the BTE location for sows. For boars, the C style equipment was 100% effective when applied to the FRO and BTE location, but the P style equipment was not effective for any cranial location in boars. For stage two, treatments P-FRO, P-TEM, P-BTE and C-TEM were not included for boars, and P-FRO and P-TEM were not included for sows as they did not achieve >95% efficacy in stage one. All treatments included in stage two were 100% effective in achieving death in conscious animals.

Skull thickness

Average skull thickness for sows was numerically less than boars with a mean skull thickness of 7.12cm ± 0.87cm (range 4.5-9.5 cm; n = 35) when compared with 9.1 cm ± 0.89cm (range 7.0-10.5cm; n = 22) in boars.

Hemorrhage score

Mean (SD) hemorrhage sum values by cranial location and equipment are shown on Table 1. The final model for hemorrhage scores showed a significant interaction between cranial location and equipment type. When contrasting cranial locations by equipment, the P style gun type yielded a greater total hemorrhage score at the BTE location when compared to the TEM location ($P = 0.009$), whereas hemorrhage score was not different between locations for the C style type. Furthermore, when contrasting equipment styles, the C style produced a greater hemorrhage score compared to the P style type at the TEM location ($P = 0.019$), whereas there was no difference between equipment types for the BTE or FRO locations. Lastly, sex and having had anesthesia were not significant covariates in the model.

TBI score

Mean (SD) TBI score values by cranial location and equipment are shown on Table 2. For the final TBI statistical model, boars tended to have a lesser total TBI score compared to sows by approximately 0.61 ($P = 0.08$). The interaction between cranial location and equipment type was significant. When contrasting cranial locations by equipment, the P style equipment produced greater TBI score at the BTE location compared to both the FRO ($P = 0.04$) and TEM locations ($P = 0.01$). Similarly, the C style produced a greater TBI score at the BTE location compared to the TEM location ($P = 0.03$) but was not different from the FRO location ($P > 0.05$).

When contrasting equipment styles by cranial location, the C style produced a greater overall TBI score compared to the P style at the FRO location ($P = 0.0001$). The TBI score for the C style to equipment tended to be greater at the TEM location compared to the P style ($P = 0.08$). There was no difference in TBI score between equipment styles at the BTE location.

Table 1: Mean ± standard deviation (total number animals) sum of hemorrhage scores across four brain regions by equipment type style (Cylinder [C] and Pistol [P]) and cranial location (Frontal [FRO], Temporal [TEM] and Behind the ear [BTE])

	Cranial location		
Equipment style	**Frontal**	**Temporal**	**Behind-the-ear**
Cylinder	3.90 ± 0.31 (n = 20)	3.85 ± 0.55 (n = 13)	4.00 ± 0.00 (n = 20)
Pistol	3.67 ± 0.52 (n = 6)	3.17 ± 1.60 (n = 6)	3.92 ± 0.28 (n = 13)

Table 2: Mean ± standard deviation (number of animals) sum of traumatic brain injury scores across six regions by equipment type style (Cylinder and Pistol) and cranial location (Frontal, Temporal and Behind the ear)

	Cranial location		
Equipment style	**Frontal**	**Temporal**	**Behind-the-ear**
Cylinder	6.95 ± 1.14 (n = 20)	5.23 ± 1.58 (n = 13)	6.21 ± 1.23 (n = 19)
Pistol	4.17 ± 1.72 (n = 6)	3.83 ± 1.94 (n = 6)	5.77 ± 1.59 (n = 13)

The present study aimed to evaluate the efficacy of euthanasia at two alternative cranial locations utilizing two penetrating captive bolt gun equipment styles in mature swine. To date, the present study is the first to evaluate such euthanasia techniques in mature swine of this weight and age. Despite the limited sample size, results from the present study have provided critical information in identifying alternative euthanasia techniques that can be effectively applied on farm and produce a consistent and reliable death in mature pigs.

Based on the results from stage one, four treatments failed to achieve death in boars and two treatments failed in sows. These results are likely due to inadequate bolt penetration due to a combination of increased skull thickness and ineffective bolt gun force. When comparing sex, boars in the present study had an average approximate skull thickness of 2 cm greater than that of sows. In order for a penetrating captive bolt gun to be effective, it must produce enough force onto the skull to create a percussive effect on the brain and create skull fractures and associated intracranial bone fragments as a byproduct.[5] This shockwave of kinetic energy results in the immediate loss of consciousness followed by physical trauma to the brain tissue.[4] The increased skull thickness in boars likely prevented the bolt from effectively penetrating the brain and this idea is supported by the trend that boars demonstrated an overall lower total TBI score compared to sows.

In addition to increased skull thickness, euthanasia device type was also a significant factor. When comparing devices, the pistol style device has a smaller bolt penetration depth and cartridge load compared to the cylinder style device. The pistol style device was ineffective in achieving death when used in combination with any location in boars and both the frontal and temporal locations in sows. Individuals responsible for euthanizing large sows and boars should utilize the most powerful equipment available with the highest cartridge load to ensure maximum impact, as these equipment properties may be more effective in overcoming the anatomical challenges and variations we see in mature pigs.

In addition to the equipment, cranial location of the bolt gun directly impacted the extent of brain damage and thereby, treatment efficacy. In the present study, the BTE location consistently resulted in greater TBI scores compared to the frontal and temporal locations. This is likely due to the fact that behind-the-ear position bypasses the frontal sinus cavity and allows the bolt to directly penetrate the brain stem and cerebellum, two critical areas responsible for consciousness.

The results presented in the present study highlight the challenges associated with the euthanasia of mature swine. Several treatment combinations proved ineffective and are likely due to ineffective bolt gun penetration and greater skull thickness. Given the small sample size, future work is need to validate these alternative locations in mature sows and boars and apply these techniques on a commercial farm setting.

References

1. National Pork Board. 2009. On-Farm Euthanasia of Swine: Recommendations for the Producer.

2. Woods, Jennifer, "Analysis of the use of the "CASH" Dispatch Kit captive bolt gun as a single stage euthanasia process for pigs". 2012. Graduate Theses and Dissertations. 12706.

3. Bjarkam, C. R., Orlowski, D., Tvilling, L., Bech, J., Glud, A. N., & Sørensen, J. H. 2017. Exposure of the Pig CNS for Histological Analysis: A Manual for Decapitation, Skull Opening, and Brain Removal. Journal of visualized experiments: JoVE. 122:55511. **doi.org/10.3791/55511**

4. Anderson, K., Ries, E., Backes, J., Bishop, K., Boll, M., Branter, E., Hinrichs, B., Kirk, A., Olsen, H., Risius, B., Bildstein, C., and Vogel, K.D. 2019. Relationship of captive bolt stunning location with basic tissue measurements and exposed cross-sectional brain area in cadaver heads from market pigs. Transl. Anim. Sci. 3:1405-1409.

5. Machado C. 2010. Diagnosis of brain death. Neurology international, 2(1), e2. **https://doi.org/10.4081/ni.2010.e2**

Federal efforts to reduce the risk of African swine fever (ASF) introduction into the United States

Jack Shere, DVM, PhD

As part of its overall mission to protect American agriculture, the US Department of Agriculture's (USDA) Animal and Plant Health Inspection Service (APHIS) works to keep the country's livestock healthy. APHIS supports this mission in an international environment by collaborating with foreign partners to control pests and diseases before they can harm the United States, and by facilitating safe agricultural trade. When diseases of regulatory concern are identified in countries that were previously were of that disease, APHIS immediately takes action to impose appropriate import restrictions.

APHIS also invests in international capacity-building with foreign counterparts to build technical and regulatory skills that prevent the spread of damaging pests and diseases. As an example, APHIS is participating in a regional cooperative ASF prevention and mitigation effort in the Americas. Jointly, the United States, Mexico, and Canada have developed a framework that will support ongoing international collaboration and action in the areas of preparedness planning, enhanced biosecurity, business continuity, and coordinated risk communications. Next steps with our North American partners include developing a response foundation from this framework and moving the conversation forward toward regionalization concepts and intelligence gathering on an international scale that will inform trade policy.

The interlocking safeguards designed to prevent ASF from entering the United States extend to preventive measures at US ports of entry. Beagle detection teams conduct agriculture-related inspections and detect prohibited agricultural products that could introduce ASF. There are now 179 dogs teamed with US Customs and Border Protection at key US commercial sea ports and airports at a higher risk of encountering prohibited materials, i.e., those receiving cargo and passengers from ASF-affected countries like China. APHIS has also increased outreach and signage at key airports reminding travelers to declare prohibited meat/meat products.

Beyond port exclusionary activities, APHIS has increased messaging to industry partners to raise awareness of the direct biosecurity concerns associated with foreign visitors to domestic swine operations. In addition, since contaminated food waste fed to pigs is a risk for ASF virus transmission, APHIS is collaborating with States that allow garbage feeding to ensure these facilities understand best practices and that they undergo increased inspections. In summary, USDA continues to work with a wide range of partners—including the swine industry, US producers, other government agencies, and neighboring countries—to keep ASF out of North America. At the same time, we have response plans in place and incident management teams ready to deploy in the event that ASF does reach the United States.

For swine veterinarians: what is new in the USDA ASF Red Book?

The USDA ASF Red Book provides a comprehensive strategic response plan for the United States in the event ASF does encroach on our country. The objectives of this plan are to identify the capabilities needed to respond to an ASF outbreak in swine, the critical activities involved in responding to that outbreak, and the timeframes for these activities. In an outbreak situation, these critical activities fall under the authority of a unified Incident Command per the National Incident Management System. Emergency planners may use this strategic plan to develop a more detailed operational response plan. In an ASF outbreak, APHIS will post on its website at **aphis.usda.gov/fadprep** any response policy guidance that may change (e.g., specific virus elimination guidance, stamping-out policies, indemnity processes, etc.).

A swine veterinarian suspecting ASF on a premises should contact the office of their State Animal Health Official to discuss initiating a foreign animal disease investigation. If the investigation results in USD National Veterinary Services Laboratories (NVSL) confirmation of an ASF detection, the Secretary of Agriculture will take immediate steps to declare an Extraordinary Emergency: Issue a national movement standstill of at least 72 hours; authorize depopulation of infected premises; and authorize payment for virus eliminations at a uniform, flat rate, based on the size of the affected premises.

A national movement standstill is a complete stop in movement of live swine and swine germplasm across the United States. It is intended to allow States, Tribes, and industry time to gather critical information for a unified approach to an ASF response, establish initial control areas, conduct initial contact tracing, and increase biosecurity levels. USDA will provide implementation guidance as described in the Red Book.

Because of the long incubation period of the ASF virus and the high number of daily swine movements in the United States, practitioners should monitor for and expect to see additional cases of ASF in the days and weeks following a confirmation. These surveillance efforts may include Active Observational Surveillance (AOS). AOS relies on producers and veterinarians to track the disease status of a herd by noting and acting on clinical signs, abnormal health events (e.g., number animals treated, death loss) and production parameters (e.g., feed/water intake) that deviate from a predefined tolerance threshold. Each premises should keep a daily record of observation events, deviations from predefined thresholds, and any followup communication and action that occurred.

For effective AOS, producers/owners should create a set of guidelines that outline specific production parameters expected at that individual site with corresponding investigation triggers related to abnormal health events and decreased production parameters. Ideally, this should occur before an ASF outbreak occurs in the United States. It is also not too early for swine veterinarians to encourage their clients to maintain strict biosecurity protocols and assist biosecurity managers in identification of enhanced biosecurity practices that could be implemented in case of an outbreak.

APHIS will initiate a coordinated awareness campaign to engage and leverage Federal, State, Tribal, local, and stakeholder relationships to provide unified public messages for all audiences. Key communication messages to consumers and producers will emphasize that ASF is not a public health threat and is not a food safety issue.

Testing expectations and protocols in control areas and free areas

Effective and appropriate sample collection, diagnostic testing, surge capacity, and reporting are critical in an effective ASF response. These activities may require additional resources in the event of an ASF outbreak. Swine veterinarians may be asked to provide training and oversight for sample collection and submission. Surge capacity is also needed for diagnostic laboratory testing.

The NVSL Foreign Animal Disease Diagnostic Laboratory (FADDL) and National Animal Health Laboratory Network (NAHLN) laboratories have significantly enhanced testing capacity to meet the expected surge and have established procedures so that test reagents, kits, and supplies are available when needed.

- The number of NAHLN laboratories approved to test for ASF has been more than quadrupled since the Fall of 2018.
- The list of approved sample types has expanded from one option (blood) to four (blood, spleen, lymph node, tonsil) and pooling of up to 5 blood, spleen or tonsil samples has increased NAHLN testing capacity to 200,000 pigs/day. NVSL-FADDL continues to evaluate more sample types to offer more options for both individual animal and herd-level diagnostics. Samples currently being evaluated include spleen swabs, blood swabs and oral fluids.
- Five commercial ASF polymerase chain reaction kits have been evaluated and FADDL/NAHLN are prepared to deploy two of these in case of an outbreak.
- FADDL has stockpiled additional ASF proficiency tests and controls.
- FADDL and NAHLN are working with manufacturers to develop strategies for continuous-supply or rolling inventories of reagents, kits, and testing supplies.

Surveillance protocols

Passive surveillance is the voluntary reporting of suspect cases by producers, veterinarians, and laboratory personnel. It is conducted continuously in the United States through foreign animal disease investigations. In the event of an ASF detection, passive surveillance is intensified through rapid and clear communication to all producers, veterinarians, and other relevant stakeholders. In addition, there will be enhanced outreach emphasizing increased infection risk and the need for biosecurity. Specific morbidity and mortality levels that signal the need for investigation will vary by operation/production type. Herds within the Control Area that experience ASF-compatible clinical signs, abnormal health events (increased death loss/animals treated) or decreased production parameters (feed/water consumption or rate of gain) should be reported, investigated and sampled as rapidly as possible.

Animal health officials will immediately investigate and sample any new reported cases of ASF-compatible clinical signs or abnormal mortality events and premises with epidemiologic links to an infected premises. USDA will prioritize sample collection on a premises using targeted surveillance (selecting animals to sample by focusing on those animals more likely to have disease caused by ASF virus) as follows:

a. Abnormal mortalities (sudden death, found dead);
b. Sick animals with clinical signs compatible with ASF;
c. Animals with high probability of disease: animals that have decreased production/feed intake, appear unhealthy, are under high stress (i.e., recent movement), or may have an exposure risk (being housed near entry/exit points), but not necessarily showing clinical signs consistent with ASF; and
d. Apparently healthy animals.

Under current protocols, based on studies by the Center Epidemiology and Animal Health and the University of Minnesota—where initial surveillance modeling used individual animal samples targeted at sick and dead animals—17% of farms testing with intent to move animals during an outbreak would ship infected but undetected swine. Future addition of ASF oral-fluids testing will not significantly reduce the risk of shipping infected but undetected pigs. Given this, additional safeguards tied to biosecurity and truck washes are needed to accompany animal movements during an outbreak. Regardless of testing method used (individual samples vs. oral fluids) there is a timeframe after initial disease introduction to a farm when testing will not detect disease; this highlights the importance of biosecurity and other safeguards.

What is needed to restart trade, and realistic timelines

The OIE does not grant official recognition for ASF-freedom, but as a member of the OIE, the United States can self-declare the entire country, a zone, or a compartment free of ASF. Please refer to the most current OIE *Terrestrial Animal Health Code* for provisions on ASF freedom in a country or zone, compartment, establishment of a containment zone free from ASF, recovery of free status, and recommendations for importation of swine and swine products from countries not free of ASF. Trading partners will evaluate, on an individual or multilateral basis, all self-declarations from the United States for ASF-freedom both after an incident and for any regionalization activities that may occur during an incident. Trading partners determine whether to lift or modify trade restrictions.

Should an outbreak of ASF occur in a previously free country or zone, its status may be restored three months after the disinfection of the last infected establishment, provided that: a stamping-out policy has been implemented and a robust surveillance program in accordance with OIE standards has been carried out with negative results. In the case where ticks are suspected or known to be involved in the epidemiology of the infection, stamping-out was followed by the use of sentinel pigs in the infected establishments for two months.

In August 2020, APHIS executed a Cooperative Agreement with investigators at Iowa State University (ISU) to support a pilot project to develop a US Swine Health Improvement Plan, modeled after the National Poultry Improvement Plan. The pilot will allow State and Federal agencies, diagnostic laboratories, and industry stakeholders to collaborate directly to establish regulatory and diagnostic priorities, creating a more efficient and engaged process and fostering greater engagement in national swine disease control efforts. The pilot also has the potential to provide assurances to trading partners and consumers about US animal disease status.

Table 1: Recommended sampling scheme for an ASF outbreak response by zone and premises types

Zone/ Premises designation[1]	**Number of premises to sample within a zone**	**Within herd prevalence to detect**	**Frequency of sampling**	**Sampling duration**
Infected zone IP, CP, ARP, MP	All	10%	Every 3 days for 2 samplings, then every 6 days	Duration of quarantine
Buffer zone CP, ARP, MP	All	10%	Every 6 days[2]	Duration of quarantine
Surveillance zone FP	1% zone-level prevalence	10%	Within 15 days of first ASF detection, then every 15 days or as new zones are designated[3]	Duration of quarantine
Network CP	All	10%	Every 6 days	Duration of quarantine

1 Premises designation: Infected Premises (IP), Contact Premises (CP), At-Risk Premises (ARP), Monitored Premises (MP), Free Premises (FP).

2 Frequency of sampling is subject to change and can be adjusted based on observed incubation periods, likely route of disease introduction, feasibility of sampling and disease introduction risk.

3 If feral swine are present, more frequent sampling throughout the quarantine period is recommended.

USDA ASF Red Book: **https://www.aphis.usda.gov/animal_health/emergency_management/downloads/asf-responseplan.pdf**

Table 2: Sample sizes[1] for five prevalence values and a range of group sizes. Sample sizes achieve 0.95 probability of detection using a 95 percent sensitive test[2]

Group size[3]	**Sample size at 0.5% prevalence**	**Sample size at 1% prevalence**	**Sample size at 5% prevalence**	**Sample size at 10% prevalence**	**Sample size at 15% prevalence**
40	40	40	33	22	16
50	50	50	41	23	18
75	75	61	50	27	18
100	100	100	47	27	18
200	200	164	54	28	19
300	300	199	56	29	20
400	327	222	58	29	20
600	399	248	59	30	20
800	444	262	60	30	20
1000	474	272	60	30	20
> 5000	594	305	62	30	20

1 Sample sizes computed using Cannon, 2001. These are not appropriate for pooled or aggregate samples.

2 If pooling is performed at the lab, the number of individual samples collected on a premises may increase to account for different test sensitivity. The number of tests run by the lab will decrease if five sample pools are performed.

3 Group size refers to the size of the population being sampled. The same calculations would apply to individual animals or premises.

USDA ASF Red Book: **https://www.aphis.usda.gov/animal_health/emergency_management/downloads/asf-responseplan.pdf**

Producer and veterinary-led initiatives to prevent ASF introduction on US pig farms

Wesley Lyons, DVM
Pipestone Veterinary Services, Sycamore, Illinois

Introduction

African swine fever is one of the largest existential threats to the US pork industry. Knowing and understanding the production losses, export losses, and time to rebuild from a domestic break has been top-of-mind for producers and veterinarians since China first broke with the disease in 2018, decimating their sow numbers. As the US industry has learned more about routes of transmission, experts believe this virus has likely made the trip to our shores already, luckily failing to find its way into a pig. As the knowledge base of transmission through feed ingredients[1] and frozen and processed pork[2] has grown, veterinarians and producers have put science into action and bolstered their defenses against this catastrophic disease.

Eliminating pork products on farms

The United States is a net exporter of pork, however, in 2019, there was 945 million pounds of pork imported into the country. This included pork products from ASF-positive countries. It is proven that ASF persists in raw meat that is stored chilled or frozen for periods of 80+ days in fresh and even longer in frozen meat. ASF also persists longer than the curing process in specialty meats and survives within their shelf-stable life. This risk combined with the ever-present risk of smuggled products into cultural markets throughout the US has led producers to eliminate the entry of pork products into farms for employee meals.

Bolstering biosecurity through standardization

The Secure Pork Supply program quickly became a priority for producers everywhere in the event of a stop movement or compartmentalization of the industry. This program tasked producers and veterinarians alike to standardize thought processes and implementation of biosecurity across all sites. In lessons learned from HPAI breaks, producers also ensured adequate premise identification with state entities.

Mitigating risk of feed ingredient transmission- producer

The highest risk of introduction of ASF into the US swine herd lies within imported feed ingredients. Studies have shown that this virus survives in a trans-pacific shipping model in several key feed ingredients that the US imports from high risk countries.[1] Members of the Minnesota Pork Producers followed the Pipestone Veterinary Services' "Responsible Import" model to write policy that was introduced at Pork Forum receiving 100% approval. This producer-led initiative poses key questions and considerations to be taken anytime a feed ingredient is to be imported from an at-risk country.

- Necessity: is importation of the ingredient an absolute necessity?
- Alternatives: can the ingredient be obtained from a country free from foreign animal diseases?
- Virus: which virus is causing the concern?
- Viral half-life: is there published information on the half-life of the virus in the designated ingredient?
- Transport time: what is the projected time for delivery of the ingredient from the source to its destination?
- Viral load: are there safe products that can be added to the ingredient to reduce viral load during transport?
- Storage period: is there published information on storage time and temperature that will eliminate residual virus from the ingredient prior to use?

Producers representing 31 of the top pork-producing state associations also sent a signed letter to Secretary of Agriculture Dr. Sonny Perdue asking for regulatory action on high-risk organic soybean imports from positive countries.

Mitigating risk of feed ingredient transmission- veterinarian

Pipestone Applied Research has pioneered investigation of in-feed mitigation of viruses through a series of projects collectively called "Big Mig." A variety of products have been tested for efficacy in feed intentionally spiked with infectious doses of PRRS, PEDV, and SVA. Feed was fed through natural feeding behavior in pigs and testing was done on the complete feed as well as in the pig as a bioassay. The outcome has been a list of products that can effectively mitigate live viruses in feed and prevent infection of pigs being fed that feed. Discussion on inclusion in pre-mixes, vitamins, or complete feed is ongoing, but tools for this ever-present risk are now available.[3]

Summary

ASF continues to be a real risk for the US swine herd. Producers and their veterinarians have led the charge on protecting our industry and livelihoods through science-based research and recommendations to our government leaders.

References

1. Dee SA, Bauermann F, Niederwerder MC, et al. Survival of viral pathogens in animal feed ingredients under transboundary shipping models. *PLoS One* 2018;13:e0194509.

2. Mazur-Panasiuk, N., Żmudzki, J., & Woźniakowski, G. (2019). African Swine Fever Virus - Persistence in Different Environmental Conditions and the Possibility of its Indirect Transmission. Journal of veterinary research, 63(3), 303–310. **https://doi.org/10.2478/jvetres-2019-0058**

3. Dee, S, Shah, A, Cochrane, R, et al. Use of a demonstration project to evaluate viral survival in feed: Proof of concept. *Transbound Emerg Dis*. 2020; 00: 1– 5. **https://doi.org/10.1111/tbed.13682**

Gaps that exist to prevent African swine fever introduction into the United States

Emily Byers[1], DVM; Lisa Becton[2], DVM, MS; Rebecca Robbins[1,3], DVM, PhD
[1]National Pork Board ASF Task Force; [2]National Pork Board, Des Moines, Iowa; [3]Seaboard Foods, Guymon, Oklahoma

Introduction

African swine fever (ASF) is a contagious, viral disease of pigs, that can cause up to 100% death loss in domestic pigs, wild boar and feral swine. The ASF virus is the sole virus of the genus Asfivirus and within the family Asfaviridae.[18] The ASF virus was first identified in Kenya in the early 1920s. While there is a mild form of the virus found in Africa, the highly lethal, virulent and more common form of ASF has spread to many different areas in Africa, Europe and Asia. In 2007, the virus was identified in the republic of Georgia and has subsequently spread into parts of Russia and the Caucasus. This strain, called the 2007 Georgia strain, was identified in China in 2018 and rapidly spread into other Asian countries. Since then, ASF has continued to spread to many eastern and central European countries including Poland, Ukraine, Lithuania and Romania. In September 2020, ASF was identified in Germany's wild boar population along the Polish border. Currently, the United States, Canada and Mexico are free of the devastating disease.

If ASF were to strike the United States, options for managing a positive ASF herd would be limited. Unlike other foreign animal diseases that infect swine, such as classical swine fever and foot-and-mouth disease, there are currently no effective vaccines commercially available for use against ASF. Likewise, there are no approved ASF treatments or antivirals available for use. Measures for both prevention and management of ASF include adherence to strict biosecurity and sanitary measures on both local and national level. However, even with biosecurity measures, the spread of ASF globally has not been stopped. Measures to eliminate the virus if it were found in a US herd are outlined in the USDA's Redbook guidelines for ASF.[2] Such measures include a complete halt of all animal movements for a period of time, culling of infected animals, and extensive cleaning and disinfection along with an extended downtime for affected farms.

Positive identification of ASF in the United States would cause significant economic hardship for many different sectors of pork production and likely even extend into other livestock species. With nearly 30% of all US pork produced exported, the immediate economic impact from the identification of ASF in the US will be the closure of export markets. Even countries that are considered positive for ASF would likely stop importation of pork from the US as evidenced by actions post-identification of ASF in Germany's wild boar population.[4] A recent, updated study by Carriquiry, et al., from Iowa State University, Center for Agricultural and Rural Development, has shown that many different sectors of pork production and other farm animal species, would be impacted by ASF.[5] Estimated revenue losses to the industry are likely to be as high as $50 billion over 10 years with immediate loss of international markets that will take a lot of time and effort to recover.[6]

For many years, preparedness for a potential foreign animal disease introduction into the United States has been a high priority and remains a key component for maintaining business continuity and viability for production agriculture. However, with the more recent introduction and spread of ASF in Asia and Europe, prevention of ASF incursion into the United States is of the highest priority. There have been many different organizations, agencies and governments actively involved in prevention and preparedness activities. However, additional work remains in order to continue to prevent this virus from harming US agriculture.

2013 Border Protection against Foreign Animal Disease Gap Analysis[9]

The 2013 analysis was conducted during a workshop hosted by the National Center for Foreign Animal and Zoonotic Disease Defense Center (FAZD) in collaboration with the National Pork Board (NPB). Participants involved in the analysis included FAZD, NPB, Department of Homeland Security (DHS), US Department of Agriculture (USDA), state government officials, academicians, and swine industry representatives. The broad summary conclusions and recommendations of this analysis are found in Table 1. A long list of vulnerabilities was defined in the report for specific gaps and needs associated with respect to movement of foreign visitors, international surveillance and legal importation of foreign pork products, border control inspections and illegal importation of foreign pork products, domestic surveillance programs and diagnostic capacity, and feedstuff importation, among other concerns. A critical key finding within the report was a lack of resource allocation to meet these needs citing, "The biggest gap identified was that a broad increase is needed for resources at all levels... There is a maximum capacity at which agencies and the private sector can work, and at some point it becomes impossible to continue to 'do more with less'."

It is important to note that this consortium met just twelve days following the first confirmation of porcine epidemic diarrhea (PED) virus, a transboundary disease, introduction into the United states.[16] As the gap analysis report was being developed and finalized, the US swine industry suffered tremendous losses, and federal and state agencies struggled to respond or assist in the control efforts of a highly important, economically significant transboundary disease that would become endemic to the US swine herd in rapid fashion. This clearly exemplifies the potential for FADs, especially ASF, to do the same. While the introduction of multiple PED viruses and later porcine deltacoronavirus into the US swineherd were devastating events, many lessons have been learned since 2013, and the swine industry has benefitted from this knowledge. These informative experiences and resulting scientific knowledge should be used to help guide the direction of the industry and continued efforts to prevent future transboundary and foreign animal diseases.

Current industry opinions on ASF prevention

In preparation for this discussion, a brief and informal survey of industry representatives familiar with ASF was conducted. A total of sixty-seven (n = 67) responses were received from the following self-reported industry roles: academia (n = 11), allied industry (n = 6), government (n = 13), production (n = 5), veterinarian (n = 29) and other (n = 2). Using a sliding scale, participants were asked to quantify their level of concern of ASF introduction into the US from Not at all concerned (0) to Extremely concerned (100). The level of concern ranged from scores of 49 to 100 with a median concern score of 88 and a mean concern score of 84. Using a sliding scale numerically representing values from Poor (0) to Excellent (100), participants were asked to quantify their confidence in the ability of the US to mitigate the risks of ASF introduction. Confidence scores ranged from 0 to 100 with a median confidence score of 39 and a mean confidence score of 42. The remainder of the survey allowed for open-ended responses to the following question: "What are the 3 most important gaps that you believe exist to prevent ASF introduction into the United States?" One hundred ninety responses were received to this question and are categorized in Table 2.

Table 1: 2013 Border Protection against Foreign Animal Disease Gap Analysis[9]

Summary conclusions and recommendations[10]
Improve coordination, communication and information sharing between federal and state agencies, and between government and industry
Strengthen international relationships, especially in neighboring and high-risk countries
Improve message clarity and utilize easy to understand, visual outreach tools whenever possible to overcome language barriers
Gain industry trust and promote industry support in both the US and partner countries to help ensure success
Support development of new assays, models, surveillance methods and technologies to assist in border screening and investigations
Work across species/commodities to implement effective, consistent strategies for communicating
Support policy initiatives to clearly identify processes to be implemented in the event of FAD introduction and communicate those processes/policies across commodities and levels of government

Table 2: Current industry opinions on ASF prevention when asked, "What are the 3 most important gaps that you believe exist to prevent ASF introduction into the United States?"

Importation biosecurity: CBP limitations, contaminated supplies, inputs and/or feedstuffs from ASF countries	45
Travel biosecurity: customs & border patrol (CBP) limitations/enforcement, product smuggling and/or tourist inspection	43
Current biosecurity and transportation practices (including implementation/compliance)	16
Bioterrorism	12
Lack of active surveillance and diagnostic testing capability/capacity/validation	11
Lack of coordinated prevention / response plans, risk assessments	11
Importation biosecurity: pork products from ASF countries	8
Lack of confidence in leadership: urgency, resources, and funding availability	8
Communication/Eeducation/awareness: on-farm, backyard producers, general public	5
Depopulation / mortality management	5
Delayed detection potential of ASF	4
Garbage feeding programs	4
Continued international spread of ASF, including risk to Mexico/Canada	4
Lack of effective vaccine	3
Lack of adequate feral swine control programs	3
Communication/collaboration between producers and state, federal officials	2
Scientific lack of understanding	2
US Pork industry representatives traveling between US and ASF countries, especially with swine contact	2
Transparency of international data	1
Accredited veterinarians	1

Although this survey was limited in a number of ways and potentially included a sampling bias, the concerns reflected by the participants are consistent with the previously published gap analysis, risk assessments, and peer-reviewed literature. In general, the survey reflects concern surrounding the likelihood of ASF introduction into the US and a low level of confidence in the ability of the US to mitigate the risks of ASF introduction. By far, the most consistently expressed concerns regarding ASF prevention gaps involved surveillance, detection, and enforcement limitations associated with importation biosecurity and international traveler biosecurity. Together, these two categories represented 46% of the responses, and are consistent with a statement published by Brown and Bevins in 2018 noting, "The legal or illegal movement of live animals or their products, byproducts, or animal feed...comprise the routes of highest concern for ASFV introduction into the United States."[4] The concerns expressed in the survey may reflect knowledge of international risk reports including retrospective analysis of initial ASF entry into Georgia near a major port,[8] Australia's detection of ASF in illegally imported meat products,[14] Japan's risk assessment[11] and eye-opening publications from American researchers[7,8,13] highlighting feedstuff importation risks. Additional notable concerns included limitations of current industry biosecurity and transportation practices (including implementation and compliance), agroterrorism/bioterrorism, lack of adequate active surveillance programs, and a lack of coordinated prevention/response plans and risk assessments.

Discussion

Much has been learned since 2013 that can be used to enhance ASF prevention strategies in the US. The list of funded research, peer-reviewed publications, interagency and international collaboration projects, and other work is extensive, especially when one incorporates knowledge gained and lessons learned through experiences of other emerging diseases in the industry, such as PED and COVID-19. Many industry-led task forces and working groups have been established and/or expanded to further response planning and progress, such as: ASF Crisis Team, Swine Health Information Center, National Swine Disease Council, Packer ASF Business Continuity Group, NPB/AASV Depopulation Working Group, AASV Committee on Emerging and Transmissible Diseases, and a number of state and national swine emergency and rapid response teams.

Presentations during the Foreign Animal Disease seminar at the 51st AASV Annual Meeting in 2020 highlighted a substantial amount of progress that has been made in terms of ASF preparedness.[1] These included a wide range of topics from pig movement shutdown planning, euthanasia and disposal, Secure Pork Supply, vaccine development efforts, and validation of practical sampling and diagnostic modalities. Commitment toward continued regulatory and industry coordination in preparedness planning efforts was exemplified by through high-level and multi-state tabletop exercises in 2019.[1] While all of the knowledge and preparedness gained should be celebrated, it should not overshadow the critical fact that enhanced efforts to prevent ASF introduction in the first place is the single most important area of progress that is still needed. Industry-led groups that are prioritizing the area of prevention include but are not limited to: The National Pork Producer's Council, Swine Health Information Center, NPB ASF Task Force, National Swine Disease Council, and joint Feed Safety Taskforce.

Communication, collaboration and cooperation between pork producers, veterinarians, and regulatory officials has significantly improved since 2013, which is a tremendous accomplishment towards bridging the remaining gaps. Sundberg et. al. described during the 51st AASV Annual Meeting a number of important advances that have been achieved as a result of collaborative efforts.[14] This presentation showed that collaborative partnerships between industry and regulatory groups continues to advance; however, there remains opportunity to make continued improvements in all sectors. While the progress made is positive, there will always be gaps of knowledge in an industry that conducts business and promotes innovation at the rapid pace of the vertically integrated swine industry. Simply advancing knowledge will never be enough to prevent ASF introduction into the US. Much work remains to apply the knowledge into continued tangible, practical implementations which serve to increase barriers of ASF introduction pathways and further protect the US swineherd. All entities must continue to diligently place a sense of urgency, fiscal priority, and mutual collaborative effort into continued progress. Industry stakeholders, swine veterinarians and regulatory agencies must continue to strive for advances that result in collaborative ACTIONS that promote prevention of ASF introduction into the US.

Finally, while beyond the scope of this presentation, it is also important to note that key components of FAD preparation and response should simultaneously remain an important focus in the US. Like PED introduction, market disruptions associated with the COVID-19 pandemic allowed for a real-world exercise and practical application of response plans, which further enhanced understanding of gaps and challenges associated with implementing a future nationwide FAD response. Aside from the direct impacts to the swine industry through market disruptions, the overall damage caused in the US by the 2020 global human Coronavirus pandemic (COVID-19) is an important example to learn from. Many swine veterinarians and producers express a concern that failure to prevent an ASF introduction into the US will result in similar havoc.

Lessons will continue to be extrapolated from this experience that are likely predictive of challenges the animal protein industries will face in the event of ASF introduction. Of notable concern are the parallels to the drastic overwhelm of national diagnostic laboratory capabilities, insufficient national stockpile of medical supplies and personal protective equipment and overwhelm of the frontline healthcare system resources and personnel. The strikingly clear and obvious immediate messages from this experience that the swine industry and regulatory agencies must recognize are: (1) prevention of ASF is critical for sustainability of the industry; and (2) if prevention measures fail, a lack of cohesive response planning and preparedness with immediately available access to sufficient resources will result in dramatically poor outcomes for the industry.

References

1. AASV. Foreign Animal Disease Seminar. Proceedings. 51st Annual AASV Meeting (2020). Available online: **https://aasv.org/library/proceedings/get.php?2020%2FSeminar11.pdf.**

2. AASV. News Archive. USDA Strengthens Partnerships and Protections to Keep African Swine Fever Out of the Country. March 6, 2019. Available online: **https://aasv.org/news/story.php?id=11478.** African Swine Fever Response Plan. The Redbook. FADPReP Unites States Department of Agriculture. **https://www.aphis.usda.gov/animal_health/emergency_management/downloads/asf-responseplan.pdf.**

3. ASF Confirmed in Germany. Implications for the Global Pork Trade. Market Intel September 17, 2020. **https://www.fb.org/market-intel/african-swine-fever-confirmed-in-germany.**

4. Brown, VR and SN Bevins. A Review of African Swine Fever and the Potential for Introduction into the United States and the Possibility of Subsequent Establishment in Feral Swine and Native Ticks. *Front Vet Sci*. 2018;5:11. Published 2018 Feb 6. Available online: **www.ncbi.nlm.nih.gov/pmc/articles/PMC5808196.**

5. Carriquiry M, Elobeid A, Swenson D, and Hayes D. Impacts of African Swine Fever in Iowa and the United States. Working Paper 20-WP 600. March 2020. Center for Agriculture and Rural Development. **www.card.iastate.edu.**

6. Dee, S. A., Bauermann, F. V., Niederwerder, M. C., Singrey, A., Clement, T., de Lima, M., & Diel, D. G. (2018). Survival of viral pathogens in animal feed ingredients under transboundary shipping models. PLoS One, 13(3), e0194509. **https://doi.org/10.1371/journal.pone.0194509.**

7. Dee, S. A., Niederwerder, M. C., Patterson, G., Cochrane, R., Jones, C., Diel, D., Brockhoff, E., Nelson, E., Spronk, G., & Sundberg, P. (2020). The risk of viral transmission in feed: What do we know, what do we do? Transboundary and Emerging Diseases. **https://doi.org/10.1111/tbed.13606.**

8. Food and Agriculture Organization of the United Nations. Emergency Prevention System for Animal Health. African Swine Fever in Georgia, 2007. Available online: **www.fao.org/docs/eims/upload/230205/EW_ASF_Georgia_Jun07.pdf**

9. Gap Analysis Workshop Report. Border Protection Against Foreign Animal Disease. National Center for Foreign Animal and Zoonotic Disease and National Pork Board. Crystal City, Virginia. May 29-30, 2013. v1.0. Updated August 20, 2013.

10. Herrera-Ibatá, DM, et. al. Quantitative approach for the risk assessment of African swine fever and Classical swine fever introduction into the United States through legal imports of pigs and swine products [published correction appears in PLoS One. 2018 Nov 20;13(11):e0208065]. *PLoS One*. 2017;12(8):e0182850. Published 2017 Aug 10. **www.ncbi.nlm.nih.gov/pmc/articles/PMC5552331.**

11. Ito, S., Jurado, C., Sánchez-Vizcaíno, J. M., & Isoda, N. (2020). Quantitative risk assessment of African swine fever virus introduction to Japan via pork products brought in air passengers' luggage. Transbound Emerg Dis., 67(2), 894-905. **https://doi.org/10.1111/tbed.13414.**

12. Patterson G, Niederwerder MC, Spronk G, Dee SA. Quantification of soya-based feed ingredient entry from ASFV-positive countries to the United States by ocean freight shipping and associated seaports. Transbound Emerg Dis. 2020 Oct 16. doi: 10.1111/tbed.13881. Epub ahead of print. PMID: 33064921.

13. Smith, Aidan. Farm Weekly. Border Checks Detect African Swine Fever. 2019. Available online: **www.farmweekly.com.au/story/6253800/border-checks-detect-african-swine-fever.**

14. Sundberg P, et. al. Preparing for the next transboundary or emerging disease: What the industry needs to do and what it is doing to prepare. Proceedings of the 51st AASV Annual Meeting (2020.) Available online: **https://aasv.org/library/swineinfo/item.php?AASVPCS/2020/04-003_Sundberg.pdf.**

15. United States Department of Agriculture. APHIS. Technical Note: Porcine Epidemic Diarrhea (PED). Available online: **www.aphis.usda.gov/animal_health/animal_dis_spec/swine/downloads/ped_tech_note.pdf.**

16. United States Government Accountability Office. Actions Needed to Improve Information on Facilities and Capital Planning at Land Border Crossings. Public release July 11, 2019. Available online: **https://www.gao.gov/products/gao-19-534.**

17. Wang, T., Sun, Y. & Qiu, HJ. African swine fever: an unprecedented disaster and challenge to China. Infect Dis Poverty 7, 111 (2018). **https://doi.org/10.1186/s40249-018-0495-3.**

What the state veterinarians' African swine fever working group has done to help improve preparedness

Mike Neault

No paper submitted

Industry-led efforts to develop better tools to detect and respond to an ASF outbreak

Paul Sundberg, DVM, PhD, DACVPM
Executive Director, Swine Health Information Center, Ames, Iowa

Introduction

The recent spread of ASF in Russia, Europe, and Asia is a stark reminder of the importance of protecting the US pork industry from FADs. The economic repercussions of the introduction of FADs into the US are enormous, making it imperative to protect the industry. Much progress has been made during the last two years. Without review of that progress, some of the more recent projects and activities to help prevent, prepare and respond to an ASF outbreak in the US are presented.

It is important to point out that the "industry-led" efforts and tools are not possible without the broad collaborative effort among pork producers, practicing veterinarians, allied industries, university and government academics, state animal health officials and federal animal health officials. The pork industry helps to lead, as do these collaborators in their best manner. This is a group effort, and its success depends on everyone working together to do the best possible to keep foreign animal diseases out of the US but also to be as prepared as possible should it enter.

Prevention

National pork industry biosecurity risk assessment

The National Pork Board and the Swine Health Information Center (SHIC) are funding a project, with the collaboration of the National Pork Producers Council and the American Association of Swine Veterinarians, to identify gaps in US pork industry national biosecurity. It will point out potential pathways through which FADs can be introduced, spread, and affect the domestic pork industry. The goal is to identify and act on biosecurity gaps to prevent entry of foreign animal disease (FAD) into the country.

International collaboration

Swine Innovation Porc, a non-profit corporation in Canada, facilitates research in the Canadian swine sector. Their goal is enhancing the profitability and sustainability of the Canadian pork industry. In 2019, Swine Innovation Porc developed a Coordinated African swine fever (ASF) Research Working Group and SHIC was invited to participate. Over six months, the ASF Working Group created an ASF-related research priorities document which was completed in December 2019.

This outline resulted in collection of information regarding research gaps in addressing the threat of ASF which was provided to the ASF Research Working Group, a value to the US pork industry, as well. The Group facilitates coordination between scientific experts, industry leaders, and Canadian federal agency partners.

Potential feed risk of introduction and dissemination

Dr Chad Paulk, Kansas State University, is doing a comprehensive look at production, importation and uses of soybean meals. This will include their possible role in introduction and/or dissemination of swine viral pathogens, monitoring for fecal contamination indicators and 'Best Practices' for the importation of organic soy products, which is often difficult to procure from domestic sources and has been reported to retain infectious virus longer than many other ingredients.

Studies have shown the theoretical ability for swine viruses to be transmitted through feed and feedstuffs. Ongoing research caused SHIC, the National Pork Board, the National Pork Producers Council, and the American Association of Swine Veterinarians to offer updated and revised information for feed holding times to mitigate virus transmission. Feed component holding times could be one tool used in conjunction with other mitigations to enhance feed safety. Sourcing feed components from regions or countries of the world that don't have ASF, or other diseases foreign to the US, will also negate potential risk.

Preparedness and response

National Swine Disease Council

The National Swine Disease Council is an opportunity for pork producers, swine veterinarians, allied industry and state and federal animal health officials to discuss and cooperatively respond to an emerging swine disease. Included in its mission is providing recommendations to animal health officials and industry stakeholders to mitigate threats and negative impacts to the US pork industry from diseases of concern.

The Council also works to protect international trade and interstate commerce of pigs, pork and pork products. An additional objective is to promote capacity-building to rapidly detect diseases of concern and limit the scope of a disease outbreak. It is supported by the professional staffs of SHIC, NPB, NPPC, AASV and the North American Meat Institute.

Feral swine

This year, USDA estimates there are 6 million feral swine in the US creating issues for traditional livestock production, natural resources, and other species. To address these ongoing issues, including domestic and foreign disease surveillance priorities of feral swine, USDA convened a technical working group consisting of swine industry representatives, state and federal animal health officials, university, and wildlife experts.

In an article published in the *Journal of Animal Science* (2020, Vol. 98, No. 8, 1-3), "Perspectives on the past, present, and future of feral swine disease surveillance in the United States," progress in management of feral swine is detailed. The first priority for the revised program was function as an FAD surveillance system. Specifically, it focuses on CSF (serologic) and ASF (morbidity and mortality) as well as FMD (syndromic). Another priority was for the framework to be nimble enough to respond to emerging threats.

Feed mill biosecurity

Monitoring for Enterobacteriaceae (EBAC) within feed mills is a demonstrated way to predict the risk of Salmonella spp. also being present. Viral pathogens in swine feed are a high interest health issue but there is little information available on how the presence of EBAC correlates with viral pathogen presence, especially on farms or in feed mills. The first step is to evaluate the efficacy of mill or farm level feed biosecurity. The purpose of a study conducted by Kansas State University was to identify EBAC presence in the feed manufacturing facilities and farms of a multi-farm system.

Compliance with biosecurity protocols had a substantial impact of EBAC prevalence and distribution throughout the feed mill. As facilities begin to transition biosecurity from the farm to the feed mill, using environmental monitoring to evaluate risk for biosecurity gaps, as well as success in their mitigation, will be useful and necessary.

Oral fluids

Recent unpublished, observational research indicates oral fluids (OF), when collected by rope as an aggregate sample, could be a good sample for rapid detection of ASF virus. However, as the viral load, as measured by PCR CT comparison, was much lower in OF than individual tissue samples, the study indicated false negatives do occur. This limits the potential use of swine OF as an official monitoring and surveillance tool in the event of an ASF outbreak in the U.S. without further study and refinement.

A call for proposals to develop methods to improve the detection of low levels of nucleic acid in OF through enhancements to sample pre-extraction treatment(s) or through improved extraction methodologies compatible with the high throughput testing currently done in NAHLN laboratories was published. Two proposals were selected for collaborative SHIC and National Pork Board funding and the work has begun.

The use of oral fluids for monitoring and surveillance potentially in concert with mortality testing is also being investigated and considered. Added onto the Vietnam oral fluids project, described later, is a project in Romania to collect a variety of field samples as soon as possible on farms after an ASF outbreak. It will help add to the body of knowledge needed to evaluate how oral fluids might be incorporated in surveillance in the US post outbreak and it will help inform USDA outbreak models to make them more accurate so they can continue to improve ASF surveillance plans.

USDA-Foreign Ag Service funded Vietnam ASF control project

With support from NPPC, SHIC has been funded by the USDA-Foreign Ag Service to enhance the capability of Vietnam to respond to ASF and to study ASF in field settings. The outcome will be to enhance the US industry's opportunity for pork exports to that country and to learn the lessons needed to be prepared to effectively respond if the virus arrives here.

Strengthening capacity for response

The USDA grant, "Building capacity to support the control of African swine fever (ASF) in Vietnam," has two sections. The first section focuses on strengthening Vietnamese veterinary services capacity for mitigating ASF impact. To achieve this goal, the project will create a capacity-building program to train veterinarians, laboratory workers, and/or farm advisors or managers on topics described by the OIE as necessary for assurance of functional national veterinary services organizations, with a focus on ASF prevention and control. The program will involve delivery of three training courses using a blend of mentored online delivery and face-to-face workshops.

Field projects to learn lessons needed for response and control

The second section "Implementation of field projects, and collection and analysis of samples" will help with industry preparedness and response, in Vietnam and should ASF reach the US. Contracts have been written and work has begun on a variety of ASF priorities. Project titles and objectives follow.

ELISA

"Full validation of two commercial ELISA assays for the detection of antibodies to African swine fever" and "Evaluation of the diagnostic performance of a ASFV serum/oral fluid antibody ELISAs under field conditions in Vietnam." The objectives of these two projects include evaluating the performance of ASF serum and/or oral fluid ELISAs for use in the surveillance and monitoring of ASF outbreaks in commercial farms in Vietnam. More broadly, fulfillment of the project aims could bring the development of a comprehensive integrated swine surveillance system ("no single best diagnostic approach") based on the combination of molecular and antibody methods.

Rodents

"Potential of rodents to be a vector in the transmission of African swine fever in two commercial farms in Vietnam with differing biosecurity levels." The objectives are to determine if ASF virus can be detected in mice and rats, and if so, which tissues are the best to sample; to determine the impact of farm biosecurity level on rodents' ability to carry the disease; and to measure mouse-to-mouse transmission of ASF in a controlled laboratory situation.

Baking for disinfection

"Time and temperature required for complete inactivation of African swine fever virus." Objectives are to determine the optimal baking time and temperature required to completely inactivate ASFv in aluminum surface contaminated organic materials. This research is designed to simulate the sanitation protocols currently used to disinfect animal trailers.

Epidemiological analysis of pathways of entry

"Identifying pathways of entry of ASF virus onto farms to enhance information for improving biosecurity in Vietnam." Objectives are to use the rapid response outbreak survey of SHIC's Rapid Response Program to investigate ASF pathways on Vietnamese farms and to test an electronic format for completing the outbreak investigation forms.

"Determining the pathways for ASF introduction into boar studs and risk of ASF transmission via semen movements during an ASF outbreak." This project will determine the risk of introducing ASF to a sow farm as a result of semen movement from apparently healthy boar studs located in an ASF disease control area. A pro-active risk assessment will be performed to systematically evaluate the potential risk of semen movements during an outbreak to investigate the pathways of ASF introduction into boar studs, the pathways of ASF introduction associated with semen movements into sow farms, the simulated spread of ASF in a sow farm in which multiple sows are simultaneously exposed to ASF via the AI process, and what surveillance is needed and how long should semen be held to increase the likelihood of ASF detection.

Composting

"Validating the Composting Process for ASF Virus Inactivation." Funded by the National Pork Board, this project will assess the effectiveness of the swine carcass composting process to inactivate the ASF virus by empirically estimating the degradation curve for ASF virus using composting as a management tool; will determine if regionally sourced carbon materials affect the composting process and subsequently the inactivation of ASF virus; and will determine if ASF virus survives in bone marrow during the compost process

Targeted test and remove

"Validating protocols for the targeted removal of individually housed sows when infected to move the herd to negative status." The objective of this project is to test sufficiency of current "Test and Remove" methodology for ASF elimination in Vietnam sow herds. Also, these field samples will be used to evaluate diagnostic performance of 4 commercially available ASF point-of-care (POC), pen side, assays for detecting ASF in ASF-suspect and non-clinical neighboring animals.

Pen-side tests

"Evaluate the diagnostic performance of pen-side tests for ASF detection." Three different pen-side tests will be evaluated and compared using both whole blood and oral swabs as collected tissues. This project will work to determine the time from infection to the earliest detection using pen-side tests and the sensitivity and specificity of the pen-side tests for detection of ASF in the field.

Feed transport and mill biosecurity

Initially proposed as "Investigating methods for decontamination of interior surfaces (cabs) of transportation vehicles" as those decontamination methods apply to the Vietnamese feed industry in that environment, a modified project has been funded by the National Pork Board. It will develop a model to evaluate methods of cleaning and decontamination of transportation vehicles at facilities located at (KSU) using surfaces representative of real-world conditions and evaluate multiple strategies using PEDV in BSL-2 conditions at KSU. Those results should then be able to apply lessons learned and build capacity to mitigate feed and delivery risk in Vietnam and the US, should the virus enter this country.

Oral fluids

"Field evaluation of oral fluids as a convenient, aggregate sample for early detection of African swine fever." The objective is to conduct a field evaluation in Vietnam of oral fluids in conjunction with other sample tissues for early detection of African swine fever virus. A collaborative project of the Canadian Food Inspection Agency and USDA, the outcome will help evaluate the scientific appropriateness of using swine oral fluids as a monitoring or surveillance tissue for PCR analysis and inform US surveillance plans through better understanding ASF disease outbreak dynamics.

Veterinary perspective on managing herds within a control area

Marisa Rotolo
Pig Improvement Company, Hendersonville, Tennessee

What would you do if ASF hit tomorrow?

If African swine fever (ASF) is detected within the United States, the swine industry as we know it will drastically change. No one should be surprised by that statement; the real question will be what will change and how can we manage the damage.

Since the detection of ASF in China in August 2018, the US swine industry has increased surveillance and preparedness. The United States Department of Agriculture (USDA) began to test for ASF as outlined in the national swine hemorrhagic fevers surveillance plan. Tabletops and exercises have occurred across the country in an effort to identify gaps and offer solutions to challenges such as the mass depopulation of a site. In April 2020, the USDA released the African Swine Fever Response Plan: The Redbook. This 97-page document provides a reference guide for how an ASF outbreak will be handled. The Redbook allows for producers and veterinarians alike to begin to prepare for what may come.

What you need to know

From the Redbook, there are four phases of an ASF outbreak; 1) a national stop movement and traceback investigation to identify all exposed and potentially exposed pigs, 2) establishment of control zones, 3) sampling for business continuity, and 4) eradication. While each phase is important, special attention should be paid to control areas and what requirements farms will need to satisfy to maintain biosecurity.

In the event of an ASF outbreak, control areas will be established around the infected site and any site outside of the control area with an epidemiological link to the infected premise. A control area consists of an infected zone and buffer zone. If no feral swine are present near the infected site, the infected zone will be 3 km with a 2 km buffer zone. These zone sizes would increase in the presence of feral swine. Any premises that fall within the control area will be required to collect and submit samples satisfying their specific premises designation requirements.

There are five designations for premises within a control area. An Infected premises is a premises with confirmed ASF. A Contact premises is a site with an epidemiological link to an Infected premises. A Suspect premises is a premises with ASF-compatible clinical signs. An At-Risk premises is a premises with swine but no ASF-compatible signs. The At-Risk premises are allowed permitted movement of animals within the control area. A Monitored premises is formerly an At-Risk premises; however, animals can under permitted movements out of the Control area. Each designation will have different outbreak surveillance requirements.

A group effort

Managing a herd within the control area will require increased labor, surveillance, coordination, collaboration, and patience. It is vital that premises prepare for a FAD by compiling site-specific secure pork supply plans, working with their federal and state officials to determine movement requirements, and training on farm staff for ASF sample collection in a biosecure manner. With the current protocols, veterinarians, farm staff, and federal and state officials will be needed to handle the sampling burden. In addition to the labor dedicated to collecting samples, submission and testing of the samples will need to be coordinated with an ASF-certified NAHLN veterinary diagnostic laboratory. Any personnel on premises within the control area must observe downtime before visiting premises outside the control area.

There will be multiple moving parts within a FAD outbreak, some that will be discovered in the thick of it. It is imperative that the industry continue its efforts for preparedness and response.

Packer response to an FAD outbreak

T. L. Lee, DVM, PhD
North American Meat Institute

Introduction

The occurrence of a foreign animal disease (FAD) in the US could devastate the food animal industries. Being prepared for an outbreak situation can help keep plants running or shorten the amount of time that movement or exports are stalled in the event a foreign animal disease is discovered in the United States.

In the past five years, the meat industry has increased its industry engagement regarding foreign animal diseases. The industry recognizes now, more than ever, the need for knowledge about foreign animal diseases and collaboration with others in the animal agriculture industry. As such, the North American Meat Institute and its members are now fully engaged in ongoing efforts made by the National Pork Producers' Council, the National Pork Board, the Swine Health Information Center, and the American Association of Swine Veterinarians to prepare for and possibly respond to a foreign animal disease, particularly African swine fever, in the hog population in the United States.

Response to a possible FAD detection in a slaughter establishment

Beef, pork, and lamb slaughter establishments are unique production units in FAD detection, as they operate under the oversight of two governmental bodies in the event of an FAD investigation, the Food Safety and Inspection Service (FSIS) and the Animal and Plant Health Inspection Services (APHIS). Additionally, State Animal Health Officials (SAHO) have jurisdiction in their respective states in the event of an outbreak and will facilitate response activities. If an FAD is suspected in a slaughter facility, the establishment must work cooperatively with all regulatory agencies to ensure that all necessary steps are taken in an FAD response, as well as the subsequent steps required to continue normal business operations after an outbreak is contained.

To navigate such a heavily-regulated landscape, especially in the event of an FAD outbreak, the North American Meat Institute (NAMI or the Meat Institute) formed an industry-specific working group to develop guidance on how to prepare for and handle an outbreak situation in a slaughter facility. The Meat Institute developed a specific guidance on FAD preparedness, recognizing clinical signs of FADs, and biosecurity practices in slaughter facilities. Additionally, a specific guidance document was developed to outline the steps an establishment must take if an FAD investigation is conducted in their facility. The following summarizes the information provided in the NAMI FAD investigation guidance.

If clinical signs of a foreign animal disease are detected by an establishment employee, the employee should notify establishment management immediately and if clinical signs are confirmed, management should immediately notify the FSIS public health veterinarian (PHV). If clinical signs are observed by FSIS personnel, the PHV and establishment management should be notified immediately.[1] The animals displaying clinical signs should be separated from animals not showing clinical signs. The entire lot may be held for sample collection. The PHV should notify the FSIS District Office as soon as possible, providing, at minimum, the origin of the affected lot, the time of detection, the severity of the lesions, and the number of animals in the lot. The District Office will then contact APHIS and the State Animal Health Official (State Veterinarian) and provide the information gathered by the PHV.[1]

The State Animal Health Official or the Assistant Director will determine how the case should be handled. If an investigation is warranted, a Foreign Animal Disease Diagnostician (FADD) will be sent to the establishment to investigate. FADDs should conduct a site visit within 24 hours, but experience shows that most visit within a few hours of the finding.

Meanwhile, the plant should notify their procurement, quality control, and other appropriate teams to help gather information on the lot's history. Such information should include:

- Premises ID number
- Production facility of origin
  - Name, address, and phone number of production facility
- Species of animals involved
- Number of animals in the lot
- Identification of drivers/trucks delivering the lot to the establishment
- Time of arrival at the plant
- Time at which the trucks carrying the affected animals left the establishment
- Information on other lots from the same live animal production facility arriving at the establishment
- Antemortem pen cards
- Clinical signs present/gross lesions
- Any history of the animals showing clinical signs
- The last 30 days of records for live animal movement to the establishment

When the FADD arrives at the establishment, he/she will gather the information collected (above), and collect samples from the affected animals.[2] The FADD will send samples to a National Animal Health Laboratory for preliminary testing. Samples will also be sent to the National Veterinary Services Laboratory's Foreign Animal Disease Diagnostic Laboratory at Plum Island, NY for confirmation of test results.[2] Depending on the results of testing, an establishment's outcome of the investigation may vary.[2]

If the test results are negative for any disease, all animals are released for routine slaughter, and the production facility of origin may ship animals and product as normal. If test results are negative for an FAD, but positive for a domestic disease (such as Seneca Valley virus), animals at the establishment are released for slaughter after passing FSIS antemortem inspection, and animals at the originating live animal producer facility are held until all symptoms of the disease are gone. If official lab results have been communicated to FSIS personnel assigned to the official establishment and acute signs are no longer observable in livestock at the live animal producer facility, that live animal producer facility can resume sending livestock for harvest, however, this is subject to FADD observations.[3] Finally, if test results are positive for an FAD, the FAD Incident Command (APHIS) will take over. The company's corporate office should be notified, if applicable, and should follow the direction of APHIS personnel.

Continuity of business in the face of an FAD outbreak

The above-referenced steps in an FAD investigation are well-understood within the pork packer industry, as many establishments have prior experience with FAD investigations due to SVV outbreaks in the summer months of the past few years. However, there is still much confusion about subsequent steps to be taken if an establishment receives test results confirming the presence of a true foreign animal disease such as African swine fever. Questions such as "Must pork products be destroyed?" and "Will the establishment be allowed to slaughter hogs still in lairage?" remain. Additionally, the question of which regulatory body should answer these questions remains unanswered. There is also confusion surrounding the status of slaughter facilities which may find themselves in a control zone or a surveillance zone during an outbreak. Are slaughter facilities within such zones free to operate? If so, will there be stipulations on the procurement of hogs for that slaughter facility?

Additionally, it is understood that a stop-movement order would be issued by APHIS or perhaps individual states following the announcement of an FAD finding in the United States. However, it is unclear how this will affect slaughter facilities in unaffected areas, and how movement of pigs to slaughter facilities will be permitted after a full stop movement is lifted.

Finally, resumption of trade is a large concern for the pork packer industry, as almost 30% of pork products made in the United States are exported to other countries.[4] Regionalization/zoning will be of utmost importance to slaughter facilities that wish to resume trade as soon as possible. How establishments fit into these regionalization schemes is very important to the industry, but still unclear when considering continuity of business in the face of an outbreak.

To address these and other concerns regarding slaughter establishments' ability to continue with normal business operations, the North American Meat Institute has formed a working group to help develop a plan for resumption of trade in the face of an FAD outbreak. Additionally, the industry is working with APHIS to develop a series of FAD exercises specific to the slaughter industry, to be held in 2021. Such exercises will provide a platform for much-needed discussion and collaboration between industry, the States, and the federal government agencies to ensure that slaughter establishments are as prepared as possible in the event of a foreign animal disease outbreak.

References

1. USDA FSIS Directive 6000.1, Rev. 1: Responsibilities related to foreign animal diseases (FADs) and reportable conditions. **https://www.google.com/url?sa=t&rct=j&q=&esrc=s&source=web&cd=1&ved=2ahUKEwjwpq6C1-biAhXIx1QKHQtDBgcQFjAAegQIARAC&url=https%3A%2F%2Fwww.fsis.usda.gov%2Fwps%2Fwcm%2Fconnect%2F57bfc5bf-b1ff-4b33-a831-b37c122a4272%2F6000.1Rev1.pdf%3FMOD%3DAJPERES&usg=AOvVaw0zvUSibdcY6VTzVDWUh6u5.**

2. USDA APHIS Veterinary Services Guidance 12001.2: Policy for the investigation of potential foreign animal disease/emerging disease incidents (FAD/EDI). **https://www.google.com/url?sa=t&rct=j&q=&esrc=s&source=web&cd=1&cad=rja&uact=8&ved=0ahUKEwiArOnfutjZAhWKmeAKHTVSCH8QFggpMAA&url=https%3A%2F%2Fwww.aphis.usda.gov%2Fanimal_health%2Flab_info_services%2Fdownloads%2FVSG_12001.pdf&usg=AOvVaw0XZ5SeCBMZxUNRd9cchuMF.**

3. USDA APHIS Veterinary Services Guidance 7406.2, dated 04/06/16. **https://www.google.com/url?sa=t&rct=j&q=&esrc=s&source=web&cd=3&ved=0ahUKEwjn-oCWu9jZAhVrmuAKHa6ABlAQFgg4MAI&url=https%3A%2F%2Fwww.aasv.org%2Fdocuments%2FVSG7406.2_SVA_Guidelines_4516.pdf&usg=AOvVaw1n6HoXd9dPD2AAnzlXvzQD.**

4. Pork Check Off News. March 11, 2019. Available at: **https://www.pork.org/news/u-s-pork-exports-remain-steady-2018-despite-challenges/**. Accessed November 6, 2019.

What it will take to get back into business and trade: industry perspective

Russell A. Nugent III[1], PhD
Dogwood Ag Services, LLC Lowell, Arkansas

Introduction

Over the last two plus years, I have had repeated exposure to experts in live swine production, slaughter, meat processing, sales, state, federal and international political entities, state and national swine organizations, swine health organizations, and biosecurity experts on strategies for African swine fever (ASF) and implications to our industry. I realize I do not understand all the details and intricacies of these ASF risk topics. My comments need to be taken as such, but I will contend herein that we can plan better and move much faster to action plans than we have so far as an industry. Subject matter experts in charge of these areas need to sort through my ideas and flesh out those of value. Where we are today in terms of ASF preparedness in the US is not acceptable from an industry perspective as it relates to potential loss of export markets.

It is November 2020 as of this writing and things could be very different in March 2021. Any errors borne out by time are mine alone. I speak herein as a US pork producer. My goal is to help us examine our current approach and consider changes to improve our processes, now, before we need them. I challenge the reader to think more philosophically about the points I am making and why. I know we as an industry have worked on these topics, but I contend we are not thinking innovatively enough and are taking way too long to be prepared.

My goal is to regain as much US pork export market volume as possible in the face of an ASF outbreak as quickly as possible. We need to re-engineer our plans now because getting "back into business" entails not really losing these markets in the first place or losing them for as brief a time as possible. The timeframe needs to be so short that ASF negative sow herds are not eliminated because they cannot maintain cash flow. Once significant ASF negative sows are depopulated due to market price collapse, it will be too late.

Background

The relevant concepts are: the experience of human contact tracing for COVID-19, the current state of affairs in Germany with ASF in wild boars, and ASF likely not being immediately clinically obvious in our pig herd.

The US live production technical infrastructure is much more suited to mass clinical observation and testing for ASF in pigs than we are for COVID-19 in humans. This is a big strength of our country to leverage towards export customers. We have learned a lot from COVID-19. We need to be ready to identify premises that have had potential contact with ASF positives. But, we also need to be prepared to test BEYOND known contact tracing epidemiology. COVID-19 taught us that. There will be positive premises with no known contact that we need to identify. I will discuss below the need now to take advantage of this industry wide testing capability very specifically. We must seize this potentially competitive advantage of US pork in the international market place.

Germany has known that ASF was in wild pigs across the border in Poland for some time. Wild pigs do not recognize geopolitical borders and ten meters to either side of the Polish/German border makes no difference to them. But it makes a world of difference to us when it comes to export eligibility. At some point, an ASF infected wild boar made it across into Germany and it was some additional amount of time later before the first dead ASF positive carcass was found in Germany. So the key event was not an ASF positive wild boar within the German borders, but rather a wild pig carcass being found and subsequently tested for ASF. Consequently, the announcement of positive results led to Germany losing many export countries for their pork very quickly. No export of pork meat is allowed to those countries, even for pork from German commercial pig herds hundreds of kilometers away from the Polish border. Thus it is the perceived risk of the situation more so than the situation itself that can negatively impact meat price for a country that relies on exports for market price strength. Unfortunately, meat that is wanted for purchase in a foreign country is being banned from export even though ASF is not known to be present in Germany's domestic herd. This is not a human health issue or even a presence of ASF in pig meat sold for export. It is rather a traditional political reaction to ASF presence inside a political boundary. Furthermore, it does not appear to matter how far pig herds are from the known positive situation; the entire country loses export eligibility, at least initially. The fear is that somehow the domestic pigs in an "ASF free zone" in an infected country will have ASF, that meat will be exported to another country and ASF from the imported meat will find a way into the importing country's pig herd. Which in some cases, the importing country already has ASF in their domestic pig herd! But, interestingly, this all or none fairly immediate "drop of imports from Germany" response is not the stance of all export customers of Germany. There are better ways that we should all be reacting to a new ASF presence in a country. I agree that it is good to be conservative to protect pig health in a country by being careful about imports. But this is a self-imposed prohibition of sale of perfectly safe pork to people who need the protein and are willing to pay for it. Wild pigs with ASF in Poland ten kilometers from Germany have no effect on German exports, but, to switch geographic reference, if ASF was only in Maine, US, for example, it would potentially stop at least some exports from other US states a thousand kilometers or more away from Maine. How can we change these globally imposed regulations for the greater good of us all? People came up with these strategies at some point, so as a people, we surely can correct the unintended negative implications we feel today.

Hypothetically, if ASF does find its way into commercial pigs in state "A" in the southeast US and there is excellent data supporting the conclusion that no ASF is west of the Mississippi River,

[1] The views expressed herein are my own and not those of the National Pork Board.

excellent logistical controls are in place and audit results are consistently favorable, why shouldn't we be exporting that western US ASF free pork to countries that need that protein? Or, if ASF is clearly only in wild pigs, why can't domestic pig meat be exported? Why can't meat from northwest Germany currently be exported to all previous export partners under certain conditions? Could we keep the necessary current restrictions, but remove any potentially counterproductive pieces that unnecessarily stop safe and viable meat from export? Our industry should be talking with export customers now about modernizing these concepts. We need government officials asking each potential export country to partner with us on defining specific data needs and processes to allow importation of our pork in a partial ASF outbreak in the US. We need to be doing this now. Changing our ways AFTER ASF gets to our country is too late. Let's try not to lose even one export partner country. We have regionalization agreements with some countries and efforts ongoing with others. That is fantastic! Let's strive to have agreements in place with all export markets that would treat ASF and exports commensurate with true risk to all involved parties.

Finally, ASF does not clinically present itself like vesicular foreign animal disease; we will not immediately recognize what we are looking at visually at a pig farm. The clinical signs of ASF are too similar to endemic pig diseases we see every week. When we get the first indication from a veterinary diagnostic lab that we may have an ASF positive, we will already have moved pigs from the index farm. We will be looking back and saying, "Oh no! That wasn't PRRS, that was ASF I saw in my pig farm starting last week!"

Thoughts to consider

US pork production in 2020 is much different from pork production in 1985, let alone 1945. But, many of our health containment approaches are legacy strategies that worked well in eras gone by when pigs grew out and were slaughtered much closer to where they were born. Pork production today is also very different from US egg or broiler production. As it relates to export implications and/or regional control functionality, an Exotic Newcastle Disease outbreak situation or High Path Avian Influenza control analogy is just not relevant for large scale multi state US pork production. Our current US pork industry entails the movement of tens of thousands of pigs across state lines for grow out every day.

The common large scale broiler "complex" model consists of a genetic multiplication pyramid, hatchery, feed mill, grow out, slaughter plant, and further processing all being in a "circle" with a radius of 60 kilometers and is effective from a regional disease control standpoint. Therefore, governmental lessons learned from controlling poultry disease impacts are usually not applicable to US pork and thinking otherwise carries a lot of risk. That poultry model allows an export relevant disease outbreak in say, California, to often not disrupt export from a production entity in say, Mississippi, at all or for only a matter of hours or days. US pork production is very connected across multiple state lines and the task at hand will be much more like COVID-19 contact tracing than anything that has happened in the large scale US poultry industry. Only real time national pork production data on premises IDs, animal movements, clinical disease signs, and veterinary diagnostic lab testing results VISUALLY available to state and federal officials will allow for the best day to day decisions to contain ASF to manageable levels, especially in the first 72 hours post known ASF presence.

This database will be essential to any attempt to preserve a large geographic portion of US pork production for export eligibility. This database does not exist for poultry and has not been needed to recover from export losses or to preserve exports due to that industry's geographical structure. That will not be the case for US pork. Any previous experience from poultry outbreaks that provides us comfort as it relates to pork foreign animal disease export impact should be critically evaluated because that "comfort" is false.

This database will be not be easy to maintain accurately, but if we want to keep pork exports, we need to try. We must realize that time and cost of constant update of animal movement data in the face of no outbreak will make it difficult for our industry to maintain focus, energy and spending in these areas over time. The database system must be simple, consistent from state to state, efficient to make it viable for producers to keep information updated real time and ready for immediate use by government officials in the face of a sudden unanticipated potentially widespread outbreak. It did not take long for ASF to spread hundreds of kilometers in China. Anything less than a high percentage of US pork producing premises and real time movement data in the database will be too little, too late. Soon after exports are lost, ASF negative sows in the national herd will be depopulated. It will not be easy, but acknowledging the actual logistical infrastructure of the US pork industry as it is today will be paramount to having an effort in place that will allow return to export channels very quickly. State pork associations, state animal health officials, and National Pork Board need to agree on one database system, across all states, now. Producers will then need to provide real time premises and movement data on an ongoing basis to that system.

There will be a need for slaughter plants and further processing facilities that are sourcing only "certified" ASF negative pigs and pork to be eligible for export. Conversely, there will need to be some slaughter and processing facilities that handle the ASF positive raw material. We need to think now as a meat packing industry how that might look. I realize that will depend greatly upon where and how widespread ASF is in our country but I think the logistical effort can be modeled and practiced now to prepare. Any post ASF industry reconfiguration on the slaughter side will need to be quick and efficient to minimize time of lost exports. The packing industry needs to plan ahead.

If ASF is contained in some sub region of the US, we need to think and plan now how they will be moved, fed, slaughtered, rendered, etc. Can feed mills feed positive and negative herds? What about effluent handling? What about cull and dead movements? Many smart people have worked on these issues already, but we do not have a consensus plan to quickly save some or most of US exports. We still have work to do here and the clock is ticking. AASV has the functional member base to plan these strategies in conjunction with allied industry.

Let's finish this section talking about additional testing that will be needed IMMEDIATELY upon thinking that ASF may be in our national herd. I mentioned above that we will not immediately recognize ASF for what it is. That means ASF will have potentially spread a far distance from where we first discover it. Our inclination, because it follows our training and experience, will be to draw a circle around the infected farm, stop movements and test very heavily in a concentrated geographic area. We will quickly learn there are too many sites, not enough human resources to do the sampling and, most importantly, ASF might already be outside the geographic range we anticipated. If that is true, we risk greater country wide spread in the meantime (even with stop movement in place) before knowledge allows us to contain it.

We need to consider that to truly preserve any semblance of exports, we will need to do testing that at first will seem counter intuitive. Traditional epidemiological testing will be too slow, too concentrated geographically and will be like chasing a ghost. Remember the initial PEDV spread in the US? Instead we need to test broadly and non-statistically beyond known animal movement premises links during the first 72 hours to see where we can find it. It would be great if we could do this and the traditional focused epidemiological tracing simultaneously but I do not think we will have enough time and manpower to do both.

The first 72 hours post realization of ASF in our country will require us to change our methods and will require us to be very careful about how we interpret, announce and explain the test results and implications of what we find. We need to come out of the national 72 hour stop movement with a very broad perspective of WHERE WE CAN CURRENTLY FIND ASF. That will help us make informed decisions on risk management of what to do and where at hours 73 plus. It will feel very strange, but I believe it's necessary and we need to plan now.

Immediately upon suspecting ASF is on a US farm, we need to test pooled blood from several large slaughter plants across a broad geographic area. I understand there will be no tracing value and there will be risk of dilution leading to false negative. This in itself will not help exports at all. A negative test will mean nothing. But a positive test will mean a lot. Example 1: on day 2 of knowing we have an ASF positive on farm in Wyoming, we have positive tests at packing plants in North Carolina, Indiana, and Minnesota; all other packing plants yielded negative tests but again, that means nothing. But we do know now, quickly, that the virus somehow already exists way beyond Wyoming. That will be much more useful to devise a plan to save exports than only testing within a circle in Wyoming. Example 2: we know it is on a farm in Wyoming and a quick broad test of several packing plants shows only one positive and that's the small plant in Wyoming that kills sows from the positive farm. That is a huge piece of knowledge for hour 73 decision making. We are not saying it hasn't spread beyond Wyoming but it's certainly a different scenario to plan for than the outcome of Example 1 above. This will help greatly with intra and interstate movement plans at hour 73 plus. This will greatly improve chances of partial containment and thus formation of US regions hopefully eligible for export much sooner. USDA will need to create official ASF geographic zones; this will help that effort be accurate more quickly.

Similarly, our training and experience will lead us to want to test 95:5 or something similar in those first 72 hours. That will be way too much resources focused on too small a geographic range of farms. Instead of twenty or ten or even five pen samples in a hog house, we need to do one or two pooled samples per site, maximum. As above, negative test results mean nothing. But a positive will mean a lot. And since this method allows us to greatly expand the geographic range and number of farms tested in the first day or two, again we get much more useful information to estimate the severity of spread. This allows for a much better strategic plan past hour 72. We cannot let the impossible pursuit of perfection get in the way of achievable good when it comes to knowing the extant of the spread quickly. USDA has experts in this area; I am asking for help in managing risk by deploying resources differently than we would strive for in a perfect world.

Please do not get hung up on exactly the details of what I say above. My point is that if we want to preserve exports at all, we need to move quickly and broadly and differently than what we envision today. We need to test broad and fast beyond known epidemiological contact tracing links looking only for positives and then narrow down to focused and detailed smaller regions as we get smarter and smarter on where it is and hopefully is not. The broad and fast philosophy is what I hope you embrace. The experts devising the ASF strategy may differ in exact details, but they must embrace some close version of this concept or ASF will be spread over large swaths of our country before we know exactly what we are dealing with and no significant pork producing region will be ASF free.

I could discuss other topics, but I think you get the point. This paper has been a success if the common theme of my concern has struck a note with you and inspired you to think outside the box to enhance our current strategies and greatly reduce the time to delivery of FUNCTIONAL plans we currently do not have, but desperately need. The National Swine Disease Council has an excellent portfolio of members. I challenge them to embrace these type of concepts and lead an effort to aggressively eliminate any historical paradigms that cause potentially ineffective strategies in our approach. The goal is to maintain and or quickly resume exports. There is work for all of us. No entity is free from potential improvements in our plans. We can be better positioned if we act together as a team now to shore up all fronts.

Conclusion

There is nothing in the above comments that has not been stated before elsewhere. My take home message is that we have two choices: **1)** create better working plans now that can be executed immediately upon ASF entering our country that may allow preservation and/or quick resumption of significant export volume or **2)** we lose all exports, we spread ASF over a large part of our country; we lose over 30% of our sales volume essentially overnight and a corresponding proportion of the herd depopulates due to bankruptcy and we spend five to ten years building back a domestic meat herd. I like the first scenario better. We are smart enough to develop plans NOW that are realistically functional. I'm sure they will not be perfect. I'm sure there will be unforeseen challenges. But we must have a sense of urgency to get these plans in place. ASF broke in China August 8, 2018. We have not built an export viable US plan in those two plus years. Daylight is burning, we must do better. We will not "get back into business and trade" in a way that is truly viable for our industry without a viable plan in place.

What happens in the US if we have to learn to live with African swine fever?

Keith Erlandson, MS, MBA, DVM, DABVP (SHM)
Zoetis, Florham Park, New Jersey

Introduction

This paper and presentation will be somewhat short on scientific references and long on anecdotes and extrapolations made from my experience in dealing with African swine fever (ASF) while working for an integrated swine production company in China. ASF was first reported in China in August of 2018 and I worked on my first outbreak approximately one month later on a sow farm that was located within 10 km of the index case.

Since that first case, I have had multiple opportunities to see the disease in both sow farms and finishing farms. The course of the disease, diagnosis of the disease and the methods to control or eradicate the disease were more complex than I believed they could be prior to experiencing it firsthand. I hope that the musings in this paper and presentation will assist you in thinking about how to prepare your clients to keep the disease out of the US, or deal with it if it should appear here in the United States.

Diagnosis

I had previously believed that an infection with ASF would be apparent as soon as I walked into the barn and saw the tell-tale clinical signs listed in any edition of Diseases of Swine. Diagnosis, I thought, would be the easiest part of dealing with ASF. I found this not to be the case, however. We fight diseases on pig farms every day that can lead to increased mortality and have clinical signs that are shared with ASF.

There are occasions where the initial clinical signs in affected animals were off feed sows and some vomiting, which could easily be mistaken for infection of those animals with PRRS, PED, SIV (as well as PRV or CSF in areas where those disease are common) or any of several other diseases.

Increased death loss in the sow farm or the finisher may go undetected in the early stages of the disease, when few animals are infected. It is difficult to determine if the mortality on any given day is the result of the introduction of a deadly disease or normal day to day variation. By the time the death loss is high enough to trigger a reaction, there are likely enough animals infected on the farm that the only option is a complete depopulation.

Gross lesions are frequently seen, but animals may not have all of the lesions of ASF infection, and those lesions are also associated with other diseases. This can lead to confusion as to the root cause of mortality early in the infection on a farm.

In the early days of ASF taking a foothold in China, there were no laboratories to send confirmatory diagnostics for ASF. This left the practitioner to rely on ruling out other diseases that could be tested for and assessing clinical signs and gross lesions to make the diagnosis.

Vigilance

Finding the infection early in its course is critical to stopping the spread of the disease within the farm, within the system and within an area. This fast response depends on the caretakers in the farm, as they are the critical first line of defense for detection of problems in the farm. Initial training and subsequent training emphasized the importance of detecting and reporting animals that exhibit clinical signs consistent with ASF infection.

In areas where ASF is known to be, farms I worked with were required to report the number of animals showing clinical signs that were consistent with infection by ASF to the herd veterinarian daily. Many farms have invested in bench-top PCR equipment and will automatically sample and test animals exhibiting clinical signs. Those samples are also sent to a Veterinary Diagnostic Laboratory for independent confirmation. Having the PCR onsite is advantageous in that it allows the farm to initiate on-farm protocols for ASF should any sample be detected as positive.

Dealing with outbreaks

Depopulation

Early in the epidemic of ASF in China, it was believed that the only way to clean up a farm from an ASF infection was by total depopulation. After the depopulation, farms were cleaned and disinfected several times and allowed to sit empty for 4, 5 or six months or even longer. I worked with numerous farms developing and executing clean-up after the farm had lost the battle with ASF. In the early stages of ASF in China, there was not much practical field experience to draw on to develop the plan for cleaning and repopulation, thus plans tended to be over conservative on washing requirement and downtime requirements.

Partial depopulation or "tooth extraction"

It soon became apparent that the disease was going to become endemic in China and complete depopulations of farms was not a viable long-term solution to the ASF problem.

I am not sure where the terminology "tooth extraction" came from, but it is now widely used in the swine industry in China to describe a targeted test and removal program to eliminate ASF from a herd. The theory is that by removing the bad tooth (infected animal) one can save the rest of the teeth (other animals on the farm) from infection.

A critical component to successful partial depopulation is access to diagnostics with a rapid turn around time. In the early stages in the infection on a farm when few animals are infected, the virus appears to move slowly and we can take advantage of this by removing animals that are infected, suspected of being infected or exposed to infected animals.

In most cases, once a diagnosis of ASF is found (or even a suspicion of ASF), the farm follows these steps to stop the spread and eliminate the virus:

1. All non-essential movement on the farm is stopped for animals and people. Breeding and weaning are discontinued.
2. People movements are eliminated except for essential functions such as feeding.
3. Animals that tested positive are quickly removed from the farm as well as a ring of animals around them that may have been exposed. The number of animals that need to be removed varies depending on several factors:
   a. The number of animals that are showing clinical signs or have tested positive
   b. The location of those animals within the farm
   c. The likelihood of the spread from infected animals to animals nearby
4. Once positive and suspected animals have been removed, the entire herd is tested for ASF. Any animals that test positive are removed along with animals that have been exposed to the positive animals. After the removal, the whole herd is again tested. If there is confidence that the infection is limited to a specific area of the farm, that area can be tested in lieu of a whole herd test. I have seen good success using a swab to collect oral fluids from individual sows. While this method may not be quite as good as testing serum, it has several advantages.
   a. There is no need to snare animals, making repeated tests easier to perform in the farm should it be necessary
   b. There is no risk of transmission of the virus through contaminated snares
   c. Very little training is required for workers on the farm to become competent at collecting the sample.

Partial depopulations that I have been involved in have been mostly successful, if you define success as anything other than the complete depopulation of the farm. Some farms have removed less than 10% of the breeding herd in order to stop the spread of the virus and others have been over 50% depopulated.

Partial depopulation on finisher farms

Most cases of ASF that I have had to deal with on Finisher farms were the result of the virus being drug back from a slaughter plant. In these cases, the farm should be depopulated as soon as possible. In farms where the pigs were not yet large enough to go to slaughter, the spread of the disease can be stopped by removing the pen exhibiting clinical signs along with the surrounding pens. This method has worked, but success is not guaranteed.

Most cases of ASF I have dealt with in nursery age pigs are infections that have gone undetected in the sow farm and have been carried with the pigs to the nursery or wean-to-finish farm. In these cases, the only option is to depopulate.

Biosecurity

Obviously, keeping any disease out of a farm is far superior to dealing with an outbreak of disease. Farms I worked with that experienced ASF outbreaks frequently had excellent biosecurity. Some farms contracted ASF despite being in 100% lockdown, leaving me to conclude that there is an "area spread" component to the transmission of the disease.

ASF changed the manner in which farm visits were conducted. While strict downtimes were in place before the introduction of ASF into China, those downtimes and isolation requirements were dramatically increased after the virus arrived. In no case did we visit multiple sites in one day after the outbreak in the country.

Transportation

One of the most important risk factors for farms that become infected seems to be selling loads to slaughter or selling cull sows. Systems I worked with implemented transfer points to create a gap between the farm and the slaughter plant. Multiple trucks were purchased so that a truck that delivered animals to a slaughter plant was never allowed with 2 km of a farm. A dedicated truck picks the animals up from the farm and delivers them to the transfer point and leaves. Once the delivering truck has departed, then the truck that will deliver the animals to the slaughter plant is allowed to arrive at the transfer point and load the animals.

Many farms constructed a holding bin away from the barns so that feed delivery trucks could be kept away from the buildings housing animals. In other instances the farm purchased a dedicated feed truck that stayed on the farm. This truck takes delivery of feed from the feed mill truck and then carries the feed the final kilometers to the farm.

Testing

Rapid diagnostics are essential to quickly identifying an outbreak of ASF on a farm. The quick diagnosis allows us to eliminate infected animals before they can infect too many other animals, thus stopping the spread of the disease on the farm.

In general, samples from animals that are showing clinical signs consistent with ASF and animals in close proximity to those affected animals are tested individually by PCR. Animals outside the "zone of contamination" are pooled in groups of 5 and tested. A positive test of a pooled sample triggers an immediate test of each of the samples in that pool individually.

African swine fever as a cost of doing business

Despite increased awareness, training and significant capital investment in biosecurity, farms still become infected with ASF. Were the US to reach a state where the virus became endemic, it must then be assumed that a certain proportion of farms will break with the virus each year.

Conclusion

I am aware that some statements I have made in this paper and presentation may appear contradictory. My experience has led me to believe that in order to combat African swine fever producers and veterinarians must be able to hold two mutually exclusive facts to be true at the same time. Namely, that ASF is a disease that is difficult to transmit and ASF is a disease that is easy to transmit. When dealing with cases of ASF, we must be able to select which of these two statements applies and run with it to solve the problem at hand with the knowledge that the next case may require us to think of the disease in completely contradictory terms.

Acknowledgments

My thanks to Dr. Yang Li, Dr. Xi Zhaofang, Dr. Chen Yang, Dr. Lei Ting, Dr. Li Shengbin and Dr. Richard French and Dr. Gary Stoner for their insight, ideas, encouragement and friendship as we learned about battling African swine fever together.

Economic impact of COVID-19 on macro and swine economies

L. Schulz, MS, PhD
Department of Economics, Iowa State University, Ames, Iowa

Global markets were roiled by COVID-19 and the hog and pork market was no exception. The market upheaval generated intense interest among policy makers, the media, producers and allied industry. Agricultural economists were thrust into the spotlight to help explain the supply and demand shocks, answer questions about price implications, and estimate economic damages. Here is a non-exhaustive list of that applied research and outreach.

Hart, C.E., D.J. Hayes, K.L. Jacobs, L.L. Schulz, and J.M. Crespi. The Impact of COVID-19 on Iowa's Corn, Soybean, Ethanol, Pork, and Beef Sectors. CARD Policy Brief 20-PB 28. April 2020. **https://www.card.iastate.edu/products/publications/pdf/20pb28.pdf.**

Hayes, D.J, L.L. Schulz, C.E. Hart, and K.L. Jacobs. 2020 A Descriptive Analysis of the COVID-19 Impacts on US Pork, Turkey, and Egg Markets. *Agribusiness: an International Journal.* **https://onlinelibrary.wiley.com/doi/10.1002/agr.21674.**

Lusk, J.L., G.T. Tonsor, and L.L. Schulz. 2020. Beef and Pork Marketing Margins and Price Spreads during COVID-19. Applied Economic Perspectives and Policy. **https://doi.org/10.1002/aepp.13101.**

Schulz, L.L. Hog producers limited in short-run to change production. Ag Decision Maker Newsletter. April 2020. **https://www.extension.iastate.edu/agdm/articles/schulz/SchApr20.html.**

Schulz, L.L. Hog numbers: What counts, what doesn't for USDA? Wallaces Farmer. July 2020. **https://www.farmprogress.com/hog/hog-numbers-what-counts-what-doesnt-usda.**

Schulz, L.L. How to interpret USDA's Hogs and Pigs reports. Wallaces Farmer. October 2020. **https://www.farmprogress.com/hog/how-interpret-usdas-hogs-and-pigs-reports.**

Tonsor, G.T. and L.L. Schulz. Assessing Impact of Packing Plant Utilization on Livestock Prices. Kansas State University Department of Agricultural Economics Extension Publication. AgManager.info. April 2020. **https://www.agmanager.info/livestock-meat/marketing-extension-bulletins/price-risk/assessing-impact-packing-plant-utilization.**

Tonsor, G.T. and L.L. Schulz. COVID-19 Impacts on the Meat Processing Sector. CAST Commentary: Economic Impacts of COVID-19 on Food and Agricultural Markets. QTA2020-3. June 2020. **https://www.cast-science.org/wp-content/uploads/2020/06/QTA2020-3-COVID-Impacts.pdf.**

Tonsor, G.T., L.L. Schulz, and J.L. Lusk. Meat Availability and Shortages Overview. Kansas State University Department of Agricultural Economics Extension Publication. AgManager.info. April 2020. **https://www.agmanager.info/livestock-meat/marketing-extension-bulletins/trade-and-demand/meat-availability-and-shortages**

Weersink, A., M. von Massow, J. Ifft, J. Maples, K. McEwen, M. McKendree, C. Nicholson, A. Novakavic, A. Rangarajan, T. Richards, B. Rickard, J. Rude, M. Schipanski, G. Schnitkey, L. Schulz, K. Schwartzkopf-Genswein, M. Stephenson, J. Thompson, K. Wood. COVID-19 and the Agri-Food System in the United States and Canada. *Agricultural Systems.* **https://www.sciencedirect.com/science/article/pii/S0308521X20309008.**

Lessons learned from plant closures, testing employees, and modeling outbreaks

Kimberly VanderWaal[1], PhD; Lora Black[2], MSc; Judy Hodge[3], DVM, MPH, Dipl ACVPM; Sasidhar Malladi[1], MS, PhD; Addisalem Bedada[1], DVM, MS; Scott Dee[4], DVM, MS, PhD, Dipl ACVM
[1]University of Minnesota; [2]Sanford Health; [3]Hylife, Ltd; [4]Pipestone Veterinary Services

Introduction

Pork processing plants were apparent hotspots for SARS-Cov2 in the spring of 2020. As a result, the swine industry was confronted with a major occupational health, financial, and animal welfare crisis. A recent paper from the CDC reported that almost 9000 workers tested positive for SARS-COV2 among US food manufacturing (meat and poultry) and agriculture workers from March through May, 2020.[1] While these outbreaks have been attributed to prolonged close contact with coworkers, shared transportation, and frequent community contact among workers, data related to this issue have thus far been reported in an aggregated manner,[1,2] thus limiting insights about how these outbreaks unfolded at individual plants. To better understand how workplace outbreaks can be controlled or prevented, screening/testing protocols have been proposed to identify infected and asymptomatic individuals; however, the population-level effectiveness of different screening/testing protocols has not been rigorously tested in this context. Therefore, the objective of this work was to characterize the epidemic curves across four pork processing plants, summarize epidemiological data related to PCR and antibody testing at individual plants, and develop mathematical models that are calibrated to this epidemiological to simulate transmission in these plants. We used the model to estimate the outbreak's reproduction number (R) during the early phases of workplace transmission, and calculated time-dependent reproduction numbers (R-TD) that allow the effective R to vary through time to measure the influence of interventions over time.[3,4] Finally, we used the model to evaluate a range of PCR-screening scenarios to test their effectiveness at minimizing SARS-CoV2 circulation.

Materials and methods

Data were available from four plants located in four different states. These plants ranged in size from 750 workers to 2400 workers. Plants participated in this study on condition of confidentiality. Each plant had two types of data available: daily incidence of new self-reported SARS-CoV2 cases and company-initiated antibody or PCR testing data. Daily incidence of self-reported cases was summarized as the number of new workers per day, and testing data was summarized as percent positive per testing event.

Second, we developed a mathematical model that accounts for asymptomatic, pre-symptomatic, and background "community" transmission. By calibrating this model such that simulated outbreaks resembled the observed epidemiological data, we were able to estimate the reproduction number (R) of the virus during the initial phases of each outbreak. In addition, we estimated effective R-TDs for each plant from the observed epidemic curves.[5]

We evaluated the effectiveness of PCR-based screening (every 3, 7, 14, 28 days) in reducing the number of clinical cases relative to a temperature-screening only baseline. For each testing frequency, we also evaluated the effect of delays in testing results, where infected workers remained at work until results were received (after 1, 3, or 5 days). Finally, we also evaluated each scenario if used proactively to prevent disease introduction, or reactively during the early, peak, post-outbreak stages. We also considered an outbreak stage (35%-Imm.) in which testing is used to prevent re-introduction to an already partially immune population (35% of workers start as recovered to resemble the epidemiological situation of Plants B and D). Based on a sensitivity analysis, model projections are sensitive to the assumed workplace transmission rate, R, and background "community" transmission rate, b, thus all scenarios were run for both high and low transmission parameters.

Results and discussion

In this study, we described the epidemiological situation across four pork processing facilities in the United States. All plants experienced SARS-CoV2 infections in their workforce beginning in March and April 2020, a time period in which SARS-CoV2 was rapidly spreading throughout the country and scientific understanding of its epidemiology, transmission, and appropriate protective and mitigation measures was rapidly evolving. Cumulative incidence of clinical (PCR-confirmed) disease plateaued at ~2.5% to 25% after two to six weeks. For larger outbreaks, antibody prevalence was approximately 30% to 40%.

Across plants, R generally ranged between 2 and 4 during the initial phase, indicated that about 2 to 4 additional cases resulted from each infected worker (asymptomatic, pre-clinical, or clinical). However, effective R's subsequently declined to ~1 after two to three weeks, most likely due to implementation/compliance with biosecurity measures in combination with population immunity.

Using the calibrated model to simulate a range of possible scenarios, we show that the effectiveness of routine PCR-screening at minimizing disease spread was far more influenced by testing frequency than by delays in results, workplace R, or background community transmission rates. Testing every three days generally averted about 25% to 40% of clinical cases across a range of assumptions, while testing every 14 days typically averted 7 to 13% of clinical cases. However, the absolute number of additional clinical cases expected and averted was influenced by whether there was residual immunity from a previous peak (i.e., routine testing implemented after the workforce had experienced an initial outbreak), and assumptions about transmission rates. In contrast, when using PCR-screening to prevent outbreaks or in the early stages of an outbreak, even frequent testing may not prevent a large outbreak within the workforce. This research helps to identify protocols that minimize risk to occupational safety and health and support continuity of business for US processing plants.

References

1. Waltenburg, M., Rose, C., Victoroff, T., Butterfield, M., Dillaha, J., Heinzerling, A., Chuey, M., Fierro, M., Jervis, R., Fedak, K., et al. 2021 Coronavirus Disease among Workers in Food Processing, Food Manufacturing, and Agriculture Workplaces. Emerging Infectious Disease journal 27. **doi:10.3201/eid2701.203821**

2. Waltenburg, M.A., Victoroff, T., Rose, C.E., Butterfield, M., Jervis, R.H., Fedak, K.M., Gabel, J.A., Feldpausch, A., Dunne, E.M., Austin, C., et al. 2020 Update: COVID-19 Among Workers in Meat and Poultry Processing Facilities - United States, April-May 2020. MMWR Morb Mortal Wkly Rep 69, 887-892. **doi:10.15585/mmwr.mm6927e2**

3. Obadia, T., Haneef, R. & Boelle, P.Y. 2012 The R0 package: a toolbox to estimate reproduction numbers for epidemic outbreaks. BMC Med Inform Decis Mak 12, 147. **doi:10.1186/1472-6947-12-147**

4. Alkhamis, M.A. & VanderWaal, K.L. 2016 Spatial and temporal epidemiology of Lumpy Skin Disease in the Middle East, 2012-2015. Frontiers in Veterinary Science 3, 19.

5. Wallinga, J. & Teunis, P. 2004 Different epidemic curves for severe acute respiratory syndrome reveal similar impacts of control measures. Am J Epidemiol 160, 509-516. **doi:10.1093/aje/kwh255**

What could veterinarians do to help human medical doctors in response to COVID-19

T. Loula[1], DVM; D. Bomgaars[2], DVM; P. Yeske[1], DVM, MS
[1]Swine Vet Center, St. Peter, Minnesota; [2]RC Family Farms, Orange City, Iowa

Introduction

As the pandemic began, we all learned a host of new terms such as "social distancing" and "shelter in place," just to name a few. Once these regulations were put in place many farms worried their greatest concern would be if they would have enough staff to get the pigs fed and watered, let alone for normal production activities. Many farms took measures to ensure this such as splitting work crews to AM/PM, alternating days, and some even alternating weeks. Also among new procedures were temperature monitoring and recording, increased office hygiene, and split breaks. The swine industry is set up for social distancing since many activities are spread out and workers could establish good open space. Barns are well ventilated already for animals. Farm staff have learned to deal with population disease control with many different pathogens over the years. They also have a good working knowledge of epidemiology and biosecurity, so these things were not necessarily new to them. They never imagined that slaughter capacity would soon be the major concern that they would face.

As the pandemic continued and more regulations were put into place, one of the challenges no one expected from COVID 19 was shut down of the packing plants. Packing plant staff became infected from their communities and contacts resulting in reductions in plant staffing making it difficult for plants to maintain the daily numbers harvested that they were accustomed to. Farms started to feel the pinch as loads were cancelled and delayed.

This quickly became a risk as local health officials started to look at these plants and their ongoing operations as potential risks to their communities even though they were deemed essential businesses. They applied pressure to local health officials and boards to close these essential businesses due to concerns that there would be more spread within the local community. With all the publicity, we soon started to see some plant employees afraid to go to work because they were concerned that they would become sick. This only escalated the problems at the plants for harvest capacity. The snowball gained momentum very quickly in a short amount of time.

Once the first plant closed it became like a domino effect, affecting all the plants in the region. Trying to share the capacity and get customer loads to different plants only quickly compounded the problem to more producers. Local health officials were having greater concerns because other plants being closed only put more pressure on them to take the same action.

One of the things that happened as one plant was closed, and other plants were looking for more employees, was that employees started moving between plants to keep working, in some cases commuting on buses and in vans. This caused a mix of different social communities which did not help with spreading concerns and again only complicated and compounded the problem.

Then, the Easter holiday came around with families still wanting to get together, resulting in another surge of cases. The problems just continue to mount. At the same time more plant employees continued to be infected and were fearful of going to work, only further complicating the problem.

One of the biggest challenges as plants were closed was that there was no idea of when they would be able to be reopened. It seemed like the rules continued to change not only with available employees but how many could be asked to work on the line. With social distancing being implemented at the plants where traditionally people worked shoulder to shoulder, it dramatically reduced the plant capacities. There appeared to be no apparent plan on how these plants would re-open.

As the plants closed, this very efficient "just in time production system" suddenly had nowhere for market pigs to go. The sow farms continued to make pigs every day and needed to have a place to move them offsite, but finishers were not available since there was no place to harvest pigs. Once all spaces were full, things started to back up very quickly. It felt like we were on a long escalator of people with no one moving once they got off ... people just kept getting pushed into the pile.

Results and discussion

Producers and veterinarians such as Dr Loula quickly identified the problem and saw that it was only compounding. These farms have become very good at logistics and had everything dialed in tightly, they could read the tea leaves very well and saw huge problems starting. Unfortunately, their perceptions were very accurate. One of the advantages veterinarians like Dr Loula and others in the swine industry had was that they had connections and knew what was happening across the entire industry and region. These veterinarians quickly organized "at hoc" conference calls to discuss what was going on and what could be done. There was much discussion about euthanizing pigs to get them out of the way until plants were able to reopen. Much discussion had occurred in ASF practice exercises on how this could be done but really had not ever been tested, at least in the US system and at the size and scale of the industry.

Dr Loula soon recognized that the best thing we could do was keep the plants running and get those closed back in action as soon as possible. This was really the only efficient way to handle the numbers necessary. In the conversation with other veterinarians, he talked with Dr Dave Bomgaars who was very connected to the local hospital CEO and suggested collaboration with Sanford Health in Sioux Falls, that was located in the area in the center of where problems were occurring. Connections were made with now Chief Medical Officer Dr Jeremy Cauwels and the collaboration began in earnest as he was extremely receptive in gaining population medicine insight. One of the challenges that the doctors had was they are accustomed to working with one patient at a time in a one-on-one situation. Now they not only had to do this but also had to work with populations such as in the plants, communities, etc.

Drs Loula and Bomgaars related this to some of the regional control projects and what we had used every day in the past for population medicine...working more on diagnostics to not only assess the individual but how to assess what was going on in the population. They used both the antigen tests (PCR) and antibody tests to make a combination assessment of not only what is going on with the infection but with immunity as well. By encouraging whole population testing there were 2 goals with this project; to understand and assess what was going on within the plants and to build confidence in the employees that there was a plan or road map in place to manage the infections.

Through this collaboration of veterinarians, epidemiologist, and health care providers; the outcome was the development of a testing algorithm (Figure 1) that defined the status of employees and provided a road map on how to get them back to work. It also showed who could safely work together and what the criteria were for this to happen. There was much discussion about the deep nasal swabs and use of a new spit test. The veterinarians shared the very positive results we have had with the use of oral fluids and encouraged the development of these technologies to help alleviate compliance issues for those having to be potentially tested at frequent intervals.

Unfortunately, only limited use of whole population testing (both antigen and antibody) was done. However, lessons were learned and later plants that experienced problems had this information to use and keep as many employees working as possible. This is a good example of the "One Health initiative" and how using the collective knowledge of the group resulted in faster learning and response to the problem.

Figure 1: COVID-19 testing algorithm for packing plants and other businesses.

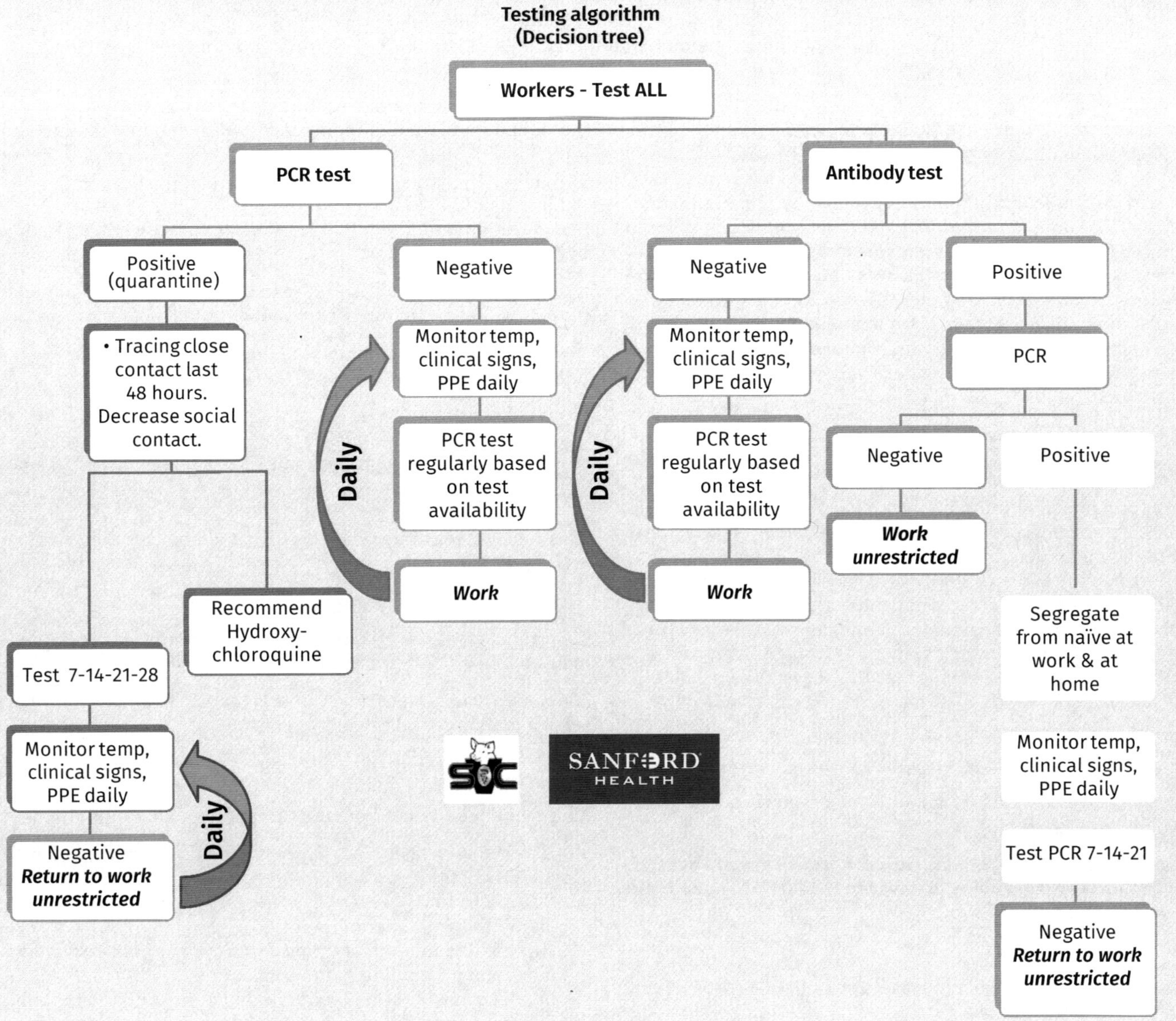

Talking points for the testing algorithm

- **Assumes adequate availability of tests, both PCR and Antibody/Elisa (SNAP)**
- **Identify a communication person for each language group of workers to educate, explain testing program and goal (preferably prior to issues developing at the plant).**
- **Consider hiring a medical director to coordinate education, testing and contact tracing.**
- **Goal: Identify the 3 populations**
  - Naïve (Negative PCR, Negative for antibodies)
  - Immune (Positive for antibodies)
  - Positive PCR (Shedding)
- **Workers would be managed differently according to health status to decrease health risk and promote worker safety**
  - Negative PCR population and Immune population
    - *Could work together*
    - *These workers and their families could be housed in hotels/dorms to stay safe*
  - Positive PCR with positive antibody test (very low risk of shedding if no clinical signs)
    Segregate in plant or work different shifts (second shift)

How to handle COVID-19 down on the farm

Larry Coleman, DVM

As a swine veterinarian, I can say without any hint of exaggeration that my work experience during 2020 has been defined by the COVID-19 pandemic. As a result, I have had to provide safeguards at the farm level to keep employees safe and feeling secure, so that barn staffs were maintained and animal care was not compromised. The novelty of the virus and consequential changes in the understanding of how it was spreading added complexity to determining sound recommendations. However, the general guidelines for COVID-19 virus prevention were to avoid the 3 C's (crowds, confined spaces, and close contact) as well as wear face masks and stay home if you are sick. These measures have been implemented on many of the swine farms that I work with.

In order to determine the degree to which COVID-19 health practices have been instituted across various North American swine farms, I conducted an informal survey of several swine veterinarians. This survey involved 15 veterinarians, 30 systems, 385 sow sites, approximately 1.1 million sows, and an estimated 4,000 employees. For the purpose of my survey, I asked the veterinarians to limit their comments to farrowing facilities, which typically have larger staffs and potentially have been more proactive with COVID-19 safeguards. Table 1 details the safeguards that may or may not have been initiated on farms and how those safeguards changed over three different time periods from the start of the COVID-19 pandemic to November 1, 2020. I have also included data concerning how COVID-19 has impacted employees at the farm level.

The survey results are shown in Tables 1 and 2.

I also solicited comments from the surveyed veterinarians on this question:

"What is the number one take-home you have learned about disease this year from being involved in the COVID-19 pandemic and swine production?"

Here are their comments:

"We all think we are experts in biosecurity, but COVID-19 proved we have serious gaps in biosecurity compliance."

"Thank goodness this virus wasn't zoonotic in nature."

"In integrated swine systems, health policies are easy to put in place, but compliance is determined by leadership at the farm level."

"Population medicine is easier to apply in animals than people."

"Employees were open to COVID-19 health practices initially, but by fall, fatigue set in."

"Some employees will find excuses to stay home while other employees will come to work sick, even when told not to."

"Swine employees understand the implications of their actions because of biosecurity training."

"For the 1st time, due to the health department contact tracing, we are finding out which employees are living with other farm employees accurately as opposed to PRRSV outbreak investigations."

"COVID-19 has negatively affected our ability to staff farms."

"It is easier to mandate swine biosecurity protocols than employees' biosecurity protocols."

"Compliance reflects company culture."

"Overtime, employees became comfortable with a lack of compliance on human biosecurity policies."

"Compliance to biosecurity protocols (people or swine) is directly related to a perceived threat."

Conclusions

My general take-home from the survey is as follows:

1. The swine industry initially responded well with standard COVID-19 safeguards.
2. Safeguard practices have lessened since the start of the pandemic to the fall period.
3. To date, the number of hospitalizations has been relatively low among front line swine employees.
4. All veterinarians surveyed reported difficulties getting employees to comply with the COVID-19 safeguards at work and at home.

Table 1: Percentage of sites implementing COVID-19 health practices

	Mar-May	Jun-Aug	Sep-Oct
Daily employee temp check	40%	24%	24%
Daily health questionnaire	35%	36%	38%
Break room social distancing	42%	25%	22%
Hand-sanitizing protocols	50%	30%	30%
Face masks	37%	19%	19%
Break room sanitation	39%	37%	30%
Split break times	57%	37%	34%
Split shifts	4%	2%	2%
Restricted supervisor visits	16%	10%	7%
Restricted veterinarian visits	17%	11%	6%
Restricted visitor/vet student	32%	36%	21%

Table 2: COVID-19 employee impact percentages/numbers

	Mar-May	Jun-Aug	Sep-Oct
Contact tracing quarantines*	29%	29%	25%
COVID-19 positive tests**	9%	14%	12%
COVID-19 hospitalizations***	4	3	2
COVID-19 deaths***	0	0	2

* Percentage of sites that had workers stay home because of quarantines

** Percentage of sites with at least one employee testing positive for COVID-19

***Total number of employees from all sites combined

US Department of Agriculture (USDA) response to COVID-19: What did we learn to be better prepared?

Jack Shere, DVM, PhD

USDA views preparing for and responding to foreign animal diseases (FAD)—such as highly pathogenic avian influenza (HPAI) and foot-and-mouth disease (FMD)—as critical actions to safeguard the Nation's animal health, food system, public health, environment, and economy. USDA's Animal and Plant Health Inspection Service (APHIS) maintains several programs that plan, prepare, and conduct exercises for such events. The HPAI outbreak that devastated poultry producers during the winter and spring of 2015 was the worst animal disease in US history. Subsequently, much of our funding and many of our efforts focused on HPAI preparedness, although not exclusively. Unsurprisingly, our 2018 and 2019 functional exercises for FMD and African swine fever (ASF) demonstrated a series of gaps in the Nation's readiness to respond to a large animal disease, particularly concerning depopulation and disposal methods.

By March 2020, COVID-19's effects were becoming increasingly evident in the United States, and packing plants were experiencing outbreaks of the virus and workforce absenteeism. The situation reached its nadir in late April to early May, when US pork processing plants sustained losses of 60 percent of their processing capacity. The effects were multifold: not only was less pork reaching grocery stores, but millions of pigs remained on farms and could not move as scheduled to slaughter or to their next production site.

APHIS and our other USDA and Federal partners reviewed our respective authorities to determine how we might assist producers caught up in the predicament of having to cull thousands of hogs. The authorities Congress granted us must be used for specific purposes, however, and we must work within those parameters. For example, the Animal Health Protection Act does not allow APHIS to expend funding unless we demonstrate the presence of a disease or pest of livestock. And the Animal Welfare Act does not apply to livestock on farms or those used for food or fiber purposes, nor does it include a funding mechanism.

In April, we in APHIS and industry representatives joined an existing ASF working group comprised of State Animal Health Officials, which had expanded its focus to include COVID-19; we later brought other Federal agencies into discussions centering on finding ways to alleviate the backlog in the pork supply chain. Although working group participants initially thought the Department of Homeland Security's Federal Emergency Management Agency (FEMA) may be a funding source, FEMA also found this particular issue outside the scope of its mission.

Through its Environmental Quality Incentives Program (EQIP), USDA's National Resources Conservation Service (NRCS) was able to help qualifying producers with animal disposal costs. NRCS also worked with our officials to expand the disposal methods its program would cover. The agency made these changes quickly enough to help some affected producers who were using aboveground burial as a disposal method.

Additionally, USDA leadership was able to provide a little over $2 million in discretionary funds. Portions of that funding went to cooperative agreements with two States (Iowa and Minnesota) to aid with disposal equipment and processes. The bulk of it went to the National Veterinary Stockpile for the purchase of a Horizontal Grinder and Mobile Inert Gas Depopulation System for future use.

We also reviewed another potential funding source: the National Animal Disease Preparedness and Response Program (NADPRP). Established by the 2018 Farm Bill, the NADPRP allows APHIS to collaborate with animal health partners throughout the Nation to boost efforts to keep high-consequence animal diseases from entering and spreading in the United States. While NADPRP could not provide direct, immediate grants, the window for submitting 2021 proposals was open during this timeframe. The program increased its available funding from $5 million to $10 million and specifically solicited proposals for depopulation and disposal projects.

For the most part, individual producers managed their supply chains by culling piglets or conducting depopulation for animal welfare reasons. Funding allocated via the Coronavirus Aid, Relief, and Economic Security (CARES) Act also provided support to several States for humane depopulation processes and disposal of thousands of pig carcasses. There is no question that States made great strides in testing and developing equipment and methods that we can rely upon in the event of an ASF or other FAD outbreak.

Stakeholders showed tremendous resilience and a spirit of collaboration and innovation in managing the swine processing slowdowns due to COVID-19. And there are many more examples, such as renderers working together to share their operational capacity to handle additional carcasses; animal health laboratories expanding their services to include human COVID diagnostic testing; and, private organizations developing novel online live pig replacement sites, to name but a few.

These lessons learned remind us that no single entity can carry out the entire range of preparedness or response activities needed to manage an emergency. While only the Federal Government can perform certain activities, such as test validation and development of government-to-government trade agreements, our State, Federal, and industry stakeholders and collaborators also have critical roles to play in an effective emergency response. State governments can work proactively on agreements regarding pre-movement testing and permitting of animals, as well as their own response planning and implementation. Large producers can work with us in advance to ensure their growers have correct premises identifications and designations on file. They may also have capital to fund research and development on depopulation and disposal methods. In addition, those on the farm should learn about and how to implement proper biosecurity measures and herd management plans, and we also encourage them to work with environmental officials on disposal options appropriate for their land and local requirements. Industry and academia, too, may have resources to conduct innovative research on depopulation, disposal, and virus elimination methods.

By keeping these lessons in mind and continuing to work collaboratively and proactively to identify and close any remaining gaps, we can reach a common state of readiness for the next threat.

American Association of Swine Veterinarians
830 26th Street
Perry, IA 50220-2328
USA

ISBN 978-1-7138-2952-2
9 781713 829522

CURRAN ASSOCIATES INC.
proceedings.com